"Ancient civilizations across the world started building the powerful mathematical toolkit powering today's science but failed to discover the most important number of all: zero. This well-written book is a fascinating interdisciplinary expedition to unearth zero's best-kept secrets."

– Professor Max Tegmark, Massachusetts Institute of Technology, USA, author of *Our Mathematical Universe* and *Life 3.0*

•••

"Was everything born from Nothing, or has Nothingness always been impossible? Trying to understand Zero might help us to answer such questions. Zero has been central to battles in mathematics, in philosophy, in religion, and in the sciences. Those battles, and the cultural backgrounds to them, are described in this fascinating volume."

– John Leslie, Professor Emeritus of Philosophy, University of Guelph, Canada, co-editor of *The Mystery of Existence: Why Is There Anything At All?*

•••

"This book is a fascinating compilation of reflections on Zero – the digit, the symbol, the concept – from a plethora of perspectives: mathematical, scientific, philosophical, theological, spiritual, historical, linguistic, artistic. Once again proving that there is a great deal to say about nothingness, and that there's always a new angle to find no matter which page of the book you choose to open."

– Vinod Subramaniam, President, University of Twente, The Netherlands

•••

"To the ancient Greeks, a number was the ratio of commensurable quantities. How then could one divide a distance by a time? Time and distance are incommensurable. It took the human race over a thousand years to figure out how to divide distance by time and thus arrive at the concept of velocity. We could finally outrun Zeno's tortoise. In another three hundred years we had celestial mechanics and a few hundred more we walked on the moon. Zero is much harder. Zero is the number of elements in the empty set, the number of things that are not equal to themselves. What is that number?

Depending on the consistency of our arithmetic, the empty set contains either nothing or everything. Is our arithmetic consistent? Thanks to Gödel's incompleteness theorem we know that we will never know. At least we know that we can stop asking. But we can't stop wondering how we got here, not knowing whether everything is nothing or not. This book may not let the fly out of this bottle but it will do something better, it will create wonder. Wonder is after all a wonder. It's over the moon. Enjoy this book."

– Roger M. Cooke, Emeritus Professor of Applied Decision Theory at the Department of Mathematics at Delft University of Technology, The Netherlands and Chauncey Starr Senior Fellow at Resources for Future in Washington

•••

"The classical Chinese rice bowl was painted with the image of a tiger. The inept forms show that none of the painters had ever seen the actual animal. They imagined a tiger. The dictionary says that imagination is the faculty of making mental images of things that are not present. Seeing the Sun setting over the ocean, we imagine that we are riding an immense spinning sphere, even though our senses tell us that this is absurd. But we can even imagine things that have no image, that are not even things – such as numbers, and especially the number zero – as is shown in this fascinating book. Even more than our image of the cosmos, zero is an ultimate product of imagination. Imagination and creation are twins. In mathematics and in art, something exists as soon as it has been imagined. A flying tiger, everlasting love, the square root of a negative number; even theorems are conceived before they are proven, or proven to be wrong. And zero, symbolic for science and for art. We can make visual images of things that are not present. In French, zero is *l'Oeuf*, the egg-shape, pronounced as 'love' in the game of tennis. Zero is love, the love for zero in this volume celebrates imagination."

– Dr Vincent Icke, Professor of Theoretical Astrophysics and visual artist

∵

The Origin and Significance of Zero

# Value Inquiry Book Series



VOLUME 395

The titles published in this series are listed at *brill.com/vibs*

# The Origin and Significance of Zero

*An Interdisciplinary Perspective*

*Edited by*

Peter Gobets
Robert Lawrence Kuhn

BRILL

LEIDEN | BOSTON

Funding towards the cost of publication received from Mr S. A. Norden, Dr A. Seeley, The Kuhn Foundation, Drs M. Oort, Mr M. Freriksen, Ms S. Lodaya, Ms A. van Dongen, Mr W. Saleh, Mr P. Gobets and NXP Semiconductors Netherlands. All such data are published by law on the Zero Project website annually (About Us tab): https://www.thezeroproject.nl/about-us/



Cover illustrations: 4,000 years of the numeral zero – the earliest and latest zeros in history (Mesopotamia, ~2000 BCE, and computer chip technology, ~2000 CE) (1) Placeholder text showing a record of deliveries of silver, dated to the second year of the reign of Ibbi-Suen of the dynasty of Ur III (2027 BCE in the Middle Chronology). Catalogue number: YPM BC 16534 (formerly YBC 1793). Collection: Yale Peabody Museum, Babylonian Collection. Photography: Klaus Wagensonner. Courtesy of Jim Ritter. (2) Computer chip, NXP Semiconductors, The Netherlands © 2023 NXP B.V. Reprinted with the permission of NXP Semiconductors.

The Library of Congress Cataloging-in-Publication Data is available online at https://catalog.loc.gov
LC record available at https://lccn.loc.gov/2024930950

Typeface for the Latin, Greek, and Cyrillic scripts: "Brill". See and download: brill.com/brill-typeface.

ISSN 0929-8436
ISBN 978-90-04-69155-1 (hardback)
ISBN 978-90-04-69156-8 (e-book)
DOI 10.1163/9789004691568



This book is printed on acid-free paper and produced in a sustainable manner.

*This volume is a tribute to the life and legacy of Peter Gobets, whose vision, commitment, creativity, ingenuity, insight, persistence and passion transformed a phantasmal dream into reified reality – apropos, perhaps, of zero itself, from empty placeholder to axial number. Peter was our Commander Zeronaut (as Peter called us explorers of the truth behind the discovery or invention of zero), and, as fate would have it, he departed Mothership Earth on 5 March 2024, just days before publication. Peter's laser-sharp focus and dedication towards the study of zero, his charm in engaging scholars, scientists, artists, entrepreneurs and captains of influence, all combined to actualize the project, realizing the holding of various zero events and the publishing of this zero monograph – the culmination of Peter's 50-year vision and accomplishment.*

*In addition, it is dedicated to the memory of René Samson, chairperson of the Zero Project Foundation from its launch in 2015 until his untimely passing in 2019. René was not only an accomplished scientist with wide professional interests, including in mathematics and physics, co-authoring two of the chapters in this book, but he was also a gifted composer of modern classical music (https://renesamson.nl/en/).*

∵

# Contents

## PART 2
## *Zero in the Arts*



## PART 3
## *Zero in Mathematics and Science*

# Foreword

Next to one, zero is the most common symbol in today's digital world, which often uses symbols, signs and concepts without knowing their true meaning and origin. Such a situation is not unique to today's society; similar phenomena have occurred many times in the history of civilization. We are simplifying the world by studying it more and more deeply and thoroughly with technologically advanced analytical tools and instruments, basing our conclusions on evidence and repeatable observations, of which there are never enough, and as a result, the intensity of research and the scope of the new knowledge we acquire is increasing significantly. At the same time, for several decades already, society has not been able to keep up with and absorb all of this newly generated knowledge, nor are scientists, who make their discoveries in narrow fields of knowledge, able to do so.

This knowledge can no longer be acquired through traditional methods and techniques. It is possible that a new field theory and support from artificial intelligence will soon be needed to extract the necessary knowledge from the overflowing container of scientific research. However, such generalization is lagging behind. Likewise, a fundamental reassessment of scientific research strategies themselves is also lagging, often without recognizing the diversity of intellectual traditions around the world, which are rooted in different cultures with different views on the meaning and consequences of human action. In this context, the Zero Project is a kind of cross-cultural encounter that enables experts from different cultures to carry out something like a 'syncing of the clocks' regarding the concept of Nothing.

An essential problem is that scientists, in their single-minded focus, find it difficult to accept the significance of the discoveries and research priorities of their colleagues in other fields. From this point of view, alongside intercultural dialogue, it is equally important to support interdisciplinary approaches, and this is also where the Zero Project shows initiative.

The aim of this monograph is to present different arguments, approaches and perspectives of scholars from different points of knowledge and experience in research. Such an approach has been very rare in recent decades, and it is therefore important that The Zero Project Foundation has taken note of the need for this perspective and provided its support.

At the same time, the *Expertise Center on Zero* is a wonderful initiative for the future development of the discussion forum, and we hope to involve doctoral students and young scholars from a variety of disciplines in its audience, thus giving them the opportunity to directly assess their knowledge and gain motivation and inspiration for future research, including targeted research on the invention/discovery of zero.

*Professor Valdis Segliņš*
Vice-Rector, University of Latvia

# Preface

Zero should need no introduction. It is meaningful in its representation of Nothing and significant in humanity's mathematical and philosophical development. This volume, *The Origin and Significance of Zero: An Interdisciplinary Perspective*, exemplifies the Zero Project, which explores the origin or origins of zero, and the role and impact of zero, in world intellectual history.

I come to zero via Nothing.[1] Since a child, I have been haunted by Nothing. When I was 12, in the summer between seventh and eighth grades, a sudden realization struck such fright that I strove desperately to blot it out, to eradicate the disruptive idea as if it were a lethal mind virus. My body shuddered with dread; an abyss had yawned open. Six and a half decades later I feel its frigid blast still.

Why not Nothing? What if everything had always been Nothing? Not just emptiness, not just blankness, and not just emptiness and blankness forever, but not even the existence of emptiness, not even the meaning of blankness, and no forever. Wouldn't it have been easier, simpler, more logical, to have Nothing rather than something?

The question would become my life partner, and even as I learned the rich philosophical legacy of Nothing, I do not pass a day without its disquieting presence. Here we are, human beings, conscious and abruptly self-aware, with lives fleetingly short, engulfed spatially and temporally by an ineffable, unimaginably vast, seemingly oblivious cosmos.

Call all-that-exists, anywhere, anytime, and in any form, 'Something.' Why is there Something rather than Nothing? Why is there anything at all? Of all the big questions, this is the biggest. The ultimate puzzle. The mystery of existence.

This biggest question seems unfathomable, impenetrable, uncrackable. But are there ways to probe? Can we learn about Nothing? Can we learn from Nothing?

Some scientists claim that the universe came from Nothing. But what's the nature of their kind of Nothing? That's where the confusion lies.

A milestone on my journey to Nothing was engaging with the philosopher John Leslie, who for decades had focused on Nothing and whom I had come to know through our discussions on *Closer To Truth*, the US public television

1 This Preface is derived, in part, and developed further, from R. L. Kuhn, "Levels of Nothing," *Skeptic Magazine*, Vol. 18, No. 2, 2013, and R. L. Kuhn, "Why Not Nothing," in J. Leslie and R. L. Kuhn (eds.), *The Mystery of Existence: Why is there Anything At All?* (Chichester: Wiley-Blackwell, 2013).

series that I created and host.[2] Together, we co-edited a book of readings and commentaries on our favorite ultimate question.[3] *The Mystery of Existence: Why is there Anything At All?* – long in gestation – presents the ideas of contemporary thinkers as well as others throughout intellectual history, grouped under five possible 'solutions' to the 'Why-is-there-Something-rather-than-Nothing?' puzzle: (1) a blank is absurd; (2) no explanation needed; (3) chance; (4) value/perfection as ultimate; and (5) mind/consciousness as ultimate.

Nothing has entered public discourse.[4] But how to progress on Nothing? I set a limited goal: clarifying Nothing. What do we mean by Nothing? When scientists, philosophers, people in general, speak about Nothing, to what are they referring? Are they referring to the same kind of No-Thing? Again, that's where the confusion lies.

My approach is to distinguish what I call 'Levels of Nothing,' and to use these Levels of Nothing to standardize discussions about Nothing.

Start by lumping together everything that exists and might exist – physical, mental, Platonic, spiritual, God, other non-physicals. As for the physical, include all matter and energy, space and time, all the fundamental forces of physics, and all the laws and principles that govern them (known and unknown); as for the mental, imagine all kinds of consciousness and awareness (known and unknown); as for the Platonic, gather all forms of abstract objects (numbers, logic, forms, propositions, possibilities – known and unknown); as for the spiritual and God, embrace anything that could possibly fit these non-physical categories (if anything does); and as for 'other non-physicals' – well, I just want to be sure not to leave anything unclassified.

Lump together literally everything contained in ultimate reality. Now call it all by the simple name 'Something.' Why is there 'Something' rather than 'Nothing'?

Note that Nothing is not an odd kind of Something, called 'Nothing,' that is existing. Rather, Nothing is defined by negation: No Thing, No Something.

---

2 Why is there Something rather than Nothing?, the ultimate question, is a continuing theme of *Closer To Truth* (www.closertotruth.com), the public television/PBS series that I created and host, and is co-created, produced and directed by Peter Getzels. For many *Closer To Truth* television episodes and videos on Nothing, see https://closertotruth.com/search-results/?searchwp=Nothing.

3 Leslie and Kuhn, *The Mystery of Existence.*

4 Jim Holt, *Why does the World Exist: An Existential Detective Story* (New York: Liveright, 2012); Michael Shermer, "Much Ado About Nothing," *Scientific American* (May 2022); Lawrence Krauss, *A Universe from Nothing: Why There Is Something Rather Than Nothing* (New York: Free Press, 2012); David Albert, "On the Origin of Everything," *The New York Times* (March 23, 2012); Ross Andersen, "Has Physics Made Philosophy and Religion Obsolete," *The Atlantic* (April 24, 2012).

So, why Not Nothing?

From this point, what guides me, I must admit, is more gut feeling, less clever reasoning, which is why no argument has ever dissuaded me from continuing to think, following Leibniz,[5] that Nothing, no world, would be simpler and easier than any world, that Nothing would have been the least arbitrary and 'most natural' state of affairs.

As I have continued to think about Nothing, I have continued to think that Nothing 'should,' in some sense, have obtained, and the only reason I accept the fact that Nothing does not obtain is not because of any of the arguments against Nothing,[6] but because of the raw existence of Something – because in my private consciousness I am forced to recognize that real existents compose Something.

In other words, an a priori weighing of Nothing vs. Something (from a timeless, explanatorily earlier perspective) would, for me, tip the balance heavily to Nothing, but for the fact of the matter. Thus, since I have no choice but to recognize that there is Something, I have no choice but to conclude that, if I eschew brute fact, there is some foundational force, selector, productive principle or, more likely, a type of necessity – some deep reason – that brings about the absence of Nothing.

None of this seems credible. Perhaps embrace brute fact?

I reject the argument that because there is an infinite number of possible worlds of Something, and only one possible world of Nothing, therefore the probability of a Nothing world is precisely zero (one divided by infinity). I reject this argument because it assumes that the prior weighting of a Nothing possible world is equal to that of *each* of the infinite number of Something possible worlds. To me, on the contrary, the prior weighting of a Nothing possible world exceeds the sum of *all* of the infinite number of Something possible worlds.

I cannot rid myself of the conviction that Nothing would have obtained had not something special somehow superseded or counteracted it. Yes, I know that seems circular – and many well-regarded philosophers say, 'So there's a world not a blank; what's in any way surprising about that?' 'There has to be something or other.'[7] But I just can't help feeling that those who do not take

5 Gottfried Leibniz, *The Principles of Nature and Grace* (1714).

6 Arguments against Nothing include asserting that Nothing is unimaginable, nonsensical, meaningless or absurd, or as soon as something is possible it must exist somewhere. Some would have God's necessary existence as proscribing Nothing.

7 See Bede Rundle, *Why there is Something rather than Nothing* (Oxford: Oxford University Press, 2006).

Nothing seriously are passing right over the problem most probative of ultimate reality.

Defining 'Nothing' may seem simple – no thing, not a thing. But what's a 'thing'? I invoke the term 'thing' in the most general possible sense, and therefore, given some possible notions of Nothing, it is no contradiction to find 'things' that compose these different kinds or levels of Nothing. Teasing apart these constituent things, as if scaffolds or sinews of Nothings, may help enrich understanding of the nature of Nothing, yielding a taxonomy that arrays opposing kinds of Nothing that could be conceived and might have existed.

This taxonomy is structured as a deconstruction or as a dissection, as it were, a reverse layering, a peeling, a progressive reduction of the content of each Nothing in a hierarchy of Nothings. As such, this taxonomy takes its heritage from the so-called Subtraction Argument, which seeks to show that the absence of all concrete objects would be metaphysically possible. (Stated simply, the Subtraction Argument works by imagining a sequence of possible worlds each containing one fewer concrete object than the world before, so that in the very last world even the very last object has vanished. It is no surprise that complexities emerge.[8])

Developing this way of thinking, there might be nine levels of Nothing, with a general progression from Nothing most simplistic (Nothing 1) to Nothing most Absolute (Nothing 9). (Nine is not a magic number of levels of Nothing; cut differently, there could be more levels, or fewer.)

Each level of these Nothings can be criticized. My point here is not so much to argue the legitimacy or conceptual contribution of any one kind of Nothing, but rather to construct an exhaustive taxonomy of all potential or competing Nothings, and a taxonomy in which those Nothings are mutually exclusive. If so, then one level of Nothing must be the most correct in some fundamental sense, even if we cannot adjudicate among them (i.e., adjudicating among levels 5 to 9; we can reject levels 1 to 4 immediately as they are obviously not scientifically sufficient).

8 For the Subtraction Argument, see the following: T. Baldwin, "There might be nothing," *Analysis* 56 (1996): 231–38. G. Rodriguez-Pereyra, "There might be nothing: the subtraction argument improved," *Analysis* 57 (1997): 159–66. G. Rodriguez-Pereyra, "Metaphysical nihilism defended: reply to Lowe and Paseau," *Analysis* 62 (2002): 172–80. Alexander Paseau, "The Subtraction Argument(s)," *Dialectica* 60(2) (2006): 145–56. For the opposing view, that it is metaphysically not possible that there would be no concrete objects, see the following: D. M. Armstrong, *A Combinatorial Theory of Possibility* (Cambridge: Cambridge University Press, 1989); David Lewis, *On the Plurality of Worlds* (Oxford: Blackwell, 1986); E. J. Lowe, "Why is there anything at all?" *Aristotelian Society Supplementary* Volume 70 (1996): 111–20. See also, Roy Sorensen, "Nothingness," *Stanford Encyclopedia of Philosophy* (2009).

Following are the nine levels of Nothings.

1. Nothing as existing space and time that just happens to be totally empty of all visible, observable objects (particles and energy are permitted) – an utterly simplistic, pre-scientific view.
2. Nothing as existing space and time that just happens to be totally empty of all matter (no particles, but energy is permitted – flouting the law of mass-energy equivalence). Another easy reject.
3. Nothing as existing space and time that just happens to be totally empty of all matter and energy.[9]
4. Nothing as existing space and time that is by necessity – irremediably and permanently in all directions, temporal as well as spatial – totally empty of all matter and energy.
5. Nothing of the kind found in theoretical formulations by physicists, where, although space-time (unified) as well as mass-energy (unified) do not exist, pre-existing laws, particularly laws of quantum mechanics, do exist. And it is these laws that make it the case that universes can and do, from time to time, pop or 'tunnel' into existence from 'Nothing,' creating space-time as well as mass-energy. (It is standard physics to assume that empty space seethes with virtual particles, reflecting the uncertainty principle of quantum physics, where particle-antiparticle pairs come into theoretical or mathematical being, and then, almost always, in a fleetingly brief moment, annihilate each other.)
6. Nothing where not only is there no space-time and no mass-energy, but also there are no pre-existing laws of physics that could generate space-time or mass-energy (universes).
7. Nothing where not only is there no space-time, no mass-energy, and no pre-existing laws of physics, but also there are no non-physical things or kinds that are concrete (rather than abstract) – no God, no gods, and no consciousness (cosmic or otherwise). This means that there are no physical or non-physical beings or existents of any kind – nothing, whether natural or supernatural, that is concrete (rather than abstract).
8. Nothing where not only is there none of the above (so that, as in Nothing 7, there are no concrete existing things, physical or non-physical), but also

9 As an example of an objection to a kind of Nothing, some would resist the idea that there could be space and time that had been emptied of existing things. The 'relational' theories of space and of time assume that emptying space and time of existing things is impossible, because space is the system of spatial relations between things, and time is the system of temporal relations between things.

there are no abstract objects of any kind – no numbers, no sets, no logic, no general propositions, no universals, no Platonic forms (e.g., no value).

9. Nothing where not only is there none of the above (so that, as in Nothing 8, there are no abstract objects), but also there are no 'possibilities' of any kind. (It is recognized that possibilities and abstract objects overlap in that possibilities are an abstract object, though allowing that the two concepts can be distinguished, such as in the possibility of there being abstract objects.)

Nothings 1 through 7 progressively remove or eliminate existing things, so that a reasonable stopping point – a point at which we might well be thought to have reached (what I hesitatingly call) 'Real Nothing' or 'Absolute Nothing,' the metaphysical limit – would be Nothing 7, which features no concrete existing things (no physical or non-physical concrete existents) of any kind.

Nothings 8 and 9 go further, eliminating non-concrete objects, things, existents and realities. Do they go too far? Many philosophers assert that neither Nothing 8 nor Nothing 9 is metaphysically possible, arguing that the claimed absence of abstract objects and/or possibilities would constitute a logical contradiction and hence abstract objects and/or possibilities exist necessarily.

This necessity of abstract objects and/or possibilities could be important because, as John Leslie points out, among the realities which aren't concrete things, or which do not depend on the existence of concrete things, and thus cannot be eliminated, there may be some realities that are plausible candidates for explaining the world of concrete things (i.e., value, Platonic good).[10] In this way of thinking, the crucial distinction is between realities that seemingly can be eliminated and realities that seemingly cannot be eliminated, rather than any particular way of distinguishing between levels of nothingness or particular ways of defining nothingness.

I like to point out that among all these levels of Nothing, one of the 'lesser Nothings' – that is, a kind of Nothing with more 'things' in it – is the Nothing of physicists, Nothing Level 5.

What physicists contemplate – the sudden emergence or 'tunneling' of universes from 'Nothing' – is fascinating and indeed may be cosmogenic, but the tunneling process or capacity itself is not Nothing.[11] The Nothing of physicists

10 See sections 3.7, 3.8, and 3.9 in the taxonomy of possible generators or creators of the universe, in my essay, R. L. Kuhn, "Why This Universe?" *Skeptic Magazine*, Vol. 13, No. 2 (2007): 36.

11 That the universe may have popped or tunneled into existence via some sort of cosmic spontaneous combustion, emerging from the 'nothing' of empty space (i.e., vacuum energy generated by quantum fluctuations, unstable high energy 'false vacua'), or from 'quantum tunneling' (Alex Vilenkin, *Many Worlds in One: The Search for Other Universes*

is thick and rich with the complete set of the laws of physics, and so between physicists' Nothing and Absolute Nothing lies a vast, unbridgeable gulf.

On this taxonomic scale, physicists' Nothing, as Nothing Level 5, is barely halfway to utter, Absolute Nothing. If physicists' Nothing were in reality Absolute Nothing (i.e., bedrock ultimate reality, with no lower Nothings), the laws of quantum physics (or whatever might turn out to be the most fundamental physical laws underlying quantum physics) would have to be either impossible to remove (meaning that eliminating them would involve logical contradiction) or a brute fact about existence beyond which explanation would be meaningless. Few would argue that the ultimate laws of physics are logically necessary, not contingent, and I doubt I could ever get over the odd idea that something so intricate, so involved, so organized and so accessible as the complete laws of physics would be the ultimate brute fact as the furniture of Absolute Nothing.

As a separate consideration, philosophers of religion argue that (if there is a God) God is a 'necessity,' meaning that it would be impossible for God not to exist – God must exist in all possible worlds – thus precluding Nothing 7 (which has no non-physical concrete things such as God but still has abstract objects) and crowning Nothing 6 (which has no space-time, no mass-energy, no laws of physics, but still has God and other non-physical things) as the metaphysical limit of what is to be explained.[12]

---

(New York: Hill and Wang, 2006), may be the proximal cause of why we have a universe in the first place, but cannot be the reason, of itself, why the universe we have works so well for us. Universe-generating mechanisms of themselves, such as unprompted eternal chaotic inflation or uncaused nucleations in space-time, can only address the fine-tuning problem of our universe by postulating innumerable universes, perhaps an infinity of universes, a vast multiverse, in which the laws of physics must reset randomly in each universe, and must be, in some sense, primordial and foundational. Nor can vacuum energy or quantum tunneling or anything of the like be the ultimate cause of the universe, because, however hackneyed, the still-standing, still-unanswered question remains 'from where did those laws come?'.

12 The question of whether God, assuming God exists, would be 'necessary'– which means that God would exist in all possible worlds – has beset philosophers and theologians for centuries. The much-debated now commonly refuted Ontological Argument for the existence of God, which defines God as 'a being than which no greater can be conceived,' leads to the claim that God is necessary because necessity is a higher perfection than contingency. Richard Swinburne originally asserted that God is a 'factual necessity' but not a 'logical necessity' in that the non-existence of God would introduce no logical contradiction (*Closer To Truth*). Swinburne later strengthened God's necessary existence to a kind of 'metaphysical necessity' where God is both the necessary and sufficient cause of God's own existence, and so gives a probability of 1 to the probability of the effect. Swinburne said he was not convinced of this, but it did seem more plausible than any other answer to 'Why is there anything at all?'; it provides a genuine intermediate possibility between

I find the move of imputing necessity to God, especially logical necessity, challenging. Moreover, based on the levels of Nothing in this taxonomy, it would seem less of a leap to imagine a world without God (Nothing 7) than to imagine a world without abstract objects (Nothing 8). For the traditional God, that won't do.[13]

---

God being contingent and God being logically necessary. He developed this idea (including explaining the sense in which 'cause' is being used analogically) in R. Swinburne, *The Coherence of Theism: Second Edition* (Oxford: Oxford University Press, 2016), chapter 14 (private communication). Timothy O'Connor defends God's necessity in his monograph on the topic. Timothy O'Connor, *Theism and Ultimate Explanation: The Necessary Shape of Contingency* (Oxford: Blackwell, 2008).

13 The relationship between God and abstract objects is particularly troublesome for those who believe that God created and sustains all things and who privilege above all else God's absolute sovereignty (aseity). The reason is that abstract objects, many philosophers believe, exist necessarily, which means that it would be impossible for abstract objects not to exist, which further means that it makes no sense for even God to have created them. What would it take to create the idea of the number 3 or the truth that 1 + 2 = 3 or the reality that squares are not round? How could such ideas, truths, realities even conceivably be created? Peter van Inwagen calls abstract objects 'putative counter-examples' to the thesis that God has created everything. But if abstract objects do exist necessarily, then wouldn't God's mental life be encompassed by blizzards or swarms of infinities of infinities of abstract objects, not only which God would not have created but also over which God could exercise no control? The problem posed by abstract objects for a God whose sovereignty must be absolute is complex and requires metaphysical analysis. Consider two of the more general ways to defend God's sovereignty (aseity): 1) Deny that abstract objects are real, in that numbers, universals, propositions and the like are mere human-invented names with no correspondence in reality (nominalism, fictionalism); and/or 2) claim that abstract objects are thoughts in the mind of God. Van Inwagen rejects both ways; he must therefore defend the position that there are besides God other uncreated beings and he thus prefers to restrict God's creation of 'all things visible and invisible' to 'objects that can enter into causal relations' (which excludes abstract objects). Peter Van Inwagen, "God and Other Uncreated Things," in Kevin Timpe (ed.), *Metaphysics and God* (London: Routledge, 2009). On the other hand, William Lane Craig rejects the view that 'there might be things, such as properties and numbers, which are causally unrelated to God as their Creator.' Craig says that 'Abstract objects have at most an insubstantial existence in the mind of the Logos,' adding, 'If a Christian theist is to be a Platonist, then, he must, it seems, embrace absolute creationism, the view that God has created all the abstract objects there are.' However, Craig himself resolves the conundrum by espousing nominalism and fictionalism, by judging Platonism to be false – so that those pesky abstract objects no longer exist and thus no longer undermine God's sovereignty. See William Lane Craig, *God and Abstract Objects: The Coherence of Theism: Aseity* (New York: Springer, 2017) and his more popular version, William Lane Craig, *God Over All: Divine Aseity and the Challenge of Platonism* (Oxford: Oxford University Press, 2018). Richard Swinburne argues that abstract objects, which seem to contradict his concept of God, are fictions; the only things that are true or false are human sentences

Cosmic visions are overwhelming, but I am sometimes preoccupied with another conundrum. How is it that we humans have such farsighted understanding after only a few thousand years of historical consciousness, only a few hundred years of effective science, and only a few decades of cosmological observations? Maybe it's still too early in the game. Maybe answers have been with us all along. This is a work in process and diverse contributions are needed.

Setting aside my taxonomy and consulting my gut, I come to only two kinds of answers. The first is that there can be no answer: Existence is a brute fact without explanation. The second is that at the primordial beginning, explanatorily and timelessly prior to time, some thing was self-existing. The essence of this something necessitated its existence such that non-existence to it would be as inherently impossible as physical immortality to us is factually impossible.

Various things or substances could conceivably contain this deeply centered self-existing essence, from the most fundamental meta-laws of physics to diverse kinds of consciousness, one of which could be God or something like god. Perhaps even these explanations are too mundane and bedrock is so bizarre that abstract objects or pure possibilities somehow harbor generative powers.

Why is there Something rather than Nothing? Why is there anything at all? Why Not Nothing? If you don't get dizzy, you really don't get it.

This is the mystery of existence and this is the foundation of *Closer To Truth*, our web resource and US public television series on cosmos, life, consciousness, and meaning.

•••

The Zero Project's working hypothesis is that to account for the 'otherwise inexplicable emergence of the mathematical zero in the historical record relatively recently,' the invention/discovery of zero may have been facilitated by the prior evolution of a sophisticated concept of Nothingness or Emptiness (as it is understood in non-European traditions); and conversely, inhibited by the absence of such a concept of Nothing, or an active aversion to it, in the West. While the Zero Project's working hypothesis is supported by a majority of the authors in this volume, I myself am 'passionately neutral.'

I am perennially skeptical of all 'just so stories' of untestable hypothesis, such as in evolutionary psychology. While I would find the philosophical nexus

---

(*Closer To Truth*). Also, Matthew Davidson, "God and Other Necessary Beings," *Stanford Encyclopedia of Philosophy* (Stanford, 2009).

between mathematical zero and philosophical Nothingness intellectually satisfying, my tastes are no substitute for definitive research.

So, this is my question, and my challenge, to the Zero Project: Where is the evidence that the philosophical Nothingness or Emptiness in the East was a precursor – or more challenging, a driver – of the mathematical zero that emerged? More specifically, what might such evidence to support this working hypothesis even look like? Moreover, how to falsify what I might consider the default, base case, that mathematical zero emerged via the less lofty midwife of, say, 'mathematical agriculture'?

No doubt, the concept of zero is both a critical event in intellectual history and a milestone in the development of mathematics, science, and technology. The exploration of zero's origin, culturally and linguistically as well as mathematically and philosophically, could elicit novel ideas and new ways of thinking. Thus, reversing the explanatory arrow of putative causation, could the broader philosophical significance of zero reveal transcendental ideas of Nothing, Emptiness, Void, Blank as features of reality to be apprehended and appreciated and perhaps applied to entirely new categories of thought? Indeed, as described above, 'Nothing' is a prime *Closer To Truth* leitmotiv and driving theme.[14]

Nothing also commands global fascination. Well over half of *Closer To Truth* audiences are outside of the US and it is satisfying to see viewers from across the globe, from dozens of diverse countries, engaging with the profound philosophical issues and implications of Nothing.

In this sense, the diverse, contemporary global interest in Nothing parallels the diverse, historical global development of zero. Zero is humanity's treasure, cutting across religions, regions, races, ethnicities, genders, ages, educational levels, income levels, and social class. Perhaps recognizing the broad origin and ubiquitous impact of zero can play a small part in catalyzing human harmony in a fractious world.

*Closer To Truth* co-creator and producer/director Peter Getzels and I are pleased to collaborate with the Zero Project, to provide our wholehearted endorsement, in bringing the origin and impact of zero to global audiences.[15] We will interview several of the contributors to this volume for a mini-series on the Zero Project to be streamed on the *Closer To Truth* website

14 See footnote 2.

15 'The Zero Project – International Conference/Workshop on Zero,' October 3, 2021, video, https://youtu.be/W59_tyUHfK0. Closer to Truth, 'Is Zero More Than Nothing? Introducing the Zero Project,' October 8, 2021, video, https://youtu.be/bUVN7E1wiBs.

(www.closertotruth.com) and *Closer To Truth* YouTube channel (over 580,000 subscribers as of August 2023).

The Zero Project is a tribute to the commitment of its board of trustees – 'Team 0,' as they like to call themselves – and to the vision of its founder and organizer, Peter Gobets, co-editor of this volume, whose ingenuity, insight, passion, and persistence transformed a phantasmal dream into reified reality – apropos, perhaps, of zero itself, from empty placeholder to axial number.

Our appreciation to the outstanding scholars and thinkers in this volume. Their interdisciplinary nature characterizes and distinguishes the thrust of this volume, and of the entire Zero Project, from prior explorations of zero.

The Zero Project is both for, and not for, Nothing.

*Robert Lawrence Kuhn*
*Closer To Truth*
December 2023

# Acknowledgements

The Zero Project would not have got off the ground without the generous financial support of, among others, Mr S. A. Norden, Dr A. Seeley, Dr R. L. Kuhn, Drs Marianne Oort, Mr M. Freriksen, Ms S. Lodaya, Ms A. van Dongen and Mr W. Saleh. Ms S. B. Cowsik provided valuable fundraising expertise. Support-in-kind was supplied by internet provider Combell and website development and maintenance by Eastern Enterprise.

A vote of sincere thanks goes to the contributors for their patience and perseverance throughout the years leading up to publication and to all who extended support in one way or another to bringing the book to fruition.

We are indebted to Professor Marcis Auzinsh and Professor Kaspars Klavins, University of Latvia, and their in-house UL Press, who made a vital contribution at a key stage of the publication process.

In addition, our sincere thanks go to UL Vice-Rector for Natural Sciences, Technology and Medicine, Professors Valdis Segliņš, for the Foreword endorsing the Zero Project volume and *Expertise Center on Zero*.

We are also obliged to Professor Dr Sharda Nandram, Vrije Universiteit, Amsterdam. Readers will find the chapter she co-authored with colleagues in the spawning grounds of contiguous fields of the Humanities which touches on the 'immaterial' significance represented by zero.

Wende Wallert, Vrije Universiteit Art Curator and her assistant, Maria Miccoli, organized an art exhibition on the theme of Zero (https://youtu.be/c_pPMPVFOsg).

Much of the credit for the final form of the book goes to the meticulous copy editor, Mary Davis, based in the UK, who whipped the content we provided into shape.

It has been a pleasure working with editors Erika Mandarino and Helena Schöb of Brill who have been the source of expert guidance and enthusiasm in equal measures.

Seasoned videographer, Kevin Lee of Lee Pictures, was in charge of recording the presentations by the *Monograph on Zero* contributors during the online event from October 2021 to May 2022, the list of links to which are included in the book for ready reference.

Drs P. H. M. Keesom not only provided invaluable legal advice on issues related to the establishment of the Zero Project Foundation in 2015 and subsequently, but as sworn translator/interpreter rendered the Foundation's Articles of Association from the original Dutch into English.

Notary Aniel Autar had generously drawn up the Foundation's Articles of Association by way of support-in-kind.

We would be remiss not to mention the seminal role played by all the members of our philosophy club (FWF) founded in The Hague over 30 years ago. The Zero Project traces its genesis to the stimulating annual gatherings held in well-nigh Epicurean environs.

Elizabeth den Boer framed the first academic research proposal in 2014, which she presented at an international conference.

Annette van der Hoek, organized and led two Camp Zero workshops in 2016 and 2017.

Erik Hoogcarspel advised us and contributed a chapter on the origin of zero included in the book. Jildi Mohamad Sjah provided media intel and moral support.

Bhaswati Bhattacharya's chapter provided the methodology to be observed throughout.

Our appreciation, too, goes to another member of the philosophy club, Sander J. Cohen, who, proofread the chapters to ensure requirements were met with respect to form and content.

Victor van Bijlert, one of the founding members of the philosophy club, would in due course join the board of trustees of the Zero Project and provide invaluable advice.

The youngest member in the company, virtuoso violinist Satyakam Mohkamsing, composed and performed the Zero Project's signature tune as musical score to a number of promotional videos as well as to the International Conference/Workshop on Zero online event.

Finally, we especially wish to express our sincere appreciation to Dr Robert Lawrence Kuhn for taking an interest in the work of the Zero Project – as indeed the Zero Project takes in his own quest for Ultimate Questions. His involvement gave the Zero Project the boost needed to complete the daunting task of shedding light on the origin of the mathematical number zero and the underlying concept of Nothingness.

# Figures and Tables

## Figures

## Tables

# Notes on Contributors

*Debra G. Aczel*
has over 40 years in educational program management, including as program manager at Massachusetts Institute of Technology's Terrascope Program – an interdisciplinary environmental program working to solve pressing global issues. She is co-founder and current co-director of the Amir D. Aczel Foundation for Research and Education in Science and Mathematics, supporting cultural and educational projects in Cambodia.

*Miriam R. Aczel*
is a Postdoctoral Scholar at the California Institute for Energy and Environment (CIEE) at University of California, Berkeley, working on the Oakland EcoBlock pilot project. Dr Aczel is a Research Fellow, Sustainability Transitions and Justice, United Nations University Institute for Water, Environment and Health (UNU-INWEH). She is also an Honorary Research Associate at the Centre for Environmental Policy (CEP) at Imperial College London, and a Fellow of the Royal Society of Arts. She is co-founder and current co-director of the educational non-profit Amir D. Aczel Foundation for Research and Education in Science and Mathematics.

*Djamil Aïssani*
is President, University of Bejaia, History of Muslim Mathematics, Algeria. He started his career at the University of Constantine in 1978. He received his PhD in 1983 from Kiev State University (Soviet Union). He has been at the University of Bejaia since it opened in 1983/1984 in the following roles: Director of Research, First Head of the Faculty of Science and Engineering Science (1999–2000), Director of the Research Unit LAMOS (Modeling and Optimization of Systems, http://www.lamos.org), Scientific Head of the Doctoral Computer School (2004–11), President of the Learned Society GEHIMAB (History of Sciences in the Maghreb and Mediterranean area) http://www.gehimab.org, as well as teaching in many other universities. Professor Aïssani was the president of the National Mathematical Committee (Algerian Ministry of Higher Education and Scientific Research, 1995–2005).

*Ravi Prakash Arya*
is a Chair Professor of Maharshi Dayanand Saraswati Chair (UGC) at Maharshi Dayanand University, India. Professor Arya has to his credit 75 research papers and 70 books running into 84 volumes on the various aspects of Vedas, Vedic

Sciences, Vedic Exegesis, Vedic Philology, Vedic Philosophy, Religion, Indian History and Culture. He is the Chief Editor of a quarterly research journal, Vedic Science, dealing with the scientific interpretation of Vedas and allied literature and ancient Indian traditions. He is also the editor of the annual World Vedic Calendar. He has convened several national and international conferences/seminar/workshops on Vedic Sciences, Ayurveda, Yoga, Spirituality and Indian knowledge system. Contact: vedicscience@gmail.com, https://.vedic-sciences.com.

*Marcis Auzinsh*
is currently a Professor and Head of the Chair of Experimental Physics at the University of Latvia. From 2007 till 2015 he served as the Vice Chancellor of the University of Latvia. His main professional interests lie in the field of quantum physics. Professor Auzinsh has co-authored more than a hundred research papers published in leading research journals and several hundred conference reports. He is also a co-author of two monographs published by Cambridge University Press and Oxford University Press and both appeared in several editions. During his academic career he worked in many countries – China, Taiwan, the United States of America, Canada, the United Kingdom, Israel and Germany.

*Bhaswati Bhattacharya*
is a Fellow of the University of Göttingen, Germany and a historian affiliated with the Centre for Modern Indian studies, Göttingen University as a senior research fellow. Her recent publications include a monograph on the Indian Coffee House, Much Ado over Coffee: Indian Coffee House Then and Now (Social Science Press, 2017). She is the co-editor of Politics of Advertisement and Consumer Identity: The Making of the Indian Consumer (Routledge, in press). Dr Bhattacharya is working on another monograph on the marketing and consumption of coffee in India.

*Sudip Bhattacharyya*
is a full Professor of Astrophysics at the Tata Institute of Fundamental Research (TIFR) in India. He has published more than 90 peer-reviewed research and review papers, and received National Aeronautics and Space Administration (NASA) Space Science Achievement Award due to his research. He is also the current Principal Investigator of the Soft X-ray Telescope aboard AstroSat, the first dedicated Indian astronomy satellite. Before joining TIFR, Professor Bhattacharyya was a Research Associate in the University of Maryland at

College Park and NASA in USA. In addition to science papers, he has published the peer-reviewed philosophy paper entitled 'Zero – a Tangible Representation of Nonexistence: Implications for Modern Science and the Fundamental'. Contact: sudippdb@gmail.com.

*Joseph A. Biello*
is a Professor of Mathematics at the University of California, Davis. He began his career studying astrophysical fluid dynamics, but gradually moved to fluid dynamics of a more mundane nature – the Earth's atmosphere. His mathematical expertise lies at the interface of partial differential equations and asymptotic methods, and his perspective on the concept of zero is informed by ideas of dominant balance that arise from asymptotics. He lives with and is raising his two teenage daughters in California. While on sabbatical in 2012, Dr Biello and his family, by stroke of serendipity, found themselves renting an apartment above Dr René Samson. From this have sprung many collaborative explorations into fluid dynamics and mathematics, including a contribution to the present book.

*Puneet K. Bindlish*
is Assistant Professor of Hindu Spirituality at the Faculty of Religion and Theology, Vrije Universiteit, Amsterdam, The Netherlands. Earlier, he taught at the Department of Humanistic Studies, Indian Institute of Technology (Banaras Hindu University), Varanasi, India. He brings a blend of academic, consulting and entrepreneurial experience across healthcare, telecom, technology, banking and insurance, education, sports, public sectors. His fields of interest are: Spirituality, Leadership, organizational behavior, entrepreneurship and integrative intelligence. Contact: p.k.bindlish@vu.nl.

*Johannes Bronkhorst*
was Professor of Sanskrit & Indian Studies at the University of Lausanne (Switzerland) until his retirement in 2011. He began his studies of Sanskrit and Pali at the University of Rajasthan, India, and was awarded a PhD by the University of Pune, India, and a doctorate by the University of Leiden, The Netherlands. Professor Bronkhorst has written more than 200 research papers, 19 books and numerous reviews. He is editor of a number of collective volumes, various book series, sits on several editorial boards, and is a member of the Royal Netherlands Academy of Arts and Sciences (KNAW). Contact: johannes.bronkhorst@unil.ch.

*Jonathan J. Crabtree*

is Historian of Mathematics, Melbourne, Australia. Having initially graduated in economics at the University of Melbourne, he is now an autodidact, studying the history of mathematics since 1983. His first paper was on the history of Euclid's definition of multiplication while his next paper explored the writings of Descartes and Newton. Another paper presented in Hungary spanned the writings of Diophantus, Cardano, Euclid, Liu Hui and Brahmagupta. Having reviewed original writings in Latin, Greek, Arabic and Sanskrit, Crabtree has also written magazine articles on mathematics history and been a guest lecturer on the history of zero at Indian universities.

*Vishwanath Dhital*

is Assistant Professor at Department of Humanistic Studies, Indian Institute of Technology (Banaras Hindu University), Varanasi, India. Post Doctorate: ICPR fellow, SVDV, BHU, Varanasi and Doctorate, Masters, Bachelors: Sampurnanand Sanskrit University, Varanasi. Early Sanskrit education: Mahesh Sanskrit Gurukulam, Devghat, Nepal. He has a traditional background in Indian Philosophy especially in Nyaya-Vaisheshika Philosophy. He is honored by the President of India for his contribution to Sanskrit Language. He has published about 10 Books and Papers. Contact: vdhital.hss@iitbhu.ac.in.

*Esti Eisenmann*

studies Jewish thought over the generations, and especially the philosophy of the Middle Ages. Her research into medieval thinkers deals with the concept of divinity, biblical exegesis, and the interpretation of Jewish tradition in light of Greek philosophy, with special attention to the Jewish contribution to the development of the sciences. She has published sections of the fourteenth-century Hebrew scientific encyclopedia Ahavah ba-Ta'anugim. Compiled by an otherwise unknown savant named Moses b. Judah, it summarizes Aristotelian physics as it had come down through the Arabic tradition, with some nods to Christian scholasticism. Dr Eisenmann is currently investigating the commentaries on Aristotle's Physics written in Hebrew by a famous Jewish scholar of the fourteenth century, Levi b. Gershom (Gersonides). She teaches courses on Jewish philosophy, biblical exegesis, and Jewish mysticism at the Open University of Israel, the Hebrew University, and Herzog College.

*Paul Ernest*

is Emeritus Professor of the Philosophy of Mathematics Education at Exeter University, UK. He established and ran the doctoral and masters programme specialisms in mathematics education at Exeter University. His research concerns questions about the nature of mathematics and how it relates to teaching, learning and society. He continues to contribute to developing social constructionism as a philosophy of mathematics. His concern to locate philosophical issues in broader social, cultural and historical contexts leads to new ways to address questions about how mathematics relates to epistemology, ontology, aesthetics and ethics including social justice, and he continues to work on these areas. Professor Ernest's publications include The Philosophy of Mathematics Education (Routledge, 1991) (with over 3,000 citations), and Social Constructivism as a Philosophy of Mathematics (SUNY Press, 1998) (with over 1,000 citations). He founded and edits the Philosophy of Mathematics Education Journal, now in its thirty-second year of publication, a free web journal accessed via http://socialsciences.exeter.ac.uk/education/research/centres/stem/publications/pmej/.

*Fabio Gironi*
was Humboldt Research Fellow at the University of Potsdam, Germany. He works on the history of American philosophy, epistemology and contemporary realism. He has published numerous articles on Wilfrid Sellars, realism in contemporary continental philosophy and the philosophy of science. Dr Fabio Gironi is the author of Naturalizing Badiou: Mathematical Ontology and Structural Realism (Palgrave, 2014) and editor of The Legacy of Kant in Sellars and Meillassoux: Analytic and Continental Kantianism (Routledge, 2017).

*Peter Gobets*
emigrated to the United States from Paramaribo, Surinam, in the sixties. He was awarded a BSc in Organic Chemistry, San Jose State College, and started a Chemistry PhD program with the University of Berkeley, CA. He travelled extensively as 'Road scholar' throughout the USA, Europe and Asia and settled in the Netherlands in 1977. He describes himself as an independent autodidact, with a particular interest in Eastern and Western philosophy, linguistics and the history of mathematics. He has published two books of poetry and a philosophical novel, he also worked as translator/interpreter and media advisor until his retirement in 2013. It was in the early 1990s that Peter drew up the blueprint for the Zero Project and it was officially launched in 2015.

*Hab Touch*
spent his early childhood under the Khmer Rouge regime that ruled Cambodia (1975–79). In early 1980, Touch (pronounced Tooch) attended the School of Fine Art in Phnom Penh, specializing in sculpture and ceramics. He received a scholarship to study at the Nicholas Copernicus University in Torun, Poland.

He obtained a Master's Degree in 1993 on the field of Conservation and Restoration of Works of Art and then worked as an Assistant Professor at the University. Returning to Cambodia in 1995, he began work at the Ministry of Culture and Fine Arts in Phnom Penh and was appointed Deputy Director of the National Museum of Cambodia responsible for the conservation, documentation and exhibition programs. He was promoted to the directorship of the museum in 2007. In 2011, he was appointed Director General of Heritage responsible for museums, antiquities and ancient monuments. Since 2014, he has been the Director General for Intangible Cultural Heritage (General Department of Technical Cultural Affairs) at the Ministry of Culture and Fine Arts.

*Friedhelm Hoffmann*
studied Egyptology, Latin and German at the universities of Würzburg and Oxford. He was awarded his PhD in Egyptology in 1994 and Habilitation in 2001. Since 2010, he has been Professor of Egyptology at Ludwig-Maximilians-Universität, Munich. His main research interests include Demotic (late Egyptian) and Egyptian scientific and magical texts.

*Erik Hoogcarspel*
is a philosopher, working in the Netherlands. He studied mathematics, philosophy, phenomenolgy, Buddhology, Indology and religious anthropology at the Universities of Groningen and Leiden. He lectured in different institutions including the University of Nijmegen and the Internationale School Voor Wijsbegeerte. As well as a number of articles, Dr Hoogcarspel published The Central Philosophy: Basic Verses (2005) which is an annotated translation from Sanskrit into English of the second century text called the Mūlamadhyamakakārikāḥ, written by the Indian Buddhist philosopher Nāgārjuna; a Dutch translation, called Grondregels van de filosofie van het midden (2005); Phenomenal Emptiness (2016), suggesting the use of phenomenology as a method for better understanding and developing Buddhism in the West. He is currently researching into the origins and peculiarities of the concept of self in the history of Western philosophy.

*Anupam Jain*
is presently Professor of Mathematics in the School of Data Science and Forecasting and Director, Centre of Ancient Indian Mathematics, Devi Ahilya University, Indore, India. Previously he served as Professor and Head in the Department of Mathematics and Controller-Exams in Government Holkar (Autonomous) Science College, Indore. He was awarded his PhD by Meerut

University (now CCS University) in 1992 on the topic 'Contribution of Jaina charyas to the Development of Mathematics'. He has written 26 books and over 150 research papers on History of Mathematics, and over 55 on various aspects of Jainism, together with 185 general articles. He has been honored by various awards since 1993, including the Shastri Parishad Journalism Award (1993).

*Ankur Joshi*
is Assistant Professor at FMS-WISDOM, Banasthali, Vidyapith, and Visiting Professor at Vrije University, The Netherlands. He is a Fellow of MDI-Gurugram in Public Policy and Governance, where he did doctoral research on Gurukul education system. At Banasthali, he contributes to research, training and teaching through the OM RISE Research Group (a collaboration with researchers from Buurtzorg Nederland and Praan Group), the National Center for Corporate Governance (supported by NFCG, New Delhi), and National Resource Center (setup by Ministry of Education, Government of India). He has developed online modules, conducted various training sessions for teachers, civil servants and students. He has published about 20 papers and book chapters. Contact: ankurjoshi87@gmail.com.

*Kaspars Klavins*
is a professor in the Department of Asian Studies at the Faculty of Humanities of the University of Latvia specialising in Korean Studies. He has written on Middle Eastern history; Islamic civilisation; East Asian and Korean history, culture and spirituality; medieval history; German studies and Baltic history. Recently he has published a book on East–West intercultural relations, The 'Other' and the 'Self': Supplement to East–West Cross-Cultural Studies (University of Latvia Press, 2020). Klavins has been a visiting professor at Roosevelt University (USA) through the Fulbright Program, an adjunct research associate at Monash University (Australia), a research associate in the Department of Archaeology at the University of Reading (UK), a visiting professor at the University of Münster (Germany), a professor of history and cross-cultural management at the Emirates College of Technology in Abu Dhabi (United Arab Emirates), a professor in the Faculty of Engineering Economics and Management at Riga Technical University (Latvia) and a visiting professor at Pusan National University (South Korea). Klavins is currently the head of the Department of Asian Studies at the University of Latvia (UL), a member of the Scientific Council of the UL International Institute of Indic Studies and the project director at the UL Centre for Korean Studies.

*Robert Lawrence Kuhn*
is creator, host, writer and executive producer of Closer To Truth (TV), the PBS/ public television series on cosmos (cosmology/physics), consciousness (brain/ mind) and meaning (philosophy of religion, critical thinking) that presents leading scientists, philosophers and creative thinkers discussing fundamental questions of existence and sentience (www.closertotruth.com). Dr Kuhn has written or edited over 30 books, including The Mystery of Existence: Why is there Anything At All? (with John Leslie); *Closer To Truth: Challenging Current Belief; Closer To Truth: Science, Meaning and the Future; How China's Leaders Think* (featuring President Xi Jinping); *The Man Who Changed China: The Life and Legacy of Jiang Zemin* (China's best-selling book in 2005 and again after his death in December 2022); and *Xi Jinping's New-Era China: The Inside Story*. Dr Kuhn is a recipient of the China Reform Friendship Medal, China's highest award for foreigners. He chairs The Kuhn Foundation. He has a BA in Human Biology (Johns Hopkins), SM in Management (MIT), and PhD in Anatomy/ Brain Research (UCLA).

*Alexis Lavis*
is Associate Professor of Philosophy at Renmin University of China (RUC; Beijing). He is a specialist of comparative philosophy (Europe, India, China). He was previously lecturer in philosophy and Asian religions in Paris Institute of Political Studies (Sciences Po); lecturer in history of philosophy in University of Normandy. He received his PhD from University of Normandy (Sanskrit Buddhist Studies and Comparative Philosophy) and obtained the Agrégation in philosophy from the French Ministry of Education and Research. He is editorial director of collection "Asian Studies" (Cerf Publishing House), research director at the Collège International de Philosophie (CIPh), member/ researcher of the Husserl Archives-Paris, member of the French Asian Society, and of the International Society for Chinese Philosophy (ISCP).

*Beatrice Lumpkin*
is Associate Professor at Malcolm X College, Chicago City Colleges, USA. She became a mathematics teacher after years of work as an electronics technician, technical writer and labor organizer. She earned an MS in Mathematics Education at Northeastern University and an MS in Mathematics at Illinois Institute of Technology. As a mathematics professor at Malcolm X College and with high school teaching experience, Lumpkin has consulted with schools and authored books on the multicultural history of mathematics.

*John Marmysz*

teaches Philosophy at the College of Marin, Kentfield, California, USA, and holds a PhD in Philosophy from SUNY Buffalo. His research focuses on the issue of nihilism and its cultural manifestations. Dr Marmysz is the author of Laughing at Nothing: Humor as a Response to Nihilism (SUNY Press, 2003), The Path of Philosophy: Truth, Wonder and Distress (Wadsworth, 2011), The Nihilist: A Philosophical Novel (No Frills Buffalo, 2015), Cinematic Nihilism: Encounters, Confrontations, Overcomings (Edinburgh University Press, 2017) and he is coeditor (with Scott Lukas) of Fear, Cultural Anxiety and Transformation: Horror, Science Fiction and Fantasy Films Remade (Lexington Books, 2009).

*Parthasarathi Mukhopadhyay*

is Associate Professor of Mathematics at Ramakrishna Mission Residential College, India. He is a Gold Medalist MSc in Pure Mathematics, did his MPhil in Topological Algebraic Structures and PhD in Algebraic Theory of Semirings, from the University of Calcutta. He has published several research papers and visited a number of Universities and Institutes in India and abroad delivering over 150 invited lectures on his areas of interest, including the History of Mathematics. He has jointly authored several text books on school Mathematics, on Abstract Algebra, he has edited a book on Linear Algebra and translated The Man Who Knew Infinity in Bengali.

*Sharda S. Nandram*

is Full Professor Hindu Spirituality and Society at the Vrije Universiteit Amsterdam and Full Professor, Business and Spirituality at the Nyenrode Business University in the field of Business and Spirituality. She has earned two Bachelors and two Masters at the University of Amsterdam: one in Psychology and the other in Economics, and was awarded a PhD in Social Sciences from the Vrije University in Amsterdam. She is an adjunct professor at Banasthali University in Jaipur, India. Her fields of interest are: Hinduism and Spirituality, Business and Spirituality, Entrepreneurial behavior, organisational innovation and integrative intelligence. Contact: s.s.nandram@vu.nl and s.nandram@nyenrode.nl.

*Upasana Neogi*

is working as a Senior Software Developer in an IT company and a Guest Lecturer in History of Astronomy in M. P. Birla Institute of Fundamental Research. She has an MSc in Electronic Science from Dinabandhu Andrews College, affiliated to University of Calcutta; worked with Megatherm Electronics Pvt Ltd, Kolkata, as a Research and Development Engineer. She pursued a post-graduation Diploma in Astronomy and Planetarium Science and did a six months project on Radioastronomy at the Giant Metrewave Radio Telescope, Khodad. She also worked in Archeo-astronomy and did a project at Tata Institute of Fundamental Research, and also worked as a Research Assistant at Jawaharlal Nehru University, Delhi.

*Andreas Nieder*
is Director, Institute of Neurobiology, University of Tübingen, Germany. He studied Biology at the Technical University Munich, Germany and, in 1999, he received his PhD from the Rheinisch-Westfälische Technical University in Aachen, Germany. He then moved to the Massachusetts Institute of Technology, Cambridge, USA, to carry out postdoctoral research on the neural basis of numerosity judgments in monkeys. From 2003 to 2008, he worked as a junior research group leader at the Hertie-Institute for Clinical Brain Research & Department of Cognitive Neurology of the University of Tübingen, Germany. Since 2008, he has been Professor of Animal Physiology at the Department of Biology at the University of Tübingen, Germany, where he is also the director of the Institute of Neurobiology. He is interested in how higher brain centers of humans, monkeys and corvids enable intelligent, goal-directed behaviors in general, and numerical competence in particular.

*Jeffrey A. Oaks*
received his PhD in mathematics in 1991 and since 1992 has been a Professor of Mathematics at University of Indianapolis. He began studying Arabic mathematics in 1999 and has focused on the history of algebra from Diophantus to Viète. He is co-author with Mahdi Abdeljaouad of the book Al-Hawārī's Essential Commentary: Arabic Arithmetic in the Fourteenth Century (Max Planck Institute for the History of Science, 2021) and with Jean Christianidis of the book The Arithmetica of Diophantus: A Complete Translation and Commentary (Routledge, 2023).

*Alberto Pelissero*
is Full Professor of Indology at the Department of Humanities at Torino University, Italy, where he is referent for the Indological curriculum of the PhD program in Humanities and teaches Philosophies and Religions of India & Central Asia, and Sanskrit Language and Literature. Most recent publications in English: 'Metaphysical Landscape, Interior Landscape: Two Variants of Mythical-Religious Geography within Ancient Indian Worldview' (Journal of Oriental Research, 2017), 'The Epistemological Model of Vedantic Doxography

according to the Sarvadarśanasaṃgraha for the Study of Indian Philosophy' (Annali di Cà Foscari, 2020).

*Venkata Rayudu Posina*
has a long-standing commitment to the development of the science of consciousness, which propelled him to pursue wide-ranging experimental and theoretical investigations of the brain, mind, and cognition at Harvard Medical School (Boston), National Brain Research Centre (Delhi), Salk Institute for Biological Studies (La Jolla), and National Institute of Advanced Studies (Bengaluru). His research findings were published in Nature, Mind & Matter, Neuron, Neuroscience, Proceedings of the National Academy of Sciences *USA*, and Progress in Brain Research, among other scientific journals. Currently, in joint work with Professor Sisir Roy, he is characterizing the geometry and algebra of consciousness. Using the category theory of mathematical modelling of mathematical objects, he is building a mathematical representation of consciousness.

*T. S. Ravishankar*
is former Director of the Archaeological Survey of India, Mysore. His MA (Sanskrit), MA (Ancient History & Archaeology) and PhD were awarded by the University of Mysore. He joined the Epigraphy Branch of the Archaeological Survey of India where he eventually served as Head of the Branch, meanwhile retired as the Director (Epigraphy). He is the Chairman of Epigraphical Society of India and Vice-President of South Indian Numismatic Society. Dr Ravishankar established his expertise in deciphering Sanskrit Inscriptions, and coordinated and conducted many workshops on Paleography, Epigraphy and Numismatics at Delhi and Mysore. He has also participated in many important archaeological excavations and contributed a number of scholarly articles in English and Kannada languages to many reputed journals.

*Jim Ritter*
is a retired Professor of Mathematics and History of Science at the University of Paris. His research interests lie in the study of rational practices in Ancient Egypt and Mesopotamia and in the history of twentieth-century general relativity and unified theories.

*Brian Rotman*
is currently a Humanities Distinguished Professor in the Department of Comparative Studies at Ohio State University. He has a doctorate in mathematics from the University of London and has published articles in various scholarly

journals such as Semiotica, SubStance, Configurations, differences and Parallax as well as reviews/articles in the Guardian newspaper, London Review of Books, Times Literary and Higher Educational Supplements. His published books include Signifying Nothing: the Semiotics of Zero (Stanford, 1991), Ad Infinitum ... the Ghost in Turing's Machine (Stanford, 1993), Mathematics as Sign: Writing, Imagining, Counting (Stanford, 2000), and Becoming Beside Ourselves: the Alphabet, Ghosts, and Distributed Human Being (Duke, 2008). He is also the author of a number of stage plays and a radio drama.

*Sisir Roy*

is a theoretical physicist, Visiting Professor and Senior Homi Bhabha Fellow at the National Institute of Advanced Studies, IISc Campus, Bangalore. Previously he was Professor, Physics and Applied Mathematics Unit, Indian Statistical Institute, Kolkata (1993–2014). He worked as Distinguished Visiting Professor in many US and European Universities. His main fields of interest include Foundations of Quantum Theory, Theoretical Astrophysics, Brain Function Modeling and higher order cognitive activities. He has published more than 170 papers in various peer-reviewed international journals, twelve monographs and edited volumes. Dr Sisir Roy's recent books include: Decision making and Modeling in Cognitive Science (Springer, 2016).

*René Samson*

started his scientific career during his doctoral thesis work (1969–75) in the Chemical Physics Department of the Weizmann Institute of Science in Rehovot, Israel. The subject of his thesis was a theoretical description of the scattering of light by gases and liquids. During his post-doctorate at the University of Illinois at Urbana-Champaign and MIT (1975–78), his main subject was the theory of molecular transport phenomena (transport of heat and mass) in dispersed media (e.g., the transport of oxygen and heat in a cloud of burning fuel droplets). This was followed by work with Shell Research in Amsterdam for 30 years. He spent much time trying to understand the intricacies of Fluid Catalytic Cracking technology and trying to improve some of the hardware that is used in the process. Since retiring, Samson returned to some of the transport problems that he worked on in his MIT-time and had the good fortune to meet and start collaborating with mathematician Joe Biello of UCal. at Davis. He combined this with a second career as a composer of classical (contemporary) music but that's a whole different story (www.renesamson.nl).

*Avinash Sathaye*

is Emeritus Professor of Mathematics at the University of Kentucky, USA. He was born in India where he met his would-be professor Shreeram Abhyankar.

Under his inspiration, Avinash proceed to do a PhD in Algebra and Algebraic Geometry at Purdue University. He had studied Sanskrit as a child and developed an interest in Classical Indian philosophy as well as Mathematics. He has been working at the University of Kentucky, Lexington, Kentucky, USA for the last 45 years.

*Manil Suri*
is a Professor of Mathematics at the University of Maryland, Baltimore County. His field of research is finite element methods. He has been involved in several outreach projects to raise the cultural profile of mathematics, including co-writing the play, The Mathematics of Being Human, which was staged at different venues in 2014–16. He is also the author of the novels, The Death of Vishnu, The Age of Shiva and The City of Devi, which form a trilogy exploring the present, past and imagined future of India. His fiction has won several awards and has been translated into 27 languages. As a contributing opinion writer for The New York Times, he has written several op-eds on mathematics, India and gay rights. His 2023 book, *The Big Bang of Numbers: How to Build the Universe Using Only Maths*, explains mathematics to non-mathematicians.

*Esther Freinkel Tishman*
has published extensively on Petrarch, the Renaissance poet sometimes thought to have influenced the Tarot. An award-winning teacher, Zen Buddhist minister and certified mindfulness educator, she retired from her position as humanities professor and dean at the University of Oregon to begin work as an interfaith chaplain at Sacred Heart Medical Center in Springfield, Oregon. Dr Tishman's publications include *Reading Shakespeare's Will: The Theology of Figure from Augustine to the Sonnets* (Columbia, 2002), *Mindful Tarot: Bring a Peace-Filled and Compassionate Practice to the 78 Cards* (Llewellyn, 2019), and numerous articles on a broad range of topics from fetishism to usury and early modern encounters with Buddhist Asia.

*Solang Uk*
obtained his Bachelors and Masters degree at Louisiana State University in Baton Rouge in 1966 and 1968, respectively, and his PhD in insecticide toxicology. In 1972, he joined the Ciba-Geigy Agricultural Aviation Research Unit (AARU) based at Cranfield Institute of Technology (CIT), England. He has lectured on the aerial control of the migratory locust in Nairobi, Kenya (1975) and in Karachi, Pakistan (1976), organized by the FAO Locust Control and Emergency Operations, Plant Protection Division, Rome, and attended by crop protection specialists of the ministry of agriculture of Algeria, Egypt, Ethiopia, Jordan, India, Iran, Mali, Sudan, Tunisia, and Turkey. He retired in 1998, having

published 50 scientific articles and four books. He is the translator of Find Zero by the late Professor Amir Aczel.

*Mayank N. Vahia*
retired as Professor at the Tata Institute of Fundamental Research, Mumbai, which he joined in 1979. He spent the first 25 years building space-based telescopes for US, Russian and Indian satellites. In recent years, he has become interested in understanding the origin and growth of astronomy and science in India. He is also interested in issues related to science and society. He has led Indian efforts in the Astronomy Olympiad with consistently good results. He has published 250 research papers and written or edited five books. Contact: vahia@tifr.res.in.

*Marina Ville*
is a French mathematician, working at CNRS and attached to Université Paris-Est Créteil. Before that she belonged to other French universities, among them the University of Tours. She also spent several years as a visitor at the Hebrew University of Jerusalem and several years in Boston, where she taught at Boston University and Northeastern University. Dr Ville's work is in geometry and topology, her current interests being minimal surfaces and knot theory. She enjoys explaining mathematics to non-specialists, mainly through history. Her PhD is from Université de Paris and her habilitation à diriger des recherches from Paris Saclay. Contact: villemarina@yahoo.fr.

*Célestin Xiaohan Zhou*
is an Assistant Researcher at the Institute for the History of Natural Sciences, Chinese Academy of Sciences, Beijing, China. He was awarded his PhD from Laboratoire SPHERE (Sciences, Philosophy, History, UMR 7219), University Paris Diderot, France in 2018. At present, Dr Zhou works mainly on the continuities between mathematical writings from the thirteenth century to the sixteenth-century in ancient China and he is also interested in the historiography of sciences since the nineteenth century in terms of mathematics in ancient China.

# Introduction

*Peter Gobets*

To date, the inexplicable emergence of the number zero in the historical record some 2,000 years ago has never been satisfactorily addressed. There has been speculation, but no in-depth, systematic and above all interdisciplinary research program focusing specifically on this key issue.

## Partners and Projects

April 2020 marked the start of a lively discussion between the Zero Project and Dr Robert Lawrence Kuhn, creator – host of the popular PBS TV series *Closer to Truth*, on the shared quest to address what Robert refers to as Ultimate Questions – first and foremost, 'Why is there Something rather than Nothing?'

Eventually, this mutual interest led to close collaboration between the Zero Project and *Closer to Truth*. This culminated in a promotional video (30K+ views) on the *Monograph on Zero* book project before you and an associated online *International Conference/Workshop on Zero* (October 2021 to April 2022) – links listed *vide infra*.

The online event, co-organized by the Amir D. Aczel Foundation, was awarded Patronage by UNESCO, including use of the official UNESCO logo.[1] The academic papers from that conference are included in this volume as part of the *Retrospective* and links to the online presentations by contributors are also provided.

We were particularly fortunate that the book project also attracted the attention of world-class sculptor, Anish Kapoor, who provided selected photographs of his works inspired by the concept of Nothingness and Void.

Additionally, *Closer to Truth* plans to produce a series of videos based on interviews with select contributors to the monograph, to be posted on their YouTube channel. The channel enjoys over half a million subscribers and tens of millions of minutes viewed annually, thus ensuring greater professional and

1 'After consulting my science colleague I am happy to say that we think that the Zero Project concerns a rich discussion, both mathematical as well as philosophical, with both scientific and heritage components.' Marielies E. W. Schelhaas, Secretary-General, Netherlands Commission for UNESCO, translated from email correspondence in Dutch, January 27, 2021.

 | DOI:10.1163/9789004691568_002

public awareness of the issue of zero's central role in society generally and the sciences in particular.

Agreeing to much more than close collaboration, Robert Kuhn accepted our invitation to be the book's co-cover editor, while his long-time collaborator, co-producer and award-winning documentary maker, Peter Getzels of Getzels Gordon Productions, will produce a popular commercial edition on the subject of zero's emergence from an anthropological and artistic perspective.

The Zero Project's objective to shed light on the origin of the mathematical number zero and its signification of the concept of 'Nothing' (no-thing) appears to be an excellent way to explore further the Ultimate Question of why there is Something rather than Nothing; a world, a universe, rather than nothing for eternity and no one to notice.

## The Invention/Discovery of Zero (IDZ)

Given the ubiquitous praise heaped upon the *invention/discovery of zero* (hereafter IDZ), it is remarkable, to say the least, that not a single academic or group has ever even proposed that a systematic, interdisciplinary research project should be conducted to find fresh evidence of zero's origin.

All authors without exception rehash known data on the subject or speculate on new serendipity finds that happen to surface. Proactive targeted research as such is never initiated. Most books and articles on the history (or philosophy) of mathematics make an obligatory reference to zero but tend to gloss over the IDZ issue, often without further reference to zero in their work.

Very likely the interdisciplinary nature of such an academic exercise as proposed by the Zero Project and the associated research project present a serious challenge to academics since the prevailing academic culture is one of specialization, while the nature of the question of IDZ requires an interdisciplinary team of experts, apparently not easily assembled.

## The Zero Project's Objective

The independent, unaffiliated Zero Project Foundation based in the Netherlands was established in 2015 for the very purpose of finding fresh evidence of zero's invention/discovery and has meanwhile organized two Camp Zero workshops by way of a feasibility study – as well as launch the aforementioned book project and associated online event.

Several hundred scholars, on whose expertise the Zero Project as organizers rely, were invited to participate – of whom some 40 signed on. Together they represent the main ancient civilizations that played a role in the invention/discovery and/or transmission of zero: Mesopotamia, Egypt, China, India, the Islamic world and Europe. The reader will find brief bio-data and synopses of these scholars included.

Not for wont of trying, the Zero Project was unable to involve a specialist in Hellenist mathematics/astronomy. This is a hiatus given claims (and counterclaims) that zero was invented by Hellenist mathematicians in Alexandria in the early centuries of the Common Era,[2] as well as a tantalizing reference by the contemporaneous third-century CE Syrian Neoplatonic philosopher – mathematician Iamblichus, who alludes to zero as a number.[3] It is a subject that we expect to revisit in due course.

The Mayans of Central America, known to have innovated their ingenuous base-20, vigesimal numeration system, including zero, are also conspicuous by their absence as no scholar could be identified and invited to contribute a chapter.

## Contents

The reader will find the following main themes covered in the book, with individual chapters listed in the Contents.

- Part 0: Zero in Historical Perspective
- Part 1: Zero in Religious, Philosophical and Linguistic Perspective
- Part 2: Zero in the Arts
- Part 3: Zero in Mathematics and Science

The summaries at the beginning of each Part will serve as a 'road map' in guiding the reader's interest. The chapters are written by specialists, who provide varying levels of detail and sophistication, and provisional context. The authors also present and discuss their material online, links for which can be found in Appendix 2.

We stress at the outset that the Monograph on Zero book project is not intended as exhaustive treatment of the history of mathematics as a whole, but focuses solely on aspects relevant to zero. Inevitably, however, some authors

2 Email correspondence in 2018 with Jan Hogendijk (on al-Khwārizmī, Ptolemy); Alexander Jones (on Oxyrhynchus) and Nathan Sidoli (on Hypsicles) in 2021.

3 See Chapter 24, Kaspars Klavins, 'Challenges In Interpreting The Invention Of Zero'.

range further afield in their chapters precisely because little is known about the emergence of zero. The dearth of tangible evidence is the main challenge throughout and it is this that the Zero Project aims to address.

Furthermore, as you will see, authors present their own perspective on zero and its underlying concept in the related interdisciplinary fields of philosophy, linguistics and the arts. We applaud new thinking, such as cross-referencing the various disciplines to ascertain connectivity among them.[4]

Plotting geographically early appearances of zero is another strategy being explored in the hope and expectation to home into local epicenters of zero's origin as well as to shed light on possible transmission/diffusion channels among cultures.

Neither mathematics – nor any other cultural discipline for that matter – ever evolved in a vacuum. Robert Kuhn's own project that led him to interview leading thinkers on these and related interdisciplinary fields of expertise in the context of his quest to answer his Ultimate Questions thus enriches the Zero Project considerably.

## Context

In the words of the late Frits Staal, Professor of Philosophy and South Asian Languages at the University of California, Berkeley, himself trained in mathematics and physics, referring to our *modern* zero in these terms:

> It is widely believed that zero originated in Indic Civilization but the evidence in support of that belief is only meager; it is almost zero. No place or time, let alone the name of a discoverer or inventor, has ever been suggested. How can we handle such a problem? We must start from the beginning ... Fortunately, it is not the end of the story since those who look for origins of something, even if it is zero, must look beyond the domain where it is customarily located, even if it is absent.[5]

Consider also the reference by Søren Brier, Department of Management, Society and Communication, Copenhagen Business School, Denmark, quoting none other than C. S. Peirce:

4 See Chapter 2, Mayank Vahia, 'Connecting Zeros'.

5 F. Staal, "On the Origins of Zero," in: C. S. Seshadri (eds.), *Studies in the History of Indian Mathematics* (Gurgaon: Hindustan Book Agency, 2010).

> If we are to proceed in a logical and scientific manner, we must, in order to account for the whole universe, suppose an initial condition in which the whole universe was non-existent, and therefore a state of absolute nothing … We start, then, with nothing, pure zero. But this is not the nothing of negation. For not means other than, and other is merely a synonym of the ordinal numeral second. As such it implies a first; while the present pure zero is prior to every first. The nothing of negation is the nothing of death, which comes second to, or after, everything. But this pure zero is the nothing of not having been born. There is no individual thing, no compulsion outward nor inward, no law. It is the germinal nothing, in which the whole universe is involved or foreshadowed. As such, it is absolutely undefined and unlimited possibility – boundless possibility. There is no compulsion and no law. It is boundless freedom.[6]

Or Robert Kaplan's surmise:

> If you look at zero you see nothing; but look through it and you will see the world. For zero brings into focus the great, organic sprawl of mathematics, and mathematics in turn the complex nature of things. From counting to calculating, from estimating the odds to knowing exactly when the tides in our affairs will crest, the shining tools of mathematics let us follow the tacking course everything takes through everything else – and all of their parts swing on the smallest of pivots, zero.[7]

## Working Hypothesis

Given that a strictly historical account of zero's emergence within the field of mathematics itself has not led to a satisfactory explanation, we look beyond mathematics to the wider cultural context – in particular, religion, philosophy and linguistics – to explore whether such a project may produce plausible results.

In doing so, we hit upon what may well constitute a fundamental lacuna in Western philosophy within the field of ontology. The strength of the central

6 Quoted in S. Brier, "32: Pure Zero," in Torkild Thellefsen and Bent Sorensen (eds.), *Charles Sanders Peirce in His Own Words: 100 Years of Semiotics, Communication and Cognition* (Berlin, Boston: De Gruyter Mouton, 2014), p. 207.

7 R. Kaplan, *The Nothing That Is: A natural History of Zero* (Oxford: Oxford University Press, 2010).

argument advanced here – yet to be worked out in detail – is based on the unprecedented semiotic power exhibited by the mathematical zero as conferred upon it by the underlying ontology of Nothingness – or, if you will, to coin the neologist term, 'nonism' – as suggested by the our study of the IDZ.[8]

If zero proved to be so undeniably useful in mathematics and the sciences, then it warrants taking its philosophical and linguistic, semiotic, context seriously. That is to say, zero's unprecedented significance as *closure* of the numeration system concerned.

And since one's language and concepts strongly influence or even determine thought, in accordance with Sapir-Whorf, nonism is required – rather than monism and/or dualism that still hold sway in our time.[9] Nonism, the missing ontology in the West, which had only functionally adopted the numeral zero severed from its philosophical moorings.

Nonism, signifying the concept of the world, the universe, not made of *any* essence, substance or principle – which are still the *raison d'être* of mainstay philosophy. That innate bias may in part account for zero's late emergence in the historical record.

## In-Depth, Systematic, Interdisciplinary Approach

The *Expertise Center on Zero*, which was launched in 2022, is set to delve deeper into the crux of this typically interdisciplinary issue linking philosophy, linguistics and mathematics at a foundational level. Zero, and all that it implies, has remained a virtual enigma to this very day, while zero-based mathematics made progress by leaps and bounds, no longer hobbled – at least in practice – by the limited ontologies of pluralism, dualism and monism.

As mentioned, the stated objective of the Zero Project is to find fresh evidence of the IDZ. The book project and online event are two of the stepping stones to conducting the long overdue, in-depth academic research project required. It was to this end that the Zero Project launched the *Expertise Center on Zero* in 2022, with nine universities across Europe, Asia and North Africa endorsing it and several more expected to follow suit.[10] The signatories will

8 'Nonism', the missing ontology, in contrast to Monism, Dualism, Pluralism, refers to what does not exist, no essence or substance.

9 B. A. S. Hussein, "The Sapir-Whorf Hypothesis Today," *Theory and Practice in Language Studies*, Vol. 2, No. 3, (March 2012): 642–646.

10 Model Collaboration Agreement appended. Universities and Research Institutes as well as individual academics in their personal capacity are invited to join the XCZ. Contact us at: www.TheZeroProject.nl.

formulate targeted PhD programs in collaboration with the Zero Project to attain common objectives.

### Anachronistic Projection?

The collection of some forty monograph chapters by reputed academics from practically all continents and fields of expertise, rather than proposing definitive answers, raise a number of questions related to what may, in the first place, be our *tacit assumptions about the origin of the number zero as viewed from our modern perspective*.

In other words, anachronistic projection of modern concepts upon ancient traditions that may have confounded the matter of zero's origin in the literature for generations on end requires urgent review. The reader may like to bear this in mind while going through the chapters to examine the implicit or explicit assumptions regarding zero's status.

The survey that the Zero Project ran among contributors following their presentation at the online event showed opinions roughly equally divided between the two options, invented or discovered? At the same time, most agreed that the underlying concept of Nothingness is profound rather than trivial.

As Georges Ifrah argues:

> The zero originally meant 'void', an empty column on the counting board. When and how did it become enriched by acquiring the meaning of 'nothing', as in '10 minus 10'? That question is one of the most interesting in the history of science, but unfortunately we cannot answer it in the present state of our knowledge.[11]

Was zero as *number* indeed invented/discovered, or could it be the case that what today we consider to be the number zero slowly evolved over the centuries from the *placeholder* zero (empty or sign)? It looks increasingly likely that the slow accretion of interrelations between the placeholder and the other numbers in the notational system concerned (be these decimal, vigesimal or sexagesimal) converged into what we today see as zero's dual role – both mathematically as well as algebraically. Is the placeholder zero 'just' a special case of the number zero (or vice versa)? Grist to the mill of the *Expertise Center on Zero*.

---

11 Georges Ifrah, *From One to Zero: A Universal History of Number* (New York: Viking, 1985), p. 459.

The issue of zero's emergence in human history may perhaps be better considered in terms of the respective contributions made by all ancient traditions that eventually culminated in zero's pivotal role in modern axiomatic mathematics as we know it today, say roughly from the seventeenth century onwards. As such it covers a relatively brief span of time considering the evolution from simple counting (one-to-one correspondence with objects) going by the earliest evidence of tally marks on stone, bone, wood and other substrates tens of thousands of years ago, to eventual cipherization some five thousand years ago and beyond.

What is uncontested in the literature is that our *modern* zero may be traced to seventh-century India from where it spread across the globe via the Islamic world and Europe. The fact that the *concept of zero* integral to the Hindu – Arabic numeral system was eventually adopted by practically all countries and peoples worldwide is testimony to its efficiency and ease of handling.

However, the other positional notation systems with the zero concept in human history (Mesopotamian, Chinese, Mayan) are equally evidence of the *invention/discovery of zero* as such, even though over time these would be displaced (barring a few exceptions such as some sexagesimal units, minutes/ astronomical degrees and nostalgic use of serial Roman numbers).

But if our avowed aim is to find fresh evidence of zero's invention, these cultures are no less rich traditions as they potentially offer us an opportunity to learn the cultural context that gave rise to what has been hailed as among the greatest inventions/discoveries in human histories.

## Methodology

In this pursuit the methodology to which the Zero Project subscribes is that of 'connected histories' and 'circulation of knowledge', where both independent co-invention and/or transmission may have played a role in zero's genesis and, as such, must be further explored for possible leads under the *Expertise Center on Zero*.[12]

As an aside, an interesting – and possibly important – observation may be that the inescapable empty space between even tally marks, and later written numbers (or for that matter the silence between word numerals) came to be recognized as itself integral to numbers (read: as *placeholder in positional*

12 See Chapter 1, Bhaswati Bhattacharya, 'Viewing The Zero As A Part Of Cross-Cultural Intellectual Heritage'.

*notation*).[13] It is uncontested that this breakthrough was first achieved in Mesopotamian mathematics in the late centuries of the third millennium BCE).[14] Roughly contemporaneous ancient Egyptian mathematics was, and remained, non-positional by comparison.[15] Chinese and Indian mathematics followed suit centuries later in variations on the theme.[16]

It was the Zero Project's intention that there should be a 'joint chapter' to cap the monograph, written by a subgroup of the scholars involved, representing all the major cultures that played a role in the emergence (or transmission) of zero. This has been postponed for the time being as they were unable to agree on a single, unique, distinguishing feature of the *number* zero versus the *placeholder* zero accounting for a skeptical rather than naïve approach to IDZ. This and related questions will be the subject of future research under the aegis of the *Expertise Center on Zero*. It is one of several stratagems to home in on zero's epicenter.

The IDZ may even serve as a 'marker' in human evolution, when Nothing became Something within the field of mathematics – as argued from the standpoint of the cognitive sciences presented in the Monograph.[17]

It also accounts for Robert Kuhn's abiding interest in the context of *Closer to Truth*'s many thousands of interviews over the past twenty-five years across a wide-range of interdisciplinary academic disciplines. Furthermore, it suggests the corollary to the Zero Project's working hypothesis: namely that the IDZ required an underlying philosophical concept of Nothingness to be facilitated in the context of mathematics.

## Corollary

The corollary to this working hypothesis referred to above holds that via a 'reverse semiotic mechanism of transmission' – that is, the 'sign' (0) preceding the underlying concept lacking in recipient cultures (in other words, a sophisticated concept of Nothingness) – zero as number began to dawn in those

13 G. G. Joseph, *The Crest of the Peacock*: *Non-European Roots of Mathematics,* 3rd ed. (London: Tauris, 2011), note 11, p. 243.

14 See Chapter 3, Jim Ritter, 'Babylonian Zeros'.

15 See Chapter 4, Friedhelm Hoffmann, 'Aspects of Zero In Ancient Egypt' and Chapter 5, Beatrice Lumpkin, 'The Number Zero In Ancient Egypt'.

16 See Chapter 6, Célestin Xiaohan Zhou, 'On The Placeholder In Numeration And The Numeral Zero In China' and Chapter 8, Parthasarathi Mukhopadhyay, 'Genesis Of Zero: An Eternal Enigma.

17 See Chapter 33, Andreas Nieder, 'The Unique Significance Of Zero: A Sense Of "Nothing"'.

recipient cultures through familiarity with, and practical application of, the sign zero (0).

The corollary first suggested itself in the earlier work of one of the contributors to the Monograph.[18] The book thus includes novel documented examples of this surprising phenomenon and these may also shed light on related issues of transmission to account for what otherwise may be interpreted as independent co-invention or simply uncanny coincidence of the invention/discovery of zero.

Supporting evidence may be attested by the fact that, for centuries on end, recipient cultures of zero in the Islamic world (late eighth-century CE),[19] Jewish world (twelfth-century)[20] and Europe (early thirteenth-century CE),[21] which had not themselves invented/discovered zero, *explicitly denied* that the placeholder was a number.

Esther Freinkel Tishman provides us a rare glimpse into premodern Europe as zero surfaced in popular culture alongside the Roman numerals in the context of 'divination' – the immaterial aspect of zero deliberately overlooked in our scientific age.[22]

Thus Marcis Auzinsh points out the likely knock-on effect of the significance of zero's underlying philosophy on cutting-edge physics.[23]

The Iamblichus passage referred to in a footnote above suggests a 'bridge' that zero crossed between philosophy, linguistics and mathematics – and, as such, it may tentatively be adduced as supporting evidence of the Zero Project's working hypothesis to be pursued in future research. The concept of Nothingness itself may be necessary but not sufficient to lead to the IDZ.[24]

The jury is still out on that and related issues, hopefully to be taken up by the *Expertise Center on Zero* in due course.

---

18 See Chapter 16, Brian Rotman, 'On The Semiotics Of Zero'.

19 See Chapter 13, Jeff Oaks, 'The Medieval Arabic Zero'. An expanded version of the chapter, acknowledging the original chapter in the Monograph, has meanwhile been published by the *British Journal for the History of Mathematics*: J. Oaks, "Zero and Nothing in Medieval Arabic Arithmetic," *British Journal for the History of Mathematics* (2022), DOI: 10.1080/26375451.2022.2115745.

20 See Chapter 29, Esti Eisenmann, 'The Concept of Naught in Jewish Tradition'.

21 Fibonacci, *Liber Abaci*.

22 See Chapter 15, Esther Freinkel Tishman, 'The Zero Triumphant: Allegory, Emptiness And The Early History Of The Tarot'.

23 See Chapter 37, Marcis Auzinsh, 'Nothing, Zeno Paradoxes And Quantum Physics.

24 See Chapter 18, Alexis Lavis, 'The Influence of Buddhism on the Invention and Development of Zero' and Chapter 19, Fabio Gironi, 'Zero And Śūnyatā: Likely Bedfellows', among others.

### Theory and Practice

Even so, denials notwithstanding, all these ancient cultures managed implicitly or explicitly to apply the *placeholder concept* properly in their calculations either by algorithm or with the aid of some sort of device. Put another way: *The issue of the distinction between the placeholder and number zero appears to be a semantic rather than a syntactic one*. It is not what mathematicians say about zero in theory that matters, but how they use it in practice.

Several contributors to the Monograph contest the Zero Project's working hypothesis and corollary mentioned above, asserting rather that only a written notational system is required, whereafter zero appears naturally. The Zero Project was pleased to include dissenting views and suspends judgement in the absence of incontrovertible evidence, given the significance of zero's invention/discovery in human history.[25]

### Conclusion

In sum, the issue of zero's origin and its significance has never been resolved satisfactorily. The Zero Project hopes to contribute in small measure to future targeted research on the invention/discovery of zero that may conceivably take many years, if not generations, to achieve.

On behalf of the Board of Trustees:



Peter Gobets
Secretary/Research Coordinator

The Zero Project Foundation
The Hague, the Netherlands
February 2024

---

25 See Chapter 26, Johannes Bronkhorst, 'Much Ado About Nothing, Or How Much Philosophy Is Required To Invent The Number Zero?' and Chapter 27, Erik Hoogcarspel, 'From Emptiness To Nonsense: The Constitution Of The Number Zero, For Non-Mathematicians'.

# PART 0

## *Zero in Historical Perspective*

# Introduction to Part 0

The exploration of zero from an interdisciplinary standpoint begins with a historical framework, delving into the origins and evolution of the numeral zero. In 'Viewing The Zero As A Part Of Cross-Cultural Intellectual Heritage', BHASWATI BHATTACHARYA argues that despite many single-nation claimants to the invention of zero, research along the lines of 'connected history' or 'circulation of knowledge' yields a different, border-crossing biography of zero. Cultures react very differently to the concept of zero and, in 'Connecting Zeros', MAYANK N. VAHIA illustrates this by considering the example of Indo-European cultures and their way of life in their respective natural environments post-separation. The wider context of the concept of zero and the relation between its different forms of expression are explored.

JIM RITTER's chapter, 'Babylonian Zeros', explores the origins and development of two aspects of zero in Ancient Mesopotamian number writing: as a placeholder in positional notation and as a 'nothing' in numerical operations, showing how the material and bureaucratic constraints in its scribal practice shaped their development. The following two contributions – 'Aspects of Zero in Ancient Egypt' by FRIEDHELM HOFFMANN and 'The Zero Concept In Ancient Egypt' by BEATRICE LUMPKIN – look at zero in Ancient Egypt. Hoffmann examines the concept of zero from the perspective of 3,000 years of Egyptian cultural history – reviewing language and number writing, and a critical assessment of the word *nfr*. Expressions for 'nothingness', missing objects, the absence of entries in bookkeeping and of numbers and placeholder signs and later on the importance of astronomical texts in the Mesopotamian sexagesimal notations before and after the Hellenist period are also explored. Lumpkin evaluates the use of an additive numeral system in Ancient Egypt, without positional value and no need for placeholders, is compared with ancient cultures of pre-Columbian Peru and Central America. An analysis of the uses of zero in Ancient Egypt, and the importance of non-Western contributions to classical Greek mathematics.

An analysis of zero in China and in India follows. 'On The Placeholder In Numeration And The Numeral Zero In China' by CÉLESTIN XIAOHAN ZHOU analyzes the existence of a decimal system of numeration in ancient China, evidenced in oracle bone scripts, coinage and counting rods. Precise vacant positions and their signs in numbers were emphasized and said to have been strictly respected by the ancient practitioners, deemed to have understood zero as number. In 'Reflections On Early Dated Inscriptions From South India', T. S. RAVISHANKAR re-examines early-dated inscriptions from recent archaeological excavations and from epigraphical explorations, connecting them with

 | DOI:10.1163/9789004691568_003

existing hypotheses and views, seeking to confirm the presence and usage of decimal writing and the use of zero as a placeholder in South India. Embracing the paradigms of claims and counterclaims about the origin of zero, 'From Śūnya To Zero: An Enigmatic Odyssey' by PARTHASARATHI MUKHOPADHYAY, provides a nutshell version of the unparalleled journey of zero, from a concept to a number in its own right, and perhaps the most significant creation of the human mind ever. In 'The Significance Of Zero In Jaina Mathematics', ANUPAM JAIN argues that references to zero in ancient mathematical texts are explored and analyzed, revealing the use of zero and the decimal place value system in the Jaina tradition during the early centuries of the Common Era. JONATHAN J. CRABTREE argues that Brahmagupta's definition of zero and India's symmetric and scientific zero failed to migrate West which led to the emergence of a misunderstood mathematical concept of zero in 'Notes On The Origin Of The First Definition Of Zero Consistent With Basic Physical Laws'.

In a revised and updated version of an opinion piece from the *New York Times* – 'Putting A Price On Zero' – MANIL SURI examines a survey among university students in the context of hypothetical royalty allocation for the invention of zero, revealing how the provenance of zero is laden with deeply held feelings that provoke strong reactions. In 'Revisiting Khmer Stele K-127', DEBRA G. ACZEL, SOLANG UK, HAB TOUCH AND MIRIAM R. ACZEL analyze and interpret the inscription on a Khmer stone stele (K-127), which was discovered in 1891, rediscovered in 2013 and shows the date including zero in numeral form, taken to be the world's oldest tangible evidence of zero.

JEFFREY A. OAKS investigates how authors of Medieval Arabic books on arithmetic conceived of nothing and zero, in the context of their understanding of number as an amount or measure of some material species in 'The Medieval Arabic Zero'. He argues that all the authors reviewed deny explicitly that zero is a number – a finding unique in the literature on the subject that came to light unexpectedly during the course of the Zero Project. In 'Numeration In The Scientific Manuscripts Of The Maghreb', DJAMIL AÏSSANI analyzes the beginning of Islamic mathematics, emphasizing the influence of Indian arithmetic, the specificity of the digits and the symbolism by the Muslim West, and the role of the Algerian city of Bejaia in popularizing Arabic numerals in Europe via Fibonacci's *Liber Abaci*.

ESTHER FREINKEL TISHMAN analyzes zero from the perspective of the fifteenth-century Italian game of Tarot, or '*trionfi*' in her chapter, 'The Zero Triumphant: Allegory, Emptiness And The Early History Of The Tarot', where the Fool card – numbered 0 – systematically subverts the logic of play and anticipates the vision of Folly in later philosophy and literature in premodern Europe.

CHAPTER 1

# Viewing the Zero as a Part of Cross-cultural Intellectual Heritage

*Bhaswati Bhattacharya*

**Abstract**

Currently, there are different claimants for the honor of inventing the zero as a digit, a part of the decimal system. These claims are made on the basis of evidence found in one particular country. None of the claims have, however, been able to convincingly rule out the rival claims. This chapter suggests that research along the terms of cross-cultural history and or connected history linking the threads of information coming from different civilizations regarding the knowledge of the zero may yield a different, border-crossing biography of the zero.

**Keywords**

ancient history – border-crossing – circulation – connected history – cross-cultural connections – the zero

## 1 Introduction

The historiography of the knowledge of the zero, one of the most important findings in the history of science, and a digit as essential in simple arithmetical calculations as in scientific computations, is often reconstructed in the context of national history. Currently, scholars coming from different nation states claim that the relevant knowledge was first developed as part of the ancient script/knowledge in their respective country before it was gradually transmitted to other parts of the world.[1]

Thus, it has been suggested that the Chinese performed decimal calculations with the help of the Chinese counting rod system as early as the fourth

1 Here I am not concerned with the question if the respective scholarly claim with regard to zero as the placeholder, or zero as a digit on its own merit.

 | DOI:10.1163/9789004691568_004

century BCE. The rods 'gave the decimal representation of a number, with an empty space denoting zero' (Hodgkin, 2005). Lam Lay Yong has also argued that the Hindu-Arabic numerals including the zero originated in the Chinese counting rod numerals. Indians, in her view, 'transcribed' the rod numeral system to a written version, for which they used the Brahmi script (Yong, 2004). According to some, the symbol o used in the astronomical data in ancient Greece alludes to its Greek pre-history (O'Connor and Robertson, 2000).

Other scholars have traced the zero east of India on a stone Khmer inscription in Cambodia (Aczel, 2015). Although it is not known who invented the zero, the position paper written for the Camp Zero-2 workshop, akin to the title of the Zero Project, underlines the widespread belief that the zero had its beginning in India. Within India again, there is an old claim that the concept of the zero was part of the ancient classical Indian astronomical and mathematical knowledge system written in Sanskrit.

All these studies deal with the subject of the knowledge of the zero as developed within one culture in isolation from the rest of the world. In addition, supposedly there was a hierarchy in the knowledge system in question: it was developed in one culture – naturally superior to the rest – from where the knowledge was transmitted to other civilizations that were passive recipients of that knowledge. While this point of view boosts cultural chauvinism, so far no scholar has been able to present any conclusive proof that confirms the claim of a specific nation as the finder of the zero as irrefutable in the face of similar counterclaims (Suri, 2017). The question of cultural nationalism apart, it also evokes a skepticism in the academic community at large about such cultural nationalist claims, usually seen as part of a political agenda.

## 2 Cross-cultural Legacy and Connected Histories

The purpose of this chapter is to question the prevalent presupposition and underline that it is possible to think of the invention of the zero as a cross-cultural legacy in view of the concept of 'the circulation of knowledge' or even as a part of the process of 'connected histories'. The revolutionization of communications technology emerging in the late twentieth century makes it difficult to comprehend how information was transmitted across spatially dispersed communities in ancient times, and how ancient communities exchanged information and knowledge about the world they experienced.

It is absolutely necessary to keep in mind here that while nation states and national boundaries are of recent origin, civilizations and cultures are known to have been connected to the outside world and engaged with one another since the earliest times for trade, religion, and other purposes (Pierre Chaunu, 1969). *The Periplus of the Erythraean Sea* illustrates the navigation and trade from Egyptian ports to the ports of the Red Sea, along the coasts of Northeast Africa, Sind and Western India (Schoff, 1912). Indeed, as the seals found in Sumer testify, commercial and cultural exchanges between Babylon and India – two ancient civilizations and major contenders for the claim of having invented the zero – predated *The Periplus* (Parpola, 2015).[2]

One of the most important sources of knowledge of ancient Central Asia and India is the account left by the Buddhist monk Hiuen Tsang (Xuanzang), who went on a pilgrimage to India via the Silk Road in the seventh century CE. He spent a considerable time in India, traveling extensively, studying at Buddhist monasteries, and collecting manuscripts. While his collection, meticulously copied and translated after he went back, made preservation of lost Indian Buddhist texts from translated Chinese copies possible, the detailed accounts of political and social aspects of places visited by him enabled reconstructing the little-known history of regions such as Bengal (Li, 1995).

As far as the knowledge of the zero is concerned, Amartya Sen has drawn our attention to the evidence substantiating the existing claim that the symbol for the zero as it is currently used was taken from Babylon via the Greeks to India, although the concept of *śūnya* existed in India prior to that date. The idea of the import of the concept of the zero from Babylon is not oblivious to the contribution of India to the process of the evolution of the zero. For, Sen adds: 'It is, however, also clear that the combination of zero with the decimal system was a particularly fruitful consolidation, and it is in exploring the nature and implications of this integration – critically important for the use of a decimal system – that Indian mathematicians seem to have played quite a decisive role in the early and middle parts of the first millennium' (Sen, 2005). Thus, conceding the concept that the zero was not invented in India does not does not take away the glory of Indian mathematicians. For, they further explored the possibilities of the digit by integrating it into the decimal system in a form globally current now.

Cultural, ideological, and material exchange and encounters across ethnic, linguistic, and territorial boundaries, crucial to the formation of knowledge, are as old as human history. Ideas circulate through active human agents and are often reinterpreted in a different context. Recent research has shown how astronomical and astrological knowledge circulated among different communities of scholars in multiple ways over time and space, and the multifarious

2 I am grateful to Dr Ophira Gamliel, Glasgow University for drawing my attention to this publication.

ways any received knowledge was continuously questioned, absorbed, or adapted to suit existing sets of local knowledge and practices at the point of receipt. Offering examples of such crisscross traffic in knowledge from Mesopotamia, Egypt, the Greco-Roman world, India, and China, John Steele argues that knowledge in the field circulated in numerous ways: in the form of written texts, through oral communication, 'the conscious assimilation of sought-after knowledge and the unconscious absorption of ideas to which scholars were exposed'. Circulation took place among scholars of two or more spatially and temporally segregated and distinct cultures, such as a late adaptation in reworking of an earlier text (Steele, 2016).

Circulation of knowledge thus presumes that the development of knowledge may have resulted from the circulation and exchange of ideas and information – and not a monocausal and mono-directional transmission – in which there is no one-way import of knowledge in its fully consolidated form from one culture to another, but both, or more parties contributing to it and adapting/transmitting the information further as part of intellectual exercise and exchange.

Naturally, the received knowledge is, as mentioned above, often adapted in a way to suit the local context (Ragep, Ragep and Livesey, 1996). The knowledge and practices of Buddhism are an excellent example, showcasing the extent of circulation from their original homeland to other distant parts of the world, where, in combination with local beliefs and practices, they gave birth to distinct forms of art (Gandhara or Indonesian), lifestyle (non-vegetarian food), and philosophy (Zen). At a time when Buddhism is almost non-existent in India, it flourishes in Southeast and East Asia. It has even been suggested, and there seems to be a consensus about the fact that the short version of the Mahayana text of the Heart Sutra, one of the the most famous of Mahayana scriptures, is a Chinese product, brought back to India to be expanded into the larger version of the Heart Sutra (Nattier, 1992).[3] Although originating in India, the rituals of everyday Hinduism practiced in Bali too are different from the multiple regional varieties come across in India.

The concept of connected histories is based on the hypothesis that there are overlaps between civilizations. Eurasia being a landmass as seen above, there is a long history of connections at different levels among the countries in this vast area. Human movement, cultural and economic exchanges between the Indian subcontinent and the Mediterranean world through Central and West Asia, for example, date at least to the Persian Achaemenid Empire, predating

3 I owe this reference to Dr Victor van Bijlert of the Department of Religious Studies, The Free University of Amsterdam.

the Baghdad Caliphate by at least a millennium. It is thus essential to overlook territorial boundaries delineated by empires and states to focus on connections, movements, exchanges, influences, relations, and continuities over a long period of time to understand the processes at work. To quote Sanjay Subrahmanyam, a distinguished scholar of early modern Eurasia, 'connected histories' is an approach in which the historian acts to some extent like an electrician, repairing the continental and intercontinental connections made oblivious by national historiographies in an attempt to render their borders impermeable. Connected histories may shed light on modes of interaction between the local and regional level, and, at the same at a supra-regional level that can also imply the global (the micro and macro processes, respectively) (Subrahmanyam, 2005, p. 299). Subrahmanyam's study based on Portuguese, Persian, Turkish, and South Asian sources points to a 'millenarian conjuncture [that] … operated over a good part of the Old World in the sixteenth century'. Not only commodities, currencies, gems, flowed between Europe and the Indian Ocean; information on aspects of society including fears about portents, omens and signs also crisscrossed as part of the circulation of human agents in this period. For example, groups of Portuguese mariners and South Asian merchants and courtiers interacting with each other shared a common set of apocalyptic beliefs and concerns. Considered from this point of view, the concept of the zero may have emerged in different contemporary ancient cultures that may or may not have been directly or indirectly linked with each other but formed part of intellectual exchange at the macro level.

Focus on the life stories of individual 'border-crossers' in the early modern and modern periods of history underscores a cosmopolitanism inculcated by strangers – persons who crossed territorial, religious, and linguistic boundaries in sync with the exploratory and scientific spirit of their time, adapting or self-fashioning their identities as part of the complicated reality they lived in (Subrahmanyam, 2011; Alavi, 2015). Is it possible to think of the concept of the zero in the terms of a 'border-crossing' intellectual concept in the course of its life from that of a space-holder to the one of a digit in its own right? The trans-civilizational life history of the zero has received some attention in the works of scholars of the zero, such as Robert Kaplan, who acknowledged the contribution of diverse cultures from the Mayans to the Romans for paving the way for the evolution of the zero from a placeholder to an indispensable tool of modern mathematics. In his opinion, the zero appeared for the first time on the cuneiform tablets of Mesopotamia in a place value system between 700 and 300 BCE to travel to Greece before it landed in India where it was transformed into the mathematical zero we are familiar with in modern times (Kaplan, 2000).

## 3 New Vistas

In lieu of a conclusion, I put forward the idea that exploring cross-cultural interconnections at hitherto unknown levels may help a better understanding of the development of the phenomenon called the zero. By tracing the conduits, trajectories facilitating the circulation of cultural ideas and connected developments in different cultural areas of the ancient world that possibly led to the integration and consolidation of the knowledge of the zero, it is conceivable to offer fresh perspectives for research on the question relating to the zero.

## References

Aczel, Amir. (2015). *Finding Zero: A Mathematician's Odyssey to Uncover the Origins of Numbers*. New York: St Martin's Press.

Alavi, Seema. (2015). *Muslim Cosmopolitanism in the Age of Empire*. Harvard University Press.

Chaunu, Pierre. (2005). *Conquête et exploitation des nouveaux mondes*. Paris: PUF, 1969.

Hodgkin, Luke. (2005). *A History of Mathematics: From Mesopotamia to Modernity*. Oxford: Oxford University Press.

Kaplan, Robert. (2000). *The Nothing That Is: A Natural History of Zero*. New York: Oxford University Press.

Li, Rongxi (tr.). (1995). *The Great Tang Dynasty Record of the Western Regions*. Berkeley, Ca.: Numata Center for Buddhist Translation and Research.

Nattier, Jan. (1992). The Heart Sutra: A Chinese Apocryphal Text? *The Journal of the International Association of Buddhist Studies*, 15(2) 153–223.

O'Connor J. J., and E. F. Robertson. *A History of Zero*. http://www-history.mcs.st-andrews.ac.uk/HistTopics/Zero.html.

Parpola, Asko. (2015). *The Roots of Hinduism: the early Indians and the Indus Civilization*. Oxford: Oxford University Press.

Ragep, F. J., Ragep, S. P. and S. Livesey. (1996). *Tradition, Transmission and Transformation*. Leiden: Brill, 1996.

Sen, Amartya. (2005). *The Argumentative Indian*. London: Allen Lane.

Schoff, William H. (1912). *The Periplus of the Erythraean Sea: Travel and Trade in the Indian Ocean by a Merchant of the First Century*. New York: Longmans, Green, and Co.

Steele, John (ed.). (2016). *The Circulation of Astronomical Knowledge in the Ancient World*. Leiden, Boston: Brill.

Suri, Manil. *Who Invented "Zero"?* (2017). https://www.nytimes.com/2017/10/07/opinion/sunday/who-invented-zero.html, 2 October 2017.

Subrahmanyam, Sanjay. (1999). *Connected histories: Notes towards a reconfiguration of early modern Eurasia. In Beyond Binary Histories: Re-imagining Eurasia to c.1830*, Victor Lieberman (ed.), pp. 289–316. Ann Arbor: University of Michigan Press.

Subrahmanyam, Sanjay. (2005). *Explorations in Connected Histories: From the Tagus to the Ganges*. OUP India.

Subrahmanyam, Sanjay. (2011). *Three Ways to Be Alien: Travails and Encounters in the Early Modern World*. Waltham, MA: Brandeis.

Yong, Lam Lay. (2004). *Fleeting Footsteps: Tracing the Conception of Arithmetic and Algebra in Ancient China*, Revised Edition, Singapore: World Scientific.

Lam Lay Yong. (2004). *Fleeting Footsteps: Tracing the Conception of Arithmetic and Algebra in Ancient China*, Revised Edition, Singapore: World Scientific.

CHAPTER 2

# Connecting Zeros

*Mayank N. Vahia*

## Abstract

As various articles in this volume demonstrate, zero has many forms and finds expression in a whole host of contexts, each as mystifying as the other. Each culture seems to have a very specific reaction to the number, and the reaction varies significantly from culture to culture. From the great hatred in Europe to the near obsession of their cousins in the Indian subcontinent, the entire spectrum of reactions can be found. Egyptians, Sumerians, and the Chinese managed to do some interesting algebra while circumventing zero completely. In this chapter we will investigate why cultures even with deeply common roots reacted so differently to zero. We take the example of the Indo-European cultures and their diametrically opposite reactions to zero. We suggest that this may be related to their experiences of nature post-separation, which were further amplified by their way of life. We then study the wider context of zero in its different manifestations and explore the relation between its different forms. We show that music seems to make a connecting link between these different aspects.

## Keywords

zero – zero in different forms – zero of grammar, music and art – zero of science, mathematics – interrelation of zero in various forms – zero, the centrality of the idea in different aspects of human creativity

## 1 Introduction

Zero is a strange entity. At one level, in the decimal representation of numbers, it means null value at the location where it appears. This is straightforward and humans have used this format since at least the rise of the first human civilization on the banks of the Euphrates and Tigris rivers. But when you expand the idea to physical sciences, religion, philosophy, advanced mathematics, grammar, etc., this null entity becomes problematic. They are intensely debated in

 | DOI:10.1163/9789004691568_005

various cultures, and different cultures reacted differently to this idea of an entity that has no physical value – null is without form or structure in every sense of the word. Zero grams has the same value as zero meters. Once an entity reaches the value of zero, the units disappear.

The contrast is particularly interesting in the context of Indo-European cultures, which have a common root language, and the exchange of ideas between the two cultures was deep (McEvilley, 2002). Despite this, their response to the idea of null was diametrically opposite. We study this contrast and attribute it to the different geological environments they encountered and their lifestyles that evolved after they separated.

We also explore the interconnectivity among different and apparently unrelated forms of zero and show that music and meditation provide an interesting interconnectivity between these otherwise disparate zeros.

## 2 Development of Perspective to Zero

The contrast in human reaction to zero is nowhere seen than between the cultural group generally called Indo-Europeans. The close similarity of Latin and Sanskrit suggests that Europeans and Indians had a common origin in Central Asia, from where they probably separated about 5,000 years ago. They went on to form what is broadly called Vedic culture in India and the culture rooted in Latin in Europe. As the cultures became separated, they formed their world view according to their environment, perspectives, and the way their leaders perceived reality.

The parameters that defined their attitude to themselves and their response to the challenges of existence and intellectual advancement have been differently driven. While in both cultures similar ideas crisscrossed in their great struggle to understand themselves and their environment, their responses to the same questions about life and the universe have been both similar and different. McEvilley (2002) and others have documented the similarity and contrast in their intellectual pursuits. This path is largely through the Arab world. The primary differences between Indians and Europeans are in their approach to nature and its study.

Indians embraced the idea of the world being complete and the universe having been born of nothingness (*creatio ex nihilo*), as exemplified in the Nasadiya Sukta in the Rigveda. The idea of void and emptiness has been the guiding principle of Indic thought. They are also humbler in their confidence in understanding the workings of nature. To them, God or Brahman is not a great account-keeper of the good and evil of human beings, nor is He even human-centric. Indeed, to them humans are not necessarily the most perfect and refined product of creation. Humans are just one form among many vying for attention. All living entities have a soul and the whole story of life is about the transmigration of this strange, formless idea. All living beings are born of a formless Brahman and return to him on their death. The creation therefore is his prerogative. But clearly, there are patterns and harmonies in this creation. Brahman has chosen to be consistent and logical. It is possible to *inductively* interpret and understand bits and pieces of it, like the movement of heavenly bodies, the idea being that this incomprehensible universe can be understood in part, whose existence can be objectively understood through association of similarities: that idea has fascinated all religions and philosophies that have arisen in India.

In contrast, European thought, diverted by the fact that God sent down his son, whose preoccupation was to ensure that humans follow a path of virtue and a good life, held a completely different perspective. Not diverted by the eternal reminder of the Great One, whose universe we are only in a position to comprehend in parts, they have been far more aggressive in their investigation of nature. Their *deductive* approach (compared to the gentler *inductive* approach of the Indians) has been a very aggressive call to nature, demanding that she explain herself to us. It has also been enormously successful. This patterning then, which Indians only gently wondered about, constituted the very foundation of European thought.

Yet the success of this basic mindset in Europe is of relatively recent date. In its earlier Greek form, it had limited success. It required the fund of enormous analysis of the workings of nature by Indian and Arab scientists that heralded the Renaissance a millennium and a half later for Europeans' more aggressive approach to bear fruit. There is therefore a symbiotic relation between the contributions of different cultures in this great human adventure called science.

But the deductive approach to the working of nature pioneered by the Greeks, for all its success, often overlooked aspects that are obvious to other cultures. The invention of the mathematical zero (*shunya* in Sanskrit) is a notable example. Embraced with passion by the Indians, what originally began as the notion of positional notation soon encompassed a whole host of linguistic, philosophical, and mathematical ideas. Equally passionately, Europeans condemned the very idea of a null entity. To Europeans, in a world created by God, even the very idea of null was seen as blasphemy. For the One who is omnipresent, there is no room for the void. It was only with great reluctance and in a very gingerly manner that Europeans agreed to look at, at least, the algebraic zero, functionally adopted for its utility, severed from its philosophical moorings.

It is interesting to ask if there is anything fundamental in their experience that produced these divergences in the Indo-European cultures.

## 3 On Differences in the Two Cultures

The present archaeological data suggests that Indo-Europeans separated out from Central Asia around Anatolia. One group traveling along the Steppes of South Central Asia entered the subcontinents via the Khyber Pass and another group came to South Asia via Iran and the Bolan Pass. Against that, the passage to Europe would have been faster and easier through present-day Turkey. The former group a landscape best exemplified by its image of barren, largely unstructured Steppes where nothing was the most common entity (Figure 2.1). Trees and bushes or large bodies of water would be hard to find: a landscape that we commonly call the cold desert. The featurelessness of this environment must have made a deep impression on the people who spent more than a thousand years in these landscapes (Figure 2.1) before arriving in the lush green of South Asia fed by the Indus and other rivers. Those coming from Iran and carrying the roots of the first religious ideas of Avesta came via the Bolan pass. Their merger with the local populations of the subcontinent in the decaying phase of the Harappan civilization (Vahia and Yadav, 2011) created the belief system that we find in Vedic literature where the primal gods – the Sun as a golden egg, Prajapati (god of all life), Pashupati (god of animals), the Avestan gods (see, for example, the Wikipedia entry on Avesta), and more complex gods all find their place. In this the impressions of the wilderness in earlier epochs and a nomadic lifestyle must have also contributed significantly to these ideas, which find their reflection in, for example, poems such as *Nasadiya Sukta* in the Rigveda and *Kaala* in Atharvaveda. This sense of void pervades much of the belief system of the Vedic people.

Against this, the Europeans encountered a rich and varied environment. Apart from relatively cold winters, the landscape was lush and green full of a large variety of variety of plant and animal life and sources of food. Their thoughts therefore reflected little of the barren, lifeless landscape of their South Asian cousins.

Clearly, this difference must have mattered significantly in their perception of life and reality. Zero would certainly not be very natural for them to accept.

FIGURE 2.1 A typical landscape in Central Asia

## 4 Divergence

The result of these varying experiences of the communities that went East and West was that there were important differences and divergences between the Europeans and Indians. Indians incorporated an idea of null as real.

Vedic Indians had no permanent places of worship – oblations were offered at a place and then people moved on. Their own god was formless and could be reached through the sacrificial fire. They had also assimilated ancestor worship, all its abstractions. Indians were comfortable with the idea of visualizing a non-existent entity. To them it represented an amorphous, undefined entity that could be experienced in absence. The null entity represented something that did not exist, or had ceased to exist. This was an interesting idea that came up in a variety of contexts:

- Zero as a null entity that created the universe comes from *Nasadiya Sukta*;
- Zero as the focus of meditation from yoga;
- Zero or nulling was used extensively in the grammar of Sanskrit;
- Zero or blank verses were used in mantras (chants) with wordless rhythms;
- Ātma (the concept of soul) is treated as being a null entity unaffected by the physical universe in Vaisheshika;
- Anu, the smallest entity of matter, is similarly treated as a featureless entity in Vaisheshika;
- Zero of *Bijaganita* (algebra) from Bhāskara II;
- This zero started out as *shunya*, became *sifr* in Arabic and zero in Europe.

Indians were the first to work out the algebra of zero and opened the window to a completely new class of mathematics.

This was not true for the Europeans, to whom life without one was unimaginable. One was the natural smallest number for them. Zero made them uncomfortable. All cultures believed, in one form or another, that there exists a Great God. This was the proverbial "One". This Great God then created the universe and the many variations in life. The one therefore pervades everything and remains even when all else is gone.

In early Europe it was forbidden to study zero as it was considered unnatural and against the working of the Great One who would always be present. Fibonacci learned his mathematics from the Muslims, and learned Arabic numerals, including zero. He included the new system in *Liber Abaci*, finally introducing Europe to zero in 1202. The book showed how useful Arabic numerals were for doing complex calculations, and the Italian merchants and bankers quickly seized upon the new system, zero included. In the end, governments had to relent in the face of commercial pressure. Arabic notation was allowed in Italy and soon spread throughout Europe. Zero had arrived – as had the void. Aristotelian resistance crumbled, thanks to the influence of the Muslims and the Hindus, and by the 1400s even the staunchest European supporters of Aristotelianism had their doubts.

However, it would be legitimate to ask if the various forms of zero listed above were all independent ideas with no relationship between them, or whether a common thread existed that connected these varied ideas of null.

## 5 Interconnecting the Zeros

As we discussed, zero now pervades a whole host of human activities. The variety of zeros that we discussed above are not isolated events. Broad features of the appearance of zero are shown in Figure 2.2.

However, even if they appear as separate entities, there is an underlying set of ideas that seem to connect them to a lesser or higher degree. They seem to interconnect with each other to a larger or lesser extent. These can be quantified. One model of such interrelation is given in Table 2.1 below.

Such quantification of relation then becomes amenable to an analysis. In Table 2.1, we have given one model of the possible strengths between various parameters that use zero in one form or another (Figure 2.2). We classify these into the following six categories.

- Religion: Hindu, Jain, Buddhist, Zoroastrian, Judaism, Christianity, and Islam
- Philosophical: Formless god, Buddhist existential zero, Jain metaphysical zero, meditation, and Sanskrit grammar

**Forms of Zero in Indian context**

| PHILOSOPHY *(The Great Creator)* | GRAMMAR *(lopa)* | MATHEMATICS *(a number representing nil value)* | SCIENCES *(The Great Creator)* | COMPUTERS *(state of a circuit and a symbol of no intrinsic value)* |
|---|---|---|---|---|
| • Used first as the source of all creation it is used in the following manner: null, empty, transient, purna, defined as neti-neti.<br>• In meditation it is the blanking of the mind.<br>• In Buddhism it is also used for nirvana and the complexity of nature. | • Used by Panini to describe the sounds deleted in sandhi (merger of words) as lopa.<br>• In mantras it is a null sound. | • Zero as a counting device in location-based value systems by Babylonians, it is now recognized as the centre of the number line, a number whose operations are unique. The zero value of functions are special; a point has zero dimensions.<br>• A string of zeros is the only one with no value unless another number appears in the string preceding it. | • In the sciences zero has several meanings. These include: zero as a perfect balance of forces, null, vacuum, empty, singularity, neutral etc.<br>• Then three absolute zeros:<br>◦ Mass of a photon<br>◦ Temperature of absolute zero − 273 °C<br>◦ Charge of a neutrino | • In computers zero is the first number in the sense that registers begin with zero, voltages below a certain value are considered zero and zero is also an entity, a numeric character with no specific value in passwords. |

FIGURE 2.2 Forms of zero in an Indian context

– Physical (based on Vaisheshika): *A priori* non-existence, *a posteriori* non-existence, absolute non-existence
– Music
– Zero in science: absolute zero, relative zero
– Zero in mathematics: arithmetic zero, algebraic zero and zero of equations

Their relative values as given by us are shown in Table 2.1. In the Table, 10 denotes the strongest connection and 0 indicates no connection. The assignments are subjective and will vary depending on perspective. We therefore focus on large, relatively robust connectivities that emerge from such assignments, and we study these in Figure 2.3.

We then use the software Netdraw to plot their connectivity, as seen in Figure 2.3.

In Figure 2.3a we show the primary connectivity. In Figure 2.3b we highlight the group structures listed in the table. We merge music with meditation as they provide a bridge between physical and philosophical, religious and linguistic zero. In Figure 2.3c we re-plot the data using a computer to identify the connected groups, which are then shaded as groups. In Figure 2.3d we isolate

TABLE 2.1 Connectivity strength between various forms of zero as used in this network study

| Grp | # | Type | Title | 1 | 2 | 3 | 4 | 5 | 6 | 7 | 8 | 9 | 10 | 11 | 12 | 13 | 14 | 15 | 16 | 17 | 18 | 19 | 20 | 21 | 22 | 23 | 24 | 25 |
|---|---|---|---|---|---|---|---|---|---|---|---|---|---|---|---|---|---|---|---|---|---|---|---|---|---|---|---|---|
| 1 | 1 | Religions | Hinduism | | | | | | | | | | | | | | | | | | | | | | | | | |
| | 2 | | Jainism | 8 | | | | | | | | | | | | | | | | | | | | | | | | |
| | 3 | | Buddhism | 7 | 7 | | | | | | | | | | | | | | | | | | | | | | | |
| | 4 | | Zorastian | 6 | 2 | 2 | | | | | | | | | | | | | | | | | | | | | | |
| | 5 | | Judaism | 3 | 2 | 1 | 3 | | | | | | | | | | | | | | | | | | | | | |
| | 6 | | Christianity | 2 | 2 | 1 | 1 | 8 | | | | | | | | | | | | | | | | | | | | |
| | 7 | | Islam | 2 | 2 | 1 | 1 | 6 | 6 | | | | | | | | | | | | | | | | | | | |
| 2 | 8 | Phil | Verdic formless god Buddhist | 10 | 9 | 8 | 5 | 1 | 1 | 1 | | | | | | | | | | | | | | | | | | |
| | 9 | | Existential Zero | 9 | 8 | 10 | 3 | 0 | 0 | 0 | 9 | | | | | | | | | | | | | | | | | |
| | 10 | | Jain meta Zero | 8 | 10 | 9 | 2 | 1 | 1 | 1 | 9 | 9 | | | | | | | | | | | | | | | | |
| | 11 | | Meditation | 9 | 9 | 10 | 4 | 4 | 4 | 4 | 9 | 9 | 9 | | | | | | | | | | | | | | | |
| 3 | 12 | Linguistic | Sanskrit Grammar | 10 | 7 | 7 | 1 | 0 | 0 | 0 | 8 | 7 | 7 | 8 | | | | | | | | | | | | | | |
| | 13 | | Prior non existence | 8 | 7 | 7 | 1 | 0 | 0 | 0 | 8 | 7 | 7 | 8 | 10 | | | | | | | | | | | | | |
| | 14 | | Posterior non existence | 8 | 7 | 7 | 1 | 0 | 0 | 0 | 8 | 7 | 7 | 8 | 10 | 10 | | | | | | | | | | | | |
| | 15 | | Abs. non existence | 8 | 7 | 7 | 1 | 0 | 0 | 0 | 8 | 7 | 7 | 8 | 10 | 10 | 10 | | | | | | | | | | | |
| 4 | 16 | Music | Zero of music | 9 | 8 | 8 | 3 | 6 | 4 | 4 | 8 | 8 | 8 | 10 | 8 | 4 | 4 | 4 | | | | | | | | | | |
| 5 | 17 | Sci zero | Absolute zero | 3 | 2 | 2 | 0 | 0 | 0 | 0 | 0 | 0 | 0 | 3 | 1 | 4 | 4 | 4 | 3 | | | | | | | | | |
| | 18 | | Relative zero | 5 | 5 | 5 | 1 | 1 | 1 | 1 | 7 | 7 | 7 | 8 | 7 | 4 | 4 | 4 | 3 | 9 | | | | | | | | |
| | 19 | | Null points | 0 | 0 | 0 | 0 | 0 | 0 | 0 | 3 | 3 | 3 | 5 | 2 | 4 | 4 | 8 | 9 | 9 | 9 | | | | | | | |
| | 20 | | Singularities | 4 | 3 | 3 | 0 | 0 | 0 | 0 | 5 | 5 | 5 | 5 | 5 | 8 | 8 | 8 | 0 | 9 | 9 | 9 | | | | | | |
| 6 | 21 | Math zero | Arithmetic zero | 0 | 0 | 0 | 0 | 0 | 0 | 0 | 0 | 0 | 0 | 0 | 2 | 8 | 8 | 8 | 3 | 9 | 9 | 9 | 9 | | | | | |
| | 22 | | Algebraic zero | 0 | 0 | 0 | 0 | 0 | 0 | 0 | 0 | 0 | 0 | 0 | 1 | 8 | 8 | 8 | 3 | 9 | 9 | 9 | 9 | 9 | | | | |
| | 23 | | Zero of an equation | 0 | 0 | 0 | 0 | 0 | 0 | 0 | 0 | 0 | 0 | 0 | 0 | 3 | 3 | 3 | 3 | 9 | 9 | 9 | 9 | 9 | 9 | | | |
| | 24 | | Digital zero | 0 | 0 | 0 | 0 | 0 | 0 | 0 | 0 | 0 | 0 | 0 | 0 | 5 | 6 | 6 | 0 | 8 | 8 | 8 | 8 | 8 | 8 | 8 | | |
| | 25 | | Zero as a number | | 0 | 0 | 0 | 0 | 0 | 0 | 0 | 0 | 0 | 0 | 0 | 0 | 0 | 0 | 0 | 4 | 5 | 1 | 6 | 8 | 8 | 8 | 10 | |
| | | | | 1 | 2 | 3 | 4 | 5 | 6 | 7 | 8 | 9 | 10 | 11 | 12 | 13 | 14 | 15 | 16 | 17 | 18 | 19 | 20 | 21 | 22 | 23 | 24 | 25 |

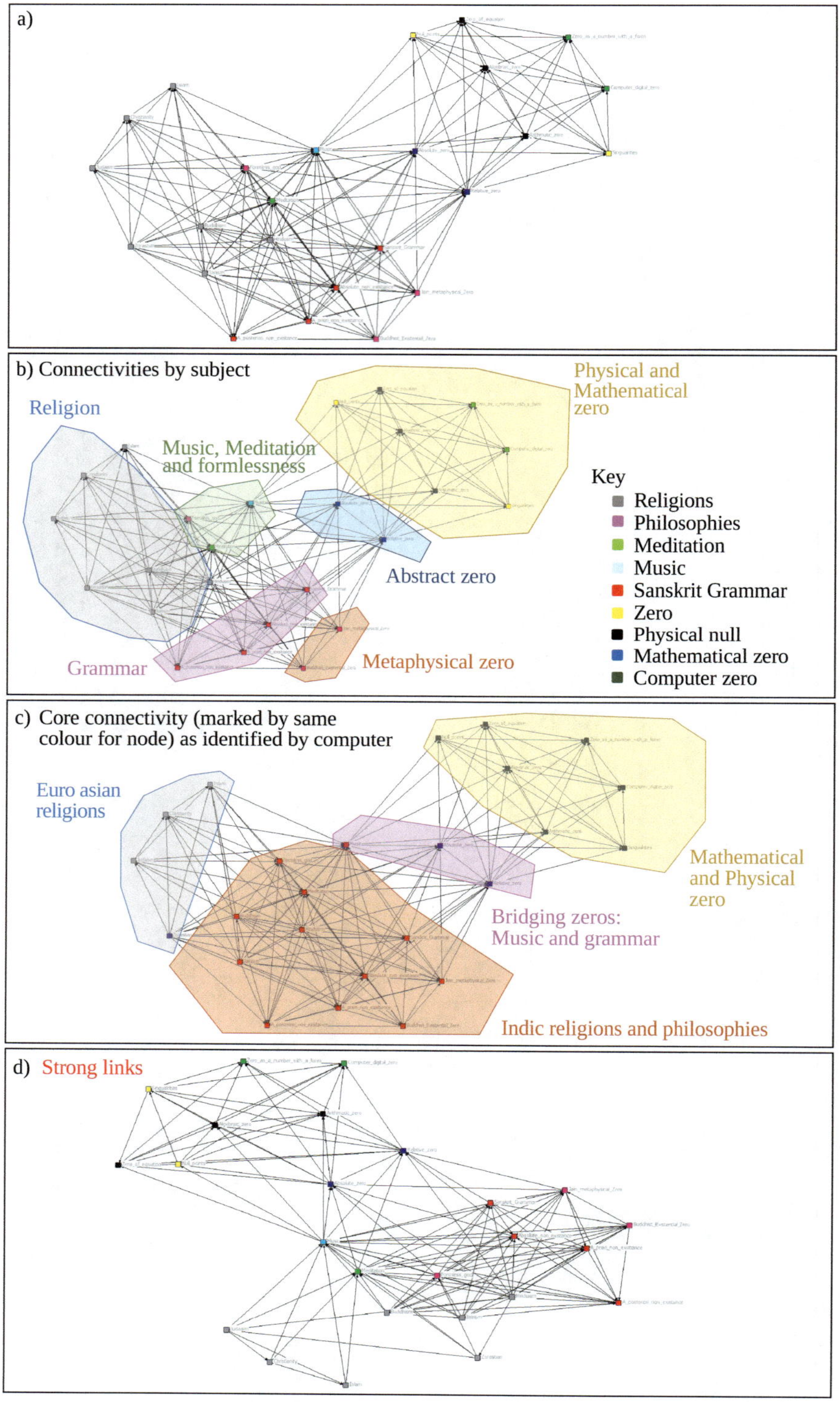


FIGURE 2.3 Network connectivity of different parameters from Table 2.1

only the strongest links – these clearly show that music is the strongest connecting link between different zeros. The result shows that in computer identification relative zero also forms a part of the bridge due to the fact that it is also used in multiple contexts.

This analysis seems logical by the fact that it naturally clusters entities that we would consider similar. Hence religions form a cluster, but within these the connection to the philosophical zeros of the Indian knowledge system connect these religious ideas to music and meditation. These in turn connect to grammar and eventually to mathematical zero and physical zero. This can be seen in Figure 2.3a, and the clustering that can be identified visually (Figure 2.3b) and through computer software (Figure 2.3b). In Figure 2.3d the isolation of strong links reinforced these conclusions. As can be seen from the figures, music and meditation form a primary bridge between the various concepts of zero.

## 6 Conclusion

Environment, experience, and perspective influence the approach of a culture to a particular situation. Zero is an excellent case in point. Comparing the response of the European cultures and South Asian cultures to the idea of zero highlights this aspect of human reaction very well. In the case of zero, the response seems to be deeply connected to the geographical environment and the lifestyle options of Indian and European subgroups of the Indo-European culture. While Europeans flourished in a relatively friendly environment, their Indo-European cousins, who came to the subcontinent via Central Asia, were not only exposed to a very harsh, featureless, cold desert (Figure 2.1) but also to a nomadic lifestyle. The latter incorporated zero naturally into a variety of creative activities (Figure 2.2) while the Europeans, happy with the One God who pervaded all existence, considered null or void to be unnatural. They resisted even the zero of mathematics until its elegance forced them to accept the idea, and still struggle to deal with zero beyond the sciences.

At first glance, the different manifestations of null and zero seem, *a priori*, to be a *post facto* association. However, human ideas and creations are interconnected. We explicitly analyze the connectivities between the various appearances of zero in different contexts well beyond mathematics. We then assign the connectivity between these ideas and their strength. We find that even an intuitive network diagram shows these ideas to be mutually connected, and a common theme that seems to connect the idea of null in religion and grammar to null in mathematics and physical sciences, seems to be the link through music and meditation. Music and to some extent meditation provide excellent links between various forms of mathematics (Figure 2.3).

Zero therefore is one of the core ideas and its appearance in a large number of contexts reveals deeper relations of human thoughts to their environment and attitude as well as to their different intellectual pursuits. Zero therefore is a thread with a capacity for providing far deeper insights into the evolution of human thought and relation to perceived reality.

## References

McEvilley, T. C. (2002). *The Shape of Ancient Thought; Comparative Studies in Greek and Indian Philosophies*. Allworth Press, New York.

Vahia, M. N. and Yadav, N. (2011). Reconstructing the history of Harappan Civilization, *Journal of Social Evolution and History*. Vol 10, No 2, pp. 87–120.

CHAPTER 3

# Babylonian Zeros

*Jim Ritter*

## Abstract

Two of the characteristics of our modern concept of zero find equivalents in Ancient Mesopotamia. First, the Babylonian sexagesimal positional notation from its invention in the late third millennium BCE required a means of indicating a 'placeholder' in a number, a missing power of the base. Secondly, the needs of mathematical astronomy, developed in the first millennium BCE, had to deal with a numerical concept of 'nothing' as a full-fledged number, capable of entering into arithmetical operations.

## Keywords

cuneiform – Babylonian mathematics – Babylonian astronomy – sexagesimal positional system – procedure texts – table texts

## 1 Introduction

> Why do we call something a 'number'? Well, perhaps because it has a direct relationship with several things that have hitherto been called number; and this can be said to give it an indirect relationship to other things we call by the same name. And we extend our concept of number as in spinning a thread we twist fiber on fiber. And the strength of the thread does not reside in the fact that some one fiber runs through its whole length, but in the overlapping of many fibers.
>
> LUDWIG WITTGENSTEIN, *Philosophical Investigations*, § 67

It is common practice when enquiring into the *origin* of mathematical or other objects or concepts to suppose somehow that the object or concept in question possesses a definite, limited and easily discernible contour. So it is with the question of the 'origin' of zero. Let the reader consult almost any general history of the question: he or she will find the same canonical story In this story, the zero existed first as a placeholder, indicating the absence of a digit in a positional number system (like in our 105, where the middle 0 marks the

 | DOI:10.1163/9789004691568_006

absence of the tens); then it evolved as a proper number in itself. We would like to tell another story here, or rather two stories, each of them rich and complex – one about placeholders, one about numbers. Neither of them easily fits the canonical story and moreover these stories were and remained distinct. We will see that there was never in ancient Mesopotamia a single sign or concept which fulfils the various roles played by the zero we now learn at school. More interestingly perhaps, I will question our aim itself; is there ever anything like a history of zero?[1]

To begin with we must situate the geographic and chronological framework of our story. Since we will be dealing with mathematical and astronomical cuneiform texts from Mesopotamia, it is important to understand that this corpus was principally restricted to a limited sub-area, namely the southern part of Ancient Mesopotamia called the 'Land of Akkad and Sumer' in the ancient texts and 'Babylonia' in modern parlance. We shall be looking at texts dating from a period stretching from the end of the third millennium until the end of the second millennium and then from the last eight centuries of the pre-Common Era.[2] We can summarize the standard periodization and the modern Assyriological names of these epochs in the following manner:

| **Ur III** | **Old Babylonian** | **Middle Babylonian (Kassite domination)** | **Under Assyrian denomination** | **Neo-Babylonian** | **Achaemenid (Persian domination)** | **Seleucid (Macedonian domination)** |
|---|---|---|---|---|---|---|
| 2300–2000 | 2000–1600 | 1600–1160 | 1160–626 | 626–539 | 539–330 | 330–63 |

## 1.1 *Numbers*

Since we will be looking primarily at numbers in mathematical and astronomical texts, it is important to have an understanding of how the sexagesimal positional system, used in both domains, operated. We begin our investigation there when the system came into existence; sometime in the closing centuries of the third millennium (Ur III Period). I remind the reader that, prior to the appearance of a positional system in Mesopotamia, numbers were written

1 For an informed and sophisticated discussion of the nature of 'origins' in mathematics see Goldstein, 1998, and for the question of 'concepts' in that discipline see Netz, 2002.

2 All dates in this chapter will be BCE.

in a half-dozen different manners, depending on which metrological system (length, area, capacity, etc.) was being utilized.[3] The symbols themselves were drawn from a small common repository of signs but their order and relative values varied from one system to another.

Starting around 2300 BCE the adaptation of the cuneiform writing system to handle the Akkadian language as well as the Sumerian language for which it had been developed, coupled with writing reforms touching sign direction and tablet formatting, led eventually to the creation of an abstract number system, in what we now call a sexagesimal positional notation, that is with 60 as its base (instead of our contemporary decimal system).[4] It used 59 symbols and since it was positional it needed no more than these to represent any number however large. For historical reasons, linked to the archaic metrological system from which it was derived, it actually made do with less than 59 distinct signs; using an additive construction, the system used in fact only 14 different signs and even these were all built out of two fundamental marks, the vertical wedge, 𒁹, and the corner wedge, 𒌋. Nine signs represented the 'digits' from one to nine, five signs the 'digits' ten to fifty by tens. Written in their Old Babylonian form, they are the following (along with their numerical values written in our modern decimal notation):

| 𒁹 | 𒈫 | 𒐈 | 𒐉 | 𒐊 | 𒐋 | 𒐌 | 𒐍 | 𒐎 |
|---|---|---|---|---|---|---|---|---|
| one | two | three | four | five | six | seven | eight | nine |
| 01 | 02 | 03 | 04 | 05 | 06 | 07 | 08 | 09 |

| 𒌋 | 𒎙 | 𒌍 | 𒐏 | 𒐐 |
|---|---|---|---|---|
| ten | twenty | thirty | forty | fifty |
| 10 | 20 | 30 | 40 | 50 |

We adopt the following *conventions* for transcribing Mesopotamian sexagesimal numbers using our modern decimal notation:

1. Every cuneiform sexagesimal digit corresponds to 2 decimal digits.
2. In the case of a sexagesimal digit in which there are no ones, we place a decimal '0' after the decimal tens place, e.g., if 𒌍 is a full sexagesimal digit it will be transcribed as '30'.

3 For an authoritative discussion of early numerical and metrological systems in early Mesopotamia see Nissen, Damerow, and Englund, 1994.

4 I have discussed this evolution in Ritter (1999) and Ritter (2001).

3. In the case of a sexagesimal digit in which there are no tens, we add a decimal '0' before the decimal ones place, e.g., if  is a full sexagesimal digit it will be transcribed as '02'.[5]
4. A sexagesimal digit in which there are no tens *and* no ones will be transcribed as '00', i.e., a sexagesimal 'zero'.

To construct any digit from 01 to 59 was simply a question of putting together first the tens if any, then the ones if any. Thus the sexagesimal *single-digit* number forty-six (our 46) is written . The sequence of symbols 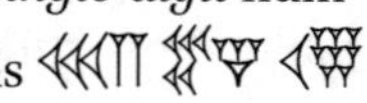 corresponds to a sexagesimal *three-digit* number; with our conventions it is transcribed as 32 54 18 (note that we have been obliged to introduce a small space between successive digits in such a multi-digit number, whether cuneiform or transcribed, for the purposes of readability).

A further point that should be stressed is that the sexagesimal system, though it could and did represent numbers having an integer part (each digit representing a certain non-negative power of 60) and a fractional part (each digit representing a certain negative power of 60), never developed a marker of the division between the two parts, the role played by the 'decimal point or comma' in the modern decimal system.[6] Related to this was the Babylonian convention that initial zeros (representing higher powers of 60 than that of the highest non-zero digit) were never represented nor were final zeros (representing lower powers of 60 than that of the lowest non-zero digit).[7]

As we shall see, the initial means of representing a missing sexagesimal digit was through leaving a blank space. This choice was an almost automatic consequence of the manner of writing cuneiform numbers. To see this we shall compare various modifications of a 3-digit sexagesimal number, say the example in the last paragraph, 32 54 18, in which different parts of the middle digit or of the preceding digit are set to zero:

5 Contrary to ordinary Assyriological transcription tradition, this decimal zero will be added in every applicable case, even where the sexagesimal digit in question is the leading digit.

6 What is essential for fixing the absolute value of a number, in any system, is to have a way of determining how many places exist between the written digits and the integer/fractional transition point. Our modern decimal system uses zeros to do this. The absence of the decimal/comma convention in Mesopotamia means that there was no way of indicating absolute values of numbers except by such manoeuvres as writing out the word equivalent for the number, etc., tricks that were very rarely used in cuneiform mathematical texts. We shall not need to consider absolute values of numbers in this chapter, only relative values.

7 Modern place notation shares the first but not the second Mesopotamian convention. There are rare cuneiform examples of explicitly written initial zeros in a few astronomical table texts in the case of fixed length numbers (though no unambiguous example of written final zeros) but we shall not deal with these here for lack of space.

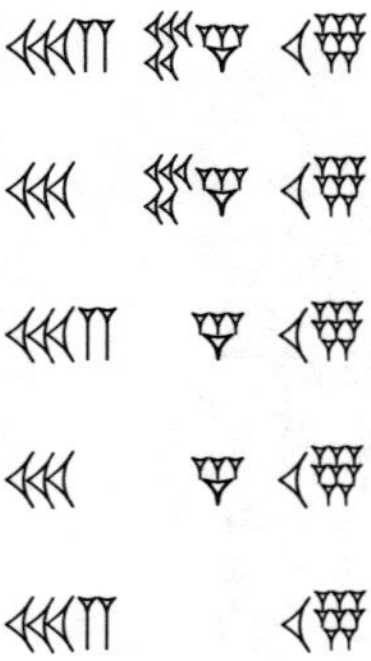

| | |
|---|---|
| 32 54 18 | |
| 30 54 18 | missing ones in the first digit |
| 32 04 18 | missing tens in the second digit |
| 30 04 18 | missing ones and tens |
| 32 00 18 | missing the whole second digit |

The small or large blanks left by the replacement of a non-zero number by zero is thus of itself a clear marker of such a missing part of a sexagesimal digit or of a complete sexagesimal digit.

### 1.2 *The Corpus*

We shall principally be involved with two types of cuneiform texts in our study. One, the older, which began around the beginning of the second millennium BCE, consists of *mathematical* tablets and falls into three main groups:

1. Procedure texts: these are comprised of anywhere from one to thirty or more solved mathematical problems, each involving numerical data and asking for a numerical answer. The solution to the problem is then developed as a step-by-step procedure (algorithm), each step invoking either an arithmetic operation (an operation which transforms one or more numbers into another) or a control command (an operation which operates a change in the information flow of the solution algorithm), with the answer to the problem given as the final result.
2. Table texts: tabular arrangements of multiples, squares and square roots, cubes and cube roots, reciprocals, useful constants, etc. constitute a second type of mathematical text. They serve to provide the numerical results of the operations called for by the problem texts or by the professional work of scholars.
3. Exercise texts: individual students' work in resolving set problems.

Our examples here will be drawn from the first two classes of documents, as the third one does not bring anything different for our restricted purpose.

The second, younger, class of texts – the *mathematical astronomical* texts – only began in the eighth century BCE and, like the mathematical texts, consists of *procedure* texts, telling the scribe how to calculate algorithmically the positions and movements of the celestial bodies, Sun, Moon and planets and *table* texts, often now called ephemerides, which give the resulting numerical results of applying the procedures to specific celestial bodies.

FIGURE 3.1 Two types of Old Babylonian mathematical text. Left: UM 29.15.503 (CDLI P254509) obverse, a multiplication table (×30). Reproduced courtesy Penn Museum, Philadelphia PA. Right: AO 6770 (Thureau-Dangin, 1936, p. 75) obverse, a procedure (solved-problem) text. Reproduced courtesy of Musée du Louvre and of the Réunion des Musées Nationaux, Paris

As we will see, the introduction of a specific sign for (a type of) placeholder zero and its later development depends in large part on the formal and stylistic differences between tables and procedure texts. These differences are clear from even a cursory view of the tablets themselves, as shown in the Old Babylonian examples of (Figure 3.1).

It should also be stressed from the outset, as this will equally be relevant for our purpose, that there is no extant procedure text that provides an example of the technique for the explicit carrying out of an arithmetical operation: addition, multiplication, reciprocal or root extraction. So, unlike some other cultures we cannot look for a direct example of how, for example, multiplication by a number with an intermediate zero digit was carried out. All we have are tables which resume the results of such operations; clearly the Babylonians knew how to handle such calculations, if only to construct the tables, but just how they did so is not clear.[8]

However, in order to take stock of the state-of-the-art before the types of texts of our main corpus existed, we will begin each of our two stories with a few examples of the Ur III period.

## 2 Placeholders

Most of the texts from the Ur III period we have are concerned with book-keeping and as such only display a variety of metrological numerical systems. In the standard story, as explained, zero starts out as a placeholder within a positional notation – in our case of Ancient Mesopotamia, its base-60 place notation. However we are in luck. Though we have only two texts exhibiting sexagesimal abstract numbers at their earliest appearance in the Ur III period, one of the texts has a number including an empty digit (Figure 3.2).[9] The section of the tablet shown in Figure 3.2 shows the first seven lines of the text, of which the first four lines contain one 3-digit sexagesimal number each and, following a blank line, the total quantity of a delivery of silver, expressed in terms of the Ur III weight system.[10]

The arrangement of the sexagesimal numbers in clear columns, the fact that the fifth inscribed line begins with the word 'total', the labelling of the entry as 'individual deliveries', all imply that it is a question here of four individual deliveries of silver whose sexagesimal sum, 01 33 27 50 in our transcription, when converted into the ordinary Ur III weight system yields precisely the total indicated (with the 50 in the fourth, least significant, place ignored as is often the case). Note that the metrological system here uses unit-names (***ma-na, gín, še***) which are of course absent from the abstract sexagesimal number system

8 The modern reconstruction of such ancient Babylonian calculational techniques has been attempted by a number of authors, often based on the analysis of numerical errors in the texts, but the question remains open. For a summary of such attempts in the case of multiplication see Proust, 2000.

9 Though the tablet was published in 1919 (Keiser, 1919, pl. LXVIII) its importance in our context was first realized by Marvin Powell (1976, p. 420).

10 Our convention for the transliteration of texts: *italic* represents a word in the Akkadian language; ***bold italic*** represents a word in the Sumerian language or a Sumerian logogram (abbreviation of an Akkadian word); SMALL CAPITALS represents the name of a sign whose pronunciation in the immediate context is not known; [...] square brackets represent lost or forgotten signs by the scribe.

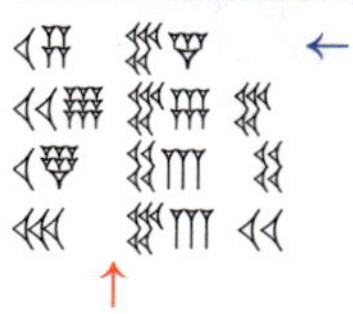

14 54 00
29 56 50
17 43 40
30 53 20

***šunigin*** 𒁹 𒈦 ***ma-na*** 𒐈 𒈦 ***gín***
***lá*** 𒐌 ***še kù-a***
***mu-ku**$_x$*(DU) ***didli***

Total 1½ ***mana*** 3½ ***gín***
less 7 ***še*** of silver.
Individual deliveries.

FIGURE 3.2 YPM BC 16534 (YBC 1793) obv I 1–7. Record of deliveries of silver, dated to the second year of the reign of Ibbi-Suen of the dynasty of Ur III (2027 BCE Middle Chronology), possibly from Ur. Top: photo of the upper left of the tablet (CDLI P142357). Reproduced courtesy of the Yale Peabody Museum, Babylonian Collection. Photograph by Klaus Wagensonner. Bottom left: Transliteration of the text in the photo. Bottom right: Translation of this text

and that the repertoire of numbers used is different as well, including fractions ('½') and subtractive writing ('less 7 *še*').[11]

The careful columnar arrangement of the sexagesimal entries in the first four lines of text yield inevitable blank spaces where, for a given digit, there is either no value present (the blue arrow in Figure 3.2, Left) or there are no ones or no tens in the digit (in our case, no ones; the red arrow in Figure 3.2, Left). In our decimal 2-digit transcription of these numbers, the first case is represented by a double zero '00' and the second by a single zero in the appropriate place. In what follows we shall call the first kind of indicator of absence a 'sexagesimal zero', denoted by '$0_S$', and the second a 'demarcation zero', denoted by '$0_D$'.[12]

The blank space inevitably left by a $0_D$ inspires a certain resistance to possible encroachment. Were the space to disappear it would in some cases allow a misreading due to a possible amalgamation of the digit in question with the following one. With a single missing blank, (45 03) might be mistaken for the single sexagesimal digit (48), or (20 15) mistaken for (35). This danger is particularly true in the specific case where there a missing ones value *and* an immediately following absence of a tens value in the next lower digit, e.g., (50 06) might well risk being misread (56), a type of error of which there is ample trace in the cuneiform record.[13]

It is not our intention to deal in any detail with the use of blanks as placeholders. However, let us comment on the use of a blank as a final zero in the first line of Figure 3.2, as it is historically unique. Figure 3.2 is a product of the fact that the four numbers are arranged specifically to carry out an arithmetical operation, addition in this case, and such an explicit arrangement in view of an operation is simply not extant in any other text.

In ordinary use, even in tables, final zeros, like initial zeros, are simply not written.[14] The study of blank placeholders is thus one of their use in

---

11 The arrangement of the ones and tens involved in numbers in the metrologically-bound writing can also differ from the abstract system arrangement, compare the metrological number '7' in line 5 with the abstract sexagesimal '7' used in line 3. Note also the use of two forms of the abstract '4' in line 1.

12 This 'demarcation zero' has no relation to any purported decimal system in Mesopotamia but is simply an indication that the absence is in a ones or tens place only. Though there is evidence of a base-10 system in Mesopotamia, it is totally distinct from the sexagesimal system under examination here. See Friberg, 1990, p. 537.

13 Such a misreading is not possible in our particular tablet in line 4 of Figure 3.1, since (30 53) could not possibly be read '83', there being no grouping of 8 tens allowed in a sexagesimal system. But, as we shall see, even where there can be no risk of misreading, an explicit placeholder $0_D$ is often present.

14 This use has been excellently treated in an article that is unfortunately almost never cited, probably because it was written in Russian, by the Soviet historian of mathematics Mark Jakovlevič Vygodskij (1959).

*intermediate* positions and thus necessarily in 3-digit or larger numbers. This restriction has an immediate consequence due to the frequency with which 3-digit or larger numbers occur in the cuneiform texts. In the Old Babylonian period, 3-digit or larger numbers make up a very small percentage of the total quantity of numbers.

We have already mentioned the distinction in format and disposition between table and procedure texts. This distinction was to play an important role in the *evolution* of placeholders in Mesopotamian mathematics in the Old Babylonian period. Aside from the difference in size (but this can vary and depends on the number of entries in the table text and the number of problems in the problem text), the two classes of texts differ principally in the *arrangement* of the signs on them. The tables have entries arranged in columns with the leading (leftmost) numerical signs usually lined up to form visual columns, leaving much free space (see, for example, Figure 3.1, Left, and Figure 3.5) while the problem texts, with their problems and solution algorithms written in ordinary prose, have a continuous, linear disposition of signs arranged to more-or-less uniformly fill space (see, for example, Figure 3.6). In this latter context, numbers in particular are written with little or no spaces between the successive digits, thus rendering difficult or impossible the use of blank spaces to indicate absences of units or tens, as was done in earlier texts, as seen above in the Ur III example.

| **Table texts** | **Procedure texts** |
|---|---|
| signs arranged in columns | signs arranged linearly (space-filling) |
| 'free' choice of precision (number of places) | 'free' choice of parameters used in problems |
| blanks as placeholders stable | blanks as placeholders subject to compression |

When 3-'digit' or larger numbers with missing tens or ones appear in table texts, they can and do retain the blanks that the construction of sexagesimal numbers automatically generate, as we have seen in Section 1 above. But such numbers almost never appear in problem texts; with their freedom of parameter choice, they need only show how the solution works, such texts can simply avoid large numbers. Indeed, I know of only one second millennium problem text in which a 3-digit number appears whose value should require an internal sexagesimal zero: it is a very late Old Babylonian or a Middle Babylonian problem text, AO 17264 (Thureau-Dangin, 1934, p. 63, obverse, lower edge 2). The

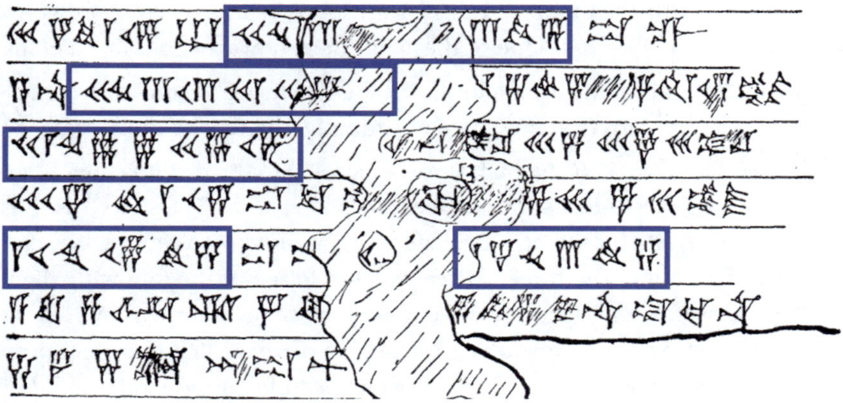

Square 34 41 15 – 20 𒀹 03 [13 21 3]3 45 you will see.
To 20 𒀹 03 13 21 33 [45 …] add 01 05 55 [0]4 41 15:
21 𒀹 09 08 26 15 [you will see]. W[ha]t is (its) square root? ⌜3⌝5 37 30 is the square root.
Add 34 41 15, your square, [to 3]5 37 30:
01 10 𒀹 18 45 you will s[ee]. W[hat] should I multiply [by] 14 𒀹 03 45,
the false area, which will give me [01 10 1]8 45?
05 you will put down. 05 is the upper breadth.

FIGURE 3.3 S$^{b}$ 13934 + 13935 (Bruins and Rutten, 1961, Text XXIV + XXV) obv 15–21. Mathematical procedure text. Late Old Babylonian Period, Susa. Hand copy courtesy of Éditions Geuthner, Paris. Top: copy of the obverse of the tablet with the selected numbers in blue rectangles. Bottom: Translation with selected numbers in blue

number in question is the square of 𒐖 𒌋𒌋𒐛 (02 27) which leads to 06 00 09, but the square is written 𒐋 𒐝, as though it were simply '06 09' with no empty space of either the $0_S$ or $0_D$ kind.

Dating from the end of the Old Babylonian period, we have a number of mathematical texts from Susa, far to the east on the Iranian plain but written in Akkadian cuneiform and, like their counterparts in Mesopotamia proper, consisting of tables and procedure texts. One of the tables, with the usual carefully disposed entries, uses a blank for a sexagesimal placeholder $0_S$ (S$^{b}$ 13091, Bruins and Rutten, 1961, Text v, rev 39) but the procedure texts show a new development. An example is the following portion of a Susa mathematical procedure text (Figure 3.3).[15]

15 Following standard Assyriological notation, missing text is indicated by square brackets '[ ]' with text restored by the modern editor placed within these brackets. Mathematical texts, tables or procedures, are by their very nature often easy and safe to restore. But though the restoration of numerical values is relatively sure, nothing can be said about the way in which missing numbers, etc. were actually written. Since our analysis hangs

Of the nine distinct numbers in this excerpt, five include what should be a placeholder of the form $0_D$. None of these numbers however contains a blank placeholder; four of them have, in its place, a nonnumerical cuneiform sign 𒃵 (called GAM in modern Assyriology). For example, the first of them reads 𒎙 𒃵 𒐈 𒌋𒐈 𒎙𒁹 𒌍𒐈 𒐏𒐊 which since, according to the text it is the square of 34 41 15, can be restored as the number 20 03 13 21 33 45. The sign 𒃵 is thus an explicit substitute for the blank placeholder, situated between the two corner wedges for 20 and the three vertical wedges for 03 in the following digit, allowing the twenty and the three to be written close together without fear of being read '23'. GAM is therefore an early example of an explicit representation of the demarcation placeholder. The same sign is to be found in all the other numbers in the text where $0_D$ is to be expected, signaling either a missing tens entry (line 17: 21 𒃵 09 08 26 15 and line 19: 14 𒃵 03 45) or a missing ones entry (line 19: 01 10 𒃵 18 45).[16] Note that the number 01 05 55 04 41 15 where at least two possible places for a $0_D$ are found (01 05 … and … 55 04 …) do not use 𒃵. The sign is thus neither systematic nor obligatory.[17]

GAM is not the only explicit placeholder sign developed in the Old Babylonian period. In the Diyala region to the north-east of central Mesopotamia, in particular in the town of Me-Turan (modern Tell Ḥaddād), a number of mathematical tablets were found in 1979 during Iraqi excavations; most have remained unpublished until comparatively recently. One procedure text in particular is pertinent to the discussion here; it bears the Iraqi Museum number IM 121512 (Figure 3.4). The tablet presents another distinct nonnumerical sign 𒂊 (E) with the same use as 𒃵, i.e., a $0_D$, as can be seen in the selection on the obverse where it indicates a missing ones place in the first sexagesimal digit of 40 𒂊 50, as well as (twice) in 40 𒂊 50 𒂊 15. The selection from the reverse shows the same sign serving as an indicator of an empty tens place in 06 𒂊 06 40. Once again, the following line on the reverse shows that the use of E is no more obligatory than that of GAM, since it is not present in 07 06 40.

The reason why these two signs were selected to indicate an explicit $0_D$ is not clear. E, but not GAM, was used in another context in Old Babylonian mathematical problem texts, to separate two numbers written one directly

precisely upon the material traces of numbers, we will not use any such restored numbers in our analysis, here or in what follows.

16 The same sign is to be found serving the same purpose in another of the procedure texts from Susa: Sb 13922 obv. ll. 4, 11, 12 (Bruins and Rutten, 1961, Text XII).

17 Yet another use of 𒃵 as a $0_D$ is known to me from the Old Babylonian period. This is from a small tablet, MS 2731, with a squaring exercise(?) and a use of GAM to indicate a missing tens place in the answer (Friberg 2007: 40). The tablet, of unknown provenance, is from a private collection about which serious ethical issues have been raised; see NESH, 2005.

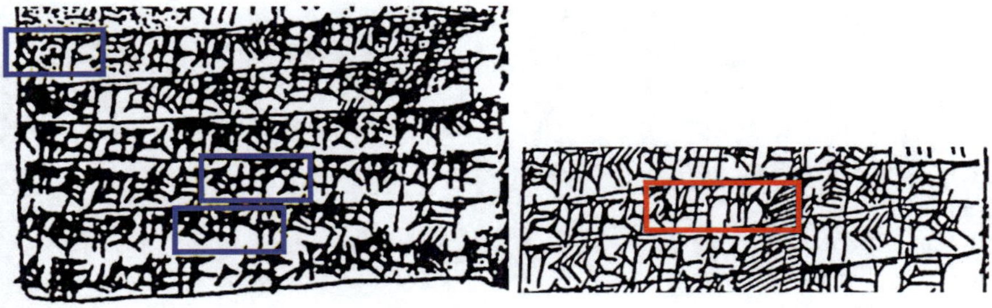

(…) Add 15 to the field **40** **50**, that your head retained, – **40** **50** **15** you will see. Extract the square root of **40** **50** **15** – 49 30 you will see. (…)

Add 01 to **06** **06 40**, that your head retained – 07 06 40 you will see.

FIGURE 3.4 IM 121512 (Friberg and Al-Rawi, 2016, pp. 254–255). Mathematical procedure text. Old Babylonian Period, Me-Turan. Left: obv I 44–46. Copy and translation with selected numbers in blue. Right: rev II 7–8. Copy and translation with selected number in red

after another. This was often the case when the result of one arithmetical operation is then to have its square root taken, e.g., '15 to 14 30 you will add – 14 30 15 29 30 is (its) square root.' (BM 13901 obv I 7, Neugebauer, 1937, p. 1) which is to be understood as '15 to 14 30 you will add – 14 30 15. (Then) 29 30 is (its) square root.'

GAM also had a separative function; in mathematical table texts it served as a separation sign, usually to separate two consecutive entries when, for reasons of space or otherwise, they were written on the same physical line. Outside of this specific mathematical use, it was frequently employed to indicate a gloss – a translation from a word or phrase from one language to another or a commentary on a difficult passage.[18]

From the end of the Old Babylonian period towards 1600 BCE until the seventh century BCE we possess a few mathematical texts in which it is clear that the use of a sign for $0_D$ continues, though sometimes using a modified form of .[19] But in the seventh century, during the Neo-Babylonian empire, we find the first extant text in which an explicit sign is used, not only for $0_D$, but also for $0_S$, the full *sexagesimal* placeholder. It appears moreover in a table text, not a procedure text, a table of squares from Kish (Figure 3.5). The novelty here, in comparison to Old Babylonian period tables of squares, is the 2-sexagesimal-digit precision, there where the older texts had only 1-digit precision – the

18 For these non-mathematical uses, one may consult Krecher, 1971, p. 433. There unfortunately is no detailed analysis of the use of GAM or E in mathematical texts.

19 The use of both and (TAR) for $0_D$ in the Middle Babylonian (circa 1200 BCE) mathematical procedure text HS 245 is a good example (Oelsner, 2006).

| | | |
|---|---|---|
| [15] | times 15 | 03 45 |
| ⌜15⌝ 30 | times 15 30 | **04 𒌍 15** |
| 16 | times 16 | 04 16 |
| 16 30 | times 16 30 | 04 𒌍𒐖 15 |
| 17 | times 17 | 04 49 |

| | | |
|---|---|---|
| [42] | times 42 | 29 24 |
| [42] 30 | times 42 30 | **⌜30 𒌍⌝ 06 15** |
| [43] | times 43 | 30 ⌜4⌝9 |

FIGURE 3.5 Ashm 1924.796 + 1924.2194 + 1931.38 (Robson, 2004, Text 28). Table of squares: Neo-Babylonian Period, Kish. Left: obv 29–33 (first half of line). Copy and translation of selected text with selected number in blue. Right: rev 18–20 (first half of line). Copy and translation of selected text with selected number in red. Handmade copies courtesy of the author and SCIAMVS

integers from 1 to 59. Now the writing of the squares of these latter numbers do not require a zero sexagesimal digit. But in the case of 2-digit precision numbers – from 01 00 to 59 30 – there are precisely 6 numbers whose squares require such a zero intermediate sexagesimal digit:[20]

$(15\ 30)^2 = 04\ 00\ 15$ $(24\ 30)^2 = 10\ 00\ 15$ $[(35\ 30)^2 = 21\ 00\ 15]$
$(39\ 30)^2 = 26\ 00\ 15$ $(44\ 30)^2 = 33\ 00\ 15$ $(59\ 30)^2 = 59\ 00\ 15.$

All but the entry for 35 30 are extant on the tablet and of the remaining five cases all but 59 30 have the sign 𒌍 in the place of the intermediary missing digit (00 in our transcription). We have translated the first of these cases (obv, line 2 of the excerpt) in Figure 3.5 (Left); here 𒐉 𒌍 𒌋𒐊 should represent 04 00 15 (as the square of 15 30), not *04 30 15 nor *04 45 (as a direct reading of this number might suggest). This, and the other three examples above are the first indications I know of a special sign for a missing sexagesimal digit, a $0_S$. On the other hand, the example given above in Figure 3.5 (Right), line 4, states that $(42\ 30)^2 = 30\ 06\ 15$, written 𒌍 𒌍 𒐋 𒌋𒐊, the first 𒌍 representing the number '30' while the second 𒌍 is a placeholder, here a $0_D$. Of course, in the vast majority of cases in this text, as in the first digit of the last example, the sign 𒌍 stands simply for '30'.

In the following centuries, the signs used for intermediate placeholders, both $0_D$ and $0_S$, remain infrequent but varied. The sign 𒑱 appears around 600

20 If one assigns an absolute value of sexagesimal units to the leading digit (the simplest case) the following 30 for those entries which have it is to be understood as 30/60 = ½. That is, this is a table of squares of integers and half-integers.

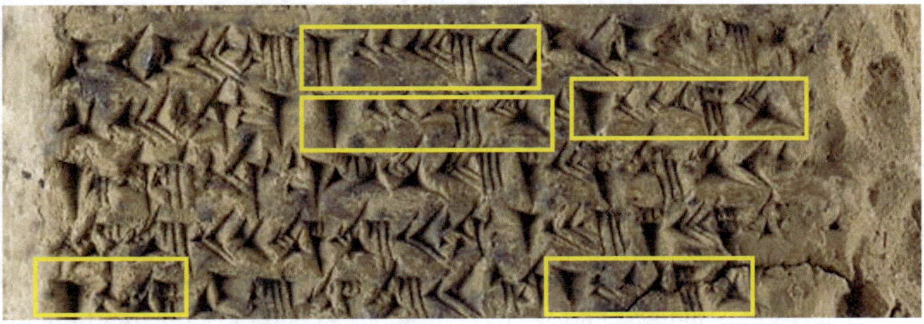

An *igû* and its *igibû*: 02 𒑲𒑲 33 20. [How much are] the *igû* and its *igibû*? [02 00 00 33 20]
by 30 multiply – 01 𒑲𒑲 16 40. 01 𒑲𒑲 16 40 by [01 00 00 16 40 multiply – 01 00 00 33 20 04 37 46 40].
01 from it decrease – the remainder is 33 <20> 04 37 46 40. How much [by how much should I multiply (to obtain) 33 20 04 37 46 40?]
44 43 20 by 44 43 20 multiply – 33 <20> 04 37 46 [40. 44 43 20 by 01 00 00 16 40 increase – ]
01 𒑲 45 is the *igû*. 44 43 20 from 01 𒑲 𒑲 16 40 decrease [ – 59 15 33 20 is its *igibû*].

FIGURE 3.6 AO 6484 (Neugebauer, 1935, pp. 96–107) rev 10–14. Mathematical procedure text; Seleucid Period, Uruk. Top: Photograph of reverse of tablet with problem (CDLI: P254387). Reproduced courtesy of Musée du Louvre and of the Réunion des Musées Nationaux, Paris. Bottom: Translation of the selected text

in a table for clepsydra weight-time equivalents (BM 29371; Brown, Fermor and Walker, 2000, pp. 144–148) as a marker for $0_S$,[21] while the numerical sign 𒌋 is to be found in an Achaemenid table of squares as a symbol both for $0_D$ and for $0_S$ (CBS 1535; Neugebauer and Sachs, 1945, No. 33, p. 34).

But it is in the mathematical and astronomical texts of the Achaemenid and especially Seleucid periods (sixth to first centuries BCE) that a final decision as to the form and the manner of use of placeholder zeros seems to have been made. The choice devolved on a non-numerical sign, 𒑲, one which, though composed of two corner wedges, has them arranged quite differently than the numerical '20', 𒎙. The sign 𒑲 was used systematically and generally in all astronomical and mathematical texts, both procedures and tables, of the period, such as the mathematical procedure text AO 6484 (Figure 3.6).[22] Though the right half of the tablet is broken off, what remains is sufficient to restore the whole problem and, in any case, shows clearly in what remains the use of the separation sign 𒑲 as a marker for the sexagesimal placeholder $0_S$.

21 The sign 𒑳 is originally a 'separator sign' like 𒑱; indeed they are often used in the same mathematical texts to separate two or more entries placed on the same line (see for example the neo-Babylonian reciprocal table MM 86.11.410 in Friberg, 2005: No. 76, p. 15). Later, in the Seleucid period, 𒑳 is also used to represent the number '9', replacing the older form 𒐝.

22 In this text, an *igû* and an *igibû* are mutually reciprocal numbers, i.e., *igibû* = 1/*igû*.

It appears singly in the first number in the last line of the excerpt, 01 𒑲 42 (i.e., 01 00 45) and doubly in the numbers 02 𒑲 𒑲 33 20 and 01 𒑲 𒑲 16 40, representing 02 00 00 33 20 and 01 00 00 16 40 respectively. The sign 𒑲 is the final form of $0_S$ in cuneiform, generally used in both mathematics and mathematical astronomy and remaining unchanged in form for some 300 years.

Of course the principal use of 𒑲 is to be found in the astronomical texts since it is here that many-digit numbers, constrained by empirical observations, are to be found and it was no longer possible to deliberately avoid intermediate zero sexagesimal digits as had been the practice in earlier purely mathematical procedures. Astronomical texts have been found almost exclusively in two sites only; Babylon, in the North of Babylonia, and Uruk, in the South. The two sites represented separate astronomical schools distinguished not only by some of the procedures used to predict the future position of the Moon, stars and planets but also, more pertinent for us, by the fact that, though 𒑲 serves as a *sexagesimal* placeholder in both, it retains its *demarcation* placeholder role in Uruk alone.

We can resume the results of this first strand of our double history by Table 3.1.

TABLE 3.1 Placeholder signs other than blanks

| Approximate date (BCE) | $0_D$ | $0_S$ |
|---|---|---|
| 1700 |  (Susa)<br> (Me-Turan) | — |
| 1200 |  | — |
| 650 |  |  |
| 600 | — |  |
| 500 |  |  |
| 300 | 𒑲 (Uruk)<br>— (Babylon) | 𒑲 |

The em dash '—' signifies that no non-blank sign is yet known for the period in question

In short, we see an uneven development in which local traditions and experiments no doubt played a major role in the choice and pattern of use of placeholder zeros. Coming into existence in the late Old Babylonian period, at first in outlying areas of cuneiform culture, an explicit marker for a missing place was originally used exclusively in the case of missing tens or ones in the construction of sexagesimal digits and only some thousand years later was it also employed for missing entire sexagesimal digits.

## 3 Babylonian 'Nothings'

The other strand of our story is that of the Mesopotamian ways of indicating 'nothing' in a numerical context, e.g., as a result of the subtraction of a number from itself. As in Part 2, before studying the texts of our main corpus, let us start with some Ur-III-period administrative texts. We can pick out two ways of representing this kind of 'nothing' in such texts.

The first case, HUJI 1701, a receipt for the delivery of sheep and goats by a herdsman, dates from the forty-eighth regnal year of Šulgi (2077 BCE in the Middle Chronology), and was found in the archives in the city of Girsu (Figure 3.7):

| | | | |
|---|---|---|---|
| 13 | ewes | | |
| | adult sheep | | |
| 20 | male sheep | | |
| 05 | shorn lambs "for the chosen ewes" | | |
| 05 | unshorn lambs | | |
| | lead goats | | |
| delivered. | | | |
| | ewes | 02 | male sheep |
| expended | | | |

FIGURE 3.7 HUJI 7100 obverse. Ur III administrative text. Šulgi 48, Girsu. Left: Photograph (CDLI P332525) with yellow arrows marking blanks. From the collections of the Institute of Archaeology, The Hebrew University of Jerusalem. Right: Translation (Conlan, 2013)

The numbers of different varieties of sheep (and goats) is here indicated with blank spaces left (marked in the photo by yellow arrows) in the case where no animals of the given kind were actually delivered (ewes or lead goats) or expended (ewes). The use of a fixed sequence of animal types means that a 'nothing' quantity must be included where applicable and, as originally in the case of a missing place in the sexagesimal positional notation, an uninscribed blank space was used to indicate it.

Another contemporary indication of the way to handle 'nothings' in administrative texts is to be found in the large Ur III tablets, called ***šà-bi-ta*** (commonly translated as 'balanced accounts').[23] These accounts, produced by the Ur III state administration, register material and financial resources made available to a merchant against which the turnovers realized by the merchant are offset, all values being given in silver equivalents. The difference produced between the former and the latter is then indicated in the last column on the reverse. If the amount advanced is greater than the turnover, a 'deficit' is indicated, if less, a 'surplus' is noted. Examples from the case of the Umma merchant Ur-Dumuzida, an important merchant in the city of Umma are:

- NBC 11448, dated to Amar-Suen year 7 = 2040 BCE (Snell, 1982, #10, pl. XVIII)
  Total: 2 ***ma-na*** 10⅙ ***gín*** 13½ ***še*** of silver.
  Expended.
  Deficit: 2 ***ma-na*** 15¼ ***gín*** less ½ ***še*** of silver.
- AO 5680, dated to Amar-Suen year 2 = 2045 BCE (Genouillac, 1922, pl. XVII)
  Total: 5⅚ ***ma-na*** ½ **gín** 7½ ***še*** of silver.
  Expended.
  Surplus: 14⅚ ***gín*** 13 ***še*** of silver.

The case where the sum advanced is equal to the expenditures is rarer but does exist:

- YPM BC 2908, dated to Amar-Suen, year 9 = 2038 BCE (Snell and Lager, 1991, #123)
  Total: 14⅓ ***ma-na*** 9⅓ ***gín*** 10½ ***še*** of silver.
  Expended.

Here the fact that the expenditure is equal to the quantity received – that there is neither a deficit nor a surplus – is expressed not by a term or symbol meaning 'nothing' but by silence of a blank line.

As the second millennium unfolds, one notes what is clearly a conscious attempt to avoid just such equal subtractions in mathematical texts. In the

23 The details of the exact nature of these 'balanced accounts' is subject to some debate in Assyriological circles. A balanced account of this 'balanced accounts' debate can be found in Neumann 1999.

case of *table* texts this is not surprising because the standard subjects of such texts – multiplication, reciprocals, squares and cubes and the corresponding roots – involved precision of one or two digits during the Old Babylonian period and at this precision, none of the tabulated values requires such a 'zero'. At the same time, since mathematical *procedure* texts are unconstrained in their choice of initial parameters, it is not difficult to avoid any need to express 'nothing'. There is one exception. This isolated example is to be found in a procedure text – from Susa once again – where a mathematical problem involving the determination of the dimensions of a rectangle contains the following set of instructions:[24]

(...) 30 *ù* 20 ***gar*** 5-*a-dù a-na* 4 *ri-*[*ba-ti*] ***sag***
*i-ši-ma* 20 *ta-mar* 20 ***sag*** 30 *a-na* 4 *ri-ba-*[*ti* ***sag***]
*i-ši* 2 *ta-mar* 2 ***gar uš*** 20 *i-na* 20 ***zi***
*ù i-na* 2 30 ***zi*** 1 30 *ta-mar* (...)

(......) Put down 20 and 30. Multiply 05 by 04, the fourth of the width, –
you will see 20, the width. Multiply 30 by 04, the fo[urth of the width]:
you will see 02. Write down 02, the length. Subtract 20 from 20,
and from 02 subtract 30: you will see 01 30. (......)

FIGURE 3.8 S$^b$ 13093 (Bruins and Rutten, 1961, #7, pl. 14) obv 5–8. Mathematical procedure text. Old Babylonian Period, Susa. Top: transliteration; Bottom: translation

The final two lines of our section involve a pair of subtractions: the first is 20–20 and the second 02–30 (to be understood as 02 00–00 30). Only the second operation is actually carried out, the (correct) result, 01 30, being indicated; the result of the first subtraction however, which should be 00, is simply passed over in silence, as we have seen to be the case for a similar situation in the earlier Ur III balanced accounts.

It is in fact not in the strictly mathematical tradition but in the much later mathematical astronomy tradition that the question of 'nothing' became unavoidable.[25] Particularly is this so in a new *observational* textual genus, the so-called 'Astronomical Diaries' and related texts, of which the oldest extant is dated to the sixteenth regnal year of the neo-Babylonian king Šamaššumukin

24 The interpretation of this problem contains numerous difficulties, though not for the two subtractions that interest us here. For a recent analysis of the problem see Høyrup, 1993, pp. 246–254.

25 For reasons of space, we will not enter into any explanatory technical detail beyond that strictly necessary to understand the role of 'nothings' in the astronomical tablets.

(652 BCE). Here were inscribed, for each year, astronomical observations and, for the 'Diaries,' also such events as epidemics, ominous monstrous births, meteorological information, prices and important political events. In a tablet of lunar observations for the year 7 of the reign of the Achaemenid ruler Cambyses II (524 BCE) there is the following entry:

| | | |
|---|---|---|
| ⸢*še* 30 | 15⸣ 30 | Month XII, (the 1st of which was identical with) the 30th (of the preceding month, NA$_1$:) 15 30. |
| 12 | 10 30 ŠÚ | The 12th, ŠÚ: 10 30. |
| *ge*$_6$ 13 | 05 20 ME | Night of the 13th, ME: 05 20. |
| 13 | ŠÚ *u* NA ***nu tuk*** | The 13th, there was no ŠÚ and NA. |
| *ge*$_6$ 14 | 10 GE$_6$ | Night of the 14th, GE$_6$: 10. |

FIGURE 3.9 BM 33066 (Hunger, 2001, #55, pp. 164–172, pl. 21) III 12′–16′. Lunar Six observation text, Cambyses II 7, Babylon. Left: transliteration. Right: translation

The tablet treats of the daily values of time, for the middle of each Babylonian lunar month, of what modern Assyriologists call the 'Lunar Six', that is, six parameters, named:

- KUR = time from last *visible* moonrise before sunrise to sunrise (near New Moon)
- NA$_1$ = time from sunset to first *visible* moonset after sunset (near New Moon)
- ŠÚ = time from moonset before sunrise to sunrise (near Full Moon)
- NA = time from sunrise to first moonset after sunrise (near Full Moon)
- ME = time from last moonrise before sunset to sunset (near Full Moon)
- GE$_6$ = time from sunset to first moonrise after sunset (near Full Moon)

and measured in units of UŠ (a 4-minute time interval). Their values are given in this excerpt for the twelfth month of the Babylonian year (corresponding to the month of February/March). For the thirteenth it is noted that there was neither an interval from moonset to sunrise nor one from sunrise to moonset. The expression used is 'ŠÚ *u* NA ***nu tuk***' where ***nu*** is the Sumerian logogram for 'not …' followed by the verb thus negated. Unfortunately, the Akkadian value of the Sumerian verb ***tuk*** is not completely known in the astronomical context; there are several possibilities. The meaning of the phrase though is quite clear; it corresponds to our 'nothing' or 'zero'.

---

ŠÚ *u* NA *ana* ***dù**-ka* 18 ***mu-meš*** *ana* ***egir**-k*[*a* ......]
*šal-šá-šú-nu* ***giš-ma ta*** NA *šá* ***murub***$_4$ BAR [......]
*ù šal-šú šá* ŠÚ *u* NA KI ŠÚ TAB-*m*[*a* ......]
*ki-i šal-šú šá* ŠÚ *u* NA ***ta*** NA ***zi***-[*ma* ......]
*eš-še-tú* E *ù mim-ma šá ina* ***ugu na*** [......]
*tak-mu-ru* ***lal***-⸢*u a*⸣-*na* NA šá MU-*k*[*a* ......]
*la i-ti-qu* ⸢ŠÚ *u* NA⸣ *šá* ***igi*** *šámaš* [......]
ŠÚ *u* NA *la i*-⸢*šu u šal*⸣-*šú* [......]

---

In order for you to calculate ŠÚ and NA, [you will return] 18 years behind you [... you will add ŠÚ and NA together,]

you will take one-third of them, and [you will subtract] it from the NA of the middle of month I, [and what remains of NA you will predict as the NA of your new year;]

and you will add one-third of ŠÚ and NA to the ŠÚ, and [you will predict it as the ŠÚ of your new year.]

If you subtract one-third of ŠÚ and NA from the NA [and it goes beyond the NA, (whatever goes beyond the NA,) as the ŠÚ]

[of your] new [year] you will predict. And whatever [it goes] beyond the NA [(of your old year), whatever] (this result) is less [than ŠÚ and NA which]

you have added [to each other, you will predict it] as the NA of your [new] year. [If it]

does not go beyond [the NA (of your old year)], ŠÚ and NA in front of the sun ... [...]

it has no ŠÚ and NA; and one-third [of ŠÚ and NA ...]

FIGURE 3.10 BM 42282 + 42294 (Brack-Bernsen and Hunger, 2008) rev 11′–18′. Astronomical procedure (Goal-Year) text. Achaemenid Period, unknown provenance. Top: transliteration; Bottom: transliteration

The determination of the Lunar Six was not only an observational question; the prediction of their values for a given date was also the subject of a number of astronomical procedure texts. One early class of such texts, called 'Goal-Year' texts in modern Assyriology, made use of the known 18-year periodicity (the 'Saros' cycle) in basic lunar positions. The oldest known example of such a text, most likely, like the preceding example, from the Achaemenid period, contains one section on the determination of ŠÚ and NA based on a correction added to or subtracted from their recorded values from the corresponding month one Saros cycle earlier (Figure 3.10).

The expression of interest here, similar to that in Figure 3.9, is 'it has no ŠÚ and NA' but the verb is in Akkadian here: *lā išû*, the negative (*lā*) of the verb 'to have, to own' (*išû*). It serves exactly the same purpose as in the previous example, that of picking out the zero values of the Lunar Six parameters ŠÚ and NA. Though only the last line contains the expression which interests us here, I have cited the full section of the tablet so that the algorithmic pattern might be clear; in particular that the broken-off ending of the last line of the excerpt is to be continued, as in lines 2–4 of the extract by 'of ŠÚ and NA', that is specifically here one-third of 'nothing'.

The use of the Akkadian expression *lā išû* for 'nothing' reposes the question of the Akkadian reading of the Sumerograms ***nu tuk***, because in a number of the Akkadian-Sumerian lexical lists, the learning of which made up a good part of the education of a scribe in Mesopotamia, one of the Akkadian equivalents for ***tuk*** is precisely *išû* (and even more specifically *ul išû* for ***nu tuk***).[26] This suggests that the expression ***nu tuk*** and *lā* (or *ul*) *išû* are equivalent.[27]

There is one final manner of expressing 'nothing' used in the astronomical texts. Rare, I know of only one appearance in the astronomical literature, occurring in a procedure text, one that also contains tables that are meant to be utilized in carrying out the algorithms on the tablet. The procedure which interests us begins in the following way:

⌜*ki*⌝-*i* 26 21 ***kur*** *ki-i an-na-a* ***tab*** 5 ***tab*** *šá zib*
5 ***tab*** *šá* 5 ***ḫun*** *bi-rit* ***tab*** *ana* ***tab*** *ja-a-nu*

If on day 26, the KUR is 21, the addition is the following: the addition for Pisces is 05, the addition for Aries is 05, the difference from addition to addition is *jānu*.

FIGURE 3.11 BM 36722 + 37205 + 40082 (Ossendrijver, 2012, #52) rev I 8–9. Lunar procedure text with tables. Early Seleucid Period (reign of Philip III Arrhidaeus?), Babylon. Top: transliteration; Bottom: translation

The indeclinable verbal form *jānu* represents '(there) is not', often used in informal sources such as personal letters. Here clearly it refers to the 'nothing' that is the result of the subtraction of 05 from 05.

26 The choice of which of the two Akkadian negative adverbs, *lā* or *ul*, is used in a given context is a question of syntax.

27 Note that this was apparently the choice of Otto Neugebauer who, in a list of lunar eclipses (AO 6485 + 6487), understands a writing of '*ul* ***tuk****-ši*' (Neugebauer, 1955, p. 477 s.v. *išû*) to represent '*ul išī*'. On the other hand, Mathieu Ossendrijver, citing the same source, reads '*ul ibašši*' with an equivalence ***tuk*** = *bašû*, 'to exist' (Ossendrijver, 2012, p. 18, n 100).

From the same tablet comes another 'nothing', this time in the tables used in calculating the Lunar Six parameters, specifically from that used for ME and $GE_6$:

| | | | | |
|---|---|---|---|---|
| ***pa*** | 06 | ***tab*** | 15 | ***zi*** |
| ***maš₄*** | 04 | ***tab*** | 14 | ***zi*** |
| ***gu*** | ***nu tuk*** | ***tab*** | 12 | ***zi*** |
| *zib* | 04 | ***tab*** | 10 | ***zi*** |
| ***ḫun*** | 08 | ***tab*** | 13 | ***zi*** |
| (…) | | | | |
| | | | | |
| *Pabilsag* | 06 | addition | 15 | subtraction |
| Capricorn | 04 | addition | 14 | subtraction |
| Aquarius | nothing | addition | 12 | subtraction |
| Pisces | 04 | addition | 10 | subtraction |
| Aries | 08 | addition | 13 | subtraction |

FIGURE 3.12 BM 36722 + 37205 + 40082 (Ossendrijver, 2012: #52) obv I 16–20 (2nd half of lines). Lunar procedure text with tables. Early Seleucid Period (reign of Philip III Arrhidaeus?), Babylon. Top: transliteration; Bottom: translation

This text is an excerpt from the 12-line full table listing the numerical values to be added (col. 2) or subtracted (col. 4) from a parameter value during the period when the planet is be found in the sector of the sky attributed to the zodiacal sign in col. 1. Here we find, not *jānu* as in the procedural part of the tablet (Figure 3.11) but, as in Figure 3.9, ***nu tuk*** for the zero value of the amount that is to be added; a difference that might be understood from a functional point of view as the distinction between a 'nothing' – *jānu* – that is the result of an operation (of subtraction) and a 'nothing' – ***nu tuk*** – which serves as the argument in an operation (of addition). We shall return to this question at the end of this section.

A similar but even more explicit role is played by ***nu tuk*** in an astronomical procedure text from Babylon which unfortunately cannot be securely dated but is probably Seleucid. Even more unfortunately it is in fragmentary shape with many signs missing both on both sides of the remaining text. Once again we present several lines to indicate context:

[...] ⌜***me***⌝ 10 ***ninda zi-šú nu tuk ù*** 10 ***ninda*** ⌜GAR⌝.[...]
[...] ⌜***me***⌝ 12 30 ***zi-šú*** 10 ***ù*** 12 30 [...]
[...] ⌜4⌝5 GAR.GAR-*ma* 17 45 PAP.PAP ***ta*** [...]

[...] ⌜'days'⌝ 10 ***ninda*** is its displacement. *Nothing* and 10 *ninda* you will ad[d ...]
[...] ⌜'days'⌝ 12 30 is its displacement. 10 and 12 3[0 ...]
[...] 45 you will add – 17 45 is the total. From [...]

FIGURE 3.13 BM 36680 (Ossendrijver, 2012, #25), obv 7′–9′. Astronomical procedure text for Jupiter (System A). Seleucid Period (?), Babylon. Top: transliteration; Bottom: translation

The point of the procedure is to determine the synodic motion of the planet Jupiter, that is the motion of the planet from the point of view of an observer on Earth. The displacement of the planet is measured in time units (here in ***ninda***, a unit equal to 1/60 of an UŠ, that is 4 min/60 = 4 sec). What remains of the first line of our excerpt is that 'nothing', ***nu tuk***, is to be added to 10 ***ninda*** and a numerical result is expected (as shown in the case in line 3 of the excerpt), the result itself having alas been lost.

There are a half dozen further examples, in every case with ***nu tuk***, of 'nothing' being indicated in an astronomical text: table, procedure or 'diary'. Nothing further would be gained by giving them explicitly here and all are later than Figure 3.9 given above. We can thus produce a small table with the earliest examples of our three (or is it two?) forms of 'nothing'.

TABLE 3.2 Earliest occurrence of 'nothing' in astronomical texts (BCE)

| **Earliest occurrence in astronomical texts (BCE)** | **'Nothing'** |
|---|---|
| Old Babylonian – Neo-Babylonian periods | — |
| 524 | ***nu tuk*** |
| 500 | *lā išû* |
| 300 | *jānu* |

If the story of the Babylonian placeholders was one of a long period of experimentation with a unified and systematic set of rules appearing only very late, the history of 'nothings' in the mathematical astronomical tradition started at the time of, and within, the astronomical genre. It knew fewer attempts – two or perhaps three (depending on the identity or not of ***tuk*** and *išû*) – to settle on a formulation and a much more restricted use of the terms involved. In part this was due to the fact that the occasions in which operations on astronomical parameters would lead to a value of zero were comparatively rare; it was even the case that a special algorithm would sometimes be invoked to avoid just such a situation.

## 4 Conclusion

To summarize, two of the functions we attribute now to zero, a place holder in our positional system and the result of an operation, never converged in Mesopotamia. Their histories had their own separate developments, often local ones, over two thousand years. As seen in the difference between tables and procedural texts, the material questions of display is a key feature of these developments.

Even if the canonical story is right about the placeholder role in the sexagesimal system being played initially by empty spaces, the introduction of specific signs to replace the blank is both earlier and qualitatively different from that version. We have seen that the avoidance of sexagesimal zeros is quite clear in the early mathematical textual tradition and the earliest signs in a placeholder context arose not to indicate *sexagesimal* but rather *demarcation* zero placeholders. The need to disambiguate two consecutive sexagesimal digits in the case where the first had no 𒁹s and/or the following digit had no 𒌋s led to the creation and utilization of non-numerical separator signs to play this role.[28] This continued as the dominant use of placeholders until the second half of the first millennium BCE and the extension of the demarcation sign to represent a placeholder for sexagesimal missing digits – but which in Uruk at least continued besides to play its demarcation role as well until the very end.[29]

The other strand of our story, that of the Babylonian concept of 'nothing', is quite distinct. Though the method of expressing 'nothing' in some Ur III administrative texts is to use a blank, like the early placeholder method, the issue is simply avoided in the mathematical domain until the advent of mathematical astronomy in the first millennium forced a confrontation with the problem. As we have seen, there are two or three means of representing 'nothing' in the astronomical texts. And these 'nothings' serve precisely as the candidates for the existence of a concept of a 'number zero' in Ancient

28 Cuneiform separator signs would merit further study; they seem generally to have played the role of typographical signs in modern print culture.

29 Among the various choices made over time of signs for this function, particularly puzzling to a modern are the use of numerical signs to represent placeholders; a choice that seems to invite ambiguity. The final adoption of 𒑱 avoided the difficulties of earlier choices and its universal adoption in Babylonia does appear both to ancient Babylonians and to us to have been a good one.

Mesopotamia. The essential question is then: what would serve as evidence of such a concept? In the absence of any formal definition of mathematical concepts in Mesopotamia one must rely on the usage that can be determined in the texts, that is, on the functional role played by the various appearances of 'nothings'. Their roles in the astronomical texts (we recall that they do not appear in the mathematical corpus) are threefold:

– to indicate that a zero numerical value to be attached to a parameter (Figures 3.9 and 3.10) – *lā išû*, ***nu tuk***[30]
– to indicate that the result of a given arithmetical operation is zero (Figure 3.11) – *jānu*
– as one argument in an arithmetical operation (Figures 3.12 and 3.13) – ***nu tuk***.

The first point, though suggestive, does not imply that 'nothing' was conceived as having an *operational* role in arithmetic. The second and third of these points however do so imply; 'nothing' can be the result of an arithmetic operation or a parameter to be operated on in an arithmetic sense. Exactly how one would wish to characterize this – in particular, the extent to which one is willing to say that this counts as evidence for the existence of a 'concept of the number zero' in late Mesopotamia – depends on what one wishes to understand as the 'number concept.'

This question brings us back full circle to the epigraph at the beginning of this chapter. Wittgenstein's 'thread' of overlapping fibers provides, I think, a good starting point for talking about the question: Did the Babylonians have a concept of zero? We have seen that two aspects of what we today think of as zero – placeholder and 'nothing' – were extant in ancient Mesopotamia but as two separate practices. In the thread of modern zero they form, if you will, two overlapping but distinct fibers; they neither combine nor is one the simple prolongation of the other. In general, our modern concept of zero contains many fibers: zero as the neutral element of an additive group, as the cardinality of the empty set, as the starting point in the recursive definition of non-negative integers, etc.[31] There is no event in history at which one may point and say: There the true zero was invented/discovered. This is true of course for all concepts of a certain degree of complexity, even in a domain so apparently

30 To these two can be added a small number of other, very similar, examples not discussed here, all using the expression ***nu tuk*** – cf. for tables, Neugebauer, 1955, #101 obv XII 9; #120 obv VIII 5, VIII 11; #200 obv II 25, rev II 12; #211 obv 12, and for procedure texts, Ossendrijver, 2012, #22 obv 8; #34 obv(?) I 6′; #38 obv(?) 11′; #53 obv II 25′, rev II 14 (part of a table); #74 obv 7′, 13′.

31 In this respect it is suggestive that in Peano's famous creation of the axioms of natural numbers, his initial formulation left out zero and began the construction with 1 (Peano, 1889, p. 1).

formalized as mathematics. And one must keep in mind that in the historical creation of mathematical concepts, not only can fibers be created, they can also be lost. From this point of view our zero is, like that of the Babylonians and other civilizations, complex and multiple.[32]

## References

Brack-Bernsen, Lis and Hunger, Hermann (2008). BM 42282+42294 and the Goal-Year Method. *SCIAMVS* 9, pp. 3–23.

Brown, David, Fermor, John and Walker, Christopher (2000). The water clock in Mesopotamia. *Archiv für Orientforschung* 46/47, pp. 130–148.

Bruins, Evert M. and Rutten, Marguerite (1961). *Textes mathématiques de Suse* (Mémoires de la Mission archéologique en Iran 34). Paris: Librairie Orientaliste Paul Geuthner.

CDLI. Cuneiform Digital Library Initiative, https://cdli.ucla.edu [organized by CDLI number].

Conlan, Christopher (2013). Note on an Ur III administrative tablet. *NABU* 2013/4, pp. 104–105.

Friberg, Jöran (1990). Mathematik [in English]. In D. O. Edzard (Ed.). *Reallexikon der Assyriologie und Vorderasiatischen Archäologie*. Volume 7. Berlin: Walter de Gruyter, pp. 531–585.

Friberg, Jöran (2005). Nos. 72–77. Mathematical texts. In I. Spar and W. G. Lambert (eds.). *Cuneiform Texts in The Metropolitan Museum of Art* I: *Literary and Scholastic Texts of the First Millennium BC*, New York: Metropolitan Museum of Art, pp. 268–314.

Friberg, Jöran (2007). *A Remarkable Collection of Babylonian Mathematical Texts* (Manuscripts in the Schøyen Collection: Cuneiform Texts 1). New York: Springer.

Friberg, Jöran and Al-Rawi, Farouk N. H. (2016). *New Mathematical Cuneiform Texts*. Cham: Springer.

Genouillac, Henri de (1922). *Textes économiques d'Oumma de l'époque d'Our* (Textes Cunéiformes du Louvre 5). Paris: Paul Geuthner.

Goldstein, Catherine (1998). À la recherche des origines : contenus, sources, communautés et histoires. In Catherine Goldstein, Jeremy Gray and Jim Ritter (eds.). *L'Europe Mathématique : Histoires, Mythes, Identités*. Paris: Éditions de la Maison des sciences de l'homme, pp. 15–32.

Høyrup, Jens (1993). Mathematical Susa texts VII and VIII. A reinterpretation. *Altorientalische Forschungen* 20, pp. 245–260.

32 For a set of examples of this multiplicity of modern zeros see Vignes, 1987.

Hunger, Hermann (2001). *Astronomical Diaries and Related Texts from Babylonia* Volume V. *Lunar and Planetary Texts*. Vienna: Verlag der Österreichische Akademie der Wissenschaften.

Keiser, Clarence E. (1919). *Selected Temple Accounts of the Ur Dynasty* (Yale Oriental Series 4). New Haven [CT]: Yale University Press.

Krecher, Joachim (1971). Glossen A. In sumerischen und akkadischen Texten. *Reallexikon der Assyriologie und Vorderasiatischen Archäologie*, Vol. 3, pp. 431–440. Berlin: Walter de Gruyter.

NESH (2005). Statement from the National Research Ethics Committee for Social Sciences and the Humanities regarding research on material of uncertain or unknown origin. (English translation) https://www.forskningsetikk.no/globalassets/dokumenter/4-publikasjoner-som-pdf/statement-regarding-research-on-material-of-uncertain-or-unknown-origin-2005.pdf.

Netz, Reviel (2002). It's not that they couldn't. *Revue d'histoire des mathématiques* 8, pp. 263–289.

Neugebauer, Otto (1935). *Mathematische Keilschrift-texte*. Vol. I. Berlin: Julius Springer.

Neugebauer, Otto (1937). *Mathematische Keilschrift-texte*. Vol. III. Berlin: Julius Springer.

Neugebauer, Otto (1941). On a special use of the sign 'zero' in cuneiform astronomical texts. *Journal of the American Oriental Society* 61, pp. 213–215.

Neugebauer, Otto (1955). *Astronomical Cuneiform Texts*, London: Lund Humphries for the Institute for Advanced Study. (Reprinted 1983 by Springer, New York).

Neugebauer, Otto and Sachs, Abe (1945). *Mathematical Cuneiform Texts*, New Haven [CN]: American Oriental Society.

Neumann, Hans (1999). Ur-Dumuzida and Ur-Dun. Reflections on the relationship between state-initiated foreign trade and private economic activity in Mesopotamia towards the end of the Third Millennium BC. In J. G. Dercksen (ed.). *Trade and Finance in Ancient Mesopotamia*. Leiden: Nederlands Historisch-Archaeologisch Instituut te Istanbul, pp. 43–53.

Nissen, Hans, Damerow, Peter and Englund, Robert (1994). *Archaic Bookkeeping. Techniques of Economic Administration in the Ancient Near East*. Chicago: University of Chicago Press.

Oelsner, Joachim (2006). Der 'Hilprecht-Text': Tafel HS 245 und die Paralleltexte Sm 162 Rs. (CT 33, 11) sowie Sm 1113 (*AfO* 18, 393f.). *Archiv für Orientforschung* 51, pp. 108–124.

Ossendrijver, Mathieu (2012). *Babylonian Mathematical Astronomy: Procedure Texts*. New York: Springer.

Peano, Giuseppe (1889). *Arithmetices principia, nova methodo exposita*. Turin: Bocca.

Powell, Marvin (1976). Antecedents of Old Babylonian place notation and the early history of Babylonian mathematics. *Historia Mathematica* 3, pp. 417–439.

Proust, Christine (2000). La multiplication babylonienne: la part non écrite du calcul. *Revue d'histoire des mathématiques* 6, pp. 293–303.

Ritter, Jim (1999). Metrology, writing and mathematics in Mesopotamia. In Jaroslav Folta (ed.). *Calculi 1929–1999*. Prague: Prague Studies in the History of Science and Technology, pp. 215–241.

Ritter, Jim (2001). Les nombres et l'écriture. In Yves Michaud (ed.). *Université de tous les Savoirs IV : Qu'est-ce que l'Univers ?* Paris: Odile Jacob, pp. 114–129.

Robson, Eleanor (2004). Mathematical cuneiform tablets in the Ashmolean Museum. *SCIAMVS* 5, pp. 3–65.

Snell, Daniel C. (1982). *Ledgers and Prices: Early Mesopotamian Merchant Accounts* (Yale Near Eastern Researches 8). New Haven: Yale University Press.

Snell, Daniel C. and Lager, Carl H. (1991). *Economic Texts from Sumer* (Yale Oriental Series, Babylonian Texts, vol. 18). New Haven: Yale University Press.

Thureau-Dangin, François (1934). Une nouvelle tablette mathématique de Warka. *Revue d'Assyriologie* 31, pp. 61–69.

Thureau-Dangin, François (1936). Textes mathématiques babyloniens. *Revue d'Assyriologie* 33, pp. 65–84.

Vignes, Jean (1987). Zéro mathématique et zéro informatique. *Cahiers du Séminaire d'Histoire des Mathématique* (1) 8, pp. 25–42.

Vygodskij, Mark Jakovlevič [Выгодский, Марк Яковлевич] (1959). Происхождение знака нуля в Вавилонской нумераци [Origin of the sign for zero in Babylonian numeration]. *Историко-Математические Исследования* 12, pp. 393–420.

CHAPTER 4

# Aspects of Zero in Ancient Egypt

*Friedhelm Hoffmann*

## Abstract

Investigating zero in ancient Egypt raises different questions and covers about 3,000 years of cultural history. Starting with the Egyptian language (which was easily able to express the idea of non-existence), I first provide some general information about Egyptian number writing, followed by multiplicative number writing and then a critical assessment of the traditional opinion that the Egyptian word *nfr* was 'zero'. After that are considered, in turn, Egyptian expressions for 'nothingness' and missing objects (from the third millennium BCE), the absence of entries in bookkeeping (from the second millennium BCE), the absence of numbers and placeholder signs (from the second half of the first millennium BCE). Of particular importance was *ἰwty/ἰwṯ* which was incorporated into Greek (in the first century CE) and finally into Arabic (in the eleventh century) astronomical texts in sexagesimal notations.

The Egyptians did not understand zero as a numerical value. They did not need it for expressing any number and they did not calculate using zero. The question is finally raised as to which criteria one should accept as a firm and definite proof for the existence of the concept of zero as a numerical/mathematical value in a culture, not only the Egyptian.

## Keywords

ancient Egypt – non-existence – 'which does not exist' – absence of numbers – non-existing dimension in geometry – number writing: general, positional, multiplicative, sexagesimal – *nfr* – placeholder

## 1 Introduction

Writing about ancient Egyptian numbers means dealing with roughly 3,000 years of written sources. During these millennia many things changed and several developments can be observed. Ancient Egypt is not the uniform and stable block that it is sometimes portrayed as. Of course, only a very short

 | DOI:10.1163/9789004691568_007

Egyptological summary of Egyptian conceptions of non-existence, nothingness, and zero mainly in a mathematical context can be given on these pages.[1]

Our sources come from all areas of the Egyptian culture: religion, literature, scholarship, science, administration, private documents, and many more, partly on papyrus, partly on ostraca (pottery sherds or limestone chips), partly in inscriptions. This variety of written evidence gives good, although fragmentary, insights into the Egyptian ideas of zero and related concepts. Due to its long existence and the political vicissitudes throughout its history, Egypt became both the recipient of foreign ideas from other peoples and in turn was able to transfer its own concepts to foreign countries. Most important here were the contacts with the Mesopotamian and the Greek scholarly traditions.

## 2 Non-existence

The non-existence of something could be easily expressed in Egyptian by the Old Egyptian negation *n*(*y*) (> *nn* later in this function). One can find, e.g., *wn*=*ṯ n*(*y*) *ṯn wnn*=*ṯ n*(*y*) *ṯn* ('Did you (fem.) exist or did you not (exist)? Do you exist or do you not (exist)?') (Pyramid Spell 415 § 738c; ca. 2300 BCE) or *nn wn*=*i* ('I will not exist') (Papyrus BM EA 10754 A,10 = Discourse of Sasobek; ca. 1800 BCE).

One could also use *tm wn* ('not to exist'), formed from the verb *wn* ('to exist') preceded by the verb of negation *tm*:[2] (the Egyptian king) *ir s*(*t*) *m tm wn* 'who makes them (= the foreign countries) as (= into) non-existing ones (Beth-Shan Stela of Ramesses II, line 12 = Kitchen, 1979: 151,4; ca. 1267 BCE). As one can see, this combination can be used to characterize things or persons that/who do not exist, are destroyed, or have never come into being. Several further examples are found in the *Book of Overthrowing Apophis* (P. British Museum EA 10188; fourth century BCE): the evil god Apophis is made *m tm wn* ('as one non-existing').[3] In the same papyrus, there is also a long litany declaring that the evil god Apophis and everything belonging to him will not exist. Each invocation begins with *nn ḫpr*=*k* ('You will not exist'), making use of the negation

1 Cf. already Hoffmann (2004/05) [online at http://archiv.ub.uni-heidelberg.de/propylaeumdok/volltexte/2013/1959]. See there for additional references.

2 Perhaps there are connections between *tm* ('to come to an end'), etc. (Wb V 301,4–302,3), *tm* ('to be not') (Wb V 302,5–303,11), and *tm* ('to be complete') (Wb V 303,12–304,16). For the explanation of the name of the god of creation Atum (Egyptian (*i*)*tm*), interpretations like 'the Complete One', to whose perfection the not-yet-existing and the development out of nothing belong, have been proposed (cf. Kákosy, 1975, p. 550; Myśliwiec, 1979, pp. 78–81).

3 P. Bremner-Rhind = P. BM EA 10188 27.9, 27.12 (late fourth century BCE; ed. Faulkner [1933]).

*nn* and the verb *ḫpr* ('to become, to exist').[4] The only kind of existence left over for this foe is *ḫpr=k m tm wn* ('You exist as someone non-existent.').[5] It is interesting to see that non-existence was considered as a kind of existence.

In leaving these more philosophical ideas and coming closer to mathematical problems, I would nevertheless like to stress that mathematical expressions are derived from general words and concepts of a culture. In ancient Egypt, this development can be followed over millennia.

While in Old Egyptian grammar *n(y)* ('not to exist') was used as a predicate in sentences of non-existence and *nn* in Middle Egyptian (corresponding to *nn wn*), a new word *mn* became normal in Late Egyptian and Demotic,[6] the old negation(s) *n(n)* now being used for negating other types of sentences.

But in our search for something like zero another expression became more relevant. The Egyptian language had not only a relative pronoun/converter *nty*, fem. *nt.t* 'who (is)', but also a negative relative pronoun/converter *ỉwty*, fem. *ỉwt.t* 'who (is) not'.[7] The feminine can be used for the neuter, so *ỉwt.t* could also mean '(that) which (is) not'. This word, which played a major role in later periods, is attested from the Pyramid Texts, the earliest Egyptian religious text corpus, first written down in the Old Kingdom in the middle of the third millennium BCE. Here already, the opposition between *nt.t* and *ỉwt.t* becomes apparent: Spell 510 § 1146c[8] characterizes the creator as 'who says what is (= exists) (*nt.t*), who creates what is not (= does not exist) (*ỉwt.t*)' (here again, not to exist is a particular ontological status resulting from a process of creation and complementary to existence). By the Middle Kingdom (*c.*2119–1794/93 BCE), *nt.t ỉwt.t* 'that which exists and that which does not exist' has become a common phrase for 'everything'.[9]

The word *ỉwt.t* also played a role in bookkeeping from the Old Kingdom. Objects that should be there, but are lost or destroyed, together with their quantity could continue to be kept in the records under the rubric *ỉwt.t*.[10]

4 P. Bremner-Rhind 31.4–9.

5 P. Bremner-Rhind 31.4.

6 The Egyptian language existed for about 4,000 years. Since it changed considerably over time, several phases can be distinguished: Old Egyptian mainly in the third millennium BCE, Middle Egyptian in the first half of the second millennium, Late Egyptian from the second half of the second millennium to the first half of the first millennium, Demotic from the seventh century BCE to the fifteenth century CE, and Coptic (Egyptian written in the Greek script) from the third century CE.

7 Wb I 46,1–47,3; Edel, 1964, pp. 551–54 §§ 1065–72; Gardiner, 1957, pp. 152–53 §§ 202–03. Peust, 2023, discusses a possible etymological origin from 'who (is) coming' > 'who (is) not (yet)'.

8 Attested in the pyramid of Pepi I (around 2300 BCE) for the first time.

9 Gardiner, 1957, p. 153 § 203,4; Wb I 47,2.

10 For examples see, inter alia, Posener-Kriéger, 1976, p. 666.

Although *iwt.t* and the numbers of the things that are missing – normally written in red ink – remind us of negative numbers, *iwt.t* does not play a role in Egyptian mathematical texts. In the extant corpus, there are no exercises that make use of zero, and neither did one need a sign like '0' for writing numbers.

## 3 Egyptian Number Writing – Including Some Less Well-Known Aspects

The traditional Egyptian number[11] signs comprised signs for the units, tens, hundreds, thousands, ten thousands, hundred thousands, and millions that were written, e.g., twice for writing 200: 𓍢𓍢.[12] 3,020 would be 3 times the sign for thousand + 2 times the sign for ten: 𓎆𓎆𓆼𓆼𓆼. There was no need to express the fact that there are no hundreds and no units in our example. Since every power of 10 had its own sign and no positional decimal notation was used[13] (like 75, 7.5 or 0.75), no sign for zero was used or needed, nor for fractions. 1/2 1/4 (✕𓂋) would be the Egyptian way of expressing 3/4 = 0.75.[14] These principles also remained the same in the cursive hieratic and Demotic scripts, attested from the third millennium BCE and the middle of the seventh century BCE respectively. One should add that in the cursive writings most groups of units, tens, hundreds, and thousands resulted in special ligatures.[15] Nevertheless, the Egyptians were able to cope with any calculation within the four basic mathematical operations: addition, subtraction, multiplication, and division.

Occasionally a dot was used to separate a number from a preceding word. Examples can be found, e.g., in P. Rhind problems 65, 68, 71, 75, 78,[16] and in Demotic before the number 1,[17] probably to differentiate a simple stroke at the end of a word from the number 1.

There was, however, also the possibility of writing numbers in an ornamental way – visual poetry as one might call it.[18] This is a common phenomenon in hieroglyphic inscriptions of the Graeco-Roman period and it is not restricted to numbers. The Egyptians used these writings to add visual associations with

11 Sethe (1916) is still the best overview of numbers in the Egyptian language and culture. The articles by Bibé (1984) on the basics of number writing and (1989) on fractions give an overview of Egyptian numerology from a mathematical historical perspective and are mostly based on Egyptological secondary literature.

12 All Egyptian examples are given in the direction of writing from right to left.

13 Cf. Imhausen (1996).

14 Neugebauer (1926); Waerden (1938); and Reineke (1993) for Egyptian fractional arithmetic.

15 Möller, 1909–12, nos 614–649; Erichsen, 1954, pp. 694–703.

16 For the texts, see Peet (1923) and Imhausen (2003). 'P.' = 'papyrus'.

17 Cf. Hoffmann, 2014, p. 88 n. 42.

18 See, e.g., Morenz (2008).

the content of the text to the signs used for writing it.[19] Instead of writing, e.g., '8', one could use, for example, or or or , as well as other signs.[20] I cannot enumerate these writings here. They did not have a practical benefit and were not used in mathematical texts, but only in religious inscriptions and papyri.[21]

The Egyptians of the Graeco-Roman period had also a special way of expressing day numbers in a date by writing these as fractions of the month. Each Egyptian month had 30 days. Thus 1/5, e.g., could be used for 'day six'. These writings, too, were restricted to religious inscriptions. I mention them here because only from the context – in a date – does it become clear to which quantity the fractions refer and of which they are a part.

In combination with the principle of visual poetry mentioned earlier, e.g., the five strokes of a normal writing for 1/5 ( ) can be arranged as a star ( ), thus resulting in a writing for 'day 6' as .

To conclude this introductory chapter on less widely known aspects of writing numbers in ancient Egypt it should be added that there was already in the third millennium BCE a strong tendency to write month and day numbers horizontally (e.g., instead of ).[22] While in hieroglyphic writing, this habit does not have any particular effect, the horizontally written numbers resulted in completely new ligatures in the cursive Egyptian scripts.[23] In addition, special ligatures were created for writing fractions of the *arura* (a measure of area) and still others for fractions of measures of capacity.[24]

Sometimes the form of a sign can influence that of a following number. The scribe of P. Vienna G 19818 verso (covering the years 91/92–96/97), for example, even uses three different forms of the sign for 2.[25]

The special ligatures of fractions of the corn measure led to signs that were later interpreted as parts of the eye of the god Horus ( ).[26]

19 An extreme example is the cryptographic hymns in the temple of Esna from the Roman period (Sauneron, 1982, nos 103 and 126; Leitz, 2001 and 2022).

20 Wit, 1962, p. 278.

21 For a list, see Wit (1962).

22 Edel, 1955, pp. 180–81 §§ 415–17.

23 Möller, 1909–12, nos 656–666; Erichsen, 1954, pp. 707–712.

24 Möller, 1909–12, nos 695–719; Erichsen, 1954, p. 706. Cf. also Pommerening (2005) and Zauzich, 1987, pp. 464–471.

25 Hoffmann, 2014, p. 86 n. 20.

26 Gardiner (1957), but cf. Ritter (2002) and Pommerening, 2005, p. 140.

## 4 Multiplicative Number Writing

When I said that there was no positional use of ciphers in writing numbers, this is not strictly true. There are some exceptions.

Multiplicative writing was common for the highest numbers. The sign for 1 million (𓁨, Egyptian *ḥḥ*) was mainly used, for example, in 'historical' and religious texts for expressing the idea of a really high number. But in accounting and in mathematics, the sign for 100,000 was the usual highest number sign.[27] For example, '6 million' was written with the sign for 100,000 and the sign for 60 below it. This stood for 60 times 100,000. As one can see, 100,000 was understood as a kind of unit that could be counted. Only the context would make it clear (or plausible) whether 60 times 100,000 or 100,060 was meant.[28] Multiplicative writing for higher numbers was also usual with 10,000s in the cursive scripts.[29] In hieroglyphs occasionally 1,000 was used as a base.[30] These multiplicative writings of higher numbers were probably related to the realization of numerals in the spoken language. For example, 3,000 was 'three thousands' in Egyptian.[31] Only rarely do expressions like 'four twenties', i.e., 80 – cf. French *quatre-vingt(s)* – occur in Coptic, the latest phase of the Egyptian language.[32]

Administrative routine tried to reduce the amount of scribal work. So with the *heqat* measure, which was typically used for corn, the high numbers were reduced by using a base of 100 *heqat* measures. In the hieratic (cursive) writing, the signs for 1 and *heqat* ([33]) mean 100 *heqat*, while the signs in the sequence *heqat* and 1 () mean 10 *heqat*; the *heqat* sign and a dot () mean

27 There are, however, exceptions, where two million signs are used for writing 2,000,000, e.g., in a hieratic cloth list in a non-royal tomb of the twenty-third century (Blackman, 1924, pl. 20 no. 6; cf. Edel, 1955, p. 166 § 385) or in a hieroglyphic inscription of Osorkon I in the early first millennium BC (Lange-Athinodorou, 2019, pp. 573 and 576 fragment I 2; probably also p. 573 and p. 575 fragment C 2).

28 Cf. Gardiner, 1957, p. 191 § 259; Quack (2002); Hoffmann, 2018, pp. 16–18.

29 Möller, 1909–12, no. 653; Erichsen, 1954, p. 703.

30 Imhausen, 2016, p. 21; Hoffmann, 2018, pp. 16–17.

31 Edel, 1955, pp. 171–76 §§ 397–403 (Old Egyptian); Till, 1961, p. 23 § 102 (Coptic). During the periods in between, almost no numerals were written out as words.

32 Till, 1961, p. 23 § 104. It should also be noted that the normal hieratic sign for 70 can be analyzed as a multiplicative writing 10 × 7 (Meyrat, 2014, p. 272; I think the sign for 10 is standing, not lying, before the Roman period [cf. Möller, 1909–12, no. 629]).

33 The examples here are hieratic in the original and have only been transcribed into hieroglyphs through convention.

1 *heqat*[34] (cf. also the next section). In a similar way, the notation of areas can also be simplified: stands for 40 *aruras*, for 4 *aruras*.[35] Again, it makes a difference whether the number is written before or after the abbreviated word.

Finally, one could also write fractions in a simplified way, because the signs for fractions are normally arranged according to their value. A sequence like 1/2 1/4 8 is, of course, not a writing for 8 3/4 (this would be 8 1/2 1/4), but for 1/2 1/4 1/8 = 7/8. The indication that 8 is a fraction (in hieroglyphs by , in hieratic and Demotic reduced to a mere dot over the number sign) could simply be left out, since the fractions had to be arranged in a descending order.[36]

All these phenomena are at least loosely connected with what could be called a positional writing system of numbers, as the context and the position of a number tell us the intended value. But this kind of writing was only used under specific circumstances and did not form a coherent general system.

## 5 Indicating Zero Amounts in Corn Measure Units

The different usages of dot and stroke gave rise to an interesting corn measure notation during the Ramesside period (1292–1069 BCE; New Kingdom), in which the non-existence of the number of the higher unit was implicitly noted. Ostracon DeM 213[37] from the twelfth century BCE may serve to illustrate this: was an abbreviation for the *khar* measure, which consisted of four quadruple *heqat* (= *oipe*) in this period. The number of *khar* was indicated by strokes, the number of *oipe* by dots. These consistently different number signs made it unnecessary to write the word *oipe* when it was meant: thus is '1 *khar*, 2 (*oipe*)' (line 6), and consequently stands for '(0) *khar*, 2 (*oipe*)' (line 5). Probably because dots can also occur in other contexts, one refrained from writing the dots alone ( ), but a clear connection with the corn measure was kept by starting the notation with .

34 In original hieroglyphic writing, multiples of the *heqat* were obviously expressed with the ordinary numbers (Gardiner, 1957, p. 198 § 266). For the Demotic writings, cf. Zauzich, 1987, pp. 462–464, and for the historical development Vleeming (1981).

35 Gardiner, 1957, p. 200 § 266.3.

36 Cf. the discussion in Jordan, 2015, pp. 343–364.

37 For the text see also Kitchen, 1983, p. 590, 10–14 and cf. Pommerening, 2005, p. 155.

## 6 *nfr*

In the large accounting papyrus Boulaq 18 from the Theban royal court of the 13th Dynasty (1794/3–1648/45 BCE),[38] several times one finds the sign 𓄤, e.g., in fragments XVIII (= Scharff, 1922, p. 5** no. 12; Allam, 2019, vol. a: 5+5a, vol. b: 50) or XXVII (= Scharff, 1922, p. 12** no. 41; Allam, 2019, vol. a, p. 15+15a, vol. b, p. 64) as well as in other places.[39] In the accounts, incoming and outgoing goods are listed. In the last line the balance ('rest', i.e., what remains) is given. This is expressed by the normal numbers. But if there is nothing here, because income and expenditure are identical, *nfr* is used. In XVIII 13, e.g., one reads as the amounts of the balance of bread, diverse deliveries, jugs of beer, baskets of dates, baskets of *ḥnw*-vegetables, bunches of *iꜣq.t*-vegetables, *ḥr.t*-cakes, and other cakes: '240, 203, 16, *nfr*, *nfr*, *nfr*, *nfr*, *nfr*.' Line 15 of fragment XXVII has for the same eight goods '200, 338, 21, 2, *nfr*, *nfr*, *nfr*, *nfr*'. It is true that *nfr* stands next to numbers and seems to be a sign indicating the number 0. Of course, if income and expenses are identical, the remainder is 0. For that reason, Gardiner (1957: 266 § 351 note 8a) and other authors[40] take the *nfr* sign as a symbol for zero – this goes too far, for *nfr*, which Gardiner rightly etymologically connects with *nfr* 'good', can also simply be an indication that income and expenses are equal. The account is 'good', the matter 'fits'. This is certainly the reason, why Scharff (1922, p. 5**, 12**) and elsewhere does not write '0' in his transcription of the text, but 𓄤, and why Allam (2019, vol. b, pp. 50 and 64) translates *erfüllt* ('fulfilled', 'accomplished', 'carried out'): *nfr* simply indicates that the accounts of the relevant items are settled and that there is no remainder.[41]

Another word or another use of *nfr* should be briefly mentioned here, namely *nfr* denoting the baseline in architecture.[42] A foundation or a wall can be below it or higher. This in itself is not necessarily a proof for something like 'zero', since the etymological meaning of *nfr* is simply 'good'. Also the negation *nfr* does not imply nothingness. It could simply be a similar use of 'good' as, e.g., in German *Jetzt ist aber gut!*, meaning *Jetzt ist aber Schluss!*: 'Stop!', 'No!'. In

38 Are the papyri or at least some of them to be specifically dated to the reign of Sebekhotpe II? Cf. *imny*[…] *sbk-ḥtp* (= a variant *of imn-m-ḥꜣ.t sbk-ḥtp*?) in Sch 52 line 2 and perhaps *sḫm*[*-rʿ ḫw-tꜣ.wy*] in Sch 3 line 12 which would be the prenomen of the same king.

39 Cf. the word index in Allam, 2019, vol. a: p. 19.

40 E.g. Spalinger, 1985, pp. 181–188, who uses '0' in his translation.

41 Cf. Imhausen, 2016, p. 20.

42 Arnold, 1991, pp. 17–18; cf. also Figure 1.8 on p. 12. Another example is Dobrev, 2005, p. 175, Figure 2. Dobrev calls it the 'horizontal 0 level of the monument'.

a similar way, the Egyptian *tm* ('to be complete') is related to the negative verb *tm* ('not to do').[43]

Finally the use of *nfr* in the negations *nfr-n*[44] and *nfr pw* should be mentioned here. But unlike Gardiner (1923 and 1957, pp. 266–67 § 351) and Brose (2009, p. 4), who suggest a derivation from *nfr* ('zero'), one can understand this meaning on the basis of 'good': the perfect state of something prohibits its change and thus produces prohibitive usages of the word 'good'.[45]

## 7 Nothingness/Absence of Numbers/Placeholder

I would like to come back to the notation of nothingness and zero.[46] We had *iwt.t* ('which is not') as an expression of things missing in inventory lists of the Old Kingdom. Also 𓈖 exists in the Old Kingdom, which is either a writing of the negation *n*(*y*) ('there is not')[47] or – less probable – a shortened writing of *iwt.t*.[48] The use of the negation 𓈖 as a mark before some calendar dates is used in the New Kingdom (1550–1070/69 BCE) to make it clear that the day marked in this way was not to be considered, that something did not take place or similar – unlike the days marked with a simple dot in the same lists. It is significant that this use of the mark 𓈖 is only found in continuous lists of days. The system of an uninterrupted list necessitated an entry for *every* day. Without the basis of a continuous list, it would have been enough to note down

43 Gardiner, 1923, p. 81 unconvincingly tried to explain the use of *nfr* as a negation in the following way: 'It seems quite easy to discern a sort of satisfactoriness about zero, this being the point where mathematical and other troubles cease'.

44 The doubts of Bibé, 1984, pp. 5 and 11, about the correct reading are unfounded, since in nearly all attestations the *n* is clearly written. Rare cases with the negating arms sign are conventional Egyptological transcriptions of hieratic *n* with a space filler on top of it.

45 *nfr-n* is a typically Old Egyptian negation, cf. Edel, 1964, pp. 588–94 §§ 1130–40.

46 Contrary to Gillings, 1972, pp. 227–28, it is not expressed by a blank space. What Gillings presents is his own reconstruction of what he thinks was the way the Egyptian scribe had to perform his calculation in writing. Gillings argues that in one of the steps there must have occurred a '0'. But in the original (P. Reisner I H24) only the result is given (without any zero or blank space in its writing), not the calculation (Simpson, 1963, pl. 14 and 14A). Thus there is no proof whatsoever for the use of a blank space as an indication of zero in Ancient Egyptian number writing. Blank fields do, however, occur in tables, if there was no entry.

47 Cf. Hoffmann, 2004/05, p. 52.

48 Posener-Kriéger, 1976, pp. 218–19; *iwt.t* is treated by Posener-Kriéger, 1976, p. 197.

only the days on which something *did* happen. The necessity of considering days with and without a particular event made a sign for 'not (at this day)' necessary.

In a similar way, Demotic  (from hieroglyphic ) *wš*, literally 'hole', 'without', is used in uninterrupted day lists. While, e.g., in ostracon ODL 134 in the Louvre different numbers of bunches are given for most days, there are entries that simply say *wš* 'without', meaning there was nothing to record for this particular day.

In the early Ptolemaic period, *ỉwt.t*, the old negative relative pronoun, surfaces again, this time in the originally masculine form *ỉwty*, in Demotic *ỉwṱ*. We find it in the context of a mathematical problem. In P. Cairo JE 89127–30+89137–43 verso S8 (c. fourth century BCE),[49] it occurs in respect of a pyramid that has a base of 500 by 500 cubits *wbꜣ ỉwṱ r pꜣ ḥꜣt.ṱ* 'against *ỉwṱ* toward the heart (= apex)'. In spite of some uncertainty, the general meaning seems to be clear: we are dealing with a real pyramid, not with a frustum. One should note that the text does not say '*ỉwṱ* cubits' or something like this for '0 cubits', but simply 'without'/'nothing'. This clearly shows that *ỉwṱ* is not the number zero, but an expression for saying that there is nothing, meaning that there is no extension at all.[50]

Another example of the use of Demotic *ỉwṱ* comes from the famous field register of the Horus temple at Edfu. The text is dated to the reign of Ptolemy X Alexander I (107–88 BCE), but says that it covers donations back to the time of Darius I (522/1–486/5 BCE) and even before that.[51] To understand the text and its relevance for our subject it is important to realize how the Egyptians used to give the measurements of fields. They were taken as four-sided areas – rectangular or irregular. The opposing sides were added together, the totals were halved and the two halves multiplied by each other. The calculation was only correct if the area was rectangular. But the approximation was easy and applied in real life. The scribes used a simplified notation like:

$$1\ 1/2\ \frac{2\ 1/4}{2\ 1/4}\ 1\ 1/2$$

49 Parker, 1972, p. 51; Jordan, 2015, p. 398.

50 Another example of *ỉwṱ* occurs in verso R13 (Parker, 1972, p. 49; Jordan, 2015, p. 389). The context (a triangle) is badly broken.

51 Meeks, 1972, pp. 134–35 thinks of Necho II (610–595 BCE).

which means a rectangular field such as this:[52]

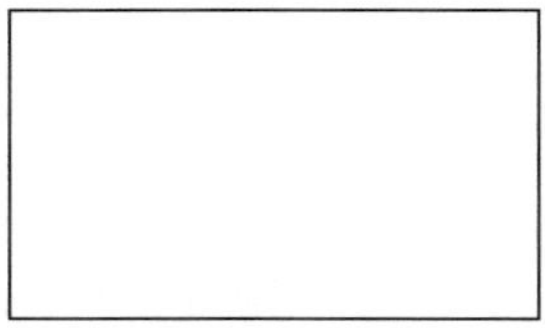

Even triangular fields were calculated in a similar way:

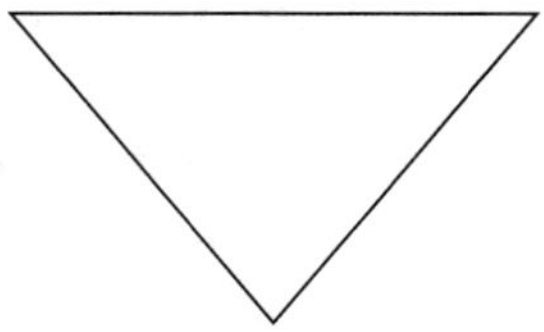

One simply assumed that one of the four sides did not exist. The opposing side was halved and multiplied with the half of the sum of the other two sides. In this case, the notation used by the Egyptian scribes looked like this:

2 1/4

1 1/2 —— 1 1/2

What was taken to be a non-existing side was simply left empty.

In using this notation around a thick stroke that symbolized the field, one could leave a position without a number. One might compare the zero *khar* measure in the New Kingdom, where also the writing of nothing at a certain position suffices to unambiguously indicate the omission of a number.[53] But how was this pronounced or transformed into words? This is where the Ptolemaic inscription in the temple of Edfu becomes relevant. The text has, for example:

52 Parker, 1972, p. 71. The text of problem 64 makes clear that really opposing sides 'the south and the north', 'the east and the west', are meant. The doubts of Gericke, 1996, vol. 1, pp. 59–60, are unfounded.

53 A really impressive example of representing something by writing no sign at all might be mentioned here: in this case it was an unwritten space that was left intentionally in an inscription on a wooden writing palette of the New Kingdom. If one accepts that the unwritten space allows the reader to see the wooden material and if this is *mrw*-wood, one can include the name of the material into which the inscription is incised as a rebus writing for a form of the verb *mri* ('to love'), which can be expected here on the basis of parallels in normal orthography (Seidlmayer, 1991, pp. 323 f. In contrast to Seidlmayer, I would prefer an active interpretation: 'who loves myrrh').

> (fields); their dimensions ...: 5 to 8, 19 to 19, makes 1[2]3 1/2 (Meeks, 1972, p. 29 [24*, 1])

This is an irregular four-sided field. And (5 + 8) / 2 × (19 + 19) / 2 = 123.5. But the immediately preceding entry has:

> *ỉwty* to 5, 17 to 17, makes 42 1/2 (also Meeks, 1972, p. 29 [24*, 1])

and thus refers to a triangle, for (0 + 5) / 2 × (17 + 17) / 2 = 42.5. In this case, the idea that it is triangular and that one side of a quadrangle is 'missing' is expressed by *ỉwty* ( is the most frequent writing;[54] only occasionally is an extra land sign added). By the use of this word, it is clearly stated that one side is non-existent. The text does not say that the length of one side is zero cubits or anything like this. Therefore, *ỉwty* is used to express the non-existence of a side, not the length of it, although the formulation suggests at first sight an expression for '0'. But the text can – and should – be understood as 'a non-existing side to 5, 17 to 17, makes 42 1/2'.

The same word was also used to indicate a missing number. In the hieratic papyrus Berlin P 23057 a–j (fourth century BCE) and the Demotic papyrus Vienna D 12006 (first half of the first century CE?), a system of three numbers was used for a kind of divination. Each position of numbers can have the values from 1 to 9. In addition, *ỉwṱ* was used. '1 1 2' or '2 5 7' are straightforward examples. But if one of the digits was absent, the empty position had to be clearly indicated. This was done by [55] in the hieratic text, in Demotic by *ỉwṱ* , which directly derived from , e.g., '2 *ỉwṱ* 2' in 3.26 or '3 1 *ỉwṱ*' in 3.30. In the Vienna papyrus, *ỉwṱ* served as a marker for the absence of a number at a position. Again the requirements of the system made it necessary to indicate the absence of a number.

Very similar is the use of *ỉwṱ* in Demotic astronomical lists, which make use of the sexagesimal notation known from Mesopotamia for degrees and minutes. The entries in P. Carlsberg 32 (second century CE) consist of four positions and have, e.g., '*ỉwṱ* 32 43 42' or '*ỉwṱ* 38 10 59'.[56] *ỉwṱ* has the form in this text. In the recently published P. Monts.Roca inv. 314 (first century CE), another

---

54 Not simply , as Reineke 1980, p. 1241; Gericke, 1996, pp. 59 and 287; or Imhausen, 2016, p. 20, state.

55 Hieratic in the original and transcribed into hieroglyphs through convention.

56 Edited for the first time by Parker (1962), later included in Neugebauer and Parker, 1969, pp. 240–41 and pl. 79 B.

Demotic astronomical table, the same method of notation is used for degrees and minutes.[57]

Finally, I would like to mention that in my opinion the Demotic sign *iwṯ* ( , , , etc.)[58] was taken over into Greek astronomical texts in the form , , , , and similar.[59] This Greek sign occurs from the first century CE onwards. As far as I can see, it remained restricted to sexagesimal tables and was taken over by Arabic scientists in exactly this usage, namely as a zero in sexagesimal astronomical number writing.[60] Earlier forms (eleventh to sixteenth centuries CE) are, e.g., . To the best of my knowledge, the sign only occurs in Arabic astronomical manuscripts within a system of writing sexagesimal numbers by letters. In this system, a sign for the absence of a number was necessary, but not for zero, as, e.g., '20' was written with the twentieth letter of the old Semitic alphabet, i.e., ك (*k* in Arabic).

## 8 Conclusion: the Place of Ancient Egypt in the History of Zero

The overall development in Egypt can be summarized as follows: Already in the third millennium BCE, there were words for 'non-existence', 'absence', 'missing', 'emptiness', etc. During the Old Kingdom, there were words for the notation of lost/not-existing things in inventory lists. From the New Kingdom, the possibility of marking whole entries in lists as not applicable can be found. From the middle of the first millennium onwards, *iwty*/*iwṯ* was used to note the non-existence of values or numbers. Thus, that whose non-existence could be written down became more and more abstract. But all expressions, including *iwty*/*iwṯ*, remained negations or placeholders and the step to a *number* 0 was not made. The Demotic sign for *iwṯ* as a placeholder was taken over by Greek astronomers and finally handed down to medieval Arabic astronomy.

I would like to raise the question as to what one can accept as a firm criterion for distinguishing a placeholder sign from zero as a mathematical value. Words for concepts like 'lack', 'nothing', 'emptiness', 'non-existence', etc., cannot be accepted as firm proof for the existence of the concept of zero as a numerical/mathematical value.

---

57 Ed. Escolano-Poveda (2018/19). In this papyrus, a secondary use of the *iwṯ* sign is found, by which the digits corresponding to the degrees and minutes are separated from each other, even if there is no ambiguity (Escolano-Poveda, 2018/19, pp. 26–27). The reason for this use is still unclear.

58 Cf. Erichsen, 1954, p. 25.

59 Jones, 1999, p. 62; cf. Irani, 1955, p. 11.

60 Irani, 1955, pp. 11–12.

A writing like '204' in itself would only show that it is a writing that follows a positional system, but it does not imply with certainty that 0 was also a number. Even higher numbers are open to interpretation. For example, 2,030, could be the result of a multiplication of 203 × 10, but neither the factors nor the result nor their writings are an incontrovertible proof that the person who performed this mathematical operation had any idea of zero as a numerical value. An abacus could be a simple, but effective tool for such an operation.[61] And no concept of zero as a number would be needed. In 10, 203, and 2,030, the cipher 0 could simply act as a placeholder in writing numbers.

Even the occurrence of mathematical problems like 55 × 0 = 0 or 72 + 0 = 72 would not provide clear evidence for 0/zero as a numerical value, since the sign 0 could stand for nothing (72 plus nothing remains 72). Written Egyptian mathematical problems are styled as texts (rhetoric) and do not make use of any operator symbols like + or × or of formulas; they present the solutions as a sequence of instructions (they are algorithmic); finally they are numerical, i.e., they do not use variables, but only specific numerical values and 'nothing'.[62] But zero as a numerical value does not occur.

Combinations like '0 bread' or '0 kg' (or in words, of course) could prove that zero existed as a value on the same conceptual level as other numbers. There is no such evidence from ancient Egypt. So, one can be as good as certain that no concept of zero as a numerical value existed.

## Acknowledgement

I would like to thank Cary Martin for correcting my English in this article.

## References

Allam, Schafik. (2019). *Hieratischer Papyrus Bulaq 18*: 2 vols. Urkunden zum Rechtsleben im alten Ägypten 2. Tübingen: [published by the author].

Arnold, Dieter. (1991). *Building in Egypt. Pharaonic Stone Masonry*. New York; Oxford: Oxford University Press.

Bibé, Celia. (1984). Numerología Egipcia, I. *Aegyptus Antiqua* 5, pp. 5–12.

Bibé, Celia, (1989). Numerología Egipcia, II. *Aegyptus Antiqua* 6–7, pp. 2–8.

61 Ifrah, 1992, p. 97, takes it for granted that abaci, which normally have 8 to 12 rows, can be enlarged to have up to 30 or more rows, which would allow calculations of up to $10^{30}-1$.

62 Imhausen, 2003, p. 13; Imhausen, 2016, pp. 70, 72.

Blackman, Aylward Manley. (1924). *The Rock Tombs of Meir. Part 4. The Tomb Chapel of Pepi'onkh the Middle Son of Sebekḥotpe and Pakhernefert (D, No. 2)*. Archaeological Survey of Egypt, Memoirs 25. London: Egypt Exploration Fund.

Brose, Marc. (2009). Die mittelägyptische *nfr-pw*-Negation. *Zeitschrift für ägyptische Sprache und Altertumskunde* 136, 1–7 [DOI: 10.1524/zaes.2009.0002].

Dobrev, Vassil. (2003). Builders' Inscriptions from the Pyramid of King Pepy I (Sixth Dynasty). In Hawass, Zahi and Lyla Pinch Brock (eds), *Egyptology at the Dawn of the Twenty-first Century. Proceedings of the Eighth International Congress of Egyptologists. Cairo, 2000*: vol. 3, pp. 174–177. Cairo; New York: American University in Cairo Press.

Edel, Elmar. (1955). *Altägyptische Grammatik* I. Analecta orientalia 34. Roma: Pontificium Institutum Biblicum.

Edel, Elmar (1964). *Altägyptische Grammatik* [II]. Analecta orientalia 39. Roma: Pontificium Institutum Biblicum.

Erichsen, W. (1954). *Demotisches Glossar*. Kopenhagen: Ejnar Munksgaard.

Erman, Adolf and Hermann Grapow (eds.). (1926–31). *Wörterbuch der aegyptischen Sprache im Auftrage der Deutschen Akademien* (5 vols). Leipzig: J. C. Hinrichs.

Escolano-Poveda, Marina. (2018/19). Astronomica Montserratensia I: A Demotic monthly almanac with synodic phenomena (P. Monts.Roca inv. 314). *Enchoria* 36, pp. 1–36.

Faulkner, Raymond O. (1933). *The Papyrus Bremner-Rhind (British Museum no. 10188)*. Bibliotheca Aegyptiaca 3. Bruxelles: Fondation Égyptologique Reine Élisabeth.

Gardiner, Alan H. (1923). A hitherto unnoticed negative in Middle Egyptian. *Recueil de travaux relatifs à la philologie et à l'archéologie égyptiennes et assyriennes* 40, pp. 79–82.

Gardiner, Alan (1957). *Egyptian grammar being an introduction to the study of hieroglyphs*, 3rd, rev. ed. Oxford: Oxford University Press.

Gericke, Helmuth. (1996). *Mathematik in Antike und Orient* (2 vols in 1). Wiesbaden: Fourier.

Gillings, Richard J. (1972). *Mathematics in the time of the pharaohs*. Cambridge, Massachusetts; London: MIT Press.

Hoffmann, Friedhelm. (2004/2005). Astronomische und astrologische Kleinigkeiten IV: Ein Zeichen für „Null", in P. Carlsberg 32? *Enchoria* 29, pp. 44–52 [online at http://archiv.ub.uni-heidelberg.de/propylaeumdok/volltexte/2013/1959].

Hoffmann, Friedhelm. (2014). Doppelte Buchführung in Ägypten: Zwei Wiener Abrechnungen (P. Wien G 19818 verso and 19877 verso). In Depauw, Mark and Yanne Broux (eds), *Acts of the Tenth International Congress of Demotic Studies: Leuven, 26–30 August 2008*, pp. 83–114. Leuven; Paris; Walpole, MA: Peeters [online at http://archiv.ub.uni-heidelberg.de/propylaeumdok/volltexte/2020/4773].

Hoffmann, Friedhelm. (2018). Einige Bemerkungen zu Zahlen im Buch vom Fajum. In Donker van Heel, Koenraad, Francisca A. J. Hoogendijk, and Cary J. Martin (eds.),

*Hieratic, Demotic and Greek studies and text editions: Of making many books there is no end. Festschrift in honor of Sven P. Vleeming*. Payprologica Lugduno-Batava 34, pp. 15–18. Leiden; Boston: Brill.

Ifrah, Georges. (1992). *Die Zahlen. Die Geschichte einer großen Erfindung*. Frankfurt; New York: Campus.

Imhausen, Annette. (1996). Das Zahlensystem der Ägypter – (k)ein Dezimalsystem? *Discussions in Egyptology* 36, 49–51.

Imhausen, Annette. (2003). *Ägyptische Algorithmen: Eine Untersuchung zu den mittelägyptischen mathematischen Aufgabentexten*. Ägyptologische Abhandlungen 65. Wiesbaden: Harrassowitz.

Imhausen, Annette. (2016). *Mathematics in ancient Egypt: A contextual history*. Princeton; Oxford: Princeton University Press.

Irani, Rida, A. K. (1955). Arabic numeral forms. *Centaurus* 4(1), pp. 1–12.

Jones, Alexander (ed.). (1999). *Astronomical papyri from Oxyrhynchus (P. Oxy. 4133–4300a)* (2 vols in 1). Memoirs of the American Philosophical Society, held at Philadelphia for Promoting Useful Knowledge 233. Philadelphia, PA: American Philosophical Society.

Jordan, Birgit. (2015). *Die demotischen Wissenstexte (Recht und Mathematik) des pMattha*: 2 vols. Tuna el-Gebel 5. Vaterstetten: Patrick Brose.

Kákosy, László. (1975). Atum. In Helck, Wolfgang and Eberhard Otto (eds) *Lexikon der Ägyptologie*: vol. 1, cols. 550–552. Wiesbaden: Otto Harrassowitz.

Kitchen, Kenneth A. (1979). *Ramesside Inscriptions: Historical and Biographical*, II. Oxford: Blackwell.

Kitchen, Kenneth A. (1983). *Ramesside Inscriptions: Historical and Biographical*, V. Monumenta Hannah Sheen dedicata 3. Oxford: Blackwell.

Lange-Athinodorou, Eva. (2019). Der „Tempel des Hermes“ und die Pfeile der Bastet: zur Rekonstruktion der Kultlandschaft von Bubastis. In Brose, Marc, Peter Dils, Franziska Naether, Lutz Popko, and Dietrich Raue (eds.), *En détail – Philologie und Archäologie im Diskurs: Festschrift für Hans-Werner Fischer-Elfert*: vol. 1. Zeitschrift für ägyptische Sprache und Altertumskunde, Beihefte 7/1, pp. 549–585. Berlin; Boston: De Gruyter.

Leitz, Christian. (2001). Die beiden kryptographischen Inschriften aus Esna mit den Widdern und Krokodilen. *Studien zur Altägyptischen Kultur* 29, pp. 251–276.

Leitz, Christian. (2022). *Esna-Studien II. Einleitung in die Litaneien von Esna* (3 parts). Studien zur spätägyptischen Religion 38. Wiesbaden: Harrassowitz.

Meeks, Dimitri. (1972). *Le grand texte des donations au temple d'Edfou*. Bibliothèque d'étude 59. Cairo: Institut français d'Archéologie orientale.

Meyrat, Pierre. (2014). The First Column of the Apis Embalming Ritual. Papyrus Zagreb 597–2. In Quack, Joachim Friedrich (ed.), *Ägyptische Rituale der griechisch-römischen Zeit*. Orientalische Religionen in der Antike 6, pp. 263–337. Tübingen: Mohr Siebeck.

Möller, Georg. (1909–12). *Hieratische Paläographie. Die ägyptische Buchschrift in ihrer Entwicklung von der fünften Dynastie bis zur römischen Kaiserzeit* (3 vols). Leipzig: Hinrichs.

Morenz, Ludwig D. (2008). *Sinn und Spiel der Zeichen: visuelle Poesie im Alten Ägypten*. Pictura et Poesis 21. Köln: Böhlau.

Myśliwiec, Karol. (1979). *Studien zum Gott Atum II. Name – Epitheta – Ikonographie*. Hildesheimer Ägyptologische Beiträge 8. Hildesheim: Gerstenberg.

Neugebauer, Otto. (1926). *Die Grundlagen der ägyptischen Bruchrechnung*. Göttingen: Dieterich.

Neugebauer, Otto and Richard A. Parker. (1969). *Egyptian Astronomical Texts III: Decans, Planets, Constellations and Zodiacs* (2 vols). Brown Egyptological Studies 6. Providence, RI: Brown University Press.

Parker, Richard A. (1962). Two Demotic astronomical papyri in the Carlsberg Collection. *Acta Orientalia* 26(3–4), pp. 143–147.

Parker, Richard A. (1972). *Demotic mathematical papyri*. Brown Egyptological Studies 7. Providence, RI: Brown University Press.

Peet, T. Eric. (1923). *The Rhind mathematical papyrus: British Museum 10057 and 10058*. Introduction, transcription, translation and commentary. London: University Press of Liverpool; Hodder & Stoughton.

Peust, Carsten 2023. Was bedeutet *ntt jwtt*? In Bayer, Christian, Henning Franzmeier, Oliver Gauert, and Regine Schulz (eds), *Dem Schreiber der Gottesworte. Gedenkschrift für Rainer Hannig*. Hildesheimer Ägyptologische Beiträge 57, 171–181. Hildesheim: Gerstenberg.

Pommerening, Tanja. (2005). *Die altägyptischen Hohlmaße*. Studien zur Altägyptischen Kultur, Beihefte 10. Hamburg: Buske.

Posener-Kriéger, Paule. (1976). *Les archives du temple funéraire de Néferirkarê-Kakaï (Les papyrus d'Abousir): traduction et commentaire* (2 vols). Bibliothèque d'étude 65. Cairo: Institut français d'Archéologie orientale.

Quack, Joachim Friedrich. (2002). A Goddess Rising 10,000 Cubits into the Air …: Or Only One Cubit, One Finger? In Steele, John M. and Annette Imhausen (eds.), *Under One Sky: Astronomy and Mathematics in the Ancient Near East*. Alter Orient und Altes Testament 297, pp. 283–294. Münster: Ugarit-Verlag.

Reineke, Walter F. (1980). Mathematik. In: Helck, Wolfgang and Wolfhart Westendorf (eds.), *Lexikon der Ägyptologie*: vol. 3, cols 1237–1245. Wiesbaden: Harrassowitz.

Reineke, Walter F. (1993). Zur Entstehung der ägyptischen Bruchrechnung. *Altorientalische Forschungen* 19(2), pp. 201–211.

Ritter, Jim. (2002). Closing the Eye of Horus: The Rise and Fall of 'Horus-eye Fractions'. In Steele, John M. and Annette Imhausen (eds.), *Under One Sky: Astronomy and Mathematics in the Ancient Near East*. Alter Orient und Altes Testament 297, pp. 297–323. Münster: Ugarit-Verlag.

Sauneron, Serge. (1982). *L'écriture figurative dans les textes d'Esna*. Esna 8. Cairo: Institut français d'Archéologie orientale.

Scharff, Alexander. (1922). Ein Rechnungsbuch des königlichen Hofes aus der 13. Dynastie (Pap. Boulaq Nr. 18). *Zeitschrift für ägyptische Sprache und Altertumskunde* 57, pp. 51–68 with supplement 'Umschrift des Papyrus Boulaq No· 18' 1**-24**.

Seidlmayer, Stephan Johannes. (1991). Eine Schreiberpalette mit änigmatischer Aufschrift (Städtische Galerie Liebieghaus / Frankfurt a.M. Inv.-Nr. IN 1899). *Mitteilungen des Deutschen Archäologischen Instituts, Abteilung Kairo* 47, pp. 319–330.

Sethe, Kurt. (1916). *Von Zahlen und Zahlworten bei den alten Ägyptern und was für andere Völker und Sprachen daraus zu lernen ist: Ein Beitrag zur Geschichte von Rechenkunst und Sprache*. Schriften der Wissenschaftlichen Gesellschaft in Straßburg 25. Straßburg: Trübner.

Simpson, William Kelly. (1963). *The Records of a Building Project in the Reign of Sesostris I. Papyrus Reisner I. Transcription and Commentary*. Boston: Museum of Fine Arts.

Spalinger, Anthony. (1985). Notes on the Day Summary Account of P. Bulaq 18 and the Intradepartmental Transfers. *Studien zur Altägyptischen Kultur* 12, pp. 179–241.

Till, Walter C. (1961). *Koptische Dialektgrammatik mit Lesestücken und Wörterbuch*, 2nd ed. München: C. H. Beck.

Vleeming, Sven. (1981). The artaba, and Egyptian grain-measures. In Bagnall, Roger S., Gerald M. Browne, Ann E. Hanson, and Ludwig Koenen (eds.), *Proceedings of the Sixteenth International Congress of Papyrology: New York, 24–31 July 1980*, pp. 537–545. Chico: Scholars Press.

Waerden, Bartel Leendert van der. (1938). Die Entstehungsgeschichte der ägyptischen Bruchrechnung. In: *Quellen und Studien zur Geschichte der Mathematik, Astronomie und Physik: Abt. B, Studien*: vol. 4, pp. 359–382.

Wb – see Erman, Adolf and Hermann Grapow (eds). (1926–31).

Wit, Constant de. (1962). À propos des noms de nombre dans les textes d'Edfou. *Chronique d'Égypte* 37(74), pp. 272–290.

Zauzich, Karl-Theodor. (1987). Unerkannte demotische Kornmaße. In Osing, Jürgen and Günter Dreyer (eds.), *Form und Maß: Beiträge zur Literatur, Sprache und Kunst des alten Ägypten: Festschrift für Gerhard Fecht zum 65. Geburtstag am 6. Februar 1987*. Ägypten und Altes Testament 12, pp. 462–471. Wiesbaden: Harrassowitz.

CHAPTER 5

# The Zero Concept in Ancient Egypt

*Beatrice Lumpkin*

## Abstract

Throughout history, cultures around the world have independently developed distinct number systems and distinct systems of recording numerals. In pre-Columbian Peru and Central America, astronomers and mathematicians developed positional-value systems of numerals that used different methods to indicate an empty or zero-valued position. Others, as in Ancient Egypt, used an additive numeral system, without positional value and no need for placeholders. Ancient Egyptian mathematics did use zero as a number, a magnitude, in bookkeeping records containing zero balances. Also labels on construction guidelines, still visible at pyramid and mastaba construction sites, show use of a system of integers including zero. The concept of infinity was also explored in the papyri and may have developed in the course of extensive Egyptian exploration of series of numbers. The ancient Egyptian numerals and zero concept are part of the non-European foundations of classical Greek mathematics and can serve as useful examples in the mathematics classroom. However, non-Western achievements are little known due to European and North American rewriting of history to justify slavery and colonial occupations.

## Keywords

multicultural mathematics – Maya zero – Inca Zero Concept – Ancient Egyptian Zero – Egyptian numerals – zero at the Pyramids – multinational origins

## 1 Introduction

Reconstructing the history of the zero concept is an important part of reconstructing human history. That process has been affected by the rise of racist apologetics as a cover for centuries of the African slave trade and European and North American capitalist colonization of countries in Asia, Africa and the Americas. The conquered peoples were labeled as inferior, people who could not possibly be founders of civilization. Criticizing this racist concept, W. E. B. DuBois wrote, 'that the science of Egyptology arose and flourished at

 | DOI:10.1163/9789004691568_008

the very time that the cotton kingdom reached its greatest power on the foundation of American Negro slavery' (DuBois, 1965). Before this period, according to Martin Bernal, the earlier European historians respected the Ancient Egyptian civilization as foundational for the later Greek civilization. The revised historiography, that Bernal called 'Aryan', denied an Egyptian foundation for Greek mathematics. Instead, the Aryan, or Eurocentric model claimed, 'Greek civilization was the result of cultural mixture following a conquest from the North by Indo-European speaking Greeks' (Bernal, 1991).

Eurocentrism claimed that civilization began in Greece, then spread to the rest of the world. This contradicted the classic Greek scholars themselves, who credited Ancient Egyptians as their teachers in mathematics, law and other fields. Some devaluations of the Egyptian contribution have been extreme. For example, the respected historian Otto Neugebauer wrote, 'The role of Egyptian mathematics is probably best described as a retarding force upon numerical procedures.' Continuing in this vein, he also wrote, 'Ancient astronomy was the product of a very few men; and these few happened not to be Egyptian' (Neugebauer, 1969). Morris Kline adds the Babylonians to his put-down: 'The mathematics of the Egyptians and Babylonians is the scrawling of children just learning how to write, as opposed to great literature' (Kline, 1962).

Facts presented in this chapter will show developments of significant mathematics more than a thousand years before the classical Greeks. As George Gheverghese Joseph has shown, 'practically all topics taught in school mathematics today are directly derived from the mathematics originating outside Western Europe before the fifteenth century AD' (Joseph, 1991).

From the viewpoint of mathematics education, Eurocentric bias in the curriculum has had a negative, restricting effect. A truthful, multicultural presentation of the history is urgently needed to raise the level of mathematics education worldwide. If students are misled to believe that mathematics is alien to their culture, they may be discouraged from pursuing the study of mathematics. It is also a disservice to students of European descent to teach them an elitist, imperialist view of the origin of mathematics. For all children, portrayal of mathematics as knowable only to a 'chosen few' can turn many away from the study of mathematics.

For Ancient Egyptian mathematics, there have been a number of serious studies of the mathematical papyri, including Richard J. Gillings' *Mathematics in the Time of the Pharaohs* (1972). Gillings provides an insightful analysis of the mathematical papyri. But the problems posed in the few surviving mathematical papyri did not involve the zero concept. However, a bookkeeping papyrus, translated as early as 1922, did show zero remainders but may not have been known to historians of mathematics. Other evidence of use of a zero concept was sketched on the walls of construction projects. These were independent

Egyptian applications of a zero concept. There were also other countries where an early, independent development of a zero concept took place.

## 2 Zero Concept in Pre-Columbian Peru

When the Spanish conquistadores reached the Andes mountains in Peru, they saw pyramids about as old as those in Egypt. The later Inca rulers presided over what was then the largest empire in the world. Obviously, such an empire would need a method of keeping track of their vast commerce and the tribute due to the state. But there was no form of writing, as we know it. However, there were *quipus* (or *khipu*), complex bundles of knotted strings hanging from a main cord that seemed to record something. The conquistadores burned the quipus; the few hundred quipus that escaped are now the subject of modern research. The knots were found to represent a base ten or decimal system of numerals with 'place value'. Among these researchers, Marcia Ascher and her archaeologist husband Robert Ascher also studied the use of color coding as it relates to the number zero (Ascher and Ascher, 1981).

> On some quipus, cord positions within a group are reinforced by color coding. That is, when for example, colors 1, 2, 3, 4 appear in the groups in positions 1, 2, 3, 4 respectively. This redundancy enables a distinction to be made between a value of zero and the absence of a category. Consider a set of data that is an inventory of the number of pairs of sandals owned by a community of people. Say that the numbers are recorded on pendants categorized into families by spatial groups and, within each group, categorized into men, women and children as represented by cord colors 1, 2, 3. A cord of color 3 with no knots would have a value of zero and mean that the children of a family have no sandals. The absence of cord color 3 in the cord group would mean that the family had no children … The fact that the distinction is made is important because it corroborates that a blank cord, in fact, represents what is symbolized by our zero. (Ascher and Ascher, 1981, pp. 89–90)

## 3 The Maya Zero Placeholder

It is widely accepted that the Maya of Central America had a zero 'placeholder' and a well-developed, accurate calendar. Some diminish the importance of the Maya zero because 'Europeans didn't know about it'. However, the development of Maya mathematics was an important factor supporting the advanced

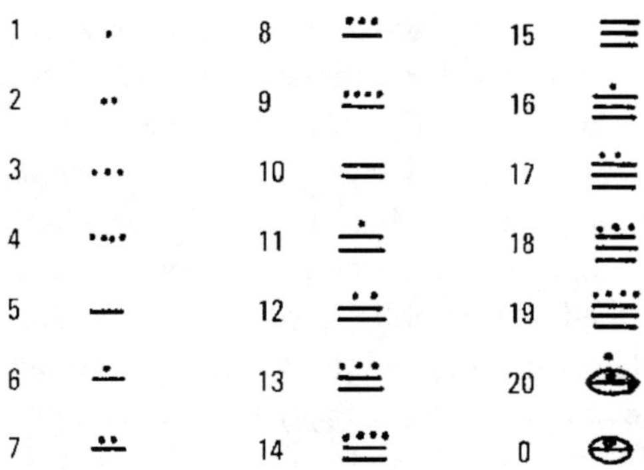

FIGURE 5.1
Maya Numerals
FROM *MATEMÁTICA MAYA*, LEONEL MORALES, WITH PERMISSION (MORALES, 1994)

Maya technology, including metallurgy. The large quantities of silver and gold, forcefully taken from the Americas to Europe, fueled the European Industrial Revolution. This new wealth made it possible for European colonists to invade and exploit the people of other continents. At the same time, the corn, beans and potatoes of the indigenous peoples of the Americas improved European nutrition, allowing European nations to support larger populations. So yes, the Maya zero, and the developed Maya economy it helped produce, had a profound influence on Europeans.

The Maya used a base-20 number system with positional or place value. They wrote their numerals vertically. In the lowest place were the units, or $20^0$. The next level above was for the twenties; $20^1$. The next higher level was $20^2$ and so on. For the values 1, 2, 3, or 4, they used dots. But for 5 they used a bar: —; four dots was the maximum allowed in one place. They continued this way up to the value 19. See Figure 5.1. For 20, they did not add another dot to the units' place. Instead, they emptied the units' place and went up one level to put one dot in the 20s place. But how could the Maya be sure this was one 20 and not one unit? About 1,000 years ago, the Maya realized they needed a zero-placeholder. Their zero symbol is thought to represent a shell or a fist. As shown in Figure 5.1, the value, 20, is written with one dot in the 20s place and a shell as a zero-placeholder in the empty units' place.

Maya numerals are easy to model in the classroom, using lima beans for dots and small sticks such as toothpicks for the bars. They provide an effective teaching tool to achieve a basic understanding of place value and the importance of the zero placeholder.

## 4 Applications of the Zero Concept in Ancient Egypt

Thousands of years of mathematical activity in Africa preceded the development of the zero concept in Egypt. We can go back beyond written mathematics, to the evidence from ancient mathematical artifacts from around the world. These artifacts show that all peoples developed mathematical thinking, that mathematical thinking is a universal human trait (D'Ambrosio, 1997).

Geometric thinking is certainly very ancient. Rock paintings of people and animals, showing an understanding of similar figures, are said to go back 65,000 years in Australia (Morell, 1995). Even older examples of abstract, geometrical thinking have been unearthed in East and Central Africa. Partially undercut outlines for tool blanks, dated to *c.*200,000 BP (Before the Present), have been discovered on a cliff face in Kenya (Gutin, 1995). Saw-toothed harpoon bits, excavated on the Semliki River near Lake Rutanzige (Edwards) in the Democratic Republic of the Congo, have been dated to *c.*90,000 BP (Yellen and Brooks et al., 1995). There may be future adjustments in these dates as techniques are refined. However, it is already clear that evidence for mathematical thinking goes far back into the prehistoric period. Ancient humans visualized the tools they needed and retained abstract geometric images of the tools. Then they made plans to produce such tools, sometimes on a large scale.

A 33,000-year old flint mine in the Nile valley, with bell pits and underground galleys, is further evidence of planning and perhaps the beginning of social division of labor as well as trade (Vermesh et al., 1984). It appears that by the time modern humans began to emigrate from Africa to other continents, they already had the ability to think abstractly and possessed an advanced tool kit. A study of human languages reveals some of the early history of mathematical ideas. According to Karl Menger, every human language contains the structure and vocabulary needed to develop mathematics. These language essentials he listed as vocabulary for natural numbers and words for the logical concepts of 'and, or' (Menger, 1972). The earliest records of numbers were made with tally marks, one tally for 1, two tallies for two, etc. The oldest known numerical record, dated 37,000 BP, comes from Border Cave in the Limpopo mountains of Southern Africa. The record appears on a fossilized baboon bone inscribed with 29 tally marks. The number of tallies suggests the number of days in the lunar cycle.

Alexander Marshack describes similar tally records from many parts of the world (Marshack, 1991). For some of these ancient records, he has found a correlation between the arrangement of tally marks with the number of days for phases of the lunar cycle. He finds such correlations in the markings on an African fossil bone from Central Africa. This fossil has been popularized by Claudia Zaslavsky as 'the Ishango bone' (Zaslavsky, 1999). The Ishango bone, carved 20,000 BP to 25,000 BP, shows levels of complexity beyond the older tally record from Border Cave. Numerals inscribed on the Ishango fossil appear to show some patterns. Counting the tallies, one edge shows: 3, 6, space; 4, 8, space; 10, space; 5, 5, space; 7. Turning the bone shows 11, 13, 17, and 19. On the other edge, we see 11, 21, 19 and 9. It appears that more is going on than random tallies.

## 4.1 *Egyptian Numerals*

Humanity has moved from tally records to our modern numerals in several stages. Each breakthrough has required new, daring concepts. Some have been invented more than once. The three key concepts of modern numerals are what Carl Boyer calls, 'cipherization', the use of a single, abstract symbol to stand for a given number of tallies) positional value and the zero concept (Boyer, 1944).

The cipherization of numerals in Ancient Egypt was at least as early as the development of writing. For example, the artifact known as the Metropolitan Museum knife handle, crafted *c.*5100 BP, showed one lotus blossom representing the value, 1000 (Williams and Logan, 1987). One cipher was used, the lotus blossom, instead of 1000 tally marks. But cipherization was only partial in hieroglyphic numerals. From 1 up through 9, tallies had to be repeated. For 10, there was a new cipher, the 'upside-down U'. Different symbols were used for each power of 10 up to 1,000,000. To reach the desired value, these symbols for powers of 10 were grouped and repeated.

In practical use, hieroglyphic writing was used mainly for monuments. For everyday records and calculations, the scribes used a faster, cursive form of the hieroglyphs, known as 'hieratic'. Hieratic numerals, unlike the hieroglyphic, were completely cipherized and required fewer characters. Hieratic numerals used distinct symbols (or ciphers) for 1 to 9, for 10. 20, 30 etc. to 90, for 100, 200 to 900 and so on. Compared to the hieroglyphs, hieratic numerals were much more compact but required memorizing more symbols. Arrangement in columns by powers of 10 promoted ease of calculation

Egyptian numerals were written right to left. Starting at the right in Figure 5.2, one tally mark, | , stands for 1. For 3, we would need three tally marks | | | . The 'upside-down' U is the cipher for 10. The 'backwards' C is 100. Next is the lotus blossom for 1000, bent finger for 10,000, tadpole for 100,000 and the figure with upraised arms used to mean 1,000,000. But we'll meet him later.

The following example from problem 79 of the Rhind Mathematical Papyrus illustrates the advantage of hieratic numerals (Chace, 1979). Only four hieratic numerals were needed to represent 19,607 but 23 hieroglyphs were required, as

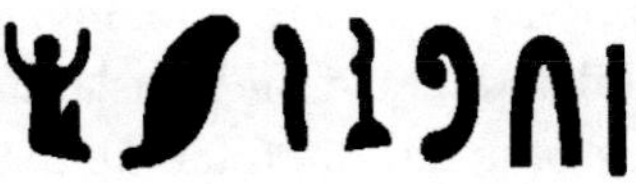

FIGURE 5.2
Egyptian hieroglyphic numerals

7 + 600 + 9000 + 10,000 EGYPTIAN HIEROGLYPHS | 7 + 600 + 9000 + 10,000 EGYPTIAN HIERATIC | $5 \times 60^2 + 26 \times 60^1 + 47 \times 60^0$ BABYLONIAN CUNEIFORM

FIGURE 5.3 Egyptian and Babylonian Numerals for 19,607

in Figure 5.3. The Babylonian numeral for 19, 607 is also shown for comparison purposes; 24 characters are needed. Babylonians wrote in clay with a stylus. A vertical triangle mark was one unit; the 'curved' horizontal mark stood for ten. Babylonian numerals used another basic principle of modern numerals: positional or place-value. For the value 19,607, the Babylonian numeral used many more characters than the hieratic. Their numeral system was based on 60 while Egyptian and Inca numerals were based on ten and the Maya base was 20. In India, cipherization, positional value and the zero placeholder were brought together to create modern numerals.

Egyptian numerals were additive, without place value. Therefore, they had no need for a zero 'placeholder.' In Figure 5.3, simply add the values of the Egyptian ciphers, in any order. The sum is 19,607. But the Babylonian numerals for 19,607 do have place value. In Figure 5.3, reading from left to right, the place values are $60^2$, $60^1$, $60^0$ since their base was 60. When a place was empty, sometimes the lack of a zero-placeholder caused confusion. But for most examples, the value was clear from the context of the problem. After the Greek conquest, they used a special symbol to 'hold' an empty space in the middle of a numeral (Neugebauer, 1969). However, the empty-space symbol was not used to represent zero as a value.

### 4.2 *Zero as a Number, a Magnitude*

Most of our knowledge of Ancient Egyptian mathematical thinking comes from surviving papyri, architects' diagrams and guidelines inscribed on walls of tombs and pyramids. Papyrus Boulaq 18 was a daily account book containing balance sheets of income and expenditures for the Pharaoh's court as they traveled. At the end of each day, bookkeepers totaled the receipts and expenditures. Then total expenditures were subtracted from total receipts. Balances were forwarded to the next day's balance sheets.

Boulaq 18 was recorded about 3,900 years ago and may be the oldest extant record of the use of zero as a number, the result of an arithmetic operation on two other numbers. Several subtraction examples with zero remainders appear on each of pages 59 and appendices 5**, 8** and 12** of the translation to German by Alexander Scharff (1922). In these examples, the total number of items disbursed equals the total number of items received. Subtracting disbursements from receipts leaves zero remainders that were entered in the bottom line of the balance sheets.

The following list from Boulaq 18 shows rows 4 to 15 of one of these balance sheets. Rows 4, 5, 6 and 7 record incoming deliveries and their sum is shown in Row 8, total incoming deliveries. Rows 9 through 13 record disbursements to officials of the royal house. For example, Row 9 shows disbursements of 625 loaves of daily bread, 60 jugs of regular beer, 2 cakes, 1 sweet baked goods of some type, 52 jugs of special beer, and 100 bundles of sticks, perhaps firewood. Row 14 shows total disbursements and the last row shows balances or remainders left after subtracting Row 14 from Row 8.

The symbol used by the scribe for the four zero remainders in the bottom row of the above table was the triliteral hieroglyph, n'f'r (Gardiner, 1978). (Vowels were not shown.) N'f'r also means good, beautiful (Faulkner, 1976). Scharff translated all the non-zero hieratic numerals in the above table as modern numerals. The zero remainders he transcribed from hieratic to hieroglyphs but did not translate. Why? Of course, Scharff knew that 1–1 = 0, that 52–52 = 0,

TABLE 5.1 Balance sheet from 13th dynasty

| Row no. | Bread daily | Bread extra | Beer regular | Cakes | Sweets baked | Beer special | Vegetables | Bundles of sticks |
|---|---|---|---|---|---|---|---|---|
| 4 | 1680 | | 135 | 2 | 1 | 52 | | 200 |
| 5 | 200 | | 2 | | | | | |
| 6 | 100 | | 10 | | | | | |
| 7 | | 938 | 90 | 7 | | | 7 | |
| **8 total in** | **1980** | **938** | **237** | **9** | **1** | **52** | **7** | **200** |
| 9 | 625 | | 60 | 2 | 1 | 52 | | 100 |
| 10 | 630 | | 61 | | | | | 50 |
| 11 | 525 | | 38 | | | | | 50 |
| 12 | | 310 | 35 | 5 | | | 7 | |
| 13 | | 290 | 22 | | | | | |
| **14 total out** | **1780** | **600** | **216** | **7** | **1** | **52** | **7** | **200** |
| Balance | 200 | 338 | 21 | 2 | 𓄤 | 𓄤 | 𓄤 | 𓄤 |

Adapted from Alexander Scharff (1922, p. 59)

that 7–7 = 0 and that 200–200 = 0. But he was writing in 1922 when the existence of an Egyptian zero symbol was widely denied. However, Anthony Spalinger's 1985 economic analysis and translation of Boulaq 18 to English does show the zero remainders as '0' (Spalinger, 1985).

In discussing the Mesopotamian mathematics of that time, George Ifrah wrote, 'that the notion of nothing was not yet conceived as a number.' He adds, 'In a mathematical text from Susa, the scribe, obviously not knowing how to express the result of subtracting 20 from 20, concluded in this way: '20 minus 20 … you see.' And in another such text from Susa, at a place where we would expect to find zero as the result of a distribution of grain, the scribe simply wrote, 'The grain is exhausted.' (Ifrah, 1985).

### 4.3 *Zero in the Set of Integers*

As early as 1,000 years before Boulaq 18, the nfrw hieroglyph was used at Ancient Egyptian construction sites to label a zero-reference point at or near ground level. Construction of large tombs and pyramids required deep foundations to support the massive weight. To locate the exact level, architects created a system of integers to use as vertical coordinates. These coordinates gave distances above or below ground as the number of cubits below or above nfrw or zero level.

Such an example can still be seen at Meidum, in the foundation trench of Mastaba 17, a massive tomb with base 52.5 m × 105.0 m. Starting at ground level and going down to the bedrock, thirteen horizontal lines were drawn on the wall of the trench, as shown in Table 5.1. These lines were spaced one cubit apart. Egyptologists believe that these horizontal lines were leveling guidelines, to help the stonemasons level each course of the huge stones. A vertical line crossed the horizontal lines; some intersection points were labeled. Labels that can still be read include 'nfrw' for ground or zero level, then 5 lines lower, '5 cubits below nfrw' and 3 lines further down, '8 cubits below nfrw.'

This provides a good multicultural example for the classroom to help beginning algebra students internalize the non-intuitive subtraction of signed or directed numbers: $a - (-b) = a + b$. Take 'above nfrw' as positive, 'below nfrw' as negative and translate 'nfrw' as zero. Then what is the distance from 8 cubits below zero to 5 cubits above zero? Or, $5 - (-8) = ?$ The answer is obviously 13 cubits. For those who would rather translate 'nfrw' as 'good', the above question would be rephrased as, 'What is the distance from 8 cubits below "good" to 5 cubits above "good"?' The answer is still the same, 13 cubits. The mathematical content is the same, of subtraction on the set of integers, (… −4, −3, −2, −1, 0, 1, 2, 3, …).

The numbering system of these construction lines, including the use of a zero-reference level, was well known to the early Egyptologists who studied the pyramids. In 1931, George Reisner described zero leveling lines at the Mycerinus (Menkure) pyramid at Giza, built *c.*4620 BP. He gave the following list, including terms collected earlier by Borchardt and Petrie from their study of Old Kingdom pyramids (Reisner, 1931).

| | |
|---|---|
| nfrw | zero |
| m tp n nfrw | zero line |
| hr nfrw | above zero |
| md hr n nfrw | below zero |

The above examples of the Ancient Egyptian concept of zero may not have been known to historians of mathematics. However, some historians have found additional examples of the use of zero as a number. Fortunately, they were not deterred by widely accepted claims that 'Egyptians had no zero' (Bunt, Jones, Bedient, 1976). Carl Boyer referred to a deed recorded some 1,500 years after Ahmose copied the 'Rhind' Mathematical Papyrus. This deed gave a formula for the area of a quadrilateral as the product of the averages of opposite sides. 'Inaccurate though the rule is,' Boyer wrote, 'the author of the deed deduced from it a corollary – that the area of a triangle is half the sum of two sides multiplied by half the third side. This is a striking instance of the search for relationships among geometric figures as well as an early use of the zero concept as a replacement for a magnitude in geometry' (Boyer, 1944, p. 18).

In the Reisner Papyrus, calculation of volume involved multiplication of lengths measured in cubits, palms (1/7 cubit) and fingers (1/4 palm). A measurement such as 3 cubits, 0 palms, 2 fingers was written as 3 cubits, space, 2 fingers. Gillings commented, 'In the papyri, a blank space indicates zero.' But then he added, 'Of course zero, which had not yet been invented, was not written down by the scribe or clerk' (Gillings, 1972, p. 228). Despite Gillings' many contributions to uncovering the achievements of Ancient Egyptian mathematics, he may not have been aware of Boulaq 18 or the zero reference points for vertical coordinates at construction sites.

In Ancient Egypt, as in Ancient India, much of the mathematical knowledge was expressed in the form of rules and guides for construction of monuments and religious structures (Joseph, 1991, p. 228). In 1930, Clarke and Engelbach published an architect's diagram and called it 'of great importance' (Clarke and Engelbach, 1990). The diagram was sketched on a stone ostracon found at the Saqqara step-pyramid built *c.*4750 BP.

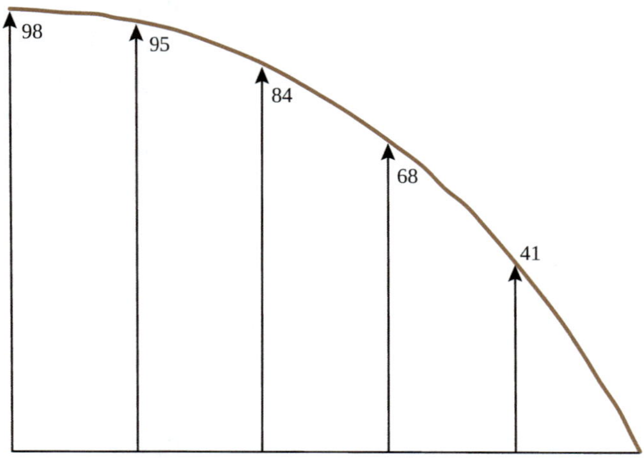


FIGURE 5.4 Coordinates for an arc

This diagram, shown in Figure 5.4, provides an example of the early Egyptian use of rectangular coordinates. The curve matched a section of a nearby temple roof. Notice that the curve turns down toward the base line where the vertical coordinate would be zero. Sadly, that corner was broken off.

It is assumed the vertical line spacing is 1 cubit apart, giving horizontal coordinates of 0, 1, 2, 3, 4, cubits. Converting to fingers gives us coordinates: (0, 98); (28, 95); (56, 84); (84, 68); (112, 41); the last point is broken off. Plotting on square-grid graph paper and smoothly connecting the points, gives us an arc very close to what the architect sketched by hand.

Ostracons, such as used for this sketch, were disposable stone scraps, meant to be thrown away after the construction. And that's what happened to this sketch. It was tossed when no longer needed. Fortunately, it was dug up in modern times, with the sketch preserved by the dry climate. Otherwise, we might not have known about the use of rectangular coordinates to define a roof curve at an early pyramid complex.

Still, we should not disregard what Joseph calls, 'Maths From Bricks: Evidence From the Harappan Culture' (Joseph, 1991, p. 221). The construction itself can often tell a lot about the mathematics used by the architects. Calculating from the data given by I. E. S. Edwards in his study of the Great Pyramid at Giza, the error in the right angles at the base of the Great Pyramid at Giza is only one part in 27,000 (Edwards, 1947, p. 118). Such accuracy implies the mathematical knowledge of the pyramid architects, even if written evidence is missing.

Systemic racism in Europe and North America led many to believe that Africans could not have built pyramids, that they didn't have the knowledge. To this day, some think that Aliens from another planet came to Earth and built the pyramids! Others, misled by some school textbooks, believe that Egyptians are 'Mideastern', not African. No doubt, it was to counteract this misinformation that led Chicago's Field Museum to hang a big sign over the entrance to their Ancient Egyptian exhibit, proclaiming, 'EGYPT IS IN AFRICA'. New York's Metropolitan Museum of Art is even more explicit, with its exhibit, 'The African Origin of Civilization.'[1]

## 4.4 *Series in Ancient Egyptian Mathematics*

The Egyptian method of multiplication gave the scribes a powerful tool for computation and reveals some of their insights into subjects such as power series, field properties of integers, and inverse operations. Suppose they wanted to multiply 134 × 17. First the commutative property of multiplication was used to select the smaller factor as the multiplier. For example, it's quicker to multiply 17 × 134 than the equivalent 134 × 17.

Starting with 1 × 134 taken once, multiplier and multiplicand were then successively doubled, probably by adding to itself (Robins and Shute, 1987, p. 16). That process produced the partial products whose sum made up the total product. The distributive property was used much as we do today except that the multiplier was partitioned into powers of 2 instead of powers of 10. To get 17 × 134, we add the lines for 1 134, and 16 2144.

| | |
|---|---|
| 1 | 134 √ |
| 2 | 268 |
| 4 | 536 |
| 8 | 1072 |
| 16 | 2144 √ |
| 17 | 2278 |

$$17(134) = 1(134) + 16(134) = 134 + 2144 = 2278$$

In this method, the multiplier was expressed as a base 2 number: $17 = 2^0 + 2^4$. After studying the Egyptian method, Lorenzo Curtis concluded that the Ancient Egyptians were masters of the geometric progression, and constructed their entire system of basic arithmetic operations around it (Curtis, 1978). It is in the work with series and geometry in the Ahmose (Rhind) papyrus, and the second-degree equations and geometry in the Moscow and Berlin papyri, that we begin to get a glimpse of the advances made by Ancient Egyptian mathematicians.

1 https://www.metmuseum.org/exhibitions/listings/2021/african-origin-of-civilization, accessed January 9, 2022.

In Ahmose Problem 64, the scribe asks for a division of 10 measures of grain among 10 men so that there is a constant difference of 1/8 between portions (Lumpkin, 1987). Ahmose solves the problem by using the equivalent of the modern formula for the sum of an arithmetic series. The Ahmose method also has the virtue of making the derivation of this formula more intuitive. He finds the average share, then adds to it the product of the common difference times half the number of differences. This gives him the last, or highest share. The remaining shares are found by subtraction of the common difference. Ahmose's Problem 40 asks for a division of 100 loaves of bread among five persons so that there is a common difference. Also, the sum of the highest three must be 7 times the sum of the lowest two.

The scribe also shows expert knowledge of geometric series in the famous Ahmose problem 79. Some think this problem was the forerunner of Mother Goose's, 'As I was walking to St. Ives, I met a man with seven wives.' It is very interesting that Fibonacci, also known as Leonardo of Pisa, returned to Italy from North Africa with a similar series about the year 1200 (Boyer, 1965, p. 281). The problem was theoretical and/or recreational, rather than the purely practical type that some claim was the limit of Egyptian mathematics. The problem was given in tabular form:

| | | | |
|---|---|---|---|
| houses | 7 | 1 | 2801 |
| cats | 49 | 2 | 5602 |
| mice | 343 | 4 | 11204 |
| spelt (grain) | 2401 | Total | 19607 |
| hekats (1/8 bushel) | 16807 | | |
| Total | 19607 | | |

The wording, 'The sum according to the rule,' introduces a multiplication that Gilling believes shows deep knowledge of geometric series. In this example, the multiplier or common ratio, is a = 7. Notice that 2801 is 1 more than 7 + 49 + 343 + 2401. Gillings describes an inductive process that the Ancient Egyptians could have used to get the equivalent of:

$$a + a^2 + a^3 + a^4 + a^5 + \dots + a^n = a(1 + a + a^2 + a^3 + a^4 + a^5 + \dots + a^{n-1})$$

## 4.5 *Concept of Infinity*

Given the interest in series, it should not be surprising that Ancient Egyptian mathematics developed a concept of infinity. At a 2013 lecture on 'Dream Analysis in Ancient Egypt' at the University of Chicago, Egyptologist Robert K. Ritner pointed to the hieroglyph 'heh' and translated it as 'infinity', instead of one million.

FIGURE 5.5
The hieroglyph, 'heh'

In response to questions, Ritner said that the hieroglyph meant millions and millions and more millions past that, in effect, infinity (Ritner, 2013). He explained the development of 'heh' as the hieroglyph for infinity.[2]

> You are correct that the word 'heh' originally meant 'million'. However, its usage in actual counting fell into disuse before the New Kingdom and was replaced by multiples of 'hefen' (100,000). (Gardiner, 1978, p. 191) As a result, the word 'heh' became a term attached to objects signifying a great number without end, 'unendlich grosse Zahl/unendlich/unendlich langen Zeit' (Erman and Hermann, 1971). This is often inexactly translated as 'many' or 'millions and millions'.

There have been other hints of Egyptian exploration of the concept of infinity. A papyrus from Kahun was described by the translator Francis LI. Griffith, as, 'one I fear is beyond hope: it was beautifully written in columns, and still contains the most tantalizing phrase, "multiply by 1/2 to infinity"'. This reminded Griffith of Zeno's paradox which ends up with a denial of motion, because one can always travel half the remaining distance but there will still be the other half left to cover (Griffith, 1898). Unfortunately, the papyrus breaks off and we do not know the findings of the Egyptian mathematician who let n increase indefinitely (approach infinity) in the sequence $(1/2)^n$. Did that infinite series ever reach its limit of zero? The finite sequence 1/2, 1/4, 1/8, 1/16, 1/32, 1/64 was of great interest to the Egyptians who named them 'Horus-eye' fractions and used them to measure grain.

2 Robert K. Ritner, Professor of Egyptology, The Oriental Institute, The University of Chicago.

## 5 Conclusion

The zero concept played an important role in the development of systems of numerals, and for calculations and measurements. Expressions of the zero concept in India, Egypt, Central America, South America and elsewhere show independent development. As early as 5,000 years ago, geographically widely-separated civilizations developed distinct number systems, using different bases and different numeral systems. There is also evidence of early exploration of the concept of infinity.

Examples cited in this paper confirm the multinational, multicultural, multiracial development of mathematics. The facts of the history of mathematics refute any claims of European intellectual superiority. Such claims have adversely affected the study of the history of mathematics that developed outside of Europe.

In recent years, there has been some progress in addressing racist distortions of history. For example. there is an increasing interest in exploring the multicultural history of the zero concept. Further study of the existing mathematical artifacts and the use of new technology to find still buried ancient documents will deepen our understanding of earlier societies and the contributions that these societies have made to the total body of human knowledge.

## References

Ascher, Marcia and Ascher, Robert. (1981). *Mathematics of the Incas: Code of the Quipu*. Ann Arbor, MI: University of Michigan Press.

Bernal, Martin. (1991). Black *Athena*. Vol. 2, New Brunswick, NJ: Rutgers University Press, 1–2.

Boyer, Carl B. (1944). 'Fundamental Steps in the Development of Numeration.' *Isis*, 35: 158.

Bunt, Lucas N. H., Jones, Philip S., Bedient, Jack D. (1976). *The Historical Roots of Elementary Mathematics*. Englewood Cliffs, NJ: Prentice Hall, 7.

Chace, Arnold B. (1979), originally 1929. *The Rhind Mathematical Papyrus*. Reston, VA: National Council of Teachers of Mathematics, 137.

Clarke, Somers and Engelbach, Reginald. (1990). originally 1930. *Ancient Egyptian Construction and Architecture*. New York: Dover, 52–3.

Curtis, Lorenzo, J. (1978). 'Concept of the exponential law prior to 1900,' in American *Journal of Physics*, 46 no. 9, 896.

D'Ambrosio, Ubiratan. (1997). 'Ethnomathematics and its Place in the History of Mathematics.' In *Ethnomathematics, Challenging Eurocentrism in Mathematics*

*Education*. Powell, Arthur and Frankenstein, Marilyn (eds.) New York: State University of New York, 17.

DuBois, William E. B. (1965) originally 1947. *The World and Africa: Inquiry Into the Part Which Africa Has Played in World History*. New York: International Publishers, 64.

Edwards, I. E. S. (1979). *The Pyramids of Egypt*. originally 1947. New York: Penguin Books, 118.

Erman, Adolph and Grapow, Hermann. (1971), originally 1926–1931. *Wörterbuch der Ägyptischen Sprache*, III, im Auftrage der Deutschen Akademien (on behalf of the German Academies). Reprint, Berlin, 158.

Faulkner, Raymond O. (1976). *A Concise Dictionary of Middle Egyptian*. Oxford: Griffith Institute, 132.

Gardiner, Alan A. (1978). *Egyptian Grammar*. Oxford: Griffith Institute, 266, sec. 351, 574.

Gillings, Richard J. (1972). *Mathematics in the Time of the Pharaohs*. Cambridge: MIT Press.

Griffith, Francis Ll. (1898). 'The Hieratic Papyri,' In *Ilahun, Kahun and Gurob* by Sir W. M. Flinders Petrie, Warminster, England: Aris and Phillips and Joel L. Malter, 48.

Gutin, JoAnn. (1995). 'Do Kenya Tools Root the Birth of Modern Thought in Africa?' *Science* 270: 1118–19.

Ifrah, Georges. (1985). *From One to Zero: A Universal History of Number*. New York: Viking, 382.

Joseph, George Gheverghese G. (1991). The Crest of the Peacock, Non-European Roots of Mathematics. London: Tauris, 71.

Kline, Morris. (1962). *Mathematics, a cultural approach*. Reading, MA: Addisson Wesley, 14.

Lumpkin, Beatrice. (1987). *African and African American Contributions to Mathematics*, a Baseline Essay. Portland, OR: Portland Public Schools. See also Gillings, 175.

Marshack, Alexander. (1991). *The Roots of Civilization*. Mt. Kisco, NY: Mayer, Bell.

Menger, Karl. (1972). Lecture, History of Mathematics, Illinois Institute of Technology.

Morales, Leonel. (1994). *Matemática Maya*. Guatemala, Guatemala: Editorial la Gran Aventura.

Morell, Virginia. (1995). 'The Earliest Art Becomes Older – and More Common.' Science 267: 31 1908–09.

Neugebauer, Otto. (1969) originally 1957. *The Exact Sciences in Antiquity*. New York: Dover, 80, 91.

Reisner, George A. (1931). *Mycerinus: the Temples of the Third Pyramid at Giza*. Cambridge: Harvard University Press, 77.

Ritner, Robert K. (2013). Email message to author, 10 June 2013, reprinted with permission.

Robins, Gay and Shute, Charles. (1987). *The Rhind Mathematical Papyrus*. New York: Dover, 16.

Scharff, Alexander. (1922). 'Ein Rechnungsbuch des Königlichen Hofes aus der 13. Dynastie (Papyrus Boulaq Nr. 18.' *Zeitschrift für Ägyptische Sprache und Altertumskunde* 57, 51–68 and appendix 1**–24**.

Spalinger, Anthony. (1985). 'Notes on the Day Summary Accounts of P. Bulaq 18 and the Intradepartmental Transfers.' *Studien Zur Altägyptischen Kultur* 12: 179–241. http://www.jstor.org/stable/25150093.

Vermeersh, Pierre M. et al. (1984). 33,000-year-old chert mining site and related *Homo* in the Egyptian Nile Valley. *Nature* 309: 24 342–44.

Williams, Bruce and Logan, T. J. (1987). 'The Metropolitan Museum Knife Handle and Aspects of Pharaonic Imagery Before Narmer.' *Journal of Near Eastern* Studies 46: 4 245–48.

Yellen, John E., Brooks, Allison et al. (1995). A Middle Stone Age Worked Bone Industry from Katanda, Upper Semliki Valley, Democratic Republic of the Congo. *Science* 268: 553.

Zaslavsky, Claudia. (1999). *Africa Counts*. 3rd edition. Chicago: Lawrence Hill Books, 17–20.

CHAPTER 6

# On the Placeholder in Numeration and the Numeral Zero in China

*Célestin Xiaohan Zhou*

## Abstract

In China, a decimal system of numeration is reflected in the number of oracle bones scripts between the fourteenth and eleventh centuries BCE. The place value system is clearly shown in the numbers on coinage from the eighth century BCE to the third century BCE. The decimal place value system was completely established in counting rods numeration, a description of the use of which is recorded in *Mathematical Canon of Sun Zi* of the late third century and the beginning of the fourth century; the earliest written counting rods numbers exist in a Dunhuang manuscript from around the tenth century. The procedures of multiplication, division or extraction of roots show that precise vacant positions in numbers are emphasized and strictly respected by the ancient practitioners. In thirteenth century mathematical works, the written symbol ○ is systematically used as a placeholder in numbers. The 'procedure of the positive and the negative' is usually regarded as a general rule of addition and subtraction between positive, negative numbers and zero, but this procedure should be interpreted in the context of the mathematical method *fangcheng*. The general rules concerning the process of elimination numbers in which zero is involved are clearly expressed and adopted, even though there is no explicit declaration of the existence of the numeral zero.

## Keywords

numeration in China – decimal system – place value system – counting rods number – placeholder – numeral zero – procedure of the positive and the negative (*zhengfu shu*)

## 1 Introduction

Since the first half of the nineteenth century, European sinologists or historians of mathematics have been interested in numeration in China on the basis of the limited mathematical works to which they have access; they introduced

 | DOI:10.1163/9789004691568_009

their findings to Western readers (Libri, 1838, pp. 201–203; Biot, 1839).[1] At the beginning of the twentieth century, the Japanese scholar Yoshio Mikami (三上義夫 1875–1950) systematically first wrote on the development of mathematics in China and Japan in English (Mikami, 1913), which served as the main basis of the contents concerning mathematics in China written in Florian Cajori's influential *A History of Mathematics* (Cajori, 1919) and David Smith's *History of Mathematics* (Smith, 1923, 1925). Mikami included chapters on Indian and Arabic influence on mathematics in China and assumed the employment of the symbol ○ in Chinese mathematical works was derived from the Indian source (Mikami, 1913, p. 59). During the twentieth century, Western historians have made more detailed descriptions of numeration in ancient China and further compared them with those discovered from other old civilizations (Needham and Wang, 1959, pp. 5–17; Martzloff, 1997, pp. 179–209) Some of the research has been absorbed into the popular history of numbers (Ifrah, 2000, pp. 263–296). A new revelatory and insightful research on this topic is Chemla (2022, pp. 13–70), which challenges the tacit assumptions in the former historical accounts.

Because of the limited newly discovered historical records or excavated objects, mutual influences among Chinese numeration and those from other civilizations are still mainly assumptions; an assertion of a clear route of the transmission of the mathematical knowledge still requires more exploration and proof. However, some new discoveries concerning the use of zero in Chinese historical literature and more reasonable interpretations of the related texts of the classics need to be introduced, which constitutes the main aims of this chapter.

A clarification of the topic, purpose, and scope of this chapter should be made before making any statement concerning '0' or 'zero' in China. First, this chapter does not aim to discuss directly the history of the so-called 'Chinese zero/0', nor incline to use expressions that combine the adjective 'Chinese' with an Arabic symbol or an English word. To some extent, this kind of expression is not precise enough to allow one to interrogate the question under study and might lead to confusion about the dynamic concept and its diverse uses in different contexts. For example, if one uses the expression 'Chinese zero' or 'Chinese 0' to refer to a Hindu-Arabic number, an elliptic symbol used in a context similar to modern mathematics or scientific fields, written or printed by a Chinese person in the geographical China, the history of the Chinese zero

1 This chapter is a product of my research project, 'Translations, Interpretations, and Understandings since the Nineteenth Century of Mathematical Works in Ancient China' (20CZS081) and it is supported by the 'Youth Innovation Promotion Association, CAS', (E2292G01).

is relatively short and such historical narrative will exclude much meaningful material from earlier periods. This symbol systematically appears in works as late as the beginning of the seventeenth century, during which European missionaries cooperated with the Chinese scholar-officials to translate Latin mathematical works into Chinese. However, the symbol standing for zero in these works was usually a round circle rather than an ellipse. Besides, the other nine Hindu-Arabic numerals were not copied from those Latin works, but were transcribed as Chinese characters for numbers, (Li and Ricci, 2013, pp. 11–96). It was not until the late nineteenth century that the ten Hindu-Arabic numerals (including the numeral zero) were introduced into China and adopted in some translations of European mathematical works or introductions to arithmetic in written form.[2] In the 1900s, many public schools around the country used original modern mathematics textbooks in English and other Western languages, or their translations.[3] Thereafter, the ten numerals including the symbol zero used in scientific fields in China have had less and less discrepancy from those used anywhere else in the world.[4]

As for the transmission of numerals between India and China, and that between the Islamic world and China, which happened before the seventeenth century, several pieces of evidence have been found in ancient literature or in excavated antiquities. At the beginning of the eighth century, the President of the Astronomical Board and Director of the Royal Observatory of the Tang dynasty (618–907) was Gautama Siddhārtha (or Gotama Siddha, in Chinese 瞿曇悉達, and Qutan Xida in Romanized Pinyin of Chinese), who was born in the Tang dynasty's capital city and was of Indian descent. He translated some astronomical material based on the Navagraha ('nine planets') system under the Chinese title *Jiuzhi li* (九執曆 *Jiuzhi li Navagraha Calendar*), which was included as the 104th volume of [*Astronomico-astrological*] *Divination Canon of Kaiyuan* (*713–741*) *Reign* (*Kaiyuan zhanjing* 開元占經) (Gupta, 2019, pp. 592–593; Yabuuti, 1979; Bo, 2003).

At the beginning of this volume, Siddhārtha introduced the Indian numerals and method of calculations. Unfortunately, the extant editions of *Navagraha Calendar* do not include any symbols for these Indian numerals when the

2 According to Yan Dunjie, in 1885 a book published in Shanghai entitled *Introduction of Western Arithmetic* (*Xisuan qimeng* 西算啟蒙) introduced the zero written as a circle, which is listed at the end of the series of nine Hindu-Arabic numerals (Yan, 1957).

3 For more about educational reform in this period, see Qian Baocong (Qian, 1964, pp. 344–345) Tian Miao (Tian, 2005, pp. 218–221), and Zhang Dianzhou (Zhang, 1999, p. 10).

4 The process of introducing Hindu-Arabic numerals from Europe into China is relatively clear and is evidenced in much historical literature. The scholars Li Yan (Li, 1998) and Yan Dunjie (Yan, 1957, 1982) have contributed detailed evidence related to this topic.

Chinese text indicates 'the shapes of the numerals for mathematics (*suanzi* 算字) from India are shown in the right column'. However, the Chinese text does give a short description of the use of a dot in the space between digits:

> For any digit [in one position which] that has reached up to ten, [it should be] carried to the previous position. [In] the place of each empty position, [the practitioner[5]] consistently places a dot. [If] there is space [between digits], [the practitioner] always marks [the space with such a dot]; [if] there is no [such case] by which [space is offered], [the practitioner] thus [uses the nine digits] alternatively.[6]
>
> 凡數至十，進入前位。每空位處，恆安一點。有間咸記，無由輒錯。(Siddhārtha, 1792, *juan* 104)

In India, a small circle has been used as a symbol for zero. Historians have also found that such zero represented by a dot had already appeared in the seventh-century Indian Bakhshali manuscript and in the Cambodia inscriptions under Indian influence (Gupta, 2019, p. 593).[7] However, most historians agree that Gautama's introduction of Indian numerals did not influence the way Chinese practitioners used numbers in mathematical science (Qian, 1933, p. 6; Needham and Wang, 1959, p. 12).

The Arabic numerals including a symbol circle representing the numeral zero appear in three magic squares engraved on an iron board, a stone board of the thirteenth or fourteenth century, and a jade pendant of the sixteenth century, respectively. Historians have pointed out these numerals are similar to Eastern Arabic numerals except for the symbol for zero, which in the latter system is often written as a dot, while in these magic squares is a circle (Tong, 2013; Wang, 1985). These numerals engraved on excavated antiquities might signify the transfer of a commodity, religious brief, or mathematical

5 Considering the topic of this chapter is about the numerals, in translations of this chapter I have followed the referee's suggestion to avoid confusion between the word 'one' representing the 'people in general' and the numeral 'one', and I use 'the practitioner' to designate the historical actor who performs the procedures described by the text.

6 A referee has pointed out that the character *cuo* in the quotation should be interpreted as 'alternatively' in this historical context, instead of the meaning 'mistake' that appears in a later period and is not common in the classical style of writing. According to the context of this quotation, another interpretation of this character is 'putting (using) [the nine digits]', which corresponds to the last word (mark, *ji*) of the previous clause.

7 An interesting phenomenon is that a common Sanskrit name for this dot is *bindu*, whose etymological meaning is 'a drop' (from the root *bhid*), while the written Chinese character (零 *ling*) used for zero also means 'a dewdrop' (Martzloff, 1997, p. 208; Wang, 2003).

knowledge between the Islamic world and China; but similar to the introduction of Indian numerals of the eighth century, these individual, fragmented pieces of evidence or the brief introduction of numerals show that before the seventeenth century, Hindu-Arabic numerals made few echoes in extant Chinese mathematical works.

After clarifying the issues concerning the Hindu-Arabic numerals occasionally appearing in China before the seventeenth century, but without much influence, and after knowing those numerals used in the mathematical science works systematically introduced, which lead to the modern science gradually established in China, a question formulated more precisely arises: In ancient China, how were elements of the functions of numeral 0, as a placeholder of numeration and as a numeral in calculations in modern mathematics, achieved in mathematical works or practices without a clear external influence? Temporarily suspending the issue of mutual influence of mathematics among civilizations, of which there is only a little evidence, and considering only the ages before the missionaries came, which had a large influence on mathematical sciences in China, mathematics in China could be regarded as an entity that has continuedly evolved (or developed) since the fourteenth century BCE, attested to by the numeration on the oracle bones script and, has formed a relatively complete system covering various branches of knowledge – not only practical use in administrative affairs or in daily commercial activities, but also theoretical studies. This chapter explores that these historical literatures related to mathematics show the process of the gradual emergence of a placeholder in numeration and its representation in the procedures of calculation and in written form. It also discusses that in a mathematical context where the numeral symbol for zero is not clearly given, how general rules of calculation in which '0' is involved are expressed, understood, and adopted by the ancient practitioners.

## 2 The Decimal System and Place Value System of Numeration in China

### 2.1 *The Decimal Numeration Reflected in the Oracle Bone Script (Jiagu wen* 甲骨文*)*

An examination of the placeholder in numeration must begin with an introduction to the decimal place value system used in ancient China. An embryonic form of decimal numeration appears in the divinatory texts engraved on the surface of oracle bones and tortoiseshells between the fourteenth and eleventh centuries BCE. The practitioner engraved the times of making divinations or numeral sequences (*zhaoxu* 兆序) besides the artificial cracks on the bones or carapaces. Since the end of the nineteenth century, tens of thousands of inscribed bones and tortoise carapaces have been discovered through systematic excavations or found accidentally during construction and agricultural works. The small numbers showing the sequence of divination are abundant. Besides, there are also large amounts of numbers in the main divinatory texts, among which the largest number recognized is three thousand.

The numbers in the oracle bone script mainly consists of nine numerals from one to nine and characters of ten (*shi*), hundred (*bai*), thousand (*qian*), myriad (*wan*, ten thousand). The basic symbols for these numerals and characters were given in works by Needham and Wang (1959, pp. 6, 14) and Martzloff (1997, pp. 180–181), but more variations of these symbols are given in *A History of Science and Technology in China, Volume of Mathematics* edited by Guo (2010, p. 17) – see Figure 6.1.

For these numeral symbols, it is clear that from one to four, which are represented by one, two, three, and four horizontal strokes respectively, their forms demonstrate a probable relation to the configurations formed by counting rods. The origins of other symbols are still in debate. Historians suggest some of the symbols (like those of six and seven) were initially borrowed to represent the digits because of the similar pronunciation between the borrowed characters and those numerals the ancient people pronounced conventionally. The symbol that stands for myriad (ten thousand) seems to be a pictogram of a scorpion, as its newborn offspring evoke an immense number to people (Martzloff, 1997, p. 181). However, it is hard to have a convincing answer that relies only on the shape of a symbol to deduce its origin and the reason why it was used to represent a certain digit. For example, H. Sakai suggests that the pictogram for a hundred is related to that of a toe (Martzloff, 1997, p. 181); nevertheless, this identical symbol is interpreted as a cypress cone (Needham and Wang, 1959, p. 5).

As for the notations for tens, it is clear from Figure 6.2 (Guo, 2010, p. 19) that the numbers 20, 30, and 40 are denoted by repetitive addition of the sign of ten, and the signs of ten in these numbers were linked by a curve or by two joined strokes under these vertical signs. Martzloff explained that after 40, the following tens, hundreds, thousands and myriads were denoted using a 'multiplicative principle' (Martzloff, 1997, p. 181). In fact, this multiplicative principle corresponds to two ways of forming a character in the oracle bone script from the perspective of paleography. A combined character (*hewen* 合文) refers to the combination of two symbols as an entity, in which the strokes of these symbols might be linked together or crossed. A symbol for 2,000 is usually shown as two strokes representing 'two' engraved on the top of a symbol representing

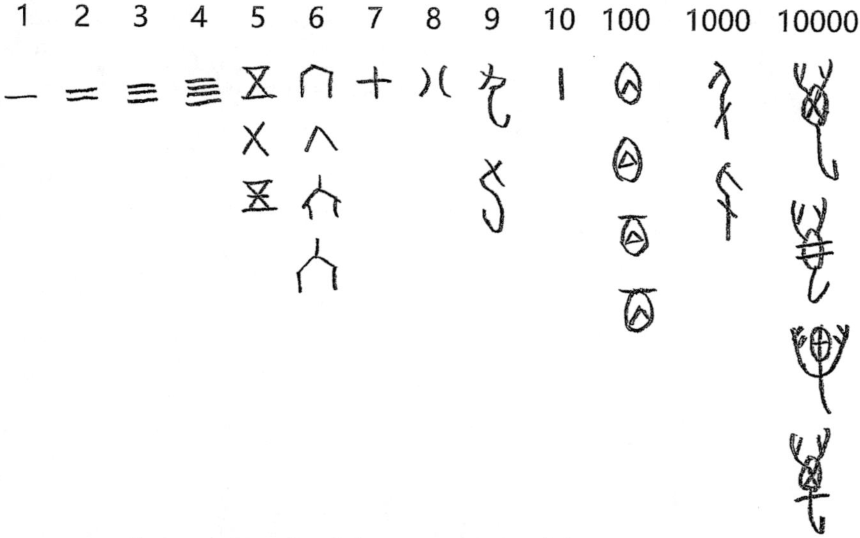


FIGURE 6.1 The basic numerals and characters in the oracle bone script
Note: There are variations of symbols representing a numeral, for example, the symbols for 'hundred' could be carved as a combination of a horizontal stroke at the top representing 'one' and an oval-like symbol with jointed short strokes inside, the interpretation of which varies from scholar to scholar (see discussion below). However, whatever the inside component is carved like those in the figure above, the oval-like symbol is the key element representing the numeral 'hundred', since this numeral could also be represented by this symbol without a horizontal stoke. The two sorts of symbols (with the stroke at the top or without one) for the numeral 'hundred' possibly exist in two different systems of numerations in oracle bone script, and they also correspond to the variations of numbers of hundreds in Figure 6.2. The previous analysis also holds true for symbols for thousand, in which the oblique stroke representing 'one' crosses over one stroke of the symbol representing 'thousand'. Figure 6.2 shows that one way of representing two thousand, three thousand, etc., is accumulating the oblique stroke on the same stroke of the symbol 'thousand', the total number of oblique strokes of which represents the value multiplied by one thousand. The other way is putting horizontal strokes at the top of the symbol for the combination of 'one thousand', in which one oblique stroke has been added; thus the horizontal strokes at the top indicate the values should be multiplied by one thousand.

'thousand'. The symbol for 3,000 is carved by means of replacing the two strokes of the symbol for 2,000 by three strokes.[8] The other way of forming numeral characters is called 'writing separately' (*xishu* 析書), by which the symbols representing digits are separated from those representing tens, hundreds, and thousands. Historians conjecture that people from antiquity first expressed

8 For a detailed description of the symbol, see Note to Figure 6.1.

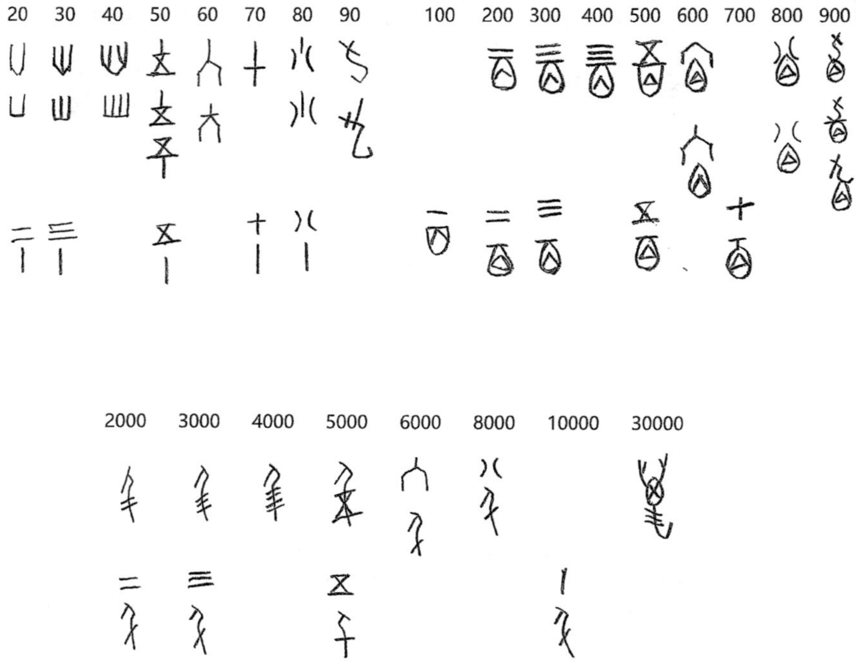


FIGURE 6.2 Tens and hundreds, thousands and myriads (ten thousands) in oracle bone script

numbers by means of multiplication and separating the individual symbols, and thereafter these certain numerals (such as 200, 300, 2,000, etc.) in separated form were then gradually developed into the combined form (Guo, 2010, p. 19). But from the perspective of mathematics, the probable earlier numerals in separated form are more meaningful when considering the characteristic features of Chinese numeration, for they are more closely related to the place value system, a discussion to which will return – see Figure 6.2.

As for the intermediate numbers, Needham (1959, p. 14) and Martzloff (1997, p. 182) have given many examples clearly showing how the additive and multiplicative principles are used simultaneously. In some cases, the character '*you* (and)' is added between the units, hundreds, and tens, but it is not mandatory, for example (Chen, 1978, p. 276) – see Figure 6.3.

On the basis of the symbols of numerals listed in Figure 6.1 and the multiplicative and additive principle, it is not difficult to know the symbols are '2,656 people', even though these symbols are not engraved in the same horizontal or vertical line.

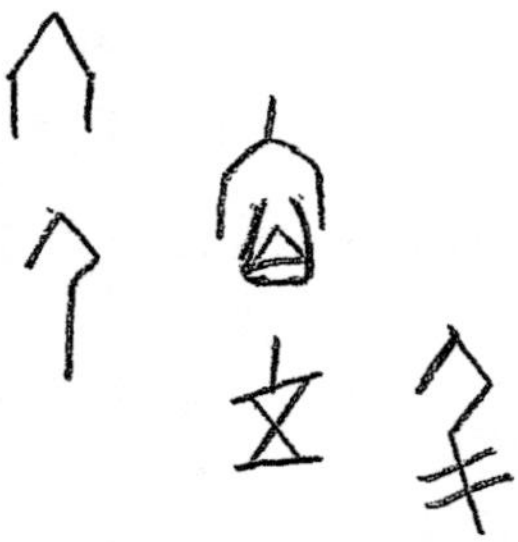

FIGURE 6.3
A copy of handwritten script '2,656 people' on one piece of oracle bone

The discussion above is already sufficient to show a decimal system for numeration in the oracle bones script. From numeral one, every time it is multiplied by ten, there is a special symbol created to represent the number. By relying on only the nine numbers and symbols for ten, hundred, thousand and myriad (ten thousand), most of the numbers carved on the bones or carapaces are expressed. Even in the number including the combined character (*hewen*), one can also find the 13 basic symbols above as the crucial components of the symbol.

### 2.2 *The Place Value System of the Numbers on Coinage*

During the period from the tenth to the third century BCE, texts carved or cast on the inner surface of bronze vessels have provided abundant characters for historians, the numerals among which demonstrate a close link to those engraved on oracle bones or tortoiseshells. Most symbols of the basic numerals in the bronze inscriptions (*jinwen* 金文) are similar to those in oracle bones script (Needham and Wang, 1959, p. 7), except that the bronze inscriptions contain more variations for the numerals four, five and ten (Guo, 2010, p. 20). Besides, the character 'and (*you*)' appears more in the number of bronze inscriptions, showing the additive principle was commonly adopted during this period. This character is used not only between the digits of the hundreds and tens, but also between the digits of the tens and the units (Chen, 1978, pp. 275–276).

The frequent use of the character 'and' might reflect a more formal and official appearance of a number or form of written language. Apparently, the bronze inscription numeration combing the symbol 'and' symbols representing ten, hundred, and thousand, etc., adopting the additive principle and multiplication, seems to imply less possibility of generating a place value system, in which the position of a digit in a number determines its value, namely, the powers of ten in decimal numeration.

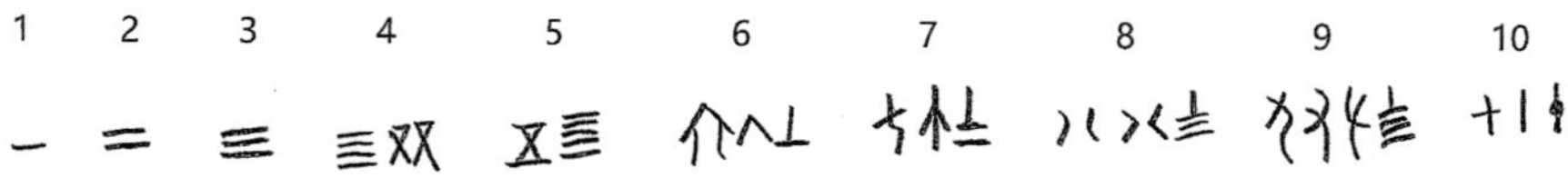


FIGURE 6.4 Common numerals on the coins from the eighth to the third century BCE

FIGURE 6.5 Numbers on the coins from the eighth to the third century BCE
Note: I appreciate that one referee has pointed out the inconsistency that emerged in some numbers involving digit five in Figure 6.5. These inconsistencies probably resulted from the different forms of the numeral five. Considering the evidential similarities and quite possible continuities between the numerals in the oracle bone script and those on coinages, readers could also refer to the variations of the numeral five in Figure 6.1 to recognize the digit five in the numbers in Figure 6.5, in which it could be in the form of '×'. Rotation of the symbol ⧗ by 90 degrees and the symbol without the bottom stroke could also be a possible symbol for digit five. Further explorations pertaining to the use and the evolution of the numeral symbols need to be carried out.

However, in the numbers found on the coinage dating from the eighth to the third century BCE, the place value system of numeration is relatively obvious. These numbers are used to indicate the order of making models of the mintage, on the basis of which it is possible to know the way of expressing numbers by ordinary artisans of casting coins. Most of the numbers have two digits and the largest number is 2,000 (Chen, 1978, p. 278). There are variations of numerals in the number on coins, and the most common shapes are listed in Figure 6.4 (Chen, 1978, p. 278).

Using the numerals above, the numbers on coins adopt a place value system that is clearly demonstrated by the examples in Figure 6.5 (Chen, 1978, p. 279). Comparing with the numbers on the oracle bones, a prominent feature of the numbers on coinage is that nearly all the symbols for ten in the number of two digits disappear; furthermore, even the symbol for hundred in the number of three digits disappears. For example, in the example of 154, the three digits are arranged horizontally, and the digit one in the position of hundreds is arranged on the left and the digit four in the position of units is arranged on the right. The symbols for hundred and ten are all omitted. Only the position taken by the digit determines the powers of ten, by which the digit in that position will be amplified. This way of expressing the numbers is very close to numeration by the configuration of counting rods. But in these numbers on coins, no solid evidence has been discovered that supports a placeholder having been used – see Figure 6.5.

### 2.3 *Counting Rods Numeration*

The use of counting rods as a calculation tool has been reflected on in the text of *The Nine Chapters on Mathematical Procedures* (*Jiuzhang suanshu* 九章算術 also translated as *The Nine Chapters on the Mathematical Art* (Shen, Crossley, and Lun, 1999), hereafter *The Nine Chapters*), which dates from to the first century BCE to the first century CE, but a clear description of the method of using counting rods to represent a number is recorded in the *Mathematical Canon of Sun Zi* (*Sunzi suanjing* 孫子算經), completed at the end of the third century or the beginning of the fourth century. At the beginning of the discussion about the method of multiplication and division, the text reads:

> For all the methods of mathematics, [the practitioner must] first know the position [where a digit is put]. The ones (units) are [expressed by a] vertical [model], and the tens are [expressed by a] horizontal [model]. The hundreds are [expressed by a model in which counting rods representing first several small digits, like one to four] stand, while thousands are [expressed by a model in which counting rods representing first several small digits] lie down. [The models of expressing] thousands and

tens look [the same] with each other, [and the models of expressing] the 'ten thousands' and the hundreds are corresponding to each other.

[The practitioner] does not [express] number six by accumulating [six counting rods]; and [the practitioner] does not [express] number five by a single [rod].[9]

凡算之法，先識其位。一從十横，百立千僵，千、十相望，萬百相當。
……六不積，五不隻。[10] (*Mathematical Canon of Sun Zi, juan* one, Folio 2, b)

According to the description above and other clues acquired in this book and other mathematical works, one can deduce the nine digits are expressed by arrangements of counting rods in the following two ways. In way I, a vertical counting rod represents digit one; two vertical counting rods represent digit two. Digits bigger than five use a combination of a horizontal counting rod representing five with vertical rod(s) below. The list below shows numerals from one to nine in way I:

| | 1 | 2 | 3 | 4 | 5 | 6 | 7 | 8 | 9 |
|---|---|---|---|---|---|---|---|---|---|
| Way I (vertical mode *zongshi* 縱式): | 𝍠 | 𝍡 | 𝍢 | 𝍣 | 𝍤 | 𝍥 | 𝍦 | 𝍧 | 𝍨 |

9 Since the classic Chinese is very concise, and because of the pursuit of harmonious sounds for the sentence, from the perspective of modern English grammar, components of the sentence of classic Chinese are usually omitted. Many omitted components need to be added into the translation to make the English sentence meaningful and precise in terms of the mathematical context for modern English readers. Needham's translation (1959, p. 8) and Lam and Ang's translation (2004, p. 193) give a general idea of this paragraph, and they are clear and direct in modern English, but they might not contain enough supplements of the sentence and do not completely fit each character of the original text. Translation in this chapter is based on interpretations of each Chinese word in the paragraph; besides, some added components in square brackets are not reflected from the text translated but are permeated by modern historians' received interpretation of the whole text from which the paragraph is extracted. The components in round brackets are an explanation of the word or sentence previous to them.

10 In *Mathematical Canon of Xiahou Yang* (*Xiaohou Yang suanjing* 夏侯陽算經) from the second half of the eighth century, a variation of the last sentence reads '[In the process of accumulating counting rods to express numbers, when the number is] above six, [the single counting rod representing] five is above [the other rods]. [One] does not [express] number six by accumulating [six] counting rods; [and the rod representing] five is not placed individually (it must be used with other rods to present numbers larger than five). (滿六以上，五在上方；六不積算，五不單張。) (Ji, 1999, p. 274).

In way II, a horizontal rod represents digit one; two horizontal counting rods represent digit two. Digits six, seven, eight, and nine use a combination of the horizontal counting rod(s) and a vertical rod above representing five.

In a number, the digits represented by the two ways appear alternatively. According to the description in *Mathematical Canon of Sun Zi*, the digits in the positions of units, hundreds and ten thousands use way I, and the digits in their neighboring positions use way II. However, numbers represented by these ways do not appear in the extant *Mathematical Canon of Sun Zi* or any other mathematical canons printed in the thirteenth century. The earliest counting rods numbers are recorded in a manuscript titled *Mathematical Canon of Ready-made [Rules]* (*Licheng suanjing* 立成算經) around the tenth century, which was founded in Dunhuang.

| | 1 | 2 | 3 | 4 | 5 | 6 | 7 | 8 | 9 |
|---|---|---|---|---|---|---|---|---|---|
| Way II (horizontal mode *hengshi* 橫式): | 𝍩 | 𝍪 | 𝍫 | 𝍬 | 𝍭 | 𝍮 | 𝍯 | 𝍰 | 𝍱 |

In this manuscript, the last number of each piece of 'pithy formula (i.e., short sentences in concise Chinese with a certain rhythm composing a set of instructions for calculation that could be recited or memorized easily)', which is the product of multiplication between two digits, is represented by the configuration of counting rods in Figure 6.6 (see the first extracted sentence and its translation in Figure 6.6). The numbers of the summation of all the products one gets after each pithy formula are also calculated and represented by counting rods (see the second extracted sentence and its translation in Figure 6.6). For example, 36 is written as 𝍫𝍥, and 196 is written as 𝍠𝍱𝍥. Martzloff (1997, pp. 206–208) and Chen Liangzuo (1977, p. 176) have pointed out that Li Yan's transcription (1963, pp. 36–39) of this manuscript slightly modified the formation of the rods numeration. For example, in his transcription of the second extracted sentence in Figure 6.6, Li adds a space between the 𝍠 and 𝍧 in the number 108. In one of the subsequent sentences, he adds a space after the 𝍪 in the number 120. In fact, the cursive handwriting in the manuscript does not strictly make use of space or any other marks to avoid the ambiguous representation of counting rods. The number 18 (𝍩𝍧) is distinguished from 108 (𝍠𝍧) in the counting rods numeration by means of the different modes adopted for digit one of tens and that for digit one of hundreds. According to the text of *Mathematical Canon of Sun Zi* and the variation of it placed at the beginning of *Mathematical Canon of Ready-made* [*Rules*], the digits of tens are in horizontal mode (see Way II) while the digits of hundreds are in the vertical mode (see Way I). Considering that the characters in a pithy formula are conventionally

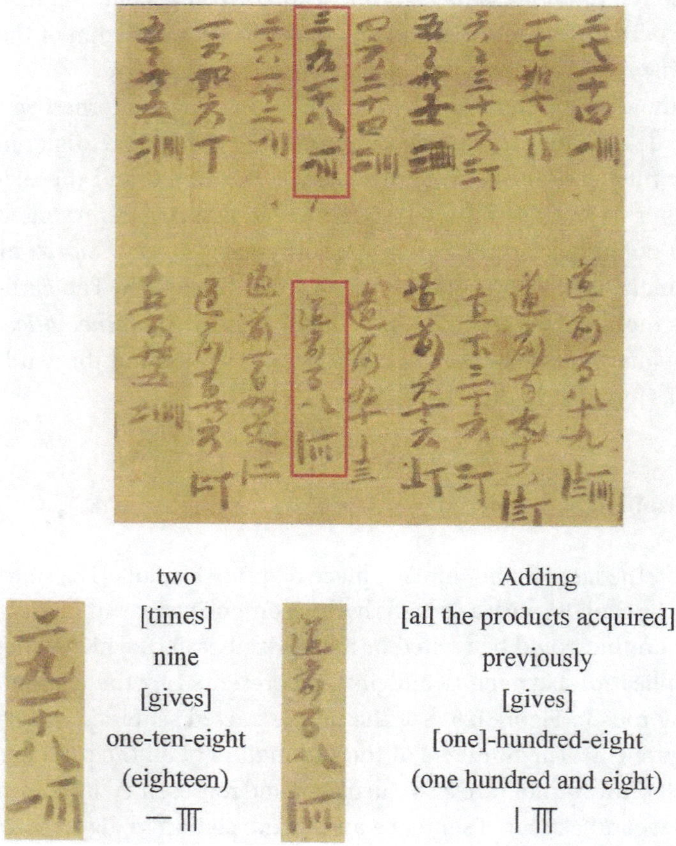


FIGURE 6.6 Parts of manuscripts from *Mathematical Canon of Ready-made* [*Rules*] (No. S. 930) from Dunhuang

written in a vertical line and, on the contrary, the representations of digits are listed in a horizontal line, there is not much space left between the digits of a number. Thus, the change of the mode of digits in turn plays an important role to avoid ambiguity in certain cases.

## 3 A Placeholder in the Numeration and the Calculations with the Placeholder

If there is no space or any mark between the digits of a counting rods number, the change of mode of digits is not sufficient to avoid ambiguity in cases such as representations of the numbers 18 and 1008. In fact, these representations of numbers without space between the digits do not faithfully reflect the actual physical appearance of counting rods on a counting board. Space between digits must exist and be carefully treated on the counting board. This point of view can be verified by the procedure carried out on the basis of the description of the method of multiplication or division.

An analysis of division is shown in the example below. According to the text of *Mathematical Canon of Sun Zi*, the operands dividend (*shi* 實), divisor (*fa* 法), and the quotient acquired are placed in three rows, and all the digits represented by counting rods should be strictly placed in certain positions relative with each other. For instance, the position of hundreds of the dividend and that of the quotient are aligned vertically. And in the whole process of division, the position of the first digit of the divisor is moved from left to the right to help practitioners determine the position of the digit of quotient in a certain partial division.

In the example of dividing 100 by 6 taken in the introduction to the method of division in *Mathematical Canon of Sun Zi*, the author says:

> [When the practitioner] divides one hundred by six, [the practitioner] should advance [the digit six] two grades (positions) [to the left] and make [it] precisely below [the position of] hundreds [of the dividend]. [In this position,] digit one is divided by six, thus [this is a case in which] the divisor [of this position] is lager while the dividend [of this position] is smaller. [The practitioner] could not make a division [in this position]. Thus [the practitioner] should retreat [digit six rightward] to the position [below that of] tens [of the dividend]. [In this position, the practitioner] divides the dividend by divisor [and one gets the quotient one in the position of tens, thus the practitioner] calls digit one [of the quotient in the position of tens] multiplies [the divisor] six, and deducts [the product sixty] from the one hundred, [the result of which] is forty. Thus [this process shows that] division [in this position] is possible.
>
> 以六除百，當進之二等，令正在百下。以六除一，則法多而實少，不可除，故當退就十位，以法除實，言一六而折百爲四十，故可除。(*Mathematical Canon of Sun Zi, juan* one, folio 3, a)

According to the explanation above, the dynamic configurations of counting rods on the counting board of the whole process of division are demonstrated below, and a transcription of them in Hindu-Arabic numbers is placed as follows.[11]

11 In Lam and Ang's transcription of the configuration of counting rods (LAM, ANG, 2004, p. 64), the symbol of zero is added while there is no clue in the text showing that such a physical placeholder appears on the counting board.

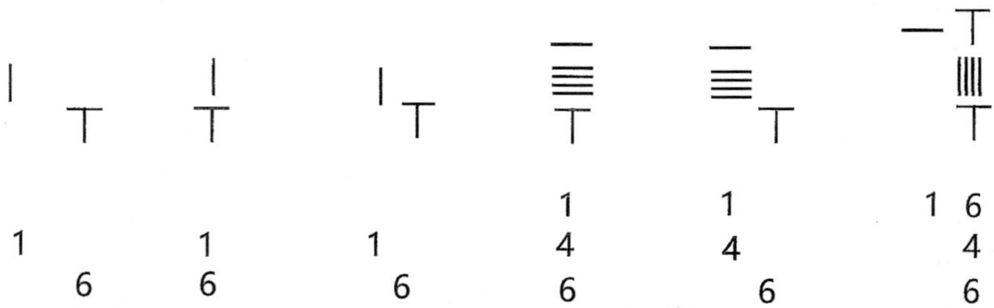

FIGURE 6.7 Dynamic configurations of counting rods of the process of a division and their transcription in Hindu-Arabic numbers

In the process of division conducted, one could naturally perceive that vacant positions in certain positions are necessary, no matter whether the vacant position is located at the end of a number or between digits of the number. In commentary by the seventh-century commentator Li Chunfeng (李淳風 602–670), the word '*kongjue* (空絕 literally 'absolute empty')' was used to describe the case in which there is no digit in a position between two digits of a quotient. Li further stated that when this case has happened, the counting rods representing divisor have been moved two positions to the right.

In the process of calculation with counting rods, a considerable emphasis on the accuracy of the positions of the digits has been made not only in the text of *Mathematical Canon of Sun Zi*, but also in the procedure of the extractions of a square root and a cubic root in *The Nine Chapters*, from which one could also conclude that empty positions must have appeared in numbers represented by counting rods during calculation.[12]

Taking advantage of the decimal and place value system of numeration in China, it is clear that even though no particular object or mark is used to represent the digit zero in a number, the vacant position between digits sometimes plays the role of zero in a mathematical calculation.

In the thirteenth century, a symbol for zero frequently appears in numbers in Qin Jiushao's (秦九韶 1208–1261) *Writing on Mathematics in Nine Chapters* (*Shushu jiuzhang* 數書九章 1247) and Li Ye's (李冶 1192–1279) *Sea-mirror of the Circle Measurement* (*Ceyuan haijing* 測圓海鏡 1248). For example, in the latter

12 Chen Liangzuo even made a further assumption that the counting board should have grids to help practitioners place digits represented by counting rods in the correct positions. When numbers with many digits are involved in a division, a counting board without grids might lead to great confusion. Illustrations of counting board with grids in seventeenth and eighteenth century Japanese mathematical works show that grids exist on a counting board. Chen suggested that at least the marks of the order of vertical columns should have appeared on the counting board in China (1977, p. 171).

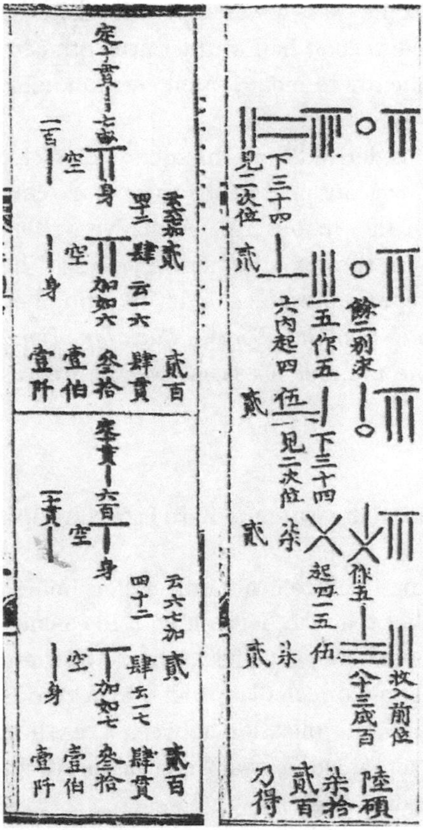

FIGURE 6.8
○ or the character '*kong* (empty)' in written procedures for multiplication and division in *Mathematical Methods of Yang Hui*

work, the integer part of a number, 3,056, was written as ≡ ○ ≣ ⊤ (Li, 1998, p. 386). In mathematical procedures in Zhu Shijie's (1249–1314) *Jade-mirror of Four Elements* (*Siyuan yujian* 四元玉鑒 1303), if the numbers corresponding to the coefficients of an equation from the perspective of modern mathematics are zero, their positions are also taken by a circle symbol ○. An example is shown in the configuration

|○太○|
○○○○○
○○|○○

which corresponds to equation $x^2 + y^2 + z^2 = 0$, in which the coefficients before $x, y, z, xy, x^2y, xy^2, x^2y^2, xz, x^2z, xz^2, x^2z^2$ are all zero (Yan, 1982, p. 37).

In the surviving fifteenth-century printed edition of *Mathematical Methods of Yang Hui* (*Yanghui suanfa* 楊輝算法 1274, 1275), one can find that ○ or the character '*kong* (空 literally empty)' is used in the procedure for multiplication or division by numbers represented by counting rods in Figure 6.8. The symbol ○ was also used in the books concerning pitch pipe or calendar calculation. In the *Daming Calendar* (大明曆 1180), this symbol as a placeholder is written between characters for numbers, but in some cases it was also omitted. In the *Shoushi Calendar* (授時曆), used since the second half of the thirteenth century, the symbol ○ was used more frequently to record numbers (Yan, 1982, pp. 36–37) – see Figure 6.8.

Historians suggest that the symbol ○ is derived from the square character written as □, which is used to represent a vacant position because of obscure or missing characters on a book page or in a manuscript. A cursive writing of the square could lead to a circle seen in those mathematical works.[13] The square itself was also used directly to represent a placeholder in a number, a well-known example of which is seen in *Complete Book of Pitch Pipe* (*Lülü chengshu* 律呂成書 *c.*1300), in which the number 10,636,863 was written as 一□⊥|||⊥𝍧⊥||| (Yan, 1982, p. 36).

## 4 The Rules of Calculations in Which the Numeral Zero Is Involved

From the perspective of modern mathematics, zero is not only a placeholder, it is also regarded as an individual number that can be dealt with in calculations such as addition, subtraction, multiplication, and division, etc. And there are rules that declare explicitly the principle of calculation in which zero is involved, such as the rules that the result of multiplication between a zero and any number is always zero, the result of zero divided by any number is always zero, and zero should never be taken as a divisor, etc.

Before the end of the sixteenth century, in Chinese mathematical literatures that do not show a clear influence by Western mathematical works, their authors did not directly state that there is an invisible number as a placeholder in the vacant position of the counting board. Neither did they state that a mark, like ○, in the written counting rods number system is a numeral. Furthermore, a clear declaration of the principles in which zero individually involved in general calculation is not to be found in mathematical texts. For example, in different kinds of multiplication table found in bamboo strips dating back to the Warring States period (475–221 BCE), in the *Mathematical Canon of Sun Zi*

13 Yan Dunjie explained that the user of an ink brush would draw a square in a way such as 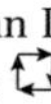, a quick drawing of which will lead to a circle, but this assumption is in question because it is difficult to prove the ancient writer conventionally preferred to wiring the □ in such way. On the contrary, in a writing-copy version of *Mathematical Methods of Yang Hui* in the *Special Collection of Wanwei* (*Wanwei biecang* 宛委別藏) at the beginning of the nineteenth century, the symbol □ representing the missing characters was not written in this way, and the unlinked strokes of the square demonstrate that they were probably written as the character 口 (mouth) by three non-continuous strokes.

from the end of the third century or the beginning of the fourth century, or in tenth-century mathematical manuscripts, the pithy formulas do not contain the case in which zero is involved into any multiplication with other number. In the thirteenth century, a series of pithy formulas for divisions appeared in Yang Hui's work, on the basis of which a division by any number could be calculated; however, the divisors of these pithy formulas do not include a numeral zero either.

However, in a mathematical method titled *fangcheng*, the practitioner must have dealt with the cases in which a zero is involved from the perspective of modern mathematics; the ancient practitioners have summarized the rules for these calculations.

The *fangcheng* chapter of *The Nine Chapters* deals with problems related to solving what today are known as simultaneous sets of linear equations.[14] A problem in the chapter *fangcheng* of *The Nine Chapters* sets out:

> Suppose that if one sells two oxen and five sheep to buy 13 pigs, they still have 1,000 coins, that if one sells three oxen and three pigs to buy nine sheep, they have just enough coins, and that if one sells six sheep and eight pigs to buy five oxen, they are short of 600 coins. One asks what the prices of ox, sheep and pig are.
>
> 今有賣牛二、羊五，以買一十三豕，有餘錢一千；賣牛三、豕三，以買九羊，錢適足；賣六羊、八豕，以買五牛，錢不足六百。問牛、羊、豕價各幾何。(Chemla and Guo, 2004, p. 634)

14 Translations of the title *fangcheng* and the ways used to interpret its contents vary from scholar to scholar. For example, Wang Ling and Needham translate the title as 'The Way of Calculating by Tabulation' (Needham and Wang, 1959, p. 26), and describe its contents as: 'The counting-rods, in Chinese style, were placed in different squares in a table so as to represent coefficients of the different unknowns, and the constant terms. The Chiu Chang (*Jiuzhang*) (Chapter 8) has numerous problems requiring the solution of equations of the type ax + by = c, a′x + b′y = c″ (Needham and Wang, 1959, p. 115). Shen Kangshen et al. (Shen, Crossley, and Lun, 1999, p. 386) translate it as 'Rectangular Arrays' and explain the procedures as 'the coefficients are laid out in an array and then the coefficient matrix of the array is reduced to triangular form using elementary matrix operations.' Roger Hart (2011, pp. 67–84) thought that the 'fangcheng procedure' is in many ways similar to modern Gaussian elimination except for the system for writing numbers and the orientation of the matrix. Karine Chemla leaves it untranslated, but in the glossary interprets this title as 'mesures en carré' [measures (cheng 程) in square (fang 方)] (Chemla and Guo, 2004, pp. 599, 922), which is approved by the author of this chapter. The reason of this interpretation is seen in (Chemla, 1994; Chemla, 2000; Chemla, 2019, pp. 106–107).

A solution to this problem in modern mathematics is shown below, in which the prices of one ox, one sheep, and one pig are $x$, $y$ and $z$ respectively.

$$\begin{cases} 2x + 5y = 13z + 1000 \\ 3x + 3z = 9y \\ 6y + 8z = 5x - 600 \end{cases}$$

To solve this problem, the practitioner needs to place the counting rods on the counting board in the configuration in Figure 6.9. One possible way is that a slash over the counting rods represents the number whose property is opposite to that of the rods number without a slash. The former number corresponds to a number one called a negative number and the latter corresponds to today's positive number. However, there are subtle differences between the calculation between those numbers.

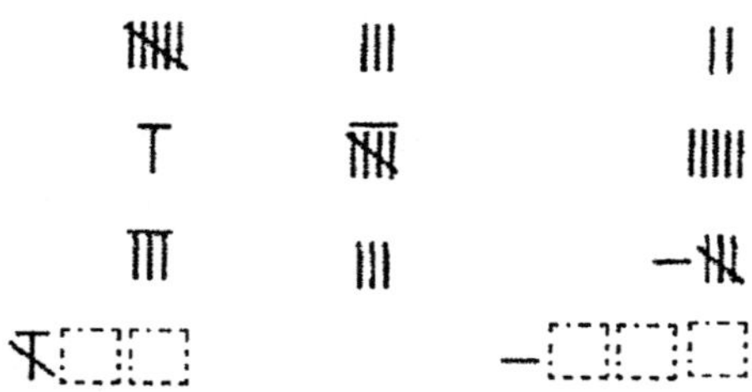

| | | |
|---|---|---|
| -5 | 3 | 2 |
| 6 | -9 | 5 |
| 8 | 3 | -13 |
| -·600 | | 1000 |

FIGURE 6.9 Configuration of numbers represented by counting rods and its transcription in Hindu-Arabic numbers

A 'procedure of the positive and negative' is recorded in *The Nine Chapters* to solve this kind of problem. In this procedure, two rules concerning the calculation of zero appear. Because the rules are very concisely expressed, interpretations of the sentences of rules (1-C and 2-C) vary from person to person. A translation by Martzloff of the two sentences are as follows; nevertheless, the chapter below suggests that the explanation in the round brackets is problematic to some extent, and this chapter will give a detailed translation that makes up the missing parts according to the context where the two rules are extracted.

> 1-C: [If a] positive [rod] does not have a vis-à-vis, it is made negative (a positive number subtracted from nothing becomes negative); 正无入負之，
>
> [If a] negative [rod] does not have a vis-à-vis, it is made positive (a negative number subtracted from nothing becomes positive). 負无入正之。(Martzloff, 1997, p. 203)

> 2-C: If a positive [rod] does not have a vis-à-vis, it is made positive (a positive number added to nothing is left positive); 正无入正之，
> If a negative [rod] does not have a vis-à-vis, it is made negative (a negative number added to nothing is left negative). 負无入負之。 (Martzloff, 1997, p. 203)

As Martzloff's explanations in the round brackets, historians generally thought that these two sentences were a part of sign rules (Shen, Crossley, and Lun, 1999, p. 405; Guo, 2009, pp. 333–335), which correspond to the following formulas:

1-C: $0 - a = -a$
$0 - (-a) = a$
2-C: $0 + a = a$
$0 + (-a) = -a$

However, Mei Rongzhao (1984), Chemla (Chemla and Guo, 2004, pp. 625–629), Zhu Yiwen (2010), and Zhou Xiaohan (2020) do not agree that these four sentences are general rules of signs for addition and subtraction in which zero is involved. Sentences of 1-C were placed after the sentences that Martzloff translated as follows:

> 1-A: [Rods of the] same name [=sign] [are] mutually reduced. 同名相除，
> 1-B: [Those with] different names [are] mutually increased. 異名相益。

And the latter sentences of 2-C were placed after the sentences that Martzloff translated as follows:

> 2-A: [Rods] with different names [are] mutually reduced. 異名相除，
> 2-B: [Those with the] same name are mutually increased. 同名相益。

In fact, none of the sentences above could be interpreted individually without the context of the *fangcheng* method. They are all used to guide practitioners in dealing with numbers with different signs in certain steps of the whole *fangcheng* procedure. As for sentences 1-C and 2-C, focused on in this chapter, they deal with the procedure in which zero is involved from the modern perspective, but they are different to some extent from that of today.

A major process of the *fangcheng* procedure is to eliminate numbers in certain positions of the configuration of numbers, such as in Figure 6.9. After a series of eliminations, only the numbers of one certain object and its corresponding price (in the context of finding the price of each object) are left on the counting board. By division between the two numbers, one can find the price of the object and other prices are then deduced on the basis of the price already found.

How to eliminate the numbers on the counting board without changing the relationships between the numbers? Two ways are recorded in *The Nine Chapters*. One way is to multiply, for example, *b* in a certain position by number *a* in its corresponding position of the configuration and then carry out *b* times of deduction, in each of which *a* is deducted from the product *ab*; thereafter the position *b* originally takes is empty. Even though the other way was introduced in *The Nine Chapters*, it was popular after the thirteenth century according to surviving historical literature. This way uses mutual multiplication between *a* and *b* in corresponding positions of a configuration and uses a one-time elimination between the two numbers (*ab*) in two positions to make the number in one of the positions empty, achieving the aim of elimination of numbers. No matter which way of elimination, the numbers in the same column with *a* and *b* need to be multiplied, added to, or subtracted from each other. According to the context of the question, when the nature of numbers is opposite with each other (such as 'remaining money' and 'money that one lacks'), the numbers are marked as positive and negative. When these numbers with opposite signs are to be dealt in *fangcheng* procedure, the 'procedure of the positive and negative' is applied.

The whole context of *fangcheng* procedure and the commentaries by the third-century commentator Liu Hui all support an interpretation of the sentence as follows:

> 1-A and 1-B: [If the numbers in the position which is to be eliminated are] the same *ming* (名 lit. name i.e., sign),[15] [the numbers with the same *ming*] are eliminated from each other, [the numbers with] different *ming* are added to each other.

15 The first clause of the procedure always implies a conditional, letting the readers know in which cases they should choose the rules of series 1 or series 2. This is the reason why translations such as 'different *ming* are eliminated from one another' or 'opposite signs subtract' (Shen, Crossley, and Lun, 1999, p. 404; Dauben, 2007, p. 280) is not adopted. The translation is modified on the basis of Karine Chemla's translation: 'If the different names are eliminated from one another, the same names are increasing each other. (1994, p. 10)'; 'Si des nombres de noms différents sont éliminés l'un de l'autre, les nombres de même nom s'augmentent l'un l'autre. (Chemla and Guo, 2004, p. 627)' (異名相除，同名相益).

> 2-A and 2-B: [If the numbers in the position which is to be eliminated have] different *ming*, [the numbers with different *ming*] are eliminated from each other, [the numbers with] the same *ming* are added to each other.

An issue that deserves great attention is that, in the text of The Nine Chapters and some works that were influenced by this canon, what 'positive' and 'negative' refer to are not equivalent to what we call 'positive numbers' and 'negative numbers' in modern mathematics.[16] In these works, 'positive' refers to the number that was marked by a character written as 'zheng 正' (positive), and 'negative' refers to the number that was marked by a character written as 'fu 負' (negative).[17] It was the numbers without any marked sign, rather than a 'positive number' or a 'negative number' as treated in modern mathematics, that were operated and calculated.

In *Mathematical Methods Explaining in Detail The Nine Chapters* (*Xiangjie jiuzhang suanfa* 詳解九章算法 hereafter *Mathematical Methods*, 1261), a configuration of numbers could support this interpretation. In the extraction of the configuration (Guo, 1993, p. 967) transcribed on the left of Figure 6.10, the two number sixes in the middle positions are marked as positive and negative respectively. Since they have different names (signs), rules 2-A and 2-B are applied to each pair of numbers in corresponding position in the two columns. The 'six of positive' and 'six of negative' are not added with each other as one expects from the perspective of modern mathematics; on the contrary, the columns where the numbers belong to are called 'minuend (原數)' and 'subtrahend (減數)' in terms of a 'subtraction', which is also different from modern 'subtraction' (Zhou, 2020, pp. 51–55). The 'subtraction' precisely corresponds to the term 'elimination (*chu* 除)' in 2-A of *The Nine Chapters*.

However, there are some cases that are not covered by sentences 1-A, 1-B, or 2-A and 2-B. In these cases, there is not any number in a position of configuration of *fangcheng*. This vacant position corresponds to the number zero from

---

16 For research on the concepts of 'positive number' and 'negative number' in early Chinese history, see Zou (2010), where, on the basis of the earliest Chinese mathematical texts of the first millennium BCE written on bamboo or wooden slips, the subtle meaning of these concepts and the process of their emergence are analyzed.

17 This understanding of 'positive number' and 'negative number' could be supported by the use of counting rods of different color or of different shape as calculation tools. The opposite color or opposite shapes are marks of the nature of 'positive' and 'negative'. Only the number of counting rods was calculated. For a detailed discussion of this topic, see Zhu (2010).

| Positive and negative is deducted [to acquire] this number | Minuend | Subtrahend |
|---|---|---|
| 3 Negative [sentence 2-A] | 5 Negative | 2 Positive |
| Empty [sentence 2-A] | 6 Positive | 6 Negative |
| 10 Positive [sentence 2-B] | 8 Positive | 2 Positive |

| Remove the number through deduction | Minuend | Subtrahend |
|---|---|---|
| Ox Empty [sentence 1-A] | 2 Positive | 2 Positive |
| 11 Negative [sentence 1-B] | 6 Negative | 5 Positive |
| 15 Positive [sentence 1-B] | 2 Positive | 13 Negative |
| 1000 Negative Positive has [no number] to enter, to be negative | No [number] to enter | 1000 Positive |

FIGURE 6.10 Transcription of parts of configuration of the fangcheng procedure in *Mathematical Methods*
Sentences marked with 2-A, 2-B, 1-A and 1-B in this configuration are variations of their corresponding rules from *The Nine Chapters*. The differences are not discussed in this chapter and the corresponding sentences are marked with the same label.

the perspective of modern mathematics; however, for the ancient practitioners of *fangcheng*, there is not a number in this position, thus a character '*kong* 空 (empty)' is taken in its place. The positive and negative signs could never be marked on a place without any number, therefore sentences 1-A, 1-B, or 2-A and 2-B cannot deal with this case.

There are different reasons that might lead to a case in which there is no number in a certain position. For example, the original condition in the question might state that there is not any object, reflecting on the counting board in the first configuration of numbers, a certain position is empty. There is also the case that after deduction of numbers according to sentences 1-A or 2-A, a number in the column that is called 'minuend' is totally deducted, thus there are no counting rods in that position, while there are still counting rods left after the deduction in the corresponding position in the 'subtrahend' column.

In these cases, sentences 1-C or 2-C are used and a revised translation of them are as follows:

> 1-C: [If] the positive has no [number] to enter (*wuru* 無入 or *wuren* 無人, i.e., the other operand is equal to zero), [one makes it] to be negative; [if] the negative has no [number] to enter, [one makes it] to be positive.
> 2-C: [If] the positive has no [number] to enter, [one makes it] to be positive; [if] the negative has no [number] to enter, [one makes it] to be negative.

In the extraction of the configuration (Guo, 1993, p. 968) transcribed on the right of Figure 6.10, the ten-thousand of positive in the right column (subtrahend column) has no corresponding number in the middle column (minuend column) to enter. And since number two in the top position of the minuend column has the same *ming* (which is positive) with number two in the subtrahend column, the rules 1-A and 1-B are applied to the numbers in the two columns. Simultaneously, sentence 1-C should be applied to the case in which ten-thousand of positive has no number to enter, thus one makes it to be negative.

The 'procedure of the positive and the negative' clearly demonstrates the case in which even though the numeral zero is not introduced by Chinese practitioners of mathematics, general rules in terms of a certain mathematical methods *fangcheng* were still precisely expressed and applied to address the question that must be solved by rules of addition and subtraction with the numeral zero from the perspective of modern mathematics. This procedure was recorded in *The Nine Chapters*, which dates back to the first century BCE, and it was inherited by commentators and learners of this book in the subsequent millennium.

## 5 Summary

Scholars have researched the communication and mutual influence among the numeration of China and that of other civilizations. On the one hand, materials from after the end of the nineteenth century are abundant, and they clearly demonstrate the process of gradual adoption of Hindu-Arabic numerals in the modern mathematical sciences established in China. On the other hand, before European missionaries came to China at the end of the sixteenth century, the limited and fragmentary historical records concerning the interaction of numeration between China and other civilizations still hinder scholars from concluding a clear path of such influence from China to another place, *vice versa*.

This chapter focuses on the question that in ancient China, how elements of the functions of numeral '0', as a placeholder of numeration and as a numeral in calculation in modern mathematics, are achieved in mathematical works or practices. To examine Chinese numeration, this chapter introduces the decimal system reflected in the numbers on oracle bones script and demonstrates the place value system used in the numbers on coinage. The decimal place value system was fully established in counting rods numeration. A description of the use of counting rods to express numbers is found in *Mathematical Canon of Sun Zi* and the earliest written counting rods numbers exist in a Dunhuang manuscript from around the tenth century.

Because there are two models of series of counting rods numerals appearing in turn in a number, which could avoid the ambiguous case in which one digit is zero, the counting rods numbers in the manuscript do not contain a placeholder. However, through conducting the procedure of multiplication, division, or extraction of roots according to the textual description of these mathematical methods, no one could deny that precise vacant positions in the number are emphasized and strictly respected by the ancient practitioners, even though there is probably no physical object or written symbol representing zero as a placeholder. In the thirteenth-century mathematical works, the written symbol ○ is systematically used as a placeholder in number.

In modern mathematics, zero is not only a placeholder, but also an individual number that can be calculated. Furthermore, a prominent demonstration of zero being regarded as an individual number is a clear declaration of the rules of calculation in which zero is involved. This chapter interprets the 'procedure of the positive and the negative', which is usually thought of as general rules of addition and subtraction between positive numbers, negative numbers and zero. In fact, this procedure is used in the context of the mathematical method *fangcheng*, but general rules concerning the process of elimination numbers in an array in which zero is involved are clearly expressed and adopted. This case, together with the fact that counting rods numbers with vacant positions as placeholders are precisely calculated, show that in a mathematical context in which the numeral zero has not been widely and clearly adopted, the elements of the functions of zero could be achieved by their own means.

## Acknowledgements

I greatly appreciate the discussions by email with Peter Gobets, for they shaped the thesis of this chapter. I would also like to pay tribute to two referees for their critical and yet benevolent reading. Their feedback on my text helped me to improve my ideas and writing. I would like to deeply thank Louise Bolotin and Mary Davis for polishing the chapter and editing the format of the bibliography. I take full responsibility for the remaining shortcomings.

## References

Biot, Edouard. Note sur la connaissance que les chinois ont eue de la valeur de position des chiffres [J]. *Journal Asiatique*, 1839(12): pp. 497–502.

Bo, Shuren 薄樹人. (2003.) Astronomico-astrological Divination Canon of Kaiyuan (713–741) Reign: An exotic book in the history of Chinese culture (*Kaiyuan zhanjing: zhongguo wenhuashi shang de yibu qishu* 開元占經——中國文化史上的一部奇書) [A]. In Collected Works of Bo Shuren (*Bo Shuren wenji* 薄樹人文集) [C]. Hefei: University of Science and Technology of China Press, pp. 189–196.

Cajori, Florian. (1919). *A History of Mathematics* [M]. New York, London: Macmillan.

Chemla, Karine. (1994). Different concepts of equations in *The Nine Chapters on Mathematical Procedures* and in the commentary on it by Liu Hui (third century) [J]. *Historia Scientiarum*, Presentation in NHCS, 1994, 4(2): pp. 113–137.

Chemla, Karine. (2000). Les problèmes comme champ d'interprétation des algorithmes in *Les Neuf Chapitres Sur Les Procédures Mathématiques* et leurs commentaires. De la résolution des systèmes d'équations linéaires [J]. *Oriens-Occidens: sciences, mathématiques et philosophie de l'Antiquité à l'Âge classique*, 2000(3): pp. 189–234.

Chemla, Karine. (2019). Different clusters of text from ancient China, different mathematical ontologies [J]. *Hau: Journal of Ethnographic Theory, Special Issue, Science in the Forest, Science in the Past*, 9(1): pp. 99–112.

Chemla, Karine. (2022). Cultures of Computation and Quantification in the Ancient World: An Introduction in Chemla, Keller and Proust (eds.) Cultures of Computation and Quantification in the Ancient World, Springer, 2022.

Chemla, Karine and Guo, Shuchun 郭書春. (2004). Les neuf chapitres: le classique mathématique de la Chine ancienne et ses commentaires [M]. Paris: Dunod.

Chen, Liangzuo 陳良佐. (1977). The vacant places in the rod-numeral system, zero, and some related questions (*Zhongguo chousuan zhong de kongwei ling jiqi xiagnguan de yixie wenti* 中國籌算中的空位 零 及其相關的一些問題) [J]. *Dalu Zazhi* (大陸雜誌), 54(5): pp. 1–13.

Chen, Liangzuo 陳良佐. (1978). The development and influence of pre-Qin mathematics (*Xianqin shuxue de fazhan he yingxiang* 先秦數學的發展及其影響) [J]. Bulletin of the Institute of History and Philology (*Zhongyang yanjiuyuan lishi yuyan yanjiusuo jikan* 中央研究院歷史語言研究所集刊). Part 2 (49): pp. 263–320.

Dauben, Joseph. (2007). Chinese Mathematics [G]. In *The Mathematics of Egypt, Mesopotamia, China, India, and Islam, A sourcebook*. New Jersey: Princeton University Press, pp. 178–384.

Guo, Shuchun 郭書春 (1993). (ed.). Complete Collection of Chinese Classical Works on Sciences And Technology, 1st Volume of Mathematics (*Zhongguo kexue jishu dianji tonghui shuxuejuan yi* 中國科學技術典籍通彙 數學卷 一) [G]. Zhengzhou: Henan Education Press.

Guo, Shuchun 郭書春. (2009). Translation and annotation of The Nine Chapters on Mathematical Procedures (*Jiuzhang suanshu yizhu* 九章筭術譯註) [M]. Shanghai: Shanghai Ancient Works Publishing House.

Guo, Shuchun 郭書春 (2010). (ed.) A History of Science and Technology in China, Volume of Mathematics (*Zhongguo kexue jishu shi, shuxue juan* 中國科學技術史數學卷) [M]. Beijing: Science Press.

Gupta, Radha Charan. (2019). Indian Mathematical Sciences in Ancient and Medieval China [A]. In K. Ramasubramanian (ed.) *Gaṇitānanda, Selected Works of Radha Charan Gupta on History of Mathematics* [C]. Singapore: Springer: pp. 575–604.

Hart, Roger. (2011). *The Chinese Roots of Linear Algebra* [M]. Baltimore: The Johns Hopkins University Press.

Ifrah, Georges, Bellos, David et al. (2000). (transl). *The Universal History of Numbers* [M]. John Wiley & Sons.

Ji, Zhigang 紀志剛. (1999). Guide to the Reading of Mathematical Canon of Master Sun, Mathematical Canon of Zhang Qiujian, Mathematical Canon of Xiahou Yang (*Sunzi suanjing Zhang Qiujian Xiahou Yang suanjing daodu* 《孫子算經》《張邱建算經》《夏侯陽算經》導讀) [M]. Wuhan: Hubei Education Press.

Lam, Lay Yong, and Ang, T. S. (2004). *Fleeting Footsteps: Tracing the Conception of Arithmetic and Algebra in Ancient China* [M]. Revised edition. Singapore: World Scientific.

Li, Yan 李儼. (1963). Historical Source Material of Chinese Ancient Mathematics (*Zhongguo gudai shuxue shiliao* 中國古代數學史料) [M]. Shanghai: Shanghai Science and Technology Press.

Li, Yan 李儼. (1998). Chinese mathematicians' numeration (*Zhongsuanjia de jishufa* 中算家的記數法) [J]. *Bulletin of Mathematics* (*Shuxue tongbao* 數學通報) 1958(6): pp. 1–5, 20. In *Complete Collected Works of History of Sciences, Vol. 10*, by Li Yan And Qian Baocong (*Li Yan Qian Baocong kexueshi quanji dishijuan* 李儼錢寶琮科學史全集 第十卷) Shenyang: Liaoning Education Press, pp. 374–388.

Li, Zhizao 李之藻, and Ricci, Matteo 利玛窦. (2013). Arithmetical Indicator Gathering the Cultures (*Tongwen suanzhi* 同文算指): A critical edition by Ji, Zhigang 紀志剛, et al. In The Collection of Books in Chinese Concerning The Western Missionaries During The End of Ming Dynasty And The Beginning of Qing Dynasty, The Fourth Collection (*Mingqing zhiji xifang chuanjiaoshi hanji congkan disiji* 明清之際西方傳教士漢籍叢刊 第四輯) Nanjing: Phoenix Publishing & Media.

Libri, Guillaume. (1838). *Histoire des Sciences Mathématique en Italie* [M]. Paris: Chez Jules Renouard et C, Libraires.

Martzloff, Jean-Claude and Wilson Stephen. (1997). (transl). *A History of Chinese Mathematics* [M]. Berlin: Springer-Verlag.

*Mathematical Canon of Ready-made [Rules]* (*Licheng suanjing* 立成算經) [MS]. S. 930 http://idp.bl.uk/database/ooscroll_h.a4d?uid=8176139377;recnum=929;index=1.

Mathematical Canon of Sun Zi (*Sun Zi suanjing* 孫子算經) (1980). [M] in Six Mathematical Canon Printed in Song [Dynasty] (*Songke suanjing liuzhong* 宋刻算經六種) [M]. Beijing: Cultural Relics Press.

Mei, Rongzhao 梅榮照. (1984). The *fangcheng* theory of Liu Hui (*Liu Hui de fangcheng lilun* 劉徽的方程理論) [A]. *Collected Papers on the History of Sciences* (*Kexueshi jikan* 科學史集刊). Beijing: Geological Publishing House, (11): pp. 63–76.

Mikami, Yoshio. (1913). The development of mathematics in China and Japan [M]. Leipzig: Teubner.

Needham, Joseph and Wang, Ling. (1959). Science and Civilisation in China: Volume 3, Mathematics and the Sciences of the Heavens and the Earth [M]. Cambridge: Cambridge University Press.

Qian, Baocong 錢寶琮. (1933). The textual research on the way of recording numbers (*Jishufa yuanliu kao* 記數法源流考) [A]. In *Examining the origin of the ancient mathematics* (*Gusuan kaoyuan* 古算考源) [C]. Shanghai: Zhonghua xueyishe & Commercial Press: pp. 1–9.

Qian, Baocong 錢寶琮. (1964). (ed.) *The History of Mathematics in China* (*Zhongguo shuxue shi* 中國數學史) [M]. Beijing: Science Press.

Shen, Kangshen, Crossley, J. N. and Lun A. W. C. (1999). *The Nine Chapters on The Mathematical Art Companion and Commentary* [M]. Oxford & Beijing: Oxford University Press & Science Press.

Siddhārtha, Gautama (Qutan Xida 瞿曇悉達). (1792). Astronomico-astrological Divination Canon of Kaiyuan (713–741) Reign (*Kaiyuan zhanjing* 開元占經) (1792). [M]. In The Complete Library in Four Sections (*Siku quanshu* 四庫全書).

Smith, E. David. (1923, 1925). *History of Mathematics* [M]. New York: Dover Publications.

Tian, Miao 田淼. (2005). The Westernization of Mathematics in China (*Zhongguo shuxue de xihua licheng* 中國數學的西化歷程) [M]. Jinan: Shandong Education Press.

Tong, Jianhua 佟健華. (2013). Yuan Dynasty Arabic magic square discovered at Zhongdu site (*Yuanzhongdu yizhi chutu alabo huanfang zhi yanjiu* 元中都遺址出土阿拉伯幻方之研究) [J]. *Journal of National Museum of China* (*Zhongguo guojia bowuguan guankan* 中國國家博物館館刊), (03): pp. 76–83.

Wang, Yusheng. (2003). Chinese 'Ling' (o) and Indian 'Sunga' (o) Spread to China [A]. In Bag A. K. and Sarma S. R. (eds.) *The Concept of Sunya*. New Delhi: Indira Gandhi National Centre for the Arts and Aryan Books International.

Wang, Zhengshu 王正書. (1985). *Records of the graves of the Lu family in Shanghai Pudong* (*Shanghai Pudong Ming Lu shi mu jishu* 上海浦東明陸氏墓記述) [J]. *Acta Archaeologica Sinica* (Kaogu 考古), (06): 540–549+582–584.

Yabuuti, Kiyosi. (1979). Researches on the *Chiu-chih li*, Indian astronomy under the Tang Dynasty [J]. *Acta Sciences*. (36): pp. 7–48.

Yan, Dunjie 嚴敦杰. (1957). A history of the spread of Arabic numerals into China (*Alabo shumazi chuandao zhongguo lai de lishi* 阿拉伯數碼字傳到中國來的歷史) [J]. *Bulletin of Mathematics* (*Shuxue tongbao* 數學通報), (10): pp. 1–4.

Yan, Dunjie 嚴敦傑. (1982). The history of the use of numerals in China (*Zhongguo shiyong shumazi de lishi* 中國使用數碼字的歷史) [A]. In *Collection of Papers on The History of Mathematics, The Eighth Collection* (*Kejishi wenji* 科技史文集　第八輯) [C]. Shanghai: Shanghai Science and Technology Press, pp. 31–50.

Zhang, Dianzhou 張奠宙. (1999). *The Development of Modern Mathematics in China* (*Zhongguo jinxiandai shuxue de fazhan* 中國近現代數學的發展) [M]. Shijiazhuang: Hebei Science and Technology Press.

Zhou, Xiaohan Célestin 周霄漢. (2020). Preliminary Research on the Mathematical Methods and Order of Problems in Yang Hui's *Mathematical Methods* (1261 CE), *Fangcheng* chapter [J]. *Chinese Annals of History of Science and Technology*, (4): pp. 1–62.

Zhu, Yiwen 朱一文. (2010). Chinese Mathematics: Counting Sticks and Techniques Using *Fangcheng* as an Example (*Shu: shu yu shu, yi jiushu zhi fangcheng weili* 數：筭與術——以九數之方程為例) [J]. *Chinese Studies* (*Taiwan*) (*Hanxue yanjiu* 漢學研究) (台灣), 28(4): pp. 73–105.

Zou, Dahai 鄒大海. (2010). The Concepts of Positive and Negative Numbers in Early China Reflected in the Unearthed Bamboo and Wooden Slips (*Cong chutu jiandu wenxian kan Zhongguo zaoqi de zhengfushu gainian* 從出土簡牘文獻看中國早期的正負數概念). *Acta Archaeologica Sinica* (*Kaogu xuebao* 考古學報) (4): pp. 481–504.

CHAPTER 7

# Reflections on Early Dated Inscriptions from South India

*T. S. Ravishankar*

## Abstract

The study of the transition from the numerical system of writing to the decimal system of writing, including the quest to date the origin of zero as a place numeral, has actively engaged the attention of modern researchers and scholars alike. The subject is in fact so vast, gripping, as well as baffling that it will most likely continue to keep them engaged for quite a while to come.

There is no need to reiterate the fact that India is very rich in epigraphical wealth. Hundreds and thousands of inscriptions have been reported from across the length and breadth of the country, written in various languages and in varied scripts, and following different dating systems. It is because of this abundance and the sheer complexity of these inscriptions, coming from different parts of India and belonging to different dynasties, that it has become an arduous task to pinpoint when exactly the numerical system of writing gave way to the decimal system in India, and when zero first made its appearance in Indian inscriptions. Further, the chronological sequencing of the changes in the system of writing – from numerical to decimal – that had taken place over time is also difficult to establish.

In this chapter, while keeping the vastness of the subject in mind, an attempt has been made to re-examine some of the early dated inscriptions that have come to light in the last few years, both as part of the reports from archaeological excavations and from epigraphical explorations. Efforts have also been made to connect these new findings with the hypotheses of and the views expressed by earlier scholars.

The goal of this exercise was to ascertain and confirm the presence and usage of decimal systems of writing, and also the use of zero as a placeholder, in South India. Thus, the primary task undertaken has been to review some of the already known dated inscriptions from Andhra Pradesh, Karnataka, Kerala, and Tamil Nadu.

## Keywords

numerical system – decimal system – Brahmi numerals – Saka Era – South East Asia – dated inscriptions – transitional phase – Bakhshali manuscript

 | DOI:10.1163/9789004691568_010

## 1 Introduction

Thanks are due to the pioneering efforts of great epigraphists, like D. C. Sircar, B. Ch. Chhabra, Prof B. N. Mukherji, and many more for critically examining this issue and arriving at the probable earliest dated inscriptions using the decimal system and zero. It may be noted that scholars such as Richard Salomon voiced the opinion also held by others that the symbol for the numeral zero had its origin in South East Asia. A close examination of the writings of the scholars by and large reveal that the origin of the numeral zero lies around the sixth and seventh centuries CE.

After going through the available published material, I was motivated to probe further in this chapter and examine many early records, especially reported from different regions in South India, and look into the probable origin of zero. My approach was to juxtapose the inscriptions that are already available from other parts of the subcontinent and take a comprehensive view of the issue, since in recent years many more inscriptions have come to light, especially from early periods. In arriving at a probable conclusion, a review of the epigraphical record is necessary particularly from the point of view of zero.

## 2 The Epigraphical Record

It is generally accepted by scholars that the use of decimal notation in the Indian epigraphs in different parts of India to express dates becomes almost a regular feature from about the ninth CE.

To cite some examples of this from the earlier observations made on the subject, mention may be made of Mankani inscriptions of Taralaswamin of the year 346 (Figure 7.1), corresponding to 595 CE.[1] But since the record is considered spurious by some, we leave it out of consideration here.

However, there are several authentic inscriptions that can be cited to trace the beginning of the decimal system of writing in India. A very significant one is from Ashrafpur (Memoirs of Asiatic Society of Bengal, 1905, p. 85) (Figure 7.2), Dacca district (now in Bangladesh), held to be an example of the earliest known use of the decimal system in the subcontinent. Another, one of

1 The above inscription is dated in Kalachuri-Chedi era of ca. 248 CE. Keilhorn, originally proposed July 28, 249 CE as the first date of the era, but later charged this to September 5, 248 CE (Ref: Salman Richard (1998), Indian Epigraphy, New York, pp. 184 ff.

FIGURE 7.1 Spurious Mankani plates of Taralaswamin (one plate and one close-up view)

the earliest inscriptions cited by scholars, is the Mankuwar image inscription of the Buddha year 129 (Figure 7.3) (Bhandarkar, 1981).

A very significant contribution to this quest for early dated inscriptions was made by Subrata Kumar Acharya, who attempted to identify and explain the different phases of decimal notation that prevailed in Orissa, especially the transitional stages (Studies in Indian Epigraphy, 1993). Because of the large presence of dated early inscriptions from Andhra Pradesh right from the period of the Satavahanas and the succeeding dynasties, I was led to conclude that the examples of decimal notation located in Orissa had found their way there from Andhra Pradesh.

It is also clear that from Orissa the system of decimal notation traveled to West Bengal, Bangladesh, and further to Southeast Asia, where dated inscriptions have been found belonging to the early seventh century of the Saka era (Saka years 605, 606 and 608) (Figure 7.4) (Bulletin of the School of Oriental Studies, 1930–32). As rightly remarked by Prof. B. N. Mukherji (Studies in Indian Epigraphy, 1993) and Ajay Mitra Shastri (Bag and Sharma, 2003), the occurrence of dates based on decimal notation, as early as the seventh to eighth century CE, in Sanskrit inscriptions (Studies in Indian Epigraphy, 1993) from Java (Dinaya) Bakul and Po Nagar (Champa), indicate that the areas were heavily influenced by India via Indian migrants to these Southeast Asian regions, which had adopted the decimal place value notation completely. There is broad consensus on this point.

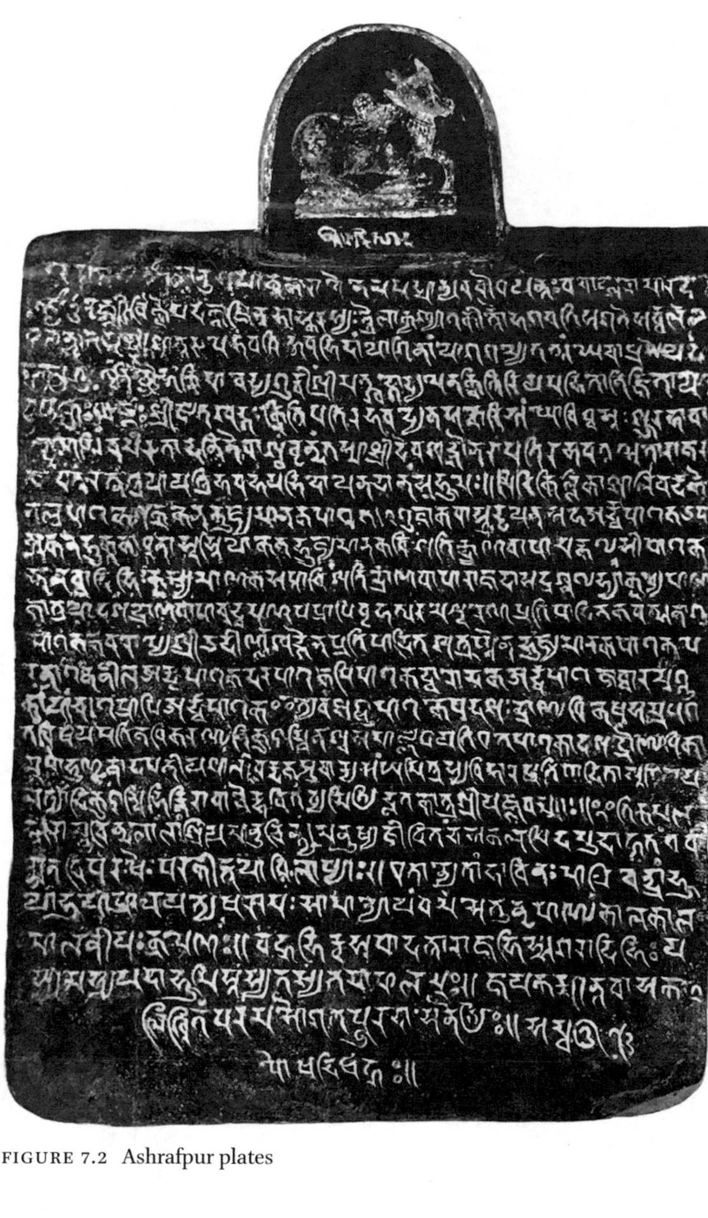

FIGURE 7.2 Ashrafpur plates

FIGURE 7.3 Mankuvar stone image inscriptions of Kumara Gupta I

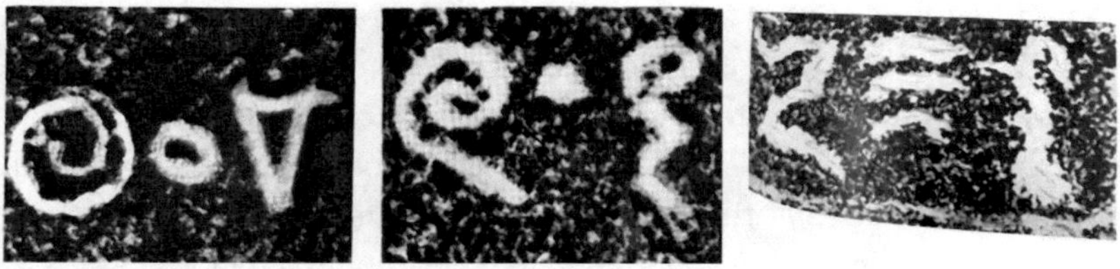

FIGURE 7.4 Three Southeast Asian inscriptions

## 3 Examples of Early Dated Inscriptions

### 3.1 *Andhra Pradesh*

In recent years many dated inscriptions have come to light here. One of these is from Chebrolu, Guntur District (Studies in Indian Epigraphy, 2018), dated regnal Year 5 (207 CE) of the Satavahana King Vijaya.

Further, excavations conducted at Phanigiri, Nalgonda district, yielded a number of Prakrit and Sanskrit inscriptions (Figure 7.5) (Epigraphia Indica, 2011–12). This inscription belongs to the eighteenth year of the reign of Ikshavaku King Rudrapurushadatta. Earlier, the oldest known regnal year of this king was Year 11, but the inscription referred to here added seven more years to his reign.

Apart from this, there are many dated inscriptions belonging to the reign of the Ikshavaku rulers. There are also many dated inscriptions, mainly using the numerical symbols, belonging to the reign of Vishnukundin, Salankayana, and many subordinate rulers.

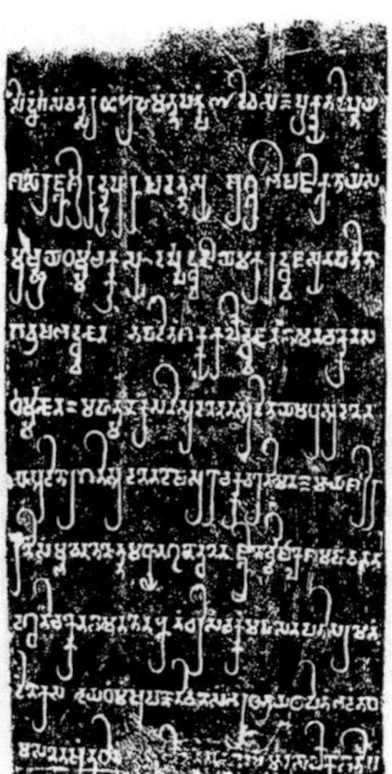

FIGURE 7.5
Panigiri inscriptions of Rudrapurusha Datta

It should be noted that in South India, Andhra Pradesh stands apart as a separate entity as far as the usage of numerical symbols is concerned. The Andhra Pradesh findings help us to closely and critically study the development of numerical symbols right from the Satavahana era to subsequent periods. It is also likely that the system of numerical symbols used widely in Andhra Pradesh had spread to and influenced the adjoining regions of Orissa.

### 3.2 *Karnataka*

One of the earliest findings of the use of the numerical system in Karnataka is on a memorial pillar inscription of Somayashas dated 10082 day (105 CE). This very interesting example is from Vadagaon Madhavpur, Belgaum District (Epigraphia Indica, 1971–73).

Among the other dated early inscriptions, mention may be made of at least six inscriptions from Kanaganalli (Sannathi) Gulbarga District, belonging to different Satavahana rulers, dated in the sixteenth regnal year of Shri Chimukha Satavahana Sri Pulumavi (36 BCE), dated in the thirty-fifth regnal year of Vasistha Putra Sri Satakarni, in the fifth regnal year (132 CE) and the eleventh regnal year (138 CE) of Sri Yagnasatakarni, in the tenth and eleventh regnal years (191 CE) and the tenth regnal year of King Madariputra Sri Pulumavi (236 CE) (Figures 7.6 and 7.7) (Poonacha, 2011).

A review of some of the early dated inscriptions reveals that apart from the inscriptions of the Satavahana rulers, there are no inscriptions dated in numerals belonging to the early Kadamba rulers. The sole exception is in the Gudnapur inscription of Ravivarman, where the numerals are used in some other context but not for mentioning the date.

Further, during the periods of the early Ganga rulers, that of the imperial Chalukyas of Badami, and even in the Rashtrakuta records, we do not find the use of numerical symbols. On the other hand, we find that the dates are expressed in words only.

FIGURE 7.6 (A and B) Dated inscription of Madhariputra Siri Pulumavi

FIGURE 7.7 Sannathi inscriptions

### 3.3 *Kerala*

Some early Brahmi inscriptions are reported from Kerala, but the use of numerals has not been found.

### 3.4 *Tamil Nadu*

Mention may be made of a well-preserved pottery inscription from Alagankulam, with only numerals incised in a fairly large size. Iravata Mahadevan has dated this to first and second centuries CE. The number is read as 408 (Figure 7.8) (Iravatham Mahadevan, 2014), the first digit resembling a cross is the symbol for 4, followed by the symbol for 8, which is incised in the reverse direction. Further, Mahadevan states that, although given that there is no accompanying text, the exact significance of the number could not be understood. It is nonetheless a unique and important discovery.

In subsequent periods, we do not find the usage of early Brahmi numerals. Some important early dated inscriptions are highlighted above only to provide some idea of the numerical symbols used in the different regions of South India.

Although till now no inscriptions have been found that could demonstrate the earliest use of decimals and of zero, future epigraphical explorations might well yield such inscriptions in decimal notation.

FIGURE 7.8
Numeral 408 incised on pottery from Alagankulam

Since there are a number of early dated inscriptions from Andhra Pradesh belonging to different dynasties, these may also yield some inscriptions belonging the transitional phase between the Brahmi numerical system and the decimal system, as found in Orissa.

## 4 Use of Numerals in Early Inscriptions Other than the Date

While examining some early dated inscriptions, I came across some that used numerical symbols in various contexts. One of these unique inscriptions is that of Nagannika (Sircar, 1965), which records gifts and honoraria given for the performance of various sacrifices. In the Gudnapur inscription of Ravivarma (Sastri, 1973), we find the use of numerals while giving an account of grants made for the worship and maintenance of the Kamajinalaya by the King. Here, the numerals 1 to 9 have been used starting from lines 18 to 22 of the inscription (Figure 7.9).

FIGURE 7.9
Eye copy of numerals found in Gudnapur

FIGURE 7.10 Mattepad inscriptions

The Mattepad plates of Damodaravarman (Epigraphia Indica, 1923–24)[2] are extremely interesting, as the eight inscribed faces are numbered and, as a result, they look like the pages of a modern book, with the numerical symbols from 2 to 8 in the left margin (Figure 7.10).

## 5 The Symbol for the Numeral 20 and Zero

It may not be out of place if we mention here that there is every possibility that the symbol for 20, which is nothing but zero with minor variations, might have influenced the process of transition from the numerical system, to enter the decimal scheme of notation. Although D. C. Sircar was tempted to read the date as 100 0 9, i.e., 109, this was not accepted as zero, but as 20 (Figure 7.3). Otherwise, this would have been the earliest instance of the use of zero (Epigraphia Indica, 1920).

## 6 About the Bakhshali Manuscript

Subjecting the Bakhshali manuscript to radiocarbon dating by an Oxford/Bodleian team and assigning an anterior date (second/third centuries CE) has triggered intense controversy, speculation, and debate in the scholarly world.[3] The two main contending groups involved are the mathematicians on the one side and the epigraphists/paleographers on the other. The crux of the matter is the appearance of the numerals, especially zero, which naturally generated tremendous curiosity among members of both groups.

For quite a long time now, scholars such Hoernle, A. F. Rudolf discussed this issue and assigned a relatively early date to the Bakhshali manuscript, which does not gel well with the paleography of the manuscript. Scholars such as Professor B. N. Mukherjee and Professor Ajay Mitra Shastri have assigned a relatively later date of the seventh or eighth century CE, keeping in mind the paleography of the script.

I, too, had the opportunity to re-examine the photographs of the Bakhshali manuscript that appeared in different media, and also some published articles. These investigations have convinced me that the manuscript belongs to a later

2 Epigraphia Indica, Vol. 17.

3 See Appendix 3 for an online petition urging the Bodleian Libraries, Oxford, UK, to take concrete steps to commission the necessary follow-up radiocarbon-dating of the Bakhshali Manuscript in the interest of scientific advancement in the field.

period, that of the eighth century CE, rather than the earlier date claimed by the mathematicians.

Two issues are involved here. The first is carbon-14 dating, the results of which do not stand up to scrutiny when set against the paleography, and the second is the script of the manuscript. However, these contradictions can be reconciled if one presumes that the creators of the Bakhshali manuscript might have used centuries-old palm leaves that had been cut and preserved for later use. Hence, while the material used for writing dates from the earlier period, the script, the numbers, and the zero used in the manuscript all belong to the later period of the eighth century CE.

## 7 Conclusion

A brief review of some of the dated inscriptions having numeral symbols found in Andhra Pradesh, Karnataka, Kerala, and Tamil Nadu, leads one to identify different scenarios at different places and at different time periods. However, one thing is certain, *viz* that they all revealed a uniform pattern of representing numerical symbols, with minor variations, followed throughout almost the whole of the South Asian subcontinent.

The study further reveals that the inscriptions from Andhra Pradesh, belonging to the Satavahana and successive dynasties, continued the trend of employing numerical symbols, to express the regnal year, and sometimes in words, too, in most of these inscriptions. This has not been seen in the inscriptions known to be from Karnataka and Tamil Nadu.

Scholars like S. K. Acharya have studied the inscriptions from Orissa in a very detailed and descriptive manner, to establish the transitional phase from numerals to the decimal system of writing. Early dated inscriptions reported from Andhra Pradesh provide the necessary backdrop for understanding and appreciating the development of numerical symbols in contiguity.

It also seems likely that from Orissa, the transitional phase spread its influence to present day West Bengal and Bangladesh. Further, it might have traveled to Southeast Asia through the Indian migrants who settled in Indonesia and elsewhere in the region, where the inscriptions were written Saka era, and as such point back to the subcontinent.

Thus, I would like to suggest, keeping in mind earlier studies made in this regard, that from the coastal belt of Andhra Pradesh, the numerals made their way into Orissa, and subsequently into West Bengal and Bangladesh and Southeast Asia. I would like to reiterate the fact, which other scholars have already endorsed, that it was only through the Indian subcontinent that the decimal system plus zero traveled to Southeast Asia.

At present, even with the available new inscriptions, it cannot be said with certainty when precisely the zero occurred in inscriptions from South India. However, future epigraphical exploration in the region may hold the key to pinpointing the first usage of zero in India.

## References

Bag A. K. and Sharma S. K. (ed). (2003). Ajay Mitra Shastry. Brahmi Numerals and Decimal Notation Nature and Evolution pp. 72 ff. In *The Concept of Sunya*. New Delhi.

Bhandarkar, D. R. (1981). Inscriptions of the Early Gupta Kings. In *Corpus Inscriptionum Indicarum* (revised), Vol. 3, New Delhi, p. 291.

Buhler, G. (1959). Indian Paleography (edited and translated by J. F. Fleet). Reprinted in *Indian Studies Past and Present*, Vol No. 1 Calcutta, p. 102.

Bulletin of the School of Oriental Studies. (1930–32). Vol No. 6 pp. 323 ff.

*Epigraphia Indica*. (1923–24) Vol. 15, pp. 146 ff and plates.

*Epigraphia Indica*. (1923–24) Vol. 17, pp. 327 ff and plates.

*Epigraphia Indica*. (1971–73). Vol. 39. pp. 183 ff and plate.

*Epigraphia Indica*. (2011–12). Vol. 43. Part I, pp. 75 ff and plate.

Mahadevan, Iravatham. (2014). *Early Tamil Epigraphy*. Chennai, p. 243.

Memoirs of the Asiatic Society of Bengal. (1905). Vol. 1, p. 85.

Mirashi, V. V. (1955). Inscriptions of the Kalahuri Chedi Era, pt. I Ootacamund. *Corpus Inscriptionum Indicarum*, Vol. 4, pp. 161–164.

Poonacha, K. P. (2011.) *Excavation at Kanayganahali* (Sannati Dist Gulbarga Karnataka). ASI, New Delhi, pp. 443 ff.

Sastri, Dr S. Srikanta. (1973). Felicitation Volume – Srikanthika, pp. 61 ff.

Sircar D. C. (1965). *Select Inscription*, Vol. 1(82) pp. 192 ff.

*Studies in Indian Epigraphy* (Bharatiya Purabhilekha patrika).(1993) Vol. 19 pp. 52 ff. Mysore.

*Studies in Indian Epigraphy*. (1993). Vol. 19. Mysore. pp. 80 ff.

*Studies in Indian Epigraphy*. (2018). Vol. 43, pp. 55 ff.

CHAPTER 8

# From *Śūnya* to Zero – an Enigmatic Odyssey

*Parthasarathi Mukhopadhyay*

## Abstract

Who invented zero? Interestingly, there is no one-line answer. And the acceptability of any attempted answer generally depends on the perception of the seeker, as the connotation of the word 'zero' can be perceived from several different but interrelated perspectives. Many ancient civilizations, including India, had their own version of zero or a zero-like concept or symbol as a representative of 'nothingness'; some as a philosophical conundrum, elsewhere some others even in a practical sense, such as a filler or a gap on their counting board; but except in India, none of these early and somewhat hesitant initial concepts did ultimately mature to its true mathematical potential. Today it is generally accepted worldwide that this peerless concept of a decimal place value system of enumeration in tandem with the true zero of our present-day mathematics, evolved in ancient India. Initially philosophically nurtured and analyzed during the early periods of oral tradition in India by several different schools of thought, eventually at some point of time the concept reflected in the Sanskrit word *śūnya* was transformed into a numeral for mathematical expression of 'nothing'. Perhaps at a later period, it bloomed into its full potential as a number, *kha*, an integer on which mathematical operations can be performed. Indeed, this unique feature makes *kha*, the Indian zero, the true progenitor of our modern mathematical zero. However, the exact time frame of this gradual evolution is still hotly debated, a recent controversy in this direction being the outcome of attempted radiocarbon dating of the famous Bakhshali manuscript by the Bodleian Library of Oxford. Going in the other direction, some scholars suggest that a trace of this concept, if not in total operational perspective, might have had a Greek origin that traveled to India during the Greek invasion of the northern part of the country in the pre-Mauryan period. A relatively recent third view professes the Chinese origin of the concept of zero as a placeholder, which might have traveled with the traders from China to the far eastern parts of Asia, to places like Cambodia, then under Buddhist influence that spread from mainland India, where it got the shape of a 'bold dot', the earliest known written form of zero. Scholars belonging to this school of thought want to credit mainland India only for 'garlanding' this concept of zero toward its modern shape, sometime around

 | DOI:10.1163/9789004691568_011

the eighth or ninth century CE. Embracing all these paradigms of claims and counter-claims about the origin of zero made till date, this article is a nutshell version of an odyssey, an unparalleled journey from a concept to a number in its own right, perhaps the most significant creation of the human mind ever.

**Keywords**

zero – *Śūnya* – *kha* – *nfr* – Babylonian zero – Mayan zero – rod numerals – *kong* – Hellenistic zero – *Śūnyavāda* – *bindu* – bold dot zero – Bakhshali manuscript – *Brāhmi* numerals – place value system

## 1 Introduction

The concept of numbers gradually evolved in early human civilizations as an abstraction of the physical process of counting tangible objects. As early as some 25,000 years ago, or even earlier, a few of our prehistoric cavemen ancestors residing in different parts of our globe seemed to have a penchant for 'counting'. This is testified by quite a few, more or less systematically, notched bones, found during different archaeological excavations, starting from the Ishango Bone found in Belgian Congo, to several others, now preserved at different natural history museums worldwide. This was possibly some kind of record-keeping, where 'counting' stood for one-to-one correspondence between the objects counted and the notch marks. However, when it came to counting the number of some tangible objects orally, which surely had occurred at a later period in time, one only then needed the linguistic supply of number names. Note that in this business, zero is never required, as we do not begin counting from zero: neither then, nor now.

Recall our first individual formal encounter with numbers along with their names during childhood. It is almost always made through some generally attractive colorful books, unfailingly soothing the abstract idea of a 'number' by putting it in association with a tangible object or other, usually chosen from our surroundings; 'one', written in both name and numerical symbol by the side of some picture, say, of an apple; 'two' might be tagged similarly with two oranges, and then three with three bananas, and so on. In that page designed for early exposure to numbers, one can never put the number zero on a similar footing. Only at a more mature stage of mental development, the idea of the number zero is introduced, usually depicted through a process of step-by-step subtraction, as a number representing 'nothingness'. Suppose you have a pen and I take it from you; then if I ask you, how many pens you have, sure enough, your answer will be 'none at all'. Observe that, while answering an affirmative question, you preferred to give a negative answer through language. However, if you want to give an affirmative answer in this case, the number that you require is zero, a number that is not usually used for counting, but without which the present prosperity of modern civilizations through various scientific achievements would never have been possible. However, its proper appreciation as a number in its own right requires a certain amount of mental maturity, when one begins to see every number as a purely mathematical entity, independent of its different linguistic names or symbols, which may vary from one language or culture to another. Amazingly, somewhat parallel was the gradual genesis of these abstract ideas in various early human societies, but only at a much longer timescale; these evolved somewhat independently, though pocketed at several different parts of the world, not always having mutual contacts or exchanges among them. While today we take only a few years' time to assimilate these ideas during our formative days, this journey from concrete to abstract, with its relevant intricacies, initially took thousands of years for mankind to assimilate.

The exact time frame, as well as geographical location of its philosophical origin as an abstract representation of conceptual nothingness, and subsequent gradual evolution to a fully bloomed number, an integer, are still hotly debated. But it is generally accepted worldwide, among a majority of scholars and common folk alike, that the decimal (i.e., base 10) place value system of enumeration in tandem with the number 'zero' (and its present universal symbol, 0, as well) – the peerless gift of Indian civilization to the world – was first thought of at some point of time in ancient India, perhaps philosophically to begin with and then mathematically, most possibly at different time epochs. The word 'zero' originated from the Sanskrit word *śūnya*, which was only one of several synonyms such as *kha*, *ākāsha*, *gagana*, *ananta*, *pūṛna* and many others used to represent the concept; a dichotomy as well as a simultaneity between nothing and everything, the zero of void and that of all-pervading fathomless infinite, hinted through the simile of the vast and limitless expanse of the sky. Initially, during the pre-Common Era days of oral tradition in India, one finds several examples of decimal place value based nomenclatures of Sanskrit number names in early Vedic Indian scriptures right from the *Ṛgveda* onward. However, these examples were without any exclusive reference to any 'number zero', since such a thing is not required in uttering the name of numbers larger than zero, until, of course, someone wanted to specifically refer to the 'number zero' itself. The earliest such known case is found in Piṅgala's *chandasūtra*, composed during the second or third century BCE, where a zero occurring as a label in a prosodic discussion steeped in mathematical spirit is referred to by the name *śūnya* in language. Zero as a numerical symbol becomes indispensable only when one starts writing numbers in a place value system. Absence of any other digit in one or more 'places' of a written number cannot otherwise be tackled successfully. Of course, one may think of the naive way of leaving a gap at the requisite place or places, but that will surely confuse readers, particularly if such requirement occurs at successive places or even at the end of the number, as the history of similar attempts in different civilizations over several centuries clearly testifies. One of the important mathematical roles of zero is that of a placeholder, where it allows the other digits, say from 1 to 9 as in the decimal place value system, to take their own 'places' in the written representation of a number, so as to be able to distinguish between, say 12, 102, and 100,002. With the passage of time this zero was eventually appreciated as an independent numeral, a number in its own right, on the same footing as any of the other digits, a number for the mathematical expression of 'nothing', with which one can operate mathematically, as we use it today, for example, $2 - 2 = 0$, or $2 + 0 = 2$. And this later phase of its evolution is purely Indian in origin.

Many ancient civilizations, including India, had their own respective version of initial zero or zero-like concept or even symbols in some cases, as a representation of 'nothing'. In India it was deeply rooted in various philosophical doctrines; while in Greece it created a philosophical conundrum; the Babylonians at one point had zero in a practical utilitarian sense, as a filler for writing numbers; early Chinese or later Romans had it as a gap on their respective counting boards; Mayans used it even as a pure number in an erroneously executed place value system in their long count calendar. But with the exception of India, none of these initial ancestors of the modern mathematical zero did ultimately, to the fullest extent, mature to its paramount mathematical potential. Today, in general, educated Indians are aware of the fact that in some remote past, Indian mathematicians discovered the number zero, as is usually mentioned in their school textbooks. But what really is meant by this discovery? It was neither invented in a laboratory, nor excavated from earth. How did the ancients manage without zero before the so-called discovery? How could they distinguish between, say, 101 and 100,001? Why is that method not being followed now?

Apart from India, there were various other glorious civilizations in the past that scaled fascinating heights in a wide variety of fields, including mathematics. What were their enumeration systems? Did they know about zero as a number or at least as a concept representing void or nothingness? If any of these civilizations had the concept of some kind of zero at all, why then today is their zero not considered as the forefather of our modern zero? To know the answers from the proper perspective, one has to scan through a rich history of various early civilizations over a period of more than 5,000 years. This engaging travel in time proceeds through the Egyptian hieroglyphics and Inca *Quipu*, via Babylonian clay tablets and Mayan glyphs and codices, followed by the great Greek civilization and the mighty Romans with their clumsy enumeration system in the West; whereas in the East we come to know about the lofty philosophy of ancient Indian seers, the clever Chinese calculation system and industrious pursuits of the wise Arabians of Baghdad. Touching on these various facts, let us try to place in a nutshell, the genesis of our modern mathematical zero in its proper historical perspective – an ode to 'the nothing that is'.

## 2 The Egyptians

The Egyptians (from around 3000 BCE) had a tradition of constructing pyramids. At a relatively later period they sometimes used the hieroglyph for *nfr* (which literally means 'beautiful') to indicate the fixation of the base level of a pyramid at somewhat higher than the natural ground level so as to save its burial chamber from recurrent inundation from the Nile. The height of the pyramid as well as the depth of its underground chambers in two opposite directions were then indicated by using the numbers one, two, etc. in cubits. Here some scholars see a tacit budding of a number line-like concept, with *nfr* playing the central role of zero. However, the Egyptians never used this zero as a number in its own right, neither in hieroglyph engraving nor in later hieratic script used in writing on *papyri*. Even though they gave definite importance to the numerical symbol for ten and its higher powers in their enumeration system, their representation of a number was additive in nature, somewhat like that of the later Romans (27 BCE–1453 CE), in the sense that the values of the numerical symbols written side by side were to be added to understand the number represented by them. In fact, in the clumsy Roman system it had even to be subtracted in some of the cases, as in writing IV to mean 5–1, i.e., 4; although in the Roman system, IC (the symbol for 1 followed by that of 100) does not stand for 100–1, i.e., 99, instead of which you have to write XCIX. These presentations required no need for zero as a placeholder, but more and more new symbols were to be introduced in the system for representing larger numbers. And since there is no largest number, the requirement for newer symbols theoretically becomes unlimited. The great historian A. L. Basham in his book, *The Wonder That Was India*, beautifully pointed out: 'Most of the great discoveries and inventions of which Europe is so proud would have been impossible without a developed system of mathematics, and this in turn would have been impossible if Europe had been shackled by the unwieldy system of Roman numerals.'

## 3 The Babylonians

A written symbol for zero in the sense of a placeholder[1] was used from the Seleucid period (312–64 BCE) in Babylon in their sexagesimal system of enumeration, as evident in their numerous clay tablets. This place value system was of base 60, and so it required 59 other number symbols as digits,[2] which were actually groups of two basic symbols, viz. that for one and for ten, thus cooking up all the numbers from 1 to 59 via the additive principle of juxtaposed symbols.[3] Note that in a sexagesimal system two numerical symbols for one, written respectively in the unit place and the 60 place, stand for 1 times 60 plus 1, representing the number 61. In the Babylonian system, the double-wedge symbol of zero was inserted not only in the median positions but also in the initial and terminal positions of a number, as many times as necessary. However, the Babylonians failed to conceptualize this zero as a number in its own right, as has been cited with specific examples by Georges Ifrah in his famous book, *The Universal History of Numbers*. Before their invention of this sign for zero to fill the vacant place in a number, the Babylonians used to leave a gap in the requisite place(s) of the number to represent occurrence(s) of zero. But that only added to the confusion of one who would have read it at a later point in time. For example, in modern decimal notations, if we see two ones side by side, but separated by some space in between, such as 1 1, can anyone ever be sure about the correct number of zeros that was meant to be inserted in the gap while originally writing it? This situation prevailed in Babylon for well over a thousand years (starting from about 1800 BCE), before some Babylonian genius came to understand that a gap-filling symbol was necessary to overcome this problem. This symbol, as we have pointed above, is considered as the oldest known written symbol for 'placeholder zero' in human civilization.

1 This symbol looked like a 'double slanted-wedge' sign, somewhat like two partially overlapping 'less than' symbols: one on top of the other.

2 We need only nine symbols for our digits named one (1) to nine (9) in the decimal system.

3 In the Indian decimal system, symbols of the basic digits happen to be distinct and single, leading to an ultimate economy of symbols for written representation of large numbers.

## 4 The Incas

The Incas (around 1500 CE), before being plundered and destroyed by the gold-hungry Spanish *conquistadors*, ruled Peru with brilliant efficiency, where rigorous arithmetical record keeping of taxes received, etc., was done using multicolored knotted strings called *Quipu*. It was a decimal (base ten) place value representation of numbers on string by means of three different kinds of knots and their repetitions judiciously spaced, where a zero in the sense of absence of other numerals was represented by keeping a gap at the required place on the string. This clearly was a primitive placeholder version of zero, which may be compared in representational aspect with the blank space of the Roman arithmetical counter board or to a certain extent of the Chinese calculation scheme with rod numerals. However, as they did not have a known system of writing, there was no need for an independent number zero either, nor was there any natural scope of evolution of any possible relevant symbol.

## 5 The Mayans

The Mayans (200 BCE–1540 CE) from the Yucatan peninsula (present-day Mexico) used the number zero in their vigesimal system of enumeration.[4] Their zero was not a mere placeholder; indeed it was a pure number in its own right. They had at least five different symbols or glyphs to represent their zero, the most common being that of a seashell, usually red-colored. They developed some very advanced calendars for timekeeping. One of these calendars (called *haab*) had 18 months (*winal*) of 20 days (*kin*) each, (followed by a phantom month *uayeb* of five days, to make up for their 365 day-long year, a *tun*) where the beginning day of each month was reckoned not as the usual day one as we do now, but as day 'zero'. However, in spite of this bold presence of zero as a counting number, which is a unique feature seen nowhere else in any civilization, we still cannot call their zero a true forebear of our present mathematical zero, since it lacked certain fundamental mathematical features as seen from their calculation in the long count calendar. In that calendar, toward recording the date of some important social or cultural events of their history, Mayans first used to count the total number of days elapsed till that event took place, starting from their understanding of Day Zero,[5] which was, according to

4 It is a place value system with base twenty, where two numerical symbols for one, written side by side, used to mean '1 times twenty plus 1', i.e., twenty one. However, Mayans used to write them differently, one above the other, to be read from top to bottom.

5 This day is seen to be equivalent to 13 August, 3114 BCE, with respect to our modern calendar.

them, the day of creation of the universe. Next they used to subdivide that total number of days (*kin*) into several groups of higher units of time, like month (*winal* i.e., 20 *kin*), year (*tun* i.e., 18 *winal*), and even further to much higher units like *katun* (i.e., 20 *tun*), *baktun* (i.e., 20 *katun*) etc. However, trying to keep the number of days in a year as close to 365 as possible, they had destroyed the pattern of a base-20 place value system by putting 18 times 20 in the third place (1 *tun* i.e., 360 *kin*), instead of as would have been mathematically required, 20 times 20 in a proper vigesimal system. This caused a serious drawback in the mathematical property of their zero. Note that if there is a fully functional mathematical zero in tandem with a proper place value system, then when one writes a zero at the extreme right of any given number (being written from left to right), the new number must be equal to the original number times the base of that system. As in our decimal (base 10) system, when we write 0 to the right of 12, the new number 120 becomes 10 times the original 12. Or, for that matter, when we express the decimal number 3 as 11 in the binary (i.e., base 2) system ($1 \times 2 + 1 = 3$), then the corresponding 110 would represent the decimal number 6 ($= 1 \times 2 \times 2 + 1 \times 2 + 0 \times 1$), which is 2 times the original 3. Similarly, in case of the ternary (i.e., base 3) 11, that stands for decimal 4 ($= 1 \times 3 + 1$), the corresponding 110 stands for 12 ($= 1 \times 3 \times 3 + 1 \times 3 + 0 \times 1$), which is 3 times 4. This is precisely where the Mayans faltered in their otherwise brilliant system. The Mayan long count number 100 is equivalent to decimal 360 $[= 1 \times 18 \times 20 + 0 \times 20 + 0 \times 1]$ rather than 400, $[= 1 \times 20 \times 20 + 0 \times 20 + 0 \times 1]$ as it should have been in a correct vigesimal place value system. Hence their zero, ingenious as it was, cannot be considered as the true predecessor of our modern mathematical zero.

## 6 China

Early Chinese civilization used a multiplicative-additive system, based on written decimal place value symbols within successive digits. For example, to write 13, they would write their symbol for one followed by that of the ten and then the symbol for three, to comprehend the number as 1 times 10, plus 3. Such a presentation did not require the placeholder zero.

Later, around 600 CE, as seen in *Sun Zi suan jing*,[6] they started calculating with the so-called rod numerals,[7] an early form of seventeenth century abacus,

6 Sun Zi's canonical Manual of Arithmetic, written by an unknown author, most likely dates from the Sui (589–618 CE) or Tang (618–907 CE) dynasties.

7 These artifacts were usually made with small bamboo sticks or bones or even ivory, depending upon the status of their users, laid out on a flat surface with apparently no lines, grooves, or material indications of any kind to distinguish between the different orders of unities of numbers.

where a gap was maintained to represent zero.[8] The first known appearance of zero in medieval China is found in a Chinese adaptation of an Indian astronomical text in 712 CE. According to Radha Charan Gupta, while writing a number, the Chinese got the idea of filling the gap representing their zero in the counting board, most likely from Gotama Siddha, a Buddhist monk who traveled to China from India. From then on they started using the 'Indian zero', which was then denoted by a thick dot (*bindu* in Sanskrit). A fourth century example of such usage of bold dot symbols in the work of a Chinese mathematician, Chu Shih-Cheih, is exhibited in a picture of a so-called Pascal triangle-like configuration[9] in the famous book, *Science and Civilization in China* by Joseph Needham and Wang Ling. However, a relatively recent claim of the Chinese origin of our present-day Hindu-Arabic numeral system, along with zero, has been put forth by Lam Lay Yong and Ang Tian Se in their book, *Fleeting Footsteps: Tracing the Conception of Arithmetic and Algebra in Ancient China,* where in line with an earlier conjecture made by Needham and Wang, they argued that zero originated as a number in China through their decimal rod numeral system and then, at a later point in time, through connections due to trade and commerce with the once Buddhism-dominated parts of Southeast Asia, where the eastern zone of Hindu culture met the southern zone of Chinese culture, it eventually traveled to India, where it got the present circular look; the nothingness of vacant space of the *Han* counting board, as Needham proposed, was thus only 'garlanded' in India. Armed by this thesis, Meera Nanda in her relatively recent book, *Science in Saffron,* remarked that the Indian origin of zero has become an article of faith in popular discourses, and criticized the attitude of analyzing India-China cultural exchange from the allegedly biased point of view that China was only a receiver. However, Martzloff pointed out: 'The representation of numbers in Chinese Buddhist literature is often borrowed from Indian culture, especially in the form of phonetical transliteration of Sanskrit words into Chinese. Conversely, Chinese mathematical terms have never been detected in Indian or Islamic technical literature ... these

8 They referred to the gap via language as *kong* ('empty') in written descriptions of rod configurations that served as numerals. However, no instances either of two successive median zeros within a number, or terminal zeros being referred to as *kong kong*, was ever found recorded, as has been pointed out by Jean-Claude Martzloff. He further observed that, at a later point in time, from the fourteenth century Ming dynasty onwards, the Chinese had regularly written *ling ling* for the same purpose. This clearly goes against the candidature of the written symbol for *kong* as a symbol for a form of zero.

9 This was about 200 years before Pascal. Incidentally, such a configuration was described by Piṅgala about a millennium and a half earlier in India, as described by most of his commentators.

aspects of the problem are passed over in silence in *Fleeting Footsteps*.' Nanda did not throw any light on the issue either. According to her, 'if the Chinese had transferred their rod numerals and the mathematical operations based upon them into writing, the result would be identical to our modern numeration and mathematical operations.' But the fact remains that the Chinese had actually replaced the counting rods by the abacus around the twelfth century and according to Lam, 'it set them back, as the step-by-step thinking that rod numerals required was replaced by rote-learning'.

Anyway, the proposition of the Chinese origin of our number system, as proposed by Lam and Ang in their book, has been strongly criticized by experts such as Martzloff, N. L. Maity, and Frank Swetz among others. Describing the book by Lam and Ang as a valuable resource for understanding early Chinese mathematics, Swetz concluded that 'a claim of a Chinese origin for the Hindu-Arabic numeral system … remain missing footsteps in the path this book has taken'. After carefully considering the candidature of *kong* as a possible predecessor of our modern zero, Martzloff rejected the idea by observing: 'If the hypothesis of a Chinese origin of our number system is to be maintained, the very important problem of the zero raises other thorny issues … According to the MS. Stein 930 from the Dunhuang caves [ninth or tenth century CE, one of the oldest known original Chinese mathematical manuscripts] the presence of blank spaces was not automatically preserved in the written version of the rod numeral system … in practice, the counting-rod system was not as perfectly decimal and positional as the description in *Fleeting Footsteps* would imply' and hinted at a possible 'Babylonian origin of Chinese zero'.

## 7 **Greece**

Greek mathematicians from at least the fifth century CE were fascinated with geometry and the concept of ratio among numbers, and thereby developed a fear and hatred of the number zero, as it threatened to destroy their much-cherished concept of ratio-based understanding of the beauty in nature and harmony of the universe, as the ratio of a number with zero was mathematically incomprehensible. So the Pythagorean brotherhood discarded zero as a number. For them, every number had to have a shape, such as triangular (e.g., 1, 3, 6, 10 etc.),[10] square (e.g., 1, 4, 9, 16 etc.), pentagonal (e.g., 1, 5, 12, 22

10 As one can arrange this many beads or pebbles in the shape of a (regular) triangle in a symmetric manner.

etc.), hexagonal (e.g., 1, 6, 15, 28 etc.) and so on.[11] They raised the question: 'If zero means nothing, how can nothing be represented by something? After all what shape can nothing have?' Moreover, the Greek philosopher Aristotle had declared, 'Nature abhors a vacuum'. He was very influential in Greek society as he had given a 'proof' of the existence of God sitting in the seventh heaven, by declaring Him as the Prime Mover, responsible for rotating the seventh, i.e., the outermost, celestial sphere, harboring the blue-black midnight sky studded with twinkling stars beyond all the other concentric spheres, one each for the seven planets that they could see with the naked eye, while the immobile earth remained stationary at the center. Mostly under the influence of his teachings, the Greeks decided that there was no zero and no infinity; everything in the universe must be finite, with the possible exception of time. This dictum was then to be followed for about 2,000 years in Europe as the Catholic Church in Europe at a later date accepted the Aristotelian doctrine and contesting Aristotle became tantamount to challenging the authority of the Catholic Church, which in medieval Europe could easily cost one's life.

Even the great Archimedes didn't use the 'number zero', as can be observed from his approach to the famous problem of the 'sand reckoner', where he devised a number scale for counting astronomically large numbers using ingenious repetitions of their basic largest unit of *myriad* over and over; but while developing this scheme he always started to count the first number 'one of a higher phase' by declaring it to be the same as the last number of the previous phase, and not from the subsequent number. However, the common Greek enumeration system was additive in nature and there were no separate numeral symbols; their usual 24 letters of the alphabet (along with three special symbols: *digamma, koppa* and *sampi*) used for representing numbers, written with a bar on top, a system where the largest number name was *miryori*, only ten thousand, denoted by the symbol M. Three blocks of nine letters each served the purpose of nine numerals from one to nine, to be put in the unit, tens and hundreds place.

At a later period, Hellenistic astronomers, while recording the angular positions of the celestial bodies, extended their alphabetic numerals into a base-60 positional system, much like the Babylonians, by limiting each position to a maximum value of 59 and including a special symbol for zero, which looked like an 'o with a bar or dumbbell on top', believed to have come from the first letter of the Greek word *ouden*, meaning 'nothing'. However, in the usual Greek alphabetic numeral system, this 'o', called *omicron*, stood for 70. The Hellenistic

11 The inclusion of 1 in all these sets indicates the trivial case.

zero was also used alone like our modern zero, more than as a simple placeholder. However, the positions were limited to the fractional part of a number (called minutes, seconds, thirds, fourths, etc.); it was not used in the integral degree-part of a number. This system was probably adapted from Babylonian numerals by Hipparchus around 140 BCE. It was then used by the Greek astronomers of Alexandria such as Ptolemy (around 140 CE), who backed the Aristotelian view of a geocentric universe through his own ingenious theory of epicycles, and by Theon (around 380 CE) and his daughter, Hypatia (arguably the last stalwart of Alexandria, who was brutally murdered in 415 CE by a Roman mob for her refusal to accept their religious faith).

## 8 India

Coming to the case of India, it is worthwhile to quote from *Mathematics in India* by Kim Plofker: 'The Indian development of place value decimal system ... is such a famous achievement that it would be very gratifying to have a detailed record of it. Exactly how and when the Indian decimal value system first developed, and how and when a zero symbol was incorporated into it, remains mysterious.' Let us try to throw some light on different aspects of this 'mystery'. In ancient Vedic India only the oral tradition (*śruti*) of transmission of knowledge prevailed in society. There was no writing at all, whence no direct written record exists that might have been called the 'history' of that period in the usual sense. This society was amazingly rich in lofty philosophical ideas and thoughts, as testified by the Vedas, *Upaniṣads* and many other texts of the Vedic corpus and also through the later Buddhist and Jaina texts. However, unlike Greece, philosophers of every school of thought in India seemed to have toyed freely with the concept of nothingness and embraced the idea of complete void. For example, in the *Tantra*, one finds the eternal divine creator *Niṣkala* (partless) *Shiva*, the undifferentiated formless omnipotence, the Lord Shiva, who creates this whole universe from the void and destroys his creation into the void again and again. From the *Nāsadiya sūkta* of the *Ṛgveda*, where the Vedic seer is philosophically trying to imagine the nature of the complete vacuum before the creation of the universe, to the concept of *māyā* (illusory non-existence in reality) in *Vedānta* philosophy, from the concept of *abhāva* (absence) of the *Nyāya* school, to the *Śūnyavāda* (doctrine of devoidness) of Nāgārjuna of the *Madhyamaka* school of *Mahāyāna* Buddhism, where the highest form of knowledge, i.e., *prajnāpāramitā*, is attributed to the perception of everything phenomenal or worldly as *śūnya*, the spiritual attainment of this sense of *śūnyata*, the pure void (the state of *catuṣkoṭi vinirmukta*) being *nirvāna* (the ultimate emancipation) as seen from the teachings of *Mulamadhyamakakārikā*.

It appears very likely that the thread of the rich philosophical and socio-academic ambiances of Indian Antiquity was quite pregnant with the immensity of the concept of *śūnya*, – a dichotomy as well as a simultaneity between 'nothing' and 'everything'. Hindu culture worshiped the supreme God *Brahma* as *eṣa śuddha śūnya śāntah* (oh thy pure, calm, void incarnate) in one of the *Upaniṣads* and in another as *anoroṇian mahatomahiān*, (minuter than the minutest, and yet simultaneously larger than the largest); the *śūnya* (zero) of void and that of an all-pervading fathomless *pūrṇa* (infinite wholeness) seemed to be merely two sides of the same coin in their philosophical outlook. In this society, where mathematics, an *aparāvidyā* (worldly knowledge), was treated with utmost importance, to be considered at the helm of all the *Vedānga śāstra*s (like the crest of a peacock or the gem on the hood of a serpent), when mathematicians felt or declared the necessity of having a numeral for *śūnya*, society at large never contradicted the idea or tried to resist it, contrary to what was the case in Greece. But to appreciate the real importance of the role of zero, it has to be considered in tandem with the decimal (*daśa*) place value system prevailing in India during the Vedic age (from around 1200 BCE) or even earlier.

Yaṣkācharya in his *Nirukta* (etymology, one of the six limbs of the Vedas) (circa 600 BCE) traced the root of *daśa* (10) to the verb *dṛś* (to see), and pointed out that the use of 10 is seen in the formation of subsequent numbers. It clearly shows that the status of the number 10 was not like the other *anka*s (digits) from *eka* (1), *dvi* (2), through *nava* (9). One early passive reference to the decimal place value system is seen in a different context in the first century CE Buddhist *Madhyamaka* school text *Abhidharmakoṣa* by Basumitra. While illustrating the theory of *dharma*s (elements or factors), he used the comparison 'like a marker (*vartikā*) in reckoning, which in the unit position has the value of a unit, in the hundreds position that of a hundred and in thousands position that of a thousand' (as translated by David Ruegg).

A strikingly similar statement, but from a completely different perspective is also found later in seventh century CE: the *Vyāsabhāṣya* of the *Yogasūtra* by Patanjali. However, if one wants to find this place value principle in an exclusively mathematical text, the earliest one known is *Āryabhaṭiya* [499 CE] by Āryabhaṭa of Kusumpura (near present-day Patna of Bihar, India). He clearly enunciated the decimal place names from unit (*eka*), 10 (*daśa*) up to the ninth power of 10 (*brinda*), stating clearly that at every stage the next higher place is 10 times (*daśagunam*) the previous one. But the previous examples clearly show that the decimal system was conceptualized and in vogue long back in ancient India. Indeed, it had gone way beyond the ninth power, as can be seen in various ancient texts through their penchant for naming larger and larger numbers in powers of 10, as if to impose a superior claim of numeracy to its audience. In *Śukla Yajurveda* we find a mention of the 12th power of 10 (*parārdha*), in a Buddhist work *Lalitabistāra*, the 53rd power of 10 (*tallakṣhna*); while in a Jaina text, *Amalasiddhi*, the 96th power of 10 (*daśa-ananta*) was mentioned. Incidentally, it is well established today, thanks to the pioneering work of Gupta,[12] that Āryabhaṭa's algorithms for extraction of the square and cube roots of a number, that he preferred to present through a language-based scheme (in *Ganitapada* verses 4 and 5), can work only in a fully functional place value system with zero. Most likely this fact, transmitted wrongly, has given rise to the common misinformation among generally educated Indian folk that Āryabhaṭa 'invented' zero, since to date we do not have any information about an exclusive and systematic Indian text per se on mathematics, written before *Āryabhaṭiya*.[13]

Amazingly, one may find about 3,000 decimal nomenclatures of number names in the *Ṛgveda* [circa 1200 BCE], which is astonishingly rich in words and phrases relating to numbers. Starting from the nine primary numbers, i.e., the digits (*aṇka*) such as one (*eka*), two (*dvi*) up to nine (*nava*), it has the first nine multiples of ten as, ten (*daśa*), twenty (*viṃśati*) and up to 90 (*navati*) and also the powers of ten up to *ayuta* (the 4th power of ten), the largest example being *navatiṃnava sahasra* (ninety[and]nine thousand). Many of these number names exhibit the 'fundamental characteristics obligatory to an orally expressed decimal system', as has been investigated at length by Bavare and Divakaran (2013). As Divakaran (2018b) points out, 'the earliest Vedic texts show in an absolutely unambiguous manner that one can implement the place value paradigm just as effectively with spoken number names … as with number symbols. The rules for forming numbers above nine – serving the same purpose as the now-universal sequence in the written symbolic notation – had to be realized by applying rules of nominal composition: in other words, rules of grammar.' However, there is no exclusive reference to the number zero to be found at this stage, as this was never required for oral treatment of number names, not even for the numbers like *sata* (hundred), *sahasra* (thousand) or, for example, a more complicated *ṣaṣṭiṃ sahasrā navatiṃnava* (sixty thousand ninety [and] nine) etc. This requirement became unavoidable only at

12 Rigorous proof of this fact was also given by Ifrah at a later period.

13 The famous Bakhshali manuscript, as we shall discuss later, is of uncertain authorship and disputed time-origin and also more of a commentary in nature.

a much later stage, when writing these numbers in a place value paradigm was necessary.

A. K. Dutta (2016) observes: 'To arrive at a written decimal notation from the above terminologies, one has to simply suppress the place names from a given numerical expression, provided one has an additional tenth numeral as a placeholder to indicate the possible absence of the nine numerals at certain places. Thus one can conclude that the structure of the Sanskrit numeral system contains the key to the decimal place value system.' This in effect indicates the polynomial structural form of Sanskrit (decimal) number names, where suppressing the place names is equivalent to considering the ordered tuple made with the coefficients of the polynomial. In the first representation, some power(s) of the variable may remain absent (just as in an oral number name [of a particular number] some place(s) may not feature at all), while in the second representation, the corresponding position(s) in the ordered tuple must be filled with zero(s) to indicate that polynomial correctly (respectively, when we have to write [that particular number], the vacant places demand the placeholder zero). And before the time of the Aśokān edicts [around the middle of the third century BCE], or the Nānāghaṭ inscription [second century BCE], no systematic written records are known to exist in India.[14] So it is clear that a written symbol for zero (as a placeholder) evolved out of a place value system (decimal, in this case) at a point when written representation of the already perfected oral system was necessary.

And it is precisely here we stumble upon a critical fact. The initial Indian written records available are almost all in the *Prākṛit* language and *Brāhmi* script, written from left to right, be it a sermon on stone inscription (for example, the Sāsārām rock edict of Aśoka from circa 250 BCE) or an inscribed date on numerous metal coins. However, some stray *Kharoṣṭhī* ones were found, mostly in the erstwhile Gāndhāra region, now in Pakistan; written from right to left, the *Kharoṣṭhī* system of writing numerals was even more primitive compared to *Brāhmi*. Basic single numerical symbols were used only for numbers 1 to 4; each of the numerals from 5 to 9 was a group of those symbols, particularly using the symbol of 4 juxtaposed with others by the additive principle (such as two symbols of 4 and then that of 1 placed side by side to represent 9); there were some separate single symbols for each of the numerals 10, 20, 100, 1,000 etc., and the whole scheme of writing numbers was that of additive juxtaposition in nature. Of course, there was no zero in it. This script had no impact

14 Leaving aside, of course, the engravings on Hārāppān seals, which have not yet been deciphered, not even to the extent of being a language at all, despite some formidable recent attempts, https://www.nature.com/articles/s41599-023-02320-7.

on the later development of written languages in India.[15] However, *Brāhmi* numerals, though basically grouped by ten and its powers (and hence decimal in spirit), were not written according to a positional (place value based) system. It was multiplicative-additive in principle and there was no numeral symbol for zero found in the system. But of course, an additive system does not need it.[16] There were nine arbitrary basic symbols for the first nine numerals 1 to 9, then separate single symbols for 10 and multiples of ten up to 100, and some special symbols for 200, 500 and 1,000 each. A multiple of higher powers of ten was expressed by attaching a symbol for the multiple to the symbol of that specific power of ten,[17] and then such groups of symbols were juxtaposed according to the additive principle of formation of a number. Some 30 instances from the time of the Sātavāhana kingdom (circa the second century BCE) of south India were found in the Nānāghaṭ inscription (written in Sanskrit using *Brāhmi* script), in the Western Ghat region of India. For example, the number 24,400 is written as three symbols side by side from left to right, the symbols being those for 20,000, 4,000 and 400. The question is, why did a civilization that had already perfected and excelled in the oral decimal (positional) place value system have an inferior additive presentation of the written numbers? In a recent article, *The Bakhshali Manuscript and the Indian Zero*, P. P. Divakaran (2018b) proposed a plausible explanation to this long-standing mystery. According to him, the *Brāhmi* numbers 'are a faithful symbolic version of the rules for forming decimal number names that had held sway from the time of the Ṛgveda, i.e., of the polynomial representation, and they result from the slavish transcription of names to symbols. They are not positional because they have not reduced a number to a list of its (polynomial) coefficients.' This approach indeed explains the missing *Brāhmi* zero, as it indicates that up to this period the written place value based notations were yet to be invented and numerals in *Brāhmi* script, while pointing to the continuing hold of the oral tradition on the then relatively recent shift to a written paradigm, played a bridging role between the existing Sanskrit oral nominal composition scheme of decimal nomenclatures and the later place value based written symbolic representation of those decimal numbers.

However, alhough a plausible reference to the number zero as *kṣudra* in the *Atharva* Veda is reported by some scholars, the most common parlance for zero

15 It is to be noted in this context that Nāgari was to evolve and develop as a written script much later; it flourished during the Gupta period.

16 Incidentally, the Brāhmi single numeral symbol for 20 looks very similar to our modern zero symbol.

17 Apparently it was quite like the early Chinese multiplicative-additive paradigm.

in ancient India was *kha*. It is interesting to note that Yaṣka [circa 600 BCE] traced *kha* to the root *khan* (to dig), and one of the earliest references to *kha* is found in *Gopatha Brāhmana* as *chidraṃ khaṃ iti uktaṃ* (*kha* is the puncture mark or hole). Now if one decides not to use the absence of testimony with regards to the lack of exclusive reference to zero (be it a placeholder or a number of its own) in early Vedic texts like *Ṛgveda* (as was deemed unnecessary in their oral system), as the testimony of the absence of the number zero at their knowledge level as well, it is very tempting to try and visualize the *chidra*, a hole, which is naturally represented by the bold dot mark or *bindu*, as the symbolic imagination of the number zero, also prevalent at that early period of time. However, *kha* literally means the sky, which is also referred to in Sanskrit as *śūnya*. This word is derived from *śūna* [+*yat*], which is the past participle of the root *śvī*, which means 'to swell' or 'to expand', and there from by semantic extension 'hollow'.[18] In the *Ṛgveda*, one may find another meaning – 'the sense of lack or deficiency'. The two different meanings were fused to give *śūnya* a single sense of absence or emptiness with the potential for growth, a womb-like hollow, ready to receive and swell. In many early Indian mathematical writings such as the Bakhshali manuscript, *Āryabhaṭiya* etc., one may find another dual role of this concept being manifest in the form that it stands for both the placeholder zero as well as the unknown solution (to begin with) of a numerical problem (somewhat like we consider the $x$ in algebra today). The Indian place value concept of the formation of a number treated the places (*sthāna*) of a number as mere locations, initially empty (*śūnya*), and eventually to be filled by one of the numerals from 1 to 9; in the case that a place is not occupied by any of these numerals it remains empty, which during the early written regime is seen to be filled by the *bindu* symbol, as is found in the Bakhshali manuscript (Datta, 1929).

## 9 The Islamic World and Europe

However, the dichotomy of the concept of *śūnya* as nothing and everything, clearly a manifestation of the philosophical and cultural ethos of Vedic India, was not properly appreciated by the Arabs, when it was transmitted to them through Kanaka in the eighth century CE via the trade route. They first received this idea of the place value based decimal number system with zero from that Indian mathematical text and literally translated it into Arabic in accordance

18 It is interesting to note that the root of the word Brahma is *bṛinh*, which also means 'to swell' or 'to expand'.

with their understanding, mostly from the linguistic or grammatical connotations of the Sanskrit terms.[19] This led to the translation of *śūnya* as *sifr*, which in Arabic bears the perspective of 'nothingness' in the sense of being empty, as can be seen from many a related word such as *safira* (it was empty), *safr* (empty), *asfara* (to empty) etc., all sharing the same stem *sfr*.[20] Later, over a long period of time, this term *sifr* as the name of placeholder zero was Latinized into various similar sounding words such as *cyfra*, *cyphra*, *zyphra*, *tzyphra* etc. In 1202, Leonardo Pisano of Italy, commonly known as Fibonacci, in his book *Liber Abaci* (The Book of Numbers) referred to this symbol as 'with these nine numerals, and with this sign 0, called zephirum in Arab, one writes all the numbers one wishes'. This was an early eye-opener for Europe where the clumsy Roman additive-subtractive system of enumeration was still in vogue. Fibonacci came to know about the new system from the Arab traders, while as a young man he traveled with his father Bonaccio to the eastern parts of Arab-dominated Africa, a place geographically located to the west of Italy. This might have influenced his coinage of the term *zephirum*, literally meaning 'light western breeze', so light that one might not even feel it. This almost negligibility clearly refers to the sense of nothingness, and his statement also clearly shows his hesitant attitude toward the status of zero in this system, as he refers to it as a mere 'sign', while the other nine symbols are called 'numerals'. The name *zephirum* gradually changed over time to *zefiro* in the local Venetian dialect, which eventually paved the path toward the evolution of the modern English (and French) word zero (and the Spanish word *cero* as well).[21]

## 10 India Revisited

Apart from *Gopatha Brāhmana*, early references to zero in Indian works are found in the *c.*300 BCE work of Bhadrabāhu as *thibuga*, which was later interpreted by Hemachandra to mean *bindu* (Aczel, 2015), as *tuccha* (trifling) in *Amarkoṣa*, a lexicon by Amarsiṃha, also as a description of the number 60

19 The translation by Al Fazari, most possibly out of a text composed by Brahmagupta, is not extant, but a later Latin translation of the Arabic work (*De Numero Indorum*) survived.

20 It may be noted, however, that it is in Brahmagupta's work *Brāhmasphuṭasiddhānta* that we find the first ever categorical statement about zero being a number in its own right, which he declared as an outcome of addition among equal positive and negative numbers (*samaikye kham*).

21 Incidentally, the first printed symbol of the mathematical zero in modern Europe is to be found in a book called *De Arithmetica Opusculum*, written by Philippi Calandri in 1491 CE.

as *ṣaṭ binduyutani* (six with *bindu*) as well as *ṣaṭ khayutani* (six with *kha*) in *Yavanajātaka* (270 CE) by Sphujidvaja (Kaplan, 1999), or as *śūnya-bindu* (zero dot) in the romantic fiction *Vāsavadattā* by Subandhu (*c.*400 CE) through a poetic allegory, referring to the shining stars in the night sky as resembling the 'zero dots', as if drawn by the creator Himself with a piece of chalk as the moon. This simile clearly refers to the possible shape or form of zero as a bold dot, as might have been prevalent at that time. Ifrah (2000) has shown numerous examples of numbers presented in Sanskrit *bhūtasaṃkhyā* (word numerals).[22] An early example of the use of this system with the decimal place value principle in tandem with word numerals for zero is found in the Jaina cosmological text *Lokavibhāga* (Parts of the Universe), composed in 458 CE, where the author claims this knowledge had been handed down to him from an even earlier period of time. Indeed, such examples are too numerous to be listed in detail here.

A tacit appreciation of the placeholder role of the mathematical zero in forming a number is seen in a statement in the Jaina text *Anuyogadvāra sūtra* (*c.*100 BCE), where a certain number, presented in the form of a product of two particular numbers, was claimed to occupy 29 places (*sthāna*). Indeed, when we calculate it in decimal notation, the described 29 digit number is seen to have a zero in its thousand place, indicating that this mathematical sense of taking into consideration the place occupied by zero (or maybe just not occupied by any of the nine digits) was already present.

Among other scholars, M. D. Pandit (2003) and Frits Staal (2010), while analyzing Pāṇini's monumental text on grammar, *Aṣṭādhyāyī*, (sometime between the sixth and fourth centuries BCE), have argued at length that at a deep structural level, the conceptual role of placeholder zero, is much akin to the Pāṇinian implementation of the rule 1.1.60 *adarśhanaṃ lopaḥ* (non-appearance or non-availability as vanishing), with reference to his grammatical reconstruction of *Prākṛit* through a minimalist's approach toward harnessing a maximum possible number of cases under a grammatical rule. This rule on *lopa* and some of Pāṇini's other linguistic rules, using some meaningless metalinguistic markers, toward a structural analysis of the Sanskrit language, pertaining to vanishing (or 'zeroing', so to speak) of an affix or infix (*morpheme* or *phoneme*) is now looked upon as a possible grammatical ancestor of the Indian zero. Or, was it the other way round? Was Pāṇini influenced by a prevailing idea of zero from the mathematical community? Of course, the answer is not known. However,

22 It is a system where every number had many equivalent names through language, handy for the metrical need of a composition using that number name; for example, *gagana*, *ambara*, *ākāsha*, *byoma*, *jaladharapatha* and many more for *kha*.

it is interesting to note that in a later grammatical text, *Jainendra Vyākaraṇa* by Pujyapāda (*c.*450 AD), the term *lopaḥ* was replaced by *kham*, (1.1.61) the most common parlance for zero in ancient India.

However, a clear cut, independent, and unambiguous reference to the number *śūnya*, this time not just a placeholder making room for other digits to occupy their places but the mathematical number zero appearing alone, can be found in the second (or third) century BCE book on Sanskrit prosody, *Piṅgalachandaḥsūtra*, by Piṅgala, where the *sūtra* numbers 29 and 30 in the eighth chapter are respectively *rupe śūnyam* and *dvi śūnye* (Emch, Sridharan, and Srinivasa, 2005; Mukhopadhyay, 2009). These two terse aphorisms, along with the *sūtras* 28 and 31 put together, as described by the later commentators, refer to a combinatorial calculation to be done in a two-phase algorithmic manner. It invokes the number zero (*śūnya*) along with the number 2 (*dvi*), used as two different labels to separate two possible cases that may occur in the first phase of the given algorithm, and they also guide through the calculation of the second phase to reach the final answer. The symbol of 2 (*dvi*) is supposed to mark the stage where the number under consideration is even, which is to be divided by 2 in the first phase (*dvirardhe*; *sūtra* 28) and where squaring (*tāvad ardhe tad gunitam*; *sūtra* 31) has to be done in the second; while the symbol of zero (*śūnya*) marks the stage where the number under consideration is odd and hence 1 is to be subtracted (*rupe śūnyam*) in the first phase and where the number at hand in that stage is to be multiplied by 2 (*dvi śūnye*) in the second. Not only that, but in the last but one step of the first phase of the algorithm one has to face the situation 1–1 as its outcome, a fact that is bound to happen here mathematically irrespective of the number you begin with; and although the symbolic label for this operation is to be taken as '*śūnya*' according to *sūtra* 29, it is not clear whether the outcome of this particular subtraction was also appreciated as the number zero by Piṅgala in that early period.[23] Or was it simply a signal to stop only when one reaches that particular operation of subtraction and begin the second phase of calculation in the reverse gear? Interestingly, the second phase has to begin compulsorily with the first counting number 1 as the number at hand (then to be multiplied by 2, according to *sūtra* 30, *dvi śūnye*, as the label of the last step must always be *śūnya*) and not the zero that would have been reached at the end of the first phase through the execution of the subtraction 1–1. Does it anyway hint toward a possible awareness of the property of the number zero under

23 As we have already mentioned, this eventuality was later categorically mentioned through a rule of addition among numbers in *Brāhmasphuṭasiddhānta* by Brahmagupta in the seventh century CE.

multiplication? Of course, this fact of mathematically reaching the number zero through subtraction is not explicitly mentioned by the commentators, but it might have been tacitly so. Again, it is a matter of fact that this *śūnya* did not take part directly in the mathematical operations indicated by the relevant rules, and this fact has been highlighted by some scholars, who are skeptical of accepting this *śūnya* as the number zero. Yes, Piṅgala could have expressed his mathematical algorithm given through the *sūtra*s 28 to 31, avoiding the labels of *śūnya* and *dvi*. Instead, he could have used the names of say, 'cow' and 'goat' respectively, to serve the same purpose. But it is absolutely important to note that he didn't. So, in a thoroughly mathematical environment, where an algorithmic recipe is seen to involve other numerals clearly by their names, it is indeed quite far-fetched logically to see one such name, *śūnya*, not to refer to a numeral as well. However, there is no direct evidence to imagine any form of a written symbol for this zero that might have been used, though there very likely was one. The spirit of *sūtra*s 20 to 35 in the eighth chapter for combinatorial calculations indicates quite strongly that they are meant to be written down … somewhere maybe on a dust board; the name of one particular group of rules, *naṣṭam*, 'lost' or 'annihilated', hints at it clearly. One rule (*ekona adhva*, that comes under *advayoga*, i.e., space measurement, as classified by all his commentators except *Halayudh*, actually calculates the amount of space required to write down the matrix bed of all possible variants of a meter with a definite number of syllables. But, he could not have done it in *Brāhmi* script as there was no zero to write the labels in the rule *rupe sunyam* or *dvi sunye*. On the other hand, we must also remember *Brāhmi* and *Kharoṣṭhī* were the scripts meant for the understanding of the common people, as attested by the Aśokān edicts, where the language was mostly archaic *Prākṛit*. So, could it have been possible that classical Sanskrit, known as *Devabhāṣā* [the language of Gods], was kept in seclusion among the venerated scholars, who could have had some other means of 'writing' as Piṅgala's rules indicate? Not much is known here. But anyway, as it stands, the absence of the symbol for zero in *Brāhmi* (and *Kharoṣṭhī* as well) is a fact and so is Piṅgala's *śūnya* as a name for the number zero. We have no known evidence of written records in India prior to the Aśokān edicts, but as the saying goes, the absence of evidence may not be evidence of an absence.

Coming to the written inscriptions of the various zero symbols known to us and discovered in different places on present-day Indian soil, we have three stone inscriptions originally deciphered by D. C. Sircar, a leading epigraphist, as reported by Michel Danino (2016) in his article, *In Defence of Indian Science*. Arranged chronologically, these are (i) a small globular symbol from the Mankuwar Buddha inscription of 428 CE, representing a placeholder zero (not in a positional context), (ii) the Dabok inscription (from Udaipur, Rajasthan) of 644 CE that displays a dot to represent zero (not in a positional system), and (iii) the Khandela inscription (Rajasthan) of 807 CE with a proper place value system, where a small circular symbol for zero was used in the engraving of the number 201. All these three pieces of evidence predate the much referred to stone inscription found in the Chaturbhuja temple of Gwalior fort, where the numbers 270 and 50 were engraved in *Nāgari*, using the positional system with a circular zero, in a charter issued by King Bhojadeva (*c.*875 CE). However, if we look at the whole Indian subcontinent, including those parts of Southeast Asia that were within India's cultural influence, a 683 CE inscription of bold dot zero was found in Cambodia and originally deciphered by the French scholar George Codes in 1931. The place was later plundered during the Khmer Rouge regime led by Pol Pot, leading to a loss of its whereabouts. Eventually it was rediscovered by Amir Aczel (2015) in 2013 from a place near Angkor Wat. This inscription in Sanskrit language, written in old *Khmer* script, depicts the number 605 in positional representation, where the zero in the middle is written as a bold dot. Indeed, many other contemporary stone inscriptions with bold dot zero were found in neighboring Southeast Asian countries, as was tabulated by Ifrah in his book, *The Universal History of Numbers* (2000). Further, if we consider the copperplate land grants as pieces of evidence,[24] there are quite a few, the oldest inscription known being the *Sankheda* (Bharuch, Gujrat) charter (596 CE).

## 11 The Bakhshali Manuscript

The first ever known mathematical writing of Indian origin that contains a symbol of written zero (in the form of a bold dot: •) in numerous arithmetical calculations, where numbers are written in the decimal place value system, is the Bakhshali manuscript (Sarasvati and Jyotishmati, 1979), which was discovered by chance in 1881. Written in ink on birch bark, it is an incomplete document consisting of about 70 partially mutilated fragments. A greater part was reported to be destroyed while it was dug up by a farmer in the village of Bakhshali, where it lay buried between stones. The site is about 50 miles from Peshawar in present-day Pakistan, known to be a part of the erstwhile Gāndhāra region and within 70 miles of the famous ancient learning center of Takṣásilā. The manuscript was sent to Dr Hoernle of Calcutta Madrasa

24 Some scholars consider them to be fake, but almost never offer concrete proof of said fraud, as observed by Danino (2016).

for examination and publication; he gave a short description of it before the Asiatic Society of Bengal in 1883. In 1902 he presented it to the Bodleian Library of Oxford, where this incredibly fragile document is presently kept. In early 2017, a team of experts from the Bodleian Library and the Oxford Radiocarbon Accelerator Unit collaborated on radiocarbon dating (C-14 analysis) of the manuscript.[25] Five samples were extracted from the birch bark manuscript, avoiding the inked portion, three of which were put to analysis, maintaining the state-of-the-art technological protocol needed for such an experiment. The result, after the statistical analysis of the outcome, revealed that the three samples tested surprisingly dated from different centuries, at least 500 years apart; Folio 16 from 224 to 383 CE, Folio 17 from 680 to 779 CE, and Folio 33 from 885 to 993 CE. Explaining this unexpected time gap, which does not conduce to the earlier proposed dating by different experts (although those estimates were nothing like unanimous, and indeed varied over a period by about a millennium), has given rise to a flurry of activities in interested quarters, as this document is clearly associated with the history of a written zero symbol in India. Kim Plofker, Clemency Montelle, T. Hayashi et al.[26] (2017) published a paper criticizing the physical evidence provided by the recent radiocarbon dating. They called it 'historically absurd' and concluded that 'the proposed division of the Bakhshali manuscript text into three chronologically distinct sections corresponding to the three radiocarbon date ranges is contradicted by the unified appearance of its content and writing.' Indeed, they have explored all the possible avenues through which the findings of the radiocarbon dating may be rejected. While one may wonder why only those three (or five) particular specimens were chosen, instead of one sample from each of the available folios, some other scholars are of the opinion that sincere efforts should be made toward reconciling the apparent contradiction by examining various ways in which these inconsistencies can be understood and explained, rather than summarily rejecting the physical evidence. In the face of this controversy, dating each and every available folio, as far as practicable, would be a welcome option.

25 See Appendix 3 for an online petition urging the Bodleian Libraries, Oxford, UK, to take concrete steps to commission the necessary follow-up radiocarbon-dating of the Bakhshali Manuscript in the interest of scientific advancement in the field.

26 An expert on the Bakhshali manuscript, Hayashi argued at length in his 1995 book that the original composition of this document was during the seventh century CE, while the manuscript copy was thought to date from somewhere between the eighth and twelfth centuries CE, in contrast to Datta's (1929) estimate, placing its origin at some time around the beginning of the Common Era.

The original authorship of this brilliant arithmetical work, containing a wide variety of problems pertaining to daily life, is not known, but on Folio 50 (recto) we find that the present work (which may be a copy along with some added illustrations and prose commentary to some of the original older work) was inscribed by a certain Brahmin mathematician who hailed himself as the king of calculators, 'the son of Chajjaka'. He claimed to have written it (the exact language is *likhitam*, which would have been more naturally *kṛtam* or *virachitam* had it been an original composition by him, as scholars believe) for the use of Vaśiṣṭḥa's son Hāsika and their descendants. The handwriting style, presentation, and other technical accounts reveal that the present manuscript had at least two different scribes, possibly more. Some scholars believe that the colophon indicates the possibility that the text was traditionally used by the lovers of calculation especially in the Northwestern part of the then India (as it was written in the *Gatha* dialect of *Prākṛt* in *Sāradā* script, once popular in the Kashmir region). Perhaps it was handed down from one generation to the next, as an important and invaluable family heirloom, that had to be refurbished again and again by replacing some worn-out, overused, older bark folios with new ones, inscribed carefully by a competent contemporary scribe as a restoration job, not just a copy of the content, but to avoid any distortion even in the presentation of it, so as to faithfully simulate the original as far as possible. This may explain the suggested different handwriting as noted in different parts of the manuscript. This may also account for differences in the ages of the bark; for example, those used in Folios 16 and 17 (as claimed by the C-14 dating), although they happen to contain essentially similar problems on the calculation of the purity of gold (*suvarṅakṣya*), solved through a weighted arithmetic mean, continued from one folio to the other. However, it is interesting to note a significant difference in presentation (*nyāsa*) of two essentially similar consecutive problems: one in Folio 16 (verso) and the other in Folio 17 (recto). The typical Bakhshali symbol + for the negative sign indicating the operation of subtraction, employed here to indicate *kṣya* (loss due to erosion) of gold (*suvarṅa*), although found present throughout in the Folio 16 (verso) sum is not found at all in the first sum of Folio 17 (recto) (Sarasvati and Jyotishmati, 1979). Anyway, the suggested differences in their age has been cited by Plofker et al. (2017) in their paper as one of the justifications to reject the physical evidence of their age.

To restrict the date of the manuscript to a narrower range than suggested by the radiocarbon dating, P. P. Divakaran in his recent scholarly article on the dating of the Bakhshali manuscript,[27] rather than relying on the usual

27 Published in the October 2018 issue of mathematics magazine *Bhāvanā*.

analysis of its language and script focused on comparing its mathematical content vis-à-vis the famous text *Āryabhaṭiya*, whose date (the end of fifth century CE) is well established. Armed with the principle of comparative history, which states that 'what comes later cannot influence what came earlier', Divakaran cited quite a few convincing instances where the same or similar topics treated elegantly in *Āryabhaṭiya* were treated rather poorly, or not considered at all, in the Bakhshali manuscript, pointing to a lesser arithmetical sophistication of its time and hence its period being expectedly earlier than that of *Āryabhaṭiya*. Again, the non-positional organizing principle of *Brāhmi* numerals, derived from the much older decimal number names, clearly predate the more advanced place value based written numbers of the Bakhshali manuscript. These considerations prompted Divakaran (2018) to conclude that the Bakhshali manuscript, with its plethora of bold dot zero in writing, might have been originally composed sometime between 350 and 500 CE. This date broadly refers to the period historically just before the Hun plundering of the *Gāndharā* region, destroying its learning centers such as Takṣaśilā forever and corroborates the appearance of the term *śūnya bindu* in literature,[28] as Divakaran pointed out.

## 12 The Islamic World and Europe Revisited

During the reign of Abbasid Caliph Harun al-Rashid (also known as Al-Ma'mun, who reigned in Baghdad from 813 to 833 CE), Arab scholars of *Bayt-Al-Hikma* (House of Wisdom) such as Al-Khwarizmi took *śūnya* and other decimal numbers from an Indian work of Brahmagupta and, through its Arabic literal translation, decimal numbers with zero were gradually introduced to Europe during the latter half of the so-called Dark Ages, from where they eventually spread throughout the world, primarily by virtue of their Latin re-translations along with various original Latin works done in the twelfth century. However, the first known reference to the Indian decimal system in Europe was made much earlier by Severus Sebokht, a Syrian monk, who wrote in 662 CE: 'I shall not now speak of the knowledge of the Hindus … of their methods of calculation which no words can praise strongly enough – I mean the system using the nine symbols.' Similar praise can be found in the *Codex Vigilanus* by the monk Vigilan (Albelda, Spain, 976 CE): 'The Indians have an extremely subtle intelligence. The best proof of this is the nine symbols with which they represent

28 The drama Vāsavadattā of Subandhu, composed sometime in the fourth century CE, has the term *śūnya bindabaḥ*, as already mentioned.

each number no matter however large.' Interestingly, these and many other works on the Indian decimal system up to the twelfth century in Europe refer to 'nine numerals' only and the absence of any exclusive reference to zero as a number at this stage is easily noted, which shows their hesitancy in accepting the role of placeholder zero in this system and also their lack of understanding or failure in appreciating zero to be a number in the same spirit as the other basic nine numerals of the decimal system. Later, Fibonacci in his 1202 CE book *Liber Abaci* named the sign zero as *zephirum*, and identified it as Arabic in origin. In 1252 CE Maximus Planudes wrote in his *Psephophoria Kata Indos*: 'There are only nine figures. A sign known as "tziphra" can be added to these, which, according to the Indians, means "nothing". The nine figures themselves are Indian, and "tziphra" is written thus: 0.' This indicates that even as late as the thirteenth century, Europe was not yet conversant with the number zero as a number in its own right. This idea dawned on them much later, as has been pointed out in recent times by Jonathan Crabtree, who investigated at length to show how this state of affairs jeopardized the initial texts of elementary mathematics in Europe.

At one point in time during the initial stage of its introduction to Europe, the Catholic Church, which embraced the Aristotelian doctrine on vacuum, was dead set against the use of the decimal system with zero, as they considered it to be Islamic in origin owing to its Arabic connection, hence marked it as anti-Christian and banned its use. For them, zero, a representation of nothingness and hence of vacuum, was Satanic in character. But a section of Italian traders continued to use this new system secretly due to the obvious ease of its calculation process compared to the Roman one. This secrecy toward using the numeral system involving the symbol of zero called *cyfra* eventually gave birth to the term 'cipher'. However, with the gradual decline of Aristotelian views about the universe as the result of the epoch-making scientific works of Galileo, Torricelli, Pascal, and others in establishing sustained vacuum in nature that culminated in understanding the concept of atmospheric pressure, and that of Copernicus, Bruno, and Kepler in firmly establishing the heliocentric model of the universe contrary to the Church's view of the geocentric Aristotelian model, backed by the Ptolemaic theory of epicycles, the grip of the Catholic Church over society as a whole gradually loosened. Meanwhile, the philosopher Descartes, who was against the concept of void philosophically, gave into his mathematician soul by putting the number zero practically at the heart of his coordinate geometry to establish that the geometric curves as given by the Greeks could be equivalently expressed through the algebraic equations of the Arabs, as if they were opposite sides of the same coin. The Church had to finally back down after stubborn resistance lasting some 500 years

(Seife, 2000). Today, use of the decimal number system with zero is practiced worldwide and its 10 basic symbols are known as the Hindu-Arabic numerals.

## 13 Conclusion

Historical reconstruction of the mathematical knowledge that was likely to be prevalent in Indian Antiquity, or in an earlier period, is much like arranging the pieces of an enormous jigsaw puzzle, many of which have gone missing. Competent historians of mathematics, all over the world, are trying to arrange the available pieces, according to their own respective stance, with the obvious intention to try and guess the picture that it may suggest. And the job is anything but easy. Preconceived ideas sometimes come in the way of scholarly acumen, subconsciously tempting one to misplace one or two pieces of the puzzle or perhaps not to place them at all, distorting the figure to accommodate one's stance. Of course, there are scholars carefully steering a middle course, analyzing objectively as far as possible, the available pieces of information or sometimes even the lack of information, trying to make a reasonable pattern out of it. With every new piece being found occasionally, as a new input to the jigsaw puzzle, one has to try and find its right place in the puzzle, sometimes destroying the existing pattern. And the journey continues.

## References

Aczel, Amir D. (2015). *Finding Zero*. Palgrave McMillan.

Bavare, B. and Divakaran, P. P. (2013). Genesis and Early Evolution of Decimal Enumeration: Evidence from Number Names in Ṛgveda. *Indian Journal of History of Science*, 48(4), pp. 535–581.

Danino, Michael. (2016). In Defence of Indian Science. In *The New Indian Express*, 13 and 14 October.

Datta, B. B. (1929). The Bakhshali Manuscript. Bulletin of Calcutta Mathematical Society, XXI, p. 1.

Divakaran, P. P. (2018a). The Mathematics of India: Concepts, Methods, Connections. Hindustan Book Agency.

Divakaran, P. P. (2018b). The Bakhshali Manuscript and the Indian Zero. *Bhāvanā*, October.

Dutta, A. K. (2016). The Decimal System in India. In The Encyclopaedia of the History of Science, Technology and Medicine in Non-Western Cultures. (ed. H. Selin (ed.). Springer.

Emch, Gérard G., Sridharan, R. and Srinivas, M. D. (eds.) (2005). Contributions to the History of Indian Mathematics. Hindustan Book Agency.

Ifrah, Georges (2000). The Universal History of Numbers. Wiley.

Kaplan, Robert. (1999). *The Nothing That Is: A Natural History of Zero*. Oxford University Press.

Lam Lay Yong and Ang Tian Se (1992). *Fleeting Footsteps: Tracing the conception of Arithmetic and Algebra in Ancient China*. World Scientific. Reviewed by Jean-Claude Martzloff in *Historia Mathematica*, 22, p. 67.

Mukhopadhyay, Parthasarathi, (2009.) Concept of *Śūnya* in Indian Antiquity. *Journal of Pure Mathematics*, 26.

Nanda, Meera. (2016). Science in Saffron: Skeptical Essays on the History of Science. Three Essays Collective.

Pandit, M. D. (2003). Reflections on Paṅinian Zero. In *The Concept of Śūnya*. A. K. Bag and S. R. Sarma (eds.)., Indian National Science Academy, Aryan Book International.

Plofker, Kim. (2009). *Mathematics in India*. Princeton University Press. Reprinted by Hindustan Book Agency, 2012.

Plofker, Kim, Keller, A., Takao Hayashi, et al. (2017). The Bakhshali Manuscript: A response to the Bodleian Library's Radiocarbon Dating. *History of Science in South Asia*, 5(1), pp. 134–150.

Ramasubramanian, K (ed.). (2015.) Ganitananda: Selected works of Radha Charan Gupta on the History of Mathematics., Indian Society for History of Mathematics.

Sarasvati, Swami Satya Prakash and Jyotishmati, Dr Usha (eds.). (1979). *The Bakhshali Manuscript*. Dr Ratna Kumari Svadhyaya Sansthan, Allahabad.

Seife, Charles. (2000). *Zero – The Biography of a Dangerous Idea*. Penguin Books.

Staal, Frits. (2010). On the Origins of Zero. In *Studies in the History of Indian Mathematics*. C. S. Seshadri (ed.) Hindustan Book Agency.

CHAPTER 9

# The Significance of Zero in Jaina Mathematics

*Anupam Jain*

## Abstract

The term 'Jaina mathematics' represents the mathematical concepts and ideas available in Jaina canonical and non-canonical literature. The literature related to the Karaṇānuyoga section of the Digambara Jaina tradition and the Gaṇītānuyoga section of the Śvetāmbara Jaina tradition have much content to hold mathematicians' interest. The classification of numbers in the three groups Saṃkhyāta (countable), Asaṃkhyāta (uncountable) and Ananta (infinite) can be found in ancient Jaina literature. The definition of Utkṛṣṭa saṃkhyāta (maximum countable) can be given only with the help of zero. Without the knowledge of numbers and zero it is not possible to define Uṭkṛṣta saṃkhyāta and other related quantities, such as Acalātma and Śīrṣaprahelikā. But we find the use of zero in ancient mathematical texts such as the Bakhshali manuscript, texts of the Digambara Jaina tradition, such as Ṣaṭakhaṇḍāgama, Pancāstikāya, Mahābandha, Lokavibhāga, Tiloyapaṇṇattī, and the Dhavalā Commentary etc., and texts of the Śvetāmbara Jaina tradition such as Sthānāṃga sūtra, Jambūdvīpaprajñapti, Anuyogadvāra sūtra, Āvaśyaka Niryukti, Bṛahatkṣetra Samāsa, etc., in the mathematical sense in many places.

All these references are explored and analyzed in this chapter to give a comprehensive picture to the academic world. The study indicates that the use of zero and the decimal place value system was popular in the Jaina tradition during the first and second centuries CE.

## Keywords

Ṣaṭkhaṇḍāgama – Anuyogadvāra sūtra – Tiloyapaṇṇattī – Lokavībhāga – Bakhshālī manuscript – Saṃkhyāta – Acalātma – *Śūnya*

## 1 Introduction

India has a glorious past and Indian scholars have made significant contributions in the development of science. Although it is an established fact that

 | DOI:10.1163/9789004691568_012

religion and philosophy were the major fields of interest, Indian scholars also significantly contributed to the field of science and technology, especially in the field of mathematics and astronomy.

India has been a religious country since antiquity and therefore much importance was given to spiritualism. The utility of a subject was proportionate to its use in understanding religion and philosophy, and the performing of religious ceremonies. Because of these very special circumstances, religious centers and texts are very important sources that reveal the development of different disciplines. The two main cultures of India were the *Śramaṇa* (ascetic) and the *Vedic* (ritualistic). The reference to *Śramaṇa* culture in Vedas indicates its presence in the Vedic era. The Jaina tradition is one of the two streams of *Śramaṇa* culture.

The concept of *Śūnya* in India has a long history in various dimensions in mathematics, philosophy and mysticism (Bag and Sharma, 2003).

In mathematical literature, *Śūnya* is used in the sense of zero having no substantial numerical value of its own but playing the key role in the system of decimal notation, to express all numbers with nine digits: one to nine and *Śūnya* as the tenth digit. The application of *Śūnya* in this system of notations was discovered in India sometime in the pre-Christian era. Its concretization in the form of a dot or a small circle and its use in the decimal place value system were first transmitted to the Middle East and then to Europe to supplement the Greek and Roman systems. This system of numeration was recognized as the most scientific system by the whole world in the course of time. Before discussing the available literature, we first look at the inscriptional evidence. It is well known that the Brāhmī script was named after the elder daughter Brāhmī of the first Jaina Tīrthaṃkara Ādinātha . Some typical Brāhmī numbers, representing the signs 10 and 7, were found in a second century BCE inscription in a Buddhist cave at Nana Ghat (Maharastra).[1]

The earliest extant physical example of decimal place value numerals is found in an inscription at the Chaturbhujanath temple of Gwalior (after 500 CE). It mentions the number 270 in Brāhmī script on the top.

Another inscriptional piece of evidence is the Gurjara Grant plate (Gurjara Dānapatra), dated 585 CE, from Sankhedā, in which *346 Vikram Samvat* is inscribed in decimal place value notation. A detailed list of relevant inscriptions with the year and other details is given by Datta and Singh (1962, pp. 40–44).

---

1 A photograph of the inscription can be seen in Plofker (2012).

FIGURE 9.1
Inscription from Gwalior
SOURCE: (PLOFKER, 2012, P. 46)

There are several other inscriptions of the King Ashoka era that show the use of numerals in *Kharoṣṭhī* but for space reasons we will not go into detail here (see Mukherjee, 2003).

Evidence shows that Jaina literature is a rich source of mathematical knowledge. The use of the decimal place value system and zero was very common during the first to fifth centuries CE in India. In Jaina tradition we find the use of zero in the following texts:

## 2 The *Ṣaṭkhaṇḍāgama* and Mahābandha (First to Second Century CE)

The *Ṣaṭkhaṇḍāgama* was written by Puṣpadanta and Bhūtabali (first century CE) on the basis of knowledge provided by Ācārya Dharasena. Thus, the contents of the *Ṣaṭkhaṇḍāgama* were based on traditional knowledge that was converted into text form in the first century CE. The initial 177 sūtras were written by Puṣpadanta and the remaining portion was composed by Bhūtabali. It has six parts (*Khaṇdas*); the final part is very big and known as the *Mahābandha*, also composed by Bhūtabali.

A remark by renowned historian of mathematics, R. C. Gupta, is worth quoting here:

> It thus appears that the background was ready in India to evolve the decimal place value notation. And it is believed that the decimal positional system of numerals was available in India before or around the beginning of the present Christian era. This is supported by relevant material found in some ancient Jaina works, namely 'Anuyogadvāra sūtra', 'Parikarma' (Kundakunda's lost commentary on the 'Ṣaṭkhaṇḍāgama'),

> 'Saṃtakamma-paṃjiya' and 'Mahābandha', dating back to 100 CE or before. (Gupta, 2003, p. 149)[2]

Here we quote some sūtras from the *Ṣaṭkhaṇḍāgama*.

> *Oghena micchāiṭṭī davvapamāṇena kevadiyā', Aṇaṃtā.*
> Ṣaṭkhaṇḍāgama, Verse 1.2.2, p. 10

> [How many are the group of wrong faith souls? Infinity.]

> *Sajogikevalī Davvapamāṇena kevadiyā, Pavesaṇena Ekko vā, Do vā Tiṇṇi vā, Uakkasseṇa Aṭṭuttara Sayaṃ.*
> Ṣaṭkhaṇḍāgama, Verse 1.2.13, p. 95

> [How many are the omniscient souls with physical bodies (Sayogi Kevalī Jīva) with respect to number? At the time of beginning, they are one or two or three and in this way maximum one hundred and eight, numerically as 108.]

> *Caduṇhamuvasāmagā davvapamāṇena kevadiyā, paveseṇa ekko vā do vā tiṇṇi vā ukkasseṇa cauavaṇṇaṃ.*
> Ṣaṭkhaṇḍāgama, Verse 1.2.9, p. 90

> [How many are the souls in the suppressor-ladder of Guṇasthāns (Upśamakajīva) in all the four stages of spiritual development (Guṇa sthāna). At the time of admission one or two or three and in this way maximum fifty-four, numerically as 54.]

> *Maṇusapajjattesu micchāiṭṭī davavapamāṇea kevadiyā. Kodākodākodīe uavari kodākodākodākodīe heṭṭado chaṇhaṃ vaggāṇa muvari sattaṇhaṃ vaggāṇāṃ heṭṭṇado.*
> Ṣaṭkhaṇḍāgama, Verse 1.2.45, p. 253

> [How many are human souls of the wrong faith among human Paryāpti? These are above Kodākodākodī ($10^{21}$) and below Kodākodākodākodī ($10^{28}$). In other words, above $2^{2^6}$ and below $2^{2^7}$.]

2 For more details of Kundakunda's contribution, see Jain and Jain (1989).

In the above sūtras, use of the words Kodākodākodī and Kodākodākodākodī and their valuations are important.

> *Khetteṇa padarassa vechappaṇṇaṃgulasayavaggapadibhāgeṇa.*
> Ṣaṭkhaṇḍāgama, Verse 1.2.55, p. 268

> [In respect to space, when the square of 256 aṃgulas divides the Jagapratara then we get the total number of the Devās of the wrong faith.]

Here the use of the number 256 and its square show that Ācāryas were well acquainted with the decimal place value system at that time.

> *Khetteṇa padarassa saṃkhejjajoyaṇasadavaggapadibhāgeṇa.*
> Ṣaṭkhaṇḍāgama, Verse 1.2.63, p. 272

In respect to space, when we divide Jagapratara by the square of countable hundred (100) yojana then we get the number of wrong faith, Vānavyantara Devās.

Here the use of 100 is important.

> *Atta choddasbhāga vā desūnā.*
> Ṣaṭkhaṇḍāgama, Verse 1.4.6, p. 166

In respect to the previous period, souls of the right and wrong faith touch only 8/14 parts.

Thus, we find that the compilers of *Ṣaṭkhaṇḍāgama* were well versed in decimal numerals, including zero. In accordance with Jaina tradition, Ācārya Dharasena provided traditional knowledge directly to Puṣpadanta and Bhūtabali, who compiled it. Therefore, we can say that knowledge of the place value system and zero existed in Jaina tradition in the first century CE.

Ancient Jaina Ācāryas were well versed in the classification of numbers in three parts:

1. Saṃkhyāta: countable or numerable
2. Asaṃkhyāta: uncountable or innumerable
3. Ananta: infinities

These were further classified in 3, 9, and 9 parts (*Tiloyapaṇṇattī*, 1986, vol. 2, chapter 4, pp. 90–101 and *Anuyogadvāra Sūtra*, Byavar Ed., 1987, pp. 409–22). Certainly, in these texts of metaphysics there was no need to give the mathematical calculations for obtaining numerical values, but they were given in other texts of the tradition such as *Tiloyapaṇṇaṭṭī* (second to seventh centuries CE) or *Anuyogadvāra Sūtra* (second century CE) or *Dhavalā*. The use of the words *Ananta* and *Kodākodākodī* is important. They show that Ācāryas were able to understand infinite quantities and countable numbers. However, they were unable to make use of this concept to explain other things due to their lack of knowledge in detail.

Time units were certainly defined on the basis of this knowledge.

Innumerable Samaya = 1 Āvali
Numerable Āvali = 12 Ucchvāsa
= 2880/3773 seconds
7 Ucchvāsa = 1 Stoka
7 Stoka = 12 Lava = $37\frac{31}{77}$ seconds
$38\frac{1}{2}$ Lava = 1 Nādī [Ghatī] = 24 minutes
2 Nādī = 1 Muhūrta = 48 minutes
30 Muhūrta = 1 Ahorātri
15 Ahorātri = 1 Pakṣa
2 Pakṣa = 1 Māsa [month]
2 Māsa = 1 Ṛtu
3 Ṛtu = 1 Ayana
2 Ayana = 1 Varṣa

After it, *Varṣa* became the base unit but the Digambara and Svetāmbara traditions differed in some respect.

According to Digāmbara traditions (*Tiloyapaṇṇaṭṭī*, Ch. 4, Verse 312, pp. 89–90):

$84^{31} \times 10^{90}$ Varṣa = 1 Acalātma
$\approx 4.49 \times 10^{149}$ Varṣa

According to Śvetāmbara traditions (Mahendra Kumar, 1969, p. 245):

$84^{28} \times 10^{140}$ Varṣa = 1 Sīrśa Prahelikā
$\approx 7.58 \times 10^{193}$ Varṣa

This Acalātma or Śīrṣaprahelikā is a countable number[3] and Uṭkṛṣta Saṃkyāta is the biggest countable number (say S). Then *Jaghanya Parīta Asaṃkhyāta* (say A) will be:[4]

$$A = S + 1$$

3 For more details of Acalātma and Sīrṣaprahelikā, see Jain, 2012, pp. 496–509; and *Anuyogadvāra Sūtra*, Ladnun Ed., 1996, ch. 10, verse 417, pp. 257–258.

4 *Tiloyapaṇṇaṭṭī* explanation after Ch. 4, verse 313, pp. 91–93.

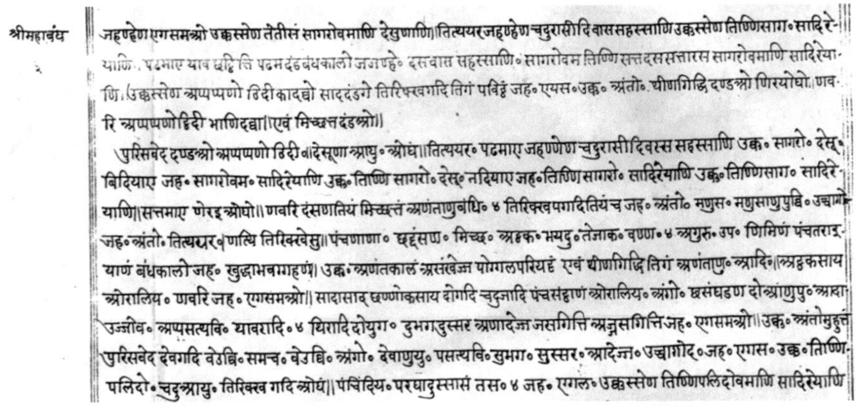
श्रीमहाबंध
जहण्णेण एगसमओ उक्कस्सेण तेतीसं सागरोवमाणि देसूणाणि॥ तित्थयर जहण्णेण चदुरासीदि वाससहस्साणि उक्कस्सेण तिण्णिसाग० सादिरे-
याणि। पढमाए याव छट्ठि त्ति पढमदंडबंधकालो जहण्णे० दसवास सहस्साणि० सागरोवम तिण्णि सत्त दस सत्तारस सागरोवमाणि सादिरेया-
णि। उक्कस्सेण अप्पप्पणो ट्ठिदी कादव्वो सादंडगे तिरिक्खगदि तिगं पविट्ठं जह० एयस० उक्क० अंतो० थीणगिद्धि दण्डओ णिरयोघो। णव-
रि अप्पप्पणो ट्ठिदी भाणिदव्वा॥ एवं मिच्छत्त दंडओ॥
पुरिसवेद दण्डओ अप्पप्पणो ट्ठिदी॥ देसूणा आयु० ओघं॥ तित्थयर० पढमाए जहण्णेण चदुरासीदिवस्स सहस्साणि उक्क० सागरो० देसू०
बिदियाए जह० सागरोवम० सादिरेयाणि उक्क० तिण्णि सागरो० देसू० तदियाए जह० तिण्णि सागरो० सादिरेयाणि उक्क० तिण्णिसाग० सादिरे-
याणि॥ सत्तमाए णेरइ ओघो॥ णवरि दंसणतियं मिच्छत्तं अणंताणुबंधि० ४ तिरिक्खगदितियं च जह० अंतो० मणुस० मणुसाणुपुव्वि० उच्चागो०
जह० अंतो० तित्थयर० णत्थि तिरिक्खेसु॥ पंचणाणा० छद्दंसण० मिच्छ० अट्ठक० भयदु० तेजाक० वण्ण० ४ अगुरु० उप० णिमिणं पंचतराइ-
याणं बंधकालो जह० खुद्दाभवग्गहणं। उक्क० अणंतकालं असंखेज्ज पोग्गलपरियट्टं एवं थीणगिद्धि तिगं अणंताणु० आदि०॥ अट्ठकसाय
ओरालिय० णवरि जह० एगसमओ॥ सादासाद छण्णोकसाय दोगदि चदुजादि पंचसंट्ठाणं ओरालिय० अंगो० छसंघडण दोआणुपु० आदा
उज्जोव० अप्पसत्थवि० थावरादि० ४ थिरादि दोयुग० दुभग दुस्सर अणादेज्ज जसगित्ति अजसगित्ति जह० एगसमओ॥ उक्क० अंतोमुहुत्तं
पुरिसवेद देवगदि वेउव्वि० समच० वेउव्वि० अंगो० देवाणुपु० पसत्थवि० सुभग० सुस्सर० आदेज्ज० उच्चागोद० जह० एगस० उक्क० तिण्णि-
पलिदो० चदुआयु० तिरिक्खगदि ओघं॥ पंचिंदिय० परघादुस्सासं तस० ४ जह० एगस० उक्कस्सेण तिण्णिपलिदोवमाणि सादिरेयाणि

FIGURE 9.2 Specimen page of the Devanāgri transcript of Mahābandha written in Kānari script, twelfth century CE, with use of zero (circle) for lopa
Transcribed by Pt. Jeevarāja Jain, Nauagava (Bansvara Rajasthan) on 30 September 1945

*Mahābandha* (*Mahādhavalā*) written by Ācārya Bhūtabali (*c.*100 CE), the disciple of Ācārya Dharasena, mentions zero in the form of a circle but it was used for void only and did not have any place value (Bhūtabali, 2007, pp. 37–39, and many other places).

R. C. Gupta wrote: 'The use of zero symbol to fill the blank space is also found in *Mahābandha*.' (Gupta, 1995, p. 58). Ācārya Pūjyapāda (fifth to sixth century CE) also uses the word *khaṃ* for the *lopa* in *Jainendra Vyākaraṇa* (sūtra VIII 4.64).[5] Hence, it is clear that Dharasena was well acquainted with the place value system and zero having positional values.

## 3 Pancāstikāya and Parikarmasūtra

Ācārya Kundakunda is the most revered Ācārya in Jaina tradition. Apart from his other books like Samayasāra etc., *Pancāstikāya* and *Parikarma sūtra* (first and second centuries CE) are more important from the mathematics point of view.

In Kundakunda's *Pancāstikāya* of we find the following two verses:

*Sattāsavvapayatthā savissarūvā aṇaṃtpajjāyā*
*Bhaṃguppāda dhuvattā aappdivakkhā havadi ekkā*

Pancāstikāya, verse 8, p. 18

5 See also Gupta (2005, p. 21).

> [[Each eternally] existing [substance] is endowed by permanence with creation and destruction [of forms], [each unit of such existence is] one … [It undergoes through] infinite number of forms.]

> *Agurulahugā aṇantātehi aṇaṃtehiṃpariṇadā saue.*
> *Desehiṃ asaṃkhādāsiyalogaṃ aavvamāvaṇṇā*
> Pancāstikāya, verse 31, p. 41

> [All souls have infinite attributes having absence-of-increase-and-absence-of-decrease property and transform in infinite forms; each soul has innumerable number of Pradeśa.]

The use of the word *Anṃanta* and *Asaṃkhādā* is also important as it shows that Ācārya Kundakunda was familiar with the difference between uncountable and infinite.

The smallest unit of time, *samaya,* is defined in Jaina tradition as 'time taken by a Paramāṇu (indivisible particle of matter) to travel from one Pradeśa (smallest unit of space) to another Pradeśa with *manda gati* (minimum speed)'.[6] From this definition of *samaya,* we define maximum countable, uncountable and infinite.

Ācārya Kundakunda is given credit for writing the text *Parikarma sūtra* (*Pariyamma Sutta*). This commentary of the first three parts of *Ṣaṭkhaṇḍāgama* gives mathematical explanation and hence a mathematical background is required to understand this text. Unfortunately, this text is not available at present.

In what follows, we quote certain references given by Ācārya Vīrāsena in his famous Dhavalā commentary. The other commentaries, written by Samantabhadra, Sāmakuṇda, Tumbulūra, and Bappadeva, are currently not available. We may hope they also have the quotes of this important text/commentary by Kundakunda.

## 4 The Dhavalā Commentary

In the Dhavalā commentary on *Ṣaṭkhaṇḍāgama* (816 CE), which is generally quoted as a standard text, we find more than 20 references to *Parikarma Sūtra* or *Pariyamma Sutta*. Of these references here are a few.

---

6 Also its samaya commentary. And see *Tiloyapaṇṇattī*, Ch. 4, verse 288, p. 82.

*Pariyamma Sutteṇa saha virujjhai*

Commentary of Sūtra, 1.2.4 in Dhavalā, Book 3, p. 36

*Taṃ Pariyamme vāuttaṃa*

Commentary of Sūtra, 1.2.15 in Dhavalā, Book 3, p. 124

*Tti Pariyamma vayaṇena*

Commentary of Sūtra, 1.5.114 in Dhavalā, Book 4, p. 390

More details of such references are available in Jain and Jain, 1989.

The *Dhavalā* commentary itself is full of references that show the use of zero. For example:

– 'Jo maccho joyaṇa sahassio'[7] has the use of zero in 1,000, and again
– 'Aṁkado vi attiyāṃ havaṁti' also has the use of zero in 29,699,103

Similarly, we can find many other references in the *Dhavalā* Books 3 and 4. It requires checking whether in old manuscripts these are written in numerals or not.

## 5 Tiloyapaṇṇattī of Yativṛṣabha (Second Century CE)

This is an important text of *karaṇānuyoga* group in which we find the structure and other details of the cosmos. In this process, the length, area and volume are discussed and many formulae are given. It was originally composed by Ācārya Yativṛṣabha in 176 CE (Shastri, 1974, p. 87; Saraswati, 1979, p. 10). However, due to many additions made in different time periods, L. C. Jain placed it as during 473–609 CE (Jain, 1958). But in any case it appears that the use of zero is not an later addition in *Tiloyapaṇṇaṭṭī*.

Zero and the decimal place value system are used everywhere in *Tiloyapaṇṇaṭṭī*. For example:

*Suṇṇa-ṇaba-Gayaṇa-Duga-Ekka-KhaTiyaSuṇṇa-Ṇava-ṆahāSuṇṇaṃ* I
*Chakkekka-JoyaṇāciyaAṃka Kame ManuvaLoya-Khettaphalaṁ* II

Tiloyapaṇṇattī, 1986, ch. 4, verse 8, p. 3

[The numerical value of the area of human world is 16,009,030,125,000 yojanas.]

Many other references are available, such as 1/123, 1/124, 2/89, 3/80, etc.

---

7 Dhavalā of *Ṣaṭkhaṇḍāgama*, Book 3, pp. 37 and 80, and other places.

## 6 Anuyogadvāra Sūtra

It is a part of Jaina canonical literature under the major section *cūlikā sūtra* and contains the knowledge of zero that was available before the Christian era. It was composed by Āryarakṣita (born 5 BCE), who rearranged the existing knowledge during the first century CE (*Anuyogadvāra Sūtra*, Ladnun Edition, 1996, Bhumikā, p. 17). There may be some modifications later on but the existing form of *Anuyogadvāra Sūtra* is from no later than the *Vallabhī* conclave (456 CE) in which all the existing canonical literature was rearranged.

The parts of *Anuyogadvāra Sūtra* that are relevant to the present topic are as follows:

1. Starting from the smallest unit of time, *samaya*, we get *Śīrṣaprahelikā*.[8]

1*Sīrṣa Prahelikā* = $(8{,}400{,}000)^{28}$ years
$= 84^{28} \times 10^{140}$ years
$\approx$ 758, 263, 253, 073, 010, 241, 157,
973, 569, 975, 696, 406, 218, 966,
848, 080, 183, 296 $\times 10^{140}$ years
$= 7.58 \times 10^{193}$ years

However, this value using decimals (Shah, 2007, p. 83) is not found in the original *Anuyogadvāra Sūtra*, but given in the commentary of Abhaideva sūri (1015–1078 CE). The use of 8,400,000 is itself important.

2. In the process of explaining counting, we find the use of Gaṇītamāna eka, dasa, sata, sahaṣsa, dasa saḥassa, saya sahassa, dasasayasasahassa, Kodi, Dasakodi, Sayakodi, Dasasayakodi. In this way it goes up to $10^{10}$, where the term Kodākodī (Shah, 2007, p. 83) used in several places, and means $10^7 \times 10^7 = 10^{14}$.

Obviously, such counting is not possible without the knowledge of the decimal place value system.

Now we quote a paragraph from *Anuyogadvāra Sūtra*:[9]

> *Maṇūsāṇāṃ Bhaṃte! Kevaiyā Orāliyasarīrā Paṇnattā? Goyamā Duvihā Paṇṇatttā Taṃ jahā baddhellayā Ya Mukkellayā Ya Tattha ṇaṃjete baddhellayā Teṇaṃ siya saṃkhejjāsiya asaṃkhejjā Jahaṇṇapae saṃkhejjā, saṃkhejjāo kodīo eguṇatīsaṃ Thāṇāiṃ Tijamalpayassa uavariṃ*

8 *Anuyogadvāra Sūtra*, Byavar Ed., 1987; *Anuyogadvāra Sūtra*, Ladnun Ed., 1996, Ch. 10, Verse 417, pp. 257–258 and 278–279.

9 *Anuyogadvāra Sūtra*, Ladnun Ed., 1996, Sūtra 490, pp. 273 and 285–286.

*Chaujamalapayassa, heṭṭhā, ahavaṇaṃ chaṭṭo vaggo, panchamvaggapaduppaṇṇo ahavaṇaṃ Chṇnaui cheyaṇagādā irāsī*

To explain the human population, the number is written with 29 digits. This number is:

$$2^{2^{6}} \times 2^{2^{5}} = 2^{64} \times 2^{32} = 2^{96}$$
$$= 79, 228, 162, 514, 264, 337, 593, 543, 950, 336$$

Here, zero has positional value. However, it is true that this value is given by the commentator. Hence it seems that the compilers and commentators of *Anuyogadvāra Sūtra*. were well acquainted with Decimal Place Value System.

## 7 Loka-Vibhāga

The present edition of *Loka-Vibhāga* is in saṃskṛta and was written by Siṃhasūṛarṣi (eleventh century CE). It is to be noted that this edition has only a language change from the ancient text under the same name written by Ācārya Sarvnandi in Prākṛta.

The Praśasti of the *Loka-Vibhāga* states (*Loka-Vibhāga*, 2001, p. 225):

*Bhavyebhyaḥ Suramānuṣorusadasi Śrīvardhamānārhatā*
*Yatproktaṃ Jagato Vidhānmakhilaṃ Jñātaṁ sudharmādibhiḥ*
*Ācāryāvalikāgataṃ Vircitaṃ tatsiṃhsūraāṣiṇā*
*Bhāṣāyaḥ Parivartanena Nipuṇaih Saṁmānyatāṃ Sādhubhiḥ.51.*

The details of the structure of the cosmos are explained by Shri Vardhaman Jinendra in the great assembly of human beings and deities (known as Samavaśaraṇa) and which are well understood by Sudharma and other Gaṇadharas, who transmitted it to the great tradition of Ācāryas. Again, that structure of the cosmos by Siṃha Sūri was written only in translation of the original language. It is respected by scholars.

*Vaiśve sthite ravisute Vṛṣbhe ca jīve*
*Rajottareṣu sitapakṣamupetya candre I*
*Grame ca Pātalikanāmanli Pāṇarāṣtre*
*Śāstraṃ purā likhitāvān Muni Sarvanandī II.52.*

When Saturn was above the Uttarāsāḍa Nakṣatra, Jupiter was above the Taurus (Vṛsa), raśi, in the bright half-moon, was above the Phalguni Nakṣatra then in the village of Pātalika of the Pāṇa nation (Rāṣtra).

It was originally written by Muni Sarvanandi.

> *Saṃvatsare tu Dvāużṃśe kā ciśaḥ Siṃhvaramaṇah*
> *Aśītyagre Śakābdānāṃ Siddha metacchattraye.53.*

This work was completed in the twenty-second year of the reign of King Simhaverma, that is in Śaka Saṃvata 380 (458 CE).

On this basis, Georges Ifrah concluded the work was composed on 25 August 458 CE (Ifrah, 2000, p. 419). I agree that many ślokas were added to the text of the *Loka-Vibhāga* at the time of Simhsuri Riṣi, but this does not affect my claim because zero was used in number of places in the text and cannot all be taken as new addition (Kṣepaka).

I quote below the *saṃskṛta* version of the relevant verse of *Loka-Vibhāga*:

> *Paṃchabhyah khalu Śūnyebhyaḥ Paraṃ Dve Sapta Cāmbaram*
> *Ekaṃ Tṛīṇi ca rūpaṃ Ca cakravālasya Pārthavam.*
> *Loka-Vibhāga, 2001, Ch. 4, verse 56, p. 79*

When we write it from right to left in accordance with Jaina tradition (***Aṃkānāṁ Vāmto Gatiḥ***), we find five voids (zeros), then two, seven, sky (zero), one, three and one. Mathematically, this is 13,107,200,000.

This is confirmative reference, but is not the only reference showing the use of zero and place value system in this text. Some other references are the given below written in the following order: chapter no. verse no./page no. – thus 1.83/9 means chapter 1, verse 83 on page 9.

> 1.7/1, 1.52/6, 1.53/6, 1.63/7, 1.65/7, 1.68/8, 1.69/8, 1.83/9, 1.86/10, 1.100/12, 1.103/12, 1.118/13, 1.114/13, 1.12515, 1.137/16, 3.40/66, 3.42/66, 3.56/68, 4.34/76, 4.36/79, 4.56/79, 6.98/101, 6.212/130, 8.33/149, 9.66/171, and 10.49/179 are important.

All these references can't be ignored, implying, thereby, that *Lokavibhāga* is an authentic source of the use of zero within a place value system. This work is contemporary to the famous *Vallabhī* Conference (454–456 CE).

## 8 Āvaśyaka Niryukti

There were two famous scholars in the name of Bhadrabāhu in the Jaina tradition. The first was Śrutakevalī Bhadrabāhu (fourth century BCE) and the second was Niryuktikāra Bhadrabāhu (fifth to sixth century CE). *Āvaśyaka Niryukti* was written by Niryuktikāra Bhadrabāhu, a contemporary of the famous astronomer Varāhamihira, in the sixth century CE.

The following verse is quoted by Jinabhadragaṇi Kṣamāṣramaṇa. (609 CE) in his Viśeṣāvaśyaka bhāṣya:

> *Thibugāgāra Jahanno, Vaṭṭo Ukkosamāyaṁo Kincī*
> *Ajahaṇnamṇukkoso, ya khelto ṇegasaṃdhaṇo.52.*[10]
> *Viśeṣāvaśyaka bhāṣya, verse 702, p. 197*

The figure of being for a minimum period has the shape of a water drop (*bindu*) and the shape of being for a maximum period is rectangular. For one that is neither maximum nor minimum it may be of any type. In Hemcandra Sūri's commentary, the word *Thibuga* is translated as zero.

H. R. Kapadia wrote: 'What is the true radical significance of the word *Thibuga* and in what sense has it been employed in the above passage?' (Kapadia, 1937). The commentator Hemcandra Sūri is of the opinion that it signifies *bindu*. Is it then the zero of the decimal numeral notation? If so, it will have to be admitted that the modern decimal place value notation was known in India in the fourth century BCE.

It seems that H. R. Kapadia had confused the author Niryuktikāra Bhadrabāhu with Bhadrabāhu-I (fourth century BCE). Needless to say, the use of zero was also popular in the sixth century CE.

## 9 Bṛhat KṣetraSamāsa of Jinabhadragaṇī (609 CE)

*Jinabhadragaṇi kṣamāṣramaṇa* is very famous for his Bhāṣyas, especially *Viśeṣāvaśyaka Bhāṣya*. *Bāhat Kṣetra Samāsa* is another important work that contains many references to zero and the decimal place value system (Kṣamāśramaṇa, 1920). Some examples are given below:

> *Veyadadha Jīvavaggo, Sattāṇaluaī sahassa Paṃca Sayā*
> *Auṇāpaṇnaṃ kodī egayālīsaṃ ca kodisayā. 68.*

10 Also *Āvaśyaka Niryukti* (Gāthā 52, p. 29); Kapadia (1937, p. xxi).

Mathematically, this represents 41,490,097,500.

> *Bharahada dh Jīvavaggo Padasayarī chacca Atta Sunnaīṃ*
> *Culle Jīvāvaggo, Dulīsa Coyāla Sunnaṭṭa. 69.*

Mathematically, this represents 75,600,000,000 and 224,400,000,000.

> *Jīvāuaggigavaṇṇā Cauvīsaṃ atta Suṇṇa Hemvae*
> *Pancahiyaṃ Sayamegaṃ mahahimave das ya sunnāiṃ.70.*

Mathematically, this represents 1,050,000,000,000.

Similarly, verses 71, 72, 90, 98, 99, 101, 102, 108, 109, 112, 113, 119, 124, 125, 126, etc. of *Bāhat Kṣetra* show clear evidence of the use of zero within a place value system.

We reproduce here a specimen page (bearing the zero symbol) of an old manuscript of *Bāhat Kṣetra samāsa* of *Jinabhadragani* with the *Vṛtti of Hemcandra Sūri* obtained from Mahāvīra Ārādhanā Kendra (Kailash Sagar Sūri Gyanbhandar) – Koba. The complete copy is available from the author of this chapter and can also be seen in Koba (Gujarat).

We now come to an important mathematical text that, however, may or may not be a Jaina text.

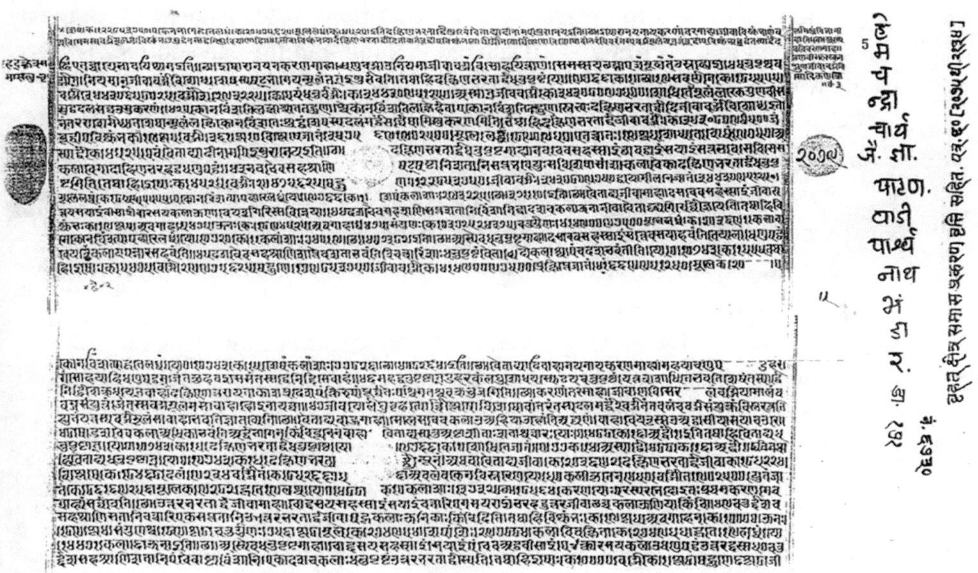

FIGURE 9.3 A specimen page (bearing the zero symbol) of an old manuscript of *Bāhat Kṣetra samāsa* of *Jinabhadragani* with the *Vṛtti of Hemcandra Sūri*
COURTESY OF KAILASH SAGAR SURI GYAN BHAṂDĀRA-KOBA

## 10 The Bakhshali Manuscript

This incomplete mathematical manuscript, excavated in 1881 by a farmer while farming his land near the village of Bakhshali about 80 km away from Peshawar (presently in Pakistan), is an important source of the mathematical knowledge in practice during the third and fourth centuries CE. Written in Prākṛta mixed with saṃskṛta it uses sāradā script and comprises 70 folios on *bhurja patra*, of which 19 are totally blank. It is presently preserved in the Bodleian Library of Oxford University, Oxford. Renowned Indologists Hornelè, Bühler, Cantor and Cajori studied and critically examined the manuscript, placed it during the third and fourth centuries CE, although G. R. Kaye and Takao Hayashi placed it during the eighth to twelfth centuries CE.[11]

The Bakhshali manuscript was written in the pattern of Pātīgaṇita, also followed by Brahmgupta, Srīdhara and Mahāvīra. Apart from other topics, it deals with decimal place value systems, numerals and zero (Bag, 1979, p. 15).

It should be noted that many Jaina texts written in the early centuries of the Common Era or before are presently not available, for example *Parikarma Sūtra* of Ācārya Kundakunda. We are tempted to agree with B. B. Datta, who said: 'As Prākṛta is the sacred language of Jainas this manuscript [not the Bakhshali manuscript] may be related to Jaina tradition' (Datta, 1936).

The language of the Bakhshali manuscript is a mixture of Prākṛta and Saṃskāta. Since most of the Jaina literature is in Prākṛta and no mathematical literature is found in Prākṛta in other traditions, it leads us to think that the Bakhshali manuscript may be related to Jaina tradition. In support of this claim, we put some further internal evidence below.

Some words common in Jaina tradition that are found in the Bakhshali manuscript:[12]

1. *Kalāsavarṇa*: A technical word used for fraction, which was previously used in Sthānāṃga sūtra.
2. *Yāvata–Tāvata*: A technical word used for linear equation, previously used in the Ṣthānāṃga sūtra.
3. *Addacheda*: Meaning $\log_2$, used in *Tiloyapaṇṇattī* of Yativṛṣabha.
4. *Sundarī*: The name of the daughter of Lord Ṛṣabhadeva, whom number script is taught as per Ādipurāṇa.
5. *Pūrva*: A part of Jaina canonical literature included in the 12 aṁgas.
6. *Siddha*: A collective name for Jaina liberated souls used in the Ṇamokāra Mantra.

---

11 See Appendix 3 for an online petition urging the Bodleian Libraries, Oxford, UK, to take concrete steps to commission the necessary follow-up radiocarbon-dating of the Bakhshali Manuscript in the interest of scientific advancement in the field.

12 For details of the verses in which these words occur, see Hayashi (1985, pp. 391–432).

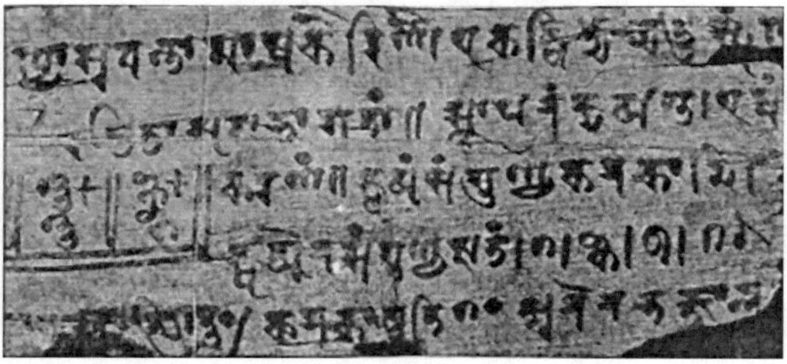

FIGURE 9.4
Folio 16, which contains data representing zeros, dates from 224–383 CE according to the radio-carbon dating results
COURTESY: INDIAN EXPRESS

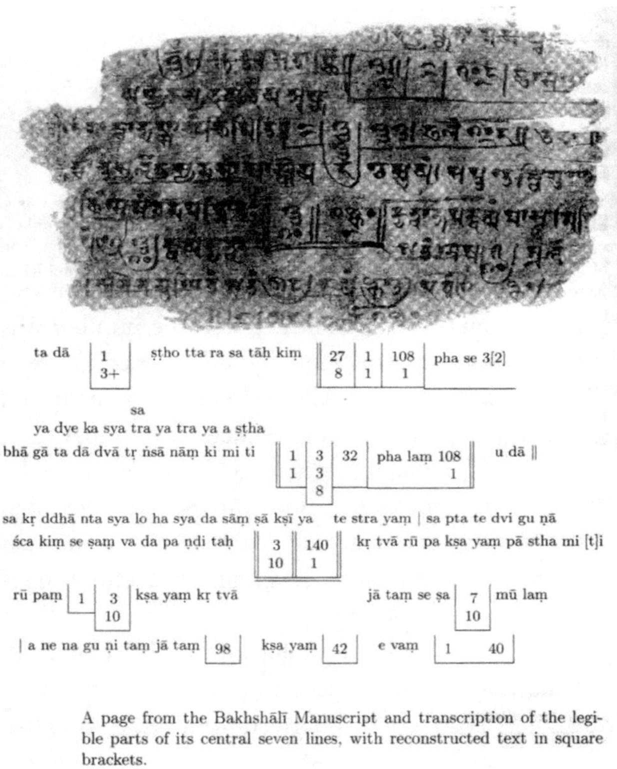


FIGURE 9.5
One page of the Bakhshali manuscript
COURTESY: PLOFKER, 2012, P. 160

Many other terms used in Bakhshali Manuscript are are very common in Jaina tradition but rarely used in other traditions. These internal proofs indicate that the Bakhshali Manuscript may have been a part of some Jaina text.

A report published in the *Indian Express* (Jaipur) on 19 September 2017 states among other things: 'In fact the Bakhshali manuscript contains material from different periods. It is actually composed of material from at least three periods some pages dating from as early as the third to fourth centuries and others dating eighth to tenth centuries, writes David Howell, Head of Heritage Science at the Bodleian Library.'

Next we examine a page from the Bakhshali manuscript and its contents, as reproduced in Plofker (2012).

It shows that zero was used in India during 224–383 CE in a form of **0** that has positional value. It should be noted that before the invention of the numeral place value system, Indian mathematicians used number names as seen in the Bakhshali manuscript, where the number

2 6 5 3 2 9 6 2 2 6 4 4 7 0 6 4 9 9 4 .... 8 3 2 1 8

is expressed in Shivkumar and Ananth (2018) as:

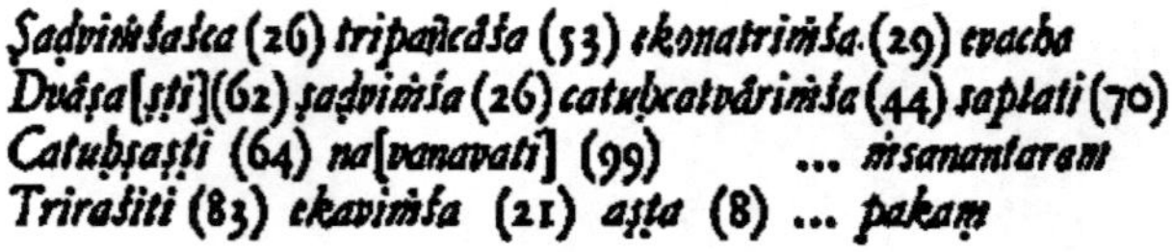

*Ṣaḍviṁśaśca* (26) *tripañcāśa* (53) *ekonatriṁśa* (29) *evacha*
*Dvāṣa*[*ṣṭi*](62) *ṣaḍviṁśa* (26) *catuḥcatvāriṁśa* (44) *saptati* (70)
*Catuḥṣaṣṭi* (64) *na*[*vanavati*] (99) ... *ṁsanantaram*
*Triraśiti* (83) *ekaviṁśa* (21) *aṣṭa* (8) ... *pakaṃ*

FIGURE 9.6 Extract from the Bakhshali manuscript
SOURCE: SHIVKUMAR AND ANANTH (2018)

The dots indicate some missing figures. This indicates that the number was expressed by grouping the digits and denoting them by their names. This system was later developed into word numerals whereby each number would be represented by names of things, or concepts, etc. Hence, the number 1 can be denoted by anything that is unique, such as the moon (*Candra*), 2 can be denoted by any pair, for example the eyes (*netra*). Accordingly, zero is denoted by words meaning void, sky, infinite, complete, etc., such as *sūnya*, *kha* and *gagana*. This system of word numerals attaches a meaning to the letters and is where the idea of zero as an empty void takes its root.

Finally, I quote R. C. Gupta: 'The Bakhshali Manuscript (A Mathematical Work) surely and explicitly used the decimal place value system with a zero symbol (dot or circle) but its date is uncertain' (as given from 200 to 1200 CE) (Gupta, 1995, p. 58).

On the Bakhshali manuscript, N. L. Maiti wrote: 'But it is evidence rather than indicative, of course circumstantial, in the (Bakhshālī) manuscript that the original MS could have been written in the late third century of the Christian era.' The analysis of a few problems, a few obscure words or terminologies found in the work, religious sects mentioned therein, etc., may shed new light on its age of composition (Maiti, 2007).

Regarding the place value system, the views of renowned historian B. B. Dutta are important. He wrote: 'The exact date of invention, however, would be nearer to the first century BCE or even earlier, because for a long time after its invention, the system must have been looked upon as a more curiosity and used simply for expressing large number' (Datta and Singh, 1962, p. 51).

His views are supported by literary and other evidence. More information on this subject can be obtained from the following sources:

1. New mathematical manuscripts (still unpublished) preserved in different Jaina Sāstra Bhaṃdārs (Jaina libraries).
2. Studies of copper plates in Jaina temples.
3. Inscriptions on the seat of Jaina idols (Pāda Pīṭha) as found in museums and Jaina temples.

We conclude with the remark that the concepts of zero and the decimal place value system were considered significant, understood well and in use during the first and second centuries CE. Jaina literature is full of such type of references. Dharasena, Kundakunda Yativṛṣabha and Āryarakṣita were probably the first people to use tḥem, but the knowledge lies in the tradition in the period before Christian era (Jain, 2005, pp. 124–37).

## References

Āryarakshita. *Anuyogādvāra Sūtra* (*Anuogadārāiṃ*). J. V. B. I. Ladnun Edition (1996).

Āryarakshita. *Anuyogadvāra Sūtra. Jināgama Granthmāla 28.* Byavar Edition (1987).

Bag, A. K. (1979). *Mathematics in Ancient and Medieval India.* Varanasi: Chaukhambha Orientalia.

Bag, A. K., and Sharma, S. R., ed. (2003). Flap, In *The Concept of Śūnya.* New Delhi: IGNCA, INSA and Aryan Books International.

Bhūtabali. (2007). *Mahābandha* (*Mahādhavala*) Book 1, Fifth Edition. New Delhi: Bhārtīya Jñānapīṭha.

Datta, B. B. (1936). A Lost Jain Treatise on Arithmetic. *Jaina Antiquary* (Arah), *1*(2), pp. 25–44.

Datta, B. B., and Singh, A. N. (1962). *History of Hindu Mathematics*, Parts I & II combined edition.

Gupta, R. C. (1995). Who Invented Zero?, *Gaṇita Bhāratī*, *17*(1–4).

Gupta, R. C. (2003). Zero in the Mathematical System of India. A. K. Bag and S. R. Sharma (Eds.), *The Concept of Śūnya*, New Delhi.

Gupta, R. C. (2005) Technology of using Śūnya in India. The concept of Śūnya. New Delhi.

Hayashi, T. (1985). *Bakhshali Manuscript.* PhD, microfilm print. Brown University.

Ifrah, G. (2000). Universal History of Numbers: from Pre-History to the Invention of Numbers, John Wiley & Sons.

Jain, A. (2012). *'Gaṇita' included in Jaina Dharma*, Paricāya, Delhi: Bhārtīya Jñānapīṭha, pp. 496–509.

Jain, L. C. (1958). *Tiloyapaṇṇattī kā Gaṇita*. Solapur: published with Jambūdīvapaṇṇattī saṃgaho.

Jain, L. C., and Jain, A. (1989). Was Kundakunda the Inventor of the Decimal Place Value System? (In Hindi). *Arhat Vacana* (Indore), *1*(3), pp. 7–15.

Jain, N. L. (2005). *The Concept of Śūnya (Zero) in Jaina canons*. Nandanvan: Parshvanath. Vidyapeeth, Varanasi.

Kapadia, H. R. (1937). *Introduction to Gaṇita Tilaka*. Baroda.

Kṣamāśramaṇa, J. (1920) *Bāhat Kṣetra Samāsa*. Bhavanagar: Jaina Dharma Prasarak Sabha (With commentary by Malayagiri.)

Kṣamāṣramaṇ, J. *Viśeṣāvaśyaka Bhāṣya*. Vaishali Edition (1972).

Kundakunda. *Pancāstikāya Vol 1*, Jaipur Edition (2014).

Kundakunda. Parikarma Sūtra (Pariyamma Sutta). Not available.

Maiti, N. L. (2007). Review of 'The History of Mathematics and Mathematicians of India' by D. Heroor Venugopal, *Gaṇita Bhāratī*, *29*(1–2), p. 140.

Manendra Kumar, M. 11. (1969). *Viśva Prahelikā, Ladnun*.

Mukherjee, B. N. Kharoṣṭī Numerals and the Early use of Decimal Notation in Indian Epigraphs, In Bag, A. K., and Sharma, S. R., ed. (2003). *The Concept of Śūnya*, pp. 87–100.

Plofker, K. (2012). *Mathematics in India*. Hindustan Book Agency.

Pūjyapāda, Jainendra Vyākaraṇa.

Puṣpadanta and Bhūtabali. *Ṣaṭkhaṇḍāgama*, Book No. 3. Solapur Edition (1993).

Saraswati, T. A. (1979). *Geometry in Ancient & Medieval India*, New Delhi: Motilal Banarsidas.

Shah, R. S. (2007) Mathematics of Anuyogadvāra Sūtra, Delhi: *Gaṇita Bhāratī*, 29(1–2).

Shastri, N. (1974). Tīthaṃkara Mahāvīra aura Unkī Ācārya Paramparā Sagar, vol. 2.

Shivkumar, N. and Ananth, M. (2018) *Zero – from nothing to everything*. Unpublished.

Siṃhasūṛarṣi. *Loka-Vibhāga* 2nd Edition. Solapur: Jaina Samskriti Samrakaka Sangh (2001).

Vīrāsena. Dhavalā commentary on *Ṣaṭkhaṇḍāgama*. Solapur Edition (1993).

Yativṛṣabha. *Tiloyapaṇṇattī*, vol. 2. Kota Edition (1986).

CHAPTER 10

# Notes on the Origin of the First Definition of Zero Consistent with Basic Physical Laws

*Jonathan J. Crabtree*

## Abstract

If mathematics is the language the universe was written in and mathematics is discovered rather than invented, then Brahmagupta's 628 CE Sanskrit text first defines a zero as old as the universe. Brahmagupta's definition of zero as a sum of equal and opposite negative and positive quantities is the first scientific definition of zero found to be consistent with laws of motion and particle physics. However, having originated in the East, the full power of India's symmetric and scientific zero failed to migrate West via the medieval Arabic world to Renaissance Europe. Thus, only a trivial mathematical concept of zero emerged, representing either nothing on its own or an arithmetical placeholder used alongside other numbers. Zero as an arbitrary midpoint for measurement purposes within a single quantity is similarly trivial, whether it be centuries BCE or CE, or a measure above or below a surveyor's foundation line as used for construction in ancient Egypt. Thus, the oldest original extant definition of zero compatible with laws of physics describing our universe originated with Brahmagupta in India 628 CE. Yet long lost in transmission and translation, the sad aftermath in classrooms today is a zero in which negatives to its left are arbitrarily less than positives to its right.

## Keywords

Brahmagupta – India – mathematics – zero – negatives – positives – physics – *śūnya*

## 1 Introduction

The entry for 'zero, n. and adj.' in the online *Oxford English Dictionary* (OED) defines zero in the sub-section 'Mathematics' as 'The absence of quantity considered as a number; nought'. This is followed by, 'The earliest example of zero considered as a number in its own right occurs in a manuscript by

 | DOI:10.1163/9789004691568_013

Indian mathematician Brahmagupta (598–668) dated to the seventh cent.' So, it seems we have our first mathematical definition of zero as a number, written in Sanskrit text as *śūnya* (pronounced shoonya) in 628 CE. For the first zero symbol as a circle, a popular meme on the internet is that Bhāskara I (*c*.600–*c*.680) the next year in 629 CE, in *Aryabhatiyabhashya*, a commentary on Aryabhata's work, was the first to use a circle for India's zero. Supposedly, Bhāskara I wrote 'nyâsaśca sthânânâm 0000000000', meaning 'writing down the places we have 0000000000' (Datta and Singh, 1962). Alas, such a simple finding for the first mathematical use of the symbol 0 is merely a mirage. More profound for mathematics students today, Brahmagupta's seventh century mathematical definition of zero (*śūnya*), led to nothing. The original mathematical zero Brahmagupta defined, as old as the universe, remains shrouded in equal parts philosophical sense and mathematical nonsense, while its symmetry continues to unlock secrets of our universe.

Just as the OED provides a trivial definition of zero, elsewhere we find definitions of zero falsely attributed to Brahmagupta such as 'the result of subtracting any number from itself' (Barrow, 2001, p. 38). This 'nothing remaining as a result of subtraction' and placeholder notion may have been an idea that reached the Arabic world on its way to Europe. Yet, as will be noted, Brahmagupta's zero is consistent with ideas such as conservation of matter and energy and Newton's third law, for every action there is an equal and opposite reaction.

On zero we read, 'We know that *śūnya* traveled from India to Europe via the *algorismus* texts, starting tenth century CE, and that the epistemological assimilation *śūnya* required some five to six hundred years' (Raju, 2007, p. 95). Yet, India's original mathematical definition of zero and its elementary applications born from empirical physical foundations were never fully transmitted from the ancient East to Western classrooms today.

Only the 'nothing' and placeholder concepts of zero made their way from India to the Arabic world and from there into Western pedagogies. For nearly all this time, zero was not considered a number mediating equal and opposite quantities in either the Arabic world or Europe. Around 300 BCE Euclid defined a number as 'a multitude composed of units' (Heath & Euclid, 1908, p. 277). So, zero was not alone in its struggle to be considered a number. For most of our Western history of mathematics via the Greeks, one was also not formally considered a number. In number theory, zero was only granted formal number status in the twentieth century. Yet accepting India's zero as both placeholder and a number in its own right still leaves zero's most important and powerful attribute, symmetry, in the dark.

In what might come to be described as a 'Black Swan' event (Taleb, 2007), Brahmagupta's original definition of zero failed to find its way out of India in time for the arrival of the printing press in the West. England, in particular, exported mathematics books to its settlements and colonies, with explanations largely built upon Greek (Euclidean) foundations. These foundations did not feature one as a number and both zero and negatives were absent. Thus, neither Brahmagupta's *zero* nor *one* came to be included in modern algorithmic definitions of multiplication and exponentiation (Crabtree, 2017a, p. 14). The aftermath has been a pedagogical Dark Age that continues to this day. Extraordinary claims require extraordinary evidence. Thus, we shine new light upon Brahmagupta's definition of zero, to see its embedded twin shadows of opposing *negative* and *positive* quantities emerge. Having a pedagogical problem with zero in 1968, the author literally began rebuilding elementary mathematics from zero in 1983. To the reader, the author's call for a major rewrite of elementary mathematics curricula might come as a shock. Yet, in hindsight, it might one day be seen as having been inevitable. Google reports hundreds of matches for the exact phrases, 'crisis in mathematics education' and 'fear of mathematics'. The social proof is real. When it comes to Brahmagupta's original mathematical definition of zero, in which negative is equal and opposite to positive (rather than less than positive), has history taught us nothing?

## 2 On Mathematics and Progress

Interviewed on the documentary *Infinite Secrets* about a manuscript of Archimedes that went missing, Dr Chris Rorres, Professor Emeritus of Mathematics at Drexel University, said: 'If we had been aware of the discoveries of Archimedes hundreds of years ago, we could have been on Mars today, we could have developed the computer that is as smart as a human being today, we could have accomplished all of the things that now people are predicting for a century from now' (Rorres, 2003).

Similarly, had Brahmagupta's definition of zero been understood and documented in the Arabic world within the writings of al-Khwārizmī and others, algebraic innovations might have occurred centuries sooner. Whether or not the absence of Brahmagupta's definition of zero in both the Arabic world and Europe held back scientific progress a century is just speculation. However, having deciphered Brahmagupta's cipher, many difficulties of great mathematicians from Diophantus to Descartes could have been avoided.

# 3 On the Pedagogical Absence of Zero and One

## 3.1 *Zero's Absence*

At school in 1968, age seven, the author knew how to count, add, and subtract. The teacher of Grade 2C, Miss Collins, said multiplication was like repeated addition. So, her explanation should have been simple, yet it wasn't. After writing 2 × 3 on the blackboard, Miss Collins asked her class, 'What is two added to itself three times?' Because I could understand that 2 added to 1 three times is seven (1 + 2 + 2 + 2), when Miss Collins chose me to answer 'What is two added to itself three times?' my answer was eight (2 + 2 + 2 + 2). Surprisingly, the English language rhetorical explanation given for 2 × 3 followed verbatim has led to eight for centuries. Miss Collins drew three 'hops' of 2 on the blackboard number line that landed on 6. However, the three hops started at India's zero. Pedagogically, explaining 2 × 3 via repeated addition requires more precise statements such as 'three twos added together' or 'two added to zero (not itself) three times', which is 0 + 2 + 2 + 2.

Many years later, it became obvious that Miss Collins' explanation of multiplication was a paraphrase of an incorrect translation of Euclid's 300 BCE definition of multiplication. This reads, 'a number is said to multiply a number when that which is multiplied is added to itself as many times as there are units in the other, and thus some number is produced' (Heath & Euclid, 1908, p. 278). Today, Euclid's definition of multiplication has been modernized (yet remains incorrect) to read 'to multiply **a** by integral **b** is to add **a** to itself **b** times' (Borowski & Borwein, 2012, p. 376). However, India's zero is needed in the definition of multiplication. Without a prior knowledge of multiplication facts, the English definition of multiplication dating back to 1570 would have us believe 2 × 1 does not equal 1 × 2. The former would be 'two added to itself once' which is four (2 + 2), while the latter would be 'one added to itself twice' which is three (1 + 1 + 1). Yet the Commutative Law (***ab*** = ***ba***) means the products of 2 × 1 and 1 × 2 must be the same. Returning India's zero to its rightful place in the English definition of multiplication fixes the anomaly, as 2 × 1 and 1 × 2 become two added to zero once and one added to zero twice respectively, both of which equal two. India's zero needs to be included in the integral definition of ***ab***. Thus, ***a*** multiplied by ***b*** is *either* ***a*** added to ZERO ***b*** times, *or*, ***a*** subtracted from ZERO ***b*** times, according to the sign of ***b***.

As we will see, Brahmagupta's scientific symmetric zero definition failed to be carried across cultures. We might think elementary mathematics has been carefully assembled and improved over time, yet the pedagogies in place today have emerged in large part due to ignorance of symmetric ideas in the

East. The writings of Bhāskara II (1114–1185) and Brahmagupta (598–668) on zero, negatives, and positives were translated into English language books too late to have any impact on pedagogies (Strachey, 1813; Colebrook, 1817; Taylor, 1816).

An author and former mathematics teacher asked:

> Do you really think that children were supposed to learn a problem like 1–2 five years after learning 2–1? The unwarranted separation of positive and negative numbers is the most glaring symptom that we merely mapped our historical misunderstanding of zero and negative numbers onto our math education, then called it a day. (Singh, 2021)

Englishmen published primary-level mathematics pedagogies built largely upon Greek foundations 1,000 years older than the writings of Brahmagupta. The British Empire then exported these out-of-date explanations of mathematics to their settlements and colonies. So, as English became the world's de facto language, the sub-optimal pedagogies sold in sixteenth/seventeenth-century England came to be disliked by customers (children) worldwide today. French pedagogies were superior, as you might expect from a country that would later champion the base-10 metric system. By way of example, around 200 years ago, an American mathematics professor wrote: 'The first principles, as well as the more difficult parts of Mathematics, have, it is thought, been more fully and clearly explained by the French elementary writers, than by the English' (Farrar, 1818, Preface).

## 4 Brahmagupta's Zero-Sum Definition

Chapter 18 of Brahmagupta's *Brāhmasphuṭasiddhānta*, was about algebra (*Kuttaka*). The section containing Brahmagupta's original definition of *śunyā* was titled *Dhanarṇa Śunyānām Samkalanam*, or 'calculations dealing with quantities bearing positive and negative signs and zero' (Prakash, 1968, p. 200). This section detailed the laws of sign for positives, negatives and zero. Within his laws of addition (*saṅkalana*), Brahmagupta defined zero as the sum of equal positive and negative (Plofker, 2009, p. 151). As an astronomer, symmetry was central to Brahmagupta's calculations. If for example, North was positive then South was negative. Zero's role as an additive identity was emphasized to the extent that even zero plus zero is zero was noted in Brahmagupta's Sanskrit Laws of addition which contain his definition of zero (Figure 10.1).

### Brahmagupta's 5 Addition Sutras

धनयोर् धनम् ऋणमृणयोः धनर् णयोरन् तरं समैक् यं खम्
ऋणमैक् यं च धनमृणधनशून् ययोः शून् ययोः शून् यम्

| | |
|---|---|
| AS1 | positive plus positive is positive |
| AS2 | negative plus negative is negative |
| AS3 | positive plus negative is the difference between the positive and negative |
| AS4 | when positive and negative are equal the sum is zero |
| AS5 | positive plus zero is positive<br>negative plus zero is negative<br>zero plus zero is zero |

FIGURE 10.1 Brahmagupta's five addition sutras

Thus, we can anachronistically depict the central theme of India's zero mediating equal and opposite line magnitudes in the diagram below (Figure 10.2).

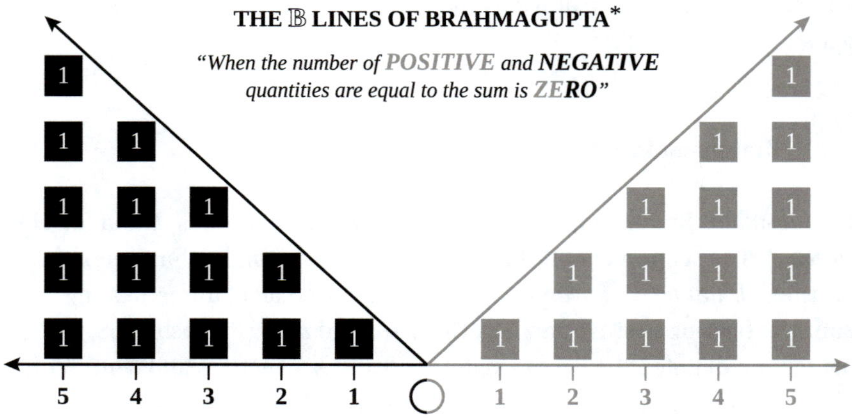


*598–668 CE INDIA

FIGURE 10.2 India's zero represented the least magnitude among equal yet opposing magnitudes

In 1039 CE, Śrīpati's treatise *Siddhāntaśekhara* also defined zero as the sum of two equal negative and positive numbers (Datta & Singh, 1962, p. 21). Thus, we read 'sunya (zero) is neither positive nor negative but forms the boundary line between the two kinds, being the sum of two equal but opposite quantities' (Joseph, 2016, p. 108). It is this zero, as old as the universe where forces exist as equal and opposite pairs, that waits to be fully incorporated into our concept of zero in mathematics classrooms today. The integer inequality symbols < and > (Oughtred, 1631) predated the first appearance of a number line (Wallis, 1685) with numbers either side of zero. Sadly, we are thus taught two negatives are greater than five negatives and told to write ${}^{-}2 > {}^{-}5$ which is contrary to the laws of physics.

China's approach was similar, yet with columns on a counting board being used to separate numbers into places; a zero symbol was not required as a placeholder. When either a positive number or negative number was to be subtracted from an empty place in Chinese arithmetic (Martzloff, 2006), the rules were as follows:

> [If a] positive [rod] does not have a vis-&-vis (i.e., a number facing it) it is made negative (a positive number subtracted from nothing becomes negative). Conceptually, this is $[\ ] - [{}^{+}3] = [{}^{-}3]$. Then, the rules continue, [If a] negative [rod] does not have a vis-&-vis, it is made positive (i.e., a negative number subtracted from nothing becomes positive). Conceptually, this is $[\ ] - [{}^{-}3] = [{}^{+}3]$.

In the second century BCE, China's negative and positive numbers were depicted with black and red rods being equal yet opposite in value (Shen, 1999). Therefore, a similar anachronistic diagram may emerge that again depicts a Chinese absence of number in a column with a conceptual midpoint between negative and positive as shown in Figure 10.3.

Following on from rod numeral arithmetic, the subsequent use of the abacus meant the 0 symbol for zero was delayed by centuries in China, first appearing in *Mathematical Treatise in Nine Sections* by Qín Jiǔsháo in 1247 CE.

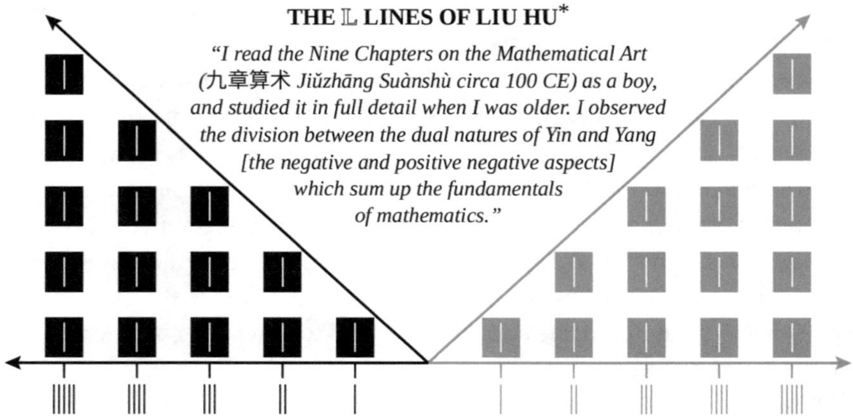


*263 CE CHINA

FIGURE 10.3 China's equal and opposite rod numeral system had empty places rather than a zero

## 5 Brahmagupta's Laws of Zero

Just as *left* and *right* cannot exist without a *center*, you cannot have the relational concept of *positive* and *negative* without a *center*. Similarly, with *above* and *below*, zero is neither term but formed from both. Brahmagupta's zero exists between positive and negative because he defined it as the sum of both (equal) positive and negative. It is this aspect, being a mathematical sum, that not only qualifies zero as a number but gives it the capacity to be the sum of all numbers in the set of real numbers. Zero has the capacity to be both the void and the infinite. As an astronomer first and mathematician second, Brahmagupta dealt with relationships between quantities. Numbers, being isomorphic to quantities, were merely tools of astronomers, as were their instruments. Zero, to the Indian scientist, acted as a reference point from which counts and measurements of quantities were made. Brahmagupta's 18 sūtras of symmetry (excluding division by zero), are shown in Table 10.1 (Dvivedin, 1902, p. 309; Plofker, 2009, p. 151).

TABLE 10.1 Brahmagupta's 18 sūtras of symmetry for zero, positive and negative

| *Addition* | *Subtraction* |
|---|---|
| 1. Positive plus positive is positive.<br>2. Negative plus negative is negative.<br>3. Positive plus negative is the difference between the positive and negative.<br>4. When positive and negative are equal the sum is zero.<br>5. Positive plus zero, is positive, negative plus zero is negative and zero plus zero is zero. | 6. A smaller positive subtracted from a larger positive is positive.<br>7. A smaller negative subtracted from a larger negative is negative.<br>8. If a larger negative or positive is to be subtracted from a smaller negative or positive, the sign of their difference is reversed, negative becomes positive and positive becomes negative.<br>9. A negative minus zero is negative, a positive minus zero is positive and zero minus zero is zero.<br>10. When a positive is to be subtracted from a negative or a negative from a positive, then it is to be added. |
| *Multiplication* | *Division* |
| 11. The product of a negative and a positive is negative.<br>12. The product of negative and negative is positive.<br>13. The product of two positives is positive.<br>14. The product of zero and a negative, of zero and a positive, or of two zeros is zero. | 15. A positive divided by a positive is positive.<br>16. A negative divided by a negative is positive.<br>17. A positive divided by a negative is negative.<br>18. A negative divided by a positive is negative. |

## 6 On Brahmagupta's Zero and the Foundations of Physics

Within Brahmagupta's quantitative laws, positive and negative co-exist as equal terms. His laws were likely derived from empirical observation and the need to solve problems, an approach similar to the 'scientific method'. Thus, as an astronomer Brahmagupta's ancient treatment of zero, negatives, and positives is consistent today with the laws of physics. Just as India's mathematics treated negative and positive as equal and opposite, today we accept Newton's third law – for every action there is an equal and opposite reaction. Thus, the zero-sum game of mathematics leads to innovations in physics. An example is Paul Dirac's predicted discovery of the positron as the antiparticle for the electron. When one positron and one electron meet, they cancel each other out, effectively summing to zero. Brahmagupta's zero-sum logic can also be associated with the physical law of conservation of matter and energy. The laws of physics and the laws of mathematics that describe quantitative relationships require a correct definition and understanding of zero. Numbers and quantities are symmetric around zero, yet the history of mathematics has essentially been asymmetric, being built upon half the system, the positive. The realization half of Brahmagupta's algebraic laws of sign (involving negative quantities) were mostly missed in the Arabic world appears to have rarely dawned on Western historians and teachers today who bear the brunt of the incomplete transmission of India's zero to the West.

## 7 The Placeholder Zero in Brahmagupta's Arithmetic

In Brahmagupta's Chapter 12 on arithmetic (*Ganita*), we read, 'He, who distinctly and severally knows addition and the rest of the 20 logistics, and the eight determinations including measurement by shadow, is a mathematician' (Colebrooke, 1817, p. 277). The content is far broader than most arithmetic curriculums today, as shown in Table 10.2 (Pingree, 1981, p. 57).

TABLE 10.2 Brahmagupta's 20 Arithmetical operations and 8 determinations

| | | | | |
|---|---|---|---|---|
| parikrama 20 fundamental operations | saṅkalita addition | vyavakalita subtraction | pratyutpanna multiplication | bhāgahāra division |
| varga square | vargamūla square root | ghana cube | ghanamūla cube root | bhāṇḍapratibhāṇḍa barter |
| pañcajātayaḥ operating on fractions (5 rules) | trairāśika rule of three | vyastatrairāśika inverse rule of 3 | and … rules of 5, 7, 9, and 11 terms (4 rules) | vyavahāra 8 determinations mixtures, series, plane geometry, solid geometry, stacks, sawn lumber, mounds of grain and shadow problems. |

It is known Brahmagupta's writings on arithmetic (or a derivative of it) made its way to the Arabic world. Zero made its way too, yet as a placeholder, not a number, and not as a *sum* of equal positive and negative. The Arabs embraced the Hindu system of base-10 numeration in which nine distinct letters (numerals) 9, 8, 7, 6, 5, 4, 3, 2, and 1 were recycled to represent different quantities in different places by virtue of zero. In an additive base-10 numeration system, III represents three, while in a multiplicative base-10 system, 111 means one hundred and eleven.

The most obvious advantage of India's zero, which enabled the base-10 multiplicative system over an additive system, such as the numerals of the Roman Empire, is compactness. For example, the number 888 in the Indian system is DCCCLXXXVIII in the Roman system, where zero as a symbol does not exist. While we read from left to right, Arabic is read from right to left. The number 6 does not change with direction, yet VI read left-right could be mistaken for 4 if written IV in a right-left context. Similar confusion may have existed with 11 written XI, being confused with IX which is 9. The single digit base-10 Hindu numeration system was the innovation the Arabs embraced – not India's number zero. As will be revealed, zero was treated as a vital placeholder enabler of the system, yet was not considered a number.

How did Brahmagupta's ideas on zero, positives and negatives reach the West? The story of mathematics has India's zero as documented by Brahmagupta being embraced by the Arabic world. Traders in Northern Africa then passed India's concept of zero onto Leonardo Pisano (Fibonacci), who helped introduce both Hindu and Arabic mathematics to Europe at the start of the thirteenth century. Alas, this story does not agree with the evidence.

Given Brahmagupta's Hindu arithmetic noted in Table 10.2 did *not* discuss the rules of operating with positives, negatives, and zero, such information appears to have not been fully appreciated in the Arabic world. As will also be discussed, it appears Hindu algebra, which featured both positives and negatives, may have been transmitted orally to the Arabic world yet al-Khwārizmī and others seemingly failed to either document it or benefit from it.

From the writings of Brahmagupta, we will explore the partial transmission of his symmetric mediating zero via several important Arabic influencers, (al-Khwārizmī, al-Uqlīdisī and Kūshyār ibn Labbān). Then we explore the influential writings of the Italian Leonardo Pisano, whose 1202 CE book *Liber Abaci* appears to have helped bring about the demise of Gerbert's abacus, which we discuss shortly. Zero the placeholder made its way West (Figure 10.4), while zero as defined and applied as the sum of equal positive and negative remained in the East.

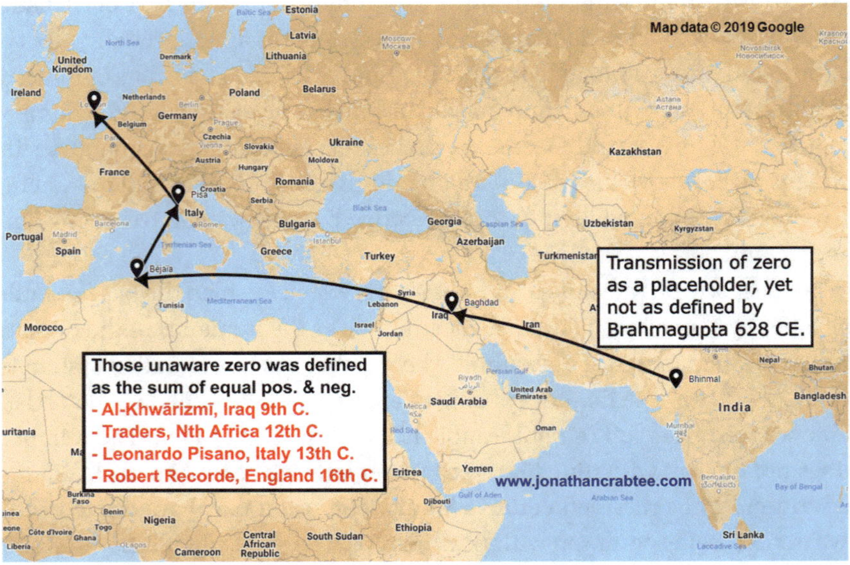


FIGURE 10.4 The transmission of zero as a placeholder from East to West

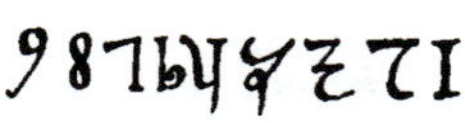

FIGURE 10.5
The oldest extant Arabic numerals to be used in Europe (without zero)

It should be noted that the above timeline depicting the westward travels of zero as a placeholder was the second introduction of (Hindu) Arabic numerals into Europe. The oldest extant Arabic numerals in Europe (without zero) are found in the *Codex Vigilanus* of 976 CE, as seen in the Wikimedia Commons image (Figure 10.5). These came via Muslims in Spain, who came in contact with the Frenchman Gerbert of Aurillac (later Pope Sylvester II) who promoted the use of Arabic numerals on European counting boards.

From the late tenth to the early thirteenth century base-10 positional calculation was performed on a counting board or abacus with Arabic numerals (Ifrah, 2000, p. 580). Given the counting boards had fixed columns, within which counters (called apices) labeled with numerals would be placed, zero as a placeholder was not required. Thus, for more than 200 years 1, 2, 3, 4, 5, 6, 7, 8, and 9 were used in Europe without zero being needed as a placeholder, as revealed by the Wikimedia Commons image Apices of the modern age (Figure 10.6).

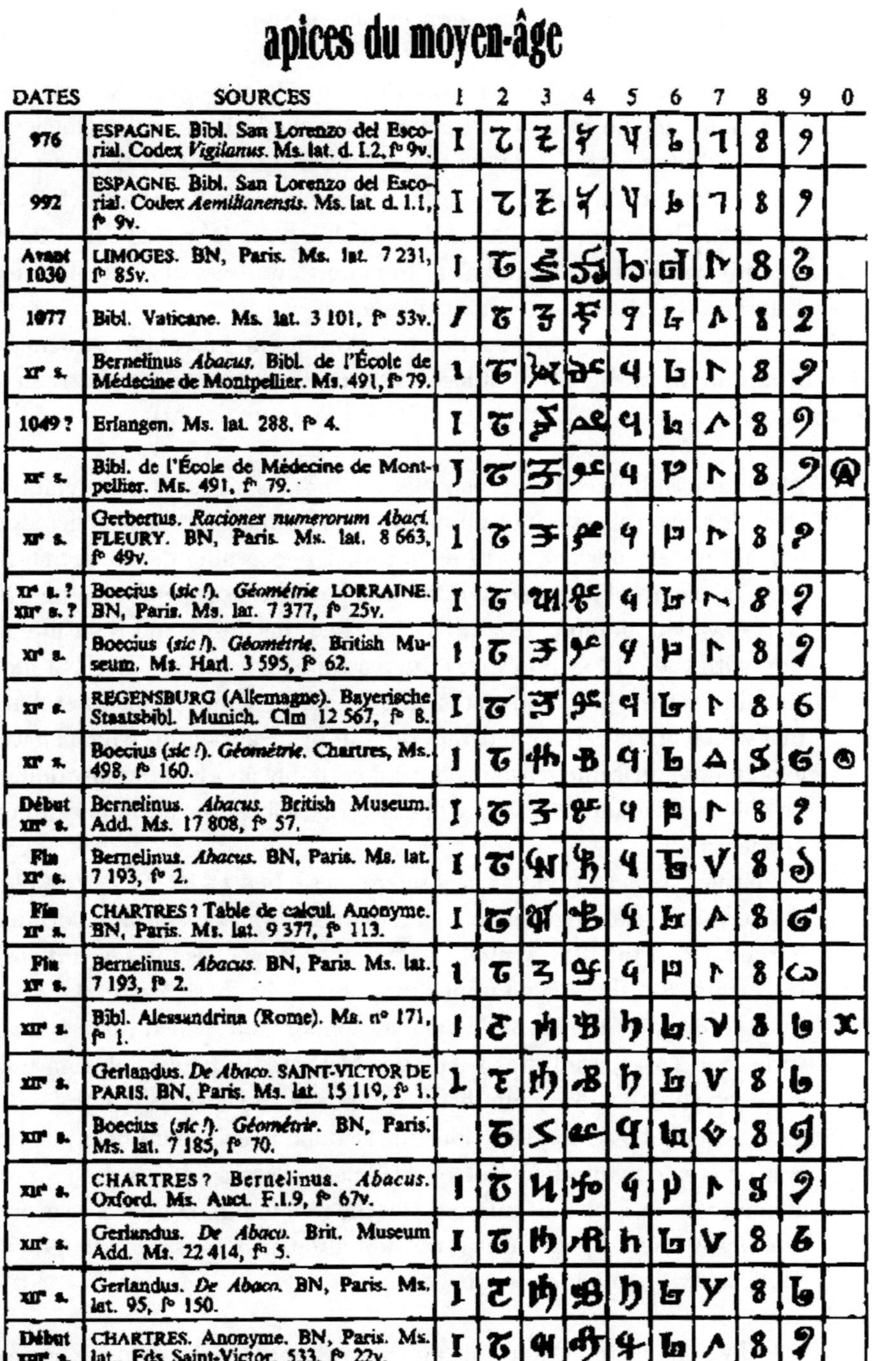

**apices du moyen-âge**

| DATES | SOURCES |
|---|---|
| 976 | ESPAGNE. Bibl. San Lorenzo del Escorial. Codex *Vigilanus*. Ms. lat. d. I.2, f° 9v. |
| 992 | ESPAGNE. Bibl. San Lorenzo del Escorial. Codex *Aemilianensis*. Ms. lat. d. I.1, f° 9v. |
| Avant 1030 | LIMOGES. BN, Paris. Ms. lat. 7 231, f° 85v. |
| 1077 | Bibl. Vaticane. Ms. lat. 3 101, f° 53v. |
| XIe s. | Bernelinus *Abacus*. Bibl. de l'École de Médecine de Montpellier. Ms. 491, f° 79. |
| 1049 ? | Erlangen. Ms. lat. 288, f° 4. |
| XIe s. | Bibl. de l'École de Médecine de Montpellier. Ms. 491, f° 79. |
| XIe s. | Gerbertus. *Raciones numerorum Abaci*. FLEURY. BN, Paris. Ms. lat. 8 663, f° 49v. |
| XIe s. ? XIIe s. ? | Boecius (*sic !*). *Géométrie* LORRAINE. BN, Paris. Ms. lat. 7 377, f° 25v. |
| XIe s. | Boecius (*sic !*). *Géométrie*. British Museum. Ms. Harl. 3 595, f° 62. |
| XIe s. | REGENSBURG (Allemagne). Bayerische Staatsbibl. Munich. Clm 12 567, f° 8. |
| XIe s. | Boecius (*sic !*). *Géométrie*. Chartres, Ms. 498, f° 160. |
| Début XIIe s. | Bernelinus. *Abacus*. British Museum. Add. Ms. 17 808, f° 57. |
| Fin XIe s. | Bernelinus. *Abacus*. BN, Paris. Ms. lat. 7 193, f° 2. |
| Fin XIe s. | CHARTRES ? Table de calcul. Anonyme. BN, Paris. Ms. lat. 9 377, f° 113. |
| Fin XIe s. | Bernelinus. *Abacus*. BN, Paris. Ms. lat. 7 193, f° 2. |
| XIIe s. | Bibl. Alessandrina (Rome). Ms. n° 171, f° 1. |
| XIIe s. | Gerlandus. *De Abaco*. SAINT-VICTOR DE PARIS. BN, Paris. Ms. lat. 15 119, f° 1. |
| XIIe s. | Boecius (*sic !*). *Géométrie*. BN, Paris. Ms. lat. 7 185, f° 70. |
| XIIe s. | CHARTRES ? Bernelinus. *Abacus*. Oxford. Ms. Auct. F.I.9, f° 67v. |
| XIIe s. | Gerlandus. *De Abaco*. Brit. Museum Add. Ms. 22 414, f° 5. |
| XIIe s. | Gerlandus. *De Abaco*. BN, Paris. Ms. lat. 95, f° 150. |
| Début XIIIe s. | CHARTRES. Anonyme. BN, Paris. Ms. lat., Fds Saint-Victor, 533, f° 22v. |

FIGURE 10.6 The various symbols used on apices (counters) for base-10 calculations

## 8 On the Absence of Brahmagupta's Zero in the Arabic World

### 8.1 *On al-Khwārizmī's Arithmetical Absence of Brahmagupta's Zero*

Popular writers on the history of mathematics would have readers believe the ideas of Brahmagupta, such as the definition of zero and the 'laws of sign' for the four arithmetical operations, were mastered by al-Khwārizmī (780–850 CE). The legend perpetuated by such writers is that al-Khwārizmī subsequently wrote a book on India's arithmetic around 820 CE, known via its Latin translation as *Algoritmi de numero Indorum*.

Notably, the translator's introduction to al-Khwārizmī's text makes no mention of zero in his introduction. It also states one is not a number. Early in a Latin translation, we read (Crossley and Henry, 1990, pp. 110–111):

> Algorizmi said: since I had seen that the Indians had set up IX symbols in their universal system of numbering, on account of the arrangement which they established, I wished to reveal, concerning the work that is done by means of them, something which might be easier for learners if God so willed. If, moreover, the Indians had this desire and their intention with these IX symbols was the reason which was apparent to me, God directed me to this. If, on the other hand, for some reason other than that which I have expounded, they did this by means of this which I have expounded, the same reason will most certainly and without any doubt be able to be found. And this will easily be clear to those who examine and learn.
>
> So they made IX symbols, whose are these: (9 8 7 6 5 4 3 2 1). There is also a variation among men in regard to their forms: this variation occurs in the form of the fifth symbol and the sixth, as well as the seventh and the eighth. But there is no impediment here. For these are marks indicating a number and the following are the forms in which there is that variation: (5 4 3 2). And already I have revealed in the book of algebra and almuqabalah, that every number is composite and that every number is put together above one. Therefore one is found in every number and this is what is said in another book of arithmetic. Because one is the root of all number and is outside number. It is the root of number because every number is found by it. But it is outside number because it is found by itself, i.e., without any other number.

Given al-Khwārizmī's translator appears so insistent 'one was outside number', it is understandable why zero did not rate a mention in his introduction. Al-Khwārizmī's legacy, whether intended or not, is zero, which acted as a placeholder, was not a number. While al-Khwārizmī covers the four basic operations of addition, subtraction, multiplication, and division, negative numbers are absent from his book on Hindu arithmetic. Yet, in Brahmagupta's laws of the four basic arithmetical operations (noted in his algebra, not his arithmetic), negatives are equally as prevalent as positives.

Comparing al-Khwārizmī's approach to Brahmagupta's, we read (Rashed, 2009, p. 77):

> Once again al-Khwārizmī differs from Brahmagupta, this time in not employing any abbreviation. Also he avoids using 'negative' numbers or simply a [larger] number subtracted from a smaller one, or from zero, whereas Brahmagupta, like other Indian mathematicians before him, does not hesitate to make use of such [negative] numbers. It is difficult to imagine that al-Khwārizmī, if he had read this chapter (i.e., chapter 18 of Brahmagupta's *Brāhma Sphuta-siddhānta*) would not have been able to profit by it, even if only to shorten the presentation of his work.

## 8.2 *On al-Khwārizmī's Algebraic Absence of Brahmagupta's Zero*

Several years later, al-Khwārizmī was said to have written his landmark book, الكتاب المختصر في حساب الجبر والمقابلة, *al-Kitāb al-mukhtasar fī hisāb al-jabr wa'l-muqābala* (The Compendious Book on Calculation by Restoration and Confrontation). (Sometimes the phrase *Completion and Balancing* is used instead of *Restoration and -Confrontation.*) The word *al-jabr* in the title gave us the word algebra while algorithm (via algorism) is derived from the name al-Khwārizmī. Writing around 200 years after Brahmagupta, it is often assumed al-Khwārizmī wrote first on arithmetic, then on algebra. However, this is not the case. Al-Khwārizmī wrote his book on algebra around 820 CE and followed this up with his book on Hindu arithmetic around 825 CE. Al-Khwārizmī refers readers to his book on algebra in his arithmetic book. Notably, al-Khwārizmī's algebra has little connection to the earlier mathematics of India.

We find possible knowledge of Hindu astronomy and definite knowledge of Hindu arithmetic in the writings of al-Khwārizmī, yet not Brahmagupta's algebraic laws, which featured negative quantities as much as positive quantities. We read (Rashed, 2009, p. 79):

> Whether we are concerned with concepts or procedures, the many divergences indicate that, even if al-Khwārizmī did know books by Āryabhaṭa and Brahmagupta, he had read them only for astronomy and, perhaps, for arithmetic. In any case, reading them had no effect on his conception of

> algebra and it exercised no influence on the techniques he employed in the discipline. The style of the mathematical reasoning that is at work in al-Khwārizmī's algebra has nothing to do with what we encounter in the work of his (Indian) predecessors.

Accepting negative numbers and the identity elements (zero and one) as numbers is critical to the development and evolution of mathematics. However, it is evident from the translations of al-Khwārizmī's text, he used zero as a placeholder yet not as a number. The Arabic world did not see one as a number (being the unit of count or measure) and did not document India's laws of sign for positive and negative numbers. (Like Diophantus of Alexandria centuries before, al-Khwārizmī was aware a positive number with a subtraction multiplied by a positive number also with a subtraction results in a positive number being added as an adjustment.)

## 8.3 *Balancing Algebraic Equations without India's Zero*

The words *al-jabr wa'l-muqābala* (literally restoration and confrontation) in al-Khwārizmī's book on algebra have been loosely translated as 'balancing an equation' (Devlin, 2012, p. 25). In an environment in which weights and scales often determined the cost of goods at marketplaces, it may have been that al-Khwārizmī had this metaphor in mind for his algebraic equations. If so, then what scale could balance against zero? Without symbols presented here for clarity, al-Khwārizmī provided solutions for linear and quadratic equations involving combinations of $ax^2$ (squares) and $bx$ (roots) and $c$ (numbers) involving positive rationals. If subtraction was involved, it would be eliminated by adding the subtracted term onto both sides to keep the equations balanced. For example, $ax^2 = bx - c$ became $ax^2 + c = bx$. Al-Khwārizmī's six standard types of balanced equations are depicted below in Figure 10.7.

From al-Khwārizmī's six types of balanced equation, his three normal forms were:

1) $x^2 + bx = c$ square + roots = number (*where roots are the side of a square*)
2) $x^2 = bx + c$ square = roots + number, and
3) $x^2 + c = bx$ square + number = roots

To popularize a little-known branch of mathematics, algebra, separate to both geometry and arithmetic is a remarkable achievement. Yet if you don't treat zero as a number, which combinations of positive and negative terms can equal, you won't solve equations of the form $ax^2 + bx + c = 0$. Al-Khwārizmī's modus operandi was to eliminate subtracted terms and confront only positive terms on opposing sides to balance equations.

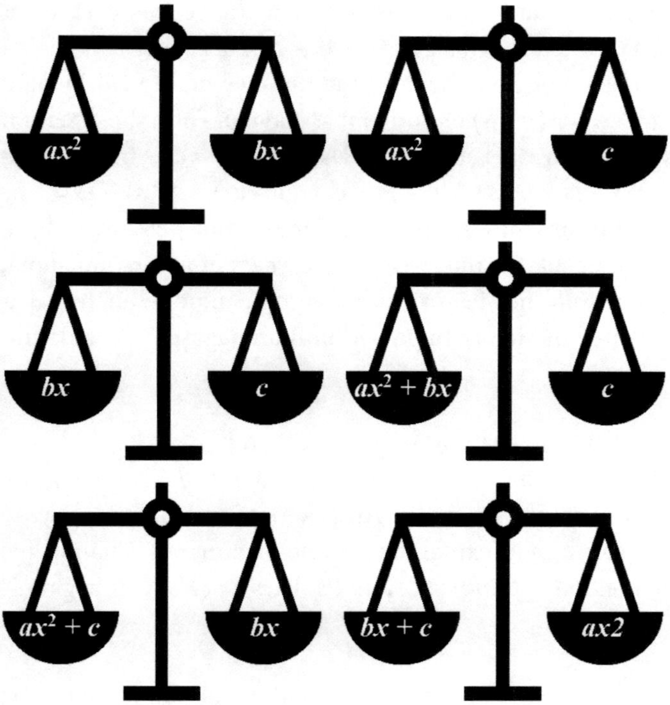


FIGURE 10.7 Al-Khwārizmī's six types of balanced equations

Thus, neither al-Khwārizmī in the ninth century nor Leonardo Pisano, who introduced Hindu Arabic mathematics to Europe in the thirteenth century leveraged India's ideas that would lead to solutions of quadratic equations in forms such as $ax^2 + bx + c = 0$ or $ax^2 - bx = {}^{-}c$. If al-Khwārizmī were to solve for $x$ in $x^2 + 2x = 15$, he would use his rhetorical formula for $x^2 + bx = c$, which today, might be written as shown below:

$$\sqrt{[(b/2)^2 + c]} - b/2$$

Solving for $x$ in $x^2 + 2x = 15$ with $b = 2$ and $c = 15$ is as follows. Two divided by two, is one, which squared, remains one. Add 15 to one and you get 16. Then take the square root of 16, which is four. Then from four, subtract two divided by two, which is one, and you arrive at $x = 3$.

Al-Khwārizmī primarily dealt with unknowns, $x$ (roots or sides of squares), their squares, and rational positive numbers. Had he been exposed to Brahmagupta's negative numbers and zero, he might have solved $x^2 + 2x - 15 = 0$. Today, we look for the factors of $x^2$ and the factors of ${}^{-}15$. The former are $x$

and $x$ while the latter are $^{-}3$ and $^{+}5$ as they sum to 2, the desired coefficient of the middle term, $2x$. So to solve for $x$ in $x^2 + 2x - 15 = 0$ we arrive at $(x + 5)(x - 3) = 0$. As Brahmagupta gave the rule any number multiplied by zero equals zero, either $(x + 5)$ or $(x - 3)$ or both, must equal 0. Thus, the equation $x^2 + 2x - 15 = 0$ is solved with $x = 3$ and $x = -5$. Using these values in the equation confirms $3^2 + 2(3) - 15 = 0$ as $9 + 6 - 15 = 0$, and $(-5)^2 + 2(^{-}5) - 15 = 0$ as $25 - 10 - 15 = 0$. Al-Khwārizmī's approach generated the positive root, 3, yet not the negative root, $^{-}5$. Despite writing 200 years after Brahmagupta, al-Khwārizmī's approach did not feature either zero or more tellingly, negatives, since they remained hidden in India within Brahmagupta's 'lost' definition of zero.

## 8.4 *On al-Uqlīdisī's Absence of Brahmagupta's Zero*

The oldest extant Arabic text on Indian arithmetic is *Kitāb al-fuṣūl fī al-ḥisāb al Hindī* (The Book of Elements on Indian Arithmetic) by al-Uqlīdisī (*c.*920–980 CE). Written in 952 CE in a textual form without numerals, Chapter 1 is titled *Justification of the Hindi (Arithmetic) and Its Whys and Hows*. It begins as follows (Saidan, 1978, p. 186):

> Here we state justifications of Hindi (arithmetic) and queries about its whys and hows; for many of the people of this craft ask saying: why and how. To every question that is asked there is an answer and if it is hit, he who asks is satisfied.
>
> One question is: Why are the Hindi letters nine, no more, no less? We say: Because the beginning of numbers from which they start is one and the last unit we pronounce is nine. Thus when we say units we mean (something) between one and nine; after that units are over, and ten comes out like one and takes its form. We add up ten to ten until we reach 90 which conforms with nine. Tens are now over and we say one hundred, coming back to one, and going up to 9. Thus we see that all places start with one and end with nine. That is why they are made nine.

And also:

> If it is said: Why do we say: units, tens, hundreds, thousands? We say: These are places. Upon them lies all the principle of Hindi (arithmetic). The first place is that of units; it may have from one to 9, and these are units only. Next comes the second place, which is the place of tens; it may have from ten to 90 and nothing else. Similarly for the third place which is the place of hundreds, and for the thousands place. No place has more

> than 9; it may have from one to 9, and thereafter we move to another place which may have the same thing again. Thus units, tens, hundreds, and thousands are repeated.

Al-Uqlīdisī ended Chapter 1 on India's number system with *So much for the nine letters*. In Chapter 4, a section titled 'Questions on Multiplication' includes the following:

> If it is said: Why is zero [multiplied] by zero equal to zero and zero by any letter zero? We say that by multiplying zero by zero the aim is only to occupy the place; the same applies for multiplying the letter by zero. We multiply the letter by zero only once, the first time, by the first letter in the upper, to occupy the place, and tell that there is a place and that it is empty.

Thus, it can be seen that al-Uqlīdisī, writing more than 300 years after Brahmagupta, clearly saw zero as being separate to the nine Indian letters (numerals) and that its purpose was to denote an empty place. Notably, al-Uqlīdisī's full name is Abū al-Ḥasan Aḥmed ibn Ibrāhīm ***al-Uqlīdisī***, which means Abū al-Ḥasan Aḥmed ibn Ibrāhīm, ***the Euclidist***. The name of the man who wrote the definitive book on Hindu arithmetic was derived from his advocacy and translations of Euclid's geometry, in which a number zero did not exist. So, by 953 CE, zero may not have been considered a number outside India, yet at least the unit one in the Arabic speaking world was.

### 8.5 *On Kūshyār ibn Labbān's Absence of Brahmagupta's Zero*

The oldest extant Arabic text that features Hindu numerals is that of Iranian mathematician, Kūshyār ibn Labbān (971–1029 CE). His *Kitáb fī usūl hisāb al-Hind* (Principles of Hindu Reckoning) was written around 1000 CE. Once again, zero was used to fill an empty space where no numbers exist (Levey & Petruck, 1965, pp. 44–46):

> Before proceeding with these principles, it is essential to have a knowledge of the symbols of the nine numerals and the place order of [any] one of them with respect to the others and the increase of [any] one [compared] to the others, and the lessening of [any] one of them compared to the others.

Then, in a section headed 'On the understanding of the symbols of the nine numerals', we read:

> In the place position where there is no number, a zero is placed as a substitute for that missing number. In the case of the ten a cipher is made to precede it in the place position of the units. Likewise the hundred is preceded by 2 zeros in the place position of the units and tens.

India's view of zero as a number defined by an equal prevalence of negative and positive numbers can be contrasted with the Arabic view (Oaks, 2018):

> I have read a few dozen medieval Arabic books on arithmetic and algebra, and there is no hint of negative numbers in any of them. Zero, too, was not regarded to be a number but was merely the placeholder for an empty place in the representation of a number in Arabic (Indian) notation. All numbers in Arabic arithmetic were positive.

Elsewhere we read, 'Like in medieval Europe, negative numbers and zero and were not acknowledged in Arabic mathematics' (Oaks, 2011, p. 2).

## 9 On the Absence of Brahmagupta's Zero in Europe

### 9.1 *Leonardo Pisano and the Liber Abaci*

Leonardo Pisano's influential 1228 CE *Liber Abaci*, (Book of Calculation) begins: (Boncompagni, 1857, p. 2)

> Novem figure indorum he sunt 9 8 7 6 5 4 3 2 1. Cum his itaque novem figuris, et cum hoc signo 0, quod arabice zephirum appellatur, scribitur quilibet numerus, ut inferius demonstratur.

This translates as (Sigler, 2002, p. 17): 'The nine Indian figures are: 9 8 7 6 5 4 3 2 1. With these nine figures, and with the sign 0 which the Arabs call *zephir* any number is written, as is demonstrated below.'

Notably, by *any number*, Pisano means any positive number. It is no surprise, we read (Sigler, 2002, p. 6). 'The zero, or zephir as Leonardo calls it, counts for nothing and serves as a place holder.' Exactly six centuries after Brahmagupta gifted the world the elements of modern arithmetic, Pisano's *Liber Abaci* neither documented the laws of sign for positive and negative, nor presented the laws of using zero as a number in the four arithmetical operations.

Pisano is feted for having helped introduce Hindu mathematics to Europe. He did, yet like al-Khwārizmī, also credited with transmitting India's mathematics 400 years earlier, Brahmagupta's symmetric definition of zero as a sum of opposing quantities was long lost in transit, if it even ever left India. On the treatment of equations, as Brahmagupta implied, if the product of two or more factors is zero, then at least one of those factors is zero. Some 200 years later, around 820 CE, al-Khwārizmī missed the idea of 'balancing with zero' to solve equations. Pisano also missed this insight and used the same six standard forms of linear and quadratic equations developed by al-Khwārizmī. Had Brahmagupta's negative numbers and zero been embraced westward, modern equations would most likely have evolved sooner.

As many may remember (or wish to forget), the equation $x^2 + 2x - 15 = 0$ can also be solved with the quadratic formula for determining roots or $x$-intercepts.

$$x = \frac{-b \pm \sqrt{b^2 - 4ac}}{2a}$$

Solving for $a = 1$, $b = 2$ and $c = {}^{-}15$ in $x^2 + 2x - 15 = 0$ is as follows:

$$\frac{-2 \pm \sqrt{2^2 - 4(1)(-15)}}{2} \text{ which is } \frac{-2 \pm \sqrt{2^2 - (-60)}}{2} \text{ then } \frac{-2 \pm \sqrt{64}}{2} \text{ and } \frac{-2 \pm 8}{2}$$

so $x = 3$ and $x = {}^{-}5$.

With $x^2 + 2x - 15 = 0$, India's zero multiplication rule means either factor $(x + 5)$ or $(x - 3)$ must equal zero. This is simpler than both the formula of al-Khwārizmī for $x^2 + 2x = 15$ which is $\sqrt{[(b/2)^2 + c]} - b/2$ and the quadratic formula above.

So, it took around 1,000 years for Brahmagupta's zero to truly be given the same algebraic status as a number in its own right, that combinations of other numbers could equal. Notably, the first to apply the zero definition of Brahmagupta to arrive at the general formula for quadratic equations was India's Śrīdhara (*c.*870–*c.*930 CE).

## 9.2 *How Positives Became Negative and India's Symmetric Zero Was Assumed Understood*

You may not have noticed that equal and opposite negatives opposing positive quantities have been missing in books on the evolution of Western mathematics. The reason is we often read about negative numbers (that are not) in the writings of mathematicians throughout history. The Greek Diophantus of Alexandria (circa 250 CE), wrote the following in his *Arithmetica*: Λεῖψις ἐπι λεῖψιν πολλαπλασιασθεῖσα ποιεῖ ὕπαρξιν, λεῖψις δέ επί ὕπαρξιν ποιεί λεῖψιν (Heath, 1910, p. 130). In Latin, this was correctly translated as (Bachet de Méziriac, 1621, p. 9) *Minus per minus multiplicatum, producit Plus. At minus per plus multiplicatum, producit minus.* Heath gave this translation of Diophantus: 'A minus multiplied by a minus makes a plus; a minus multiplied by a plus makes a minus.'

Importantly, on the nature of $(\boldsymbol{a} - \boldsymbol{b}) \times (\boldsymbol{c} - \boldsymbol{d})$, in Heath we read: 'To be emphasized is the fact that in Diophantus the fundamental algebraic conception of negative numbers is "wanting".' *In* $2x - 10$ *he avoids as absurd all cases where* $2x < 10$. In $(\boldsymbol{a} - \boldsymbol{b}) \times (\boldsymbol{c} - \boldsymbol{d})$, the terms $\boldsymbol{b}$ and $\boldsymbol{d}$ are not negative terms. Despite the fact $\boldsymbol{b}$ and $\boldsymbol{d}$ are positive terms being subtracted, we read comments such as 'Diophantus formulates for relative numbers the following rule of signs: a negative multiplied by a negative yields a positive, whereas a negative by a positive yields a negative' (Bashmakova & Silverman, 1997, p. 6). Diophantus did not write about positive and negative numbers related by zero and neither did al-Khwārizmī, yet you may not have noticed.

## 9.3 *How Explanations of* al-jabr *Misled Educators*

As mentioned, al-Khwārizmī eliminated subtracted terms and only confronted positive terms on opposing sides to 'balance' his equations. Yet through modern eyes, subtracted terms often magically morph into negative terms, which implies zero mediates the positive and negative. Al-Khwārizmī's algebraic equations never contained negative terms, yet we read:

> In mathematical language, the verb [jabr] means … when applied to equations, to transpose negative quantities to the opposite side by changing their signs. The negative quantity thus removed. (Rosen, 1831, p. 178)

and

> The usual meaning of jabr in mathematical treatises is: adding equal terms to both sides of an equation in order to eliminate negative terms. (van der Waerden, 1985, p. 4)

and

> Al-jabr means 'restoration' or 'completion', that is, removing negative terms, by transposing them to the other side of the equation to make them positive. (Devlin, 2012, p. 53)

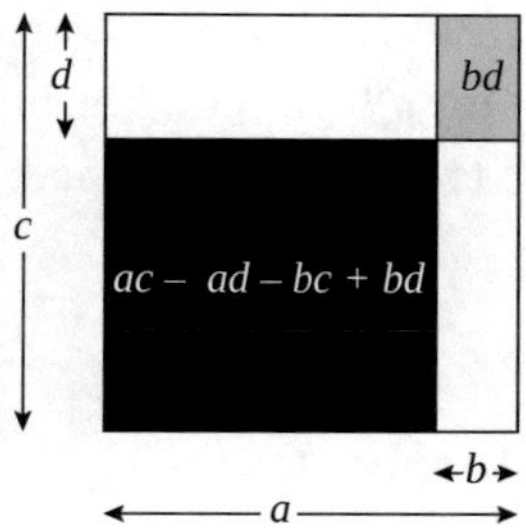


FIGURE 10.8
The geometric instantiation of *'minus* × *minus* = *plus'* given false equivalence to *'negative* × *negative* = *positive'*

Positive numbers being subtracted are not the same thing as negative numbers, yet they are often conflated. In the equation $a - b = c$ (where $b < a$) the term $b$ is not a negative number. Mathematicians before and after Brahmagupta, such as Diophantus and al-Khwārizmī, have given geometric explanations for $(a - b) \times (c - d)$ provided $a > b$ and $c > d$. Yet, all the terms were positive, with $b$ always less than $a$ and $d$ always less than $c$. Today teachers might discuss the expansion of $(a - b) \times (c - d)$ which becomes $ac - ad - bc + bd$ to say we define *negative multiplied by negative as positive* in order to preserve the distributive property of multiplication. Looking more closely at $(a - b) \times (c - d)$ the term $bd$ is a positive term being added back after having been subtracted once too many times as shown in Figure 10.8.

With $(a - b) \times (c - d)$ in the form $ac - ad - bc + bd$, from $ac$, we subtract $a \times d$, (the white horizontal strip from left to right) and subtract $b \times c$, (white vertical strip from bottom to top) to get $ac - ad - bc$. Yet we *twice* subtracted $bd$ (the gray shaded area resulting from overlapping white strips), so we add back $bd$ in the corner once, to get $ac - ad - bc + \underline{bd}$. With $(10 - 2) \times (10 - 3)$ as the example for $(a - b) \times (c - d)$, we know the answer is $8 \times 7$, which is 56. So, by distribution over subtraction we get 100 (i.e., $ac$) – 30 (i.e., $ad$) – 20 (i.e., $bc$). We have arrived at $100 - 30 - 20$ which is 50, yet we must make an adjustment as we twice subtracted the gray area $bd$. Therefore, we add back $bd$ once to arrive at $ac - ad - bc + \underline{bd}$ and get $100 - 30 - 20 + (2 \times 3) = 56$. Notably, every term in this explanation is positive, regardless of whether it is added or subtracted. Subtracted terms have for so long been anachronistically described as negative terms, that many people have thought negative algebraic terms were a feature of al-Khwārizmī's equations. Similarly, we have long seen discussions about positive and negative exponents with $a^{+b}$ and $a^{-b}$ where $a$ is the base and $b$ is the exponent. Yet the signs of the exponent are misleading. They might be better written $a^{\times b}$ and $a^{\div b}$ as the exponents are a count of the number of times 1 is either multiplied by $a$ or divided by $a$. For example, $2^{+3}$ more clearly means $2^{\times 3}$ as it becomes $1 \times 2 \times 2 \times 2$ which is 8 and $2^{-3}$ more clearly means

$2^{\div 3}$ as it becomes $1 \div 2 \div 2 \div 2$ which is 1/8. The problematic $a^0$ (which often gets dumped into the index laws) then becomes simple. With $a^0$ the number of times 1 is either multiplied by $a$ or divided by $a$ is zero, so 1 remains unchanged, thus $a^0 = 1$.

## 10 How Brahmagupta's Zero Could Have Helped Diophantus and Descartes

The author's analogy, inspired by fond memories of playing with cubic bricks, is just as a wall is composed of a number of bricks, we do not consider a single brick a wall. Yet where might a cubic mud brick have come from? The answer is, of course, a hole in the ground. Before the brick and hole were created, we had ground-level zero, treating height above and below ground level as an implicit vertically aligned number line. Consistent with the physical law of conservation of matter, for every brick made another hole is made as zero is split and rearranged into opposite quantities in the spirit of Brahmagupta's definition.

Diophantus called the equation $4 = 4x + 20$ absurd as it would result in a negative value for $x$, which he thought was impossible. Yet, what if Diophantus had played with Brahmagupta as a child? Brahmagupta might have said he had four bricks, which was the same as Diophantus' 20 bricks combined with four things. After a bit of fun, they would have realized the four mystery things were holes, each four bricks deep. In the game a brick would be a positive unit and a hole a negative unit, while obeying both India's laws of mathematics and the basic laws of physics. Having imagined taking Brahmagupta back to the third century to play with Diophantus, what if Brahmagupta had played with René Descartes in the seventeenth century?

Like others in the West before him, Descartes only considered line segments to be positive. After all, how can you travel a distance less than nothing? Importantly, borrowing from an ancient idea of Euclid, Descartes came up with a model of multiplication involving line segments where the product was another line, rather than an area (Descartes, 1637, p. 298). A modernized comparison of the area model of multiplication with Descartes' idea of using similar triangles is shown in Figure 10.9 (Crabtree, 2017b, pp. 94–96).

Descartes never extended his line segments backward onto the opposite side of the origin (zero), yet Brahmagupta might have suggested he do just that. After all, Brahmagupta surely knew if from his origin he walked 100 steps south then 100 steps north, the net distance he would have traveled from his starting point to his endpoint would be zero. Had Brahmagupta told Descartes negative lines were simply equal and opposite to positive lines on the other

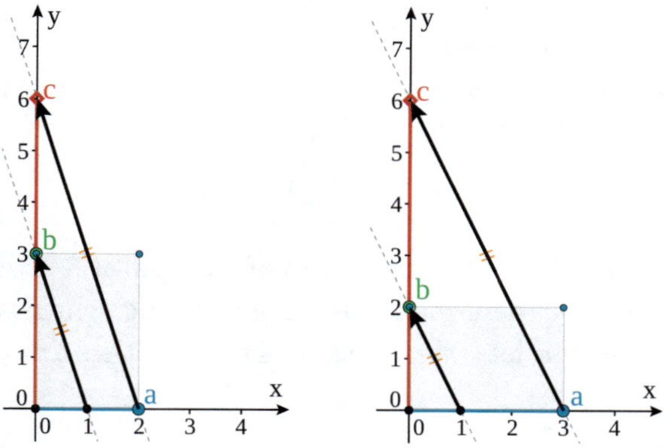


FIGURE 10.9
Standard area models for $2 \times 3$ and $3 \times 2$ alongside modernized versions of Descartes' original proportional approach

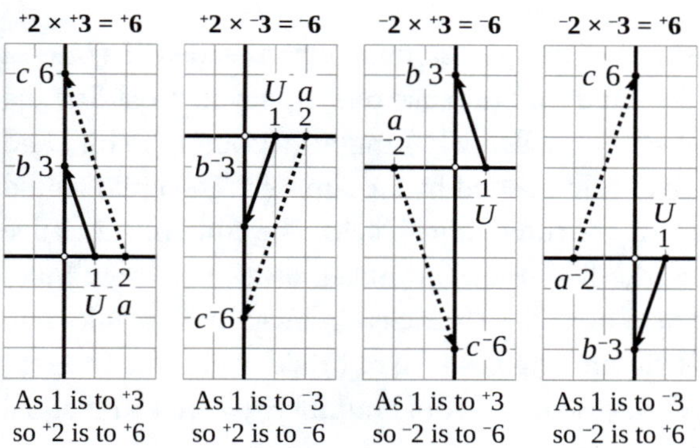


FIGURE 10.10
Instantiations that might have emerged had Brahmagupta met Descartes

side of zero, children might now be taught laws of sign via similar triangles (Figure 10.10).

## 11 Rebuilding Elementary Mathematics from Zero

From the time of the ancient Greeks, 'number theory' has been developed on the natural numbers where N = {1, 2, 3, …}. For example, the first axiom in the highly influential late nineteenth century Dedekind/Peano axioms stipulated 1 to not be the successor of any number, implying zero was not a number. Joseph Peano eventually included zero in his axioms (Peano, 1902, p. 8) yet it had little effect on the academic status of zero as we continue to read 'there exists no number whose successor is 1' (Landau, 1966: 2). For all the time scholars have invested in number theory, there has been no pedagogical benefit in lower-level classrooms. Bad English ideas from centuries ago remain just as bad in classrooms today. For example, in 1685, the English mathematician John Wallis

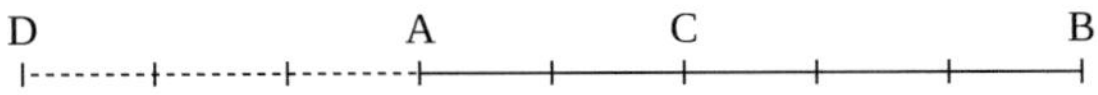


FIGURE 10.11 John Wallis's diagram with which he said a movement from A to D was less than no move at all

[Wallis, 1685] drew a diagram in which a movement west of an origin at A with a magnitude of three from A to D was described as 'less than nothing', which is why three negatives are still said to be less than 0 negatives today (Figure 10.11).

## 12 Teaching 2 – 5

So, how might a child understand 2 – 5? Being familiar with the idea of a bucket and spade, children can understand that to make one brick you must first dig one hole. Now, in the role of a brick seller, what might happen if a child had only two bricks for sale yet five were needed by a customer? The child would simply dig three more holes to make three more bricks. The child now has five bricks and three holes. Once the five bricks are taken away, the three holes remain. Simply assign the idea of positives to bricks and negatives to holes and the pedagogy of 2 – 5 is self-evident. The absence of bricks and holes can represent a nothing zero while the presence of equal numbers of brick and equal numbers of holes of equal dimension can represent an Indian zero.

In a physical sense, ${}^{+}2 - {}^{-}5$ cannot be resolved as you only have positive quantities and so cannot subtract negative quantities. A solution is to add a Brahmaguptan zero-pair in the form of ${}^{+}5 + {}^{-}5$. The equation then becomes ${}^{+}2 + {}^{+}5 + {}^{-}5 - {}^{-}5$ which is ${}^{+}7$. Alas, ${}^{+}2 - {}^{-}5$ and ${}^{-}5 - {}^{+}2$ is taught five years after 5 – 2 because India's zero definition as a sum of equal and opposite quantities, empirically consistent with science, failed to be transmitted to the West.

So, if God made the integers, the devil is in the detail. We cannot reclaim lost progress. Yet will the world be prepared to rebuild its curriculums upon Indian foundations that featured both zero and negative quantities rather than Greek foundations that did not? Only time will tell. In the meantime, further pedagogical research based on visual instantiations of Brahmagupta's writings on zero, negatives and positives consistent with physical laws (Crabtree, 2018) is suggested.

## References

Bachet, Claude. (1621). *Diophanti Alexandrini Arithmeticorum Libri Sex*. Lutetiae, Parisorum.

Barrow, John D. (2001). *The Book of Nothing*. Vintage, London.

Bashmakova, Izabella G., and Joseph H. Silverman. (1997). *Diophantus and Diophantine Equations*. Mathematical Association of America, Washington, DC.

Boncompagni, Baldassarre. (1857). *Il Liber Abbaci Di Leonardo Pisano*. Fiorentina, Badia.

Borowski, Ephraim, and Borwein, Jonathan. (2012). *Collins Dictionary of Mathematics*. Glasgow, UK: Harper Collins.

Colebrooke, Henry T. (1817). *Algebra with Arithmetic and Mensuration from the Sanskrit of Brahmegupta* (sic) *and Bhascara*. Murray, London.

Crabtree, Jonathan J. (2016). The Lost Logic of Elementary Mathematics and the Haberdasher Who Kidnapped Kaizen. *Proceedings of the Mathematical Association of Victoria Annual Conference 53*, Melbourne, Australia. Retrieved March 2019 from www.mav.vic.edu.au/files/2016/2016_MAV_conference_proceedings.pdf.

Crabtree, Jonathan J. (2017a). Fun in Podo's Paddock: New Ways of Teaching Children the Laws of Sign for Multiplication and Division (and more!). *The Mathematics Education for the Future Project: Proceedings of the 14th International Conference Challenges in Mathematics Education for the Next Decade*, Balatonfüred, Hungary. Retrieved March 2019 from www.bit.ly/TeachingLawsOfSign.

Crabtree, Jonathan J. (2017b). How mathematics teachers can explain multiplication and division in the manner of René Descartes and Isaac Newton. *Proceedings of the 26th Biennial Conference of the Australian Association of Mathematics Teachers* 90–98, The Australian Association of Mathematics Teachers (AAMT) Inc., Adelaide, South Australia.

Crabtree, Jonathan J. (2018). *The Relevance of Indian Mathematics*. Presentation for the Department of Mathematics, Jadavpur University, Kolkata, India. Retrieved March 2019 from www.bit.ly/NewMaths.

Crossley, John N., and Henry, Alan S. (1990). *Thus Spake Al-Kẖwārizmī: A Translation of the Text of Cambridge University Library Ms. Ii. Vi. 5*. Historia Mathematica.

Datta, Bibhutibhusan, and Singh, Avadhesh N. (1962). *History of Hindu Mathematics*. Asia Publishing House, Bombay.

Descartes, R. (1637). *La Géométrie in Discours de la Méthode*. I. Maire: Leyde.

Devlin, Keith. (2012). *The Man of Numbers*. Bloomsbury Publishing, London.

Dvivedin M. Sudhakāra. (1902). *Brāhmasphuṭasiddhānta and Dhyānagrahapadeṣādhyāya*. Printed at the Medical Hall Press, Benares.

Farrar, John. (1818). *An Elementary Treatise on Arithmetic, Taken from the Arithmetic of S. F. Lacroix*, Printed by Hilliard & Metcalf, at the University Press, Cambridge, New England.

Heath, Thomas, L. and Euclid. (1908). *The Thirteen Books of Euclid's Elements, Vol II*. Cambridge: At the University Press.

Heath, Thomas L. (1910). *Diophantus of Alexandria: A Study in the History of Greek Algebra*. Cambridge University Press.

Ifrah, Georges. (2000). *The Universal History of Numbers*. John Wiley, New York.

Joseph, George G. (2016). *Indian Mathematics: Engaging with the World, from Ancient to Modern Times*. World Scientific, London.

Landau, Edmund G. (1966). *Foundations of Analysis: The Arithmetic of Whole, Rational, Irrational and Complex Numbers*. Chelsea Publishing, New York.

Levey, Martin and Petruck, Marvin. (1965). *Principles of Hindu Reckoning*. University of Wisconsin Press, Madison.

Martzloff, J.-C. (2006). *A history of Chinese mathematics*. New York: Springer.

Oaks, Jeffrey A. (2018). Personal communication with Dr Jeffrey Oaks, Professor of Mathematics on Medieval Arabic algebra and the mathematics of Greece and medieval Europe.

Oaks, Jeffrey A. (2011). *Al-Khayyām's Scientific Revision of Algebra*.

Oughtred, W. (1631). *Arithmeticæ in numeris et speciebus institutio: quæ tum logisticæ, tum analyticæ, atque adeo totius mathematicæ, quasi clavis est, etc*. Apud T. Harperum: Londini.

Peano, Giuseppe. (1902). *Aritmetica Generale E Algebra Elementare*. (General Arithmetic and Elementary Algebra) Ditta G. B. and Paravia C., Torino.

Pingree, David. (1981). *Jyotiḥśāstra: Astral and Mathematical Literature*. Wiesbaden: Harrassowitz.

Plofker, Kim. (2009). *Mathematics in India*. Princeton University Press, Princeton.

Prakash, Satya. (1968). *A Critical Study of Brahmagupta and His Works*. Indian Institute. of Astronomical and Sanskrit Research, New Delhi.

Raju, Chandra K. (2007). *Cultural Foundations of Mathematics: The Nature of Mathematical Proof and the Transmission of the Calculus from India to Europe in the 16th c. CE*. Pearson Longman, Project of history of Indian Science, philosophy and culture, Delhi.

Rashed, Roshdi. (2009). *The Beginnings of Algebra*. Saqi, London.

Rorres, Chris. (2003). *Infinite Secrets*, www.pbs.org/wgbh/nova/transcripts/3010_archimed.html.

Rosen, Fred. A. (1831). *The Algebra of Mohammed Ben Musa*. J. L. Cox, London.

Saidan, Ahmad S. (1978). *The Arithmetic of Al-Uqlídisí: The Story of Hindu-Arabic Arithmetic As Told in Kitab Al-Fusul Fi Al-Hisab Al-Hindi*. Reidel, Dordrecht.

Shen, K., Liu, H., Crossley, J. N., and Lun, A. W.-C. (1999). *The Nine Chapters on the Mathematical Art: Companion and Commentary*. Oxford: Oxford University Press.

Sigler, Laurence. (2002). *Fibonacci's Liber Abaci: a Translation into Modern English of Leonardo Pisano's Book of Calculation*. Springer, New York.

Singh, S. (2021). *Chasing Rabbits: A Curious Guide to a Lifetime of Mathematical Wellness*. Impress, San Diego.

Strachey, Edward. (1813). *Bija Ganita: Or the Algebra of the Hindus*. Glendinning, London.

Taleb, N. N. (2007). *The Black Swan: The Impact of the Highly Improbable*. London: A. Lane.

Taylor, John and Bhāskarāchārya. (1816). *Lilawati: Or, a Treatise on Arithmetic and Geometry*. Printed at the Courier Press, by S. Rans, Bombay.

Van der Waerden, Bartel, L. (1985). *A History of Algebra, from Al-Khwārizmī to Emmy Noether*. Springer, Berlin.

Wallis, John. (1685). *A Treatise of Algebra, both Historical and Practical*. Printed by John Playford, for Richard Davis, Bookseller, in the University of Oxford.

CHAPTER 11

# Putting a Price on Zero

*Manil Suri*


## Abstract

Many cultures contributed to the invention of zero, but how to apportion credit? This article explores a new way to do so: imagine royalties were due for every usage of zero, and then figure out how these should be divided among present-day countries. We report on what such a divided allocation might look like, and give reactions, both positive and negative, to it. We also explore how such ideas might be relevant in terms of potential exploitation of the Indian cyber market by foreign corporations.



## Keywords

zero – patent – royalties – Hindu – reparations – discovery – invention – non-European


## 1 Introduction[1]

In the Fall semester of 2017, I was teaching a course on the history of mathematics to a class consisting entirely of mathematics majors. News of the carbon dating of the Bakhshali manuscript came just as we were discussing the origins of zero. To motivate the class, I had them go through an exercise: suppose royalties had been imposed on every use of zero (like for a patentable discovery), how should the money be divided? It was a cute exercise and gave rise to some interesting larger questions, so based on the feedback, I decided to write an opinion piece for the *New York Times*.

The piece was published soon after our class discussion. Right afterwards, I started receiving a barrage of comments by Twitter and email in response. This was a touchier subject than I'd imagined – it seemed to have struck a nerve for some readers. It was also a great teaching moment, since I was able to discuss the comments with the class. In the next section, I present the article,

1 This chapter was first published in the *New York Times* on 8 October 2017, under the title 'Who Invented Zero?'.

 | DOI:10.1163/9789004691568_014

and in the final section, very briefly summarize some reader comments and my class's responses to them.

## 2 Opinion Piece

Carbon dating of an ancient Indian document, the Bakhshali manuscript, recently placed the first written occurrence of the number zero in the third or fourth century CE, about 500 years earlier than previously believed. While the news has no practical bearing on the infrastructure of zeros (and ones) underlying our hi-tech civilization, it does remind us how indebted we are for this invention. But to whom is this debt owed? And how should it be repaid?

Chauvinistic politicians might loudly trumpet India's role (as they have, more controversially, in the case of the Pythagorean theorem), but the history of zero remains unsettled enough to still be the subject of continuing quests. The Babylonians used it as a placeholder, an idea later developed independently by the Mayans. The Chinese, at some point in time, indicated it by an empty space in their counting-rod system. Some claim the Greeks flirted with the idea but, finding the concept of the void too frightening in their Aristotelian framework, passed it on to the Indians. The Hindus are generally acknowledged as being the first to formulate it as an independent number – the key to using it in mathematical calculations or representations of binary code. What's clear is that this history is dominated by non-European civilizations. Truly an alt-right nightmare.

Obviously, there were no intellectual property rights in force back then. Had there been a patent office, it might have ruled, as courts do now, that mathematical advances uncover pre-existing knowledge rather than create anything new – and are hence unpatentable. The conundrum of whether mathematics is discovered or invented is as old as Plato. Certainly, zero displays this duality: The void is as old as time, but it was a human innovation to harness it with a symbol.

In recognition of this innovation, and ignoring all practicalities, suppose someone, somehow, had figured out how to put a price tag on zero. The royalties generated would be staggering – imagine the tab for just your personal use alone! This might lead to a significant redistribution of wealth, most of it going to the developing world.

One difficulty is splitting the payments, since no one could claim exclusive 'ownership' of zero's creation. I asked my History of Mathematics class to come up with an exact breakdown based on zero's provenance, something that, coincidentally, we had just discussed when the carbon dating news broke.

Not unexpectedly, India fared best, with 42% of the proceeds, though students directed it be split with neighboring countries – after all, the manuscript was found in what is now Pakistan (I can already hear the Indian ministers howl their protests).

Babylon ended up with 18%, which if allotted to Iraq, the present-day country of its location, might be just compensation for the years of war endured. Greece came next, with a surprising 15% – perhaps my class felt the country was getting shortchanged for all its other mathematical contributions. The Mayans raked in 14%, which means Mexico would be rolling in so much money from its share that it might be the one clamoring (and paying) for a wall. My class's most left-leaning group declared it wouldn't disburse the money at all, 'so as not to encourage capitalism.'

Of course, the exercise was pure fantasy for many reasons; any compensatory scheme would be dead on arrival based on the mention of 'reparations.' And yet it highlighted the fact that there were cultures and peoples that parented zero, whose descendants may not be doing as well now. If not financial recompense, is at least some enhanced ethical responsibility toward them owed?

If so, the primary onus might fall on tech companies, arguably the biggest users of this resource. Right now, their prize target is India, with Microsoft, Google, and Facebook all vying to bring its enormous population online. These giants might point out that they're already being altruistic by offering free connectivity, through schemes that will plug in rural areas, vitalize the economy, and transform the country – and just happen to add hundreds of millions of potential customers to their rosters for a variety of ads and e-products. Could it be a coincidence that Microsoft, for instance, has also been investing heavily in future cloud services, cybersecurity, and e-commerce for India?

Think of it. The companies will use the indigenously developed resource of zeros and ones (the Arabs got their numerals from India, after all), package them into new services and products, and sell them back. India has lived through such irony before. The British Empire took her raw cotton and sold it back as finished garments, destroying the local textile industry and helping lower India's share of the world gross domestic product to 3% from 23%.

Fortunately, the parallel flounders. The finished e-products will mostly be manufactured in India, even if backed by foreign investment. Also, the country is wiser: It will not succumb easily to a new cyber-colonialism. In 2016, under a broader 'net neutrality' decision, the government banned Facebook's 'Free Basics' plan, which offered free Wi-Fi but only to websites of the company's choosing (Facebook, undeterred, is already marketing a replacement). A year later, it also declined a bid from Microsoft to offer connectivity through old television bands. Instead, under pressure from Indian cellular operators, the bands were to be auctioned off.

Whether Indian tech companies will prevail remains to be seen. What's clear is that vigorous market competition is underway to control all those zeros and ones. Despite my liberal student group's disapproval, zero encourages capitalistic forces, after all.

In fact, zero is essential to much of human endeavor; it has become a fundamental part of our legacy, too seemingly immutable for any kind of compensatory reckoning. And yet the Bakhshali manuscript reminds us that zero wasn't always at hand. Rather, it was the intellectual product of cultures perhaps far different from our own, of peoples and regions that may have subsided but could once again rise to dominance.

## 3 Reader Comments

A number of the emails were appreciative, often accompanied by anecdotes revolving around zero. At least four readers contributed poems, one of them a haiku. Some readers wanted clarifications of various points – for instance, one wrote that she'd always been taught 'zero was a digit, but not a number,' and when had that changed? Other readers pointed out alternative sources that contradicted some of the historical 'facts' I had quoted (which can be found, for example, in popular books by Seife, Kaplan, and Teresi).

Predictably, there was resistance to the idea that Indian royalties should be shared with Pakistan. As one reader put it, the current country of Pakistan had little in common with what existed at the time of the Bakhshali manuscript, so why should it be rewarded? My students pointed out that this was just following the premise I'd laid out – in fact, the same question could be raised about, for instance, allotting Babylon's share to Iraq.

In this regard, several people on Twitter 'liked' the phrase, 'The Hindus are generally acknowledged as being the first to formulate it as an independent number.' This was soon being tweeted with 'Hindus' replaced by '#Hindus' to tie it to a popular platform for posts related to Hindu identity and Hinduism. Of course, my use of the word 'Hindu' was in the historical sense, i.e., indicating geographically coming from the Indian subcontinent, and not in the sense of those practicing Hinduism. Were all these retweets a sign of religiously motivated chauvinistic appropriation?

My students didn't have a problem with it. They pointed out that several popular accounts of the history of zero we'd read in class credited early Hindu religious concepts for having created the conditions needed to recognize zero as a number. The #Hindu tweets were just a manifestation of the same notion.

The idea of considering, even as a purely theoretical exercise, how zero's discovery should be compensated, given all its value to humanity, elicited the most hostility. One particularly industrious reader posted a 27-part tweet, sprinkled with epithets and ending with the verdict, 'Wrong question. Ignorant answer. Missed opportunity.' Others bristled at the idea that non-European civilizations deserved most of the credit or were owed anything. A letter posted over a month after the article appeared was particularly unvarnished in terms of racial rancor, referring to my 'particular shade of brown,' accusing me of propagating a 'self-victimizing birthright' to collect royalties, and charging India with 'the most racially biased naturalization laws on this planet.' The class, though unpersuaded by the arguments, was shocked by the tone.

Perhaps this was the most valuable lesson for my students – a glimpse into how antagonistic the world outside their protected academic environment can be, how even hypothetical thought exercises can give rise to such vitriol. The question of zero's provenance is laden with deeply held feelings that provoke strong reactions. Historians working on it might want to take note.

## References

Suri, M. (2017). Who Invented Zero? *New York Times*, 8 October 2017.

CHAPTER 12

# Revisiting Khmer Stele K-127

*Debra G. Aczel, Solang Uk, Hab Touch and Miriam R. Aczel*

## Abstract

A stone stele code-named K-127 was discovered in 1891 on the east bank of the Mekong River in Sambaur district of Kratié province, Cambodia by the French governor of the province. A recent search to relocate the ancient temple where the stele was found showed only brushes and a piece of plain rock of similar colour to K-127. French epigraphist, George Cœdès translated its inscriptions in 1930 and was astonished to find that it bears the number 604 of the śaka era which is equivalent to 682 CE. This is the first Khmer inscription that shows the date in numeral form. The top right corner of the stele had been broken when it was found, with the first eight lines of the inscriptions partly broken off on the right side. The zero on K-127 is two centuries older than the internationally accepted oldest Indian zero found in the city of Gwalior. In 1931 Cœdès published an article to debunk the belief that our modern decimal number system with the numeral zero as place holder originated in Europe from the Greeks. The origin of the Indian zero comes from the Buddhist logic of *catuskoti* or tetralemma. To underline the logic, a brief discussion between the wandering ascetic Vacchagotta and Buddha is presented. During the Cambodian civil war in the 1970s, the short-lived Cambodian republican government moved many ancient artefacts to warehouses of Angkor Conservation in Siem Reap for safe keeping. When the Khmer Rouge came to power in April 1975, nobody knows what happened to K-127. It was presumed lost. In 2013, an American mathematics professor who had been travelling the world in search of the origin of the numeral zero in our decimal number system, learned about K-127. He set off again on his search through Southeast Asia, and eventually found K-127 in a shed of Angkor Conservation. He organized with the ministry of Culture and Fine Arts to have the stele brought back to the National Museum in Phnom Penh. In July 2018, the Bodleian Library at Oxford University in England announced in newspapers and through YouTube the results of their C-14 analysis of the Bakhshali manuscript found in 1881 in Bakhshali village near Peshawar in today's Pakistan. The age of the manuscript from the C-14 test spans from the third to the tenth century CE. The results were immediately rejected by five experts in Indian mathematics, historians of Indian math, and a specialist in Indian palaeography. So long as the age of the Bakhshali manuscript remains unsettled, K-127 can rightfully claim to bear the world's oldest zero.

 | DOI:10.1163/9789004691568_015

### Keywords

Stele K-127 – Mekong River – Kratié Province – George Cœdès – Catuskoti – Bakhshali manuscript

## 1 Introduction[1]

In 1891, French orientalist, poet and politician Adhémar Leclère, governor of the province of Kratié, Cambodia from 1890–1894, was working in the ruins of Trapaing Prei village in Sambaur district.[2] He found a stone stele which was later code-named K-127 (Figure 12.1).

The numeral zero is represented as a dot • between 6 and 4. *604 was favored by Damais (1952, 1), over the original 605. Cœdès himself later doubted his translation.

In 1930, renowned French epigraphist in ancient Khmer, George Cœdès translated its inscriptions (Cœdès, 1942). The following is an English translation combining the original French translation by Cœdès and the English translation by Jenner (2007):

(lines 1–3) – *List of servants.*

(lines 4–5) – *The year 604 of the Śaka era, the fifth* [*day*] *of the fortnight of the waning moon of* [*the month of*]…., *lunar mansion of.* ….

(line 6–8) – *The lord Vidyākīrti has given dā* …[*the servants*] *to My Holy High Lord Śrī Amareśvar.*

(lines 9–10) – *Warder* (*aṃraḥ*): *male* servants (*vā*)*: vā Tvāl; vā Kvan; dn* ……*; vā Knoc; Female servants* (*ku*)*: ku Vñau* [*and*] *child*(*ren*) *ku Kaṃpañ;*…….[*child*] *vā Tal Tol; ku Croñ; ku Anaṅ* [*and*] *her children vā Tlos, vā Utpala a suckling* □□□□□ṅ*: 10*□*. Five pairs of cows. Respectable people* (*pādamūla*) *in residence: one substitute varī; a special guard, the poñ*□□□.[3]

(line 10) – *Allowance of* [*white*] *rice for their sustenance at* [*the sanctuary: 2 liḥ.*

(line 10–13) – *Clerics* (*paṃjuḥ*) *shall be responsible for the tours of duty of the servants of My Holy High Lord. He who overcomes Ignorance on these premises – he shall pass up to heaven* [*and*] *there remain. Miscreants who would seek to*

1 This chapter is in memory of the late Professor Amir D. Aczel, author of the book, *Finding Zero.*

2 See Appendix 3 for an online petition urging the Bodleian Libraries, Oxford, UK, to take concrete steps to commission the necessary follow-up radiocarbon-dating of the Bakhshali Manuscript in the interest of scientific advancement in the field.

3 □□□ denotes unidentified letters of the inscriptions.

*steal these servants who have been given and this rice for their provision at [the sanctuary of] My High Lord Śrī Amareśvar shall know the sufferings of hell [with] all their kin for as long as the sun and moon [shall shine].*

(lines 14–15) – *The servants given by the lord Iśvaravindu to My Holy High Lord of the golden linga* and *My Holy High Lord Maṇiśiva who shall share [their] use (upabhoga) with My Holy High Lord Śrī Amareśvara.*

(lines 16–17) – *Warder ('aṃraḥ) Vidhival; warder Phal; male servant (vā): vā Pa'em; vā Kma; vā Ksān; vā Tkah; vā 'Añes; vā Kañces; vā Cap Mān; vā Sudatta; vā Ratnapāla; vā Kansiṅ; vā Klaṅ 'Asa; vā Caṭaka.*

(line 18) – *Female servants (ku): ku Vrau [and] 4 children; ku Syam [and] 1 child; ku Nak Dai [and] 2 children; ku Utpala [and] 4 children; ku Kaṃvut [and] 2 children; ku Vilāssā [and] 3 children; ku Pkāy [and] 3 children; ku Nanāṅ.*

(line 19) – *11 oxen; 4 uncastrated [bulls]; 17 water buffaloes.*

(line 20) – *This is what [he] has given to My Holy High Lords of the golden liṅga and Maṇiśiva who are to share [their] use with My Holy High Lord Śrī Amareśvara.*

(line 21) – *Persons who wreak damage on Śiva --[their] fathers [and] mothers onto the seventh generation shall fall into the five Great Raurava hells.*

In 1931, he published an article in the *Bulletin of the School of Oriental Studies* (Cœdès: 1931, 332) to counter English science historian G. R. Kaye who had propounded the idea that India's mathematical knowledge came from the ancient Greeks by way of Persia (Kaye: 1919). However, Cajori (1919) and Gâñguli (1929) stated that there was not a shred of evidence in old manuscripts or inscriptions to support the Greek origin of our numerals.

The numeral zero is one of the most important human inventions, for it enables the use of place-based number representation. This percolates through almost all of our daily lives in commerce, science, and technology, which mostly depend on it to function. Life as we know it today could not be lived without it.

## 2 Ancient Concept of Nothingness

The original concept of zero is 'nothingness', that the Babylonians represented by leaving a blank space and later as  in their sexagesimal (base 60) system. The ancient Hindus and Buddhists in particular, called it *sunyata* meaning 'emptiness'. It arises from the Buddhist logic *catuskoti* or tetralemma propounded by Buddhist philosopher Nagarjuna (*c.*150–*c.*250 CE). In his writing

FIGURE 12.1 Upper part of K-127 showing the date śaka year 604*

in the *Mūlamadyamakakārikā,* he based his analysis on four basic propositions (Anonymous: 2019, p. 3):

1. Affirmation
2. Negation
3. Double affirmation
4. Double negation

Nāgārjuna rejects all firm standpoints and traces a middle path between being and nonbeing. *Catuskoti* had been expounded in *Aggi-Vacchagotta Sutta* – the Fire Discourse to *Vacchagotta* (English translations by Thanissaro, 1997, p. 483; Tan, 2003, p. 483) on the discussions of the wandering ascetic named *Vaccagotta* with *Gotama* Buddha. The discussions are well known in the Buddhist world. For example, there is a painting on the wall of Mogao Grottoes in Dunhuang on the ancient silk road in North Western Gansu Province, Western China. The following is an excerpt of one of the discussions adapted from Aczel's book *Finding Zero* (Aczel, 2015, p. 141):

> Q: Does *Gotama* believe that the *Arahat* (enlightened Bhikkhu) exists after death and that this view alone is true and other false?
> A: Nay *Vacca, Tathāgata* (the way Buddha called himself) do not hold that the Arahat exists after death and that this view is true and every other false.
> Q: Does *Gotama* believe that the *Arahat* does not exist after death and that this view alone is true and other false?
> A: Nay *Vacca, Tathāgata* do not hold that the *Arahat* does not exist after death and that this view is true and every other false.
> Q: Does *Gotama* believe that the *Arahat* both exists and does not exist after death and that this view alone is true and other false?
> A: Nay *Vacca, Tathāgata* do not hold that the *Arahat* exists and does not exist after death and that this view is true and every other false.
> Q: Does *Gotama* believe that the *Arahat* neither exists nor does not exist after death and that this view alone is true and other false?
> A: Nay *Vacca, Tathāgata* do not hold that the *Arahat* neither exist nor does not exist after death and that this view is true and every other false.

In short Buddha said no to everything *Vaccagotta* asked. In the last conversation with Buddha, *Vaccagotta* said '*Gotama*, you said no to everything I asked. Now I am very confused. It negates all my belief in you from our previous discussions.' Buddha said: '*Tathāgata* had rid myself of all theories, greed, attachment, anger, excitement, etc.' Buddha then asked *Vaccagotta* some questions while commenting that it is difficult for people from another belief (*Vaccagotta* was a Brahman) to understand the Buddha's view. At the end of the discussion, *Vaccagotta* understood Buddha perfectly and asked to be initiated into the community of monks. He later reached the *Arahat* status.

The logic of *catuskoti* and the concept of *sunyata* (nothingness) lead to the general acceptance that the ancient Hindu introduced the decimal place holding number system. Nevertheless, the Brahmi numerals (ca. first CE) that all subsequent numerals in the East from Burma to Southeast Asia were derived from, had only nine symbols representing the numbers 1 to 9 and no symbol for zero. Two famous Indian mathematicians and astronomers, Aryabhatta (476–550? CE) and Brahmagupta (*c.*598 to *c.*668 CE) were credited with having used the concept of zero in their arithmetic operations (Dani, 2012). Their use of a symbol for zero was not clear. They mainly referred to zero as '*sunya*' which means empty, or '*kha*' which means space. They wrote their mathematical work mainly in verses that used some accepted words to represent numbers in order to maintain the strict rules of Sanskrit poetic rhymes.

In Cambodia, the earliest known record that followed the Indian method of using word symbols to represent the date is on stele K-151 from Prasat Robaing Romeas (Kampong Thom province). It bears the words *kha, dvi, shara* literally translated as *kha* = space = 0; *dvi* = two = 2; *shara* = arrows = 5. This corresponds to units (0) + tens (2) + hundreds (5) which is 520 śaka year equivalent to 598 CE (Cœdès and Parmentier, 1923, p. 6).

The universally recognized oldest Indian numeral zero is in the number 270 inscribed on the wall of *Chaturbhuj* Temple at Gwalior City in India, dated back to 876 AD (Casselman, 2018). As mentioned earlier about Aryabhata and Brahmagupta, the reason that there was no actual numeral zero in India earlier than the Gwalior zero, is that mathematical works were written in verses. Numeral signs cannot rhyme as 0, 1, 2, 3, etc., unless they are written down as *sunya, kha, vyoma*, for zero; *eka, âdi:, soma*, for one; *dvi:, netra:, . paksha:*, for two; *loka:, kâla:, Haranetra:*, for three, etc.

## 3 The Mystery of K-127

Further to the east of India, George Cœdès discovered in Cambodia the numeral zero inscribed on stone stele K-127. Prior to the Cambodian civil war in the 1970s, K-127 was kept at the National Museum in Phnom Penh. In 1968, the government foresaw the risk of the war in Vietnam spreading into Cambodia and moved K-127 along with other priceless ancient Khmer artefacts to the enclosure of Angkor Conservation in Siem Reap for safe keeping. Eventually, the Khmer Rouge (KR) took control of Cambodia on 17 April 1975 and ruled until 7 January 1979 when Phnom Penh fell to the advancing Vietnamese troops. No one knew what happened to K-127 or its whereabouts. Under this cruel KR regime, there had been looting and sheer destruction of a large number of Khmer archaeological treasures, and K-127 was presumed lost or destroyed.

In 2013, armed with his knowledge of the excellent epigraphic work of George Cœdès, but with scant information on K-127, American mathematician Professor Amir Aczel embarked on a journey through Bangkok (Thailand), Hanoi (Vietnam), Luang Prabang (Laos), Siem Reap (Cambodia) in search of K-127. In his search through Southeast Asia, Aczel spoke to historians, antique dealers, mathematicians, Buddhist monks, and politicians. Finally, he found the missing K-127 stele tucked between hundreds of other ancient steles in a shed of Angkor Conservation's enclosure in Siem Reap. Later, he arranged with the Cambodian Ministry of Culture and Fine Arts to have K-127 brought back to the National Museum in Phnom Penh (Aczel, 2015) where it is currently on display since 2017.

## 4 Social Implication of K-127 on Seventh-Century Cambodia

While ancient stone inscriptions throughout Cambodia used both Sanskrit and vernacular old Khmer language, K-127 is of special interest as it is the first inscription in old Khmer that bears the date in local numerals. It was definitely intended for the populace who could not read Sanskrit. Following the list of gifts of servants, land, and animals, the end of the inscriptions gave stern warnings of punishments to those who steal or desecrate the temple.

The fact that many inscriptions used numerals for the lists of servants, lands, and animals in temples, could allude to some form of mathematics being taught in schools by Buddhist monks as Charles Seife conjectures (Seife, 1997, p. 6). Fillon (1925) suggested that the whole development of arithmetic had been largely a question of notation that eventually evolved from additive, *viz.* the roman arithmetic to the positional notation (the Babylonian sexagesimal system and the decimal system). Faraut (1910) mentioned that astronomy was known to the Khmers since ancient time as they had learned it from India, even before the era of the powerful Khmer Empire of Angkor. Stone inscription K-50 from Vat Prei Voar, Prei Veng Province records the moment of the installation of a statue of Harihara (Visnu and Śiva combined in a single body) in a horoscope expressed in a Sanskrit stanza. It specifies the year in the śaka era, the day of the month, the position of the seven planets in the sign of the Zodiac. The date of this event was equivalent to 667 CE. Another stone inscription of the same year from Kampong Speu Province refers to the installation of a linga by means of a horoscope also expressed in a Sanskrit stanza. Both stone tablets are on display at the National Museum in Phnom Penh. Zhou Daguan, the Chinese diplomat who lived in Angkor then known as Yaśodharapura for almost a year in 1296–97, mentioned that in this country there are also people who know astronomy. They can calculate solar and lunar eclipses. (Uk and Uk: 2016, 61). However, documents that were written on animal hides or palm leaves mentioned by Zhou Daguan did not survive the warm and humid tropical climate for us to study today. Zhou Daguan visited Cambodia at the time that the country was at war with neighbouring Thailand, and was somewhat restricted in places he could explore. Often, he relied on information from local Chinese residents.

## 5 Influence of Ancient India

Cambodian culture as well as that of other Southeast Asia states (Champa in present day Central Vietnam and Srivijaya (in present day Sumatra) were heavily influenced by ancient India. George Cœdès (1964) went so far as to call these civilizations *Les Etats Hindouisés d'Indochine et d'Indonésie* (*The Indianized States of Indochina and Indonesia*), but upon close examination we find a great deal of originality and innovation in the region. An excellent example is the discovery that Cambodians used a dot as a symbol for the numeral zero two centuries earlier than the firmly attested Indian zero at Gwalior.

Anthony Diller (1996) of the Australian National University, quoting Datta and Singh (1962), stated that ancient Indian Sanskrit inscriptions were almost invariably in verse. The poetic constraints of Sanskrit verse forms necessitate the use of word symbols to represent numbers, and the use of direct number symbols was rare.

While Sanskrit stone inscriptions used word symbols in letters for dates, stone carvers in Southeast Asia including seventh century Cambodia, rarely carved numbers in word form in their own vernacular language (Ifrah, 1988). To record specific dates in the *śaka* era, the dates of a king's accession to the throne, a conquest, or other important historical events, they used decimal numerals, strictly adhering to the place holding value system so that the common populace could understand the inscription.

Dominique Soutif (2009), in his doctoral thesis (Organisation Religieuse et Profane du Temple Khmer du VII^e^ au XIII^e^ Siècle. Volume I: Les biens du dieu), made a detailed study of the Khmer ancient numerals. He quoted Cœdès as saying the numerals used in Cambodia were sometimes very different from those in India. No complete series of Indian numerals corresponding to those in Khmer had been identified. Despite the difficulty in determining their origin and date of their introduction, it remains certain that the Khmer numerals, just as the Khmer alphabets, were directly burrowed from India during the sixth *śaka* era (6–7 CE) at the latest. The necessity of representing *sunyata* (emptiness) in the decimal place holding system of numeration, a shape has to be given to *sunyata*, for example to distinguish 11 from 101. The Khmer pre-Angkor and Angkor periods used a dot to represent *sunyata*. (Vong Sotheara, 2003).

## 6 The Bakhshali Manuscript

Until July 2017 the numeral zero in stele K-127 had been recognized internationally as the oldest known recorded zero in the world. On 3 July of the same year, the Bodleian Libraries at Oxford University, England, published the radioactive isotope C-14 dating of the Bakhshali Manuscript (BMS) – a manuscript that had been discovered in 1881 in the Bakhshali village in Peshawar, in today's Pakistan. The manuscript contains valuable ancient arithmetical treatises, but as it was written on birch bark, it was extremely fragile and heavily damaged at its discovery. The late Professor Aczel was impatiently hoping for a C-14 dating study, but because of its extremely fragile state, the Bodleian Libraries did not allow radiocarbon dating analysis on the manuscript for many years. Eventually, the dating work was carried out, possibly as a result of the publication of Aczel's *Finding Zero* in 2015. Three small pieces of the manuscript were taken from three folios (folios 16, 17, and 33).

The results of the analyses were as follows (Howell, 2017):

Folio 16: age 224–383 AD
Folio 17: age 680–868 AD
Folio 33: age 885–993 AD

The age of the BMS folio 16 greatly predates the K-127 zero and could be said to push the Khmer stele into oblivion. However, the Khmer stele bears the name K-127, which is an auspicious prime number as remarked by the late Amir Aczel.

## 7 Issues with the Dating of the BMS

Soon after the news broke in the newspapers, a group of five professors of mathematics, historians of ancient Indian math, and a historian-linguist in Sanskrit, Pali, and Prakrit, from institutions in five different countries (Plofker, et al., 2017) published an article in the *Journal of History of Science in South Asia*, refuting the findings of the Bodleian Libraries C-14 dating. They criticize the sampling method of clipping a clean, un-inked portion of the birch bark, for the C-14 dating tests. Takao Hayashi who wrote his PhD thesis on the BMS at Brown University in 1985 (Hayashi, 1995), commented that considering the palaeographic evolution of Indic scripts, the scripts written in BMS remain consistent throughout, despite the age difference of the three tested folios that spanned over seven centuries. He mentioned that the scribe on folio 17 is no different from that of folio 16 and the content on the reversed side of folio 16 continues on the obverse of folio 17. The Bodleian $C\text{-}_{14}$ results show that the ages of the two folios are separated by three, and possibly five and a half centuries. However, the written content of the two folios treats exactly the same problem on the topic of mixture of gold alloy with different impurities. The transition in handwriting and textual content for leaf 16 to leaf 17 appears to be seamless. The combination of the palaeographic consistency and continuity of content suggests that the manuscript was written as a single unified work. In that case, it follows that the date of the written zeros is the date of the scribe, which is the date of the *latest* of the folios, not the earliest, i.e., in the ninth to tenth century.

The group recommends the possibility of sampling the actual written characters, rather than the un-inked portion of the birch bark folios, since sampling the written area would directly prove the actual date of the writing. They express the importance of cross-checking the findings by testing more than one sample from each of the sampled folios. The group also suggests that more consideration should be given to the ancient techniques of preparation, storage and use of the birch bark, as well as the dating of early manuscript folios, to assess the possible scenarios under which a scribe might use leaves of different dates on which to write a text. The group of mathematicians and historians suggest as follows:

> Without wishing to dampen the laudable ardour shown in this project for scientifically investigating the material characteristics of ancient documents, we urge the investigators to consider the importance of reconciling their findings with historical knowledge and inferences obtained by other means. It should not be hastily assumed that the apparent implications of the results from physical tests must be valid if the conclusions they suggest appear historically absurd (sic) (Plofker, et al., 2017).

The jury is still out regarding the status of the numeral zero in Khmer stele K-127 compared to the zero found in the Bakhshali Manuscript. As long as the results of the C-14 study by the Bodleian Libraries remain controversially unsettled, the K-127 zero remains the oldest known zero. It is currently on display for all to visit at the Cambodian National Museum in Phnom Penh.

## References

Aczel, Amir D. *Finding Zero*. New York: Palgrave Macmillan, 141. 2015. Print.

Anonymous. https://en.wikipedia.org/wiki/Catuskoti. 2019.

Cajori, Florian. The Controversy on the Origin of our Numeral. The Scientific Monthly, Vol. 9, No. 5, pp. 458–64. 1919. https://www.jstor.org/stable/6795.

Casselman, Bill. *All for Nought. By accident, it records the oldest '0' in India for which one can assign a definite date*. Monthly essays on mathematical topics. American Mathematical Society, 6/8/2018. 2018, www.ams.org/publicoutreach/feature-column/fcarc-india-zero.

Cœdès, George. A propos de l'origine des chiffres arabes. *Bulletin of the School of Oriental Studies* (University of London) 6 (2), pp. 323–328. 1931.

Cœdès, George. Inscriptions du Cambodge. Volume II [Texte imprimé] / édité et traduit par G. Coedès. Hanoi: Ecole Française d'Extrême-Orient, 1942. Out of print.

Cœdès, George. Les États Hindouisés d'Indochine et d'Indonésie. Paris: E. de Boccard. 1964. Print.

Cœdès, G. and Parmentier. 1923. *Listes générales des inscriptions et des monuments du Champa et du Cambodge*. Hanoi: *Ecole Française d'Extrême Orient*, 6.

Damais, L. C. 1952. *Bulletin de l'Ecole Française d'Extrême Orient*, tome 6, No. 1, pp. 1–106.

Dani, S. G. *Ancient Indian mathematics – A conspectus*. Resonance. Vol. 17, pp. 236–46. 2012. https://doi.org/10.1007/s12045-012-0022-y.

Datta, Bibhutibhusan and Singh, Avadesh Narayan. *History of Hindu Mathematics: a Source Book. Parts I and II*. Bombay: Asia Publishing House. 1962. Print.

Diller, Anthony. *New zeros and old Khmer*. Mon-Khmer Studies 25, pp. 125–32. 1996.

Dumoulin, Heinrich. *Zen Buddhism: a history, India and China*. New York: Macmillan Publishing, 1998. Print.

Faraut, Félix. Gaspard. *Astronomie Cambodgienne*. Phnom Penh: F. H. Schneider, 1910. Print.

Fillon, L. N. G. *'The Beginnings of Arithmetic.'* The Mathematical Gazette, vol. 12, no. 177, 1925, 401–14. JSTOR, www.jstor.org/stable/3603525. Copy.

Gâñguli, Sâradâkânta. *Notes on Indian Mathematics. A Criticism of George Rushby Kaye's Interpretation. Isis*, vol. 12, No. 1, pp. 132–45. Chicago: The University of Chicago Press. 1929.

Hayashi, Takao. *The Bakhshali Manuscript: An Ancient Indian Mathematical Treatise*. Groningen Oriental Studies, book 11. 1995. Print.

Howell, David. *Carbon dating reveals Bakhshali manuscript is centuries older than scholars believed and is formed of multiple leaves nearly 500 years different in age*. 2017. http://www.sciencemag.org/sites/default/files/Bakhshali%20Research%20Statement_13%209%2017_FINAL.pdf.

Ifrah, Georges. *The Universal History of Numbers*. New York: John Wiley & Sons, Inc. 1988. 402–21. Print.

Jenner, Phillip N. *A Manual of Pre-Angkorian Khmer*. The Dated Inscription. *C* II, pp. 89, *A/C* I, pp. 131–5, IV: 33, JENN MAUAL: *Part* I, No. 26. http://sealang.net/classic/khmer/manual/.

Kaye, G. R. '*Indian Mathematics.*' *Isis*, vol. 2, no. 2, pp. 326–56. 1919. JSTOR, www.jstor.org/stable/223883. Copy.

Plofker, K., Keller, A., Hyashi, T., Montelle, C., Wujastyk, D. The Bakhshali Manuscript: A *Response to the Bodleian Library's Radiocarbon Dating*. *History of Science in South Asia* 5.1 (2017), pp. 134–150. Print.

Seife, Charles. *Zero. The biography of a Dangerous Idea*. New York: Penguin Books. 2000, 6–18. Print.

Soutif, Dominique. *Organisation Religieuse et Profane du Temple Khmer du* VII*ème* *au* XIII*ème* *Siècle* Thèse de Doctorat. Université de la Sorbonne nouvelle – Paris III, 2009.

Tan, Piya. *Aggi Vacchagotta Sutta. The Fire Discourse to Vacchagotta*. Living Word of the Buddha SD, vol. 6, no. 15, MJ 72, 112–27, 2008. www.themindingcentre.org/dharmafarer/.../27.4-Maha-Vacchagotta-S-m73-piya.pdf.

Thanissa, Bhikkhu. *Aggi-Vacchagotta Sutta: To Vacchagotta of Fire*. Majjhima Nikaya, MN 72, PTS: M I 483, 1997. https://www.accesstoinsight.org/tipitaka/mn/mn.072.than.html.

Uk, Solang, and Uk, Beling. *Customs of Cambodia*. Holmes Beach, FL: DatASIA Press, 2016. Print.

Vong, Sotheara. *Pre-Angkorian Stone Inscriptions. 1.* (in Khmer). Buddhist Institute Printing House, Phnom Penh, 2003. Print.

CHAPTER 13

# The Medieval Arabic Zero

*Jeffrey A. Oaks*

## Abstract

Authors of medieval Arabic books on arithmetic often solved problems with 'nothing', and in calculation with Indian (i.e., 'Arabic') numerals, they frequently operated on zero while at the same time claiming that the zero is a sign indicating the absence of a number. We investigate how these authors conceived of nothing and zero from the many remarks they made in their books, and in the context of their understanding of number as an amount or measure of some material or noetic unit.

## Keywords

zero – nothing – Arabic – arithmetic – algebra – notation – decimal – sexagesimal

## 1 Introduction[1]

Today, when compelled to perform calculations without the aid of some electronic device, most people begrudgingly pull out paper and pencil and resort to the one manual way that they learned in school: calculation with Arabic numerals. Even when we do not need to work out the calculations ourselves, the numbers we encounter every day are nearly always shown with the same ten figures whose positions determine their values. These numerals and the rules associated with them have a long history, one branch of which passes from their origin in India to their appropriation in the Islamicate world, and from there to medieval Europe via contact with Muslim practitioners in the Western Mediterranean. Because the Islamic world sits both geographically and temporally between India and the West in this chain of transmission, it is important to investigate their texts with regard to how the zero was understood.

1 This is a condensed version of an article that will appear in *British Journal for the History of Mathematics*. All translations from Arabic are mine unless noted otherwise.

 | DOI:10.1163/9789004691568_016

Reading through the many sample calculations made with Indian figures in medieval Arabic books, we sometimes find the zero added, subtracted, and multiplied with the other digits just as it is today, but at the same time the author will state that the zero signifies an empty place where no number resides. So on the one hand zero seems to be treated as a number, but on the other it is explicitly said to mean 'nothing'. To understand how zero was conceived in medieval Islamicate countries, and whether different authors may have had different ideas about it, we first give a general overview of Arabic arithmetic with special attention to how numbers were conceived. From there we investigate how 'nothing' was treated in problem-solving, and then look into the descriptions of zero and the wording of operations involving it in Arabic books on calculation. Then, after a final example from a book of al-Hawārī, we assess the meaning of zero (*ṣifr*) in medieval Arabic arithmetic.

## 2 The Different Arithmetics of Medieval Islam

Contrary to the way arithmetic is taught in Western schools today, in many cultures, including premodern Europe, there were several techniques of reckoning in use by different groups for different purposes. We review here the most common techniques that were employed in the medieval Islamic world.[2]

The members of several professional groups in Islamicate societies, including merchants, surveyors, and government secretaries, traditionally performed calculations mentally in which intermediate results from 1 to 9,999 can be stored by positioning the fingers in particular ways. This method of calculating was called by various names, including 'finger-reckoning' and 'mental reckoning'. Finger reckoners also worked with fractions, and by the ninth century also with irrational roots. The earliest extant book devoted to finger-reckoning is the *Book of What is Necessary for Scribes, Dealers, and Others from the Science of Calculation* by Abū l-Wafāʾ (328 AH/940 CE–388 AH/998 CE).[3] It is in this tradition of mental calculation that the various methods of arithmetical problem-solving were practiced, including the rule of three, single false position, double

2 The introductions to Abdeljaouad & Oaks (2021) and al-Uqlīdisī (1978) give overviews of Arabic arithmetic.

3 Kitāb fīmā yaḥtāju ilayhi al-kuttāb wa l-ʿummāl wa ghayruhum min ʿilm al-ḥisāb (#256 [M2]), published in Saidan (1971). '#256 [M2]' means book [M1] by author #696 in Rosenfeld and İhsanoğlu (2003). Similarly for subsequent references of this form. Also, we give dates in both the Islamic (AH) and Christian (CE) calendars. Islamic dates are roughly 600 years less than the corresponding Christian dates during the medieval period. For example, 'third/ninth century' means (roughly) the third century AH and the ninth century CE. See https://en.wikipedia.org/wiki/Islamic_calendar.

false position, and algebra. Algebra was the most sophisticated method. Where operations were performed only on known numbers in the other methods, in algebra an unknown was named and operations were performed on this name to set up and solve equations.

It was mainly astronomers who performed sexagesimal calculations. In this base-60 system the places 1 through 59 were traditionally written in what is called *jummal* notation, using the letters of the Arabic alphabet, together with a zero formed from a circle with a line over it that was copied from the Greek sexagesimal system. In some texts Indian numerals took the place of *jummal* numerals. Degrees sometimes play the role of units, so that the minutes, seconds, etc. that follow are fractional amounts. Because the techniques of performing operations in base 10 and base 60 are similar, many books describe them both. The sexagesimal zero is discussed below in Section 6.

We call our numerals 'Arabic' because Europeans learned them from Arabic sources. But in the medieval Arabic books the collection of techniques for working with these numerals was often called 'Indian calculation' (*al-ḥisāb al-hindī*). This system originated in India in the first centuries CE, and by the middle of the seventh century knowledge of its use had reached westward as far as Syria. Writing in Nisibis in 662 CE, the bishop Severus Sebokht noted the superior Indian method of calculating, which 'is done by means of nine signs' (Ginsburg & Smith, 1917, p. 368). However, it may be that Arabic calculators first learned the use of the nine figures 1, 2, 3, 4, 5, 6, 7, 8, 9 and the zero (0) later, in the second half of the eighth century through direct contact with Indian practitioners (Folkerts, 2001; Kunitzsch, 2003).

Calculation with Indian numerals was also sometimes called 'dust arithmetic' or 'board arithmetic', from the dust-board on which the calculations were commonly worked out. This was a board or other flat surface covered with fine sand on which one wrote with a stylus. One important feature of the dust-board for the rules of calculation is that digits can easily be erased and rewritten. The earliest known Arabic book on Indian calculation was aptly titled *Book on Indian Calculation*,[4] written in the first half of the third/ninth century in Baghdad by Muḥammad ibn Mūsā al-Khwārazmī (d. after 232/847). The Arabic original is lost, but we are fortunate that a reworking of a medieval Latin translation is extant. Al-Khwārazmī and later authors cover the shapes of the numerals and the rules for forming numbers, followed by rules for addition, subtraction, doubling, halving, multiplication, and division for whole

4 al-Kitāb fī l-ḥisāb al-Hindī (#41 [M1]). The redaction of the Latin translation with a German translation is published in al-Khwārazmī (1997). An English translation of roughly the first half is published in Crossley & Henry (1990).

numbers and fractions, and root extraction. Fractions in this system were usually borrowed from finger-reckoning.

Arabic mathematicians learned ancient Greek number theory mainly from translations of the arithmetical Books VII–IX of Euclid's *Elements* and Nicomachus of Gerasa's *Arithmetical Introduction.* The philosophical foundation of this theory stems from the writings of Aristotle and was elaborated by Muslim philosophers, most notably by Ibn Sīnā (Avicenna) in the early fifth/eleventh century. At the beginning of Book VII Euclid defines the unit as 'that by virtue of which each of the things that exist is called one', followed by 'A number is a multitude composed of units' (Euclid, 1956 II, p. 277). These units were regarded as being indivisible; thus there could be no fractions. And because a single unit is not a multitude, numbers consist only of the positive integers 2, 3, 4, etc. The unit, 1, was not a number, but was said to be the origin or cause of number. Rather than cover rules of calculation, these Greek authors investigated the classification of numbers into even, odd, and their sub-species; prime numbers and divisibility; amicable and perfect numbers, etc. Several of these results are applicable to practical Arabic arithmetic, such as the sieve of Eratosthenes for determining prime numbers for working with the denominators of fractions.

The transmission of knowledge in medieval Islam was largely conducted orally. Written books were regarded as being transcriptions of lectures, and indeed textbooks were typically recited to students, and books were 'published' in public recitations.[5] Authors wrote out the instructions for making dust-board calculations entirely in words, even spelling out the numbers, since notation serves no purpose to the listener. This was true for algebra as well, so that the polynomials and equations are presented rhetorically in the manuscripts. The Indian figures are shown in arithmetic books as occasional illustrations for what one should write on the board. Occasionally, though, some numbers will be written in notation in the running text, but in a way that does not interfere with recitation. Arabic is read right to left, and when these Arabic books were translated into Latin, the figures showing the numerals were not reversed, causing us to read the numbers backward. In both languages 'one hundred fifty-four' is written as 154, with the units on the right. In Arabic, it is the units that are read first. All figures reproduced in this article appear with the same orientation as shown in the manuscripts.

5 Berkey, 1992; Chamberlain, 1994.

## 3 Theoretical and Practical Numbers in Premodern Arithmetic

Modern numbers have an existence independent of whatever units they may count or measure. When I teach the course Real Analysis to undergraduate mathematics majors, we work with the axioms for an ordered field. We define a real number to be any element of a set that, together with binary operations that we call addition and multiplication, satisfies those axioms. People without this formal training often think of numbers as points lying on some imagined number line, or they might think of numbers as values that arise from counting without regard to anything counted. What these different approaches have in common is that numbers are objects in themselves.

Such was not the case in premodern mathematics. Whether one works with the discrete numbers of Euclid or the continuous numbers of the Arabic practitioners, numbers necessarily measure or count some divisible or indivisible unit. In Euclid's number theory a number is a multitude composed of intelligible, indivisible 'units', so that, for instance, any trio of these units is a 'three' (Mueller, 1981, p. 59). From the many examples in our extant Arabic texts, we know that numbers of practical arithmetic are either counted/measured in material units like men, mithqals (a unit of weight), hours, inches (*aṣābiʿ*), degrees, or dirhams (a silver coin), or they are stated in terms of intelligible, divisible 'units'. These intelligible units are often called 'dirhams', too, reflecting the focus on finance in many practical texts. So, unlike modern numbers, both Euclid's numbers and practical Arabic numbers consist of a quantity – species pair. For the number 'five and a half dirhams', the quantity is 'five and a half' and the species is 'dirhams'. The number *is* the dirhams as seen from the perspective of 'quantity'. Keep in mind that books that explain the rules for calculation rarely mention the species, since it is only the quantity that is relevant when operating on the digits.

We might have thought that we could gain a better understanding of the nature of the numbers of Arabic practitioners from the definitions given in their books. But the authors of books teaching practical calculation, whether finger-reckoning or Indian numerals, presume that students already know what a number is. Most books give no definition at all, including those of Abū l-Wafāʾ, al-Uqlīdisī, al-Samawʾal, and Jamshīd al-Kāshī,[6] to name only a few, and begin straight away with how to work with them. Others, like al-Nīsābūrī (fourth/tenth century) and Muḥammad al-Ḥaṣṣār (d. before 590/1194),[7] give

6 For the first two authors, these are the books already mentioned above. The other two are al-Samawʾal (MS) and al-Kāshī (1969); al-Kāshī (2019).

7 al-Nīsābūrī MS, p. 7.1; al-Ḥaṣṣār MS, fol. 3b.4.

their books a semblance of theoretical rigor by citing some of the characteristics of Greek numbers: that the unit is the origin of number, that a number is a collection of units, and that numbers begin with two. These characterizations, along with the implied indivisibility of those units, are then disregarded as the authors commence work with whole numbers and their fractions.

Thus, we can only ascertain what quantities were regarded as being numbers indirectly, by reading what authors wrote about them and how they worked with them. It is clear from books on arithmetic that any positive quantity that can be arrived at through calculation, including whole numbers, fractions, and irrational roots, were considered to be numbers. Negative numbers were not acknowledged. Although Arabic practitioners knew well how to work with subtracted numbers, these were not admitted as standalone amounts. Whether zero was considered to be a number will be investigated below.

## 4 Arithmetic Problems Leaving 'Nothing'

The problems posed and solved in Arabic arithmetic books are often stated in quotidian terms, such as asking how long it takes for two rivers to fill a reservoir, for the weights of gold and pearl in a piece of jewelry, how long it will take a courier to catch up with another, and how much money each man has who wants to buy a horse. Sometimes, in problems in which a man conducts business, he finds himself left with nothing. Abū Kāmil, an Egyptian author of the late third/ninth century, solves this problem in his *Book on Algebra*:[8]

> A man has some money (māl). He does business with it and profits the same amount, and he gives ten dirhams in alms. Then he does business with it and profits the same as what remained and gives ten dirhams in alms. Then he does business a third time with the remainder. He profits its same and gives ten dirhams in alms, after which he is left with nothing. (Abū Kāmil, 2012, p. 717.12)

The problem is situated in the realm of everyday experience, and it is a common occurrence that people find themselves without any money. There is no need, then, for us to investigate the meaning of this 'nothing' (*la shay'*, literally 'no thing') from a philosophical perspective, or to identify it with our number zero.

8 Kitāb fī l-jabr wa l-muqābala (#124 [M1]), published with a French translation in Abū Kāmil (2012). All translations from Arabic in this article are mine, except where noted.

Abū Kāmil solves the problem by algebra. Before giving the translation, we should explain some vocabulary. The first-degree unknown in Arabic algebra is called a 'root' (*jidhr*) or, more commonly, a 'thing' (*shayʾ*). (One can infer, then, that this 'thing' is not 'nothing'.) The name given to the second-degree unknown in algebra is *māl*, whose ordinary meaning is an 'amount of money'. In the enunciation above the word takes this non-algebraic meaning. Also, we will see below that the word is also found in many arithmetic questions with the meaning of a generic 'quantity'. Which meaning is intended is clear by the context. Units in algebraic expressions were typically counted in intelligible 'dirhams', even in problems where the units have nothing to do with money.

Abū Kāmil begins his solution by naming the amount of money that the man started with as 'a thing', and he then calculates in turn the algebraic expressions that result from the transactions. At the end of the operations he equates the final profited amount with the ten dirhams that the man last gave away, thus avoiding an equation with 'nothing' on one side. In the translation below, I insert modern algebraic expressions corresponding to those in the text in parentheses, even if the modern versions carry a different meaning.[9]

> Its rule is that you make his money a thing ($x$). He does business with it and profits its same, which is a thing ($x$), so it becomes two things ($2x$). He gives ten dirhams in alms, leaving two things less ten dirhams ($2x - 10$). Then he does business with it, profiting its same, so it becomes four things less 20 dirhams ($4x - 20$). He gives ten dirhams in alms, leaving four things less thirty dirhams ($4x - 30$). He does business with it, profiting its same, so it becomes: eight things less sixty dirhams equal ten dirhams ($8x - 60 = 10$), since he said, 'he gives ten dirhams in alms, and he is left with nothing', so what he had was necessarily ten dirhams. Simplify it, to get that the thing is eight dirhams and a half and a fourth ($(8 \frac{3}{4})$), which is his money.

This problem and most others like it that we encounter in the Arabic books are really purely arithmetical problems framed in a material setting. More often, the quotidian dressing is dropped. Abū Bakr Muḥammad al-Karajī, writing in the early fifth/eleventh century, solves this same problem in his [*Book of*] *al-Fakhrī on the Art of Algebra*,[10] but posed in these more abstract terms:

---

9 For the main differences between premodern and modern algebra, see Abdeljaouad & Oaks (2021, pp. 214ff).

10 Al-Fakhrī fī ṣināʿat al-jabr wa l-muqābala (#696 [M6]), published in Saidan (1986).

> A quantity (māl): you double it and you subtract from it ten dirhams, then you double the outcome and you subtract from it ten dirhams, then you double the outcome and you subtract from it ten dirhams, leaving nothing. (Saidan, 1986, p. 171.20)

Other abstract problems that are structured differently also terminate in nothing. For example, the enunciation of one problem in Kamāl al-Dīn al-Fārisī's *Foundation of Rules on Elements of Benefits*[11] (late seventh/thirteenth century) reads:

> If someone said, a quantity (māl): you add to it its fifth and five dirhams and you subtract from the outcome its third and five dirhams, leaving nothing. (al-Fārisī, 1994, p. 536)

Al-Fārisī gives three different solutions to this problem by three different methods: algebra, working backward, and double false position. The result of the operations stated in the enunciation is again 'nothing', in the sense that no amount is left at all. As with al-Karajī's problem, it is an intelligible, not a material, amount of dirhams that is reduced to nothing. Problems like these asking for an unknown *māl*, where the word takes the meaning of a generic 'quantity' and not an amount of money, are very common in Arabic arithmetic. Because the units are typically counted in dirhams, the association with finance would have remained in the backs of the minds of many who worked through the calculations.

I have found in all 36 problems whose enunciations end with 'nothing' – never 'zero' – among the worked-out problems in 21 books by 19 different authors ranging in time from the late ninth century to the first half of the seventeenth century.[12] The problems stem from the finger-reckoning tradition, and in fact only six of these books also explain calculation with Indian numerals. In all, 16 of those 36 problems, from the books of 11 authors, receive solutions by algebra, and in four of them the author takes the liberty to set up the equation

11 Asās al-qawāʿid fī uṣūl al-fawāʾid (#674 [M2]), published in al-Fārisī (1994).

12 These books are: #124 Abū Kāmil [M1], #179 Liber augmenti et dimintionis (a Latin translation of an Arabic book) [M1], #256 Abū l-Wafāʾ [M2], #267 ʿAlī al-Sulamī [M1], #309 al-Karajī [M2], [M3], #310 Ibn al-Samḥ [M1], #521 Ibn al-Yāsamīn [M3], #587 Ibn Badr [M1], #657 Ibn al-Khawwām [M2] (for his [M1], see al-Fārisī), #674 al-Fārisī [M2], #696 Ibn al-Bannāʾ [M6], #780 Ibn al-Qunfūdh [M1], #783 Ibn al-Hāʾim [M1], [M5], #1045 al-Qabāqibī [M1], #873 Sibṭ al-Māridīnī [M7], #924 Zakarīyā al-Anṣārī [M2], #997 al-Ghazzī [M1], #1058 al-ʿĀmilī [M1], #1066 Nūr al-Dīn al-Anṣārī [M2]. If an author repeats a problem in two different books, I only count it once.

with 'nothing' on one side. In problem II.5 from his *Book on the Fundamentals and Preliminaries in Algebra*[13] (late seventh/thirteenth century) Ibn al-Bannāʾ solves the same problem of the businessman as Abū Kāmil, but this time the man gives away one dirham in alms at each stage instead of ten. In his solution Ibn al-Bannāʾ sets up the equation after the last of the transactions: 'Then gaining its same and giving a dirham in alms leaves him with eight things less seven dirhams, and this equals nothing' (Saidan, 1986, p. 572.6). Reinterpreted in modern notation, the equation would be $8x - 7 = 0$. It might be tempting to say that because the 'nothing' occupies a place in the equation where a number would ordinarily be, that this 'nothing' must be a number. For comparison, consider how we write intervals of real numbers: $(0, 1)$ is the open unit interval, $(-10, 10)$ consists of all numbers strictly between $-10$ and $10$, etc. But we also write intervals like $(0, \infty)$ and $(-\infty, 1)$, where the signs $\infty$ and $-\infty$ occupy a place ordinarily reserved for a number. Of course, infinity is not a real number, but by a convenient abuse of notation it can take the place of a number when writing unbounded intervals. Similarly, we should consider that perhaps the 'nothing' on one side of the equations in these four problems was not regarded as a number, but conveniently expresses the 'nothing' that is left when an equal is removed from an equal. The prospect of writing such an equation must have made some algebraists feel uneasy, since most authors avoid it. Both the 'nothing' in Arabic and the 'infinity' in modern mathematics lie at a boundary of what was/is considered to be a number.

This nothing in an equation is shown in notation in one book. Ibn al-Qunfūdh (740/1339–810/1407) wrote *Removing the Veil from the Faces of Arithmetical Operations*[14] as a commentary on Ibn al-Bannāʾ's late seventh/thirteenth century *Condensed Book on Calculation*,[15] which he completed in 1370 CE. In it he solves many problems, among them the same problem just translated from Ibn al-Bannāʾ's algebra book, with only some minor differences in wording. Unlike the other books that give algebraic solutions to problems that result in nothing, this one shows the algebraic notation that was practiced in the Western part of the Islamic world beginning no later than the end of the sixth/twelfth

13 Kitāb al-uṣūl wa l-muqaddimāt fī l-jabr wa l-muqābala (#696 [M6]), published in Saidan (1986). The other three are in the books of Ibn Badr, al-Fārisī, and Ibn al-Qunfūdh. Al-Fārisī may have avoided setting up an equation with 'nothing' on one side in the problem translated above, but he set one up in another problem.

14 Ḥaṭṭ al-niqāb ʿan wujūh aʿmāl al-ḥisāb (#780 [M1]), in manuscript (Ibn al-Qunfūdh, MS).

15 Talkhīṣ aʿmāl al-ḥisāb (#696 [M1]), published with a French translation in Ibn al-Bannāʾ (1969). The entire text is included in the commentary by al-Hawārī, which is published with an English translation in Abdeljaouad & Oaks (2021).

century.[16] This is a notation that was written on the dust-board, and which was rendered into rhetorical form in the writing of books. We are fortunate that Ibn al-Qunfūdh took the pedagogical approach of showing what the student should write on the board at different stages of the solutions to problems.

In this notation the quantity for each power is shown in Indian numerals underneath the first letter of the name of the power, a *shīn* (ش) for *shay*ʾ ('thing'), a *mīm* (م) for *māl*, etc. For '=' the notation shows the letter *lām* (ل), the last letter in *taʿdil* ('equals'). In this problem Ibn al-Qunfūdh shows the notational version of the equation that is written in the text as 'eight things less seven equal nothing' (Ibn al-Qunfūdh MS, p. 304.18). We reproduce it below with some of the surrounding text together with a modern transcription. Here only the three dots of the letter *shīn* are shown:

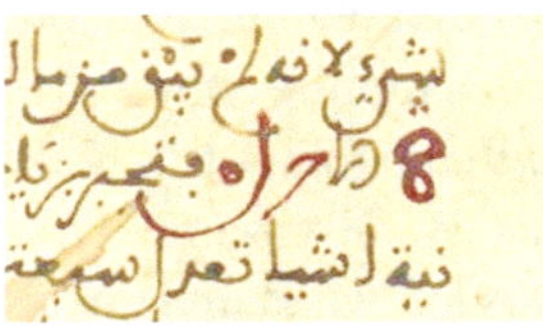

∴
8 ℓ 7 = 0

FIGURE 13.1
Notational version of 'eight things less seven equal nothing'

Reading the middle line of the image beginning on the right, you see the '8' with the three dots above it. To its left is the word *illā*, which I write in the transcription as 'ℓ' for 'less'. Next there is a '7', a large *lām* (ل) for '=', and 'nothing' is shown as a small circle, which is the Arabic zero. Ibn al-Qunfūdh needed a way to represent the 'nothing' that remains after subtracting the last dirham, and which we read in the rhetorical version of the equation. Since the notation employs Indian numerals, he chose to show this 'nothing' as a zero. This is the only instance I have seen of a zero taking a place where an amount would normally be in any Arabic text. Just what is meant by this zero will come into better focus in the next section, where we finally look into how zero was described and applied by various Arabic authors in Indian calculation.

## 5 Writing Numbers with Zeros in Indian Arithmetic

I checked 35 books by 32 different authors that explain calculation with Indian numerals.[17] One of them is the Latin redaction of al-Khwārazmī's book and

16 For an overview of this algebraic notation, see Oaks (2012).

17 The 35 books are: #41 al-Khwārazmī [M1], #218 al-Nīsābūrī [M1], #232 al-Uqlīdisī [M1], #308 Kūshyār ibn Labbān [M1], #320 al-Baghdādī [M1], #411 al-Ṣardafī [M1], #487

the rest are in the original Arabic. They range in date across a span of nine centuries, from the first half of the third/ninth to the first half of the twelfth/eighteenth century, and in place from Morocco to central Asia. Some of them were written by prominent mathematicians and scientists such as al-Samaw'al, Naṣīr al-Dīn al-Ṭūsī, and Jamshīd al-Kāshī, while others were written by lesser-known individuals. For the writing of the numerals, Ibn al-Bannā''s *Condensed Book* is an outlier. It is truly rhetorical in that it does not show the numerals at all, and no examples are given to illustrate the rules. All of the other 34 books begin by showing the nine figures for 1 through 9 and introduce the zero only later. The forms of the numerals vary across time and place and fall into two basic types. A manuscript of al-Nīsābūrī's *Attainment of Students on Truths in the Science of Calculation*,[18] copied in 843/1443, shows the eastern forms as follows:

FIGURE 13.2 Eastern forms of figures 1 through 9 from al-Nīsābūrī's *Attainment of Students on Truths in the Science of Calculation*

A manuscript of Ibn al-Qunfūdh's *Removing the Veil from the Faces of Arithmetical Operations*, copied in the fifteenth or sixteenth century CE, probably in Morocco, presents the Western forms as:

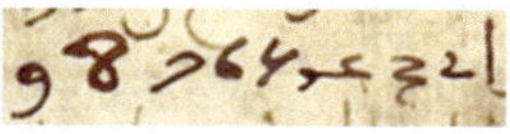

FIGURE 13.3 Western forms of figures 1 through 9 from Ibn al-Qunfūdh's *Removing the Veil from the Faces of Arithmetical Operations*
Note: al-Nīsābūrī MS, p. 20; Ibn al-Qunfūdh MS, p. 12

al-Samaw'al [M4], #521 Ibn al-Yāsamīn [M3], #532 al-Ḥaṣṣār [M1], #566 Ibn Mun'im [M1], #606 Naṣīr al-Dīn al-Ṭūsī [M17], #682 al-Abharī [M1], #696 Ibn al-Bannā' [M1], [M2], #747 al-Hawārī [M1], al-Mawāḥidī (#M176 in (Lamrabet 2014)), #780 Ibn al-Qunfūdh [M1], #783 Ibn al-Hā'im [M5], [M7], #802 al-Kāshī [M1], #815 Ibn al-Majdī [M3], #865 al-Qalaṣādī [M3], [M7], #980 Raḍī al-Dīn ibn al-Ḥanbalī [M3], #997 al-Ghazzī [M1], #1004 Taqī al-Dīn [M2], #1006 Yaḥyā al-Ru'aynī [M1], #1026 al-Sakhāwī [M1], #1045 al-Qabāqibī [M1], #1058 al-'Āmilī [M1], #1066 Nūr al-Dīn al-Anṣārī [M2], #1186 Jawād ibn Sa'd ibn Jawād al-Kāẓimī [M1], #1355 Ḥusayn al-Maḥallī [M2], the guide by the lexicographer Muḥammad ibn Aḥmad al-Khwārazmī (ed. G. van Vloten, 1895), and the anonymous Paris, BnF arabe 4441, copied in 1572.

18 Bulūgh al-ṭullāb fī ḥaqā'iq 'ilm al-ḥisāb (#218 [M1]), in manuscript (al-Nīsābūrī MS).

Both authors show several variations. The modern Arabic forms derive from the eastern versions, and the modern European forms derive from the Western versions. In the translations below I write 1, 2, 3, etc., for whatever form an author used.

Here are some examples of the ways that zero is introduced in these books. The Latin redaction of al-Khwārazmī's book begins by showing the nine numerals and explaining some variations. It is only later, as part of the description of the place value system, that the zero is brought up in the context of how to write the number ten using the numeral '1':

> So they put one space in front of it and put in it a little circle like the letter o, so that by means of this they might know that the place of the units was empty and that no number was in it beyond the little circle, which we have said occupied it.[19]

The earliest book on Indian calculation extant in Arabic is Ahmad ibn Ibrāhīm al-Uqlīdisī's *Chapters on Indian Reckoning*,[20] completed in Damascus in 341/952–3. He began the first chapter with: 'The beginner in this science should be first taught the nine letters. These are 1 2 3 4 5 6 7 8 9'. Then, after describing the place value system along with several examples, he explained the zero:

> Then he is informed that some of these places may be empty, with none of the nine letters falling in it. If this is the case, a circle is set in the empty place. This is what the people of this craft call a zero.[21]

Later in the same century al-Nīsābūrī showed the forms of the numerals, covered the place value system, and worked through the operations of doubling and halving before explaining that in each rank, 'if there is no number in it, place there a zero, which is a small circle' (al-Nīsābūrī MS, p. 16.6). The algebraist al-Samawʾal (d. 570/1174–5) introduced zero in his book on Indian calculation with: '[what] you write in an empty rank to take the place of its emptiness is called a zero, and it is this: o' (al-Samawʾal MS, fol. 3b.14). About the same time in Western North Africa, al-Ḥaṣṣār showed the nine figures and then

19 Crossley & Henry, (1990, p. 111), adjusted slightly from their translation.

20 al-Fuṣūl fī l-ḥisāb al-Hindī (#232 [M1]), published in al-Uqlīdisī (1984). An English translation is published in al-Uqlīdisī (1978).

21 al-Uqlīdisī (1978, p. 42), adjusted slightly from Saidan's translation.

explained zero in his instruction on how to write the number ten in his *Book of Demonstration and Recollection in the Art of Dust-board Calculation*:[22]

> One needs to put something in the units rank to establish the rank, so that one will know that the units rank has nothing in it. Thus the arithmeticians put a zero in it, in the form of a small circle. (al-Ḥaṣṣār MS, fol. 5a)

A few decades later, at the other end of the Muslim world, the famous Persian astronomer, philosopher, and theologian Naṣīr al-Dīn al-Ṭūsī (597/1201–672/1274) wrote a textbook on Indian calculation titled *Gathering of Arithmetic by Means of Board and Dust*.[23] About a page after showing the forms of the numerals, he wrote: 'and you put a zero, in the form of a small circle, in every rank devoid of a number' (Saidan, 1967, p. 115.1). Still later, in 830/1427, the Persian astronomer and mathematician Jamshīd al-Kāshī (d. 832/1429) wrote in his *Key to Calculation*[24] after showing the nine figures: 'A zero, in the form of a small circle, must be placed in every rank where there is no number, to avoid a gap in the ranks.'[25] Our last example is from Nūr al-Dīn al-Anṣārī's *Achievements of the Giver in the Pleasure of the Arithmeticians*,[26] completed in 1039/1629–30. Three pages after showing the nine figures he wrote: 'And the zero, spelled with an "ṣ",[27] is a sign for an empty place, and this is its figure: o' (al-Anṣārī MS, fol. 47a.24). None of the 34 books that I consulted diverge from the view that the zero is a sign for an empty place where there is no number.

At this point I should clarify what is intended by the word 'zero'. The Arabic word for zero is *ṣifr*, which ordinarily means 'empty, void, or vacant' (Lane, 1893, p. 1697). The sign for zero is described and shown in the manuscripts as a small circle and, as the texts repeat, it is a sign for an empty place. There are three parts to this relation that are better understood by comparison with an example from Greek arithmetic. Numbers in Greek were written using letters of the alphabet, beginning with 'A' for 1, 'B' for 2, and so on. So the Greek word *beta* is the name of the sign 'B', which means 'two'. Similarly, the Arabic word *ṣifr* is the name of the sign 'o' which means 'vacant' or 'empty'. Here the *ṣifr* is a technical term: the 'o' was never called by other words that mean 'nothing',

---

22 Kitāb al-bayān wa l-tadhkār fī ṣanʿat ʿamal al-ghubār (#532 [M1]), in manuscript (al-Ḥaṣṣār MS).

23 Jāmiʿ al-ḥisāb bi l-takht wa l-turāb (#606 [M17]), published in Saidan (1967).

24 Miftāḥ al-ḥisāb (#802 [M1]), published in (al-Kāshī, 1969) and (al-Kāshī, 2019), the latter covering the first, arithmetical part together with an English translation.

25 al-Kāshī (1969, p. 46.16; 2019, p. 34.15).

26 Fatḥ al-wahhāb ʿalā nuzhat al-ḥussāb (#1066 [M2]), in manuscript (al-Anṣārī MS).

27 This is an oral instruction not to misspell *ṣifr* ('zero') as *sifr* ('book of scripture').

'vacant', or 'empty', like *la shay'*, *khalā'*, or *'adam*. When I write 'zero' I mean the name of the sign 'o' that is used in Arabic books on Indian arithmetic. So far, taking our authors at face value, the zero is a sign designating a place where there is no number. It is thus not a sign for a number, and the 'nothing' that resides in a place labeled with a zero should not be thought of as a number, either.

This practice of putting a special sign where there is no number is much like the modern practice of putting a dash (–) or an asterisk (*) or some other sign in a table where data is lacking. Because we are accustomed to read 0 as a number, it might be helpful to replace it in the examples below with a different special sign, like •, at least for a while, to clear our heads of our modern reading. This way, for example, three thousand fifty-four looks like 3•54. The • indicates that there is no number in the hundreds place.

## 6 Operating on Numbers with Zeros in Indian Arithmetic

The rules for operating on numbers expressed in Indian notation call for the addition, subtraction, multiplication, doubling, and halving of the individual places, and often one or both of these places is occupied by a zero. There were different approaches to working out operations involving a zero. Most authors took the view that because there is no number in that place, there is no operation to perform. Others performed the operation with zero, and those who explained it either related that this was done just to 'keep the place' or they said what it means to operate on 'nothing'. Below we cover, in order, addition, subtraction, multiplication, and halving. Like most Arabic rules for operating on the numerals, those described below were intended for the dust-board, since they call for the erasing and rewriting of digits as the calculation progresses. In some rules one of the two numbers is transformed, one digit at a time, into the answer, while in others the answer is written on a separate line.

Ibn al-Yāsamīn (d. 601/1204) wrote in his explanation of addition with Indian numerals: 'Then you add each digit of one of the addends to its counterpart in the other. If there is no counterpart, then the answer is the addend, as if it had a counterpart' (Zemouli, 1993, p. 131.11). By this rule, no addition is performed, but an answer is obtained anyway. Al-Hawārī gave an example in his *Essential Commentary on the Condensed [Book] on the Operations of Arithmetic.*[28] He began the problem of adding 4•43 to 2685 by arranging the numbers one above

28 al-Lubāb fī sharḥ Talkhīṣ aʿmāl al-ḥisāb (#747 [M1]), published with an English translation in Abdeljaouad & Oaks (2021).

the other like this: $\begin{matrix}4\bullet43\\2685\end{matrix}$. He began the work from the units place, and after the 6 had been replaced with a 7 from carrying a 1, he wrote: 'Nothing in the upper line corresponds to the seven, so it is considered to be the sum of that rank and that of its counterpart as if it had something' (Abdeljaouad & Oaks, 2021, p. 45).

Most other authors similarly did not mention the operation of addition when there is a zero, like al-Baghdādī (d. 429/1037), who gave this instruction in his *Completion of Calculation*[29] for a rule in which the top number will be replaced with the answer: 'and if above the place of the added number there is a zero, then put the added number in place of the zero' (al-Baghdādī, 1985, p. 36.17). This was how al-Ḥaṣṣār added 65•3 to 7•2•, in which the answer is written on a separate line above. For the units, tens, and hundreds he made no mention of addition, but simply put the 3, 2, and 5 above the numbers, and then wrote, 'Then add the thousands, and that is seven and six, to get thirteen' (al-Ḥaṣṣār MS, fol. 9a.10), which he then also put in the upper line.

Al-Qalaṣādī stated the rule in a way consistent with the rules and examples above. For each place, 'if there is in one of them a zero and in the other a number, then the number which is in one of them is the answer. And if [instead] there is a zero there, then one of the two zeros is the answer' (al-Qalaṣādī 1999, p. 45.13). But then in his example of adding $\begin{matrix}5\bullet23\\25\bullet2\end{matrix}$ he explicitly added zero, without explaining what it means. Working from right to left, the answer will be written on a separate line above them. He wrote: 'Add the two to the three giving five. Put it in the units rank. Then add the zero to the two, giving two, which is the answer. Then add the five to the zero, giving five. Put it above that place …' (al-Qalaṣādī, 1999, p. 46.2).

Addition was, of course, performed place by place outside the context of Indian calculation, too. The process is so simple that it is not explained in books on finger-reckoning. To add 'three hundred seven' to 'fifty-two', for example, one easily arrives at the answer 359. The tens place of the answer retains the five of the 'fifty-two', since there is nothing in the tens rank of 'three hundred seven' to add to it. Seen this way, Ibn al-Yāsamīn's rule that 'If there is no counterpart, then the answer is the addend, as if it had a counterpart' is intuitively clear. The rule does not require any actual addition with zero, and it certainly does not imply a shift to a formal way of reading the operation. And then, rarely an author like al-Qalaṣādī will state the operation anyway. It is plausible that he understood the addition of • to 2 like the addition of nothing to a purse containing two dirhams, which leaves its contents unchanged,

---

29 al-Takmila fī l-ḥisāb (#320 [M1]), published in al-Baghdādī (1985).

or that the addition of 3 to • is like the addition of three dirhams to an empty purse, resulting in a purse with three dirhams.

Among the books I consulted, the subtraction of zero is routinely skipped. I found only one instance where the operation is explicitly performed, in al-Hawārī's *Essential Commentary*. To subtract • from 9 in one example, he wrote: 'So we subtract this nothing of the minuend from the nine of the subtrahend, leaving nine' (Abdeljaouad & Oaks, 2021, p. 50). Subtracting nothing means that nothing is taken away, so the nine remains unchanged. When a number is subtracted from its equal, we sometimes read, as in al-Qalaṣādī, 'Then subtract the two from the two, which leaves nothing. Put a zero above the two subtrahends' (al-Qalaṣādī, 1988, p. 37).

Multiplication is more interesting. Most authors stated the Euclidean definition in the beginnings of their chapters on multiplication, like Ibn al-Yāsamīn, who opened with: 'Know that multiplying numbers, one of them by the other, is that you duplicate one of the numbers by the quantity of what is in the other number in units' (Zemouli, n.d., p. 11.5). Although this definition does not cover multiplication of non-integers, it is applicable to Indian calculation, where the operation is performed on the nine digits.

The most common dust-board technique of multiplying numbers in medieval Arabic books calls for shifting and erasing, which I explain now so that the quotations given below will make sense. Here is an example paraphrased from al-Uqlīdisī (1978, p. 50) for multiplying 34 by 26. First, write the numbers with the units place of the multiplier under the highest place of the multiplicand, like this: $\begin{array}{ccc} & 3 & 4 \\ 2 & 6 & \end{array}$. Then multiply the 3 by the 2 and put the 6 above the 2: $\begin{array}{ccc} 6 & 3 & 4 \\ 2 & 6 & \end{array}$. Then multiply the 3 by the 6 in the lower line and replace the 3 as you add the resulting 18 to the 60 above it: $\begin{array}{ccc} 7 & 8 & 4 \\ 2 & 6 & \end{array}$. Next, shift the 26 to the right one place: $\begin{array}{ccc} 7 & 8 & 4 \\ & 2 & 6 \end{array}$. Then multiply the 4 by the 2 and add the 8 to the 78 above the 2: $\begin{array}{ccc} 8 & 6 & 4 \\ & 2 & 6 \end{array}$. Last, multiply the 4 by the 6, and replace the 4 with the resulting 24, to get $\begin{array}{ccc} 8 & 8 & 4 \\ & 2 & 6 \end{array}$. The answer is 884.

As with addition and subtraction, many authors did not perform multiplication by zero. Naṣīr al-Dīn al-Ṭūsī wrote this in his verbal explanation of the method just described:

> And if there is a zero in the rank of the multiplicand, we do not need to shift to it, so we shift to what is on its right and calculate the remaining ranks. And if there is a zero in the rank of the multiplier, there is no need to multiply by it. (Saidan, 1967, p. 122.18)

Then, in his first example he multiplied 2•76 by 5•3, and after multiplying the 2 by the 5, wrote: 'and it is not necessary to multiply the two by the zero, so we place a zero above it' (Saidan, 1967, p. 123.2). He may not have performed the operation, but he nevertheless put a zero where it should be.

If we read these rules in light of the standard finger-reckoning rule for multiplying numbers with more than one rank, this practice of not performing multiplications with zero makes sense. Al-Karajī stated this rule as: 'you multiply each of the numbers [i.e., ranks] of the multiplicand by all numbers of the multiplier' (al-Karajī, 1986, p. 39.9). To multiply 305 by 24, for example, only four products are necessary because there is literally nothing in the tens rank of the first number. Thus the zeros that are encountered in Indian notation can be disregarded, since they designate places where there is no number.

Al-Uqlīdisī and Isḥāq Ibn-Yūsuf al-Ṣardafī omitted multiplication by zero in some calculations and performed it in others. In his *Condensed Indian [Book] on the Science of Arithmetic*[30] al-Ṣardafī (d. ca. 500/1106) multiplied 2•2•2 by 22 by multiplying only the 2s by 2s, but later on the same page he began the multiplication of 2••••2 by 2•3••3 with: 'Multiply two by two to get four. Then multiply the zero by the two to get zero and put it above the zero …' (al-Ṣardafī MS, fol. 6b.27). Here the reason for the difference may be that in the first problem the answer 444444 has a number in each place, so there was no need to multiply by zero, but in the second problem if the zero is not placed next to the first calculated 4, and for other places as well, the answer 4•6•1••6••6 would have gaps in the ranks. This is what al-Uqlīdisī was referring to in his explanation of multiplication with zero:

> If it is said: Why is zero by zero equal to zero and zero by any letter zero? We say that by multiplying zero by zero the aim is only to occupy the place; the same applies for multiplying the letter by zero. We multiply the letter by zero only once, the first time, by the first letter in the upper, to occupy the place, and tell that there is a place and that it is empty.[31]

Two authors explained what it means to multiply by zero by recalling the duplication definition of the operation. Ibn Munʿim (d. 626/1228) explained it this way in his *Understanding Calculation*:[32] 'And know that anything multiplied by

30 Mukhtaṣar al-hindī fī ʿilm al-ḥisāb (#411 [M1]), in manuscript (al-Ṣardafī MS).

31 al-Uqlīdisī (1978, p. 190), Saidan's translation.

32 Fiqh al-ḥisāb (#556 [M1]), published in Ibn Munʿim (2005).

zero is zero, since the zero is a sign for nothing (*ʿadam*), and nothing is gathered from duplicating it' (Ibn Munʿim, 2005, p. 33.21). Ibn al-Bannāʾ covered it close to a century later in his *Lifting the Veil*: 'Multiplying a number by a zero or a zero by a number is that you zero the number or you duplicate the zero, and neither of them gives a number. So its sign is always a zero.' Note the use of the related verb 'to zero', which can also be translated as 'to empty' or 'to void'.

Other authors simply stated that the product is zero, without any explanation. Al-Baghdādī wrote in one calculation: 'Then we also multiply the five by the zero which is in the lower number to get zero, so we put a zero above the zero' (al-Baghdādī, 1985, p. 52.15), and al-Kāshī explained while working out one product, '… since multiplying the zero by whatever number gives a zero' (al-Kāshī, 1969, p. 52.13; 2019, p. 48.18).

Zero makes its appearance also in doubling and halving, and the explanations there are similar. While taking half of a particular number, al-Kāshī wrote: 'Since there is no half for zero, we put a zero underneath it'.[33]

Taking into consideration the wordings in all of the operations performed in these books, the authors did not waver from the position that zero designates an empty place where there is no number. The fact that in some instances the operation with zero was omitted, while in others it was explicitly performed, are not signs of a difference in how the zero was conceived. If it seems irreconcilable that zero can mean 'nothing' and still be operated on, it may be because we are still thinking in terms of modern mathematics. Today mathematicians define numbers to be elements of a set on which binary operations we call addition and multiplication are defined. The operations exist only in the context of that set, so adding $a$ to $b$ only makes sense if both $a$ and $b$ are numbers. But premodern arithmetic has no foundation in axioms. Numbers consist of a quantity – species pair, like the material three mithqals of grain or nine apples or the intelligible three and a third dirhams. Addition is simply the gathering together of two amounts, which is why even Euclid saw no reason to define it. To add three apples to five apples means simply to group them together into a collection of eight apples; similarly for subtraction, where one removes an amount from another. The meaning of multiplication is less immediate, so Euclid gave a definition in this case. But even there the operation only consists of a repeated duplication. Multiplication is governed not by axioms, but by the intuitive notion of accumulating together copies of a number.

Adding, subtracting, and multiplying 'nothing' make sense with this more material way of conceiving of the operations. It is commonplace in the physical world to add nothing to something or to take away nothing from

33 al-Kāshī (1969, p. 48.9; 2019, p. 38.15), Aydin's and Hammoudi's translation.

something, and duplicating nothing makes sense as well. And just as calculation with intelligible numbers remained somehow connected to the world of finance through repeated use of the word *māl* for 'quantity' and dirhams for units, the way people conceived of operations on these numbers would have likewise remained conceptually like the operations in the material world. So it is no contradiction for an Arabic arithmetician to regard zero as signifying nothing, and at the same time to operate on it. There is no evidence in any of these texts on Indian calculation, even those by prominent mathematicians, that an author thought of zero as anything other than a sign for a place devoid of a number.

At this point we can say a little more about how to read the zero in the notational version of Ibn al-Qunfūdh's algebraic equation shown above at the end of Section 3. There the zero is not a sign for an empty place. It designates the 'nothing' that the businessman was left with as well as the same 'nothing' stated in the rhetorical version of the equation. This is a new use of the zero, one that reflects the meaning of the word *ṣifr*, and which at the same time is not inconsistent with the meaning of the sign in Indian arithmetic. The algebraic notation uses Indian numerals for the quantities of the powers, and rather than write the words 'no thing', Ibn al-Qunfūdh, and probably many others whose dust-board calculations are irretrievably lost, chose to borrow the zero for this purpose. This innovation might have had the potential to lead to a more formal view of zero. But with the majority of algebraists avoiding equations with 'nothing' and with only this one notational example, we have no evidence that this happened.

## 7 The Zero in Arabic Sexagesimal Arithmetic

It should not be a surprise that the zero in base 60 is described in Arabic texts in the same way as the zero in base ten. For books in which Indian numerals were drawn to show the sexagesimal places, two digits are usually shown for each place even if the entry is less than ten, and the places are arranged vertically on the dust-board. For example, Naṣīr al-Dīn al-Ṭūsī set up the subtraction of 'three signs, eleven degrees, twenty-five minutes, and fifty seconds from two signs, eight degrees, and ten seconds' on the dust-board as:[34]

34 Saidan (1967, p. 254). The '02' is mistakenly shown as '03' in the text. The word burj designates 1/12 of the circle, or 30 degrees, and takes the meaning of 'sign' of the zodiac. It is not a problem that the minuend is smaller than the subtrahend in this example. He added 12 signs to the minuend and then performed the subtraction.

| | |
|---|---|
| 02 | 03 |
| 08 | 11 |
| 00 | 25 |
| 10 | 50 |

Here are some passages that explain zero in sexagesimal arithmetic. For calculation with *jummal* numerals, al-Bīrūnī explained: 'When zero has to be written in places lacking a number, its circle must have a line over it …'[35] For calculation in which the places are written in Indian notation, al-Uqlīdisī wrote in the context of addition: 'If a place is empty, having nothing, we put two zeros in it to indicate that there is nothing in it', and for subtraction he wrote: 'Whenever a place becomes empty, we insert a zero in it'.[36] Al-Baghdādī wrote in one subtraction problem: 'Then we subtract the two degrees from the two degrees, and we put in place of the two degrees two zeros, so there is nothing in its place' (al-Baghdādī, 1985, p. 148.6). And al-Kāshī gave this instruction for writing the numbers: 'At each position where there is no number, a zero is put to clarify any confusion.'[37] I have found no instances in sexagesimal calculations for which zero was treated differently than it was in base ten calculations.

## 8 Operating on Nothing in a Passage of al-Hawārī

There is one last noteworthy passage to review, from al-Hawārī's *Essential Commentary*, a work that supplies numerical examples and clarifying remarks to Ibn al-Bannā''s *Condensed Book*. In the chapter on algebra al-Hawārī discussed the rules given by Ibn al-Bannā' for solving the six simplified algebraic equations of degree one and two. The fifth type of equation had been written as 'some *māl*s and a number equal some roots' since the time of al-Khwārazmī (2009, p. 101.3). Rewritten in modern notation, this is the general equation $ax^2 + c = bx$. The rule for finding the 'thing' for the equation in which a = 1 is equivalent to our formula $x=\frac{1}{2}b\pm\sqrt{\left(\frac{1}{2}b\right)^2-c}$. In the special case where $\left(\frac{1}{2}b\right)^2-c = c$, we have $x=\frac{1}{2}b$ and $x^2 = c$, and one does not need to follow

35 al-Bīrūnī (1934, p. 42), Wright's translation.

36 al-Uqlīdisī (1978, pp. 87, 88), adjusted from Saidan's translations. For the second quotation, the figure again shows a pair of zeros.

37 al-Kāshī (1969, p. 104.6; 2019, p. 178.18), Aydin's and Hammoudi's translation.

through with the entire rule. Al-Hawārī illustrated this case with the example we write as $x^2 + 9 = 6x$, and then continued with the rule anyway:

> For example, suppose someone said, 'A māl and nine equal six things'. We square half of the things, giving nine, and this is equal to the number. So the number is the māl, and the half is the root. And if we continue the solution, we subtract the number from the square, leaving nothing. We take its root, giving nothing. We add nothing to the half, or we subtract it from it, leaving the half. It is the root, and its square is the number. So know it. (Abdeljaouad & Oaks, 2021, p. 113)

Here al-Hawārī explicitly operates on 'nothing' in a context unrelated to calculation with Indian numerals, and without mentioning zero. This example, together with some of the explanations of the operations on 'nothing' in Indian calculation quoted above, show that Arabic authors could conceive of 'nothing' as a kind of departed amount with which one can calculate. But it is still nothing; that is, it is not anything at all, and thus it cannot be a number of any species of unit.

## 9 Conclusion

We are fortunate that so many medieval Arabic arithmetic books are extant and that the authors took the trouble to explain what the zero is. If all we had were a few dust-board calculations with no explanatory text it would be impossible for us to understand the meaning of this zero.

Numbers in practical Arabic arithmetic consist of a quantity – species pair, like 10 dirhams, 24 starlings, $2\frac{10}{19}$ hours, $29\frac{2}{3}$ inches, or three fourths of a unit.[38] A number *is* a collection of counted objects or of something measured as seen from the perspective of 'quantity'. Fifteen dirhams is composed of dirhams, four and a half mithqals of grain is composed of grain, etc. Addition, subtraction, and multiplication take the physical meanings of gathering together, taking away, and duplicating respectively, even in the case of intelligible units. To add nothing to 15 dirhams is simply not augmenting it with anything, subtracting nothing from it means not taking anything away, and multiplying it by zero

38 Al-Hawārī wrote of 'twenty-four starlings' (Abdeljaouad & Oaks, 2021, p. 110), al-Karajī of 'two hours and ten parts of nineteen parts of an hour' (Saidan, 1986, p. 185.16), and Abū l-Wafā' of 'twenty-nine inches and two thirds of an inch' (Saidan, 1971, p. 205.11). 'Ten dirhams' and 'three fourths of a unit' are common.

or zero by it can be seen as taking none of the 15 dirhams or as duplicating nothing 15 times. Nothing is not an amount of any species, so it was not considered to be a number. It was natural, then, that most authors would not perform operations with zero, and for those who did, they could easily make sense of adding, subtracting, or multiplying nothing. The authors who explained the meaning of zero were unanimous in regarding it as a sign for an empty place, and this view is consistent with the ways they performed and explained their operations. The 'nothing' that the zero signifies does not become something because of these manipulations.

## References

### *Published Works*

Abdeljaouad, Mahdi and Jeffrey Oaks. (2021). *Al-Hawārī's* Essential Commentary*: Arabic Arithmetic in the Fourteenth Century*. Berlin: Max-Planck-Gesellschaft zur Förderung der Wissenschaften. https://edition-open-sources.org/sources/14/. (#747 [M1]).

Abū Kāmil. (2012). *Algèbre et Analyse Diophantienne*. Edition, traduction et commentaire par Roshdi Rashed. Berlin; Boston: Walter de Gruyter. (#124 [M1]).

al-Baghdādī. (1985). *Kitāb al-takmila fī l-ḥisāb*. Edited by A. S. Saidan. Kuwait: Maʿhad al-Makhṭūṭāt al-ʿArabiyah, 1985. (#320 [M1]).

Berkey, Jonathan. (1992). *The Transmission of Knowledge in Medieval Cairo*. Princeton: Princeton University.

al-Bīrūnī, Muḥammad ibn Aḥmad. (1934). *The Book of Instruction in the Elements of the Art of Astrology*. Facsimile edition of Brit. Mus. Ms. Or. 8349, with translation by R. Ramsay Wright. London: Luzac. (#348 [A2]).

Chamberlain, Michael (1994). *Knowledge and Social Practice in Medieval Damascus, 1190–1350*. Cambridge: Cambridge University.

Crossley, John and Alan S. Henry. (1990). Thus spake al-Khwārizmī: A translation of the text of Cambridge University Library Ms. Ii.vi.5. *Historia Mathematica* 17, pp. 103–131. (#41 [M1]).

Euclid. (1956). *The Thirteen Books of Euclid's Elements*. 2nd ed, translated by Sir Thomas Heath. New York: Dover.

al-Fārisī. (1994). *Asās al-qawāʿid fī uṣūl al-fawāʾid*. Edited by Muṣṭafā Mawāldī. Cairo: Maʿhad al-Makhṭūṭāt al-ʿArabīyah. (#674 [M2]).

Folkerts, Menso. (2001). Early texts on Hindu-Arabic calculation. *Science in Context* 14, pp. 13–38.

Ginsburg, Jekuthial and David Eugene Smith. (1917). New light on our numerals. *Bulletin of the American Mathematical Society* 23(8), pp. 366–369.

Ibn al-Bannā'. (1969). *Talkhīṣ aʿmāl al-ḥisāb*. Texte établi, annoté et traduit par Mohamed Souissi. Tunis: Université de Tunis. (#696 [M1]).

Ibn Munʿim (2005). *Fiqh al-ḥisāb*, edited by Driss Lamrabet. al-Rabāṭ: Dār al-Amān. (#566 [M1]).

al-Karajī (1986). *al-Kāfī fī l-ḥisāb*. Edited by Sami Chalhoub. Aleppo: Jāmiʿat Ḥalab, Maʿhad al-Turāth al-ʿIlmī al-ʿArabī. (#309 [M1]).

al-Kāshī (1969). *Miftāḥ al-ḥisāb*. Edited by Aḥmad Saʿīd Dimirdāsh and Muḥammad Ḥamdī al-Ḥifnī al-Shaykh. Cairo: Dār al-Kātib al-ʿArabī. (#802 [M1]).

al-Kāshī. (2019). *Al-Kāshī's Miftāḥ al-Ḥisāb, Volume I: Arithmetic*. Translation and commentary by Nuh Aydin and Lakhdar Hammoudi. Cham: Birkhauser. (#802 [M1]).

al-Khwārazmī. (1997). *Die älteste lateinische Schrift über das indische Rechnen nach al-Hwārizmī*. Edited, translated and commented by Menso Folkerts with the collaboration of Paul Kunitzsch. Munich: Bayerische Akademie der Wissenschaften. (#41 [M1]).

al-Khwārazmī. (2009). *Al-Khwārizmī: The Beginnings of Algebra*. Edited by Roshdi Rashed. London: Saqi. (#41 [M3]).

Kunitzsch, Paul. (2003). The transmission of Hindu-Arabic numerals reconsidered. In: *The Enterprise of Science in Islam: New Perspectives*, edited by Jan P. Hogendijk and Abdelhamid I. Sabra. Cambridge, MA: MIT, 3–21.

Lamrabet, Driss. (2014). *Introduction à l'Histoire des Mathématiques Maghrébins*. Deuxième édition revue et augmentée. Rabat: Driss Lamrabet.

Lane, Edward William. (1893). *An Arabic-English Lexicon*. London; Edinburgh: Williams and Norgate.

Mueller, Ian. (1981). *Philosophy of Mathematics and Deductive Structure in Euclid's* Elements. Cambridge, MA: MIT.

Oaks, Jeffrey A. (2012). 'Algebraic symbolism in medieval Arabic algebra.' *Philosophica* 87: pp. 27–83.

al-Qalaṣādī (1988). *Kashf al-asrār ʿan ʿilm hurūf al-ghubār*. Tunis: al-Mu'assasah al-Waṭanīyah lil-Tarjamah wa'l-Taḥqīq wa'l-Dīrasāt, Bayt al-Ḥikmah. (#865 [M3]).

al-Qalaṣādī. (1999). *Sharḥ talkhīṣ aʿmāl al-ḥisāb*. Edited by Farès Bentaleb. Beirut: Dār al-Gharb al-Islāmī. (#865 [M7]).

Rosenfeld, Boris A., and Ekmeleddin İhsanoğlu. (2003). *Mathematicians, Astronomers, and Other Scholars of Islamic Civilization and Their Works (7th–19th C.)*. Istanbul: Research Center for Islamic History, Art and Culture (IRCICA).

Saidan, Ahmad Salim. (1967). 'Jāmiʿ al-ḥisāb bi l-takht wa l-turāb, li Nasīr al-Dīn al-Ṭūsī.' *Al-Abhath: Quarterly Journal of the American University* 20: pp. 91–164; 213–92. (#606 [M17]).

Saidan, Ahmad Salim. (1971). *Tārīkh ʿilm al-ḥisāb al-ʿArabī*. ʿAmman: Jamʿiyat ʿUmāl al-Maṭābiʿ al-Taʿāwinīa. (#256 [M2]).

Saidan, Ahmad Salim, ed. (1986). *Tārīkh ʿilm al-jabr fī l-ʿālam al-ʿArabī*. 2 volumes. Kuwait: al-Majlis al-Waṭanī lil-Thaqāfah wa'l-Funūn wa'l-Ādāb, Qism al-Turāth al-ʿArabī. (#309 [M2], #587 [M1], #696 [M6]).

al-Uqlīdisī. (1978). *The Arithmetic of al-Uqlīdisī. The Story of Hindu-Arabic Arithmetic as Told in Kitāb al-fuṣūl fī al-ḥisāb al-Hindī by Abū al-Ḥasan Aḥmad ibn Ibrāhīm al-Uqlīdisī, Written in Damascus in the Year 341 (AD 952/3)*. Translated and annotated by A. S. Saidan. Dordrecht: D. Reidel. (#232 [M1]).

al-Uqlīdisī. (1984). *al-Fuṣūl fī l-ḥisāb al-Hindī*. Edited by A. S. Saidan. Sūryā: Manshūrāt Jamiʿat Ḥalab, Maʿhad al-Turāth al-ʿIlmī al-ʿArabī. (#232 [M1]).

Zemouli, T. (1993). *Mu'allafāt Ibn al-Yāsamīn al-riyāḍiyya (Mathematical writings of Ibn al-Yāsamīn)*. M.Sc. thesis in History of Mathematics, E. N. S., Algiers. (#521 [M3]).

Zemouli, T. (n.d.). *Mu'allafāt Ibn al-Yāsamīn al-riyāḍiyya (Mathematical writings of Ibn al-Yāsamīn)*.' New, and still unpublished, version of Zemouli (1993). (#521 [M3]).

### *Manuscripts*

al-Anṣārī, Nūr al-Dīn. *Fatḥ al-wahhāb ʿalā nuzhat al-ḥussāb*. Paris, MS BnF arabe 2475.2, ff.45a–99a. http://gallica.bnf.fr/ark:/12148/btv1b110024450. (#1066 [M2]).

al-Ḥaṣṣār, Muḥammad. *Kitāb al-bayān wa'l-tadhkār fī ṣanʿat ʿamal al-ghubār*. Philadelphia, MS University of Pennsylvania, Lawrence J. Schoenberg collection ljs 293. http://openn.library.upenn.edu/Data/0001/html/ljs293.html. (#532 [M1]).

Ibn al-Qunfūdh al-Qustanṭīn. *Ḥaṭṭ al-niqāb ʿan wujūh aʿmāl al-ḥisāb*. Philadelphia, MS University of Pennsylvania, Lawrence J. Schoenberg collection MS LJS 481. http://openn.library.upenn.edu/Data/0001/html/ljs481.html. (#780 [M1]).

al-Nīsābūrī, Yūsuf ibn Aḥmad. *Bulūgh al-ṭullāb fī ḥaqā'iq ʿilm al-ḥisāb*. Leiden, MS Leiden University Or. 780. http://hdl.handle.net/1887.1/item:1573840. (#218 [M1]).

al-Samaw'al. *Kitāb al-tabṣira fī ʿilm al-ḥisāb*. Berlin, MS Staatsbibliothek Glaser 40. https://digital.staatsbibliothek-berlin.de/werkansicht/?PPN=PPN646146610&PHYSID=PHYS_0007. (#487 [M4]).

al-Ṣardafī. *Mukhtaṣar al-hindī fī ʿilm al-ḥisāb*. Berlin, MS Staatsbibliothek Glaser 6. https://digital.staatsbibliothek-berlin.de/werkansicht?PPN=PPN646422162&PHYSID=PHYS_0007. (#411 [M1]).

CHAPTER 14

# Numeration in the Scientific Manuscripts of the Maghreb

*Djamil Aïssani*

## Abstract

In this chapter, we begin by recalling the beginning of the mathematics of Islamic countries, particularly by emphasizing the influence of Indian arithmetic. Next, we present the particularity of the mathematics of the Muslim West (Maghreb and al-Andalus), by revealing the specificity of the digits and the symbolism used. Third, we focus on the role played by the city of Bejaia (Algeria) in the 'popularization' of Arabic numerals in Europe, following the stay of mathematician Leonardo Pisano, or Fibonacci. The contribution of this article concerns the presentation of the numeration available in the Maghreb on the basis of the analysis of *Afniq n 'Ccix Lmuhub* (*Khizana* – the scholarly library of manuscripts of Sheikh Lmuhub). Discovered at Tala Uzrar in 1994, it is currently the only library of manuscripts cataloged in Kabylia (Algeria) (Aïssani, 2007; Aïssani, 2011).

## Keywords

Muslim numeration – mathematics of Muslim West – Arabic numerals – Huruf al-Ghubar – symbolism – Maghreb – Bejaia – Leonardo Fibonacci

## 1 Introduction

Since the twelfth century at least, it has been known that two families of figures were adopted by Muslim authors (Aïssani, 2018), both from Indian numeration. Their existence was attested as early as the twelfth century by the Maghrebian mathematician Ibn al-Yāsamīn (d. 1204), in his treatise *Talqih* (folio 7–8) (Abdeldjaouad, 2012).

The form adopted by the mathematicians of the Muslim West (Maghreb and Al-Andalus) gave birth to what are now called Arabic numerals, or Hindu-Arabic numerals. Ibn al-Yāsamīn had used a dust-covered calculation

 | DOI:10.1163/9789004691568_017

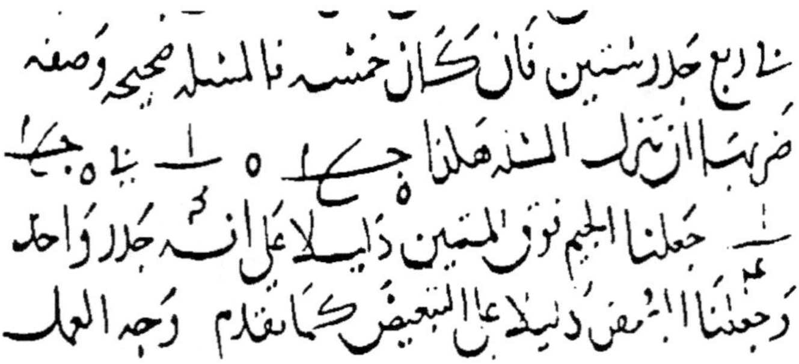

FIGURE 14.1 Talqīh al-afkār of Ibn al-Yāsamīn, folio 176.
The symbolism specific to the Maghreb was already known in the twelfth century, shown here in the Talqīh al-afkār of Ibn al-Yāsamīn, folio 176


board, *al-Lawha*, hence the name given to this arithmetic: *Hisāb al-Ghubar*. He drew the operations on this board, erased digits, copied others and continued his calculations until he obtained the results (Aïssani, 2007).

The oldest source relating to mathematics for the North Africa and Al-Andalus tradition is probably the treatise *Kitāb al-Bayān wa t-Tadhkār* of the Maghrebian mathematician al-Ḥaṣṣār (living in the last quarter of the twelfth century) (Aïssani, 2018). It is a calculation manual dealing with numeration, arithmetical operations on integers, and fractions. Al-Ḥaṣṣār used *Ghubar* numerals and the fraction line. He defined different types of fractions and reserved for each type a specific symbol, without claiming paternity, the selected notations being different from those used in the East and inherited from the Indians. When he had to represent a mixed number (that is to say, an integer added to a fraction), al-Ḥaṣṣār began by writing the integer and placed the fraction to the left of this number, for example 3/11.78, Today corresponds to the fraction 78 + (3/11). Finally, let us specify that the treatise of al-Ḥaṣṣār was known in Europe since it was translated into Hebrew by Moïse Ibn Tibbon in 1271 in Montpellier (Aïssani, 2018). Al-Ḥaṣṣār's way of representing fractions is found in use by most mathematicians of the Maghreb and even by the famous Italian mathematician Leonardo Pisano (1170–1241) (hereafter Fibonacci) in his *Liber Abaci* (1202) (Aïssani, 1994, 2003, 2014; Caianiello et al., 2012).

Indeed, it was from the city of Bejaia (Algeria) that the famous Italian mathematician Fibonacci 'popularized' in Europe, from the end of the twelfth century, the Arabic figures, the numeration system, the methods of calculation as well as the commercial techniques of the Islamic countries. He wrote in the introduction of *Liber Abaci*:

> When my father, who had been appointed by his country as public notary in the customs at Bugia (Bejaia) acting for the Pisan merchants going there, was in charge, he summoned me to him while I was still a child, and having an eye to usefulness and future convenience, desired me to stay there and receive instruction in the school of accounting. Then, when I had been introduced to the art of the Indian nine symbols through remarkable teaching, knowledge of the art very soon pleased me above all else and I came to understand it. (Aïssani, 1994, 2003, 2014)

Let us note here that mathematical activities in the Maghreb also had a different origin to the Muslim tradition. Thus, in the field of calculus science, we observe the existence since the pre-Islamic period of a calculating practice that uses what we call the 'Fez numerals'. These symbols are distinguishable from the *Ghubār* numerals, both in their number and in their form.

The use of a specific symbolism to express essential concepts is one of the main characteristics of mathematical education in the Maghreb in the Middle Ages, more than a century before the beginning of the European symbolism (with Viète). It was the Andalusian mathematician al-Qalasādī (1412–1486) who popularized symbolism in the manner of writing equations: the letter *Shin* (an abbreviation of *Shay* or thing) designates the unknown (x), the letter *Mim* (*Mal*) corresponds to $x^2$, the letter *Kaf* (*Kaab*) to $x^3$, the letter *Lam* (*Ta'dil*) represents the sign =, while the letter *Jim* (*Djadhr*) concerns the square root sign.

FIGURE 14.2
The Andalusian mathematician Al-Qalasādī (1412–1486). The Andalusian mathematician Al-Qalasādī (1412–1486), 'the last of the mathematicians' popularized the symbolism used in the Maghreb
COPYRIGHT: ABDELDJAOUAD, 2012

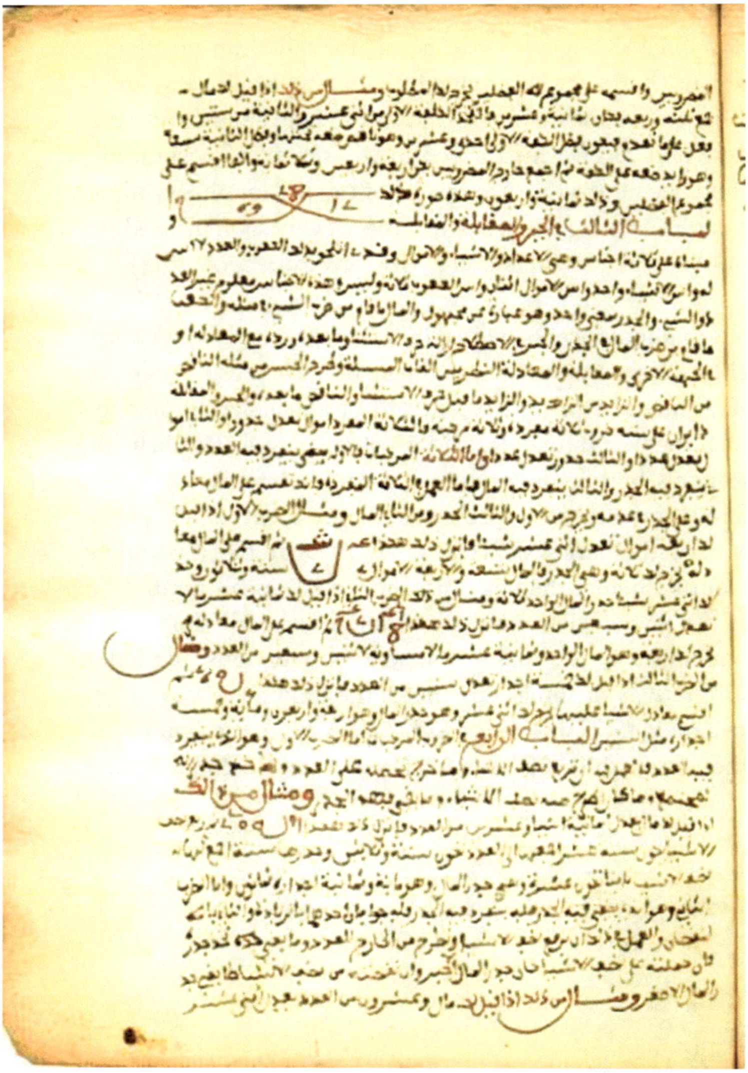

FIGURE 14.3
Figurative symbolism in the Kashf al-ʾAsrar ʿan ʿIl[m] H'urûf al-Ghûbar. Figura[tive] symbolism in the Kashf al-ʾAsrar ʿan ʿIlm H'urûf al-Ghûbar, the treatise o[n] the science of calculatio[n] Andalusian mathematici[an] al-Qalasādī
COPYRIGHT: AFNIQ N'C[CIX] LMUHUB

FIGURE 14.4
Afniq n ʿCcix Lmuhub (Khizana, scholarly library). Afniq n ʿCcix Lmuhub (Khizana, the scholarly library of manuscripts of Sheikh Lmuhub)
COPYRIGHT: AFNIQ N'CCIX LMUHUB

FIGURE 14.5
Manuscript in the Berber language (transcribed with Arabic characters)


## 2 The Mathematics of Islamic Countries

### 2.1 *Mathematics in the Muslim East*

The mathematics of the Islamic countries originated from various influences. They were constituted by the assimilation of Greek, Indian and Babylonian mathematics, then subsequently further developed. It was mainly through their translations into Arabic and their commentaries that Europe became acquainted with the works of Greek mathematicians. In Baghdad, capital of the Muslim Empire, a large-scale translation program was undertaken, first from Persian into Arabic, then from Sanskrit or Greek into Arabic. Muslim mathematicians translated Sanskrit texts on astronomy and Indian mathematics such as the *Surya Siddhanta*, the *Brahma Sphuta Siddhanta*, and the *Khandakhayaka* of Brahmagupta.

The Persian mathematician al-Khwārizmī (780–850) worked in Baghdad's *Beit al-Hikma* (House of Wisdom). Two of his treatises had a considerable impact on the European mathematics of the twelfth century. The first, *On the Calculation with Hindi Numerals*, of which only the Latin translation has been preserved (*Algoritmi de Numero Indorum*), explains the nine digits and

FIGURE 14.6 A stamp commemorating the approximate 1,200th birthday of al-Khwārizmī's (approximate), issued 6 September 1983 in the Soviet Union

FIGURE 14.7 The treatise of al-Khawrizmi (780–850)

the zero, the positional numeration, and the basic operations of written calculus. Thus began the conquest of Europe by this system, which will replace Roman numerals and abacus operations but very slowly. Indeed, penetration took place over nearly five centuries. The second treatise, *Kitāb fi al-Jibr wal Muqabala* (the book on restoration and confrontation), deals with the manipulation of equations. *Algoritmi*, the translator's rendition of the author's name, gave rise to the word *algorithm* (Aïssani, 2018).

Many numeration systems coexisted in the medieval Muslim world: the *Jummal* multiplico-additive decimal numeration system (*Ḥisāb bi'l Jummal*), the sexagesimal system of the Babylonians (which seems to have reached the Muslim world via Syria or Persia). A last system was to gradually replace the two previous ones: the positional decimal system of Indian origin consisting of nine digits and zero. One of the first books in Arabic to describe it is al-Khwārizmī's book on Indian calculus, of which only an incomplete Latin version remains.

FIGURE 14.8
Algorithm (al-Khwārizmī) in 1450. Algorithm (al-Khwārizmī) in 1450, the personification of arithmetic in the Middle Ages. The treatise 'Algorismus' by Johannes Sacrobosco, Ms. 184 Columbia University

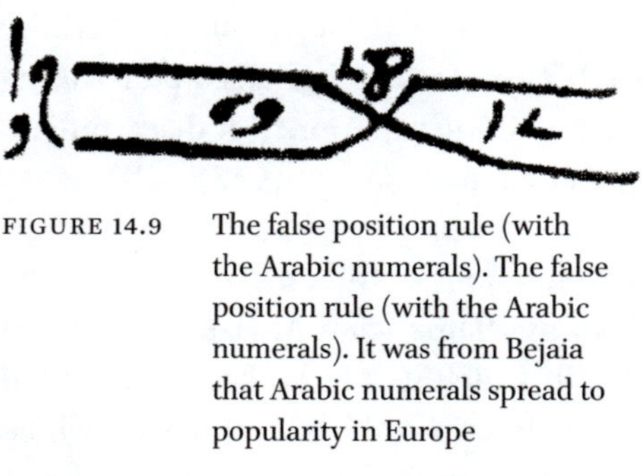

FIGURE 14.9 The false position rule (with the Arabic numerals). The false position rule (with the Arabic numerals). It was from Bejaia that Arabic numerals spread to popularity in Europe

### 2.2 *Mathematics in the Muslim West*

In the seventh century, a specificity of the mathematics of the Muslim West (Maghreb and Al-Andalus) was characterized by the appearance of a specific symbolism to represent digits (figures), fractions, roots, and algebraic expressions. Discovered in the ninth century by the first mathematicians of Baghdad, including Muhammad Ibn Mûssâ al-Khwārizmī, Indian arithmetic introduced Indian figures, the zero, and decimal position notation into Muslim science. After its introduction to the Maghreb and Al-Andalus, Indo-Muslim arithmetic had various influences through contact with the three cultures that coexisted there, Arabic, Berber, and Latin, and transformed itself by inventing in particular a specific symbolism (Abdeldjaouad, 2012).

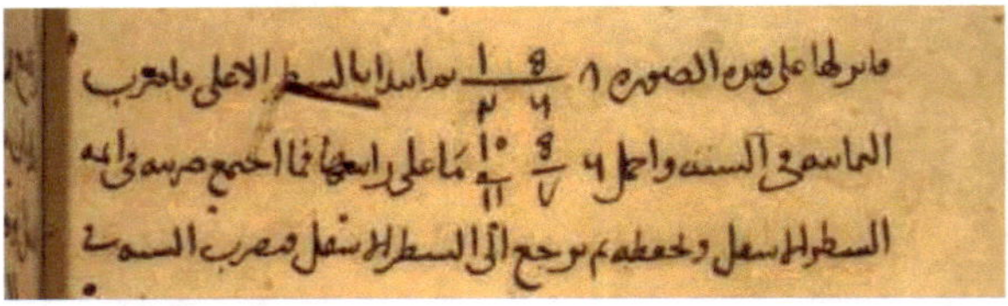

FIGURE 14.10 The Kitab al-Bayan of famous Maghrebian mathematician al-Ḥaṣṣar (twelfth century). The Kitab al-Bayan of the famous Maghrebian mathematician al-Ḥaṣṣar (twelfth century) is the first in which the fraction line appears


In the Maghreb, and especially in Kaïrouan (Tunisia), Indian arithmetic began in the tenth century during the drafting of the *Kitāb fi al-Hisāb al-Hindi* treaty, written by Ibn Tamim Abu Sahl (900–960). The enthusiasm of the Andalusian scholars for Indian arithmetic was evident in the treatise, *Tabaqat al-Ummam* (Categories of Nations) of Sāʾid al-Andalusi (1029–1070):

> Among the things that have reached us from their science of numbers, Hisab al-Ghubar, that al-Khwarizmi has simplified, it is the most concise, the most succinct, the easiest to acquire, the easiest to learn, and whose construction is the most original. It attests to the Indians a penetrating spirit, a beautiful talent for creation and the superiority of discernment, and inventive genius. (Sāʾid, 1985; Abdeldjaouad, 2012, p. 58)

The undeniable contribution to the development of Indian arithmetic in the Maghreb is the work of the Maghrebian mathematician Al-Ḥaṣṣār (living in the last quarter of the twelfth century) in which he methodically presented the decimal position notation and clearly described the operations on whole numbers illustrated in specific 'windows'. Western Arabic numerals are identifiable, and a new highly detailed typology of specific symbols used for fractions was first exhibited in Muslim literature. In addition, he explained the use of the dust board in his *Kitab al-Kamil fi Sina'at al-A'dad*: 'In our countries, calculators, craftsmen and especially scribes have become accustomed to using numbers that they have agreed between them, allowing them to express the numbers and differentiate them from each other. It's a handwriting … called *Ghubar* or even Hindi …' (Aïssani, 2018; Abdeldjaouad, 2012).

## 2.3 *Bejaia and Fibonacci*

In 1067 the town of Bejaia (Bougie in French, Bgayet in Berber, Bugia in Italian and Spanish, Buggea in Latin) became the capital of the Berber Kingdom of the

Hammadite. It would become one of the most dynamic cultural and scientific centers of the Maghreb (this town gave its name to small candles: bougies). It was in this city that Fibonacci was introduced to the numeration system, methods of calculation and commercial techniques of the Islamic countries (Aïssani, 1994, 2003, 2014; Caianiello et al., 2012).

The *Liber Abaci* by Fibonacci, its definitive version written in 1228, is an extensive book that elucidates in 15 chapters arithmetic and algebra, as well as the resolution of a number of problems that are either applications to the science of trading, or recreational or at least representing situations that are too unusual to be real.

It is not easy, within the *Liber Abaci*, to share what had been learned in Bejaia and other places, particularly in Constantinople, but also in Syria and Egypt, or in Sicily. The first chapters of the book, however, show the importance of his experience with the merchant and marine circles of Bejaia in formulating his mathematical knowledge. The contribution of the *Liber Abaci* to the Latin West, it is known, resides less in the introduction of the Arabic numerals, which were already known since the tenth century, as in the presentation of the arithmetic methods called 'Indian calculus' that use the nine digits (figures) and the zero, as well as algebraic methods.

FIGURE 14.11
Magliabechiano, National Library of Florence. Fibonacci's testament on his studies in Bejaia with an admirable teacher (exmirabili Magisterio), Liber Abaci. As seen in 'Magliabechiano', National Library of Florence

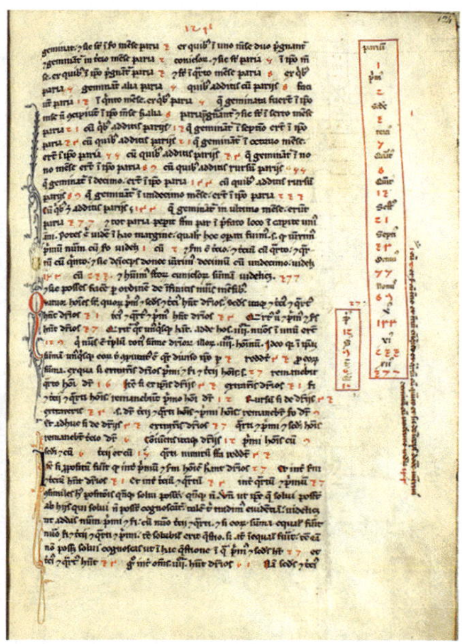

FIGURE 14.12
Arabic figures used by Fibonacci. Arabic figures used by Fibonacci (as shown in Liber Abaci)

The first pages of the *Liber Abaci* are devoted to the presentation of Arabic numerals and simple operations that they allow to achieve. The novelty lies in the possibilities of calculations offered by the system of positional notation and the use of the zero, which is the key. This allows writing operations, assigning each digit a value according to its position in the number.

The knowledge that Leonardo acquired from his master in Bejaia was not only assimilated, but immediately reformulated in Latin and with examples corresponding to his milieu, that of the merchant class. The need for this was great in a Mediterranean world that was opening up more widely and where the volume of trade, such as maritime traffic, was growing steadily. The examples that Fibonacci took to elucidate the arithmetic rules were, then, a reflection of this Mediterranean world, which he first knew in Pisa, then in Bejaia, before performing a veritable tour of the Mediterranean (Aïssani, 2003).

## 3 Muslim Numeration

Muslim writers since the Middle Ages have used several numeration systems:

1. Alphabetical numerals (*Hisāb bi'l Jummal*)
2. The *Ghubār* numerals, closer to our current Arabic numerals (or Hindu-Arabic numerals)
3. The Hindu numerals (*Al-Arqām al-Hindīa*), employed today in the Muslim East
4. The Fez numerals (known as *al-Qalam al-Fāsī*)

FIGURE 14.13 Traveler Ibn Battuta meeting the doctor of Bejaia. There were intellectual relations between the Maghreb and India in the Middle Ages. This illustration shows the traveler Ibn Battuta who meets the doctor of Bejaia, Djamel ad-Din al-Maghribi, in Indi
COPYRIGHT: GEHIMAB

### 3.1 *Ḥisāb bi'l Jummal*

In medieval times, the scholars of the Islamic countries used alphabetical numeration, that is, the use of letters of the alphabet to express digits (figures), as in all civilizations. Ibn Kẖaldūn named this system *Hisāb bi'l Jummal* in his work the *Muqqadima*, stating that it was very old. Muslim writers used it in their various writings: the science of inheritance, poetry, astronomy, astrology, divination. It can also be seen that the dates of construction of some buildings in the Islamic countries, such as mosques, palaces, graves, and schools, use this system.

The system consists of 28 letters of the Arabic alphabet. Each letter is either a number or a multiple of 10 or 100. In all, the 28 letters correspond to 58 numbers. Thus, in this ingenious system, which was well-known to the local scholars of the small Kabylia, as evidenced by the writings found in the *Kẖizāna of Sheikh Lmuhub*, an entire sentence can be digitalized, that is, evaluated by calculating the sum of the *Abjadi* values of the letters. In the table below, the two variants of this system were assembled, one used in the Middle East, the other in the Maghreb (Mechehed and Aïssani, 2012).

TABLE 14.1 *Abjadi* values of the letters

| Value | Maghreb (Kabylia) | Muslim East | Value | Maghreb (Kabylia) | Muslim East |
|---|---|---|---|---|---|
| 1 | أ | أ | 60 | ص | س |
| 2 | ب | ب | 70 | ع | ع |
| 3 | ج | ج | 80 | ف | ف |
| 4 | د | د | 90 | ض | ص |
| 5 | ه | ه | 100 | ق | ق |
| 6 | و | و | 200 | ر | ر |
| 7 | ز | ز | 300 | س | ش |
| 8 | ح | ح | 400 | ت | ت |
| 9 | ط | ط | 500 | ث | ث |
| 10 | ي | ي | 600 | خ | خ |
| 20 | ك | ك | 700 | ذ | ذ |
| 30 | ل | ل | 800 | ظ | ض |
| 40 | م | م | 900 | غ | ظ |
| 50 | ن | ن | 1000 | ش | غ |

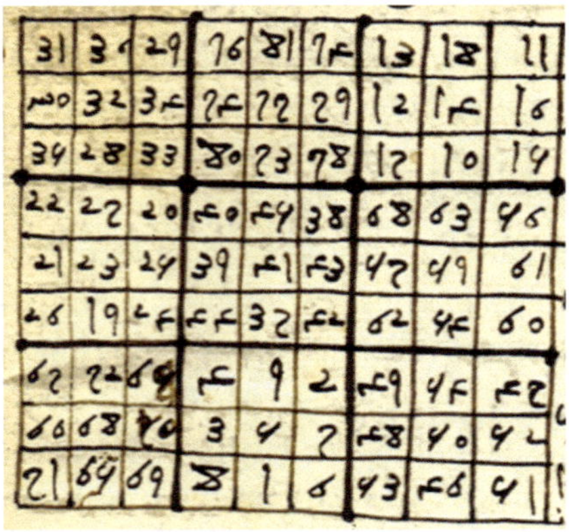

BOARD 1 (Ms ASL no. 2, Afniq n'Ccix Lmuhub)

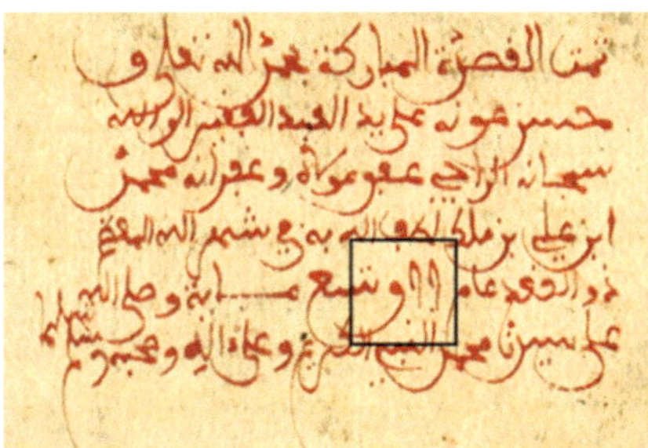

BOARD 2 (Ms LIT no 22, Afniq n'Ccix Lmuhub. Copied in 1505)

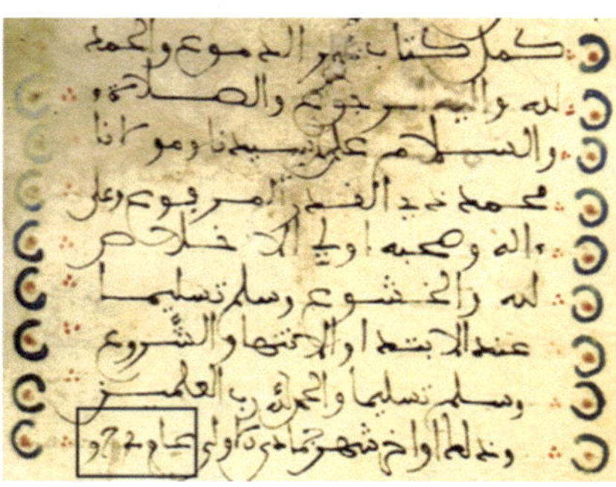

BOARD 3 (Ms TZ no 17, Afniq n'Ccix Lmuhub. Copied in 1565)

BOARD 4 (Ms SC no 14, Afniq n'Ccix Lmuhub. Copied in the nineteenth century)

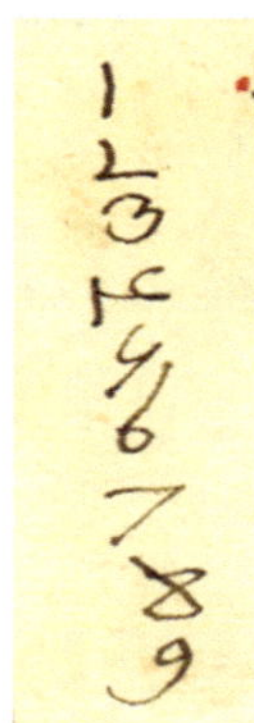

BOARD 5 (Ms SC no 14, Afniq n'Ccix Lmuhub. Copied in the nineteenth century)

FIGURE 14.14 Ghubār numerals in the manuscripts of the Sheikh Lmuhub's Library

### 3.2 Hurūf al-Ghubār (*Dustdigits*)

The Indian numeration system gave rise to two variants of numbers. One of these is the Ghubār numerals used only in the Maghreb and Al-Andalus. We find these for the first time in the book *Kitāb al-Bayān wa al-Tadhkār* (book of demonstration and recall) by the mathematician al-Ḥaṣṣār (seventh century). They were also given by the mathematician Ibn Al-Yāsamîn (*c.*1204). The *Liber Abaci* by Fibonacci, which was a great success, seems to have played a major role in spreading the *Ghubār* numerals in Europe. Indeed, at Bejaia, Fibonacci quickly realized the importance of the decimal positional notation system, with the famous zero. Over time, the Ghubār numerals changed slightly; currently they are in the form of (Arabic numerals, Hindu-Arabic numeral system):

9876543210

Board 1 represents a magic square designed with the *Ghubār* numerals. The numbers thus repeated give us a precise idea of their calligraphic forms. By comparing them with those of the other boards, there is a clear resemblance, except for the figure 1 on board 2, which is much closer to its present form, and the figure 3 on board 4.

### 3.3 *Hindu Numerals*

The modern calligraphic form of the Hindu numerals used today in the Middle East are as follows:

٩٨٧٦٥٤٣٢١٠

By way of the various dates expressed in Hindu digits inscribed in our manuscripts since the eighteenth century, it can be said that their calligraphic forms have not really changed, unlike the Ghubār numerals.

Manuscript SC no. 14 of the *Kḫizāna* of Sheikh Lmūhūb is a versified text describing the calligraphic forms of Hindu numerals (Aïssani, 2011). It is very

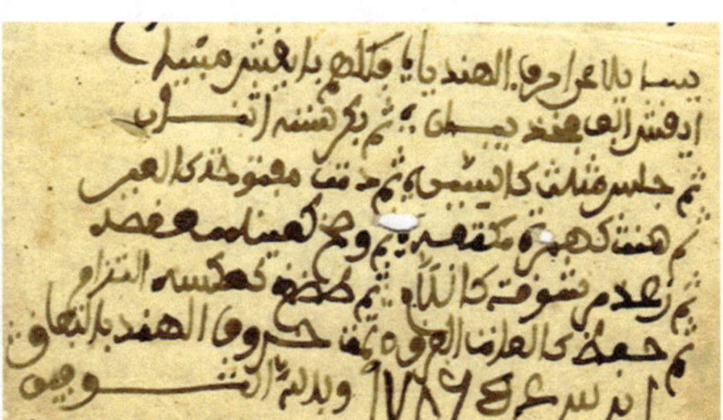

FIGURE 14.15
Poem on the calligraphic form of Hindu numerals. This manuscript, cataloged SC no. 14 Aïssani, 2011) dates from the nineteenth century

likely that this technique of memorizing by means of a poem allowed Muslim users to memorize the exact forms of the Hindu-Arabic numerals.

### 3.4 *Fez Numerals*

The practice of mathematics in the Maghreb has been inscribed in the Muslim tradition of the Muslim East. However, in the field of computational science, it has existed since pre-Islamic times, a calculatory practice that uses symbols that are called the Fez numerals. The origin of this calculatory practice is probably the West Maghreb. This practice persisted for several centuries since the famous Maghreb mathematician Ibn al-Banna (d. 1321) wrote a manual to present his principles and their use.

The Fez numerals differ from the *Ghubār* numerals and Hindu numerals, that is current digits, both by their number and by their form. This little-known, little-exploited numeration system is composed of 27 symbols.

The orientalist G. S. Colin attributed the Fez numerals to the Greeks. For his part, Ahmed Sakirj also asserted that the Fez numerals were mistakenly attributed to the inhabitants of Fez (Sakirj, 1897). Moreover, these same figures are found in the maritime registers in some Maghreb ports; Ibn Kẖaldūn designated them by the expression *Zimām* (Colin, 1933).

Manuscript SC no. 5 of *Afniq n 'Ccix Lmūhūb* is a poem by Abd al-Qādir al-Fāsī (1599–1680) on the calligraphic forms of the Fez digits. Some of them are a combination of Arabic letters. This is the case, for example, of the figure 8

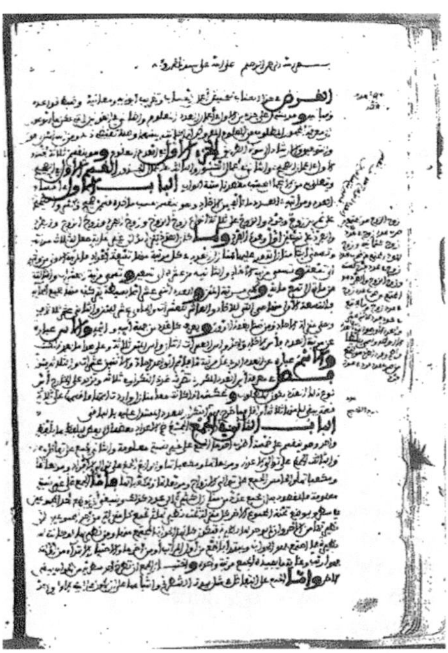

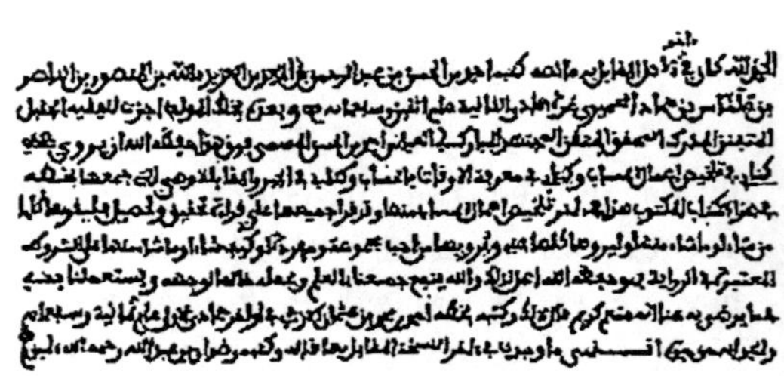

FIGURE 14.16
Ibn al-Banna's famous 'Idjaza (diploma), issued in 1308. Ibn al-Banna's famous 'Idjaza (diploma), issued to a direct descendant of the Hamadite princes in Marrakesh in 1308, appears in this copy of his treatise at-Talkhis. Left – Copyright: Afniq n'Ccix Lmuhub. Right – Copyright: De l'Escurial Library

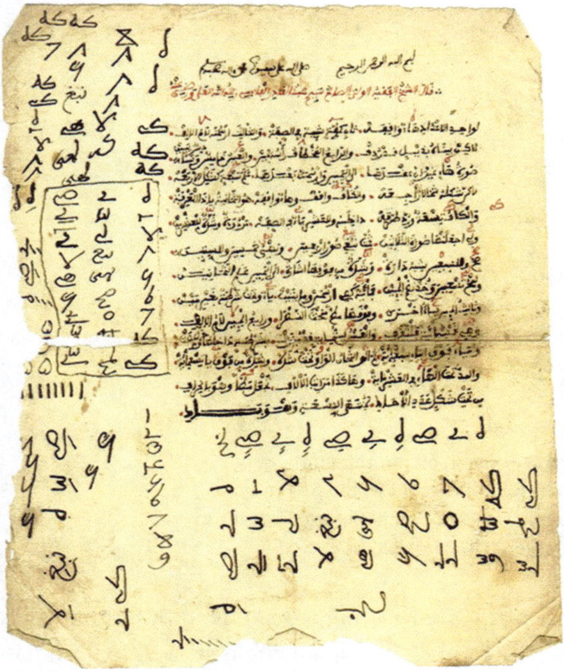

FIGURE 14.17
Description of the Fez numerals. Poem by Abd al-Qādir al-Fāsī (1599–1680). Manuscript SC no. 5 (Afniq n 'Ccix Lmuhub) (Aïssani, 2011)

which is composed of two letters – the 'ك' and the 'ه'. It is also noted that the three digits 5, 6, and 7 are expressed in *Ghubār* numerals (see Figure 14.14, Board 5).

Finally, it should be noted that a bar under a letter of the Fez numerals means multiplication by 1,000. Note here that the Romans also put a bar below the letters to denote the thousands (Monteil, 1951).

## 4 The Natural Number System

Several scholars are quoted in Manuscript AST no. 21 of *Afniq n 'ccix Lmūhūb* (Aïssani, 2011) relating to *Ilm al-Hurūf* (arithmomancy). Among them is al-Muġribī, considered by Ibn Khaldūn as a specialist in astrological and divinatory works using the 'parallel digits' known in the divinatory texts as *al Adād al-Muthḥāba* (the amiable numbers). He also cites mathematician al-Mağriṭī (950–1007) and al-Bābilī. On the last page of this manuscript is a list of 28 symbols in parallel with the letters of the *Abjadi* numeration system.

These symbols, called *al-Qalam al-Tabīʿī* (natural digits), are likely to denote numbers. Indeed, the designation *al-Qalam* is dedicated specifically to the numbers (*Qalam al-Ghubārī* for the figures of *Ghubār* numerals and *Qalam*

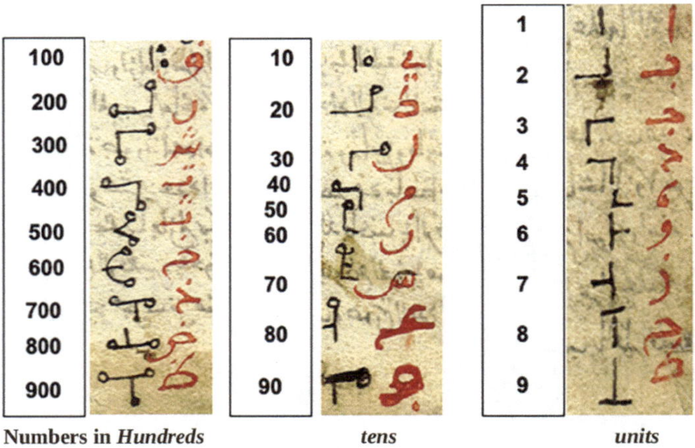


FIGURE 14.18 The 28 symbols of the system of natural numbers classified in parallel with the letters of the Abjadi numeration system. Extract of the manuscript AST no. 21 (Afniq n 'Ccix Lmuhub) (Aïssani, 2011)

*al-Hindī* for the Hindu numerals). Other elements lead us to think that these are indeed numerical values; the symbol 'o', a small circle, is used to designate tens, hundreds, and thousands.

This numeration system was reported by the orientalist Monteil, and was the subject of an article, *Cryptography among the Moors – note on some secret alphabets of the Hodh*. In this article, Monteil described the symbols in question (Monteil, 1951). It is clear that they are the same as the manuscript that we are presenting here. These symbols appear in two other manuscripts in the collection. The first is a manuscript attributed to al-Gazalī on magical squares. The second is the manuscript DL no. 40 copied in 1857 by Bachīr Ūlaḥbīb, the father of Lmūhūb (Monteil, 1951).

## 5 Luca Pacioli and the Continuation of the Work of Fibonacci

The adoption of Arabic numerals in Europe was a very slow process over several centuries. The famous Italian mathematician Luca Pacioli (1445–1517), a close friend of the universal scholar Leonardo da Vinci (1952–1519), is considered to be the inventor of accounting. He is the author of the book, *Summa de arithmetica, geometria, proportioni et proportionalita* (Venice, 1494), where in is summarized all the mathematical knowledge of his time.

The work of Luca Pacioli is considered to be the continuation of the work of Leonardo Fibonacci. Indeed, in algebra, he followed the order of the equations

FIGURE 14.19 Luca Pacioli, in his work Summa de arithmetica, geometria, proportioni et proportionalita. Luca Pacioli, in his work Summa de arithmetica, geometria, proportioni et proportionalita (Venice, 1494) follows the order of the equations in force in Bejaia (al-Qurashī, Abū Kāmil, Fibonacci)

in force at Bugia, by al-Qurashī (d. 1184), the Egyptian Abū Kāmil, and Fibonacci, and not that of al-Khwārizmī. It was at this time that the Arabic numerals had been adopted definitively.

## 6 Conclusion

The information provided in this article gives a clear idea of the variety of numeration systems used by *Ulema* in the Maghreb up to the nineteenth century.

FIGURE 14.20
'Arithmethica', G. Reisch, Margarita Philosophica, Bâle, 1508. An 'allegorical' arithmetic referees the rivalry between a holder of figures and a follower of calculation by means of tokens. 'Arithmethica', dans G. Reisch, Margarita Philosophica, Bâle, 1508

## References


Abdeldjaouad, M. (2012). The mathematical symbols specific to the Muslim West In *Les Manuscrits Scientifiques du Maghreb*, Département Expositions Ed., Ministère de la Culture, Tlemcen/Alger, August 2012, pp. 25–32. ISBN: 978-9931-361-06-0.

Aïssani, D., et al. (1994). The Mathematics in Medieval Bougie and Fibonacci. In *Leonardo Fibonacci: il tempo, le opera, l'eredità scientifica*, Pisa: Pacini Editore (IBM Italia), pp. 67–82.

Aïssani, D., and Valerian, D. (2003). Mathematics, Commerce and Society in Bejaia (Bugia) at the time of Fibonacci's stay. *Bollettino di Storia delle Scienze Matematiche*, *23*(2), pp. 9–31.

Aïssani, D., and Mechehed, D. E. (2007). The Khizana of Sheikh Lmuhub: Recovery of a Library of Manuscripts of the 19th Century. In *Les Manuscrits Berbères au Maghreb et dans les Collections Européennes: Localisation, Identification, Conservation et Diffusion*. Perrousseaux Ed., Paris, pp. 79–112. ISBN: 10-2-91-122018-8.

Aïssani, D., and Mechehed, D. E. (2011). *Manuscripts of Kabylia: Catalog of the Ulahbib's Collection*, C.N.R.P.A.H. Ed., Alger, 215 pages.

Aïssani, D. (2010). The Scientific Manuscripts of the Islamic World. In *Treasures of the Aga Khan Museum: Art of the Books and Calligraphy*. AKCP and Sabanci University and Sakip Sabanci Museum Ed., Istanbul/Geneve, pp. 200–205. ISBN: 978-605-4348-08-4.

Aïssani, D. et al. (2014). *Influence of the Algebraic Tradition of the Maghreb on the work of the mathematician Fibonacci (1170–1241)*. Educ Recherche no.7, I.N.R.E. Ed., M.E.N., Alger, pp. 58–66. ISSN: 2253-0282.

Aïssani, D. (2019). The Maghrebian Mathematics. Quaderni di Ricerca in Didattica (Mathematics), no. 3, G.R.I.M. (Departimento di Matematica e Informatica), University of Palemro, 2019, pp. 19–35.

al-Andalusi Sā'id. (1985). *Tabaqat al-Ummam*, Dar at-Taly'a Ed., Beirut.

Caianiello, E., Cifoletti, G., and Aïssani, D. (2012). Algebra in the Maghreb and its Development in Europe. In *The Scientific Manuscripts of the Maghreb*, Département Expositions Ed., Ministère de la Culture, Tlemcen/Alger, pp. 33–46. ISBN: 978-9931-361-06-0.

Colin, G. S. (1933). De l'origine grecque des chiffres de Fès. *Journal Asiatique*, pp. 193–215.

Mechehed, D. E. et Aïssani, D. (2012). The Numeration in the Manuscripts of *Afniq n 'Ccix Lmuhub*. In *The Scientific Manuscripts of the Maghreb*. Département Expositions Ed., Ministère de la Culture, Tlemcen/Alger, pp. 47–52. ISBN: 978-9931-361-06-0.

Monteil, V. (1951). *La Cryptographie chez les Maures– Note sur quelques alphabets secrets du Hodh*. Bull. de l'IFAN, T. 13(4), pp. 1257–1264.

Sakirj, A. (1897). *Iršād al Mut 'alim wa nāsī fī sifat aškāl al-qalam al-fāssī*. Lithographic copy, Bibliothèque Nationale, Rabat.

Souissi, M. (1974). *Numération arabe*. Actes du huitième séminaire sur la pensée islamique, Ministère des Affaires Religieuses Ed., Bejaia.

Yalawi, M. (1971). *Hisāb al-Ğummal 'ind al 'arb*. Review *Hawiyat* no. 8, Université de Tunis Ed., Tunis.

CHAPTER 15

# The Zero Triumphant: Allegory, Emptiness and the Early History of the Tarot

*Esther Freinkel Tishman*

## Abstract

What we now call Tarot was originally known as the game of '*trionfi*' (i.e., 'triumphs' or 'trumps'). This game emerged in fifteenth century Italy as what appeared to be an amalgamation of two different series of cards: on the one hand, a four-suited deck of playing cards brought into Europe via the Mamluk empire from the Muslim Near East; on the other hand, a deck of 22 allegorical images originating in medieval Christian iconography. The Mamluk-originating cards are numbered according to the Hindu-Arabic system, while the European Trumps cards have Roman numerals. One special trump card known as the Fool or Crazy One (*Il Matto* or *le Fol*) appears to mediate, however, between the two different series, eastern and Western. The Fool is numbered 0 – the odd one out among the Roman-numeraled allegorical cards. The Fool card subverts the logic of play in systematic ways that anticipate the vision of Folly we see elaborated in the later philosophy (cf. Erasmus) and literature of the age (cf. Shakespeare).

## Keywords

zero and the Tarot deck – zero and the Fool – playing cards (history of) – tarot (history of) – allegory – trionfi (Petrarch) – Sola-Busca

## 1 Introduction[1]

In the early modern period, what we now call Tarot was known as the game of '*trionfi*' (i.e., triumphs). The game was not imbued with esoteric import; it was just a card game, but a game embedded in the history of allegory. The Tarot emerged in fifteenth century Italy as what appeared to be an amalgamation of two different series of cards: on the one hand, a four-suited deck of playing

1 This essay draws on material from my book (2019).

 | DOI:10.1163/9789004691568_018

cards brought into Europe via the Mamluk empire from the Muslim Near East; on the other hand, a deck of 22 allegorical images originating in medieval Christian iconography. In this amalgam of straightforward number cards (the Mamluk suits) and allegorically imagined triumph cards, we find a fascinating new game that collates the arithmetic logic of counting, with the Western sublating[2] (i.e., subsuming) logic of allegory.

It was here that the zero entered the scene: one of the earliest Tarot decks (the 1491 Sola-Busca[3]) labeled the allegorical card known as the Fool with a Hindu-Arabic zero digit. This use of the Hindu-Arabic numbering system emerged despite the fact that the other allegorical triumph cards employed Roman numerals. The Sola-Busca deck thus very straightforwardly demonstrates the ways in which East meets West within the history of Tarot. More precisely, the Sola-Busca deck suggests the impact of the zero digit, and the Indian conception of *śūnyata* or emptiness that it conveys, upon Western allegorical thought.

## 2 Tarot

Playing cards were introduced in Europe in the 1300s, entering the continent via Muslim Spain. Then, as now, the regular playing card pack consisted of four suits with minimally illustrated number cards (i.e., 'pip cards' or 'pips') ranging from ace to 10, along with a series of court cards for each suit. The gambling and trick-taking games that used these early card decks followed relatively straightforward, arithmetic principles. In general, higher numbers, or more highly ranked figures like the King or Knight, beat lower numbers. However, about 50 years after these first decks appeared in Europe, fifteenth

---

2 Sublation is the movement of Hegelian dialectic (*aufheben*), where succeeding terms simultaneously negate and preserve terms that have come before. Jacques Derrida's classic discussion of Hegel's *Aufhebung* remains an indispensable exploration of this topic: 'From Restricted to General Economy: A Hegelianism without Reserve', In *Writing and Difference*, trans. Alan Bass, London, 1978, pp. 251–277.

3 Sola-Busca refers to the name of the Milanese family who acquired the deck in 1948. In 2009, the Italian Ministry of Heritage and Culture purchased the cards, and they are now held in the Pinacoteca of Brera, in Milan. Modern reproductions of the cards have been produced, including a mass market 2018 edition published by Lo Scarabeo in its 'Anima Antiqua' deck series. There is very limited reliable scholarship on the deck available in English. To date, the only significant research published is the Italian-language catalog produced for a 2012 exhibit at the Pinacoteca. The catalog explores the alchemical significance of much of the deck's symbolism.

century Italian artists and players introduced the suit of *trionfi*:[4] a word that ultimately finds its corrupted English equivalent as 'trumps'. The modified, five-suit decks operated according to somewhat different rules. If you've ever played a game like Hearts or Bridge, you'll understand immediately what the *trionfi* cards enabled: namely, a way to take a trick of cards without following suit. The triumph cards were invented in order to trump the ranking of the pips and court cards. Indeed, the fifteenth century Tarot deck invented the whole idea of trumps in game play.[5]

Originally unnumbered and unlabeled, these special *trionfi* cards included rich and immediately legible iconography: images of a vagabond, a mountebank, an Emperor, Pope, Devil, Death, etc. In game play, the value and meaning of these cards were determined not by numerical ranking – for they had no numbers or ranks – but instead by the 'divine comedy' to which they alluded. Each image, each card indicated a successive moment in the unfolding pageant of man's salvation. The cards were called 'triumphs' because they illustrated human life as a triumphant spiritual progress: each card depicted a separate stage in this progress, a victory won on the path to salvation – a new triumph on the road to heaven.

This notion of victory harkened back to the tradition of ancient Roman victory parades (*triumphi* in Latin), where a conquering general returned from battle to parade his troops, his spoils, and his captives down the streets of Rome.[6] These Roman triumphs have inspired centuries of military, religious and political spectacles. It's because of the Roman tradition that we have an Arc de Triomphe in Paris, built to commemorate Napoleon's 1806 victory at Austerlitz. But Napoleon wasn't the first latter-day ruler to imitate the triumphal imagery of imperial Rome. That honor apparently goes to Frederick II of the Hohenstaufen dynasty, who in 1237 staged a triumph in Rome to commemorate his victory over the city-state of Milan. In late medieval and early modern Europe, the iconography of the Roman triumph became a common way for rulers to celebrate their power.

---

4 The decks also apparently introduced a fourth, female court card: the Queen, positioned between Knight and King.

5 Michael Dummett pointed out that the idea of trumps might have been invented in Germany slightly earlier than the Tarot emerged in Italy. Nonetheless, it was the Italian Tarot pack – not the German game – that influenced the subsequent history of playing cards (Dummett, 1985, p. 48).

6 Tarot artist and scholar Robert M. Place offered a persuasive discussion of these ancient triumphs as essentially 'ritual reenactments' of what mythologist Joseph Campbell called the hero's journey. See Place, 2005, pp. 108–11.

At the same time in fourteenth century Italy, just a century before the invention of the Tarot deck, the iconography also made its way into literature in what would become one of the most popular and imitated poems of the Renaissance: *I Trionfi* (*The Triumphs*) by Petrarch (1374). Petrarch's poem takes the imagery of the conquering hero, and ties it to an allegory of redemption. The poem is a dream vision in which the poet, himself languishing in unrequited love to his beloved Laura, sees six interlocked visions of triumphal parades. The parades begin with the Triumph of Love, where the god of Love displays his countless victories over warriors, gods, emperors, kings, Biblical figures (etc.), leading captives who include Antony and Cleopatra, Mars and Venus, and Samson and Delilah. From there, the poet moves to the Triumph of Chastity, where Laura herself is the conquering hero over Love (because a chaste heart proves stronger than the ardor of passion). From Chastity, the poem advances to the Triumph of Death. Death is the great leveler who can overpower even the purest and most noble of souls, including chaste Laura. But in his turn, Death is conquered by Fame. The Triumph of Fame teaches us that our deeds can outlive us, just as Petrarch's writings about beautiful Laura have preserved her memory for eight centuries now. However, Fame cannot prevail forever. The next Triumph belongs to Time. Even the most famous of heroes and deeds are ultimately buried by the dust of centuries.

The sixth and final Triumph, the Triumph of Triumphs, is the Triumph of Eternity.

The poet writes:

> When I had seen that nothing under heaven
> Is firm and stable, all dismayed I turned
> To my heart, and asked: 'Wherein do you have your trust?'
> 'In the Lord,' the answer came, 'Who never fails
> His promise to one who trusts in Him. [...]'[7]

7 Translation based on *The Triumphs of Petrach*, tr. Ernest Hatch Wilkins (1962), with slight changes to reflect the original Italian more closely. The importance of Chastity in Petrarch's account is what signals the profoundly neoplatonic cast of his allegory. At root here is the logic of Platonic askesis, of the discipline that refines the human soul by denying sensual pleasures. In both the myth of the soul as chariot in *The Phaedrus*, and in Socrates' discussion of love and of Diotima's ladder in *The Symposium*, the chastity with which sexual gratification is rebuked becomes the engine for a soul's ascent from the lower forms to the highest good. I've pursued these ideas in a different context in Tishman, 2002. In a context more relevant for our purposes, see Robert M. Place's discussion of Plato's Chariot, of The Republic, and of askesis in Place, 2005.

In what can we place our trust? Each step along the way, we are chastened by the impermanence of human existence. But each step also leads us one rung closer to the final victory: the triumph of the faithful, who put their trust in God. Each conquering hero – Love, Chastity, Death, Fame, Time, Eternity – subsumes what has come before, until finally the winner, Christ, takes all. As one literary historian put it: 'Petrarch reflects all the variety of life and still unifies his pageant by the inexorable movement toward Eternity's triumph.'[8] Thanks to the sublating logic of allegory, Petrarch's poem manages to address the full range of human life, negating lesser joys on the pathway to salvation, but still preserving all within the triumph of Christ.

It has been 60 years since art historian Gertrude Moakley argued that Petrarch's poem is the source of the imagery in the Tarot trumps (Moakley, 1966). Few now would agree unequivocally with her claims. Petrarch's six triumphs do not neatly map on to the succession of 22 cards. Most significantly, not a single one of the Tarot trumps celebrates Chastity, who plays such a central role in Petrarch's account. Nonetheless, it is thanks to Petrarch that the idea of 'triumph' gets tied to the story of spiritual awakening. His poem is a conversion poem; in the final triumph, the Seeker realizes that what he has been pursuing all along lies with the One Truth in Heaven. In a way that decisively changes the history of playing cards, Petrarch marries the imagery of conquest to an allegory of the seeking heart.

Thanks to Petrarch, then, with the earliest Tarot decks we find the allegory of Christian redemption unfolding via a succession of 22 interlocking triumphs. Nearly all of the extant fifteenth century decks use the same set of 22 images (the Sola-Busca, as we will see in a moment, is the notable and telling exception). Furthermore, although there seem to be some regional variations in ordering, nonetheless consistent groupings of the cards always appear together, allowing us to discern the same three-stage path from folly to wisdom. As Tarot artist and scholar Robert M. Place describes these three stages: 'The first group [of cards] is concerned with worldly power and sensuality; the second group depicts time, death, and the harsh realities of life along with the virtues; and the third group depicts a mystical ascent through celestial bodies of increasing radiance' (Place, 1993). The succession of these *carte da trionfi*,

8 'In Petrarch a triumph is not isolated and static. In his dream-vision he portrays the successive triumphs of Love, Chastity, Death, Fame, Time, and Eternity. By providing each figure with some individuating characteristics and by having each one fall victim to the protagonist of the succeeding triumph, Petrarch reflects all the variety of life and still unifies his pageant by the inexorable movement toward Eternity's triumph. In the Renaissance this work had such stature that, like the Divina Commedia, it was considered an epic even though both poets make love, not war, the central issue' (Coogan, 1970, p. 308).

these triumph cards, is predictable, generic, and would have been immediately recognizable to any early modern game player. Indeed, that generic predictability was essential for the game to be played: the value of each card had to be immediately obvious and apparent in play.

All the same, within this generic and triumphant predictability, there is one card among the others that defies prediction. I am speaking here of the figure of folly itself: the putative starting point for the entire redemptive narrative. Each suit of 22 triumphs includes a figure of *Il Matto*: the Crazy One or Fool. The Fool card is extraordinary, literally. Within game play, it stands outside of the ordinary scope of things. The Fool's function in the game of triumphs is unique. Indeed, technically, the Fool is not a trump at all but falls out of the sequence as the 'Excuse': a card that releases players from either following suit or playing a trump.

The Fool's extraordinary status was celebrated in one of the earliest Tarot decks: the aforementioned 1491 Sola-Busca. In its overall design, the Sola-Busca may have been conceived by Ludovico Lazzarelli (1447–1500), a humanist and a poet who, like Petrarch a century earlier, sought to reconcile Roman, Christian and mystical traditions (Rowley, 2013, p. 283).

The Sola-Busca is noteworthy on a number of counts. It eschews the typical allegorical imagery in the trumps, referencing instead figures from the history of Rome. It is also the oldest Tarot deck surviving in its entirety. All 78 cards remain extant, and can easily be viewed in modern reproductions. Indeed, Pamela Colman Smith, the artist and occultist who drew the influential 1909 Rider-Waite Tarot deck, almost certainly viewed copies of the cards in the British Museum, where a complete set of photos was available. Colman Smith's artwork spawned the modern esoteric Tarot; the vast majority of Tarot packs in use today are based on the deck she produced in collaboration with Arthur Edward Waite.[9] We can see the Sola-Busca's influence most directly in Colman Smith's 3 of Swords, her 10 of Wands, and her Queen of Cups.

The Sola-Busca is also the oldest surviving printed deck: its images were produced via the art of intaglio copper engraving. It is furthermore the oldest deck with fully illustrated pip cards. It is not until the French and English occultist movements of the eighteenth and nineteenth centuries that we again find an effort to bring iconographic detail to the pips. For our purposes, however, what is most remarkable about the Sola-Busca is its Fool. The Sola-Busca provides the first image of *Il Matto*, of the crazy or foolish one, to be numbered zero. Moreover, the entire Sola-Busca deck is numbered, and incorporates an oddity

9 The 1909 collaboration between English occultists Arthur Edward Waite and Pamela Colman Smith is most commonly referred to as the Rider-Waite Tarot, in reference to Waite's direction and the earliest publisher of the deck, the company then known as William Rider & Son.

that one sometimes finds in modern decks as well: the trump cards all receive Roman numerals – I, II, III, IIII, V, etc. – while the Fool and the four suits are given Hindu-Arabic numerals: 0, 1, 2, 3, 4, 5 …

This juxtaposition of two numbering systems, Hindu-Arabic and Roman, echoes the historical moment. In the late fifteenth century, Europe was just coming to terms with the use of the zero. Hindu-Arabic numerals entered Europe through Muslim Spain and Italy in the 1200s, but only began to displace the cumbersome system of Roman numerals in the 1400s, the same century during which the Tarot was invented. Europe was suspicious of the new Hindu-Arabic numerals. In fact, in 1299 Hindu-Arabic numerals were outlawed in Florence, ostensibly because they enabled easy forgeries. If you are tabulating your goods and earnings, it's far easier to turn 50 into 5,000 than to turn L into MMMMM. The introduction of the zero as part of this number system nonetheless marked a huge and tumultuous advance for Europe, enabling such diverse phenomena as the vanishing point in perspectival painting, double-entry bookkeeping, and the rise of modern science (Rotman, 1987; Seife, 2000).[10]

But the zero digit was a terrifying oddity. Numbers help us count, add, subtract, multiply. They point to the world of things. Why have a digit – the word literally means *finger* – if you're not going to point to some thing? What is this digit that points to nothing, no thing? Even more startlingly, zero counts nothing, and yet magically adds substance to all other figures. The numeral 4 becomes 40, or 400,000, or even 400,000,000 … ad infinitum. A little nothing bends toward infinity. And thanks to zero, which itself counts nothing, the same nine numerals (1, 2, 3, 4, 5, 6, 7, 8, 9) can count everything that can ever be counted. As the placeholder that enables the decimal system underlying all of modern mathematics, finance and science, zero enables the endless expansion of number. Even though zero brings nothing to the party, it can feed everyone there. The concept of zero did not exist in the classical mathematics of the Greeks and the Romans. And it was an abomination at first to the Christian West. What use did a good Christian have for nothingness? God created *something*, not nothing.

Indeed, as this volume strives to demonstrate, the concept of zero arose from the non-Christian world, from the Hindu notion of the void, or *śūnya* (the word becomes the Sanskrit term for the mathematical zero): an emptiness that simultaneously holds the potential for everything. The zero ties together emptiness and infinity, nothingness and everything. No wonder Europe was

10 Amir Aczel's *Finding Zero* first alerted me to the Indian origin of the zero digit, and its deep metaphysical connection to a Buddhist notion of *sunyata* or emptiness.

nervous about the zero. The Christian West likes to keep *nothingness* on the side of Death, the great leveler with his mighty scythe, cutting away all that is. *Everything*, in contrast, belongs with God on the side of Eternity. Zero decisively upsets that neat dualist balance. Emptiness and infinity, nothingness and everything, instead become two sides of the same hollow coin.

Back to the Sola-Busca: In the very midst of Europe's encounter with the zero, we find a deck of cards that identifies the Fool with this transgressive new concept. Moreover, Lazzarelli's deck underscores both the 'Roman-ness' of the trumps, and the foreignness of the Fool. As we have seen, Tarot trumps derive their logic and meaning from the tradition of ancient Roman triumphs: the parades that celebrated military victory. With Lazzarelli's deck it is almost as if that triumphal logic is being disrupted by this interloping Fool. In the first place, not only does the Hindu-Arabic zero of the Fool contrast with the Roman numerals on the remaining 21 cards, but the figures on Lazzarelli's trumps are all clothed in Roman battle gear, while the Fool appears to be a Celt, bagpipe in hand and cloak fastened at the shoulder (see Figure 15.1, Sola-Busca Fool). Furthermore, the majority of the cards seem to reference periods of conflict within, and ultimately the downfall of, the Roman republic. These trumps are not so triumphant! And this crazy fool with his bagpipe manages to turn an empty bladder of wind into music.

A hundred years later, on the island associated with the Celts, another poet would show us how emptiness can indeed amount to something. In the Prologue to his 1599 *Henry V*, Shakespeare's Chorus asks whether the 'wooden O' that is the Globe theater, the great big zero on the south bank of the Thames, can possibly contain the thousand troops, horses, and firepower necessary to represent one of the greatest battles of all time: the battle of Agincourt.

> [M]ay we cram
> Within this wooden O the very casques
> That did affright the air at Agincourt?
> O, pardon! since a crooked figure may
> Attest in little place a million;
> And let us, ciphers to this great accompt,
> On your imaginary forces work.

As Brian Rotman and others have described, Shakespeare's poetry is riddled with the riddle of the zero – exploring in texts as diverse as the Sonnets and the great tragedies what it means to 'signify nothing'.[11] Rotman's reading of *King*

11 As Macbeth tells us in Act 5: 'Life's but a walking shadow, a poor player / That struts and frets his hour upon the stage / And then is heard no more: it is a tale / Told by an idiot, full

FIGURE 15.1
Sola-Busca fool. Restored image by Wolfgang Meyer and Giordano Berti. Image in the public domain
Note: https://commons.wikimedia.org/wiki/File:Sola_Busca_tarot_card_00.jpg

*Lear* is particularly noteworthy in this context, pointing as it does to the figure of Lear's Fool who serves as the 'advocate in the case of nothing' (Rotman, p. 81): the advocate for a world that resists and disrupts the sublative, winner-takes-all logic of triumph (Rotman, 1987).

Some 300 years after his Sola-Busca was printed, the father of the modern esoteric Tarot, the French author Court de Gébelin, articulated the subversive nihilism of the Fool:

> As for this Atout [i.e., Trump], we call it zero, although it is placed in the game after XXI, because it does not count when it is alone, and has only that value which it gives to others – just like our zero: thus showing that nothing exists without its folly.[12]

---

of sound and fury, / Signifying nothing.' Notably, the Sola-Busca deck identifies the collapse of triumphant, redemptive logic with the figure of the Celt; where better to find the dramatic image for this collapse than in the so-called 'Scottish play'?

12 'Quant à cet A tout, nous l'appellons Zero, quoiqu'on le place dans le jeu après le XXI, parce qu'il ne compte point quand il est seul, et qu'il n'a de valeur que celle qu'il donne aux autres, précisément comme notre zero: montrant ainsi que rien n'existe sans sa folie.'

De Gébelin was referencing the Fool's extraordinary role in game play. The Fool is a complete non-sequitur. The card cannot win tricks, cannot trump any card. It merely excuses the holder from play and at the end of the trick gets scooped back into their hand. No one can win a game of Tarot just by playing the Fool. But at the end of the game, when points are being tallied, the Fool's value is as great as the highest trump. Like the zero digit, the Fool is all and nothing.

While still extant as a card game elsewhere in France, the Tarot was practically unknown in Paris when de Gébelin wrote his account. It was relatively easy for him to posit esoteric origins and meanings for these curious cards. And while nearly all of his assertions about the Tarot were false (albeit decisive for later occult writers), his discussion of the Fool was uncannily precise. According to de Gébelin, the Fool shows us that *nothing exists without its folly*. Indeed, the Fool becomes an invitation to explore the emptiness – the folly, the blindness, the mystery, the lack of finality and substance, the interconnected open-endedness – at the center of each moment. As Brian Rotman has demonstrated, however, for an early modern audience this invitation was perhaps terrifying.

For myself, as an early modern scholar, who also happens to be an ordained Zen Buddhist, I cannot help but return to the legacy of *śūnyata* within the Prajnaparamita literature at the heart of Mahayana Buddhism – particularly relevant, perhaps, to Zen and its practice of Zazen: a sitting meditation that seeks nothing more than open-ended awareness. One of the great twentieth century Zen masters, Shunryu Suzuki Roshi, had a name for this sort of folly: 'beginner's mind'. Suzuki wrote: 'In the beginner's mind there are many possibilities, but in the expert's there are few' (Suzuki, 2011, p. 1).

A life of beginner's mind. A life of many possibilities. Such is the fundamental non-triumphant triumph of the Fool. By the time the Tarot was taken up by the likes of Pamela Colman Smith, the folly of the zero provided an opportunity for a work of occult import (see Figure 15.2, the Rider-Waite Fool). As Arthur Edward Waite wrote in 1911, in his description of the trump numbered zero:

> With light step, as if earth and its trammels had little power to restrain him, a young man in gorgeous vestments pauses at the brink of a precipice among the great heights of the world; he surveys the blue distance before him – its expanse of sky rather than the prospect below. […] The edge which opens on the depth has no terror; it is as if angels were waiting

---

See Antoine Court de Gébelin, *Le monde primitif, analys. et compar. avec le monde moderne* (The primeval world, analyzed and compared to the modern world), Paris (1774–1796).

FIGURE 15.2
The Rider-Waite Fool (1911). Public domain image from The Pictorial Key to the Tarot (London: Rider, 1911)

> to uphold him, if it came about that he leaped from the height. His countenance is full of intelligence and expectant dream. [...] The sun, which shines behind him, knows whence he came, whither he is going, and how he will return by another path after many days. He is the spirit in search of experience. (Waite, 1911, pp. 152–55)

The zero provides the Tarot with its point of origin – with, in Waite's memorable phrasing, 'the edge which opens on the depth'. Thanks to the Fool, the Tarot deck becomes a mandala of gnostic mystery. In the traditional allegories of the early modern era, a pilgrim like Petrarch places his faith in God, and in God's revealed word. Waite's pilgrim, in contrast, embodies the sheer folly of *śūnyata*. The sun shines behind him. What lies ahead, remains in shadow. He places his faith in all that has not been and cannot be revealed.

The Fool, round and open as a cipher, places his faith in the void.

## References

Aczel, Amir D. (2016). *Finding Zero: A Mathematician's Odyssey to Uncover the Origins of Numbers*. St Martin's Press.

Coogan, Robert. (1970). Petrarch's 'Trionfi' and the English Renaissance. *Studies in Philology*. Vol. 67(3), July, pp. 306–327.

Derrida, Jacques. (1978). From Restricted to General Economy: A Hegelianism without Reserve. In *Writing and Difference*. trans. Alan Bass. Routledge: London, pp. 251–277.

Dummett, Michael. (1985). Tarot Triumphant. *FMR/ America*. No. 8. New York, NY: Franco Maria Ricci International, pp. 41–60.

Freinkel, Esther. (2002). *Reading Shakespeare's Will: The Theology of Figure from Augustine to the Sonnets*. Columbia University Press.

Gébelin, Antoine Court de. The Game of Tarots. In *Le Monde Primitif, analysé et comparé avecle monde moderne* (The Primitive World, analyzed and compared with the modern world). Trans. Donald Tyson. https://web.archive.org/web/20111004232937/http://www.donaldtyson.com/gebelin.html.

Moakley, Gertrude. (1966) The Tarot Cards Painted by Bonifacio Bembo for the Visconti-Sforza Family: An Iconographic and Historical Study. The New York Public Library.

Petrarch, Francesco. (1962). *The Triumphs of Petrarch*. Tr. Ernest Hatch Wilkins. Chicago: University of Chicago Press.

Place, Robert M. (2005). *The Tarot: History, Symbolism, and Divination*. New York, NY: Penguin.

Rotman, Brian. (1987). *Signifying Nothing: The Semiotics of Zero*. St Martin's Press.

Rowley, Neville. (2013). The Sola Busca Tarocchi: Milan. *The Burlington Magazine*. Ed. Benedict Nicolson. Vol. 155 (1321), pp. 283–84.

Seife, Charles. (2000). *Zero: The Biography of a Dangerous Idea*. Penguin Books.

Shakespeare, William. (2015) *The Norton Shakespeare*. 3rd Edition. Gen. Ed. Stephen Greenblatt. WW Norton & Company.

Suzuki, Shunryu. (2011). *Zen Mind, Beginner's Mind: Informal Talks on Zen Meditation and Practice*. Shambhala.

Tishman, Esther Freinkel. (2002). Reading Shakespeare's Will: The Theology of Figure from Augustine to the Sonnets. Columbia UP.

Tishman, Esther Freinkel. (2019). *Mindful Tarot: Bring a Peace-Filled and Compassionate Practice to the 78 Cards*. Woodbury, MN: Llewellyn.

Waite, Arthur Edward. (1911). The Pictorial Key to the Tarot: Being Fragments of a Secret Tradition Under the Veil of Divination.

# PART 1

## *Zero in Religious, Philosophical and Linguistic Perspective*

# Introduction to Part 1

The next stage of the interdisciplinary exploration delves into the religious, philosophical and linguistic aspects of zero. In 'On the Semiotics of Zero', BRIAN ROTMAN expands on material in his book, *Signifying Nothing: The Semiotics of Zero,* where he presents a semiotic investigation of zero as a signifying concept in the Hindu – Arabic decimal notation and, in particular, its role in Western culture. He argues that a relation to 'nothing' – to representing it – is realized as a sign for absence of signs and challenges the notion that objects come before their names, that things are anterior to signs. PAUL ERNEST explores the mathematics, history, and philosophy of zero, looking at the development of zero purely as a sign (syntactic); its meaning (semantics); and its social contexts, roles, and uses (pragmatic) in 'Nought Matters: The history and philosophy of zero'. He demonstrates that it was not until 628 CE that the role of zero within the network of number relations in the integers was established by Brahmagupta. The stages in the development of the concept of zero are outlined, with particular reference to Indian and Mayan cultures, and Ancient Egypt.

Drawing on recent datings of the Bakhshali manuscript, in 'The Influence of Buddhism on the Invention and Development of Zero', ALEXIS LAVIS considers the way that Indian culture influenced the invention of zero – in particular, the influence of Buddhism and its role in the theory leading to the inception of the placeholder zero as well as the operative function of zero to becoming a number, with Brahmagupta. FABIO GIRONI highlights the conceptual connections between the concept of zero and the philosophical concept of *śūnyatā* in 'Zero and Śūnyatā: Likely Bedfellows'. Gironi posits the theory that the geographical congruence of both arising in the Indian sub-continent is more than a coincidence. The chapter undertakes a synthetic reading of the three 'voids' – zero, *śūnyatā* (with its central meontological significance in the philosophy of Nāgārjuna), and the 'trace' in the thought of Jacques Derrida.

In 'Indian Origin of Zero', RAVI PRAKASH ARYA gathers data from the Vedic and post-Vedic Sanskrit literature to understand the origin of zero; the philosophy behind its origin; various denominations of zero used in various phases of Indian civilization; and the philosophy and epistemology behind the symbolic representation of zero (0) in India. SUDIP BHATTACHARYYA presents 'A Philosophical Origin of the Mathematical Zero', investigating whether philosophical ideas of *nothing* could have given rise to the mathematical zero and arguing that it did not arise from a practical need, nor was a place-value number system or placeholder required to conceptualize zero. BHATTACHARYYA argues that zero and its basic operations were intellectually premised, requiring

 | DOI:10.1163/9789004691568_019

a philosophical base of *nonexistence*, and not that of *emptiness* or *void* as commonly assumed.

Within Nāgārjuna's framework of conventional reality and ultimate reality, in 'Category Theory and the Ontology of Sunyata', SISIR ROY and VENKAT RAYUDU POSINA explain the concepts of *śūnyata* and *catuskoti*, presenting category theoretic constructions that are reminiscent of these Buddhist concepts. Their elaboration of the parallels between Buddhist philosophy and category theory facilitates better understanding of Buddhist philosophy, and it brings out the broader philosophical import of category theory beyond mathematics. SHARDA S. NANDRAM, PUNEET K. BINDLISH, ANKUR JOSHI and VISHWANATH DHITAL analyze the application of numbers to the material realm as compared to the spiritual in 'Zero: An Integrative spiritual perspective with One and Infinity'. They reconsider the nature of peculiar or special numbers and ratios, numbers and concepts, their past and future possibilities – in particular, Zero, One and Infinity (ZOI) from an integrative spiritual perspective.

KASPARS KLAVINS addresses the topic of the 'Challenges In Interpreting The Invention Of Zero' which are raised by attempts to explain the importance of the invention of zero and its application, including the symbol of zero both as a scientific discovery/invention and an empirically found solution for the satisfaction of certain practical needs of humans. He considers the difficulties of synthesizing different areas of research, including scientific – technical, religious – philosophical and mathematical, with reference to medieval/early modern Europe. In 'Some More Unsystematic Notes On Śūnya', ALBERTO PELISSERO provides an overview of the themes regarding the concept of *śūnya* ('void') in Indian thought, mainly *mādhyamika*, with reference to *apoha* semantic theory, and in relation to the metaphoric use of words and the paradox of ineffability (particularly in Vedantic Brahmanical circles) – illustrating the nature of the current debate, without offering a unique answer.

JOHANNES BRONKHORST argues that the search for a philosophy that gave rise to the number zero is misguided in 'Much Ado About Nothing, Or How Much Philosophy Is Required To Invent The Number Zero?' He claims that no philosophy is required to invent this number and shows that there are good reasons to accept that Buddhism did not play a role in this invention, pointing out that the notion of number as developed in Indian philosophy had no place for zero. ERIK HOOGCARSPEL reflects on the different aspects of numbers in 'From Emptiness To Nonsense: The Constitution Of The Number Zero (For Non-Mathematicians)', with reference to Nieder, and on the precise nature of the transformation of zero from a signifier of quantity to a number as an element of a self-contained symbolic deductive system that has no sense at all. The history of zero as a number coincides with the history of abstract mathematics and the development of calculation systems, the tremendous usefulness of which is precisely based on its very meaninglessness. In 'The Fear Of Nothingness', JOHN MARMYSZ explores the genesis and influence of the Western fear of nothingness which dates back to ancient Presocratic thinkers from the Milesian School, the first school of Western philosophy. This suggested that there remains something stable, permanent, and dependable beneath it all, that is, Monism – including Thales' claim that 'all is water,' Anaximander's claim that the universe arises from *apeiron*, and Anaximenes' claim that 'all is air'.

ESTI EISENMANN considers the topic from a Jewish perspective in 'The Concept of Naught In Jewish Tradition', presenting a philosophical and mathematical consideration of zero from the Bible up to the modern age – in particular, *ex nihilo* creation, the definition of the deity as the Naught and the creation out of the Naught, and the existence of the vacuum in the material world after the Creation; plus two mathematical aspects – the Hebrew system for designating numbers (numerals) and the notion of zero as a numerical value from the Bible, Talmudic rabbinic texts and medieval philosophic literature/Kabbalah.

In 'How Does Tom Tillemans Think?', ERIK HOOGCARSPEL maintains that emptiness started as an intuition, but became in due time formalized into a concept. He argues that this has grave consequences for the interpretation of the concept of emptiness and the introduction of the zero. Hoogcarspel concludes that this interpretation has to be an historical anthropological narrative – if not it might turn out to be an oxymoron. Phenomenology is posited as a way of understanding the intuition, with the possibility of bringing the intuition back to the concept.

PETER GOBETS considers the two prevailing worldviews of an objective Reality 'out there' or of a subjective Reality 'in here' in 'Overhauling The Prevailing Worldview: An Essay'. Gobets argues that both ideologies are untenable; both have served their practical purpose but outlived their theoretical usefulness, even becoming counterproductive by eclipsing alternative worldviews. Consequently, the sciences have landed in crisis, no longer able to account for certain cutting-edge laboratory results. A viable option is offered to surmount the philosophical-logico-linguistic hurdles encountered to usher in a new worldview and the foundationless 'Nonoverse'.

CHAPTER 16

# On the Semiotics of Zero

*Brian Rotman*

## Abstract

Thirty years ago, I published *Signifying Nothing; The Semiotics of Zero* a book about zero. It was framed as a semiotic investigation of zero as a signifying concept. My interest was the concept's role in Western culture, in the Hindu-Arabic notation for decimals, and within the contemporaneous elaboration of money and painting. In each of these three codes, a relation to 'nothing' – to representing it – is realized as a sign for absence of signs. In the process the notion that objects come before their names, that things are anterior to signs, is challenged. Instead, the action of zero is to be seen as *constitutive* of the whole numbers and not merely part of the notational apparatus of 'subsequently' naming them. Moreover, the same effect holds for the codes of painting and money no less than for numbers. The extract reproduced here outlines the nature of this phenomenon for the case of numbers. In the decades since my book was published other books and essays about zero have appeared likewise revealing its extraordinary and revolutionary nature. Not least the present initiative of the Zero Project to uncover zero's origin.

## Keywords

anterior – meta-subject – closure – algebraic variable – Stevin – empty set

## 1 Introduction

The first purpose of this article is to portray the introduction of the mathematical sign zero into Western consciousness in the thirteenth century as a major signifying event, both in its own right within the writing of numbers and as the emblem of parallel movements in other sign systems.

Zero is to number signs, as the vanishing point is to perspective images, as imaginary money is to money signs. In all three codes the sign introduced is a sign about signs, a *meta-sign*, whose meaning is to indicate, via a syntax that arrives with it, the absence of certain other signs. Further, I shall argue that

 | DOI:10.1163/9789004691568_020

the shifts that take place within these codes – from Roman to Hindu numerals, from the images of Gothic art to those of perspectival painting, from gold coinage to the use of imaginary bank money – exhibit parallel patterns of semiotic disruption.

Despite its formal equivalence with the vanishing point and imaginary money, zero has a primary status, a privileged claim to attention, on several counts. First, and most simply, the introduction of zero occurred earlier than changes in the signs of painting and money and can therefore be seen in a relatively pristine state uncomplicated by historical relations it might have with these other changes. Secondly, zero is epistemologically rudimentary: not only in the sense that mathematics claims to provide objective, historically invariant, empirically unfalsifiable 'truth', thus making its signs more culturally and logically naked, transparent, and parsimonious than other signs, but in the immediately practical sense that counting is a simpler, more primitive semiotic activity than viewing a painting, and number signs are a necessary cognitive precursor of any kind of money transaction. And finally, zero's connection, obscure but undeniable, to the much older and more deeply embedded idea of 'nothing' indicates that questions about its introduction be phrased in very general terms. One can ask: what is the impact on a written code when a sign for nothing, or more precisely when a sign for the absence of those other signs, enters its lexicon? What can be said through the agency of such a sign that could not be said, was unsayable, without it?

To suggest that the Hindu number sign zero might be emblematic of fundamental changes in the signs of painting and money, and to assert that in any case it can be seen as the paradigm and prototype of these changes, is to claim for zero an exalted role: an intellectual significance and status in conflict with its banal, everyday, unspectacular appearance as a constituent of the decimal notation for numbers or of the binary system on which computer languages are based. Indeed, even to speak of zero as the origin of a large-scale change within *mathematics* seems to be at variance with how mathematicians and historians of mathematics construe the matter.

Thus, although we learn from historical accounts that zero was introduced into medieval Europe with great misunderstanding and difficulty, that it was unknown either to the Romans or the classical Greeks, that equivalent versions of it were used and obviously understood by both the Babylonian mathematicians of the Hellenistic period and the pre-Columbian Mayas, we are not presented with any sense of the semiotic difficulty – the peculiar, enigmatic, and profoundly abstract challenge that zero presented, as a *sign*. A sign, moreover, whose connection to 'nothing', the void, the place where no thing is, makes it the site of a systematic ambiguity between the absence of 'things' and the absence of signs, and the exemplar, as we shall see, of a semiotic phenomenon whose importance lies far beyond notation systems for numbers.

## 2 Special Status

Again, although mathematicians are obliged to make certain concessions to the special status of zero and its derivatives – the empty set, for example, is given a special kind of definition and is acknowledged to play a unique and privileged role as the origin of the hierarchy of infinite sets within which all of mathematics is supposed to take place – they push the matter as a question about the nature of *signs* no further. Perhaps this is because the majority of mathematicians, as well as historians of the subject, are philosophical realists, in the sense that they take mathematics to be about 'things' – numbers, points, lines, spaces, functions, and so on – that are somehow considered to be external and prior to mathematical activity; things that although formulated in terms of mathematical symbols, have nonetheless a pre-mathematical existence that cannot be accounted for solely in terms of the signs mathematicians themselves produce. In the course of this article I will reject such a view, and suggest that this insistence on the priority of certain 'things' to mathematical signs is a misconception, a referentialist misreading, of the nature of written signs; a misconception not, by any means, peculiar to mathematics, but one repeated for visual and monetary signs through the natural but mistaken notion that a painting is simply a depiction and money a representation of some prior visual or economic reality.

Focusing on the nature and status of zero, and likewise the vanishing point and imaginary money, as certain kinds of creative originating meta-signs leads inevitably to the question of origins, to the manner in which signs are produced and created. Signs, meta-signs, and the codes in which they operate do not arise and exist by themselves, they are not given as formal objects in some abstract already present space. They are *made* and remade repeatedly whether spoken, counted, written, painted, exchanged, gestured, inscribed, or transacted; they come into being and then persist, by virtue of human agency, through the continued activity of interpretation-making sign-using subjects. Such subjects are not to be identified with individuals, with persons who feel themselves to be authors and recipients of sign utterances; rather, they are semiotic capacities – public, culturally constituted, historically identifiable *forms* of utterance and reception that codes make available to individuals.

The picture presented here will, therefore, elaborate on the signifying acts initiated by certain specific semiotic subjects – the one-who-counts, the one-who-sees, and the one-who-buys-and-sells – and on the distinction made for these subjects between being an external agent and an internal recipient of signification; a distinction that although graspable as a felt, experienced difference, is notoriously difficult to convey by the subject himself. Thus, St Augustine likened the difficulty of separating signs from signs about signs, of distinguishing between talk and talk about talk, to a physical misperception:

> Discussing words with words is as entangled as interlocking and rubbing the fingers with the fingers, where it may scarcely be distinguished, except by the one who does it, which fingers itch and which give aid to the itching. (St Augustine, De Magistro in Gribble, 1968, p. 98)

Clearly, a discourse interested in illuminating the relation of signs to meta-signs has to go beyond St Augustine's image of a private self in a state of bodily confusion between inner proprioception and outer perception, and replace it by an observable relation between publicly signifying agencies, between, that is, semiotic subjects.

One consequence of employing a vocabulary of semiotic, as opposed to individual, subjects will be that the failure to distinguish between inner perception (self as author of signs) and outer perception (author as subject of signs), between words and words about words, will be seen not as an error, a confusion to be clarified, but as the inevitable result of a systematic linguistic process that elides this very difference: a process whereby meta-signs are denied their status as signs of signs and appear as mere signs; zero becomes just another number among the infinity of numbers; the vanishing point appears indistinguishable from all the many other depicted locations within a painting; imaginary money is treated as simply more money. In this naturalization of meta-signs into signs – obviously akin to figures of speech dying and becoming literal – it becomes necessary, if one is to explicate what it means for a sign to signify other signs, to retrieve the fate of the absented subject. For it is the manner of its absence, the sort of naturalization it is subjected to, that sets the stage for the emergence of a secondary formation to the original meta-sign.

## 3 Algebraic Variable

Specifically, I shall show how zero gives rise at the end of the sixteenth century to a semiotic closure of itself, namely the algebraic variable; likewise the vanishing point will be seen to engender a closure, during the seventeenth century, in the form of the multi-perspective image; and similarly imaginary money finds its semiotic closure in the emergence of paper money at the end of the seventeenth century. Moreover, we shall see that each of these new signs is itself a certain kind of meta-sign in relation to the system that spawns it; a meta-sign that requires the formulation of a new sign-using agency, a secondary subjectivity, in order to be recognized. It will be shown that this new capacity, a self-conscious subject of a subject, a *meta-subject,* is at the center of major eruptions within very different sign systems of the sixteenth and seventeenth centuries. These major eruptions were, in mathematics, the invention of algebra by Vieta; in painting, the self-conscious image created by Vermeer and Velasquez; in the text, the invention of the autobiographical written self by Montaigne; in economics, the creation of paper money by gold merchants in London.

All these signs, the original meta-signs as well as their closures, are structured around the notion of an absence, in the sense of a signified non-presence of certain signs; a notion that occurs in stark and prototypical form in the case of zero, which presents itself as the absence of *anything*, as 'nothing': the statements '1 – 1 = 0' and 'one taken away from one is nothing' are, in common parlance, read as translations of each other. Thus, in order to pursue the historical impact of zero in terms of the resistance it encountered as a sign, one has to ask questions about what 'nothing' was supposed to signify within European intellectual discourse. How, for example, was Nothing cognized within the Christian orthodoxy that dominated this discourse and in the classical Greek conception that underpinned it? If medieval hostility to zero rested on a Christian antagonism to 'nothing', since to talk of something being no-thing, to give credence to that which was not and could not be in God's world, was to risk blasphemy or heresy, then it must also be said that this antagonism was itself ambiguous and unstable. Christian adherence to the classical Greek denial of 'the void' was in conflict with its acceptance of the account in Genesis in which the universe was created out of 'nothing'.

Notions of Nothing and responses to what it – and its dialectical opposites of All, Infinity, the Cosmos – could be taken to signify in sixteenth- and seventeenth-century European thought were not, however, confined to the arena of theological and doctrinal dispute. *Ex nihilo nihil* (nothing will come from nothing) was a classical maxim that acted as a starting point for a great deal of rhetorical and metaphysical speculation that had ultimately very little to do with Christianity. Nowhere is this more dramatically the case than in Shakespeare's profoundly non-Christian play, *King Lear*: a work that not only explicitly and obviously concerns itself with a certain sort of horror that comes from nothing, but that less obviously, although as I shall demonstrate equally explicitly, locates the origin of this horror in the secular effect and mercantile purport of the sign zero.

The assertion that zero and zero-like signs permeate several, very different, signifying codes and artifacts of the Renaissance is not unexpected if one thinks of changes in these codes in historically materialist terms: the historical emergency of mercantile capitalism rode on the vector of trade, business, commerce, finance, money. And money required a system of writing, which included bookkeeping and calculation, to enable it to function as an international medium of exchange. It was precisely to meet this need that double-entry bookkeeping and Hindu numerals, both written in terms of zero, were introduced in Italy at the beginning of the Renaissance in the thirteenth century. Zero, then, was a principal element of Renaissance, that is to say mercantile capitalism's, systems of writing from the beginning.

## 4 Anteriority of Things to Signs

There is a *system* (Hindu decimal place notation, principles of linear perspective, mechanism of capitalist exchange) that provides a means of producing infinitely many signs (numerals, pictures, transactions). These signs *represent* (name, depict, price) items in what is taken to be a prior reality (numbers, visual scenes, goods) for an active human subject (one-who-counts, one-who-sees, one-who-buys-and-sells). The system allows the subject to enact a *thought-experiment* (calculating, viewing, dealing) about this reality through the agency of a *meta-sign* (zero, vanishing point, imaginary money) that initiates the system and affects a *change of codes* (gestural/graphic, iconic/perspectival, product/commodity).

I want now to deconstruct this scheme. What lies at its center, explicit in the talk of 'prior' reality, is some supposed movement into signification, some shift from object to sign, from presentation to representation, from a primary given existence to a secondary manufactured description. In each of the cases of number, vision, and money a field of entities is assumed to exist anterior to the process of assignation performed by the system. What are taken to be pre-existing numbers are given names, scenes from some supposedly pre-existing visible world are depicted, goods conceived as existing independently of, and prior to, the agency of money are assigned a price. In each case this process of assignation hinges on the meta-sign that both initiates the signifying system and participates within it as a constituent sign. And it is this double ambiguous role played by zero, by the vanishing point, by imaginary money, that ultimately destabilizes the scheme presented here and deconstructs the anteriority to signs this reality is supposed to enjoy.

In other words, the simple picture of an independent reality of objects providing a pre-existing field of referents for signs conceived after them, in a naming, pointing, ostending, or referring relation to them, cannot be sustained. What gives this picture credence is a certain, highly convincing illusion. Once the system is accepted, on the basis of a perfectly plausible original fiction, as a mechanism for representing some actuality, it will continue to claim this role, however far removed its signs are from this putative reality; so that, for example, numerals can be written that name 'numbers' that are unrealizable by any conceivable process of human counting or enumeration, pictures can be painted that depict purely imaginary, non-existent, or visually impossible 'scenes', transactions can be drawn up that price humanly unachievable relations between 'goods'.

The result is a reversal of the original movement from object to sign. The signs of the system become creative and autonomous. The things that are ultimately 'real', that is numbers, visual scenes, and goods, are precisely what the system allows to be presented as such. The system becomes both the source of reality, it articulates what is real, and provides the means of 'describing' this reality as if it were some domain external and prior to itself; as if, that is, there were a timeless, 'objective' difference, a transcendental opposition, between presentation and representation.

From one direction this deconstruction of the anteriority of object to sign can be construed, as we shall see, as the path by which the meta-sign engenders a secondary formation of itself. Each of the signs zero, vanishing point, imaginary money, will now be shown to have, in relation to the whole original system of signs, a natural closure that emerges as a new, highly potent meta-sign, important in its own right, in the codes of number, vision and money. A meta-sign whose action, at one remove from the original 'anterior' field of entities, accompanies a radically different, self-conscious form of subjectivity that I call the meta-subject.

## 5 Closure of Zero: Algebraic Variable

At the end of the sixteenth century the Dutch mathematician Stevin, advocating in his treatise, *The Dime* (Stevin, 1958) the extension of the Hindu system of numeration from finite to infinite decimals, expressed great wonderment at the creative power of zero, at the ability it gave the principle of place notation to manufacture an infinity of number signs.

Stevin rejected the classical notion of number, *arithmoi*, as a fundamental misunderstanding of the nature of numbers. For both Plato and Aristotle, *arithmoi* always, as Klein (1968) in his investigation *Greek Mathematical Thought and the Origins of Algebra* expressed it, 'indicates a definite number of definite things. It proclaims that there are precisely so and so many of these things'. And although they differed on a crucial point of interpretation, 'Plato speaks of numbers which have "visible and tangible bodies" … so that in counting dogs, horses, and sheep these processes yield dog-, horse-, and sheep-numbers' (Klein, 1968, pp. 46–7), while Aristotle, concerned to clarify what it means to say that two numbers were equal, saw them as abstractions from particular concrete collections. Both would have assented to the formula 'number as an assemblage of "units".'

For Stevin, the source of error here is not so much the formula itself but the quasi-geometrical interpretation of a 'unit' contained in the classical view that underlies it, the notion that units were abstracted, individualized 'things' that were not (obviously) pluralities and not in fact numbers at all. On the contrary, Stevin argued, the unit was a number like any other; it was not the unit that was the arche of number, but the nought. Zero was the proper origin of number, 'The true and natural beginning' (Stevin, 1958, p. 499); and just as the point in geometry generates the line, so zero, which he wanted to call *point de nombre*, gives rise to numbers.

To make zero the origin of number is to claim for all numbers, including the unit, the status of free, unreferenced signs. Not signs of *something*, not *arithmoi*, certainly not real collections, and not abstractions of 'units' considered somehow as external and prior to numbers, but signs produced by and within arithmetical notation.

In the language of Saussure's distinction, Stevin rejected numbers as signs conceived in terms of positive content, as names for 'things' supposedly prior to the process of signification, in favor of signs understood structurally, as having meaning only in relation to other signs within the sign system of mathematics. In effect, Stevin was insisting on a semiotic account of number, on an account that transferred zero's lack of referentiality, its lack of 'positive content', to *all* numbers. In so doing he overturned the belief in the anteriority of 'things' to signs that classical formulation of *arithmoi* depends on.

Stevin's primary interest, however, was not semiotic but mathematical. He proposed to extend the principles of the Hindu place notation from the whole numbers to all possible real magnitudes. This meant the creation of a system of infinitely long signifiers that relied for their interpretation on infinite summation; that is, on the process of an infinity of numbers being added together. Thus, just as 333 means 3(1) + 3(10) + 3(100), so 0.333 … ad infinitum means 3(1) + 3(1/10) + 3(1/100) + 3(1/1000) + … ad infinitum. Such non-finite addition requires an infinite limiting process; a phenomenon visualizable in geometric terms as an asymptotic approach of a line to a curve. Stevin's extension of Hindu numerals to an infinite format, which resulted in a numeralization of the real one-dimensional continuum, had far-reaching consequences for seventeenth century, and indeed for all subsequent, mathematics. Semiotically, it is the language, algebra, in which these consequences were developed, and the opposition such language offers between determinate, possibly unknown, but fixed numbers (constants) and the indeterminate non-numerical entities (variables), that is of primary interest.

Algebra, the art of manipulating formal mathematical expressions such as equations, formulas, inequalities, and identities, is co-extensive with the idea of a variable. In any algebraic expression, such as for example the identity $(x - y)(x + y) = x^2 - y^2$, the letters x and y are said to be variables, by which is meant that they denote arbitrary individual numbers in the sense that any particular number may be substituted for x and y. A variable is thus a sign whose meaning within an algebraic expression lies in certain other, necessarily absent, signs. In relation to these other signs, which constitute its 'range', a variable is a meta-sign, that is one that indicates the virtual – the potential, possible, but not actual – presence of any one of the particular signs within its range.

The essential mathematical idea of a variable, that is as an indeterminate that can be calculated with as if it were a determinate number, is due to Stevin's French contemporary, Vieta. Unlike Stevin, Vieta did not challenge the classical picture of number. His notion of variable which he called species was, therefore, formulated, as Klein observes, in referentialist terms:

> The concept of species is for Vieta, its universality notwithstanding, irrevocably dependent on the concept of *arithmos*. He preserves the character of the 'arithmos' as a 'number of …' in a peculiarly transformed manner. While every arithmos intends immediately the things or units themselves whose number it happens to be, his letter sign intends directly the general character of being a number that belongs to every possible number, that is to say, it intends 'number in general' immediately, but the things or units that are at hand in each number only mediately. In the language of the schools: the letter sign designates the intentional object of a 'second intention' (intentio secunda), namely of a concept that itself directly intends another concept and not a being. (Klein, 1968, p. 174)

For practical mathematical purposes Vieta's referentialism (repeated, as we shall see, in contemporary accounts) is of no importance. It is only significant when one asks the question what a number sign is and what it means semiotically for a variable to range over numbers. For such questions, the referentialism behind explanations of indeterminateness in terms of scholastic distinctions between first and second intentions, essentially the opposition between things and signs of things, is the point at issue: for it is precisely this opposition that mathematical indeterminacy, semiotically perceived, requires to be unmasked.

The point is central. I shall elaborate it by taking issue with the current orthodox, that is realist, explanation of a variable; a simple, purely syntactical definition within set theory that runs as follows: to say 'x is a variable ranging over the numbers N' means simply that 'x denotes an arbitrary member of the set N of numbers', where to denote a number means to point to the set constituting it. According to the set-theoretical realism behind this definition, numbers – and indeed all mathematical entities such as points, lines, invariants, geometrical figures, ratios, constructions, functions, spaces, predicates, relations, and so on – are 'things' called sets. So that mathematical discourse is entirely descriptive and referential: its signs are names, more or less explicit, more or less contained within other names, of determinate set-theoretical objects, its assertions are propositions, unambiguously 'true' or 'false' statements about some prior, extra-linguistic, universe of sets.

What, then, are *sets* as conceived by set theory? They are ensembles, collections, classes, and aggregates of members that are themselves ensembles, collections, classes, and aggregates of members: where the regression of membership is imagined to end either in some arbitrarily introduced atomic entities – ur-elements – or, as is actually the case in current practice, in the set-theoretical first cause, the equivalent of zero for collections, namely the empty set Ø. In this conception sets have precise, unambiguous criteria for membership governed by axioms; they are static, unchanging, determinate things. Theorems proved 'true' about them remain true, and so must always have been true; they are in no sense invented but discovered as timeless objects that simply exist in the Platonic (Parmenidean, in fact) universe of pure motionless being. Sets are in short the 'things', the ultimate *Dinge*, required by a nineteenth century realism immersed in the anteriority of things to signs (of things).

Conceived as sets in this way, the objects of mathematics are subjectless, unauthored referents that exist independently of human agency. By excising the subject from its conception of mathematical activity such realism offers an epistemology as well as an ontology that is semiotically incoherent: it misrepresents proof, for example, as validation of eternal truths about 'things', and misdescribes a variable as a place-holder for the members of some infinite set of such things. As *signs* variables are those that can be replaced by any sign of a given kind. In the paradigm case, that of number variables, the 'given kind' are the number signs and, since these cannot be specified except through the process of counting, it follows that a semiotic characterization of a variable can only be given in terms of the active, constructing, counting subject. Or to make the comparison in historical terms, the algebraic variable, semiotically perceived, amounts to a re-description of Vieta's *species*: one that reinstitutes the subject – the one-who-counts – into Vieta's discussion, replacing the *arithmoi* there by the purely semiotic picture of number that Stevin so marveled at.

What is the connection between zero and a variable as meta-signs? Zero functions dually: it moves between its internal role as a number among numbers and its external role as a meta-sign initiating the activity of the counting subject. So it is with the algebraic variable, except that the internal/external duality is enacted at one remove from the one-who-counts. Perceived internally, variables present their familiar appearance as manipulable algebraic objects, as signs among signs within formulas. In expressions like $x + 1 = y$, $2x + 3y - z = 0$, and so on, the letter signs enter into full arithmetical relations with number signs: multiplying them, being added to them, being numerically compared to them, being substituted for them, and generally being treated as if they *were* number signs according to a common syntax. Perceived externally, this unitary syntax comes apart, since variables are not of course number signs; on the contrary, they are signs that meta-linguistically indicate the possible, but not actual, presence of number signs.

This duality is mediated by a new mathematical subject, the *algebraic subject*, whose relation to the one-who-counts mirrors the relation between a variable and a number. Thus, the algebraic subject has the capacity to signify the absence of the counting subject, the displacement of the one-who-counts from an actual to a virtual presence. Now at certain points, when variables are instantiated by numbers, this displacement ceases to operate – the two subjects coalesce. So that, for example, the algebraic subject who reads the identity $x^2 - 1 = (x - 1)(x + 1)$ becomes the subject who checks the result of writing, say, $x = 10$; who, in other words, computes ultimately by counting – the arithmetical validity of $100 - 1 = 9 \times 11$. But this sort of arithmetical localization is extraneous to the difference between the two subjects: when variables are manipulated as algebraic objects within formal calculations any such fusion between the counting and algebraic subject is precluded; and the algebraic subject performing these calculations remains as an autonomous, arithmetically self-conscious, agent whose meta-linguistic distance from the one-who-counts is the source of the difference, when they are considered as mathematical discourses, between algebra and elementary arithmetic.

The semiotic connection between zero and the variable thus emerges as one of symbolic completion: by ranging over all number signs, that is over all possible records that can be left by the subject whose sole capacity is to repeat, the algebraic subject performs an operation of closure on the infinite proliferation of number signs that come into being with zero. In effect, the algebraic number variable is a sign for the signs that can, at least in principle, be produced by the one-who-counts.

## 6 Conclusion

Observe, finally, that zero is an origin at a very primitive, parsimonious, and minimally articulated level of sign formation. Signification codes difference, hence the need for more than one sign. How to produce, with the minimum of *ad hoc* extra-semiotic means, two 'different' signs? Answer: let there be a sign – call it 1 – and let there be another sign – call it 0 – indicating the absence of the sign 1. Of course, such a procedure *produces* the difference it appears subsequently to describe; and the use of absence to manufacture difference in this way is a viable sign practice only through the simultaneous introduction of a syntax: a system of placing signifiers in linear relation to each other in such a way that it allows signs to be interpretable in terms of the original absence/presence signified by 0.

The sign 1 can be anything. If 1 is one and 0 is zero and the syntax is the standard positional notation for numbers, then what results as the limiting case of the system, is the two-valued descendant of the Hindu decimal system. Leibniz, who spent much time formulating the rules for binary arithmetic, was deeply impressed by the generative, infinitely proliferative principle inherent in such a zero-based binarism: so much so that he refracted the binary relation between 1 and 0 into an iconic image of the Old Testament account of creation *ex nihilo*, whereby the universe (the infinitude of numbers) is created by God (the unbroken 1) from the void (cipher 0).

Again, if 1 signifies the presence of a current and 0 signifies the absence of such a current and the syntax is 2-valued Boolean algebra, then what emerges is the binary formalism within which the logic and language of all present-day computer programs are ultimately written.

Thus, to pursue zero further would be to have to say more about its role at the origin of this formalism. But such a project would require a critique of mathematical logic. In particular, one would need to unravel the assumptions behind the claim that is made for Boolean logic, with its referential apparatus of truth and falsity, to be *the* grammar of all mathematical, hence all supposedly neutral, culturally invariant, objective, true/false assertions about some prior, 'real' world. To do this would require a semiotics that went beyond zero to the whole field of mathematical discourse. A semiotics that, in order to begin at all, would have to demolish the widely held metaphysical belief that mathematical signs point to, refer to, or invoke some world, some supposedly objective eternal domain, other than that of their own human, that is time-bound, changeable, subjective, and finite-making.

## References


Alpers, S. (1983). *The Art of Describing*. John Murray, London.

Angell, N. (1930). *The Story of Money*, Cassell, London.

Bannock, G. R. Baxter and R. Rees. (1984). *Dictionary of Economics*, Penguin, London.

Baudrillard, J. (1983). *The Precession of Simulacra*, P. Foss, P. Potter and P. Beitchmun (trans.) Semiotext(e), New York.

Baxandall, M. (1972). *Painting and Experience in 15th Century Italy*. Clarendon, Oxford.

Bloom, H. (1975). *Kabbala and Criticism*. The Seabury Press, New York.

Braudall, F. (1974). *Capitalism and Modern Life, 1400–1800*. Fontana, London.

Bryson, N.(1983). *Vision and Painting: the Logic of the Gaze*. Macmillan, London.

Colie, R. (1966). *Paradoxia Epidemica*. Princeton University Press, Princeton.

Culler, J. (1981). *The Pursuit of Signs*. Routledge & Kegan Paul, London.

Cusa, N. de. (1928). *The Vision of God or the Icon*, N. Salter (trans.) Dent, London.

Davidson, J. (1983). Eurodollars, the Currency without a Country, in *Reader's Digest*, May 1983.

Derrida, J. (1976). *Of Grammatology*, G. C. Spivak (trans.) Johns Hopkins University Press, Baltimore.

Edgerton, S. (1976). *The Renaissance Rediscovery of Linear Perspective*. Harper & Row, New York.

Frege, G. (1974). *The Foundations of Arithmetic*, J. L. Austin (trans.) Blackwell, Oxford.

Foucault, M. (1973). *The Order of Things*. Vantage Books, New York.

Galbraith, K. (1975). *Money: Whence it Came, Where it Went*. Andre Deutsch, London.

Gombrich, E. (1963). *Meditations on a Hobby Horse*. Phaidon, London.

Gribble, C. (1968). *Studies Presented to R. Jakobson*. Slavica, Cambridge, Mass. p. 109.

Graham, W. (1911.) *The One Pound Note in the History of Banking in Great Britain*. James Thin, Edinburgh.

Hogan, W. and Pearce, I. (1984). *The Incredible Eurodollar*. Allen & Unwin, London.

Kirk, G. and Raven, J. (1957). *The Presocratic Philosophers*. Cambridge University Press, Cambridge.

Klein, J. (1968). *Greek Mathematical Thought and The Origin of Algebra*, E. Brann (trans.) MIT Press, Cambridge, Mass.

Mauss, M. (1969). *The Gift*, Cunnison (trans.) Routledge & Kegan Paul, London.

Miller, J. A. Suture (elements of the logic of the signifier), *Screen*, Winter, 1977–78.

Montaigne, M. (1889). *Essays*, J. Florio (trans.) Stott, London.

Needham, J. (1959). *Science and Civilisation in China, Vol. III*. Cambridge University Press, Cambridge.

Nishitani, K. (1982). *Religion and Nothingness*, J. Van Bragt (trans.) University of California Press.

Pullan, J. M. (1968). *The History of the Abacus*. Hutchinson, London.

Rice, E. (1956). *The Adding Machine*. Samuel French, New York.

Rider, F. (1973). *The Dialectic of Selfhood in Montaigne*. Stanford University Press, Stanford.

Rotman, B. Towards a Semiotics of Mathematics, *Semiotica* (in press).

St Augustine. *Confessions*, trans. Pine-Coffin. Penguin, London, 1961.

Sayce, R. (1972). *The Essays of Montaigne*. Weidenfeld & Nicolson, London.

Scholem, G. (1941). *Major Trends in Jewish Mysticism*. Schocken, Jerusalem.

Shakespeare, W. *King Lear* (Methuen, London, 1969).

Shell, M. (1980). The Gold Bug, *Genre*, vol. XIII, pp. 11–30.

Smith, A. (1976). *An Inquiry into the Nature and Causes of the Wealth of Nations*. Clarendon Press, Oxford.

Snyder, J. (1985). *Las Meninas* and the Mirror of the Prince, *Critical Inquiry*, vol. II, pp. 539–72.

Stephens, P. (1985a). *Financial Times Surveys*, 18 June, London.

Stephens, P. (1985b). *Financial Times Surveys*, 16 September, London.

Stevin, S. P. (1958). *The Principal Works of Simon Stevin*. Swets & Zeitlinger, Amsterdam.

Unamuno, M. de. (1972). *The Tragic Sense of Life* Kerrigan (trans.), A. Kerrigan and M. Nozick (eds). Princeton University Press, Princeton.

Vries, Jan V. de. (1968). *Perspective*. Dover Publications, New York.

Weyl, H. (1949). *Philosophy of Mathematics and Natural Science*. Princeton University Press, Princeton.

Wilden, A. (1972). *System and Structure: Essays in Communication and Exchange*. Harper & Row, New York.

CHAPTER 17

# Nought Matters: the History and Philosophy of Zero

*Paul Ernest*

## Abstract

This is an exploration of the mathematics, history, and philosophy of zero. It looks at zero and its development from three perspectives: purely as a sign (syntactic); its meaning (semantics); and its social contexts, roles, and uses (pragmatic). Adopting the meaning as use theory, it argues that the meaning of zero was not fully established until the time of Brahmagupta, at around 628 CE, for Brahmagupta established the role of zero within the network of number relations in the integers. Most notably, zero is the additive identity. This synthesis is analogous for number theory with Euclid's geometry. Thus, it is posited that there are stages in the development of the concept of zero. First, its predecessor serves as a placeholder in an empty column. Second, zero is understood as a sign for nothing (nought), resting on the concept of the empty void. Third, its role as additive identity is understood and accepted. The conjecture that the idea of void is a prerequisite for the development of the number zero is confirmed positively in Indian and Mayan cultures. But the case of Ancient Egypt, which has the concept of void but not zero shows that having the concept of void appears, not of itself, sufficient.

## Keywords

zero – void – nothing – Brahmagupta – semiotics – identity – place value – natural number – syntactics – semantics – pragmatics

## 1 Introduction

In his history of numeration, George Ifrah (2000) begins by correctly pointing out that the invention of the unit 1 long preceded that of zero. But whether by rhetoric or error, he spoils it by identifying zero with the void. Mathematical 'zero', the number zero, and the sign '0' have various meanings but these do not include the void, the sense of 'emptiness', whether in the unformed or becoming universe or in the hollow or detached feelings of the individual much discussed within existentialism and mysticism. These can be found in Hinduism,

 | DOI:10.1163/9789004691568_021

Buddhism, Daoism, Sufism, Judaic Kabbalah, and doubtless many other intellectual traditions. These are poetic, mythic, or even mystic concepts, at the very best analogs for the empty set, but carry much more diffuse and extraneous meanings far beyond the scope of mathematics. Thus, they are qualitative analogs, not the quantitative concept of zero. The empty set is, of course, the proper referent of zero in modern mathematics but this brings in the slippery and even more dangerous concept of set or class. This concept alone, confused with the collected referents of a predicative term, has been the ruin of more than one distinguished mathematician, including Gottlob Frege.

Even among the learned, there is confusion about the significance and meaning of mathematical terms, especially those terms with deep philosophical connotations such as zero, one and, let us never forget, infinity (sometimes linked in the ill-formed equation $1/0 = \infty$). Perhaps it is not ignorant confusion, but a deliberate attempt to amplify the significance of such concepts, to show how their echoes resonate throughout human cultures, evoking deep and lasting notes. This is all very well in a popular exposition more concerned with effect than accuracy, more concerned with the generation of excitement than the realization of precision. But mathematics, unlike poetry, must guard against the smuggling in of everyday meanings. These may well serve as epistemological obstacles (Bachelard, 1938), tripping up the more rigorous and austere chains of thought and reasoning upon which mathematics depends for its security. Indeed, such obstacles are encountered twice: first in the history of mathematics and then again in the development of the individual learner's understanding of mathematics (Vygotsky, 1978). Just as was the case in history, often a child will say 'zero isn't a number, you can't count out zero sweets'. We have to overcome the myth that zero is simply the absence of number; that it is not, in itself, a proper number.

Zero is also conspicuous by its absence from Peano's (1889) axiomatic formulation of arithmetic. He has two primitive notions or signs apart from those of logic. These are the numeral 1 and the successor function, which he designates as +1. But there is an uneasy coincidence in Peano's primitives. For 1 appears twice, first as a number and then in the operation of successor. Modern formulations of Peano arithmetic now use the numeral '0' and the successor function S as primitives. The result is a system structurally identical to Peano's and producing an equivalent set of natural numbers. However, it posits 0 as the first number and all derived numbers beginning with 1 (S0) are successors. In this formulation we see that zero is the only essential and ineliminable number, for all others are derived from 0 as their successor.

Zero is conspicuous by its absence from Peano's formulation, but its existence is necessary in the modern version in order to resolve the uneasy paradox of the ambiguous formulation that uses 1 twice, in different roles. This is the repeated story of zero. You can manage without it to a certain level, but to progress you must include it in the larger system, whether it be the formulation of numerals, the development of arithmetic, or the completion of algebraic structures such as groups, rings and fields.[1] For zero, as well as becoming itself, the fully fledged concept of zero, through the long historical and epistemological development leading up to the idea, once it has appeared it is generalized to become the all-important identity element in many structures and theories. This identity element function is vital, and indeed I shall suggest that Brahmagupta's decisive expansion of number to include zero is based on its identity properties with respect to addition. This follows on from the inclusion of negative numbers to form the integers Z, rather than just in order to fill in empty-place value columns or to represent the contents of an empty set (nothingness).

## 2 The History and Nature of Counting, Numbers, and Zero

Counting, numbers, and zero have been through a number of phases of development. Furthermore, there are several dimensions of counting and number to consider, including numerals (the signs), numbers (their meanings), and the social and historical contexts of the invention and use of arithmetic. To better incorporate each of these disparate dimensions I draw on Charles Morris's (1945) tripartite division of the field of semiotics, the study of signs, into syntactics, semantics, and pragmatics.

First, there is the technological development of counting, calculation, and numeration systems, that is, the syntactical dimension. Second, there is the philosophical and conceptual development of numbers and number relationships, the semantic dimension. Third, there is the pragmatic dimension, often neglected but equally important. This represents the human, social, and cultural needs, purposes, and practices running through the societies and civilizations of the past five millennia within which the development of number takes place. Mathematics does not appear in a vacuum, self-generated, but arises as a result of complex human needs, and zero does not appear until much of the way through all of these intertwined histories, as Ifrah (2000) indicates.

1 Brian Rotman (1987) explicitly notes this attribute of zero, and its analogy with the vanishing point in painting and paper money in finance. Each is initially unnecessary but is needed to advance the system.

## 3 Syntactics: the Technological Development of Counting and Numeration Systems

The technological development of counting, numeration, and calculation, like all other human technologies and knowledge structures, is driven by social needs and interests. In the course of history, both numbers and numerals were invented and elaborated. Although there is an important distinction between numerals (signs) and numbers (the numerical values denoted by numerals), to avoid cumbersome expressions I will not distinguish between them except where it is significant and might lead to confusion.

Key stages or milestones can be identified in the technological development of counting, numeration systems, and calculation. These are listed below in Table 17.1. This sequence or progression represents the syntax of counting and numeration, for it concerns the signs and their rules or grammar. The numbered stages listed below in Table 17.1 are approximately sequential in terms of historical emergence. Many stages are cumulative and are not superseded but remain functioning during the development of subsequent stages. Furthermore, it is more of an outline sequence of ideas than an accurate historical one and I make no claim that this account is complete.

Below, use is made of both Peirce's (1931–58) and Bruner's (1964) analyses of signs. Peirce makes the distinction between index, icon, and the purely symbolic. Bruner, himself influenced both by Peirce and Piaget, distinguishes enactive, iconic, and symbolic signs, where the last category includes spoken language, written alphabetic text, and written mathematical signs. These are explained below, following Table 17.1.

In the descriptions I use Bruner's use of the terms enactive, iconic and symbolic signs. Enactive signs operate as gestures, such as pointing left or right, to give directions. They are a form of index signs, as defined by Peirce, that indicate by proximity or adjacency. Enactive signs can also function as icons, representing meanings through movement, such as turning down fingers as one counts, or drawing a circle in the air with one hand motion. Iconic signs carry within themselves a structure resembling and partly constitutive of the meaning conveyed. Thus, iconic signs employ metaphor or analogy to embody their meaning, at least in part.

In Table 17.1, there is a degree of ambiguity in the interpretation of tally and tallies. Tallying, the process of counting with successive movements or strokes is enactive as the actions correspond (in a one-to-one manner) with the set of objects that is being counted. Tally marks are, however, iconic, for they are sign pictures that represent the domain being counted through a one-to-one correspondence.

TABLE 17.1 The syntactic development of counting, calculation and numeration

1. Verbal or oral counting, comprising spoken language signs.
2. Tallying, the action of counting with successive strokes (enactive and embodied signs).
3. Tally marks representing the actions of counting as iconic signs.
4. Numerals, incorporating tally marks for the first few numbers (iconic and symbolic sign mixes).
5. Numeral systems with different iconic signs designating different denominations (iconic and symbolic sign combinations).
6. Place differences with fixed ordering of denominational signs (e.g., Ancient Egyptian and Roman numeral notations, which retain stages 4 and 5 repetitions of signs representing tallies).
7. Algorithms for performing addition and multiplication operations with numerals.
8. Column notation representing place value, most likely derived from abacus or sand troughs used for accounting or calculation.
9. Empty column notation – zero as a placeholder.
10. Place value notation, with zero as a sign representing nil value present.
11. Place value algorithms for performing addition and multiplication operations with numerals.
12. Place value notation, with zero as a numeral of the same status as other numerals.
13. Fractional place value notation, with 'point' or other marker separating whole numerical values from parts.
14. Admission of indefinite length fractional decimal notation.
15. Admission of infinite length decimal notation.[a]

a I omit the later invention of the binary number system with just the two digits 0 and 1 used to encode computer programmes and data including numbers, text, pictures, video, music, and all of human knowledge

Symbols, the largest class of signs, are made up of arbitrary indicators associated with their meaning purely by convention. Zero (0) is a symbol because it is arbitrarily associated with its designation.[2] Table 17.1 indicates how zero plays a role of utmost significance in the development of the grammar and syntax of the sign systems of counting, numeration, and calculation, hugely

2 It is possible to interpret 0 as an iconic sign, with the oval shape enclosing an empty space.

TABLE 17.2 ASCII codes relevant to zero

| ASCII Code in decimal form | Meaning or value |
|---|---|
| 000 | value NUL (absence of sign) |
| 032 | value SP (blank space) |
| 048 | value 0 (numeral zero) |

enlarging the scope of numerical applications, leading all the way up to modern computing.

It is enlightening to reflect on the relevant ASCII (American Standard Code for Information Interchange) codes used in computing (ANSI, 1977). ASCII codes are given in decimal, octal, hexadecimal, and binary, but I shall only display the decimal value in Table 17.2.

Thus, the formulators of the internationally standardized ASCII code distinguish three codes relevant to nothing and to zero. First, there is NUL: the absence of sign. This represents the lack of any signifying function. This is a break in communication, or the lack of a signal. Second, there is SP: standing for a blank space. This is part of a signal but represents a gap between positively or actively signifying signs, except in a sign system that assigns a special place and value to this gap. Third, there is the value 0, the numeral zero, which is a numeral standing for a number or quantity: one that stands for the number zero, the empty count.

The value SP, the blank space, can be and has been employed in columns in place value notation prior to the use of the zero or empty sign. If we agree that 1 1 stands for one hundred plus one unit only, it is just as legitimate a way of representing one hundred and one as is 101.[3] With suitable and more or less identical rules, we can perform numerical computations on numerals including spaces (1 1) or designators of spaces (1–1) just as well as we can on numerals with zero notation (101), for computational algorithms only need rules for combining signs to perform their required functions. For example, ' ' (space) combined with '1' under addition gives '1', just as '-' combined with '1' gives '1', as does '0' combined with '1'. These are all equivalent operations on signs and the presence of the numeral 0 adds nothing to the performability of the operation. Over thousands of years, Mesopotamian, Ancient Egyptian, Chinese, Greek,

3 Using a space as a marker risks ambiguity because of the possible confusion between single and multiple spaces. However, using a marker for a space such as '-' avoids this risk.

and Roman arithmetics all managed to compute and calculate successfully without the number zero.

Another dimension that is woven into this historical story is the development of numbers themselves and number relationships: namely, the semantics of number.

## 4 Semantics: the Philosophical and Conceptual Development of Numbers and Number Relationships

A number of key stages in the conceptual and philosophical development of numbers and number relationships can be identified. These run in parallel to but behind the technological development of counting, calculation, and numeration systems. They are listed in Table 17.3.

TABLE 17.3 The semantic development of numbers and number relationships

1. Development of numeration systems to quantify material collections of items.
2. Extension of numeration systems so that non-present or imagined collections can be quantified, both as they are and after they have been configured and reconfigured.
3. Ontological concept of number as an independently existing entity and number mysticism (Pythagoras and his disciples).
4. Number whose meaning comprises ordinality, used in numbering sequential lists in a fixed order; cardinality representing the quantity of elements in discrete collections (such as the 'twoness' of a pair); and length as represented by the extension of geometric lines (Pythagoras).
5. The incommensurability of some lengths, which contrary to the Greek doctrine of commensurability and harmony, was known to Pythagoras and his disciples, applying to the side of a unit square and its diagonal (1 versus the square root of 2).
6. Zeno's paradoxes, that involve cutting up finite lengths to the infinitely small (length zero), which when reassembled (so it is argued) would be infinitely large. This involves the conceptualization of lines with lengths approaching 0.
7. The recognition of fractional quantities resulting from division, first regarded as derived numbers (incomplete processes), not given full ontological number status as mathematical objects (already anticipated, e.g., by the Ancient Egyptians).

TABLE 17.3 The semantic development of numbers and number relationships (*cont.*)

8. The domain of pure number distinguished by Plato and his disciples from the technological 'logistic' of applied numeration in commerce, planning and governance. There is a concomitant higher valuation of pure conceptual number and number relations over the (perceived) inferiority of technological number, calculation, and measurement (including numeration systems).
9. The technology of calculation, like that of counting, is long recognized. Numbers can be combined through the four operations by algorithms applied to numerals, preserving numerical values.
10. There are philosophies of emptiness, with nothingness conceptualized as a lack, a void, an absence, especially in Eastern philosophical and religious traditions such as Hinduism, Buddhism, Jainism, and Daoism.
11. Conceptual shift from some precursor of zero as negative sign for lack of quantity (non-existence) to zero as a positive sign for null quantity, as be found in an empty collection. This is a shift from qualitative to quantitative meaning and understanding of zero and is very likely tied in with the previous stage.
12. Invention of signed numbers or integers with negative numbers $^{-}1, {}^{-}2, {}^{-}3$ ... on a par with positive or natural numbers 1, 2, 3... (or $^{+}1, {}^{+}2, {}^{+}3$ ... in modern notation).
13. Zero as number participating in numerical relations (e.g., $^{+}1 + {}^{-}1 = 0$) and representing identity under the operation of integer addition. This is tied in with stages 11 and 12 and represents the full concept of zero, identifiably a number treated on a par with other numbers in numerical relationships.
14. Special rules for all four operations (+, −, ×, /) with 0, 1, positive and negative numbers, acknowledging that division by 0 is problematic (result undefined or infinite). This is tied in with stages 11, 12 and 13. The historical record attributes the systematic integration these stages to Brahmagupta (598–668 CE).
15. Decimal place value notation, with 'point' or other marker separating whole numerical values from fractional values.

Table 17.3 lists many of the most important advances in mathematics from ancient to modern times, with 0 playing an important role in many of the stages. The table illustrates the stages through which that which is to become zero advances from non-existence via its negative conceptual precursors such as nothingness, a lack, an absence, through positive concepts of emptiness, a pregnant void, an empty but nascent space in the process of becoming and bringing contents into being. As a result of this, nothing or the void is something, an empty collection in which the concept of the space or collection itself (an existent entity) is distinguished from the contents of that empty space or collection (non-existent, lacking objects, signs, or anything). Finally, analogous with but distinct from this qualitative concept is the number zero itself, a fully-fledged number that reflects the cardinality of an empty collection, the number that immediately precedes one.

Thus, the history leading to zero includes a shift from the world of philosophical, mystical, and qualitative concepts into the domain of properly quantitative mathematical concepts. The resultant number zero fully participates in numerical relations and, alongside positive and negative numbers, lies within the range of possible answers to algebraic equations. This last characteristic is important, for with regard to existence in mathematics 'to be' is to be the value of a variable. 'More precisely, what one takes there to be are what one admits as values of one's bound variables' (Quine, 1990, p. 26). Thus, when one accepts that solutions to equations may include zero or negative numbers, one has acknowledged their existence as possible and allowable values of a variable, say x. At this point, zero has truly arrived; it is the fully developed numerical concept with all the properties and relationships needed to join the number systems N and Z. It can join the pantheon of the other acknowledged-to-exist numbers including one, two, three, etc. Zero is a number. Zero exists on a par with all other numbers.

## 5 Pragmatics: the Social and Cultural Contexts and Roles of Number and Numeration

Pragmatics is the most important but least treated element of the history and philosophy of mathematics, asking such questions as what social forces lead to the invention and development of number systems? What social purposes do number, calculation, and indeed all of mathematics serve, both past and present? How is the social and cultural context woven in, providing limits, constraints, affordances, and encouragements for the flourishing and development of mathematical practices?

Taking a pragmatic or externalist view of mathematics, where one takes social, cultural, and personal factors into account, immediately suggests a further comment on the admission of negative and zero solutions to equations, an admission that leads toward full numberhood. If you are investigating solutions to equations, then there is a drive to accept negative and zero solutions that stem from the values of mathematicians. As we know now, and as they were finding out then, all quadratic equations point to two possible solutions.

If those that are negative or zero are disallowed, then there is a displeasing asymmetry, a contingency factor in how you treat the equations. Admitting such additional solutions as legitimate enables a fully general method for solving all quadratic equations. Indeed, versions of what we call the quadratic formula are present in Indian mathematics of the period. Mathematicians have always been susceptible to the drive toward generality, with the powerful attraction of achieving economy and elegance. These are, in part, what define mathematical values, most especially beauty (Ernest, 2015). Such a drive, coupled with the power and utility of the new methods, will likely help overcome the resistance to, perhaps for some even the repugnance of, admitting zero and negative numbers into the legitimated and accepted domain of warranted proper numbers. However, such an admission, although warranted by a few cutting-edge mathematicians, could only be the start of a broader acceptance that would take centuries to accomplish.

To provide a quasi-historical sequence for the development of mathematics from an externalist and pragmatic perspective is an immense project that exceeds the scope of this chapter.[4] However, some of the more important social aspects concerning zero are worth sketching here.

In the centuries preceding the seventh century CE, sometime after the second century BCE negative numbers were used in India to represent debts and positive numbers to indicate assets or 'fortunes' (Mattessich, 1998).[5] Thus, a lack or debt could be conceptualized as something real, alongside possession of something that whether substantial or not is real and countable. An economically and fiscally advanced culture was almost certainly necessary for the development of negative numbers as numbers on a par with natural (or positive) numbers. However, this involves a significant conceptual step beyond numbers as representing something materially present to the senses, or potentially so. For quantified ownership can be verified individually by the senses, whereas financial assets or debts can only be determined by reference to the social agreements and documents recording them. Thus, debts as negative numbers represents a significant conceptual advance in the development of number systems.[6]

---

4 For a less truncated account with greater details of the pragmatics of zero, see Ernest (2024).

5 Negative numbers appear for the first time in recorded history in the *Nine Chapters on the Mathematical Art* (Jiu Zhang Suan Shu), which in its present form dates from the period of the Han Dynasty (202 BCE–220 CE), (Wikipedia, 2021b) and in Jain mathematics, as I record subsequently.

6 Rotman (1977) stresses how the emergence of zero parallels the emergence of paper money. Likewise, the acceptance of negative numbers both parallels, and is irretrievably bound in with, the acceptance of debts as something real within the worlds of finance and governance.

Brahmagupta (598–668 CE) was familiar with negative numbers when he articulated the properties of the numeral zero as a number in its own right. He defined it as the number you get when you subtract a number from itself. He thus made (or employed) the further step in abstracting numbers from the results of counting objects materially present to the senses. Beyond having the concepts of debt and emptiness, it is still a giant step to conceptualize zero as a number on a par with other numbers. The third and most important conceptual foundation for this development is the conception of number and operations as comprising a totality: an interconnected system of relationships and meanings.

It is evident that Brahmagupta held this view because he defined rules for all four operations with the number zero, as well as for combining signed numbers (debts and assets). Thus, he accepted that both 2-3 and 2-2 are legitimate operations and that each provides a recognizable number as a solution (−1 and 0, respectively). He created an elaborate theory including all these components, thus more or less establishing the modern domain of integers (Z) as it stands today. Brahmagupta also worked in algebra and regarded zero and negative solutions as acceptable and legitimate. Indeed, as I argue above, working with equations and their solutions provides a further impetus for acknowledging and treating zero and negatives as proper numbers.

By bringing together and extending what was known about positive and negative numbers and zero and all of their relationships, Brahmagupta was making a great synthesis and a giant leap forward. Brahmagupta is in many senses the Indian analog of Euclid, as Table 17.4 shows.

Table 17.4 makes explicit the analogy between Euclid's and Brahmagupta's contributions, and it is quite striking.[7] Many dimensions tally. However, despite the strength of the analogy, there are notable differences too. Euclid's *Elements* has an elaborate logical structure in which the ordering is crucial, for later theorems depend logically on earlier ones. Brahmagupta's text on arithmetic has no such structure. The order of the rules for the most part does not matter, just as the rules of grammar are not rigorously sequenced either. Another difference is that Euclid's *Elements* has little application other than in geometry, education, and philosophy. In contrast, Brahmagupta's rules impacted on algebraic practices and on the regularization of number and numerals. Subsequent developments in extended decimals, trigonometry, and working with infinite series would not have been possible without this contribution.

7 Interestingly, Bronkhorst (2001) also makes a comparison with Euclid. He compares Panini's systematic linguistics and grammar with Euclid's elements, not Brahmagupta's contribution as I do here. Of course, there is no reason why one such comparison should invalidate the other.

TABLE 17.4 The analogy between the contributions of Euclid and Brahmagupta

| | **Euclid** | **Brahmagupta** |
|---|---|---|
| Domain formalized | Geometry | Arithmetic |
| Epistemological basis | Knowledge as based on deductive reasoning, sourced from philosophy | Knowledge as based on lists of rules defining legitimate utterances, sourced from the grammar and linguistic theory of Panini |
| Theory created | Deductive plane Geometry | Domain of integers and its rules for arithmetical calculation |
| Objects | Constructible plane figures | Integer numbers including zero and negative numbers |
| Rules of the theory | Rules for determining congruence, rules for construction | Rules for combining positive integers, negative integers, and zero using the four arithmetic operations (+, −, ×, /) in calculations |
| Additional contributions | Books of the Elements on solid figures and number | Significant contributions to mathematics of astronomy, trigonometry as well as algebra and the solution of equations (including positive, negative, and zero solutions) |
| Incorrect attribution | Inventor of deductive geometry: instead, Euclid is in fact the systematizer of geometric knowledge | Inventor of zero and integer arithmetic: instead, Brahmagupta is very likely the systematizer of an arithmetic tradition going back to Aryabhata and earlier |

Several other practices such as astronomy also depended heavily on these numerical innovations.

The word zero retains clear signs of its Hindu and Arabic roots. The Indian name for zero was *sunya*, meaning 'empty'. When the Arabs adopted Hindu-Arabic numerals, they also adopted zero, the term for which they turned into *sifr*. Some Western scholars turned *sifr* into a Latin-sounding word, *zephirus*, which is the root of the word zero.[8] Other Western mathematicians termed

8 Zephirus is also the origin of the Kabbalistic sephiroth, the tree of life. The creation myth for this begins with *Ayin* (zero, the void), from which comes *Ayin Sof* (God) who gives off an emanation of golden light *Ayin Sof Aur* that brings the tree of life into being, with its ten

zero *cifra*, which became cipher. Because of the import of zero for the new set of numbers, people started calling all numbers ciphers. This gave the French their term *chiffre*, digit, as well as giving the modern English-speaking world their name for code, cipher (Seife, 2000).

According to Seife (2000) and others, neither zero nor negative numbers were acceptable in Europe for several hundred years. Seife argues that acknowledging the void, as acceptance of zero implies, challenges Aristotle's doctrines and the medieval beliefs of Christianity built on Aristotelian roots. It was only in the twelfth century that these doctrines were rejected and not long afterwards Leonardo of Pisa (Fibonacci) introduced zero and negative numbers in his book *Liber Abaci*, published in 1202. The lesser known Nemorarius also introduced zero to Europe (Joseph, 2008). Fibonacci was thus among the first two known Europeans to accept zero and negative numbers as permitted solutions to quadratic equations. Like Brahmagupta, Fibonacci interpreted negatives as debit quantities.

Acceptance of Hindu-Arabic numerals, the related methods of computation, and new additions like zero and negative numbers took several hundred years and, as is well known, the traditional abacists (using the abacus) were trusted more than the newfangled algorists (who used the new algorithms and zero) in trade and business.

Overall, the perspective that the pragmatics of zero, and of mathematics more widely affords, enables us to evaluate the social role and function of mathematics. Adopting the perspective of the pragmatics of zero, seeing how it is enmeshed in our history and sociocultural contexts, opens up a broader and deeper history of the forces that shaped mathematics and zero and brought them into being. It also opens up new vistas on the roles of zero and mathematics in our lives, showing how powerful is their impact. It leads to questions about the legitimacy of focusing only on the internal dimensions of mathematics and the necessity of adopting a broader view that raises questions of control, power, politics, and ethics, areas traditionally excluded from mathematics and its philosophy. For we should never forget it was the interests in trade, tax, tribute, calendar, astrology, and astronomy that were the forces that brought mathematics into being and supported its development. Making the calculation systems more efficient for these purposes is what led to the invention of zero. Thus, despite the ideology of purism adopted first by the

nodes corresponding to the numbers 1 to 10. Thus, the Kabbalah, which at its very heart is numerological, is based on a creation myth that starting from 0 creates the numbers 1 to 10, and through them, creates the whole universe, seen and unseen.

Ancient Greeks, and then again by professional mathematicians in modern times, mathematics cannot and should not be viewed as divorced from its internal and external histories and contexts (Ernest, 2020). To do so is to seriously misrepresent and distort it.

## 6 How Are the Syntactics, Semantics and Pragmatics of Zero and Mathematics Valued?

What is the philosophical significance of the three stories I have told about the historical and philosophical development of zero, and how they are valued? By asking this question, I am delving further into the pragmatics of zero, as well as raising questions about its historiography, the philosophy of its history.

### 6.1 *Syntactics*

The syntactical story is about the sign system of mathematics and when zero emerged and was incorporated into mathematical notation. Histories of mathematics will discuss and trace the various numeral systems and their development. Typically, they do not explore the actual reasons why such sign systems were developed, but assume they were needed for counting and calculation without inquiring what particular social needs lay behind this technological innovation. However, some of the best modern scholars, such as Høyrup (1994, 2002) and Yadav and Mohan (2011), are meticulous in plotting both the detail of historical numeration systems, their mechanics and their social and historical contexts and usages.

Nevertheless, historians of mathematics from the nineteenth century onward often subscribe to the Eurocentric ideology that attributes the decisive conceptual innovations in mathematics to the Ancient Greeks (Bernal, 1987). This ideology interprets the contributions and advances in Ancient Egypt, Mesopotamia, India, and China as either minor or as having come from elaborations of Ancient Greek ideas. Thus, the English orientalist, Henry Thomas Colebrooke, author of *Sanskrit Grammar* (1805), undertook the task of translating classics of Indian mathematics, including Brahmagupta's *Brahmasphutasiddhanta* (Heeffer, 2011). This showed that there existed an Indian tradition in which algebraic problems were solved with multiple unknowns, in which zero and negative quantities were accepted as fully fledged numbers, as well as other innovations. In explaining this tradition, which occurred during a period when mathematics was hardly practiced in Europe and in what was (or was to become) the Islamic regions, scholars are divided into two opposing camps. Heeffer (2011) calls them the believers and the nonbelievers.

The believers showed an admiration for the ingenuity and originality of the Indian tradition. They recognized the huge intellectual advances made in mathematics before Europe became 'civilized'.

However, nonbelievers did not grant Indian mathematicians the status of original thinkers. They argued that Indian knowledge must have been passed on from the Greeks, itself the cradle of Western mathematics and civilization. A major and influential nonbeliever was Moritz Cantor (1880), who published a four-volume work *Lectures on the History of Mathematics* over the period 1880–1908. This is regarded as the first modern and comprehensive history of the subject (Swetz, 2016), but it derogates the Indian contributions to mathematics. Cantor 'takes every opportunity to point out the Greek influences on Hindu' mathematics, whether imagined, coincidental, or real. He thus denies the Indians any credit for originality of thought or for what we can see today are major advances in mathematics (Heeffer, 2011, p. 138). This is a manifestation of Eurocentrism.

In the present context, the Eurocentric ideology is comprised of two strands. These concern the questions of: who can do mathematics, and what is mathematics? The Eurocentric answer to the question of who can do mathematics is frankly racist, in that it regards the peoples of Ancient Egypt, Mesopotamia, India, China, and elsewhere (including non-Europeans in Africa and the Americas) as intellectually inferior, capable only of copying but not of originating ideas in mathematics, science and other domains of knowledge. It is no coincidence that Eurocentrism was at its height during the most financially successful years of the British Empire when India and other colonies were being exploited and looted. Eurocentrism supports the imperial project through promoting European (primarily English) exceptionalism and superiority, which justified the exploitation of what were viewed as 'inferior peoples'. The 'fact' that the minds of these subaltern peoples were incapable of the highest levels of thought confirmed that they were inferior, and not deserving of the rights and respectful treatment that is the due of Europeans and, especially, English gentlefolk.

The idea that humans and their cultures progress through distinct stages of development is an ancient one; already in the sixteenth century Montaigne (1580) criticized the assumption that civilized humanity was in some sense better than savage or barbaric peoples. Although the terms were ill-defined, in general meaning savages were tribal peoples, and barbarians were non-Christians who lived in organized societies. In the nineteenth century Darwin's theory of evolution reinforced ideas of hierarchy and development as you 'ascend' among all life forms, and especially humans: 'Victorian-era anthropologists widely accepted the terms savagery, barbarism, and civilization to describe the perceived progression of human society from the most primitive forms to the most advanced. Such distinctions, however, reinforce xenophobic views used throughout history to justify hostile acts toward dissimilar cultures.' (McNeill et al., 2010, p. 568). The idea that savage or barbaric peoples could be well advanced beyond Europeans, as Indian mathematicians undoubtedly were in what Christians call the Dark Ages, and Islamic mathematical culture in medieval times was not only anathema but inconceivable to the Victorians and indeed throughout much of the twentieth century.

The Eurocentric answer to the question of 'what is mathematics?' attributes the decisive conceptual innovations in mathematics to the Ancient Greeks. This ideology is one that regards the geometry and pure mathematics of the Ancient Greeks, centered on proofs and reasoning, as superior, and indeed as the only 'true' or 'real' mathematics (Ernest, 2009). 'Oriental mathematics may be an interesting curiosity, but [*only*] Greek mathematics is the real thing.' (Hardy 1940, p. 12, my addition in brackets). The decisive Indian and Arabic work on number, algebra, trigonometry, and analysis, which constitutes a huge leap forward from ancient mathematics, is brushed aside. Only willful ideological blindness can possibly account for this great lacuna in the history of mathematics.

It is no accident that the ideology of purism reinforces the Eurocentric perspective on mathematics. Purism values pure proof-based mathematics as significant epistemologically, pertaining to truth, wisdom, high-mindedness, and the transcendent dimensions of being. Applied mathematics and calculation are denigrated as technical and mechanical, pertaining to the utilitarian, practical, applied, and mundane; understood as the lowly dimensions of existence. Pure mathematics is the domain of philosophers, freethinkers, and gentlefolk. Applied mathematics and practical arithmetic, termed logistic by the Ancient Greeks is the domain of lower class and trades people, or even, in ancient times, slaves (Plato, 1941).

These Eurocentric and purist values have likely contributed to the scholarly separation of the history of its terminology, which I have termed here the syntactic, from the history of mathematics proper (Høyrup, 1996, p. 7). One striking example concerns prime decomposition of numbers. The Prime Factorization Theorem asserts that every number $n$ is the unique product of a fixed number of primes. The truth of this fact is one of the great foundation stones of number theory.

However, there is no significance whatever accorded to a parallel result, that might be termed the unique numerical representation theorem (in any base). In decimal form this asserts that any number $n$ is the unique sum of powers of 10.

The analogy between these two claims is as striking as is the difference in their respective valuations. In Ernest (2024), I explore both the structural similarity and disparate valuations of these two forms. Whatever their similarities and differences, it is clear that neither would be possible to manage without the sign 0.

Overall, my claim is that the syntactical dimensions of mathematics, evidenced in the symbols systems of mathematics, and especially in the decimal (or other base) place value system, are regarded as trivial, mathematically uninteresting, devoid of significant conceptual content. This is despite the fact that what I estimate to be more than 90% of the efforts expended in the history of mathematics have been involved in the invention and development of numerical representations and numerical algorithms, and well over 99% of human uses of mathematics concern nothing more than the use of number representation systems and calculations.

What this disparity of treatment and regard shows is that there is an underlying and too rarely discussed value system underlying mathematics and its history. My contention is that it is no accident that it serves the interests of power elites and hierarchies in society. These favored the free and privileged citizenry in Ancient Greece, members of a leisure class, over the humble tradespersons and slaves and their professional practices. In modern Europe these values favored the upper and middle classes (gentlefolk) over tradespersons and workers; this racist perspective regards as lowest of all aliens from colonies, conquered nations and other non-European origins. The ideology of purism cannot be separated from issues of conquest, colonialism, and empire, with its concomitant Eurocentrism and frankly racist views of subaltern and conquered peoples and their cultures as inferior. Despite the myth of neutrality, the judgments of value and worth widespread within the traditional history and philosophy of mathematics are distortions, far removed from objectivity and fairness (Powell and Frankenstein, 1997; Joseph, 2018). Such perspectives pay lip service to the Indian invention of zero as simply the invention of a sign, rather than seeing it as a cornerstone of the entire edifice of modern mathematics.

### 6.2 *The Semantics of Mathematics*

From a semantic perspective the big question is: what are the meanings of mathematical signs and symbols? The answer to this question is controversial. The traditional response is that a mathematical sign denotes a mathematical object or procedure. Thus 2 denotes the concept of twoness, an abstract concept that can be applied to any pair of objects considered as a collection in itself. Likewise 0 denotes the number zero, which stands for the value of any empty numerical descriptor such as the number of dogs currently orbiting the Earth, or the cardinality of any empty collection such as the set of real roots of the equation $x^2 = -1$. More generally, 0 is cardinality of any empty set.

In base ten, 107 is the numeral standing for and thus abbreviating the compound term $10^2 \times 1 + 10^1 \times 0 + 10^0 \times 7$, and likewise 1,000,007 is $10^6 \times 1 + 10^0 \times 7$ (with the zero terms omitted). In the representation of 107 as a sum of powers of 10 the brackets are omitted because of the associativity of addition. The property of associativity means that the value is demonstrably invariant no matter how the parentheses are inserted. However, such is the power of human understanding that the abbreviated numeral is a sign in its own right, not needing to be unpacked or reduced to its components except when operations are performed on it, such as adding it to another number also in this decimal form. Even then, we have abbreviated procedures that mean that it does not have to be fully unpacked. As millions of schoolchildren learn annually, the sum 107 + 107 can be performed as follows:

1 0 7
1 0 7 +
2 0 14
$2\ 0^1\ 4$
2 1 4

Here each column is treated by the sub-algorithm that sums each column and carries onward any excess over nine as a digit 1 in the next column. I need not spell out the obvious details.

It is important to note, in the present context, that the numeral '0' plays no essential part in this procedure. If we replace '0' by the placeholder '-' the procedure is still the same. The number 0 itself plays no part, since the algorithm operates mechanically on the signs without any recourse to their intended meanings. Thus, as remarked earlier, it is possible to have efficient computational algorithms operating on numerals prior to the 'invention' of zero, as indeed thousands of years of history attest.

Such computational procedures are of no theoretical interest to mathematicians, even though they may employ them themselves. They are typical of the computational skills taught in schools to children. The algorithms are determinate procedures with a unique output and are easily automated and performed on handheld electronic calculators or computers.[9] These computations are

9 Indeed, they were performed effectively on mechanical calculators and abacuses, which latter have no zero sign.

performed without recourse to the full expansion of numbers into the basic form of a sum of powers of 10, as illustrated above. However, the rules employed are the minimum required to respect the meaning of this underlying form.

Familiarity with multiple digit numbers leads us to see them as a single sign, just as a word is perceived as a unity even though made up of distinct letters. Psychologically, the phenomenon of treating a compound sign as a single entity in its own right is known as chunking. Thus, for example, a number (numeral in fact) such as 107 is seen as a single entity, and only needs to be opened up and examined in terms of the relationships between its constituent signs for certain purposes. Likewise, the student of chemistry will see the compound sign $C_2H_5OH$ as a single entity, namely the formula for ethyl alcohol. Chunking frees up the mind to understand and process complex signs as single entities, enabling more and more complex signs to be treated as single objects of attention. Coupled with abstraction, this facility enables us to see the composite of mathematical processes applied to mathematical objects and the end product as a unitary mathematical object in its own right. For example, 12 + 9, and $^5/_7$ can be seen both as composite signs for processes and unitary signs for individual mathematical objects.

Another effect has been remarked upon in the linguistic analysis of number. Numerals and number words 'do not refer to numbers, they *serve as* numbers' (Wiese, 2003, p. 5, original emphasis). This is an important point that contradicts the simplest referential theory: the picture theory of meaning. Numerals, number word terms, and by extension all mathematical signs need not indicate or refer beyond themselves to other objects as their meaning. They themselves serve as their own objects of meaning. Mathematical language is operative and performative, for mathematical terms create the objects to which they refer.[10] Counting creates numbers, and operations create functions. In the first instance, these are inscribed numerals and enacted operations. Repeated usage reifies and solidifies them into mathematical objects. Their currency serves as a social warrant for them, verifying their robust and legitimate existence. This is a point of deep ontological significance, for we do not have to posit a realm of being in which to locate mathematical objects outside of space, time, and material existence, the position of Platonism. Their use in social practices is what makes mathematical objects real. However, this is not to dismiss the problems of ontology, meaning, and semiotics from mathematics; it is just that

10 Mathematical language is performative in two ways, which I term inner and outer. What I describe here is the inner performativity, whereby mathematical sign usage creates mathematical objects. The outer performativity of mathematics is the way it formats the way we experience and interact with the material world (Skovsmose 2019, 2020; Ernest 2019).

no special attention needs to be paid to these characteristics of mathematical signs when they are in use. Their meanings are self-contained. Thus, it suggests that, for the most part, mathematical practices can proceed without worrying about ontology.

Historically, zero first served as a sign for an empty column, without being a number itself. Later in the development of zero it became understood to be a fully-fledged number in itself, Ironically, because mathematical signs can be used instrumentally or mechanically, it means that even when accorded the full status as a number, zero can once again be used without reference to its meaning. However, in this third stage of utilization, the meaning of zero is ignored in exactly the same way as those of all other numbers, so it remains a fully-fledged number, like the rest, although a rather important and distinguished one. There is one difference. For example, in unpacking 1,000,007 I omitted the terms $10^5 \times 0 + \ldots +10^1 \times 0$. Each of these has the value 0 and can be ignored in additive expressions as they do not affect the overall numerical value. Zero has a unique and distinguished place as the identity element of the natural numbers with respect to addition. It is the only number that can be ignored in addition operations.

Returning to the issue of meaning, it must be said that it is of great importance for mathematical signs, even if it can be temporarily ignored in some contexts. I have argued that for several reasons the referential theory of meaning fails for mathematics. Since the mid twentieth century, the Meaning as Use theory has become prominent, drawing on the work of Wittgenstein (1953). Wittgenstein offers a whole range of insights into meaning, including meaning as physiognomy, whereby meaning is partially given by the shape or inner structure and arrangement of the parts of the sign. This clearly applies to, and offers insight into, the signs of mathematics. It parallels Peirce's conception of iconic signs, whereby the inner structure of a sign in some sense is an analog of that which it signifies, representing it pictorially or by some other means. For example, the Roman numeral III represents threeness by corresponding to three tally marks. The numeral utilizes a materially given instance of threeness to designate 3.

Note that this iconic conception of meaning gives rise, at least in part, to the historical and conceptual problems of accepting zero as a number. If one stroke (|) represents 1, two strokes (||) represent 2, and three strokes (|||) represents 3, then zero or no strokes () represents 0. But that means there is no sign for zero. Therefore, in this simple sense, zero is not a number for all numbers have their signs. This is a recurring and fundamental problem of the depiction of the number zero. No sign at all to stand for no thing is in itself no sign; that is, the lack of a sign. Thus, the iconic representation of zero is paradoxical and self-negating. The only way out is to go beyond strictly iconic representations of number, which is what happens to all historical systems of numerals, or to broaden the range of icons employed. This is what the Mayans did, adopting a pictogram representing an empty turtle shell to represent zero (Coe, 1987).

However, much better known and far more influential is Wittgenstein's (1953) Meaning as Use theory. This lives up to its title, namely that the meaning of a word, sentence, or other sign is given by its usage. How a sign is used in relation to other signs and human actions is what reveals its meaning, not some indication of objects beyond itself. In the Meaning as Use the meaning of a sign is not given by a single instance of usage. One instance of use can contribute to the meaning of the sign, which is always growing as the pattern of the usage grows. Overall, Meaning as Use is a holistic theory, for meanings are not given explicitly, as when one can look up the meaning definition of a word in a dictionary. Rather it is the overall patterns of uses, the associated material activities in the form of life, and the nexus of connections and relationships into which the word, sentence or sign enters into, namely the history of its use, and expectations of its future uses that constitute its meaning. (McDermott, 2001; Quine, 1951)

The Meaning as Use theory applies to mathematics primarily via sign transformations and relationships with other signs, for such are the main usages within mathematics. One reading of this lies in Robert Brandom's (2000) inferentialism. This proposes that the meaning of a sign lies in the further signs it entails, and those that entail it. Thus, a mathematical sentence implies a whole range of other sentences and is itself implied by a further range of sentences. More generally, a mathematical sign is enmeshed with a multitude of other signs, with an array of connected signs forming a web of meaning. In my view, restricting the signs to sentences and limiting the relations between them to logical deductions, as inferentialism can sometimes appear to do, imposes too restricted a view of mathematical meaning. Mathematics contains a broader range of significant signs beyond sentences; there are also terms, concepts, diagrams, as well as calculations, proofs, methods, and theories. Likewise, mathematics contains a wider range of important relationships beyond logical deduction, including exemplification, instantiation, analogy, representation, translation, transformation, expansion (exposure of inner structure), induction, and abduction (Peirce, 1931–58). Without developing this vision any further, what can be claimed is that the meaning of a mathematical sign lies not just in the reference or direct designation of the sign, although may be included, as it is one link. It is to be found in a network of relationships making up the web of meaning surrounding the sign. This links it to a system of other signs by a complex range of relationships. So the meaning of a sign is contextual and holistic; it comprises the relationships with other signs (McDermott, 2001; Quine, 1951).

This is a complex theory requiring a much-expanded treatment and more extensive justification. However, I shall sidestep this by just focusing on the meaning of zero without developing this whole semantic and semiotic theory here.

Thus, I can now say that the meaning of zero is not just that of a placeholder in a numerical array, although that is certainly part of its use. Nor does it lie solely in the analogy with *sunya* or other concepts of emptiness, although the cardinality of any empty set is 0. Primarily, its meaning lies in its relationships within the domain of number. These relational linkages include $(^{+}1) - (^{+}1) = 0$, $(^{+}7) + (^{-}7) = 0$, $(^{-}107) + (^{+}107) = 0$, $7 + 0 = 7$, $0 \times (^{-}5) = 0$, etc. Such relationships establish zero's unique role as the identity for the operation of addition. It is Brahmagupta's great contribution. Or rather, it is one of them, in that he systematically laid out the properties and uses of zero within the systems of natural numbers and integers. By systematizing these properties and uses of zero, as well as the relationships of positive and negative numbers and rules for using the four arithmetical operations, he deserves the title the Euclid of Integer Arithmetic. The primary meaning of zero consists of its relationships and links with numbers and operations in the theory of integers, the true sentences that bind zero to positive and negative numbers via operations in equations.

It is important to remember that zero already has different sets of meanings in different contexts. Within the domain of numerals, that is number representations, zero serves as a placeholder, a sign indicating the equivalent of an empty abacus column in a place value numeral. In this sense, zero has not come into its more important full conceptual meaning. In the natural numbers zero serves as an additive identity, but within a restricted domain of subtraction wherein a larger number cannot be subtracted from a smaller one. Within the integers zero fully fulfills its role an additive identity, because every integer has its own unique inverse. Thus, for any integer there is an inverse element such that their sum is zero. Indeed, subtraction is properly defined over the integers by $i - j =_{\text{def}} (^{+}i) + (^{-}j)$, the sum of a number ($i$) with another number's ($j$) additive inverse. The meaning of zero changes still further as number systems expand to Q (the rational numbers), R (the real numbers), and finally C (the complex numbers). These expansions bring in more complications and change the meaning of zero, extending or changing its domain of use. For example, Q, the domain of rational numbers, is closed under division (the inverse of multiplication), except that division by zero is undefined (and not allowed). The same holds for R and Z.

## 6.3 *The Pragmatics of Mathematics*

The development of mathematics, arithmetic and indeed the concept and uses of zero all occur within the contexts of sociohistorical practices. I have indicated how the needs of ancient societies in terms of taxes and trade shaped the origins of numeration and calculation technologies for its first two millennia of existence. Over the following 3,000 years, mathematics continued to develop, woven into various systems of modeling and control, including mensuration, geometry, and trigonometry, especially for the purposes of astronomical and astrological calculation, control, and prediction. Royal courts, state institutions, and religious settlements trained skilled mathematicians and maintained them to fulfill these purposes.

To understand the social context of zero one must examine the Indian mathematical tradition that led up to Brahmagupta's achievements. I believe that, like Euclid, Brahmagupta pulled together centuries of work in mathematics to summarize and systematize. That tradition would have already had the concept and number zero, positive and negative numbers, the four arithmetical operations, place value numeration, and much more mathematics, including trigonometry, algebraic equations and solution methods, approximations for pi, and other constants. Thus, the huge achievement in defining the number zero and its properties and relationships belongs to a culture, a civilization, not just a single heroic figure. This is not to dethrone or denigrate Brahmagupta but to rebalance the history of mathematics and all human achievements as being social and group achievements, not those of outstanding individuals alone.

Isaac Newton acknowledged this when he wrote in a 1675 letter to fellow scientist Robert Hooke: 'If I have seen further, it is by standing on the shoulders of giants.' In his day this was already a well-known phrase dating back to the twelfth century or earlier (Merton, 1965). Thus, Newton acknowledges his huge intellectual debt to his teachers and predecessors and recognizes that his own contributions are only possible because of them. He is just one voice, albeit an important one, in the great conversation of humankind.

Joseph (1991) offers a great synoptic view of developments in Indian mathematics. In his *Chronology of Indian History and Mathematics* (pp. 313–4) he distinguishes a number of periods of development. In the period 500–200 BCE there is establishment of Indian states and the rise of Buddhism and Jainism. During this period the practice and development of Vedic mathematics continues but later declines with the ending of ritual sacrifices, with its need for the construction of mathematically precise altars. This is followed by the beginnings of Jain mathematics including number theory, permutations and combinations, the binomial theorem and astronomical work.

During the next period, which runs from 200 BCE to 400 CE, there is a triple division of India, with the Kushan dynasty (North), Pandyas (South), Bactrian-Persian (Punjab). During this time, Jain mathematics develops including rules of mathematical operations, decimal place notation, and the first use of 0. Algebra emerges including simple, simultaneous, and quadratic equations. In addition, square roots are treated as are details of how to represent unknown quantities and negative signs. Thus, zero, 0 and negative signs have already made their appearance, more than two centuries before Brahmagupta.

The following period, one that incorporates the work of Brahmagupta, stretches from 400 CE to 1200 CE. This is the period of the Imperial Guptas, reaching their height in the reign of Harsha (606–647). This period sees the flowering of Indian civilization, including developments across mathematics, science, philosophy, medicine, logic, grammar, and literature. In mathematics, this is regarded as the classical period or the golden age of Indian mathematics (Parameswaran, 1998). Notable mathematicians of this period up to Brahmagupta include Aryabhata I, Varahamihira, Bhaskara I. A whole string of important mathematical works were written. In chronological order these are the *Bakhshali Manuscript, Aryabhatiya, Panca-siddhantika, Aryabhatiya Bhasya, Maha Bhaskariya*, and *Brahmashputasiddhanta*, Brahmagupta's main mathematical work of 628 CE. The mathematical advances are too extensive to list here (see Joseph, 2016).

Brahmagupta's contributions range across much of mathematics and include algebra, arithmetic (series and zero), Diophantine analysis (Pythagorean triplets and Pell's equation), geometry (Brahmagupta's formula, triangles, Brahmagupta's theorem, pi and measurements and constructions), trigonometry (sine table and interpolation formula).

Brahmagupta lived and worked for a good part of his life in Bhillamala (modern Bhinmal in Rajasthan, India), a center of learning for mathematics and astronomy. Brahmagupta became an astronomer of the Brahmapaksha school, one of the four major schools of Indian astronomy during this period. He studied the five traditional *Siddhartha* on Indian astronomy as well as the work of other mathematicians and astronomers, including Aryabhata I and his predecessor Varahamihira. In 628 CE, at the age of 30, he composed the *Brahmasphutasiddhanta* (the improved treatise of Brahma), which is believed to be a revised version of the received *Siddhanta* of the Brahmapaksha school of astronomy. Scholars state that he incorporated a great deal of originality within his revision, adding a considerable amount of new material. The book consists of 24 chapters with 1,008 verses in the *arya* meter. A good deal of it is astronomy, but it also contains key chapters on mathematics, including algebra, geometry, trigonometry, and algorithmics, which are believed to contain new insights due to Brahmagupta himself (Wikipedia, 2021a).[11]

What this abbreviated history shows is that, first of all, Jain mathematics already includes the first use of 0 and the representation of negative signs by 400 CE, more than two centuries before Brahmagupta. Second, by his own account Brahmagupta studied the *Siddhanta* text of the Brahmapaksha school of astronomy, and his own major contribution is the much revised and extended version of the received text. However, locating Brahmagupta in the Indian historical tradition that gave rise to him, and into which he himself contributed so much, is by no means intended to belittle his contributions. Brahmagupta's understanding of the number systems went far beyond that of others of the period. In the *Brahmasphutasiddhanta* he defined zero as the result of subtracting a number from itself. He gave some further properties as follows:

> When zero is added to a number or subtracted from a number, the number remains unchanged; and a number multiplied by zero becomes zero.

He also gave arithmetical rules in terms of fortunes (positive numbers) and debts (negative numbers):

> A debt minus zero is a debt.
> A fortune minus zero is a fortune.
> Zero minus zero is a zero.
> A debt subtracted from zero is a fortune.
> A fortune subtracted from zero is a debt.
> The product of zero multiplied by a debt or fortune is zero.
> The product of zero multiplied by zero is zero.
> The product or quotient of two fortunes is one fortune.
> The product or quotient of two debts is one fortune.
> The product or quotient of a debt and a fortune is a debt.
> The product or quotient of a fortune and a debt is a debt.

11 In this paragraph and elsewhere I have replaced inflected letters by plain alphabetic letters to simplify the presentation, as scholars employing the alphabetization of Sanscrit and other Indian written scripts will notice.

Brahmagupta then tried to extend arithmetic to include division by zero. Although we reject these latter suggestions, such as *zero divided by zero is zero*, overall his work is a brilliant and groundbreaking attempt to extend arithmetic to negative numbers and zero (O'Connor and Robertson, 2000b).

I locate Brahmagupta firmly within the Indian mathematical tradition of the golden age, rather than extolling him as an individual and heroic figure on his own. My aim is to critique the 'great (hu)man theory' of history. According to this, history can be largely explained by the impact of great humans, or heroes; highly influential and unique individuals who, due to their natural attributes, such as superior intellect, have a decisive historical effect. The theory is primarily attributed to the Scottish philosopher and essayist Thomas Carlyle (1841) who claimed that the history of the world is but the biography of great men. The 'great human theory' found many supporters in the Victorian era, including William James. However, there was robust criticism of this view from Herbert Spencer (1873), who stated: 'Before he can remake his society, his society must make him.' Likewise, first Hegel then Marx proposed that history is made by large-scale forces that transcend individuals.

Ogburn (1926, p. 227) argues explicitly against this theory, stating: 'The discovery of the calculus was not dependent upon Newton, for if Newton had died, it would have been discovered by Leibnitz. And we think that if neither Leibnitz nor Newton had lived, that it would still have been discovered by some other mathematician.' Nevertheless, although *great human theories* have been widely rejected by historians in the modern era, they are still widespread in popular culture, unreformed school history teaching, and in some areas of leadership studies (e.g., Turak, 2013).

Like Euclid did for geometry, Brahmagupta provides an explicit articulation and systematic synthesis of what was known about zero, negative, and positive numbers, and how all three types combine and interact through applications of the four arithmetical operations. As the author of a synthesis of the cultural knowledge of arithmetic it is even less appropriate to apply the great human theory of mathematical advances to Brahmagupta with respect to arithmetic than to some other mathematicians, who carved out whole new fields virtually on their own, although overall it is accepted that he personally made a great number of significant contributions to mathematics.

A number of questions about the emergence of zero remain. First, what was it within the culture of Indian civilization that led to the development of zero, or, might one even say, necessitated its invention, with its full meaning and functions described here? Second, what is it that has made zero so difficult a concept and number that prevented its invention in most cultures and delayed its transmission for so many centuries when it had been invented? Third, what further support is there for the account given here of the development of zero?

## 7 What in Indian Culture Led to the Development of Zero?

The first presence of zero as a half-formed concept is as a placeholder, a marker for an empty space in a compound numeral employing place value notation. Very large numbers were employed for a variety of purposes, mostly mythological, theological, astronomical, and astrological, which provided an impetus for their development. These uses also led to a drive for optimum economy in the means of expressing number through an efficient and compressed system of numerals. In consequence, the mark of an empty column, a dot, small circle, or a circle with a dot inside is used. It signifies an empty place within a larger compound numeral represent by a string of digits whose placements indicate their value or denomination. The long elaboration of the concepts of *sunya*, *sunyata*, meaning emptiness, and nothingness, provides a philosophical backdrop for the emergence of the fully fledged concept of zero as a number.

### 7.1 *What Made Zero So Difficult a Concept?*

Numerous scholars have indicated the philosophical and theological difficulties in the network of interlinked concepts of *sunya*, *sunyata*, emptiness, nothingness, nothing, non-being, absence, void, and so on. One of the fundamental problems is having a name or concept (present) for absence. Since the name is used as a marker for that which it names, that is, is used as if it was that which it names, a paradox or contradiction arises. For the name of nothing is not nothing but something. Unless a clear distinction is drawn between a realm of objects (existents), a language (collection of names, terms, and sentences), and a metalanguage (a domain in which the names, terms, and sentences of the language are themselves treated as objects), confusion is very easy. Thus, a void is an emptiness, an absence in the domain of objects (level 0). But there is a word for void that is a linguistic term in language and is existent in the domain of language and can be used in utterance in written language or speech. It is the seventh word in the previous sentence. However, the sign 'void' is an expression in the metalanguage (level 2) denoting the linguistic object, the word in between the quote marks (void), in the level 1 domain. In the metalanguage this word is mentioned, that is named, and the convention is that quote marks make the word (void) into the name of the word ('void').

The contradiction arises because an empty state of affairs, the void, the pure phenomenon of emptiness (on level 0), meaning that there is nothing present, is referred to by an object, nothing, which is an object in the domain of language

TABLE 17.5 Distinguishing the levels of objects, language and metalanguage

| Level | Domain | Range of reference | Example |
|---|---|---|---|
| Level 2 | Metalanguage | Linguistic constants (referencing specific linguistic terms and sentences) and linguistic variables (ranging across domains of linguistic terms and sentences) | The domain of signs that ranges over the first order (level 1) universe of language and other signs: for example, 'void', 'zero', references to level 1 items such as all sentences defining nothing. Sentences applying non-existence to themselves are self-contradictory. |
| Level 1 | Language | Names of specific objects, variables ranging across domain of objects (e.g., somebody who …), sentences describing relationships between objects | The universe of language and other signs for void, *sunya*, *sunyata*, nothing, nothingness, non-being, absence, zero |
| Level 0 | Objects | None. Objects are simply themselves without referential functions (unless treated as signs in another system, where they are *not* treated simply as objects) | Experience of the pure phenomenon of emptiness (void) |

(level 1). Furthermore, to provide this analysis I need to discuss or write about both level 0 objects, contents, or states of affairs and level 1 signs that themselves refer to level 0 objects. For this I need the level 2 metalanguage domain, which can refer (write about) everything on lower levels. Level 1 enables me to *use* signs and language, but within a system with logically strict separation into levels, does not allow me to *mention* (*name or discuss*) level 1 signs.

Natural language such as I employ in writing this chapter does not include level 0 objects; these exist outside of language and are only present at one remove, mediated via referral, through their names.[12] In addition, natural

12 This differs from mathematical language, as I explained above, where terms and sentences (level 1) can become their own referents: the very objects and relationships that they name (level 0), through their (inner) performativity. This alchemy occurs because

language serves as its own metalanguage, so I can say that 'Natural language serves as its own metalanguage' is a sentence containing seven words. I can self-refer within this language. This means that level 2 collapses and everything is at level 1. This enables me to say that the term 'void' exists the same as any other term in the language; it is a normal word, but it names an emptiness, which is not the same as most linguistic objects, because its referent is non-existent. This is an antinomy or a paradox, making it hard to understand, and convincing some people it is unsound, but it is not a self-contradiction. To achieve the latter, I can make the following statement: 'This sentence is false.' This is a self-contradiction. For if I assume that it is true, then by its own statement it is false. But if instead I assume that it is false, then it is not as it states false, but true. If we accept the Law of Excluded Middle, then two negatives make a positive, so it is both true and false at the same, making it self-contradictory, and invalidating itself.

If I say, 'the void does not exist', this is not self-contradictory. Nor is 'the void exists'. For it is possible that somewhere in the universes of matter or thought there is a void, or that it can be imagined, named, or described. But if I say, 'non-existence exists', on the surface this is self-contradictory. Non-existence is a noun, and to exist is a verb. The sentence asserts that the state of non-being, that to which non-existence refers, has existence or being. This is a metaphysical or philosophical conundrum such as is found in Zen Buddhism and elsewhere, such as the existentialism of Sartre (1956) in *Being and Nothingness*. However, it is usually understood as referring to a human state of being, a sense of self or posture with regard to the world, or about the transcendence or lack of ego. It is not usually understood as meaning that there is nothing, in the sense that nothing at all exists in the universe. In my view, the latter is false empirically, because I can point to things that exist. Whether it is logically false too, because such a statement could not be true except in a universe with no consciousness, language, or objects, making the articulation of such a sentence impossible, I am uncertain.

Modern logic says that 'non-existence exists' is an illegitimate sentence because existence is not a predicate. Existence is not a property but is included in all quantification. One may say all, some, or none of some category have a given predicate or property. For example, black swans exist, meaning that

---

mathematical objects are primarily processes (applied to mathematical objects) that become reified and objectified analogously to the way verbs are transformed into nouns in nominalization. For example, the actions of counting become the numbers themselves. This is very difficult to describe because the meanings are not static but change during this process.

there are some swans that are black. One can say nothing is material. That is, no material objects exist. Or everything is material. That is, all objects are material. One can say the empty set exists. (There is an *X* such that *X* is a set, but there is no element *e* such that *e* belongs to *X*). However, in set theory one needs an axiom to make this sentence true. Even as late as 1888 the great mathematician Dedekind excluded the consideration of the empty set for reasons linked to problems of non-existence (Barton, 2020).

Descartes' ontological argument for the existence of God hinges on the use of existence as a predicate and is rejected on this basis. Descartes argues that God is the perfect being, possessing all good attributes. Existence is an attribute, and evidently existence is a better attribute than non-existence. Therefore, God has the better attribute of existence; that is, God exists.

One could continue, asking what the being of non-being means, if anything, and whether the non-being of being means the same, something different, or nothing at all. This plunges us into the domains of metaphysics, or indeed wit. Many writers have played with the ambiguities of words that mean nothing or negate existence. In Homer's *Odyssey* the hero Ulysses tells the Cyclops Polyphemus that his name is Nobody. Later when Ulysses has blinded Polyphemus the latter shouts out for help: 'Nobody is attacking me!' His cries for help are self-defeating. Through this witty ambiguity Ulysses escapes.

Overall, it is clear that there are huge conceptual, philosophical, and metaphysical difficulties in dealing with the concepts of void, emptiness, non-being, and so on (Barrow, 2000). Some of these difficulties tainted the concept of zero and prevented its acceptance and spread. As well as philosophically difficult, naming the non-existent has been seen as blasphemous, which further helps to explain the obstacles confronting the concept, and delaying its acceptance and spread for centuries. Aristotle denied the existence of the void, and in the Dark Ages Christianity based its theology on his philosophy (Seife, 2000). In medieval Europe right up until the time of Newton the idea of the void or the assertion of its existence was blasphemous because it negated the doctrine of the omnipresence of God (Barrow, 2000).[13]

### 7.2 *What Support Is There for This Account of the Development of Zero?*

Perhaps the most authoritative accounts of the development of zero are given by the historian George Gheverghese Joseph. In a series of publications including Joseph (1991), Joseph (2008) and culminating in Joseph (2016), he has not

13 To this day, nothingness, if not zero, is regarded by some as evil. "Evil is nothingness. 'Evil' is not *defined* as nothingness by Barth. Rather, evil is *identified* by Barth as *nothingness*." (Wolterstorff, 1996, p. 585. Original italics.)

only thoroughly researched and documented the origins of zero in India, but also looked at less-developed precursors in all of the great civilizations as well as the parallel emergence of the full concept of zero in Mayan culture.

According to Joseph (2016), a number of factors led to the development of zero in Indian culture.

1. The invention and use of column-based numerals with different denominations laid out in different columns, resulting in empty columns and later a sign representing an empty space in a column in compound numerals.
2. On the basis of (1) the development of a place value system for representing a number with no drawn columns and a sign representing the lack of any figure of that particular denomination.
3. The concept of void (*sunya*) or emptiness as something significant, imaginable, and representable.
4. The use of numbers in astronomy, astrology, and calendrical calculation as well as a fascination with large numbers for describing long periods of time.
5. Cohorts of specialists including astronomers, astrologers, priests, or scribes devoted to mathematics and calculation over a significant period.

Not explicitly mentioned as a possible precursor of (1) is the use of an abacus or counting troughs to distinguish numeral denominations that may well have inspired the use of columns for recording number.

Joseph (2016) describes the occurrence of zero among the Mayans as manifested in the following.

1. A focus on extensive calendar computation, including very large numbers requiring a place value numeral system.
2. Groups of priests serving as teachers and writers including mathematical skills (possibly including a high-ranking female scribe).
3. A very economical place value numeral system initially only using three symbols (for 0, 1, and 5).

Much less is known about the Mayans than developments in India, hence what one can say about the origins and numeration and zero among the Mayans as well as their culture overall is very limited. Nevertheless, the comparison raises the question: is the concept of void a necessary conceptual prerequisite for the development of zero in its full sense? Did it occur in Mayan culture?

So, the question is, did the Mayans have a concept of the void, which could be a precursor to zero? And the answer is yes, they did. The void is a part of their primordial world before anything exists, described in the opening of the Mayan sacred text *Popol Vuh*.

'This is the account of when all is still silent and placid. All is silent and calm. Hushed and empty is the womb of the sky. ... There is not yet anything that might exist.' (Christenson, 2007, p. 58). Like in many creation myths out of nothing emerges something, the empty void is pregnant, and spews forth all in the world and heavens that comes to exist. This provides confirmation for the conjecture that having the concept of the void may be a prerequisite for zero.

In comparison, Ancient Egyptian culture included the idea of 'that which does not exist' in opposition to 'that which exists', according to Hoffmann (2021). Thus, the Ancient Egyptians had the concept of the Void. However, they lacked the concept of zero in their arithmetic.

The Ancient Egyptians did have a hieroglyph that served some of the functions of zero, also standing for beauty and completeness. Its consonant sounds were 'nfr' (vowels unknown). This symbol was used to express nil remainders in a monthly account sheet from the Middle Kingdom dynasty 13 (*c.*1770 BCE). The bookkeeping record looks like a double entry accounts sheet with separate columns for each type of goods. At the end of the month, the account was balanced. When the total income matched the total disbursement this was shown by the nfr symbol (Lumpkin, 1997). One would like to think that this means income minus disbursement equals zero. But it does not mean zero (although, according to Hoffmann, some Egyptologists disagree.) Assuming that such an interpretation is incorrect, it might be described as *presentism*, seeing the past through modern eyes, or even wishful thinking. What the use of *nfr* very likely means is that the books are balanced, the results are complete, symmetrical, and beautiful.

Nevertheless, the Ancient Egyptians computed perfectly well without Zero (as also did the Mesopotamians and Ancient Greeks, for example.) So, the concept of Void is at best Necessary but appears not to be Sufficient for an advanced civilization to develop the mathematical concept of Zero.

## 8 Conclusion

I began by criticizing Seife's (2000) blurring of the distinction between the concepts of nothing, emptiness, the void, and the quantitative mathematical concept of zero. The history of mathematics is bedeviled with such confusions and they may have delayed the acceptance of zero as a number by 1,000 years in the West. Other civilizations were not so tardy in adopting it. But Seife is right to say 1 comes before 0 in historical development, and indeed it is the same in the cognitive development (Barton, 2020). From one to zero in terms of mathematical understanding took millennia. Despite its former associations, zero is now an unproblematic concept within mathematics, although its acceptance as a fully-fledged number is still an obstacle met by children learning mathematics.

In this chapter I have traced the mathematical and philosophical significance of zero, as the title promises. But also in my title I said Nought Matters. Perhaps the meaning of title is obvious but let me spell it out. One of the amazing and difficult things about zero, nothing, etc. is the ambiguity, paradoxicality, the conundrums, and indeed humor that emerge from its study and its self-negation (which impacted on its history mightily). I chose Nought Matters because it plays on three meanings:

First, the synonymous sentence: 'Nothing matters' means it is not the case that anything has significance. This says, jokily, that my enterprise (and everything) has no point.

Second, I can put this as another synonymous sentence, 'Zero counts', meaning zero and its synonyms refer to contents of importance. This surely is my message. (But it also plays with the ambiguity of the word 'counts' meaning both to enumerate and to be of importance.)

Third, as a term it is colloquially synonymous with the noun phrase 'Nothing matters', describing matters or materials relating to nothing or zero.

This last meaning, namely subject matter concerning zero, describes what comes from a study of zero and the issues it raises. It encapsulates the subject of this chapter and indeed the whole conference and project of The Zero Project.

Whitehead (1911, p. 63) said: 'The point about zero is that we do not need to use it in the operations of daily life. No one goes out to buy zero fish. It is in a way the most civilized of all the cardinals, and its use is only forced on us by the needs of cultivated modes of thought.' After the initial observation, he is of course wrong, at least today. No one in modern society can go a day without using zero. Even if they do not look at a clock or thermometer display, do not read a paper or a book, do not go shopping or handle any money, as soon as they look at any electronic display, use a mobile phone, tablet, computer, radio or television, travel in a car, bus, ship or plane, they are seeing the results of millions upon millions of zeros (and ones) controlling the system and encoding data. Zero is everywhere, throughout human society. I'm tempted to say zero or nothing means everything. But that would contradict my attempts to keep the quantitative and qualitative aspects of zero separate.

Zero is the only essential and ineliminable number. Even 1, the only other possible candidate for this title, is derived from 0 as its successor. Thus despite, or perhaps because of the controversy its adoption has caused over 3,000 years, zero with its many functions and roles is probably the most important of all the numbers. Certainly, the modern industrial revolution based on information, communication, and control technologies, could not even begin without it. Perhaps, too, the original industrial revolution depended on it. So much rests on zero and the number systems of which it is the foundation.

In coming to the end of my voyage of discovery it behooves me to give the last word to the land of its origin. *Pujyam* is the word for zero in Tamil, Malayalam, and Marathi, and it means 'worthy of worship'. As we have seen, history confirms this judgment. Beyond this, Sarma (2011, p. 211) reports a Tamil proverb that declares, 'Inside the *pujyam* (zero), there exists a *rajyam* (kingdom)'. I hope that I have opened the door to this kingdom a crack and offered a brief look inside.

## References

ANSI (1977). *American National Standard Code for Information Interchange*. Washington, DC USA: American National Standards Institute.

Bachelard, G. (1938). La formation de l'esprit scientifique. Paris: Vrin.

Barrow, J. D. (2000). *The Book of Nothing*. New York: Vintage Books (Random House).

Barton, N. (2020). *Absence perception and the philosophy of zero. Synthese* 197:3823–3850. https://doi.org/10.1007/s11229-019-02220-x.

Bernal, M. (1987). Black Athena, The Afroasiatic roots of Classical Civilisation, Vol. 1. London: Free Association Books.

Brandom, R. B. (2000). *Articulating Reasons: An Introduction to Inferentialism*. Cambridge, Massachusetts: Harvard University Press.

Bronkhorst, J. (2001). Panini and Euclid: Reflections On Indian Geometry. *Journal of Indian Philosophy*, April 2001, Vol. 29, No. 1/2, pp. 43–80.

Bruner, J. (1964). *Towards a Theory of Instruction*, Cambridge, Massachusetts: Harvard University Press.

Cantor, M. (1880–1908). *Vorlesungen über Geschichte der Mathematik* (*Lectures on the History of Mathematics*), 4 volumes. Leipzig: Trubner.

Carlyle, T. (1841). On Heroes, Hero-Worship, and The Heroic in History. London: James Fraser.

Christenson, A. J. (2007). *Popol Vuh: Sacred Book of the Quiché Maya People* (translation and commentary by Allen J. Christenson). Norman, Oklahoma: University of Oklahoma Press.

Coe, M. D. (1987). *The Maya* (4th edition). London; New York: Thames & Hudson.

Ernest, P. (2009). The Philosophy of Mathematics, Values and Keralese Mathematics. P. Ernest, B. Greer, and B. Sriraman (Eds.) *Critical Issues in Mathematics Education*. Charlotte, NC, USA: Information Age Publishing, 2009, pp. 189–204.

Ernest, P. (2015). Mathematics and Beauty. *Mathematics Teaching*, September 2015, pp. 23–27.

Ernest, P. (2018). The Ethics of Mathematics: Is Mathematics Harmful? In P. Ernest (ed.). *The Philosophy of Mathematics Education Today*. Switzerland: Springer, 2018, pp. 187–216.

Ernest, P. (2019). Privilege, Power and Performativity: The Ethics of Mathematics in Society and Education. *The Philosophy of Mathematics Education Journal* No. 35. http://socialsciences.exeter.ac.uk/education/research/centres/stem/publications/pmej/pome35/index.html. Accessed 13 October 2020.

Ernest, P. (2020). Mathematics, Ethics and Purism: An application of MacIntyre's virtue theory. *Synthese*, 2020. https://doi.org/10.1007/s11229-020-02928-1.

Ernest, P. (2024). Nought Matters: The Mathematical and Philosophical Significance of Zero. *Philosophy of Mathematics Education Journal*, no. 41 (2024). https://education.exeter.ac.uk/research/centres/stem/publications/pmej/.

Galton, F. (1883). *Inquiries into Mental Faculties*, Reprinted in Everyman's Library, London: Dent and Sons.

Hardy, G. H. (1940). *A Mathematician's Apology*. Cambridge. United Kingdom: Cambridge University Press.

Heeffer, A. (2011). The Reception of Ancient Indian Mathematics by Western Historians. In B. S. Yadav and M. Mohan (eds.), *Ancient Indian Leaps into Mathematics*, Switzerland: Springer, 2011, pp. 135–152.

Hoffmann, F. (2021). Personal communication.

Høyrup, J. (1994). *In Measure, Number, and Weight*. New York: SUNY Press.

Høyrup, J. (1996). Changing Trends in the Historiography of Mesopotamian Mathematics: An Insider's View, *History of Science*, Vol. 34, 1996, pp. 1–32.

Høyrup, J. (2002). Lengths, Widths, Surfaces: A Portrait of Old Babylonian Algebra and Its Kin. Switzerland: Springer.

Ifrah, G. (2000). A universal history of numbers: From prehistory to the invention of the computer. New York: John Wiley & Sons.

Joseph, G. G. (1991). *The Crest of the Peacock Non-European Roots of Mathematics*. London: I B Tauris, Penguin Books.

Joseph, G. G. (2008). A Brief History of Zero. *Tarikh-e 'Elm: The Iranian Journal for the History of Science*, Vol. 6, 2008: pp. 37–48.

Joseph, G. G. (2016). Indian Mathematics: Engaging with the World from Ancient to Modern Times. London: World Scientific.

Lumpkin, B. (1997). Africa in the Mainstream of Mathematics History, Ethnomathematics: Challenging Eurocentrism in *Mathematics Education*, Arthur B. Powell and Marilyn Frankenstein, eds. Albany, New York: SUNY Press, 1997, pp. 101–117.

Mattessich, R. (1998). From Accounting to Negative Numbers: A Signal Contribution of Medieval India to Mathematics. *Accounting Historians Journal*. Vol. 25, No. 2, December 1998.

McDermott, M. (2001). Quine's Holism and Functionalist Holism. *Mind*, Vol. 110, No. 440, October 2001, pp. 977–1025.

McNeill, W. H., Bentley, J. and Christian, D. (2010) *Berkshire Encyclopedia of World History*, 2nd Ed. Great Barrington, Massachusetts, USA: Berkshire Publishing Group, p. 568.

Merriam-Webster (2020). Hot button. *Merriam-Webster.com Dictionary*, Springfield, Massachusetts, USA: Merriam-Webster. https://www.merriam-webster.com/dictionary/hot%20button. Accessed 22 November 2020.

Merton, R. K. (1965). On the Shoulders of Giants: A Shandean Postscript. New York: Free Press.

Montaigne, M. de (1580). On Cannibals. *Selected Essays of Michel de Montaigne*. Translated by Donald M. Frame. New York: Van Nostrand, 1941.

Morris, C. (1945). *Foundations of the Theory of Signs*. Chicago: University of Chicago Press.

O'Connor, J. J. and Robertson, E. F. (2000a). Varahamihira. *MacTutor History of Mathematics Archive*. Accessed 8 April 2021 at https://mathshistory.st-andrews.ac.uk/Biographies/Varahamihira/.

O'Connor, J. J. and Robertson, E. F. (2000b). Brahmagupta. *MacTutor History of Mathematics Archive*. Accessed 8 April 2021 at https://mathshistory.st-andrews.ac.uk/Biographies/Brahmagupta/.

Ogburn, W. F. (1926). The Great Man versus Social Forces. *Social Forces*, Vol. 5, No. 2, December 1926, pp. 225–231.

Parameswaran, S. (1998). *The Golden Age of Indian Mathematics*. New Delhi, India: Swadeshi Science Movement.

Peano, G. (1889) Arithmetices principia, nova methodo exposita, Turin. Translated extracts in Heijenoort, J. van Ed. (1967). *From Frege to Gödel: A Source Book in Mathematical Logic*. Cambridge, Massachusetts: Harvard University Press, pp. 83–97.

Peirce, C. S. (1931–58). *Collected Papers* (8 volumes), Cambridge, Massachusetts: Harvard University Press.

Plato (1941). *The Republic of Plato* (translated and annotated by F. M. Cornford). Oxford: The Clarendon Press.

Powell, A. B. and Frankenstein, M., Eds., (1997). *Ethnomathematics: Challenging Eurocentrism in Mathematics*. Albany, New York: SUNY Press.

Rotman, B. (1987). Signifying Nothing: The Semiotics of Zero, London: Routledge.

Quine, W. V. O. (1951). Two dogmas of empiricism.

Quine, W. V. (1953). *From a logical point of view*. Cambridge: Harvard University Press, pp. 20–46.

Quine, W. V. O. (1990). *Pursuit of truth*. Cambridge, MA: Harvard University Press.

Sarma, S. R. (2011). Mathematical Literature in the Regional Languages of India. B. S. Yadav and M. Mohan (eds.). *Ancient Indian Leaps into Mathematics*. Switzerland: Springer, 2011, pp. 201–211.

Sartre, J. P. (1956). *Being and Nothingness: An Essay on Phenomenological Ontology*. London: Routledge.

Seife, C. (2000). Zero, The Biography of a Dangerous Idea. New York, Penguin.

Skovsmose, O. (2019). Crisis, Critique and Mathematics. *Philosophy of Mathematics Education Journal*, No. 35, December 2019. http://socialsciences.exeter.ac.uk/education/research/centres/stem/publications/pmej/pome35/index.html. Accessed 28 November 2020.

Skovsmose, O. (2020). Mathematics and Ethics, *Philosophy of Mathematics Education Journal*, No. 36, December 2020. http://socialsciences.exeter.ac.uk/education/research/centres/stem/publications/pmej/pome36/index.html. Accessed 16 April 2021.

Spencer, H. (1873). *The Study of Sociology*. London: Henry S. King & Co.

Swetz, F. J. (2016). *Mathematical Treasure: Moritz Cantor's History of Mathematics*. https://www.maa.org/press/periodicals/convergence/mathematical-treasure moritz-cantors-history-of-mathematics. Consulted 24 February 2021.

Turak, A. (2013). Eight Lessons from the Great Man (or Woman) School of Leadership, *Forbes*. 10 April 2013. Accessed 9 April 2021 at https://www.forbes.com/sites/augustturak/2013/04/10/8-lessons-from-the-great-man-or-woman-school-of-leadership/?sh=666f44de2b4b.

Vygotsky, L. S. (1978). *Mind in Society: The development of the higher psychological processes*. Cambridge, Massachusetts: Harvard University Press.

Whitehead, A. N. (1911). *An Introduction to Mathematics*. New York: Henry Holt & Co.

Wiese, H. (2003). *Numbers, Language, and the Human Mind*. Cambridge, UK: Cambridge University Press.

Wikipedia. (2021a). Brahmagupta. *Wikipedia, The Free Encyclopedia*. (Last revised 26 March 2021), https://en.wikipedia.org/wiki/Brahmagupta#Arithmetic. Accessed 8 April 2021.

Wikipedia. (2021b). Negative number. Wikipedia, *The Free Encyclopedia*. https://en.wikipedia.org/w/index.php?title=Negative_number&oldid=1008424119. Accessed 23 February 2021.

Wittgenstein, L. (1953). *Philosophical Investigations* (translated by G. E. M. Anscombe). Oxford: Basil Blackwell.

Wolterstorff, N. (1996). Barth on Evil, Faith and Philosophy. *Journal of the Society of Christian Philosophers*. Vol. 13, No. 4, pp. 584–608.

Yadav, B. S. and Mohan, M. (Eds.) (2011). *Ancient Indian Leaps into Mathematics*, Switzerland: Springer.

CHAPTER 18

# The Influence of Buddhism on the Invention and Development of Zero

*Alexis Lavis*

## Abstract

Since recent datings of the Bakhshali manuscript confirm our understanding that the zero is indeed an Indian creation, what is it in Indian culture that could have influenced an invention of this kind? From data relating to the Bakhshali manuscript, it is possible to recognize the influence of Buddhism and, from this, to try to see how this movement could have played a role in the theory leading to the inception of the placeholder zero. Moreover, with Brahmagupta this first form and use of zero was developed to the point of becoming a number, and thus adopting an operative role. While the writer of the *Brāhmasphuṭasiddhānta* was not a Buddhist, it is still possible to show in that document how Buddhist thinking, particularly its speculative developments in the meaning of *śūnyatā* as described by the great writers of Mahāyāna Buddhism, may provide a favorable setting for theories about the numerical concept of zero, even to the point where it is perceived as a condition of its 'thinkability' or conceptualization.

## Keywords

Buddhism – Sarvāstivāda – Brahmagupta – Bakhshali – mathematical symbolism – metaphysics – zero – ontology – point – emptiness – śūnyatā – kṣaṇa – bindu – Gandhāra – Greek mathematics – philosophy – Indo-Greek kingdoms – non-being – *Brāhmasphuṭasiddhānta* – Pratītyasamutpāda – interaction – instantaneism

## 1 Introduction

The challenge for this article is to establish the possible share of by Buddhist influence in the appearance and development of the zero in India. The Bakhshali manuscript and the *Brāhmasphuṭasiddhānta* of Brahmagupta are the basis for this study, in that they represent two founding moments in the use of zero: the first, for the positional usage it makes of it; and the second for

 | DOI:10.1163/9789004691568_022

establishing it as a number in its own right. It is first necessary to determine whether, in historical and geographical terms, Buddhist influence on these two texts can be established as fact. From there, it will be a matter of identifying correspondence between some Buddhist reflections – particularly on the moment (*kṣaṇa*) and on emptiness (*śūnyatā*) of the real – and the statuses, roles, and functions played by the zero in the two texts noted. Determining and analyzing these correspondences, between ontological speculation and mathematical usage or invention, should provide, if not proof, then at least a strong basis for the idea that Buddhism forms the epistemological field in which the arithmetical-algebraic zero appeared, thus in some way overturning the Platonic idea whereby mathematics would form the prelude to philosophy.

## 2 The Bakhshali Manuscript

The recent work undertaken at the renowned Bodleian Library, at the University of Oxford, has provided, if not final confirmation, then at least decisive evidence of the Indian origins of the concept of zero. This is the result of radiocarbon dating showing that the Bakhshali manuscript, discovered in 1881 in the Peshawar region, dates from the third or fourth centuries CE, for the oldest parts. In other words, the Bakhshali manuscript is presently the oldest known document in which the symbol appears and is used in a form, certainly already ancient, of what would become the number 0. The manuscript thus to some extent goes further than the Sumerian and Babylonian documents that, although earlier and using decimal or hexadecimal systems, developed no specific symbols referring to something like the 0.

We should also remember that the research community did not wait for the results of the Bodleian Library's tests to propose the idea that the zero was of Indian origin.[1] Much debating over two centuries, and many arguments and even proofs have now been proposed in favor of this theory. So, what does this new dating for the Bakhshali manuscript contribute? Aside from further confirmation, it does provide new direction for discussion or research. So far, it has been a matter of determining in which cultural period the zero originated. We can now be much more positive that it does indeed come from India. But in that case, what are the historical and cultural elements in India that gave rise to the invention and development of the zero? Why was this astonishing

1 See Appendix 3 for an online petition urging the Bodleian Libraries, Oxford, UK, to take concrete steps to commission the necessary follow-up radiocarbon-dating of the Bakhshali Manuscript in the interest of scientific advancement in the field.

mathematical object born in India? How can this birth and its conditions be explained and determined? So this is what the new dating for the Bakhshali manuscript allows: the context for our questions about this emergence has a much narrower focus.

## 3 The Greco-Buddhist Context

As we have noted, the Bakhshali manuscript was discovered in ruins located some kilometers from Peshawar. With reference to the information supplied by these radiation measurement tests, the manuscript originates from the ancient region of Gandhāra, also known as the Peshawar basin, at a crucial point in its history. Now, this Gandhāri cradle of zero, far from isolated within its own immutable, autochthonous cultural setting, was on the contrary a crossroads of commerce and civilization, seldom matched in history. The triangle marked by the cities of Peshawar (Puruṣapura), Taxila (Takṣaśilā), and Mardān were influenced by Achaemenid (540–326 CE), Seleucid (326–305 BCE), Maurya (305–206 BCE), Indo-Greek (206–70 BCE), Indo-Scythian (90 –/+10 BCE), Indo-Parthian (10–60 CE), and Kuṣāṇa (60–300 CE) empires.[2] As well as experiencing many incursions, such as those of the Sassanids, the White Huns or the Kidarites, these cities were also staging posts for merchants, religious travelers, and adventurers taking the Silk Road. Consequently, the fact that the Bakhshali manuscript originates from this crucial region, and at such a rich period of cultural interactions and upheavals, causes difficulties in researching the influences on this text, and hence on the origin of the zero.

It is possible, however, taking an overview of the whole period and especially of the Kuṣāṇa empire, to identify two predominant influences, running throughout the period like the thread of the weft, giving it a certain unity, along with its own unique and characteristic identity. These two factors are Hellenism and Buddhism, whose association strengthened throughout the period of influence of the Greco-Bactrian and Indo-Greek kingdoms, and from which the Kuṣāṇa empire, originating in Bactria, achieved a kind of synthesis in what could conveniently be called 'religion' and Greco-Buddhist art. Certainly Gandhāra, as we have already noted, as an economic, political, and cultural crossroads, experienced many other influences, chief among which were clearly Vedism and Brahmanism, and to a lesser extent Zoroastrianism and other features of Achaemenid Persia. However, this dual Greek and Buddhist

2 The dates shown here are only approximate, offered as guidance to a time period, rather than for precise dating.

presence seems dominant, as witnessed by the mass of evidence discovered from archaeology, scripture, numismatics, etc.

In so far as it belongs to this period and this region, it seems justified to assume that the Bakhshali manuscript and its author necessarily must have had direct contact with these Greek and Buddhist influences, all the more so as it was written in a Prakrit language then in use in the region and very common in Buddhist writings. So the task now is to determine the nature of this influence and how it was translated and spread.

## 4 Influence Difficult to Determine

It is right to ask how a contemplative tradition such as Buddhism might be likely to have any significant influence on a mathematical text of the kind in the Bakhshali manuscript. The doubt increases when looking in detail at what we have of this text, some 70 pages, with no beginning or end, the places where the author might be identified and where there would be an overview of the work, and where there is often a religious dedication that could identify membership of a particular denomination. These remaining pages only deal with arithmetic, giving a large number of problems associated with daily life, for instance three merchants, the first of whom has seven horses, the second nine ponies and the third ten camels. Each gives three animals, so that they can be equally distributed among the group, in order to make the value of the property of each one equal to that of the others. The problem is, what was the initial value of the property of each merchant and the value of each animal? This is only one example among many of a similar kind. It has to be acknowledged that such trivial questions do not fit well with the higher dimension of the practices and reflections of Buddhism. Added to this is the fact that the Bakhshali manuscript cannot be considered as arising from a key work devoted to speculation or an explanation of new mathematical theories. On the contrary, it appears to be more of a practical teaching manual, no doubt intended to train apprentice traders, as indicated by the entirely mechanical nature of the problem-solving method demonstrated, the purpose of which was to save time and effort in thinking, rather than to produce elegant or bold solutions. Such a purpose does not seem to require the aid or influence of Buddhism either.

So, at first sight it seems more reasonable to turn to the Greek part to find the essential elements, which may have influenced the arithmetic methods shown in the Bakhshali manuscript, especially as the Greeks were recognized very early on as being excellent mathematicians. Nonetheless, when looking at the text of the Bakhshali manuscript, there is no explicit, or even implicit, borrowing from Greek mathematics. The form and style of the manuscript is entirely Indian, and very close to the language (a kind of hybrid Sanskrit), the metrics (śloka), and types of reasoning appearing in the Buddhist *Abhidharma*, where a common vocabulary and logical operators are also found.[3] Nor does the notation system use anything from the Greeks. There are certainly some similarities found in problem-solving methods, but nothing indicates that these are borrowings, all the more so as the new dating for the Bakhshali manuscript now makes it significantly older than a number of Greek texts that could previously have been considered their precursors, and almost contemporaneous with the work of Diophantus, whose possible influence has, however, been invalidated by a large part of the scientific community.

Despite Kaye's statement that there is not the slightest indication in the Bakhshali manuscript of any connection with Buddhism, let us nonetheless keep trying to find some slight connection. Why should we? On the one hand, because Buddhist influence cannot be ignored, given its undisputed pervasiveness across the whole of the North-Western region; and on the other because the development of the Indian zero, of which the Bakhshali manuscript is the first known evidence, could not have arisen from nowhere, given the originality of this mathematical object. In some way it demands a philosophical basis that Buddhism, more than any other movement, is likely to have supplied.

## 5 The Prevalence of Buddhist Influence

So, while there is no explicit reference to Buddhism within the Bakhshali manuscript itself (and why would there be, given the nature and purpose of this text?), it is indisputably present in the environment in which it was present. So, let us remind ourselves of the nature of this Buddhist environment within the North-West, as a way of assessing its cultural weight in this region where the manuscript was created.

While the infancy of Buddhism was spent in the region of the Ganges, ancient Māgadha, its adolescence was certainly passed in the Indus, in other words, the Gandhāra region. It was Emperor Maurya Aśoka who instituted *Buddhadharma* in this area. It is possible, however, that small Buddhist communities settled there even before the edicts of the Great King, as indicated by the accounts of Arrian and Megasthenes that tell of debates held between Greek philosophers and *Śramaṇā* at the time of Alexander's conquests. Thus

3 Introduction of the rule by *tadā*, of the case by *sthāpana*, of the solution by *karaṇa*, and of the proof by *pratyaya*.

we know that Pyrrho, founder of skepticism, attended these meetings, which seemed to have made a great impression on him, and played a large part in the development of his school of thought. It was after Aśoka's political decision, however, that Buddhism took root and became established in the North-West. The emperor's religious activism was effective, as evidenced by the building of *stūpas* and *viharas*, by the mass distribution of monks, by a taxation policy favorable to monastic communities, and more. It also consisted of a huge communication program, mixing political messages and spiritual preaching in his famous edicts, which were engraved onto rocks or on majestic columns. On this subject, we should also note that the edicts of Aśoka seem to have been addressed mainly to merchants, who were naturally those who read routinely and who traveled, and thus were the most likely to pass on imperial messages, and spread the Buddhist teachings they expressed more widely. Let us remember also that the development of Indian mathematics also owes much to the merchant class, to which the author of the Bakhshali manuscript seems to have belonged, as indicated by the sample problems he uses, and which testify to a certain familiarity with the 'business world'. We should also remember the other major figure, to whom we will return, namely Brahmagupta, to whom we owe the development of the zero as a number, who was also vaiśya by origin.

Aśoka's edicts give us valuable information about the nature of Buddhist presence in the first part of the third century. On one of the rocks found in Kandahar, we can read the following in Greek and Aramaic: 'The conquest of Dharma is here victorious'.[4] Another passage indicates that the Greeks themselves followed the teachings of the Buddha. Besides, it was Buddhist monks of Greek origin to whom Aśoka entrusted the mission of going to preach the Dharma in the West, as reported in the *Sinhalese Chronicles*.[5] These chronicles also give astronomical figures of the number of converts and the monasteries built, in Kashmir and Gandhāra.[6] This ledger concludes rather dramatically: 'From then on, Kashmir and Gandhāra were radiant with the yellow of the monastic robes, and treasured above all the Three Jewels!'[7]

Indian culture customarily uses exaggeration, invoking the enormous numbers easily formulated in the decimal system to dazzle the imagination and symbolize grandeur and vastness. The Buddha himself, in the oldest texts, frequently used counting to express colossal quantities, in order to represent

4 Rock n°13.

5 Mahāvaṃsa, XII. See W. Geiger. (1912). *The Mahāvaṃsa or the Great Chronicle of Ceylon*, London, pp. 82, 85.

6 180,000 converts, of whom 100,000 became monks, and more than 500 monasteries founded.

7 *Ibid.*, p. 84.

the infinite in terms of the excessive. However, while the evidence from the *Sinhalese Chronicles* seems somewhat exaggerated, this use of extreme counts indicates that the establishment of Buddhism in the North-West was a massive event, with multiple and sustained effects. The mention of the Greeks and their conversion to Buddhism should also be taken in this sense. No doubt not all the Greeks of Bactria and the Gandhāra became Buddhists, but it should, however, be noted that a number of Indo-Greek kings did convert (tradition numbers seven of them),[8] including King Menander, whose dialogue with the Buddhist monk Nāgasena in 150 BCE was collected in the *Milinda-pañha*. We should also note that the renowned Menander is reputed to have achieved the supreme state of Arhat (sainthood), making him the first Western 'enlightened one'.

We mention these episodes to highlight a remarkable phenomenon that demonstrates the power of Buddhism in this region. It was not, in fact the Indians, who were converted to Greek philosophy or religion, but the Greeks, and what is more their kings, who adopted Buddhism. This phenomenon was repeated with the Indo-Scythians and the Indo-Parthians, up to the Kuṣāṇā dynasties, among the first of which was the Buddhist emperor Kaniṣka (127–163 CE), who made Peshawar not only his own capital but also that of Buddhism. Thus, despite many upheavals experienced in the North-West over nearly five centuries, Buddhism held its own at the very top of the social order, as an unbroken thread stretching between the two emperors, Aśoka and Kaniṣka, via Menander and many others. This continuity clearly demonstrates the influence exerted by Buddhism, truly a mark of identity for the entire region, since, as we know, when a king officially adopts a religion, it is never for personal reasons only, but also (perhaps above all) because this faith has a unifying power.

Another sign of the power or vitality of Buddhism, especially during the period when the Bakhshali manuscript appeared, can be seen in the creativity of Buddhism during this long Gandhāri period. Conservative attitudes actually only appear during periods of weakness or decadence, and in reality, are defensive positions. None of this happened in Gandhāra, which was the setting for one of the most innovative periods, in both doctrinal and artistic terms, at that time. It saw the emergence of one of the most original, as well as one of the boldest, schools in terms of speculative developments: the Sarvāstivāda – the area of emergence of the Mahāyāna, the Great Vehicle, and thus of the 'theory' of *śūnyatā*, especially during the reign of Kaniṣka. In institutional terms, this inventiveness is revealed through the changing state of monasteries, which

8 Menander I, Strato I, Amyntas, Nicias, Peukolaos, Hermaeus, and Hippostratos adopted the title 'Dharmikasa' (adept of the Dharma) and represented it on coins with their effigy.

transformed from retreat centers used during the rainy months, into study and training centers, in other words universities. Some actually consider, and rightly so, that the first university in history was not that of Constantinople, founded in 425 CE by Theodosius II, but that of Taxila, which legend says dates back to the time of the grammarian Paniṇi (around the fourth century BCE), and which came under Buddhist influence following the ruling of Aśoka, until its destruction in the fifth century CE. Far from being satisfied with the compilation and recitation of Buddhist texts, these monastic and educational centers, especially that of Taxila, were devoted just as much to scientific study, especially logic, medicine, mathematics, and astronomy.[9] As well as Taxila, the same region had the monasteries of Jaulian and Mohra Muradu, as well as the stūpa Dharmarajika, and the famous Takhti-i-Bahi near Mardān, as well as many more spread around Peshawar.

We must finally mention these famous Gandhāran Buddhist texts, dating from about the first century BCE to the third century CE, which were discovered in caves and are actually stored in several collections.[10] These texts, largely very badly deteriorated, testify to the strong intellectual presence of Buddhism in this region, which must be considered as the starting point of written Indian culture. Nevertheless, none of the texts found deal with mathematics, nor even with other sciences. They contain only texts from the Buddhist canon. This, however, does not mean that Buddhists lost interest in science or mathematics. On the one hand, because these manuscripts were found not in libraries but rather in reliquaries, it is the reason that they could be preserved. They were stored for devotional purposes rather than for study. It is why we have found only 'sacred' texts. On the other hand, the act of writing down had a religious function. Indians of this time did not put the various teachings in writing in order to read them, but to preserve them. The teachings of ancient India were essentially oral and based on the amazing memory of the masters who, moreover, passed on their memorization techniques to their students.[11] If the teaching of mathematics was delivered in the Buddhist monasteries of Gandhara, as some indirect sources attest, it was therefore certainly orally using the sole medium of memory and not of textbooks.

---

9 A list of secular sciences studied by monks figured in the *Milindapañha*. Among them are: calculating (finger technique) and arithmetic (II 3,7). See also Apte, *Universities in Ancient India*, pp. 8–22; Scharfe, *Education in Ancient India*, pp. 140–145; Mookerji, *Ancient Education*, pp. 475–482.

10 See Salomon, *Ancient Buddhist Scrolls from Gandhāra*, and *The Buddhist Literature of Ancient Gandhāra*.

11 See Scharfe, *Education in Ancient India*, pp. 8–28.

Thus, from very ancient times, the Buddhist monastic community formed a kind of intellectual elite. This is easily shown by the high quality of the dialogue between Menander and the monk Nāgasena. One of the characteristic traits of Buddhist monks, as well as their doctrinal commitment, of course, is their constant attention to logic and the science of reasoning. His recognition of this point is another reason why Rudolf Hoernle associated the Bakhshali manuscript with Buddhist influence, especially as mathematical science was taught in monasteries as a matter of course. He also notes that the reasoning at work in the manuscript reflects that effective in the *Abhidharmas*, where they were largely formalized. We believe that there are some other points of contact between the content of the Bakhshali manuscript and the features of speculative teaching of Buddhism, especially in this Gandhārian school that we have already mentioned, the Sarvāstivāda.

## 6 Buddhist Philosophy Favorable for the Appearance of Zero

As we have seen, although Buddhist influence is not obvious within the manuscript, it is no less certain, given its significant impact in the region at this time, that its influence must have been felt somehow. These few reminders of the history of Buddhism in the North-West are not just intended to give the reader some further historical information, but rather to redirect the question initially posed, namely: could Buddhism have had an influence on the development of the mathematical ideas present in the Bakhshali manuscript? So, in the light of the hegemony of Buddhism in the area and at the time the manuscript appeared, how could it not exercise influence on them, however that might have been affected? Given the historical data, it seems more reasonable to accept this influence, and to make it a starting point or an established fact, even though it might not be explicit within the manuscript. Thus, the question is no longer one of finding out whether or not there was influence, but of how this Buddhist influence might be revealed within this mathematics manual.

As we have said, Rudolf Hoernle had already established two similarities between the manuscript and the Buddhist texts: the language and the metrics, on the one hand, and the style of reasoning used, on the other. We believe it may be possible to detect further points of congruence, especially around the use of the zero and, in particular, its role in the manuscript.

First, let us review some points relating to the nature, function, and symbolization of the zero in the Bakhshali manuscript. There is no 0 symbol found in the manuscript that represents zero or *śūnya*, but there is a dot. This dot (*bindu* in Sanskrit) is used as one of the ten fundamental symbols in the decimal system, and shows that a quantity is null or else is unknown. In the manuscript, however, the zero-dot does not yet have the status of a number, as it will have in the *Brāhmasphuṭasiddhānta* of Brahmagupta; it is only a place-marker used to indicate an empty space or a hiatus. From this initial development, we can see that the zero is not reduced to a simple nothing, a pure deprivation of being, but more subtly refers to a 'vacant space'. The idea of 'vacancy' is much more relevant than that of null or nothing to refer to the Indian meaning of *śūnya*. So this term does not only indicate non-being, but also, and inseparably, something like availability. Now, non-being, in so far as it is open to the available, is not opposed in any way to being, but is on the contrary a kind of waiting for its coming. It is clear how understanding the nothing in this way, open to availability, makes the idea of zero as a place-marker possible, and can designate *both* the quantitative nullity and the unknown. To us, this association might go without saying, but this is not the case at all, and it was not so in Greece, where non-being had always been, since Parmenides, quite simply impossible and unthinkable.

This association of the nothing and the free, which is not in any way obvious to grasp in philosophical terms, forms one of the central points of Buddhist teaching, focused precisely around the concept of *śūnyatā*, but not only that, as we will see. Let us remember that the central role of *śūnyatā*, designating the face of the absolute itself, belongs more to the Mahāyāna. Nevertheless, this notion is found in the oldest teachings of the Buddha, as shown in the *Mahā-Suñña Sutta*[12] and the *Cula-suññata Sutta*.[13] These texts set out how contemplation of the absence of the ground of the self is precisely the entrance to *Nirvāṇa*, that is, to a state of complete freedom. The idea that the nothing liberates, or that the nothing is a complementary face of the free, a little like the two faces of the Roman god Janus, is here clearly established since the beginnings of Buddhism. Besides, the notion of *Nirvāṇa* also relies on the same idea, since it signifies both extinction *and* liberation. This is precisely why Buddhism does not depend on nihilism, since it does not so much seek annihilation as freedom, understood as complete availability, that is, openness or vacancy. The way of *Nirvāṇa* is not *destruction* but *de-obstruction* – which is not at all the same thing. This distinction is impossible while *śūnya* is considered only in the light of the nothing, of deprivation, absence or non-being, rather than that of the openness.

Let us also note that the famous dialogue between Menander and the monk Nāgasena, entitled the *Milinda-pañha*, opens precisely on a discussion of the

12 Ñāṇamoli & Bodhi (1995), *The Middle Length Discourses of the Buddha*, Boston: Wisdom Publications, pp. 971–78. [Majjhima Nikāya III, 110–118].

13 *Ibid.* pp. 965–70. [Majjhima Nikāya III, 104–109].

meaning to give to the concepts of *anātman* (non-self or non-personality) and of *Nirvāṇa*. King Menander does not understand how Buddhist teachings can affirm the non-existence of self and of the person and how *Nirvāṇa* may be rightly understood as both the annihilation of the self and its liberation. The response Nāgasena gives complies perfectly with the statements previously held, and which rely on this open understanding of *śūnya*. It is this same understanding that is at work in the positional value given to the zero by the author of the Bakhshali manuscript.

In addition to the discussions of the *Milinda-pañha*, the Buddhist school of the Sarvāstivāda, already mentioned, was developing. One of the original dogmas of this school was to determine that the *śūnyatā* does not in any way prevent the existence of reality, which this school also numbers among the 62 fundamental components (*dhātu*)[14] that are in some way established from it. This is unthinkable as long as *śūnyatā* is envisaged from the null alone, since nothing can come out of nothing, except nothing itself. But if *śūnyatā* is considered from the idea of 'empty place', that is, as indeterminate, and thus open and available to the possible, there is no contradiction in affirming both vacancy and existence, and that the first may, in some way, precede the latter. The use of zero as a placeholder, as well as the fact that it can symbolize the unknown in an equation, as it is used perfectly in the Bakhshali manuscript, agrees entirely with the metaphysics of the Sarvāstivāda, even to the point of being its most simple mathematical application.

Finally, there is one more point of contact with the views of this school, and the arithmetical appearance of zero in the Bakhshali manuscript: the use of the dot, the *bindu*, as a symbol for the latter. Another of the major speculative contributions from this Gandhari and Kashmiri school of Buddhism concerns time, of which the Sarvāstivādin have an 'instantaneist' interpretation. In other words, according to them, the truth of time is not in the duration but in the instant. They thus propose a *discontinuous* theory of time, consisting of instants, like the 'discrete' theories of matter, from ancient atomism to quantum mechanics. So this atom of time, the instant, which besides has more theoretical analogies with the famous quantum of action[15] developed by Planck than with the principles of classical atomism, has to be understood not from a simplistic idea of undivided unity, but of *limit-activity*. Vasubandhu

14 *L'Abhidharmakośa de Vasubandhu*, trans. La Vallée Poussin, vol. 1, p. 49, Louvain, 1923. Although this work dates from the fourth century CE, it includes here much older doctrinal elements.

15 In fact, the instant as the Sarvāstivāda envisages it is defined from the notion of *activity*. See *L'Abhidharmakośa de Vasubandhu*, vol. 1, pp. 228 ff.

(from Peshawar during the fourth century CE), who refers back to much earlier views, defines the instant (*kṣaṇa*) as the 'point' (*bindu*) of completion *and* of initialization,[16] or else as the union *without mediation* of birth and of perishing.[17] All the same, the instant as 'position' of termination-initializing, with the dot as metaphor, chimes strongly with the use of the zero as a placeholder in the decimal scheme in use in the Bakhshali manuscript, as well as that it is also symbolized by a dot. Finally, a further correspondence relates to the 'manifestation', that is to the moment when one passes from strict nullity, from nothing, to something – in other words, in the arithmetical world, from the 0 to 1. In the *Abhidharma* of Sarvāstivāda, this 'something' is called *rūpa* or *svarūpa*.[18] Now, the figure 1 is also called *rūpa* in the Bakhshali Manuscript.[19]

Hence, if Buddhism did indeed form a dominant element of the cultural setting in which the Bakhshali manuscript appeared, and there are correspondences in their use of zero with contemporary Buddhist reflections on the meaning of *śūnyatā* and, by extension, on that of *kṣaṇa* – which actually form part of the same field of metaphysical investigation, it might be rightly thought, without excluding it as firmly as Kaye does, that Buddhism has had a role to play from the very moment of the invention of zero in India.

What we want to show ultimately, however, is that Buddhism is as likely to have accompanied not just the creation of zero but also its development, that is, its interpretation as a 'number', as Brahmagupta established in the famous *Brāhmasphuṭasiddhānta*.

## 7 Mahāyāna *śūnyatā* and the Interpretation of *śūnya* as a Number

At the outset, let us note that there is a parallel to be drawn between mathematical developments of the zero and Buddhist reflections on the meaning of *śūnyatā*. Hoernle commented that between the Bakhshali manuscript and the *Brahmasiddhanta* of Brahmagupta, there is 'a very particular connection', a 'curious similarity'.[20] First, this affects the notation system of which the large majority was adopted by Brahmagupta. Similarly, the algebraic examples used in the manuscript are found in the *Brahmasiddhanta* chapters also devoted to

16 *Ibid.* p. 229.

17 *Ibid.* vol. 3, p. 4.

18 *Tattvasamgraha of Śāntarakṣita.* (1926). G. O. S, Baroda, p. 510, śl. 1810; *Tattvasamgraha-panjika* of *Kamalaśīla.* (1935). G. O. S., Baroda, p. 508, I, 1–25; op. cit. *L'Abhidharmakośa de Vasubandhu*, vol. IV, p. 56.

19 R. H. Hoernle. (1887). *On the Bakhshali Manuscript*, Vienna, pp. 6–7.

20 *Ibid.* p. 12.

algebra. In brief, and in the light of other analogies, we have to observe that there is a certain continuity between the texts discovered at Bakhshali and the work of Brahmagupta. We cannot talk of any 'kinship', since there is no direct proof of this, but it is certain that these two witnesses to Indian mathematical science are not entirely strangers to each other: quite the contrary. Brahmagupta's works thus develop many points from the elements in the Bakhshali manuscript, among which the zero rightly stands out, moving from a strictly positional status, to one that was much more operative, and philosophically deeper: that of a number. In parallel, it was the notion of *śūnyatā* that would experience significant conceptual development, with the advent of the Mahāyāna, whose kinship with the Sarvāstivāda school is fully acknowledged. Hence we see a correspondence that, while not proving the point is nonetheless of interest between two culturally significant advances – that of Buddhism and that of mathematics – which both take place around the same notion of *śūnya*.

The question then arises of whether Buddhist influence can be established in the work of Brahmagupta and if this influence, which probably is present in the Bakhshali manuscript, might in some way be continued with the *Brāhmasphuṭasiddhānta* author, even to the point of the development of Buddhist reflections on *śūnyatā*, leading to the zero becoming established as a number. It is a bold hypothesis, all the more so as it cannot be proven in the strict sense. Nonetheless, we would like to show that it is not entirely meaningless, since it enables some light to be thrown on the philosophical conditions from which it becomes credible that the zero could be established as a number, and thus to show in what way this institution could appear in India and nowhere else. In other words, we would like to show how Buddhism, and particularly the Mahāyāna, could become the philosophical field in which to nurture the invention of the number zero, as it is found in exemplary form with Brahmagupta. We thus begin from the principle that discoveries, whether scientific, artistic, or technical, call for a particular philosophical setting in order to happen. This philosophical setting, an area of fundamental representations, allows or forbids the appearance within it of particular ideas or practices, and thus operates as a selection space.

Thus, for philosophical reasons, the Greeks were certainly brilliant at geometry, but did not develop the idea of the number zero. We can reverse the maxim that welcomed the students of Plato's Academy, 'Let no one ignorant of geometry enter', to create the following formula: 'No one can be a student of geometry without some prior philosophy.' This philosophy was named metaphysics, that is, a fundamental way of thinking about the meaning of being, which is then passed on into all fields of knowledge. Greek metaphysics relies on two radical ideas: on the one hand, that being is absolutely opposed to non-being, so that there could not be anything like non-being in any form whatever. On the other hand, that being is to be 'something', in other words, identity is at the very basis of being. The mathematical consequences of such an idea of being apply in particular to the concept of number, which since Antiquity has been immediately understood *on the basis* of the natural integer, that is, *from* concrete magnitude, being the quantitative attribute of object, of 'something' existing as such. Hence the huge problems posed by the development of theories of irrational numbers, or the problem raised by the incommensurability of the side and diagonal of a square, which drove many Pythagoreans to suicide. Now this is nothing other than the idea that a number might express a quantitative nullity. The involvement of the zero in the *operations* of addition or subtraction is quite simply inconceivable for a Greek, that is, for a being born within the philosophical space circumscribed by Parmenides and Aristotle, even though arithmetic had accepted these strange numbers such as π, because they still remain quasi-objects, that is, 'beings'. It was not by chance that the zero was an import into the West that provoked much resistance. For instance, in 1299, the city council of Florence issued an order forbidding the use of Arab figures, especially zero. In 1494, the mayor of Frankfurt reissued the same ban, and a century later, it was the turn of the canon of Anvers. The reasons for these rejections and bans rested on religious beliefs that, associating non-being with the devil, were themselves avatars for Greek metaphysics.

It is interesting to see how the introduction of the zero in the West, that is, the object that is impossible for Helleno-Christian thought to accept, had a liberating effect, enabling a dramatic development in the concept of number. So we think for instance of Viète's introduction of symbolic notation in 1591, in the place of numerical notation, allowing the concept of the *variable* to enter mathematics. A variable is a number that does not represent any concrete magnitude or quantity. But that is nothing compared with the invention of the concept of *function*. It is to Bernoulli and Euler that we owe the great progress in this concept, which they redefined as the expression composed of a variable and any constant (real) parameters. So function allows not only magnitudes to be expressed, but also operations (that is relationships) such as those of elementary algebra, and also series and infinite products, exponentials, logarithms [$\log_{10}(1000) = 3$], what are known in mathematics as transcendental functions. Now, functions are numbers. And we can see what a demand is made on thought in envisaging the idea that a number is not necessarily the expression of an integer, but also that of combinations, that is a set of possible (sometimes infinite) situations and which are 'numbers', not because they express a magnitude, a concrete quantity, but an assembly or a set, that is, a system of variables, and not isolated or absolute units.

We can only be amazed, given the problems encountered in the West, at the ease with which Brahmagupta introduced the number zero into his work and, with almost no introduction, in the way of evidence, showed how this number behaved in various operations. Nowhere in the *Brāhmasphuṭasiddhānta* is there the least reluctance, hesitation, or awkwardness in the chapter about *śūnya*, or even any in-depth explanation about the fact that he gives it the status of a number. Let us once more note that it cannot be taken for granted that 'nothing' can be operative. However, since Brahmagupta does indeed do this, there must be some prior conditions making it acceptable, conditions that indeed are found together in Mahāyāna Buddhism.

## 8 Sūnyatā: of the Space Open to Pratityasamutpāda

Let us begin with an essential reminder: the word *śūnya* already existed before its mathematical usage. It is not then the figure that gives meaning to the word, but the word that indicates the significance of the figure. This is perfectly attested by Brahmagupta, who defines the zero first not from its value and mathematical role, but from what the common notion of *śūnya* evokes for him. So he speaks of 'space', of 'openness', and even of the 'vault of heaven'.[21] These images are in no way bland. First of all, they also appear in a number of Mahāyāna Buddhist texts to describe either the nature or the experience of *śūnyatā*. It is even mobilized, though in a less central way, in these two Pali *Sutta* mentioned above, and attributed to this notion. In this respect it is also interesting to note that when the Chinese were seeking to translate the Buddhist meaning of *śūnyatā*, they chose the term *kōng* (空), meaning free space and time, instead of *wū* (無) or *xū* (虛), meaning respectively absence and nothing.

But more radical yet, the idea of 'openness' and 'free space' appear in the etymology of the word *śūnya* itself, whose Indo-European root relates to the Greek 'κύαρ', which means 'openness', referring especially to ears or needles;[22]

21 *Brāhmasphuṭasiddhānta*, ed. Dr Sampurnananad. (1966). Indian Institute of Astronomical and Sanskrit Research, New-Delhi, t. I, pp. 88–89.

22 M. Mayrhofer. (1980). *Kurzgefasstes etymologisches Wörterbuch des Altindischen*. Heidelberg, t. III, p. 365; P. Chantraine. (2009). *Dictionnaire étymologique de la langue grecque*. Paris, p. 594.

and the symbol o, that is, the shape of a hole or an opening represents it very well – another argument in favor of the Indian origin of this symbol.

There is another interpretation besides of *śūnyatā*, not taken from nothing, or even from open space, but from the idea of *interaction* or system of interactions – in Sanskrit *pratītyasamutpāda* – thus making the operative implication of zero perfectly conceivable. The expression *pratītyasamutpāda* appears early on in Buddhism, from the second teaching of the Buddha.[23] It then indicated the concatenation of the 12 links or meshes (*nidāna*) that chain beings to this pathological regime of existence, which is called *saṃsāra*. *Pratītyasamutpāda* represents in the circumstance a specific mode of appearance by mutual causation, where the manifestation of each factor determines that of the next, which in turn retroactively reinforces the existence or the stability of its predecessor. Thus, this is a question of what is called a 'system in interaction', since the work of Norbert Wiener on cybernetics. One of the major contributions of Nāgārjuna – a founder of Mahāyāna thought, who reflected deeply on the meaning of *śūnyatā* – consisted in extending the scope of this notion of *pratītyasamutpāda*, then limited to psycho-cognitive subjectivity, to cover reality itself. The systemic nature of the real, indicated by the principle of *pratītyasamutpāda*, involves the very idea of *śūnyatā* since the interactional truth of the real assumes the non-existence of self-subsisting and self-defined entities. So, Nāgārjuna could write: 'It is in *pratītya* itself that beings exist. And that was named *śūnyatā*. For what exists in the mode of being of the *pratītya* is without its own nature'.[24] And then by way of a commentary: '*Śūnyatā* is not the privation of existence (*abhāva*), but the proper mode of being for *pratītyasamutpāda*.'[25]

We can thus see a radicalization at work with respect to ancient Buddhism in the understanding of *śūnyatā*, which here moves away from any simplistic idea of non-being to become the very principle of the idea of interactivity, that is, of mutual operativity. From such a concept, which no longer has anything to do with something like a 'nothing', the possibility of considering that *śūnya* might have the value of a number has nothing startling about it, but on the contrary appears almost taken for granted, as a 'natural' application in the field of mathematics – a field where, as we have said, Buddhists excelled since the foundation of the great monastic universities. The question that then arises is

---

23 *Mahātanhāsankhaya-sutta* and *Acela-sutta*.

24 *Vigrahavyāvartanī.* (1951). v. 22, ed. Johnson & Kunst, Chinese-Buddhist mixtures, Bruxelles, vol. 9, p. 121: 'yaśca pratītyabhāvo bhāvānāṃ śūnyateti sā proktā/ yaśca pratītyabhāvo bhavati hi tasyāsvabhāvatvam//22//.'

25 *Ibid.* p. 122.

whether Brahmagupta was likely to be influenced by Buddhist interpretations of *śūnyatā*'s meaning?

### 8.1 *Brahmagupta and Buddhism*

We certainly know the author of the *Brāhmasphuṭasiddhānta* better than the one of the Bakhshali manuscript. Born in 589 CE, we know he belonged to the merchant caste (the Cāpa family) and originated from Bhinmal, making his career in Ujjain, in the city's astronomical observatory. We do not know his religious persuasion, but we can assume he was not a Buddhist, given the dedication that opens his *Brāhmasphuṭasiddhānta*: 'I revere the unapparent primary matter …',[26] and which refers directly to the Samkhya. However, reading the rest of the dedication, his 'religion' seems to be above all that of mathematics: 'I venerate that unapparent computation, which calculators affirm to be the means of comprehension, being expounded by a fit person: for it is the single element of all which is apparent.'[27]

Thus, there is nothing to indicate he was in direct touch with Buddhism and in particular with the philosophical advances of the Mahāyāna. He was, however, reported as being very cultured and having a curious, observant mind. What is notable, however, is his geographical location. The city of Ujjain was close to Mathura, one of the greatest Buddhist centers in India, especially under the Saka (35–405 CE), mainly Buddhist, the Gupta (fifth century) and, of course, in the reign of the third great Buddhist emperor Harśa (590–647 CE). It is also on the route linking West and East, from Gandhāra to Bihār, the two regions that held the greatest monastic universities in the whole history of Buddhism. These two poles joined precisely in the region where the great mathematician lived. When the Chinese pilgrim monk Xuanzang (602–664 CE) went to Ujjain, he reported, for instance, that the sovereign showed great respect and welcome to Buddhists. He also reported that there were 20 monasteries and five Buddhist temples.[28] Of course, these institutions were not particularly devoted to mathematics and other secular sciences, but to monastic life. Nevertheless, Xuanzang reported that centers of higher learning arose in many monasteries, giving them 'university' status, such as Nālandā (Bihar) of course, but also in Kashmir despite the fall of the Taxila center, in Panjāb with Jālandara, in Uttar Pradesh with Matipura, or in Andhra Pradesh

---

26 H. T. Colebrook. (1817). *Algebra with Arithmetic and Mensuration from the Sanskrit of Brahmagupta and Bhascara*, London, p. 129.

27 *Ibid.* p. 130.

28 *On Yuan Chwang's Travels*, Vol. II, p. 167.

with the great center of Amarāvatī.[29] Xuanzang reported that these centers of high learning close to monasteries offered not only instruction in Buddhism but also in grammar, Samkhya philosophy, and secular sciences such as logic, astronomy, and mathematics.[30] According to the importance of Ujjain, which became, after the reign of Samudragupta (335–380), an important cultural and scientific center, we might think that local Buddhist institutions were also inclined to turn themselves into high learning places. A number of great Buddhist figures were also originally from Ujjain, among whom was the great master Paramārtha, born in 499 CE.

So, although Brahmagupta was certainly not a Buddhist, he lived in a place and at a time when there was indisputably a strong Buddhist presence, and at a time when Buddhist thinking was becoming more and more logical and epistemological. Moreover, for a thousand years Buddhism had exercised a profound influence on the Indian way of thinking, to the point that it had reached this metaphysical zone of which we spoke earlier. Hence Brahmagupta had no need to be a Buddhist to feel its influence, just as Euclid did not need to be a member of the Academy to be influenced by Platonic metaphysics, and by Parmenides before that. Buddhism had then become part of the ambiance of Indian thought, and this factor without doubt opened up the possibility of an operative use of zero. To express it in Kantian form: Buddhist thinking about *śūnyatā* provides the conditions of possibility for the mathematical *śūnya*, in other words, makes it 'thinkable'.

The point that has to be grasped here is that *śūnya* does not mean 'nothing'; it is not the equivalent of non-being. We owe this basic idea to Buddhism, more than to any other current of thought. It relies on discovering a meaning of being that goes much further than that in which the existent is opposed head-on to the non-existent. From this comes the possibility of a *fertile* understanding of zero, free of the sterile idea of nullity. The development of Indian mathematics has shown clearly that the zero is not in any case a nothing. And this is also shown to be true when it was introduced into the West by the Arabs. We have seen how, with Nicole Oresme, or with Leibniz, the zero is both the auxiliary then the representation of the infinite, particularly when dealing with arithmetic-geometric sequences. As a number, it is all at once real, positive, negative, and purely imaginary. In geometry, it is the dimension of a point and the symbol of the origin. In topology, it can represent a non-null dimension (the Hausdorff dimension). In probability, it is an impossibility.

29 *Ibid.*, Vol. I, p. 296, p. 322; *The Life of Hiuen-tsiang*, p. 137.

30 *The Life of Hiuen-tsiang*, pp. 68–71.

It is also an essential value in a function (Möbius function) or the only output value possible (the zero function). This conceptual fertility of the zero has thus greatly exceeded the framework of its introduction, and we can see how it is perpetuated in the field of modern mathematics, via for instance Hilbert's famous theorem of zeros, the major driver in the development of algebraic geometry throughout the twentieth century, and in the context of commutative algebra, in the notion of 'integral dependence' – an idea that in its pure principle of intelligibility is fairly close to what the Mahāyānists understood by the term *pratītyasamutpāda*, which, as we have seen, serves as the principle for the definition Nāgārjuna gives to *śūnyatā*. So, if Indian metaphysics, and as we have tried to show, more especially that developed by Buddhism, could serve in India as the basis for the development of zero, we might now consider that its mathematical development in the West will serve in return as a solid and fertile ground on which to build a new Western metaphysics.

## 9 Conclusion

Although neither the Bakhshali manuscript, nor the *Brāhmasphuṭasiddhānta* of Brahmagupta make explicit claim to Buddhism, its influence cannot be ignored. On the contrary, it is impossible these texts were not influenced by Buddhism. Externally, the geographical areas and historical periods in which they appeared were dominated intellectually, culturally, and politically by Buddhism. So it could be said that the external setting in which these texts appeared was largely Buddhist. Internally, only Buddhism, among the various philosophical fields present in India, including that of Greece, offers a non-private interpretation of emptiness (*śūnya*), one that is open, dynamic, and in a sense operational. Now, it is only if emptiness is freed from nothingness, that is, from non-being, that the zero as a mathematical object is possible and conceivable. The placement usage of zero, as well as its symbolization by the point (*bindu*) also corresponds to the philosophical proposals of the Sarvāstivāda school – originating in the same region as the Bakhshali manuscript – on the 'momentary' nature of reality. Some centuries later, Mahāyāna proposed a revolutionary interpretation of emptiness, from the concept of interaction. Shortly afterwards, in a region itself converted to the Buddhism of the Great Vehicle, appeared the great mathematical innovations of Brahmagupta, including the operational, or interactional use of zero, in other words, its interpretation as a number. So more than any other current of thought, Buddhism forms the philosophical space or the possible epistemological matrix, from which the mathematical zero is likely to have discovered its metaphysical foundations.

## References

Apte, D. G. *Universities in Ancient India*. University of Baroda Press.

Bugault, G. (1968). *La Notion de 'Prajñā' ou de sapience selon les perspectives du 'Mahāyāna'*. Paris.

Bugault, G. (1994). *L'Inde pense-t-elle?* Paris.

Burton, D. (2001). *Emptiness Appraised – A Critical Study of Nāgārjuna's Philosophy*. Dehli.

Colerus, E. (1937). *Von Pythagoras bis Hilbert – Die Epochen der Mathematik und ihre Baumeister. Geschichte der Mathematik für Jedermann*. Vienna.

Huntington, Jr, C. W. (1989). *The Emptiness of Emptiness – An Introduction to Early Indian Mādhyamika*. Honolulu: University of Hawaii.

Hsuan-tsang. *The Life of Hiuen-tsiang by the Shaman Hwui Li*, trans. Samuel Beal, London 1911, repr. 1973.

Hsuan-tsang. *On Yuan Chwang's Travels in India*, trans. T. Walters, London 1904, repr. Delhi 1961.

Kaplan, R. (2000). *The Nothing That Is: A Natural History of Zero*. London.

Katz, J. V. (ed.). (2007). "The Mathematics of Egypt, Mesopotamia, China." In *A Sourcebook*. Princeton.

Lamotte, É. (1976). *Histoire du bouddhisme indien : des origines à l'ère Śaka*. Louvain-La-Neuve.

Lavis, A. (2018). *La Conscience à L'épreuve de l'Eveil. Lecture, commentary and translation of Bodhicaryāvatāra de Śāntideva*. Paris.

Lindtner, C. (1982). *Nagarjuniana: studies in the writings and philosophy of Nāgārjuna*. Copenhagen.

Lopez, D. S. (1989). "On the relationship of emptiness and dependent-arising." *The Tibet Journal* XIV (pp. 44–69).

*Milindapañha*. (ed.) V. Trenckner, London 1889 repr. 1962; trans. T. W. Rhys Davids, Oxford 1894 (Sacred Books of the East vol. XXXVI).

Mookerji, R. (1947). *Ancient Indian Education (Brahmanical and Buddhist)*. London: Macmillan.

Nakamura, H. (1980). *Indian Buddhism*. Osaka.

Salomon, R. (1999). *Ancient Buddhist Scrolls from Gandhāra*, University of Washington Press.

Salomon, R. (2018). *The Buddhist Literature of Ancient Gandhāra: An Introduction with Selected Translations*. Boston: Wisdom Publications.

Scharfe, H. (2002). *Education in Ancient India*, Handbook of oriental studies, section II, vol. XVI. Leiden: Brill.

Seife, C. (2000). *Zero: The Biography of a Dangerous Idea*. New York.

Silburn, L. (1989). *Instant et Cause. Le discontinu dans la pensée philosophique de l'Inde*. Paris.

Streng, F. J. (1967). *Emptiness. A Study in Religious Meaning*. Madison: University of Wisconsin.

Wayman, A. (1999). *A Millenium of Buddhist Logic*. Delhi.

Williams, P. (ed.). (2005). *Buddhism: Critical Concepts in Religious Studies*, 8 vols. London/New York.

CHAPTER 19

# Zero and *Śūnyatā*: Likely Bedfellows

*Fabio Gironi*

## Abstract

The aim of this chapter is to highlight some conceptual connections between the concept of zero and the philosophical concept of *śūnyatā*. Both ideas first arose in the Indian subcontinent, and I will try to suggest that this geographical congruence is more than a coincidence. As such, this will likely be a more speculative contribution to this volume than others, but I will attempt to substantiate my proposal by stressing the conceptual advantages deriving from such a syncretic enterprise.

I will begin with an overview of the historical development of the number and concept of zero, followed by some considerations of its treatment in contemporary mathematics, as well as an examination of its less-evident semiotic properties. I will then proceed to re-examine the concept of *śūnyatā*, identifying its central meontological significance in the philosophy of Nāgārjuna. In this context, I will examine the idea of the 'trace' in the thought of Jacques Derrida, viewing it as another actor in the assemblage of concepts that I attempt in this chapter. Finally, I undertake a synthetic reading of these three 'voids' – of zero, of *śūnyatā*, and of the trace – aimed at a speculative rearrangement of these heterogeneous concepts, in a philosophical exercise that I see as both a possible and a necessary evolution of 'comparative philosophy'.

## Keywords

zero – Śūnyatā – emptiness – void – holes – Nāgārjuna – Derrida

## 1 Introduction

The problem of 'objective scholarship' and of hermeneutical procedures needs to be tackled briefly and so I would like to start by subscribing to Huntington's remarks on the topic:

> For us, meaning is necessarily embedded in the symbolic forms of our culture and our time. In response to the reader who condemns all such

 | DOI:10.1163/9789004691568_023

> attempts to interpret a text on the ground that the text itself does not employ our linguistic and conceptual structures, I can only throw my hands up in despair of ever understanding any ancient way of thinking. At some point we simply must acknowledge that no translation and no text-critical methodology can be sacrosanct. Translation and all other forms of hermeneutical activity rest firmly on the preconscious forms of linguistic and cultural prejudices peculiar to our historical situation. The most vital challenge faced by scholars is certainly summed up in their responsibility to make their … presuppositions entirely conscious and to convey through their work a sense of wonder and uncertainty of coming to terms with the original text. (Huntingdon, 1989, p. xiii)

I intend to take responsibility for my linguistic and conceptual choices, and to acknowledge my means and my ends, as my argument proceeds. That said, in order to avoid the criticism that I offer an ahistorical, decontextualized analysis, I should clarify what 'cross-cultural philosophical speculation' means. This article is not to be understood as a historical account of possible interactions between Nāgārjuna's thought and the mathematics of zero. Nor do I intend to give a historical genealogy of the social, cultural, doctrinal and political influences active on Nāgārjuna.[1] I believe (as the Madhyamikas did) that both natural and social affairs can (and should) be analyzed and explained on a *multiplicity* of ontological levels: absolute reduction is impossible. I do not claim that mine is the only possible way to understand Nāgārjuna (or zero, or Derrida's thought) but I do claim that to prioritize one (any) explanatory level over another – instead of joining them all for a more synoptic vision – is a mistake. Hence, what this chapter aims at is *precisely* to extrapolate three concepts ('emptiness', 'zero', and 'trace/void') *out* of their usual disciplinary contexts in order facilitate their 'encounter' on a philosophical ground articulated by their own mutual interaction. This is meant as neither a history of mathematics *nor* a work in Buddhist studies, *nor* as yet another meditation on Western continental philosophy.

I do not wholeheartedly accept the label 'comparative philosophy', for I do not believe that our current understanding of 'comparative' work is quite satisfactory. I hold that the 'comparative' stage is, indeed, a *stage* of a cross-cultural philosophical work, and not an endpoint. The contextualized and textually/philologically accurate recovery of the evolution of an idea in a given culture – that can then engender a constructive (albeit limited) philosophical comparison between similar concepts from other contexts – is *one* possible enterprise.

1 A more than comprehensive survey of these influences can be found in Walser 2005.

Another, different, kind of project, however, is the philosophically accurate re-employment of ideas on *new* philosophical grounds.[2] This is certainly a speculative enterprise, but it cannot be reduced to an uncritical appropriation of an alien concept in order to force it into one's conceptual scheme. To be sure, no philosopher is context-free; but my aim is not that of rekindling the fire of Western philosophy with decontextualized raw conceptual materials extracted from the Buddhist tradition, nor to steal concepts from the field of mathematics only to then clumsily rearrange them to fit my philosophical project. The goal is rather the creation of a new philosophical set of coordinates by forging new alliances between ideas.

About the problem of interpretation of ancient texts, Hayes claims that '[o]n looking at trends in twentieth century scholarship on Nāgārjuna, one can discern two fairly distinct styles, which seem to correspond to the traditional approaches known as exegesis and hermeneutics. Roughly speaking, the former attempts to discover what a text meant in the time it was written, while the latter attempts to find the meaning of a text for the time in which the interpreter lives.' (1994, p. 362)

I do not think that this distinction, one that implicitly assigns to exegesis the status of superior objectivity, can be formulated and established in such a clear-cut way, as if bracketing off one's cultural conditioning would immediately reproduce the original meaning of a text. This prejudice emerges even more clearly when Hayes comments contemptuously on deconstructionist scholarship, defining the deconstruction as 'an act of playing with the written symbols in deliberate disregard of what the author's intention may have been in first inscribing them' (Hayes, 1994, p. 347). This definition can surely be applied to something: but this would be, plainly, *bad scholarship*.[3] Quli (2009, p. 5) urges us to acknowledge that, 'this nostalgia [for original "Eastern" Buddhism], with its characteristic trope of decay and distortion, goes hand in

2 It would seem that I still take for granted the possibility of singling out a philosophical plane. For an interesting discussion around the validity of our category 'philosophy' in a cross-cultural context see the debate (regarding Chinese philosophy, but relevant to any kind of comparative enterprise) between Carine Defoort (2001, 2006) and Rein Raud (2006a, 2006b).

3 Such an understanding of deconstruction is the heritage of decades of hermeneutically lazy Derridean epigones that kept alive the myth of Derrida as an intellectually dishonest obscurantist. It would suffice to read what Derrida himself has to say on the practice of 'deconstructing texts' to understand that to accuse it of 'disregard of the author's intention' completely misses the mark: Derrida indeed claims that 'texts are not to be read according to a hermeneutical or exegetical method that would seek out a finished signified beneath a textual surface. Reading is transformational' but he immediately underlines that 'this transformation cannot be executed, however one wishes. It requires protocols of reading'. (Derrida, 1981, p. 63).

hand with the tendency to discount hybrid identities. Indeed, this tendency to reject the hybrid as inauthentic is an extension of the colonial search for pure races and pure cultures, and as such is part and parcel of what anthropology identifies as "salvage studies".'

Not only is the quest for pure and original Buddhist thought an impossible one but this theoretical nostalgia ends up denying the possibility of developing *hybrid thought*. This term indexes more or less precisely what I am trying to accomplish in this chapter: the construction of a theoretical site where different concepts can be joined together by forming new and unprecedented assemblages,[4] establishing a thought that is indescribable according to the rules of any of its previous situations/contexts. To 'always historicize' means to appreciate the contextual fabric into which texts are woven. This, however, should not translate into a practice of historical purification. The Buddhist and Western traditions will forever remain locked in their academic compartments – only to meet in sporadic 'comparative' exercises – as long as we refuse to envision the possibility of a synthesis that is not an appropriation, an encounter that is not a reduction, and the creation of a hybrid conceptual topology.

Williams observes: 'It is simply fallacious to think that because absolute objectivity is a myth, reading Mādhyamika through the eyes of Wittgenstein is no different from the attempt to understand Mādhyamika on its own terms in its own historical context. Both historical scholarship and appropriation are possible' (1991, p. 194). I agree: both are legitimate enterprises. But my approach attempts a third way, different from both text-historical scholarship (which in seeking the original context of a text's production, must recognize its own cultural emplacement) *and* hermeneutical interpretation (which should refrain from wholesale appropriation of an alien context by carefully placing its objects – texts and authors – in a coherent historical perspective). Thus, if as Keenan observes, 'the difficulty in understanding Nāgārjuna, or for that matter any ancient thinker, results from the difficulty in understanding the intertextual web of meanings within which he thought and wrote' (1985, p. 367), this difficulty cannot be overcome by employing the romanticized trope of 'objective scholarship', able to lend its ear to the faint voice of Nāgārjuna himself and to transcribe it literally. The only possibility of moving closer to the possible

4 My employment of the term 'assemblage' is informed by the Deleuzian work of Manuel De Landa (in particular, 2006). An assemblage is a construction where the single parts do not merge seamlessly under the imposition of the whole, but indeed constrain what can be built, resisting inadequate appropriations. Yet, out of such an assemblage the 'emergent property' of a new intercultural philosophical thought can proceed.

meanings of Nāgārjuna's work *for us* is precisely to pluralize it. Such a gesture does not enact historical violence, but instead remains cognizant of the historical conditions of the writer while simultaneously maintaining the possibility a live exchange[5] between two cultures and periods.

## 2 The Historical Development of Zero

I will now briefly trace the history of the concept of zero and present its various forms and uses: those of sign (digit, numeral), of placeholder (mark for an empty place) and of number (integer).

The origin of the zero lies in the invention of the positional system of mathematical notation. The introduction of this method allowed computations to be carried out at a much faster rate and with a significant economy of space. Instead of having dedicated symbols for units, tens, hundreds, and so on, the positional system allowed mathematicians to determine the value of a symbol in a string of numbers due not only to its *shape*, but also to its *position*. A clear example is to compare the way the contemporary positional notation system writes the number 111, thus repeating the same symbol while differentiating its different roles according to the position it occupies, and the way in which the Romans would have written it – CXI – using three different symbols, indicating hundreds, tens, and units. The credit for the invention of the positional system of notation has been given to several cultures as '[p]resent evidence indicates that the principle of place value was discovered independently four times in the history of mathematics' (Joseph, 1991, p. 22) in the Babylonian, Chinese, Indian, and Mayan civilizations.[6]

The use of positional notation gave rise to the necessity for a symbol to denote a 'space intentionally left blank', where one of the columns would remain empty, in order to avoid confusion (1 1 can easily be confused with 11, unless we mark the absence of a value in between with another special symbol, as in 101). While the Babylonians used two slanted wedges to signify 'nothing here', the Indians used a small dot (*bindu*). This later evolved into an empty

5 I believe that this approach allows us to avoid the danger of interpretative utilitarianism that Inada (1985, p. 220) warns us against: 'to extract concepts from the corpus of Nāgārjuna's philosophy, nay the whole of Buddhism, merely to suit one's purposes'.

6 The debate on mutual influences between these cultures (excluding the Mayan due to its geographical isolation) has been as lively as it has been inconclusive due to the relative lack of historical evidence. The most common view gives temporal priority to the Babylonian (see, for example, Ifrah 1988, p. 382).

circle and became known as *śūnya*.[7] In its first appearance, zero thus played a marginal role, being a mere placeholder, a necessary mark but without the dignity of a real number: its use was limited to the practical context of counting-board calculations. As Ifrah suggests: 'To the Babylonians the zero sign did not signify "the number zero". Although it was used with the meaning of "empty" (that is, 'an empty place in a written number') it does not seem to have been given the meaning of "nothing" as in "10 minus 10", for example; those two concepts were still regarded as distinct' (1988, 382).

But this distinction would not hold for long in India,[8] where zero matured. In the subcontinent a decimal place value notation was developed during the first five centuries CE through the adoption of three successive types of numerals: initially the *Kharosthi* (*c.* fourth century BCE–second century CE) and the *Brahmi* (third century BCE), then, as an evolution of the *Brahmi*, the *Bakhshali* (third to fifth centuries CE) and the *Gwalior* (ninth century CE), the latter being the very direct source, through the Arab manipulation and export, of modern Western numeral symbols. In the *Bakhshali* manuscript, retrieved near the village of Bakhshali in Northwest India and dated to the first five centuries CE, it is possible to find a fully developed place value system with a dot as a sign for an empty position. But the earliest appearance of the *śūnya* symbol as the empty circle, is to be found in the Gwalior inscription, found near Lashkar in central India with an inscribed date, commonly reckoned to correspond to 870 CE (see Menninger, 1968, p. 400), where in the number 270 we find the circular zero in a final position, even if slightly smaller and elevated compared to the other two digits. What was completely new here, as opposed to the Babylonian use of zero, is the evidence of the manipulation of zero not merely as a placeholder, but as an independent number. An epochal cognitive achievement.

The first certain description of zero that matches (with some imperfections, as we will see) our current understanding of zero as an integer on the number line, can be found in the work of the seventh century Indian mathematician Brahmagupta (598–668 CE). In his *Brahmasphutasiddhanta* (628 CE) the author not only deals with negative values but, for the first time in recorded history, enunciates rules for the basic arithmetical operations with zero, *śūnya*: addition, subtraction, multiplication, division, raising to powers, and extraction of roots. In these operations Brahmagupta makes a 'mistake' (according to contemporary mathematics) only concerning the division by zero, to which he essentially gives no result, tautologically claiming that the quotient of any

7 *Śūnya* was the main, but not the only name for this symbol, others being *kha* and *ākāśa*, both linked to the concept of 'sky' hence 'space', 'openness' and also 'infinity'.

8 For a comprehensive history of Indian mathematics see Plofker, 2007.

number and zero is a fraction with zero as a denominator. The correction of the problem was attempted by successive mathematicians such as the twelfth century Bhaskara II (1114–1185 CE), who interestingly proposed a $n/0 = \infty$ solution.[9] Emancipated from its role as a mere placeholder, zero had therefore now acquired the fully fledged status of a number that could be manipulated within arithmetical operations.

From Brahmagupta onwards zero as a number slowly developed until the contemporary era, traveling West with other Indian numerals in the work of Arab mathematicians. It took several centuries for these symbols to reach the West: they were first introduced into European culture only in the thirteenth century with the publication of the *Liber Abaci* by the Italian merchant and mathematician Leonardo da Pisa, better known as Fibonacci (*c.*1170–*c.*1250 CE).[10] Fibonacci was an avid reader of Arab mathematical treatises, including that of the Persian mathematician Al-Khwarizimi (780–850 CE), who in turn had studied Brahmagupta's Arabic translations at the Abbasid court of Baghdad. The very word we use today in English and in other European languages are distortions of the original *śūnya*: the Arabs translated it as *sifr*, and Fibonacci translated the Arabic word with the Italian *zefiro* – from which are derived the English 'cipher' and the French *chiffre*, which were successively deformed into *zevro*, or *zero*.

The introduction of zero into European culture was anything but painless. The feelings of awe, skepticism, and rejection toward this new number are well summarized in Menninger's vivid description:

> [w]hat kind of crazy symbol is this, which means nothing at all? Is it a digit or isn't it? 1, 2, 3, 4, 5, 6, 7, 8, and 9 all stand for numbers one can understand and grasp – but 0? If it is nothing, then it should be nothing. But sometimes it is nothing, and then at other times it is something:

9 Bhaskaracharya seemed to have known the importance of zero, not just in positional notation, but also as a number. He has special verses describing the peculiar properties of zero. He lists eight rules such as $a + 0 = 0$, $0{\cdot}2 = 0$, $\sqrt{0} = 0$, $a \times 0 = 0$ etc. The interesting aspect of this verse is the definition of infinity or *Khahara* as a fraction whose denominator is zero. In other words, $a/0 = \infty$' (Nagaraj, 2005, p. 17). In contemporary mathematics this formula is admitted only in the context of inversive geometry, where the inverted image of zero as a center is a point at infinity.

10 In the first chapter of the Liber Abaci, Fibonacci introduces the new numerals and the zero: 'Novem figure indorum he sunt 9 8 7 6 5 4 3 2 1, cum his itaque novem figuris, et cum hoc signo 0, quod arabice zephirum appellatur, scribitur quilibet numerus, ut inferius demonstratur.' Note that if the other numerals are called figuris, only the zero is merely a signo. It will also be known as nulla figura (no number), from which our current word 'null' is derived.

> 3 + 0 = 3 and 3 − 0 = 3, so here zero is nothing, it is not expressed, and when it is placed in front of a number it does not change it: 03 = 3, so zero is still nothing, nulla figura! But write the zero after a number, and it suddenly multiplies the number by ten: 30 = 3 × 10. So now it is something – incomprehensible but powerful, if a few 'nothings' can raise a small number to an immeasurably vast magnitude. Who could understand such a thing? And the old and simple one-place number 3,000 (on the counting board) has now become a four-place number with its long tail of 'nothings' – in short, the zero is nothing but 'a sign that creates confusion and difficulties' as a French writer of the 15th century put it – une chiffre donnant umbre et encombre.
>
> Thus the resistance to the Indian numerals by those who used the counting board for calculations took two forms: some regarded them as the creation of the Devil, while others ridiculed them. (Menninger, 1969, p. 422)[11]

Europe was still conceptually dependent on Aristotle – whose philosophy rejected the possibility of both absolute void (nature itself was said to abhor a vacuum) and infinity – and was preoccupied, theologically, with the fullness of being, relegating nothingness to the realm of Satan. To examine in detail the reasons for the West's aversion toward zero would lead us too far from our path, but this hostility should be kept in mind when looking back toward India: why didn't zero encounter hostility but, on the contrary, it was fully developed as a concept *there*? I will argue that it is not surprising that a culture that nourished the Buddhist doctrine of *śūnyatā* welcomed and developed this mathematical void.

## 3 Zero's Role in Mathematics

Descriptions of zero in contemporary mathematics raise some provocative problems. By definition, zero is the integer that precedes 1, it is an even

11 It is interesting to note that both the historical development of the concept of zero and its reception somewhat mirror our cognitive abilities when it comes to its use. In an interesting review of studies in developmental psychology and neuroscience, Andreas Nieder suggests that, 'The cultural hesitation to appreciate zero as a quantity and later a number is mirrored in a protracted ontogenetic development of zero relative to positive integers in children' (2016, p. 835). Nieder identifies four stages in the process of emergence and acceptance of zero, finding evidence for them in both history and in the cognitive development of the concept in children.

number, and it is neither positive nor negative. It is also a real and a rational number. Arithmetically, most of the rules firstly enunciated by Brahmagupta are still valid: zero is a neutral element with respect to addition[12] and subtraction, while the product of any number and zero is zero, collapsing any number back on itself. When it comes to division, there is an even odder result – the one Brahmagupta could not identify – for the quotient of any number and zero is considered *meaningless*.[13] Moreover, in its interaction in the decimal place value system, there are three possible roles for zero:

> It must be noted that in a fully developed place value system of numeration, the zero must be able to play all the following three roles successfully:
> – medial or internal, which is the classical role of a blank space, e.g., as in 205, 2005, etc.
> – final or terminal, which is more stringent role, e.g., in 250, 2,500, etc.
> – initial, which is a rather superfluous role ordinarily, e.g., 025 = 0025 = 25 in value. But in the computer age this role is also important. (Gupta, 2003, p. 22)

Necessary and futile at the same time, interacting in a peculiar way with other numbers, in its position at the center of the number line – squeezed on its side between the positive and the negative infinity of numbers – in mathematics zero remains the single number without repetition, neither positive nor negative, but both a beginning and an end.[14] That empty circle is all that *remains*

12 In mathematics, an additive identity occurs when an element, added to another element, remains unchanged. The number zero is the most familiar additive identity.

13 There are a few mathematical forms called undefined or indeterminate, and these – interestingly – involve zero and infinity: 'Whatever the context in which it is used, division by zero is meaningless. [Any result] is unacceptable mathematically and division by zero is therefore declared to be an invalid operation ... On the other hand, in higher mathematics we often encounter the so-called 'indeterminate forms' ... Such an expression has no preassigned value; it can only be evaluated through a limiting process ... The seven indeterminate forms encountered most frequently in mathematics are $0/0$, $\infty/\infty$, $\infty(0)$, $1\infty$, $00$, $\infty 0$ and $\infty - \infty$' (Maor, 1987, pp. 7–9). Maor here implicitly refers to L'Hopital's rule, used to convert an indeterminate form into a determinate one through the determination of limits. Without entering into the mathematical details, I should note that in these cases zero and infinity seem to run parallel and opposite to each other, carving the very theoretical limits of mathematics.

14 In geometry this property is even more striking: firstly, (0,0) are the coordinates of the origin of the Cartesian axes on a plane. The axes both begin and encounter at the point of zero, and zero is the only fixed point that enables the whole system of coordinates to have any meaning and practical value; secondly, the basic element of Euclidean geometry, the point – the basic building block of a line, in turn the building block of a 2-dimensional

when the numbers annihilate each other, and it is what made them possible in the first place. Zero stands at the same time inside and outside the number line – being anywhere where nowhere is – and it undermines its consistency. In spite of this, it is the very condition for its beginning. Zero not only indicates that nothing is present there (it is no mere vacant place, no simple omission) but indeed warns about the absence of anything at all that could be signified within the mathematical system. Rotman gives an insightful interpretation of zero's strange role:

> As a numeral, the mathematical sign zero points to the absence of certain other mathematical signs, and not to the non-presence of any real 'things' that are supposedly independent of or prior to signs which represent them. At any place within a Hindu numeral the presence of zero declares a specific absence: namely the absence of the signs 1 2 ... 9 at that place. Zero is thus a sign about signs, a meta-sign, whose meaning as a name lies in the way it indicates the absence of the names 1 2 ... 9 ... Thus, zero points to the absence of certain signs either by connoting the origin of quantity, the empty plurality, or by connoting the origin of ordering, the position that excludes the possibility of predecessors ... It is this double aspect of zero ... that has allowed zero to serve as the site of an ambiguity between an empty character ... and a character for emptiness, a symbol that signifies nothing ... In short: as a numeral within the Hindu system, indicating the absence of any of the numerals 1, 2, 3, 4, 5, 6, 7, 8, 9, zero is a sign about names, a meta-numeral; and as a number declaring itself to be the origin of counting, the trace of the one-who-counts and produces the number sequence, zero is a meta-number, a sign indicating the whole potentially infinite progression of integers. (Rotman, 1987, pp. 12–14)

For now, we can formulate the following observations about this double aspect of zero differentiating itself from all the other number signs. Zero allows us to manipulate it as a number and use it within the context of a calculation, accompanied by – and somewhat disguised among – the other, more familiar numbers. It is never quite the same as them, signaling their absence and creating the possibility for the appearance of any other number: if 2 implies the repetition of 1, hence presupposing 1, doesn't 1 presuppose 0? But if we pluck it out of the operation, and look at it, trying to grapple with its *meaning*, our

figure and so on – is a zero-dimensional object. The line acquires a measurable length out of an infinite series of points of length zero.

critical gaze falls on its openness; our categorizations encounter no ground, instead whirling down a hole, a singularity, where all shrinks to *zero*.

Zero's role seems to be consistently elusive, hybrid, neither completely *nothing* nor ever *something*. Its thin circle ambiguously divides an empty interior and an exterior vacuum, creating emptiness out of nothingness. Differently to what we observe on white pages of some books – 'This page was intentionally left blank' – the zero signals absence without claiming it, without breaking its silence. It is as if zero held a secret, a *cipher*, that can neither be kept hidden nor uttered: how to conceptualize this 'nothing' – this symbol referring to a chain of absences – that eludes reification and all conceptual capture?

## 4 Śūnyatā

The complexity of the concept of *śūnyatā* in Nāgārjuna's philosophy is best introduced by the range of possible translations of the term. We have already seen that *śūnya* meant 'zero' for Indian mathematicians, but that was not the only meaning of the term.

According to Sathyanarayana:

> Śūnya (skt.) is derivable from śūna, which is formed from the root śvi (cl.1, bhvādi, parasmaipadi) also taking the forms śū or even śvā. The root means 'to swell' (morbidly), 'to enlarge'; and by semantic extension, 'hollow'. It is in this last sense that the word śūnya has now established itself to mean empty, void, vacant, vacuity, absence, free from, nonentity, nonexistence, cipher (zero), nought and space. (Sathyanarayana, 2003, p. 264)

Therefore, *śūnyatā* is the abstract noun of *śūnya*, hence emptiness, voidness, absence, and so on. Indeed, the most common translation of this term among scholars is *emptiness*. Nonetheless, since this term has an often derogatory meaning in contemporary English usage, *śūnyatā* has been interpreted in various ways with scholars seeking to rid it from the semantic baggage of the word 'emptiness'. A small sample indicates the efforts that have been made to avoid the conflation of *śūnyatā* with emptiness: Stcherbatszky notoriously translates it as 'relativity', Sprung and Magliola use 'devoidness' while McCagney ventures 'openness'. Even if we were to consider, evaluate, and use some of these alternative translations, in the present chapter I will simply use *śūnyatā*, keeping implicit and silent the semantic dissemination of the term. As for the problem of Nāgārjuna's textual sources, my guide texts will be the

*Mūlamadhyamakakārikā*[15] (MMK) and the *Vigrahavyāvartanī*[16] (VV), the two most reliable and rich textual sources of his thought.

Let us begin with Nāgārjuna himself:

> For him to whom emptiness is clear, Everything becomes clear.
> For him to whom emptiness is not clear, Nothing becomes clear. (MMK, xxiv.14)

This passage makes clear that the first step for any Buddhist philosopher (and in fact for anyone who wants to understand the nature of reality) is to grasp the meaning of this doctrine. Thus, in this section I want to limit myself to considering the concept of *śūnyatā* both as expounded in Nāgārjuna's texts *and* as understood and re-elaborated by scholars and interpreters, proceeding in a relatively non-systematic way (a complete and exhaustive exposition of the doctrine of *śūnyatā* would require more space than I can dedicate to it here),[17] drawing its fundamental lines by grounding my observations in Nāgārjuna's texts and on interpretations of it, and keeping in mind my purpose: a synthesis of *śūnyatā* with the concept of zero and the self-voiding structure of the trace. The fundamental novelty of Nāgārjuna's anti-foundationalism, over and against the context of early and *abidharmic* speculation, is that *all* entities lack self-existence (*svabhāva*)[18] – they are therefore *śūnya*. This (negative) property characterizes everything (including the abidharmic *dharmas*), not just the self, or compounded phenomena such as King Milinda's chariot. All things *lack* objective existence, essence is nowhere to be found. To postulate a stable essence would crystallize the world into an unchanging and static reality, against the evidence of our conventional observation of an ever-changing world:

> If there is essence, the whole world will be unarising, unceasing,
> And static. The entire phenomenal world would be immutable. (MMK, xxiv.38)

15 Jay Garfield's translation (1995).

16 Kamaleswar Bhattacharya's translation (1986).

17 This task has been brilliantly accomplished by several scholars, in particular Garfield 1995, 2003; Huntington 1989; Siderits 2003; De Jong 1972, May 1978; Tola and Dragonetti 1981; and Westerhoff 2009.

18 *Svabhāva* is commonly translated as 'inherent existence'. For a thorough examination of the meanings of this term in Indian philosophy see Westerhoff, 2009.

On the other hand, *śūnyatā* is a precondition for *pratītyasamutpāda* (another concept philosophically renewed by Nāgārjuna), the universal link of everything with everything else – provided any reificationist tendency regarding *śūnyatā* itself is avoided. Indeed, *śūnyatā* is not an entity: it is simply the lack of *svabhāva* or intrinsic existence. It is the ultimate state of being of all things, but not a thing in itself, neither an immanent nor a transcendental entity – it is pure form. Given that, according to Mādhyamika thought, all things are relationally linked within the causal chain of *pratītyasamutpāda*, they all share a *devoid* nature: 'That nature of the things which is dependent is called voidness [*śūnyatā*], for that nature which is dependent is devoid of an intrinsic nature (*yaś ca pratītyabhavo bhavati hi tasyāsvabhāvatvam*)' (VV, xxii).

Therefore, *śūnyatā* is the nature of things *that consists in being devoid of (independent) nature*. It self-denies itself and the other at the same time: there is no other (since otherness, like difference, is dependent on a self-existence that is rejected) and no self (since selfhood needs to apply to an entity). It is the principle of dissolution of the ontological and cognitive illusion of an essence that was always already non-existent, and that held within itself the germ of its dissolution, in a non-place within the non-existent structure of its non-essence. In other words, *śūnyatā* is *śūnya*: the seemingly paradoxical statement of the emptiness of emptiness. To be consistent with its own claims, *śūnyatā* must negate itself, it must not present itself as an ultimate principle, nor as anything close to a self-subsisting reality; it must, therefore, denounce its own emptiness: neither a nature, nor an attribute, since all attributes of things are non-existent, not finding any self-subsistent substratum to hold on to, anywhere. Emptiness must be empty.

But this should not be mistaken with nihilism:[19] *śūnyatā* is not *nothingness*, but *emptiness*. It signals the absence of something, not the absolute[20] presence of nothing. Of course, from the ultimate standpoint, there could not be anything but void space, since nothing exists in the first place – but this

19 That of nihilism was one of the major accusations, coming from other contemporary Buddhist schools, that Nāgārjuna took pains to refute. Nonetheless, nihilistic interpretations of Nāgārjuna's philosophy have been often proposed in the history of Mādhyamika scholarship: see in particular Wood, 1994 and Burton, 1999. Moreover, the label of nihilism has been used and abused since the very first encounters the West had with the philosophy of Buddhism. See Droit, 2003.

20 In the early decades of Mādhyamika scholarship, an absolutist interpretation of *śūnyatā*, based on a Kantian reading of Nāgārjuna, was often adopted. See, in particular, Murti, 1987 and Stcherbatsky, 1989. For a critique of the Kantian interpretation see Della Santina, 1986.

constitutes no logical problem. Consider the statement 'My room is empty of winged cats': what I am predicating here is an absence. But – as far as I am aware – no winged cats can be found anywhere. The statement is not quite the same as 'In my room there is nothing', or even worse 'In my room there is nothingness'. Nihilism can be even more successfully challenged by referring to the logical tool of the two truths: if *śūnyatā* is the a-metaphysical pivot of Nāgārjuna's philosophy, the doctrine of the two truths is certainly the main a-logical one. Just as he warned about the centrality of the understanding of emptiness, so he does for the two truths:

> Those who don't understand
> The distinction between these two truths Do not understand
> The Buddha's profound truth. (MMK, xxiv.298)

The fact that *ultimately* all is *śūnya* does not correspond to a nihilistic outlook on the phenomenal world since, *conventionally*, things *do* exist. In order to understand this double level of truth, can we establish a logical priority between *śūnyatā* and *pratītyasamutpāda*? Herein lies the core of Nāgārjuna's thought. What he often seems to claim is that, being co-dependently arisen, hence caused by something else (relationally existent), things are therefore empty. And this is indeed a common explanation of *śūnyatā*. But this is a shortcut that betrays Nāgārjuna's real message. It is *because* things lack a stable, recognizable, fixed self-existence (they are empty of it) that they can conventionally be seen as arising and ceasing. *Śūnyatā* is always before: before arising, before movement, before time, before any kind of conventional designation, definition or signification. It gives conventional rise to conventional entities that enjoy a conventional existence, only to – simultaneously – deny them. Hence, *śūnyatā* always comes *after*, it is that which *remains* after we simplify and cancel out all the rest. Commenting on MMK, xxiv.18, Garfield neatly unpacks this paradoxical process:

> Nāgārjuna emphasizes here the double edge of the ontology of emptiness. Even though it is in virtue of the fact that conventional entities are constantly arising and ceasing that they are empty, their emptiness entails that they do not, from the ultimate standpoint, arise, cease or abide at all. This is an eloquent statement of the interpenetration of the ultimate and the conventional truths: The very ground on the basis of which emptiness is asserted is denied reality through the understanding of emptiness itself. (Garfield, 1995, p. 264)

*Śūnyatā*, to extend Garfield's metaphor, is indeed like a double-edged sword, *but a sword that cuts itself.* It defies the basic logical rule of a referent and a signifier, being both – or better, *neither* – at one and the same time. This is the meaning of the two truths: in the conventional field, the movement of signification – albeit frail and diaphanous – still *works*; but it is ultimately revealed to be an impossibility, because there is nothing to be signified, nor any actor capable of signifying. The articulation and conceptual mapping of reality structured into binaries – ultimate/conventional, *nirvāna/samsāra*, subject/object, and so on – is the one obstacle to be overcome by the understanding of *śūnyatā*.[21] Yet this does not mean favoring the ultimate term over the other, because this would lead into the paradoxical *ontological* primacy of *śūnyatā*, and the whole system would indeed be exposed to the accusation of nihilism; to deny the existence of an entity does not imply the affirmation of its contrary. The two truths must walk in lockstep, if in a peculiar way. The conventional world is like a phantom limb: it is absent but it feels like it is there; to perceive its conventional 'presence' correctly is to acknowledge its absence,[22] its ultimate truth, and its absence relies on the illusion of presence in an endless repetition of a swapping of roles.

In a fertile interpretation, May too comments on MMK, xxiv.18, but here he translates *upādāya prajñapti* as 'metaphorical designation',[23] so that in his view:

> [t]he term upādāya prajñapti stresses the close dependence of any signifier upon the signified: the mode of existence of signifiers, especially verbal ones, has always been felt by the Buddhists as a particularly striking example of dependent, non-absolute existence. Metaphorical designations are most strikingly empty: the very nature of metaphorical designation is emptiness. From a more metaphysical point of view, we can also say that the whole world is a metaphorical designation. It designates something; it hints at something. And at what? At its own emptiness.

21 This is indeed the core of Nāgārjuna's soteriological message. Streng (1973, p. 35) correctly notes: 'If one assumes that each opposite term refers to a different eternal quality or essence, and then desires one and hates the other, he fails to perceive that this is an empty, relative distinction.'

22 The metaphor of the phantom limb appears extremely suggestive in this context if we consider that – just like the conventional world – the main feedback that a phantom limb gives to the patient under this neurological illusion is one of pain, of suffering. See Ramachandran and Hirstein, 1998.

23 For a thorough critique of nominalist and conventionalist interpretations of MMK xxiv.18 see Berger, 2010. For a survey of other possible metaphorical meanings of *śūnyatā* see Cooper, 2002.

> He who knows how to interpret empirical existence correctly, sees everywhere its emptiness thoroughly. (May, 1978, p. 241)

The whole of reality is a conventional signifier, a signifier that signifies emptiness, being *already* in itself the same emptiness signified. The process of signification comes to a grinding halt when it realizes that it never even started. What makes the metaphorical designation possible is the one shared 'lack' that is possessed by all phenomena: *śūnyatā*. If it is true that '[u]nderstanding a metaphor consists in shuttling conceptually between two things and situations' (Steenburgh, 1965, p. 685), then the Mādhyamika philosopher will understand the metaphor of *śūnyatā* by constantly shifting between a conventional and an ultimate world.[24] And since there is nothing that 'is' not *śūnya*, and therefore nothing that 'is' not a metaphorical designation of itself, *any* attempt at signification/being will necessarily be constrained by ontological precariousness.

What kind of language can we employ in order to talk about emptiness? How can our language, based on and enclosed in basic principles of signification, say anything meaning*ful* about *emptiness*? In fact, how can it say anything meaningful *at all*? The problem of language and of meaningful philosophical statements is of course a major problem for the Mādhyamika school, a problem regarding which scholars have been debating from the Indo-Tibetan tradition of commentaries on Nāgārjuna's works (taking shape in the opposition between the *prāsaṅgika* and the *svatantrika* schools)[25] to contemporary scholarship on Mādhyamika philosophy. This was partly possible due to different interpretations of some passages of Nāgārjuna's works, such as:

> The victorious one has said
> That emptiness is the relinquishing of all views.
> For whomever emptiness is a view,
> That one has accomplished nothing. (MMK, xiii.8)
>
> I prostrate to Gautama
> Who through compassion
> Taught the true doctrine,
> Of the relinquishing of all views. (MMK, xxvii.30)

24 For an examination of the cognitive role of conventional and ultimate truth in post-Nāgārjunian Mādhyamika see Garfield, 2010a.

25 For a study on the validity of the distinction between *prasaṅgika* and the *svatantrika* see Dreyfus and McClintock, 2003.

Between these two verses the main point of contention has been concerned with whether 'all views' should be interpreted as 'all *wrong* views' or as 'all views altogether'. Given that the former interpretation coincides more exactly with the logic of emptiness, it appears to be a more compelling choice. The claim that 'emptiness is the relinquishing of all views' does not lead toward the absolute silence of an intellect mesmerized by the supreme comprehension of absolute emptiness, it isn't the nihilistic claim of a 'religion of abandonment' – the two fallacious extremes, essentialism and nihilism, that the Middle Way intends to avoid – nor again is it a skeptical claim about the non-knowability of some reality out there. On the contrary, it is an acknowledgment of the failure of the power of signification.[26] Views, which are linguistically and conceptually formulated, must necessarily refer back to something, have a ground, a starting point, a conceptual substance. Once this origin is eroded inside out by its own *śūnyatā*, nothing else stands,[27] not even *śūnyatā*, being itself *śūnya*. The rules of language are broken, or bent, to accept a discourse that is always metaphorical, groundless, fictional,[28] and hence open to an endless play of signification, arising from *śūnyatā* and shouting (or whispering) back its fall back into *śūnyatā*.

The impossibility of claims makes the problem of a *true* claim a trivial one, and in fact it brings into question the concept of truth altogether. No correspondence theory is possible: the truth of *śūnyatā* is its constant disappearance. Ontology and logic are forced to break their self-preserving partnership: the one hollowed, only provisionally *presenting* a multiplicity of conventional beings; the other dissolved, dissipated.

---

26 With this claim I seem to adopt what Huntington (1989, 30) defines as the 'linguistic interpretation' of Mādhyamika. Siderits (1988, p. 2003) shares a similar interpretation, which he defines 'semantic interpretation' as opposed to a 'metaphysical' one. For a possible application of the linguistic analysis to Tibetan Buddhism see Napper, 2003. For an explicit Wittgensteinian interpretation of Mādhyamika's use of language see Gudmunsen, 1977. For a critique of the possibility of a linguistic interpretation see Williams, 1991.

27 See MMK, iv.8: 'When an analysis is made through emptiness,
If someone were to offer a reply,
That reply will fail, since it will presuppose,
Exactly what is to be proven.'

28 I borrow Crittenden's (1981, pp. 326–327) term: 'There is nothing in the nature of fictional entities which determines which is the right logic of fiction or what fictional reality really is. Clearly there is no "essential reality" (*svabhāva*) reflected in fiction and there ought not to be any taken as reflected in literal discourse either; describing reality as fictional can be taken as calling attention to the arbitrariness of the rules of ordinary factual language.'

The paradox of a conventional/ultimate designation and an expressibility of this reality evacuated of meaning/essence and stability/safeness is exquisitely outlined by Garfield and Priest:

> We can think (and characterize) reality only subject to language, which is conventional, so the ontology of that reality is all conventional. It follows that the conventional objects of reality do not ultimately (non-conventionally) exist. It also follows that nothing we say of them is ultimately true. That is, all things are empty of ultimate existence, and this is their ultimate nature and is an ultimate truth about them. They hence cannot be thought to have that nature, nor can we say that they do. But we have just done so. (Garfield and Priest, 2003, p. 13)

Our language would then need a new form, new rules, and a grammar, wherein any existential statement – employing the verb 'to be' – must be cancelled out, but provisionally kept: 'there ~~is~~ a world', 'object *x* ~~exists~~'. Since *śūnyatā* ~~is~~ the 'nature' (and crossing out the verb here indicates consistency with the *śūnyatā* of *śūnyatā* itself) of all things, 'to be', for any-thing, is to ~~be~~ *śūnya*. 'Void' is the 'proper name' of being.

I therefore agree with Loy (1984) when he emphasizes the central thrust of Mādhyamika as being a reflection on non-duality, and who warns against reading Nāgārjuna only through a sterile 'neonominalistic' standpoint. Of course, Nāgārjuna's *śūnyatā* is not merely an aseptic tool to obtain the relinquishment of all views; his whole project is an antimetaphysical enterprise aimed toward a soteriological goal.[29] This means an informed return to conventional reality, one free from reification, subject/object distinction, and from any gesture that secretly hides cryptometaphysical presuppositions. Thus, as Garfield argues: 'Nāgārjuna replaces the view shared by the metaphysician and the person in the street, a view that presents itself as common sense, but is in fact deeply metaphysical, with an apparently paradoxical, thoroughly empty, but in the end commonsense view ... of the entire phenomenal world' (1995, pp. 122–123).

The goal of *śūnyatā* is thus freedom from (reificationist, totalizing) metaphysics and from a strictly referential theory of meaning,[30] toward the everyday reality of a phenomenal world and of the language used to describe it, that – being *śūnya* – results in having a logically indeterminate character. The

29 Siderits examines the possible connection between the 'semantic interpretation' of emptiness and the practical, everyday. soteriological goals of Mādhyamika by pointing out how *śūnyatā* frees us from a 'grand narrative' of the self and of truth.

30 See Matilal, 1973, p. 59.

multiplicity of the phenomenal world is grounded on (or better a presentation of) the emptiness at its core.

McCagney's attempt to reinterpret *śūnyatā* as 'openness', as a boundless and indeterminate arena of ever-changing events is helpful in this respect:

> The term śūnyatā functions by pointing to the incoherence of assuming that events are determinate or definable. If events were inherently one thing or another and so could be fixed in a term, they would also be unchanging and ordinary experience as well as the Middle Way and the Eightfold Path would be impossible. To assign a determinate, fixed meaning to śūnyatā utterly misses the point … Śūnyatā is like space [ākāśa],[31] there is nothing to cling to, nothing to grasp. (McCagney, 1997, pp. 95–101)

This metaphor of openness paints a fertile image of a space to be filled, an absence as a condition for any presence, a playground for any signification: freedom and 'eventual' novelty is possible only in indeterminacy. However, *śūnyatā* cannot be a play*ground*. It can be the space of signification, but not the *origin* of signification, nor the *condition* for any phenomenal entity. *Śūnyatā* is not a super-signifier, nor the sign of all signs, but (and we already gesture toward zero) we could say that it works as a meta-sign declaring the emptiness of other signs/phenomena that are always already empty. Compare the definition of zero as a meta-sign with this statement by Robinson about *śūnyatā*: 'Emptiness is not a term in the primary system referring to the world, but a term in the descriptive system (metasystem) referring to the primary system. Thus it has no status as an entity, nor as the property of an existent or an inexistent' (1967, p. 43).

On the other hand, the impossibility of thought is necessary for getting rid of any metaphysical imprint, one that might force us always to think about *something*. In this way, *śūnyatā* becomes as impossible a concept to grasp as zero and infinity are. An empty world is a state of affairs that holds as long as we do not look for an end, for a limit, because this will always elude our conceptual grasp. In the same way, it is impossible to 'think nothing' as a beginning.

This relinquishment of concepts and views does *not* mean (once again) that the emptiness of reality cannot be affirmed, accepted, and handled, or that a

31 McCagney justifies her interpretation claiming that 'Nāgārjuna has adopted this sense of ākāśa as the vast, luminous and open sky' and that he 'has adopted the more encompassing sense of śūnyatā as openness used in the Aṣṭasāhasrikā Prajñāpāramitā in which it is synonymous with space' (1997, xx; 58). The reader will recall that ākāśa was another word commonly used for zero.

nihilistic stance should be adopted. Just as we can still count even when knowing that doing so is an infinite process of enumerating empty signs, so we can talk or live with an empty language in an empty world. This 'worldview'[32] will therefore keep this contradiction within itself, being a faint *remainder* of a collapse of dualities.

## 5 Tracing Voids

It is time to pick up both the threads of the discussion so far and try to merge them into one coherent line of thought. Williams makes an excellent point when he observes: 'the appropriation of Buddhist thought might not involve only relating it to Western philosophy ... When expressed in its broadest sense, [one] not culturally determined, the principal message of Mādhyamika is to "let go of holding". To adopt Mādhyamika for the West is not the same as expressing it in terms of contemporary Western philosophy' (1991, pp. 194–195).

Even though Williams refers to a more 'religious' form of adoption (he subsequently uses the example of the penetration of Buddhism into China), this observation can usher in the crucial question: if on the one hand an objective and 'authentic' retrieval of Nāgārjuna's own thought is impossible and, on the other, the very nature of Mādhyamika is to resist any kind of assimilation, coalescence, or appropriation by other 'views', what can we say about it?

This question can be tackled through an examination of an influential contemporary trend in Mādhyamika scholarship: the deconstructivist approach,[33] a broad label that expresses any kind of interpretative move that generally follows Jacques Derrida's philosophical project. Even if this kind of interpretation does not remain immune from grandiose and frankly untenable claims – 'without Derrida it is difficult for a "moderner" to understand Nāgārjuna!' (Magliola, 1984, p. 93) – the choice of a deconstructive approach can address the *impasse* that I described. Derrida states, for example: 'Deconstruction is neither a theory nor a philosophy. It is neither a school nor a method. It is not even a discourse, nor an act, nor a practice. It is what happens, what happens today' (Derrida in Mabbett, 1995, p. 207).

32 Kakol has commented on the Mādhyamika worldview defining it as an 'open worldview' or a 'cosmological worldview': 'Closed views are complete but inconsistent whereas open views are consistent but incomplete ... Open views are either positive process-like views (creative synthesis or inclusive transcendence) or negative Mādhyamika-like "views" (negative dialectics or athesis)' (2002, p. 216).

33 This label can be reasonably applied to scholars like Magliola, Loy, Mabbett, Le Roux, Wang, and – more tenuously – Huntington.

We can see a parallel in *intention* between Derrida and Nāgārjuna,[34] since both Nāgārjuna's and the deconstructivist project, present themselves as views on reality that are not intended as new and systematic worldviews. Therefore, to read Nāgārjuna deconstructively means to collate certain terms that, in both cases, are meant to be used in a non-systematic way as a means of evaluating everyday reality. Here I will not be trying a precise point-to-point comparison. In conformity with the concerns I outlined at the beginning of the chapter I will limit myself to appropriating some Derridean terminology as a tool for expressing the connections between zero and *śūnyatā*. This does not necessarily imply that I am drawing a doctrinal parallel between Derrida and Nāgārjuna, for in fact neither of them proposed any doctrine, but rather between methods. Indeed, one point of similarity between the two philosophers is their forceful denial that they are creators of a 'new position' and further, they appear to have a shared interest precisely in the lack of *fixed* positions. Therefore, it is not a matter of 'comparing doctrines' but a question of the *methodology of deconstruction* of doctrines. In fact, more than any doctrinal convergences, what really allows us to align the two is their role as internal dissidents within the well-established metaphysical traditions that produced them. What Vattimo observed of Derrida could be applied to Nāgārjuna as well: 'If a real leap out of metaphysics it is not possible … the thought that feels itself summoned to perform such a duty has to configure itself as necessarily "parasitic" toward the tradition from which it is trying to set itself free' (2002, xiii; my translation).

One of the main Derridean concepts I will employ here is that of *trace*, best introduced in Derrida's own words:

> The trace is not only the disappearance of origin – within the discourse that we sustain and according to the path that we follow it means that the origin did not even disappear, that it was never constituted except reciprocally by a nonorigin, the trace, which thus becomes the origin of the origin. From then on, to wrench the concept of the trace from the classical scheme, which would derive it from a presence or from an originary nontrace and which would make of it an empirical mark, one must indeed speak of an originary trace or arche-trace. Yet we know that that concept destroys its name and that, if all begins with the trace, there is above all no originary trace. (Derrida, 1997, p. 61)

34 For an exhaustive enumeration of points of contact between Nāgārjuna's and Derrida's works see Mabbett, 1995. For a critique of Derrida from a Nāgārjunian standpoint see Magliola, 1984 and Loy 1987, 1993.

Derrida sketched this idea within the framework of his critique of the metaphysics of presence – a nostalgia for origins, a rooting in a unitary spring of being, a call for bright sameness, and above all of Being as presence – that is the *leitmotif* of European philosophy. Against and within this all-inclusive project, Derrida claimed, the *trace* can always be found, taking shape in any of the many external, exiled concepts that Western metaphysics has kept at bay, quarantined at the margins of its self-sustaining presence. If in the linguistic analysis operated over and against the Saussurean structuralist project or in the deconstruction of the classical texts of Western philosophy, the trace always makes its appearance in Derrida's works, often defined as the non-concrete, absent presence that threatens projects of unification; it is the locus of difference, the non-entity defying the matrix of all the binary oppositions that structure metaphysics: the opposition inside/outside. While feared as an internal stranger or underplayed as an unnecessary supplement, the *aporetic* trace is, according to Derrida, the 'original' opening of meaning, the infinite seed of an endless proliferation. Yet, the presence of the trace in the past of an origin was never a present presence. The trace is that which originates without carving its presence on the static surface of being; it is the mere and scandalous lack of being, the presence of an absence. It is impossible to trace the trace. As a matter of fact, as Derrida notes: '*The trace itself does not exist.* (To exist is to be, to be an entity, a being present, *to on*) ... Although it *does not exist*, although it is never a *being-present* outside of all plenitude, its possibility is by rights anterior to all that one calls sign' (1997, pp. 167, 62); and again: 'The trace *is nothing*, it is not an entity, it exceeds the question What is? And contingently makes it possible' (Derrida 1997, p. 75). The trace is the non-present possibility of linguistic signification (including mathematical, in the form of zero) and of every order of being.

Indeed, the structure of the trace is not to be limited, as poor readings of Derrida suggest, to textuality. As the work of Martin Hägglund (2008) has convincingly shown, the structure of the trace is a feature of reality as a whole, at work in the animate as well as in the inanimate; Derrida is offering an insight into the 'autoimmune' nature of the real, not merely providing a methodology for the deconstruction of philosophical texts. For Derrida, Hägglund clarifies, the trace is an 'ultratranscendental condition' and '[e]verything that is subjected to succession is subjected to the trace, whether it is alive or not' (Hägglund 2009, pp. 239–240): all that which is originated is always already undermined by the trace/emptiness as its own way of being.

Derrida's insistence on the ubiquity of this structure of self-voiding is what led Alain Badiou (2009a, p. 546) to describe his philosophy as motivated by a 'passion of inexistence'. Badiou offers a reading of Derrida, obviously influenced by his own theoretical preferences (and, interestingly in the context of my exposition, by his *mathematical* ontology), which effectively highlights how it is possible to employ Derrida's conceptual resources to understand emptiness as the zeroing of being. According to Badiou (2009b, p. 140) Derrida's desire is to 'locate, touch, clasp, even for less than an instant, the non-existent of a place, the vanishing of a vanishing point. Inscribe this ex-scription'. For Badiou, the absent presence of the inexistent is the claim that Derrida sets against the history of metaphysics, since:

> The metaphysical error par excellence is to have identified the non-existent with nothingness. Because the point is that the non-existent is … The non-existent is nothing. But being nothing is by no means the same as non-being. To be nothing is to non-exist in a way specific to a determinate world or place … [N]o stable opposition can really succeed in describing the precise status of the non-existent in terms of a binary opposition. (Badiou 2009b, pp. 140–141)

This inexistent, this pre-originary *différence*, this trace, is therefore what Derrida's deconstructive process attempts to 'performatively' say (for the inexistent is always below the threshold of representation), just as Badiou's mathematical ontology identifies the 'empty set' – the void always presupposed by a presented situation – with proper name of 'being'.

In its singular role as self-voiding parasite (or organizing structure, if 'conventionally' seen) of reality, the trace is a vanishing ground always to be replaced and re-said in a play of difference. It is, however, no replacement for a metaphysical principle: the trace does not *posit*. Precisely responding to a reading of his work seen as simply identifying a new set of transcendental conditions, and indeed of creating a new onto-theology 'grounded' on the trace Derrida rhetorically asked: 'Have I not indefatigably repeated – and I would dare say demonstrated – that the trace is neither a ground nor a foundation, nor an origin, and that in no case can it provide for a manifest or disguised onto-theology?' (Derrida, 1981, p. 52).

If thus correctly interpreted as the 'ultratranscendental' structuring/voiding principle of reality (a dual role that I see as mirroring the double register of a conventional/presented and of an ultimate/beyond representation level of being), the trace – I believe – is the conceptual link that allows us to connect zero and *śūnyatā*, or better to create a mirror of their own elusive nature.

We have seen how zero, as a fully fledged number on the number line, holds a particular role: it is the origin of the series of numbers and it is the opening of the possibility of denumeration. At the same time it is excluded from the possibility of full signification, kept aside, to be handled with care as the number that can implode the system of mathematics. Consider again the example of the basic arithmetical operations with zero: zero is the elusive origin and its clandestine nature within the number system is signaled when a number tries to interact with it. If in subtraction and addition zero can be considered merely a neutral element, ignored by the integrity of the number (just an empty supplement that can be ignored, as in 05), when a number needs to be multiplied – that is *repeated* zero times – it is annihilated by zero, as if the emptiness of its 'presence' were finally exposed as unable to subsist without differential repetition. Finally, to try and divide any number by zero leads to an impossible result, an undecided result. An answer within the system cannot be found. Not only does zero erase; it completely deconstructs. As Seife (2000, p. 23) remarks: 'Multiplying by zero collapses the number line. But dividing by zero destroys the entire framework of mathematics.'

When considering its double role as a placeholder and as a mark of emptiness, we can legitimately call zero a *supplement*, another term derived from Derridean/Badiouian vocabulary, if we consider that Derrida claims that:

> the supplement supplements. It adds only to replace. It intervenes or insinuates itself in-the-place-of; if it fills, it is as if one fills a void. If it represents or makes an image, it is by the anterior default of a presence. Compensatory and vicarious, the supplement is an adjunct, a subaltern instance that takes-(the)-place. As substitute, it is not simply added to the positivity of a presence, it produces no relief, its place is assigned in the structure by the mark of an emptiness. (Derrida, 1997, p. 145)

Zero would thus be the supplement within the system of mathematics, a system that, through the infinity[35] of numbers extending along the number line, merely presents a presence of being, while grounded on the void – the counting of no-thing – that zero represents.

To link together *śūnyatā* and zero we must consider the historical route of zero. It can be argued that the Indian mathematicians who developed the concept of zero did so at least three centuries after the development of the Mādhyamika school, and a millennium after the birth of the Buddha. It seems reasonable to expect high-caste individuals such as Brahmagupta and Bhaskara to be aware – even if only tangentially – of the intellectual trends and debates that were running through the subcontinent, particularly during the first five

35 See Derrida's (1997, p. 71) remark on infinitist metaphysics: 'Only infinite being can reduce difference in presence.'

centuries CE that saw the rise and expansion of Mahāyāna Buddhism. The scholastic conflicts between Buddhism and other realist Indian philosophical schools (and within Buddhism itself) must certainly have been known by the Indian educated elites, and the emphasis Buddhism placed on emptiness[36] must certainly have made its way into their intellectual milieu. The emancipation of zero in Indian mathematical thinking was possible thanks to the presence of an *existing* concept of emptiness within its cultural ground.[37]

The subsequent export of zero to the West was not simply the assimilation of an alien concept, but a difficult encounter with an external enemy that, however, was already the internal enemy of Western metaphysics, at least since Parmenides: the trace of absence. If we accept this train of thought, the connection between critiques of metaphysics of presence and Nāgārjuna's project stands out clearly: if zero, a concept developed by the mathematics of Nāgārjuna's cultural world, was only slowly, and then grudgingly assimilated by the West it is because Europeans saw it as symbol of the dissolution of the mind's very integrity.[38] And Europeans were indeed correct to fear the corrosive power of zero, since the full philosophical implication of describing something as *śūnya* implies precisely the demolition of any essentialist, reificationistic, and eternalistic ontology, or indeed onto-theology. Hence, Derrida's 'dangerous supplement' and Nāgārjuna's main concept of *śūnyatā* can be seen to function similarly conceptually, and an analysis of zero – the supplement within the mathematical system – demonstrates it, since this supplementary role is due to its hinting back to a transcendental *lack*, a peculiar character that has its historical roots in an Indian-derived thought of emptiness. To understand the real *danger* posed by such a supplement we can turn again to Derrida:

36 I am thinking here specifically of its development from the reductionism of early Buddhism and the dharmas' theory of the Abhidharmists to the radical emptiness presented in the Prajñāpāramitā *sutras*, philosophically perfectioned by Nāgārjuna.

37 What I am claiming is that – differently from other cultures – the idea of emptiness (well distinct from nihilist nothingness) had in India the legitimate status of a respectable philosophical concept. Others dare to take this further than me: '… it is quite probable that Bhāksara II's concept of *śūnya* was built up against the background of Nāgārjuna's philosophy' (Mukhopadhyaya, 2003, p. 202) and 'Although it is asserted that the discovery of zero in mathematics in India came late, probably in the fifth to sixth century AD … there are many elements of mathematical logic in Buddhist logic that antedate the above discovery by several centuries at least' (Matsuo, 1987, 34).

38 One can think, for example, of the difficulties that both Newton and Leibniz encountered, as late as the seventeenth century (several centuries after the Hindu numerals with zero were adopted in Europe) with mathematical nothingness, leading to the 'creation' of infinitesimals.

> Why is the supplement or surrogate dangerous? It is not, so to speak, dangerous in itself, in that aspect of it that can present itself as a thing, as a being-present. In that case it would be reassuring. But here, the supplement is not, is not a being (on). It is nevertheless not a simple nonbeing (mēon), either. Its slidings slip it out of the simple alternative presence/absence. That is the danger. (Derrida, 2004b, p. 112)

What was a danger to be avoided for Greek metaphysics was the state of affairs of reality as a whole for Nāgārjuna. Because *śūnya* is situated between the oppositions, between presence and absence, between being and nonbeing, Nāgārjuna placed *śūnyatā* between 'present/being' on one side and 'absence/nonbeing' on the other. In the metaphysics of presence, this becomes a hybrid, half-caste, undecidable place of void: in other words, it is unrecognizable as a term of the situation. This 'being between' however, is not *merely* an equilibrium between extremes but a complete emancipation from any polarity, an authentic *third* alternative. On a conventional level, the Middle Way is a path between extremes (recognizing but defying the law of the excluded middle) not veering in either direction (thus conventionally recognizing them). On the ultimate level the Middle Way is a complete erasure of oppositions on the ontological register (hence completely deconstructing the structure of the excluded middle), and of discriminative knowledge on the epistemological one.[39] This happens because no fixed totality (as either a self-essence or a differential principle) can be named: the logical methods employed must be consistent with a 'reality', the law of which is not the permanence and self-presence of being, but *pratītyasamutpāda*, dependent co-arising. Faure summarized this idea thus:

> Despite appearances to the contrary, Nāgārjunian logic is not the same as ours, in particular in the following respect: Instead of fixing the terms of the interrelatedness, it draws attention to their constant fluctuation.

39 What in English is commonly called the 'law of the Excluded Middle' (P v ~P) was in Latin known as the *principium tertii exclusi* (principle of the excluded third), or *tertium non datur* (the third is not given). Now, in our context, the two versions are not quite the same. We could therefore say that Nāgārjuna does refer to the law of the excluded middle when talking conventionally about his doctrine. But ultimately he rejects the principle of the *tertium non datur*. Not only does Nāgārjuna want to find a middle way between oppositions, but he also ultimately wants to destroy (deconstruct) the members of the opposition itself, finding a non-position outside the opposition, not merely in the middle of it. When we break the 1–2 binary opposition any other third option is really a multiplicity of options.

> Whereas in a tautology or an algebraic equation, the terms must turn out to be equal, in Buddhist logic, no terms ever do. The fundamental difference (always supposing that there is a 'foundation') is that Buddhism by and large values orthopraxy (correct practice) more highly than orthodoxy (correct opinion) and, in particular, ritual more highly than doctrine. (Faure, 2004, p. 96)

To value orthopraxy amounts to employing a logic that observes the everyday flow of *pratītyasamutpāda*, of dependent entities lacking any self-nature. Hence the formal, binary structure of the *alter* is broken, introducing an ever-irreducible *aliud*, whose difference is not to be found in the not-sameness of any self-existing nature of the entities under consideration, but in the universal lack of self-sameness (*śūnyatā*) that characterizes everything in the first place. If self-individuation is impossible, difference is impossible as well,[40] except if we interpret difference outside dichotomies (where 'different from' would be the basis of any proposition) as *absolute* difference.[41]

When the problem of essence and existence is at stake, when the 'question of being' is asked, the nature of the phenomena is best described by the fourth leg of the tetralemma (the most absurd and metaphysically meaningless one): beings are *śūnya*, *neither* existent *nor* non-existent.[42]

The anti-ontological (and deconstructively argumentative) spirit of Mādhyamika is non-committal[43] but, as I mentioned before, the negation of thesis A does *not* imply the affirmation of thesis B (what the Indian philosophical tradition refers to as a *prasajya* negation). In other words, we might say that the radical shift performed by the philosophy of emptiness is that from

40 The long passage (MMK, xiv.5–7) that Nāgārjuna dedicates to the problem of difference provides a wonderful example of his dialectical method and of what I will soon define as the logic of the neither/nor applied to ontological problems. Consider, as a way of comparison, Frege's statement about zero (quoted in Rotman, 1987, p. 7): 'Since nothing falls under the concept "not identical with itself", I define nought as follows: 0 is the number that belongs to the concept "not identical with itself". Zero in a logical system covers the role of an impossible nonidentity with itself, which is to say "empty of self-nature".'

41 One of Nāgārjuna's statements (MMK, vi.4) on the problem of identity seems to play on the same note of the double semantic role that Derrida gives to his word différence (to differ and to defer):
'In identity there is no simultaneity. | A thing is not simultaneous with itself. | But if there is difference, |How would there be simultaneity?'

42 I shall quickly observe how this neither/nor way of proceeding is not equivalent to the '*neti, neti*' mantra of Advaita philosophers: the problem here is not one of progressive negation of all possible positive attributes (apophatic theology) but a radical deconstruction of any possible predication, both positive and negative.

43 As Nayak (1979, p. 479) puts it: 'Nāgārjuna's critical insight is a consistent denial of all "-isms" in philosophy.'

the logical-metaphysical structure of the reified opposition 'either/or' to the empty opposition of 'neither/nor'. This is the logic of the undecidable, which does not simply stand between the poles of an opposition but rather defies its consistency, being always double-faced, unstable, unrepresentable in the closed totality of a situation. This undecidable[44] placement is what associates zero – a trace that is both the condition of possibility and of impossibility of the number line (and the numbers' power of signification) – and *śūnyatā*, the fundamental nature of things consisting in their not having any nature. These two concepts share one common focus or center of gravity – an empty one – that is never stable, but that, in their endless shifting, make the *play*[45] between them possible.

In order to clarify my discourse and to delineate more precisely the convergence of *śūnyatā* and zero – and their eluding dichotomizing thought – one more concept can be expedient: that of the *hole*.[46] Moving on three levels (metaphysical, mathematical, and linguistic) we could compare 1) the *metaphysical* juxtaposition of 'emptiness' (lack of anything at all) with 'hole' (nothing amid something) with 2) a *mathematical* one between zero as a numeral and zero as a placeholder and 3) the *linguistic* role played, in a structuralist analysis of linguistic systems, by the Derridean 'undecidable' or by a mere blank space between linguistic signs.[47] In a table:

| **Emptiness** | **Hole** |
|---|---|
| Zero as a numeral | Zero as a placeholder |
| Undecidable | Blank space |

44 Alain Badiou's own employment of the term 'undecidable' is equally applicable in this case: for Badiou an event takes place in the void of a situation precisely because it occupies an undecidable site from the standpoint of the situation itself. See Badiou, 2006: p. 181.

45 We might recall here how Derrida (2004b, p. 352) defined centered self-presence as excluding any possibility of play.

46 The discussion that we find in analytic philosophy around the metaphysical status of holes is a most stimulating one, and relevant to the central theme of this chapter. For a complete discussion about holes see Casati and Varzi, 1994; and Lewis, 1983. For an analysis of the perceptual status of holes, see Bertamini and Croucher, 2003.

47 Indeed, hole is another term employed by Derrida to indicate that trace at the center of every structure: 'If the center is indeed "the displacing of the question" it is because the unnamable bottomless well whose sign the center was, has always been surnamed; the center as the sign of a hole that the book attempted to fill. The center was the name of a hole' (Derrida, 2004a, p. 375).

Nāgārjunian *śūnyatā* represents a 'concept' that escapes the binary logic of being/nonbeing; the number zero shares this property of being a neither/nor hybrid (neither positive nor negative) and; finally, in linguistic systems those signs having impossible placement and logically aporetic role are defined 'undecidables'. On the other hand, the concept of hole, the place value zero, and the blank space in a linguistic continuum are the metaphysical, mathematical, and linguistic equivalents of a lack that *necessitates a presence*, a One, in order to be apparent, and hence whose existence is parasitical on the host system. *Śūnyatā* and zero are logical undecidables, and not merely negative counterparts (i.e., instances of nothingness), to be located within systems of binary oppositions.

If zero is a meta-sign for the absence of other signs (that ultimately have no real referent at all) it follows that it is not a symbol of nothingness, but of emptiness. Representing a lack that turns back on itself – the lack of a lack – zero is a metaphorical designation, and it is in itself empty. In the same way, *śūnyatā* underlies a multiplicity of phenomena that never add up to a totality of being, but to an infinity of de-void phenomena.

## 6 Conclusion

I would like to conclude with a summary of what I have – and what I have *not* – proposed thus far, and with a final note.

After an account of the historical development of zero, in the first section of this chapter, and an evaluation of Nāgārjuna's *śūnyatā*, in the second, I have tried to thread together these two 'concepts' with several borrowed from a (Badiouian) reading of Derridean vocabulary: trace, supplement, and undecidable. However, to describe the above as drawing 'parallelisms' (or comparisons) fails to express correctly the relation that binds these concepts: what is at stake in this interpretation is not only a new evaluation of Nāgārjuna's philosophy, but a recognition of its possible role for a contemporary attempt to *redefine* 'comparative' philosophy. To 'actualize' Nāgārjuna does not mean to read into his words our words, but to use his words to enrich the possibility for philosophy as a whole to formulate *new questions*.

What I have in mind here is the necessity of probing how our ontological commitments condition the development of any intellectual product. In other words, a 'theory' of emptiness and co-dependent truth such as Nāgārjuna's, by deconstructing the totality of a plenary being and the stiffness of a uniquely determined meaning, indexes the possibility of a new, unconstrained, and indefinable space of action for thought, beyond already charted territories. This is the liberating power of emptiness.

I have not attempted a consistent and explicit parallelism between the philosophy of Nāgārjuna and the philosophy of Derrida, nor have I tried to imply that the history of mathematics is directly indebted to Nāgārjuna for the invention of zero. The latter is at best unprovable and the former is an enterprise that, while certainly an interesting intellectual exercise that might force us to develop a better understanding of both philosophers, is often spoiled by the tendency of using the one to correct the other. Mine has been an attempt to suggest a new philosophical ground, by weaving together different concepts and heterogeneous traditions of thought and intellectual domains, trying to let them shine light on each other.

## References

Badiou, A. (2006). *Being and Event*. London and New York: Continuum.

Badiou, A. (2009a). *Logics of Worlds*. London and New York: Continuum.

Badiou, A. (2009b). *Pocket Pantheon*. London: Verso Books.

Bag, A. K. and Sarma, S. R., eds. (2003). *The Concept of Śūnya*. New Delhi: IGNCA.

Berger, D. (2010). Acquiring Emptiness: Interpreting Nāgārjuna's MMK 24:18 *Philosophy East and West*, 60(1): pp. 40–64.

Bertamini, M. and Croucher, C. J. (2003). The Shape of Holes. *Cognition* 87(1): pp. 33–54.

Betty, L. S. (1983). Nāgārjuna's Masterpiece: Logical, Mystical, Both, or Neither? *Philosophy East and West*, 33(2): pp. 123–138.

Betty, L. S. (1984). Is Nāgārjuna a Philosopher? Response to Professor Loy. *Philosophy East and West*, 34(4): pp. 447–450.

Bhattacharya, K. (1986). *The Dialectical Method of Nagarjuna. Vigrahavyāvartanī*. Delhi: Motilal Banarsidass.

Burton, D. F. (2002). Emptiness Appraised: A Critical Study of Nagarjuna's Philosophy. Delhi: Motilal Banarsidass.

Casati, R. and Varzi, A. (1994). *Holes and Other Superficialities*. Cambridge (MA): MIT Press.

Cooper, D. (2002). Emptiness: Interpretation and Metaphor. *Contemporary Buddhism*, 3(1): pp. 7–20.

Crittenden, C. (1981). Everyday reality as fiction – A Mādhyamika interpretation. *Journal of Indian Philosophy*, 9(4): pp. 323–333.

Defoort, C. (2001). Is there such a thing as Chinese Philosophy? Arguments of an Implicit Debate. *Philosophy East and West*, 51(3): pp. 393–413.

Defoort, C. (2006). Is 'Chinese Philosophy' a Proper Name? A Response to Rein Raud *Philosophy East and West*, 56(4): pp. 625–660.

De Landa, M. (2006). *A New Philosophy of Society*. London and New York: Continuum.

De Jong, J. (1972). Emptiness. *Journal of Indian Philosophy*, 2(1): pp. 7–15.

Derrida, J. (1977). *Of Grammatology*. Baltimore: Johns Hopkins University Press.

Derrida, J. (1981). *Positions*. Chicago: University of Chicago Press.

Derrida, J. (2002). *La Scrittura e la Differenza*. Torino: Einaudi.

Derrida, J. (2004a). *Writing and Difference*. London: Routledge.

Derrida, J. (2004b). *Dissemination*. London and New York: Continuum.

Dreyfus, G. and McClintock, S., eds. (2003). *The Svātantrika-Prāsaṅgika distinction: what difference does a difference make?* Somerville (MA): Wisdom Publications.

Droit, R. (2003). *The Cult of Nothingness: The Philosophers and the Buddha*. Chapel Hill and London: University of North Carolina Press.

Faure, B. (2003). Double Exposure: Cutting Across Buddhist and Western Discourse. Stanford: Stanford University Press.

Garfield, J. L. (1994). Dependent Arising and the Emptiness of Emptiness: Why Did Nāgārjuna Start with Causation? *Philosophy East and West*, 44(2): pp. 219–250.

Garfield, J. L. (1995). The Fundamental Wisdom of the Middle Way. Nāgārjuna's Mūlamadhyamakakārikā. Oxford and New York: Oxford University Press.

Garfield, J. L. (2001). Nāgārjuna's Theory of Causality: Implications Sacred and Profane'. *Philosophy East and West*, 51(4): pp. 507–524.

Garfield, J. L. (2010a). Taking Conventional Truth Seriously: Authority Regarding Deceptive Reality. *Philosophy East and West*, 60(3): pp. 314–354.

Garfield, J. L. (2010b). Buddhist Studies, Buddhist Practice and the Trope of Authenticity. Retrieved online at http://www.smith.edu/philosophy/authenticity.doc.

Garfield, J. L. and Priest, G. (2003). Nāgārjuna and the Limits of Thought *Philosophy East and West*, 53(1): pp. 1–21.

Gudmunsen, C. (1977). *Wittgenstein and Buddhism*. London and Basingstoke: Palgrave Macmillan.

Gupta, R. C. (2003). Technology of using *Śūnya* in India' in *The Concept of Śūnya*, ed. A. K. Bag and S. R. Sarma, pp. 19–24. New Delhi: IGNCA.

Hägglund, M. (2008). *Radical Atheism. Derrida and the Time of Life*. Stanford: Stanford University Press.

Hägglund, M. (2009). The Challenge of Radical Atheism. A Response. *The New Centennial Review*, 9(1), pp. 227–252.

Hayes, R. 1994. Nāgārjuna's appeal. *Journal of Indian Philosophy*, 22(4): pp. 299–378.

Huntington, C. with Wangchen, N. G. (1989). *The Emptiness of Emptiness: An Introduction to Early Indian Mādhyamika*. Honolulu: University of Hawaii Press.

Ifrah, G. (1987). From One to Zero: A Universal History of Numbers. London: Penguin Books.

Inada, K. K. (1985). Review of *Derrida on the Mend* by R. Magliola. *Buddhist–Christian Studies*, 5: pp. 218–222.

Joseph G. G. (1992). The Crest of the Peacock: Non-European Roots of Mathematics. London: Penguin.

Kakol, P. (2002). A General Theory of Worldviews Based on Mādhyamika and Process Philosophies. *Philosophy East and West*, 52(2): pp. 207–223.

Katz, V. J., ed. (2007). Mathematics of Egypt, Mesopotamia, China, India, and Islam: A Sourcebook. Princeton: Princeton University Press.

Keenan, J. P. (1996). Emptiness as a Paradigm for Understanding World Religions. *Buddhist–Christian Studies*, 16: pp. 57–64.

Lewis, D. (1983). *Philosophical Papers: Volume 1*. London and New York: Oxford University Press.

Loy, D. (1984). How Not to Criticize Nāgārjuna: A Response to L. Stafford Betty. *Philosophy East and West*, 34(4): pp. 437–445.

Loy, D. (1987). The Cloture of Deconstruction: A Mahāyāna Critique of Derrida. *International Philosophical Quarterly*, 27(1): pp. 59–80.

Mabbett, I. W. (1995). Nāgārjuna and Deconstruction. *Philosophy East and West*, 45(2): pp. 203–225.

Magliola, R. (1984). *Derrida on the Mend*. West Lafayette: Purdue University Press.

Maor, E. (1991). To Infinity and Beyond: A Cultural History of the Infinite. Princeton: Princeton University Press.

Matilal, B. K. (1973). A Critique of the Mādhyamika Position. In *The Problem of Two Truths in Buddhism and Vedanta*, ed. by M. Sprung, pp. 54–63. Dordrecht: Reidel.

Matsuo, H. (1987). The Logic of Unity: Discovery of Zero and Emptiness in Prajnaparamita Thought. New York: State University of New York Press.

May, J. (1978). On Mādhyamika philosophy. *Journal of Indian Philosophy*, 6(3), pp. 233–241.

McCagney, N. (1997). *Nāgārjuna and the Philosophy of Openness*. Lanham: Rowman & Littlefield Publishers.

Menninger, K. (1969). Number Words and Number Symbols: Cultural History of Numbers. Cambridge (MA): MIT Press.

Mukhopadhyaya, A. (2003). Impact on Indian Philosophy on Bhāskarīya Concept of Zero. In *The Concept of Śūnya*, ed. A. K. Bag and S. R. Sarma, pp. 197–205. New Delhi: IGNCA.

Mukhopadhyaya, A. (1987). The Central Philosophy of Buddhism: A study of the Mādhyamika system. London: Unwin.

Napper, E. (2005). Dependent Arising and Emptiness: A Tibetan Buddhist Interpretation of Mādhyamika Philosophy. Somerville: Wisdom Publications.

Nayak, G. C. (1979). The Mādhyamika Attack on Essentialism: A Critical Appraisal. *Philosophy East and West*, 29(4): pp. 477–490.

Nieder, A. (2016). Representing Something Out of Nothing: The Dawning of Zero. *Trends in Cognitive Science*, 20(11), pp. 830–842.

Plofker, K. (2007). Mathematics in India. In *Mathematics of Egypt, Mesopotamia, China, India, and Islam: A Sourcebook*, ed. by V. J. Katz, pp. 385–514. Princeton: Princeton University Press.

Quli, N. (2009). Western Self, Asian Other: Modernity, Authenticity, and Nostalgia for 'Tradition' in Buddhist Studies. *Journal of Buddhist Ethics* 16, pp. 1–38.

Raud, R. (2006a). Philosophies versus Philosophy: In Defense of a Flexible Definition. *Philosophy East and West*, 56(4), pp. 618–625.

Raud, R. (2006b). Traditions and Tendencies: A Reply to Carine Defoort. *Philosophy East and West*, 56(4), pp. 661–664.

Ramachandran, V. S. and Hirstein, W. (1998). The perception of phantom limbs: The D. O. Hebb lecture. *Brain*, 121: pp. 1603–1630.

Robinson, R. H. (1957). Some Logical Aspects of Nāgārjuna's System. *Philosophy East and West*, 6(4): pp. 291–308.

Robinson, R. H. (1967). *Early Mādhyamika in India and China*. Madison, Milwaukee and London: University of Wisconsin Press.

Robinson, R. H. (1972). Did Nāgārjuna Really Refute All Philosophical Views? *Philosophy East and West*, 22(3): pp. 325–331.

Rotman, B. (1987). *Signifying Nothing: The Semiotics of Zero*. Houndmills and London: Palgrave Macmillan.

Santina, Della, P. (1986). The Madhyamaka and Modern Western Philosophy. *Philosophy East and West*, 36(1): pp. 41–54.

Seife, C. (2000). *Zero: The Biography of a Dangerous Idea*. London: Souvenir Press Ltd.

Siderits, M. (1988). Nāgārjuna as anti-realist. *Journal of Indian Philosophy*, 16(4): pp. 311–325.

Siderits, M. (2003). On the Soteriological Significance of Emptiness' *Contemporary Buddhism*, 4(1): pp. 9–23.

Sprung, M., ed. (1973). The Problem of Two Truths in Buddhism and Vedanta. Dordrecht: Reidel.

Stcherbatsky, T. (1989). The Conception of Buddhist Nirvana with Sanskrit Text of Madhyamaka-Karika. Delhi: Motilal Banarsidass.

Streng, F. J. (1975). Metaphysics, Negative Dialectic, and the Expression of the Inexpressible. *Philosophy East and West*, 25(4): pp. 429–447.

Tola, F. and Dragoneti, C. 1981. Nāgārjuna's conception of 'voidness' (śūnyatā). *Journal of Indian Philosophy*, 9(3): pp. 273–282.

Tuck, A. P. (1990). Comparative Philosophy and the Philosophy of Scholarship: On the Western Interpretation of Nāgārjuna. New York and Oxford: Oxford University Press.

Vattimo, G. (2002). Derrida e l'Oltrepassamento della Metafisica. In *La Scrittura e la Differenza*, by Jacques Derrida, vii–xxiv. Torino: Einaudi.

Walser, J. (2005). *Nāgārjuna in Context*. New York: Columbia University Press.

Wang, Y. (2001). Buddhism and Deconstruction: Towards a Comparative Semiotics. Richmond: Curzon.

Westerhoff, Jan. (2009). *Nāgārjuna's Madhyamaka*. Oxford: Oxford University Press.

Williams, P. (1991). On the interpretation of Madhyamaka thought. *Journal of Indian Philosophy*, 19(2): pp. 191–218.

Wood, T. E. (1994). Nagarjunian Disputations: A Philosophical Journey Through an Indian Looking-Glass. Honolulu: University of Hawaii Press.

CHAPTER 20

# Indian Origin of Zero

*Ravi Prakash Arya*

## Abstract

Although the presence of zero has been traced to many ancient civilizations including India, the question of its origin is still a moot point among historians and mathematicians. As such, in the present chapter a humble attempt will be made to gather data from the Vedic and post-Vedic Sanskrit literature to understand the origin of zero; the philosophy behind its origin; various denominations of zero used in various phases of Indian civilization; and the philosophy and epistemology behind the symbolic representation of zero (0) in India.

## Keywords

zero – *Śūnya* – *Kha* – *Ākāśa* – *Pūrṇa* – Brahman – *Ambara* – *Chhidra* – *Śunya bindu* – *Al-Biruni* – *Āryabhaṭa* – *Ṛgveda* – *Yajurveda* – *Atharvaveda*

## 1 Introduction

The invention of zero has proved a game changer in the human history of science and technology. No scientific advancement would have been possible without the invention of zero. In fact, its invention revolutionized everything in science and technology. The famous mathematician, G. B. Halsted, and noted French mathematician and astronomer, Pierre Simon De Laplace, during the nineteenth century give credit for the invention of zero to Indians.

G. B. Halsted (1912, p. 20) observes thus:

> The importance of the creation of zero mark can never be exaggerated. This giving to airy nothing, not merely a local habitation and a name, a picture, a symbol, but helpful power, is the characteristic of the Hindu race whence it sprang. It is like coining Nirvana into dynamos. No single mathematical creation has been more potent for the general on-go of the intelligence and power.

 | DOI:10.1163/9789004691568_024

Laplace (Ifrah, 2000) observes:

> The ingenious method of expressing all numbers by means of ten symbols, (each symbol having a value of position as well as an absolute value) emerged in India. The idea seems so simple nowadays that its significance and profound importance is no longer appreciated. Its simplicity lies in the way it facilitated calculations and placed arithmetic foremost among useful inventions. The importance of this invention is more readily appreciated when one considers that it was beyond the two greatest men of antiquity, Archimedes and Apollonius.

The Vedas are the oldest ever literature in the human history of the globe. As per ancient Indian tradition, all branches of knowledge have their roots in the Vedas.

> वेदोऽखिलो धर्ममूलम्।
>
> *vedo'khilodharmamūlam*
>
> [Veda is the source of all knowledge-metaphysical, astrophysical and physical, morality and ethics.]

The entire later Vedic and post-Vedic literature speaks highly of the Vedic seers and the Vedas for their contribution to science and civilization on the globe. As such, when we search for the origin and the development of various sciences in the Indian context from the literary point of view, the Vedas are screened first to find the seed germinated in them. As in the case of other schools of thought and sciences, the science of mathematics too in India goes back to the Vedic period for its origin and development.

Algebra and Arithmetic including Geometry had their origin in India. According to Sarda (2007, p. 295):

> In mental abstraction and concentration of thought, Indians are proverbially happy. Apart from direct testimony on the point, the literature of the Hindus furnishes unmistakable evidence to prove that the ancient Hindus possessed astonishing powers of memory and concentration of thought. Hence all such sciences and branches of study as demand concentration of thought and a highly-developed power of abstraction of the mind were highly cultivated by the Indians. The science of mathematics, the most abstract of all sciences, must have had an irresistible fascination

> for the minds of the Indians. Nor are there proofs wanting to support this statement. The most extensive cultivation that astronomy received at the hands of the Indians is in itself proof of their high proficiency in mathematics. The high antiquity of Hindu astronomy is an argument in support of the still greater antiquity of their mathematics. That the Indians were selected by nature to excel all other nations in mathematics, is proved by her revealing to them the foundation of all mathematics. It has been admitted by all competent authorities that the Indians were the inventors of the numerals.

The great German critic, Schlegel (Sarda, 2007, p. 295) says that the Hindus invented 'the decimal ciphers, the honor of which, next to letters the most important of human discoveries, has, with the common consent of historical authorities, been ascribed to the Hindus.'

Professor Macdonell (1900, p. 424) says:

> In science, too, the debt of Europe to India has been considerable. There is in the first place, the great fact that the Indians invented the numerical figures used all over the world. The influence that the decimal system of reckoning dependent on those figures has had not only on mathematics but on the progress of civilization in general, can hardly be overestimated. During the eighth and ninth centuries, the Indians became the teachers in arithmetic and algebra of the Arabs, and through them of the nations of the west. Thus, though we call the latter science by an Arabic name, it is a gift we owe to India.

Sir Monier Williams (1875, p. 124) says, 'From them [Indians] the Arabs received not only their first conceptions of algebraic analysis, but also those numerical symbols and decimal notations, now current everywhere in Europe, and which have rendered untold service to the progress of arithmetical science.' Says Manning (1869, Vol. 1, p. 376), 'To whatever cyclopedia, journal or essay we refer, we uniformly find our numerals traced to India and the Arabs recognized as the medium through which they were introduced into Europe.'

W. W. Hunter (1881, p. 219) also says, 'To them (the Indians) we owe the invention of the numerical symbols on the decimal scale. The Indian figures 1 to 9 are abbreviated forms of initial letters of the numerals themselves, and the zero, or 0, represents the first letter of the Sanskrit word for *śūnya* (श). The Arabs borrowed them from the Hindus and transmitted them to Europe.'

Professor Weber (1878, p. 256) says:

> It is to them [the Hindus] also that we owe the ingenious invention the numerical symbols, which in like manner passed from them to the Arabs, and from these again to European scholars. By these latter, who were the disciples of Arabs, frequent allusion is made to the Indians uniformly in terms of high esteem; and one Sanskrit word even (uccha) has passed into the Latin translations of Arabian astronomers.

Mrs. Manning (1869, Vol. 1, p. 374) says further, 'Compared with other nations, the Hindus were peculiarly strong in all the branches of arithmetic.'

Professor Weber (Sarda, p. 298), after declaring that the Arabs were disciples of the Hindus, says, 'The same thing [i.e., the Arabs borrowing from Hindus] took place also in regard to algebra and arithmetic in particular, in both of which it appears the Hindus attained, quite independently, to a high degree of proficiency.'

W. W. Hunter (1881, p. 219) also says that the Hindus attained a very high proficiency in arithmetic and algebra independently of any foreign influence.

The English mathematician, Professor Wallace (*Edinburgh Review*, 1818, p. 147), says:

> The Līlāvati treats of arithmetic, contains not only the common rules of that science, but the application of these various questions of interest, barter, mixture, combinations, permutations, sums of progression indeterminate problems, and mensuration of surface and solids. The rules are found nearly as simple as in the present state of the analytical investigation. The numerical results are readily deduced, and if they are compared with the earliest specimens of Greek calculation, the advantages of the decimal notation are placed in a striking light.

It may, however, be mentioned that Līlāvati, of which Professor Wallace speaks, is a comparatively modern manual of arithmetic; and to judge the merits of Indian arithmetic from this book is to judge the merits of English arithmetic from Chamber's manual of arithmetic.

It may be added that the enormous extent to which numerical calculation goes in India, and the possession by the Indians of by far the largest fable of calculation, are in themselves proofs of the superior cultivation of the science of arithmetic by the Indians and their invention of *śūnya* (zero). In fact, the invention of *śūnya* (zero) was the basis of the invention of decimal system and place value notations by the Indians. That is why they used the numbers up to the endless limit. Even the *Ṛgveda* describes the name of 7 notational places from *eka* (units) to *prayuta* (millions) and the *Yajurveda* (17.2) mentions name of 12 notational places from *eka* (units) to *parārdha* (billions).

Due to a lack of knowledge of zero, decimal system, and place value notations, Arabs, Greeks and Romans could not use big numbers.

The Greeks used myriad ($10^4$) as the limit of counting while Romans defined mille ($10^3$) as the largest number. The Indians carried the decimal numeration, naming the successive power of a number (usually 10), far beyond any other civilization of the past.

Having seen this Al-Biruni (Sachau, 1964, Vol. 1, p. 174) got puzzled and criticized Bharatiyas for their passion for large numbers. He observes:

> I have studied the names of the orders of the numbers in various languages with all kinds of people with whom I have been in contact, and have found that no nation goes beyond a thousand. The Arabs too stop with thousand, which is certainly the most correct and the most natural thing to do. Those, however, who go beyond the thousand in their numeral system are the Hindus, at least in their arithmetical technical terms.

The above reference of Al-Biruni (943–1048 CE) proves beyond an iota of doubt that by his time, people of the rest of the world, including Arabs, were not aware of zero or its placement value. That is why all numbers like ten, hundred, and thousand were denoted by separate symbols. It was Indians who knew the zero and its placement value. That is why they were able to count in billions and trillions which surprised and baffled Al-Birunī.

I am sure if Al-Birunī were to write his books today, he would not have criticized the Indians for their fondness for large numbers. Al-Birunī mentioned $10^{19}$ as the largest number used by the Hindus. The Babylonians did not use numbers beyond one thousand. However, for some reason, the Arabs later on restricted themselves to 100 only.

During the first century BC, in Lalitavistara, a book on the life of Lord Buddha, in a dialogue between Lord Buddha and Arjuna, *Tallakṣaṇa* ($10^{53}$) is defined. As the story goes when Gautama reached manhood, he courted Gopa, the daughter of Kind Daṇḍāpaṇī in the tradition of *svayaṁvara* a tradition where the young men demonstrated their abilities in the presence of the bride and her family. The bride, after consultation with the family, identifies the man of her choice of marriage. Gautama demonstrated public proof of his scholarly abilities by debating (*śāstrārtha*) with Arjuna, the state mathematician of King Daṇḍāpaṇī. Arjuna asked Buddha to count numbers. Buddha defined *Koṭi* ($10^7$) and in steps of 100 defined numbers all the way up to *Tallakṣaṇa* ($10^{53}$).

The place value system is now commonly used throughout the civilized world. Without the zero and the place-value, the Indian numerals would have been no better than many other civilizations, and would not have been adopted by all the civilized peoples of the world.

*Ease of calculation*: The invention of the zero and place value system also provided ease of calculation in the Indian system. In the Indian system, at the first glance, you may recognize greater and smaller numbers according to their digits. For example, 1892 is greater than 993, because 1892 has four digits and 993 three digits.

So far as the Roman system is concerned, 100 will be written as 'C' and 99 will be written as 'IC', smaller number has greater digits. Similarly, 18 will be written as XVIII and 90 will be written as XC.

*Ignorance of zero*: The above observation of Al-Birunī clearly shows that Arabs and the rest of the world were oblivious of the concept of zero and the place value notations till the middle of the tenth century AD.

According to Mr. Clark (1929, p. 217), Arabic literary tradition, as generally interpreted, declares that the numerals with zero and place value were invented by the Indians, and they were adopted by the Arabs during the last quarter of the eighth century AD.

Indian zero was introduced by Persian scholar Abu Jafar Muh Al-Khwarizmi (780–850 CE) in the ninth century. He wrote a book in Arabic which was translated into Latin under the title '*De Numero Indorum*' (The Hindu Art of Reckoning). The manuscript of this book was copied in the thirteenth century and held by Cambridge University Library vide Ms. Ii.vi.5.. It was translated into English by John N. Crossley and Alan S. Henry in 1990 which was published in *Historia Mathematica* 17 (1990), 103–131. The following excerpt from the same translation confirms the above fact. Algorizmi [Al-Khwarizmi] said:

> I had seen that the Indians had set up 9 symbols in their universal system of numbering … for the sake of its ease and brevity, so that this work, to be sure, might be made easier for the seeker after arithmetic. … So they made 9 symbols, whose forms are these: 9 8 7 6 5 4 3 2 1 …
>
> But when X was put in the place of one and was made in the second place, and its form was the form of one, they needed a form for the tens because of the fact that it was similar to the form of one, so that they might know by means of it that it was 10. So they put one space in front of it and put in it a little circle like the letter 'o', so that by means of this they might know that the place of the units was empty and that no number was in it except the little circle, which we have said occupied it, and

> (sc. thus) it is shown that the number that is in the following place was a ten and that this was the second space, which is the place of the tens. And they put after the circle in the aforesaid second place whatever they wished from the number of tens from what is between X and XC and these are the forms of the tens: the form of X is thus (10), the form of XX [is] 20. And likewise, the form of XXX is thus (30), [Fol. 105r] and so on up to IX (sc. tens), there will be, clearly, a circle in the first place and a character pertaining to the number itself in the second place. Moreover, one must know this, that the character that signifies one in the first place, in the second signifies 'X' (10), in the third 'C' (100) and in the fourth Ī (1000). And likewise, the character that in the first place signifies two, in the second signifies XX (20) and in the third CC (200) and in the fourth ĪĪ (2000) and understand likewise about the rest.

The above statement of Al-Khwarizmi, proves two things without any shadow of a doubt.

- Arab and the Western world was unaware of zero until the time Al-Khwarizmi brought Indian zero into the limelight.
- The Indians were using place value notations from the very beginning which is why long numbers could be written.

The most glaring fact is that the Western world remained oblivious of the concept of zero till 1582 AD, until the time of Pope Gregory who amended the Julian Calendar into the Gregorian Calendar. This was the reason why the Europeans used to calculate 1 BC before 1 AD. Had they been aware of zero, they would have not calculated 1 AD after 1 BC.

Now we will make a humble attempt to find out the origin, nomenclature, and symbol of zero in Vedic, later Vedic and post-Vedic texts.

From the study of Vedic texts, one is able to glean the fact that both place value notation and zero were in ordinary use in the Vedic literature. The place value system, decimal system, and powers of ten attested in the Vedic, later Vedic, and post-Vedic texts are by no means the glaring proofs of the invention of zero by Indians in the hoary past. Now we shall take stock of the status of zero in respect of the Vedic and post-Vedic Sanskrit literature.

## 2 Powers of Ten

The concept of powers of 10 cannot be invented without the knowledge of zero. We all know the power of 10 is the number 1 followed by $n$ zeros, where $n$ is the exponent and is greater than 0; for example, $10^6$ is written as 1,000,000.

When $n$ is less than 0, the power of 10 is the number 1 $n$ places after the decimal point; for example, $10^{-2}$ is written as 0.01. When $n$ is equal to 0, the power of 10 is 1; that is, $10^0 = 1$.

The contents of power of ten in the enumeration are the Vedic invention. We find the origin of this concept in the *Ṛgveda* itself. Here we shall quote a few mantras supporting the above hypothesis. A mantra in the *Ṛgveda* (8.1.5) says:

> ये ते सन्ति दशग्विनः शतिनो ये सहस्रिणः।
> अश्वासो ये ते वृष्णा रघुद्रुवस्तेभिर्नस्तूयमा गहि॥
>
> *ye tesantidaśagvinaḥśatino ye sahasriṇaḥ;*
> *aśvāso ye tevṛṣṇāraghudruvastebhirnastūyamāgahi.*
>
> [Come quickly with those thy horses which are vigorous and fleet, and which are traverses of tens ($10^1$), or hundreds ($10^2$) or thousands ($10^3$) of leagues.]

This concept is further elaborated in the *Yajurveda*. The *Yajurveda* (17.2) illustrates how the power of *eka* (one) increases by 10 times when zero is suffixed to them.

> इमा मेऽअग्नऽइष्टका धेनवः सन्त्वेका च दश च दश च शतं च शतं च सहस्र च सहस्र चायुतं चायुतं च नियुतं च नियुतं च प्रयुतं चार्बुदं च न्यर्बुदं च समुद्रश्च मध्यं चान्तश्च परार्धद्दचैता मे ऽअग्न ऽइष्टका धेनवः सन्त्वमूत्रामूष्मिँल्लोके।
>
> *imāme'agna'iṣṭakā dhenavaḥ santvekā cha daśa cha daśa cha śataṁ cha śataṁ cha sahasra cha sahasrachāyutaṁ chāyutaṁ cha niyutaṁ cha niyutaṁ cha prayutaṁ chārbudaṁ cha nyarbudaṁ cha samudraścha-madhyaṁ chāntaścha parārdhaddachaitā me 'agna 'iṣṭakā dhenavaḥ santvamūtrāmūṣmim̐lloke.*
>
> [O Agni may these *iṣṭakā* (bricks) of *yajña* altar provide me desirous fruits like a milch-kine. They increase by 10 times like numerals when suffixed with zero; e.g., $10^0$ (one) becomes $10^1$ (ten); $10^1$ becomes $10^2$ (100); $10^2$ (100) becomes $10^3$ (1,000); $10^3$ (one thousand) becomes $10^4$ (10,000 or *ayuta*); $10^4$ (10,000 or *ayuta*) becomes $10^5$ (*niyuta* or 100,000); $10^5$ (100,000 or *niyuta*) becomes $10^6$ (1,000,000 or *prayuta*); $10^6$ (a *prayuta* or one million) becomes an $10^7$ (*arbuda* or ten million); $10^7$ (*arbuda*) becomes $10^8$ (*nyarbuda* or one hundred million), $10^8$ (a *nyarbuda*) becomes $10^9$ (a *Samudra* or one thousand millions); $10^9$ (a *samudra*) becomes $10^{10}$ (*madhya* or a

ten thousand millions); $10^{10}$ (a *madhya*) becomes $10^{11}$ (*anta* or a hundred thousand millions); $10^{11}$ (an *anta*) becomes $10^{12}$ (*parārdha* or a million million or a billion). May these bricks act as milch-kine in providing fruits in yonder world and in this world.]

In the above mantra of the *Yajurveda*, powers of 10 have been counted, from $10^0$ to $10^{12}$. It is revealed here that zero increases the magnitude of numerals by 10 times when suffixed to them. Inversely it also points out a place or positional value system, when the place or position of a number increases from right to left, its magnitude is increased by 10 times. Not only this, the place names have also been given from *eka* (unit) to *parārdha* (billion).

The above list of numerals is also reproduced in the *Taittirīya Saṁhitā* (4.4.11), the *Kāṭhaka Saṁhitā* (17.10.31), and the *Maitrāyaṇī Saṁhitā* (2.18.14).

Āryabhaṭṭa restated the same fact when he claims it to be ancient knowledge. According to him (Āryabhaṭīyam, Gaṇita Pāda, 2): if the position of a number increases from right to left, its magnitude increases by 10 times.

एकं च दश च शतं च सहस्रं त्वयुतनियुते तथा प्रयुतम्।
कोट्यर्बुदं च वृन्दं स्थानात्स्थानं दशगुणं स्यात्।।

*ēkaṁ cha daśa cha śataṁ cha*
*sahasraṁ tvayutaniyute tathā prayutam;*
*koṭyarbudaṁ cha vṛndaṁ*
*sthānāt sthānaṁ daśaguṇaṁ syāt.*

[$10^0$ (*eka*/1),
$10^1$ (daśa/10),
$10^2$ (*śatam*/100),
$10^3$ (*sahasram*/1,000),
$10^4$ (*ayutam*/10,000),
$10^5$ (*niyutam*/100,000),
$10^6$ (*prayuta*/1,000,000),
$10^7$ (*Koṭi*/10,000,000),
$10^8$ (*arbudam*/100,000,000) and
$10^9$ (*vṛnda*/1,000,000,000) are, respectively, each ten times the preceding, because of their positional/place value or notational places. With the increase in the position of number 1 from right to left, its magnitude is increased by 10 times. Inversely, zero increases the magnitude of numbers by 10 times when suffixed to them.]

Here it may be noted that Āryabhaṭa (1976, Golādhyāya, 50) does not claim this as his invention, but he says that he owes this all knowledge to the *Brāhma-siddhānta* which was in vogue in the hoary past well before his period.

> आर्यभटीयं नाम्ना पूर्वं स्वायम्भुवं सदा नित्यम्।
> सुकृतायुषोः प्रणाशं कुरुते प्रतिकंचुकं योऽस्य।।
>
> *āryabhaṭīyaṁnāmnāpūrvaṁsvāyambhuvaṁsadānityam;*
> *sukṛtāyuṣoḥpraṇāśaṁkurutepratikaṁchukaṁyo'sya.*
>
> [This *'Āryabhaṭīya'* is the same work as the ancient *Brāhma-siddhānta* (which was revealed by Svāyambhū) and as such, it is true for all times. One who finds fault with it shall lose his fame and longevity.]

## 3 The Place-Value Notations

The place value notation as pointed out above was indicated with the help of zero. In the present days also, in elementary schools in India, the student is taught the names of several notational places and is made to denote them by zeros arranged in a line. These zeros are written as ……. 000000000

The teacher points out the first zero on the right and says 'units', then he proceeds to the next zero saying 'tens' and so on. The student repeats the names after the teacher. This practice of denoting the notational places by zeros can be traced back to the time of the Vedas.

In the Vedas, we come across the names of various notational places like *eka*, *daśa*, *sahasra*, *ayuta*, etc. up to *parārdha*.

In the *Ṛgveda*, we find names of the notational places from *eka* (units) to *Prayuta* (millions). Examples are:
*Eka* (units): *Ṛgveda*, 3.29.15; 10.59.9; 1.20.7; 5.2.17
*Daśa* (tens): *Ṛgveda*, 1.53.6; 8.1.5; 10.122.13
*Śata* (hundreds): *Ṛgveda*, 1.24.9; 4.26.3; 8.32.18
*Sahasra* (thousands): *Ṛgveda*, 1.80.9; 4.26.3; 8.2.41; 8.32.18
*Ayuta* (ten thousands): *Ṛgveda*, 4.26.4
*Niyuta* (hundred thousands): *Ṛgveda*, 1.134.2
*Prayuta* (millions): *Ṛgveda*, 1.51.6

In the *Yajurveda* (17.2), as already explained above, we find names of 12 notational places from *eka* (unit) to *parārdha* (billion).

The uniqueness of the Vedic system lies in the fact that the position of a number qualifies its magnitude. Tens, hundreds, or thousands were not represented by different signs like Egyptians, Chinese, Romans, and other civilizations. They are represented by using digits in different positions. Notice that the one in the second place is 10 (ten), in the third place is 100 (hundred), and in the fourth place is 1,000 (thousand).

For example, the *Ṛgveda* (10.52.6) following the place value notation mentions 'three thousand and three hundred and thirty-nine (3339)' as the count of people in a *yajña*.

> त्रीणि शता त्री सहस्राण्यग्निं त्रिंशच्च देवा नव चासपर्यन्।
> औक्षन्घृतैरस्तृणन्बर्हिरस्मा आदिद्धोतारं न्यसादयन्त।।
>
> *trīṇi śatā trī sahasrāṇyagniṁ triṁśachcha devā nava chāsaparyan;*
> *aukṣanghṛtairastṛṇanbarhirasmā ādidhotāraṁ nyasādayanta.*
>
> [Three thousand (3,000), three hundred (300), thirty (30) and nine (9) scholars performed *yajña*, they gave oblations of ghee, they sit on the sacred grass strewn around the altars.]

In a similar manner, the *Atharvaveda* (8.8.7) mentions a hundred (100), thousand (1,000), ten thousand (10,000), and hundred-thousand (1000,000) as under:

> बृहत्ते जालं बृहत इन्द्र शूर सहस्रार्घस्य शतवीर्यस्य ।
> तेन शतं सहस्रमयुतं न्यर्बुदं जघान शक्रो दस्यूनामभिधाय सेनया ॥
>
> *bṛhatte jālaṁ bṛhata indra śūra sahasrārghasya śatavīryasya;*
> *tena śataṁ sahasramayutaṁ nyarbudaṁ jaghāna śakro dasyūnāmabhidhāya senayā.*
>
> [O Indra (lightning in clouds) your network is great, for your qualities, you are appreciated by thousands of scholars. With your help, hundreds of works can be performed. With his power, Indra (lightning in clouds) has discharged a hundred, thousand, ten thousand, and hundred thousand clouds.]

The evidence of the use of the word 'notational place' is also furnished by the *Anujogadvāra-sūtra* (100 BC) in the *sūtra* 142 (Datta, 1912, p. 12). At one place it is stated:

> a number which when expressed in terms of the denominations, *Koṭi-Koṭi*, etc., occupies twenty-nine places (*sthānas*), or it is beyond the 24th place and within the 32nd place, or it is a number obtained by multiplying the sixth square (of two) by (its) fifth square, (i.e., $2^{96}$), or it is a number which can be divided (by two) ninety-six times.

Another big number that occurs in the Jaina works is the number representing the period of time known as Śīrṣa Prahelikā. According to the commentator Hemachandra (first century AD), this number is so large as to occupy 194 notational places (*aṅka-sthānāni*). It is also stated to be $(8{,}400{,}000)^{28}$.

The first foreign proof of the use of place value notation in India was given by Al-Khwarizmi (ninth century), a Persian scholar, already discussed above.

## 4 Left to Right Notation

Here it is important to know that in a positional or place-value system, a number, represented as $x_4, x_3, x_2, x_1$ can be constructed as follows:

$$x_1 + (x_2 \times 10^1) + (x_3 \times 10^2) + (x_4 \times 10^3)$$

Where $x_1$, $x_2$, $x_3$, and $x_4$ are non-negative integers that have magnitudes less than the chosen base (ten in our case). As you may notice, the magnitude of a number increases from right to left. For example, the number 1961 will be written as

$$1 + (6 \times 10^1) + (9 \times 10^2) + (1 \times 10^3)$$

The above notation is strictly followed in the writing in Sanskrit. The numbers were written in the Sanskrit language from left to right, but numerically they were placed from right to left, as per Vedic rule described in the *Samayochita Padma Ratna Malika* (*śloka* No. 48)

> अंकानां वामतो गतिः।
>
> *aṁkānāṁvāmatogatiḥ*
>
> [Numerically the numbers are placed from left to right.]

For example, the following mantra of the *Atharvaveda* (8.2.21) follows the left-to-right notation while describing the number of years in a Kalpa period.

> शतं तेऽयुतं हायनान् द्वे युगे त्रीणि चत्वारि कृण्मः।
> इन्द्राग्नी विश्वेदेवास्तेऽनुमन्यन्तामहृणीयमानाः।।
>
> *śataṁte'yutaṁhāyanāndveyugetrīṇichatvārikṛṇmaḥ;*
> *indrāgnīviśvedevāste'numanyantāmahṛṇīyamānāḥ.*
>
> [Viśvedevāḥ (The Galaxy) takes 4,32,00,00,000 (4.32 billion) years to complete one revolution. Here *Visvedevāḥ* is representative of Galaxy.]

The above mantra states to write 2, 3 and 4 after hundred *Ayutas*. One *Ayuta* is equal to 100,000. As such, hundred *Ayutas* will be equal to one million (seven zeros). When 2, 3, and 4 are placed from left to right before a million we get the number 4,320,000,000 which is the period of a Kalpa or say the time of our Galaxy's revolution around the Supergalactic centre (*Svāyambhūva Maṇḍala*).

Later Vedic and post-Vedic Sanskrit literature is full of such examples.

## 5 Word Numerals and Place Value Notation

In addition to the above, it is pertinent to note here that a system of expressing numbers by means of words arranged as in the place-value notation was developed and perfected in India in the earliest phases of her ancient history. In this system, the numerals are expressed by names of things, beings, or concepts, which naturally or in accordance with the teaching of the *Śāstras*, connote numbers. Thus the number one may be denoted by anything that is markedly unique, e.g., the moon, the earth, etc.; number two by the eyes, the hands, etc; and similarly others. The zero was denoted by words meaning ख (*kha*), i.e., void, आकाश (*ākāśa*), i.e., space, ब्रह्म (Brahman), पूर्ण (*pūrṇa*), i.e., full or absolute, शून्य (*śūnya*), i.e., empty.

This system was used in works on astronomy, mathematics, and metrics, as well as in the dates of inscriptions and manuscripts. The ancient Indian mathematicians and astronomers wrote their works in verse. Consequently, they strongly felt the need for a convenient method of expressing the large numbers that occur so often in astronomical works and the statement of problems in mathematics. The word numerals were invented to fulfill this need and soon became very popular. They are used even up to the present day, whenever big numbers have to be expressed in Sanskrit verse.

Bhaṭṭotpala (Dvivedi, 1996, p. 163) in his commentary on the *Bṛhat-saṅhitā* has given a quotation from the original *Pauliśa-siddhānta* in which the word numerals are used. The number expressed in this quotation is:

> ख (0) ख (0) अष्ट (8) मुनि (7) राम (3) अश्विन् (2) नेत्र (2) अष्ट (8) शर (5) रात्रि / चन्द्रमा 1)

The above quotation will be numerically represented as 1,582,237,800.

Sanskrit literature is full of such examples, for the brevity of space and time, only one example has been quoted.

The words denoting the numbers from one to nine and zero, with the use of the principle of place-value, give us a very convenient method of expressing numbers by word chronograms. To take a concrete case, the number 1,230 may be expressed in many ways:

1. *kha-guṇa-kar-ādi*
2. *kha-loka-karṇa-candra*
3. *ākāśa-kāla-netra-dharā*

It will be observed that the same number can be expressed in hundreds of ways by word chronograms. This property makes the word numerals especially suitable for inclusion in meters. To secure still greater variety, the numbers beyond ten are also sometimes denoted by words.

We also find an ancient record of the place-value notation coming from Vāsumitra, a leading figure of Kaniṣka's Great Council. According to Hiuen Tsang (602–664), Kuṣāṇa King Kaniṣka (144–178 AD) called a convocation of scholars to write a book, *Mahāvibhāṣa*. Four main scholars under the chief monk, named Pārśva, wrote the book in 12 years. Vāsumitra was one of the four scholars. In this book, Vāsumitra tried to explain that matter is continually changing as it is defined by an instant (time), shape, mass, etc. As time is continually changing, therefore, the matter is different in each situation although its appearance and mass do not change. He used an analogy of place-value notation to emphasize his point. Just as the location of digit one (1) in the place of hundred is called hundred (100) and in the place of thousand (1,000) is called thousand, similarly matter changes its state (*avasthā*) in different time designations. New explanations are generally given in terms of known and established facts. Thus, the very reason Vāsumitra used place-value notation as an example established that the place-value notation was considered an established knowledge during the early Christian era. (Kumar, 2014, p. 67)

## 6 Base Ten Decimal System

Here it is important to know that the base ten (decimal) numeral system is also Vedic in origin. In fact, the words 'deca', 'deci', 'decimal', 'ten', etc are the offshoots of the Sanskrit word '*Daśa*' (दश).

The decimal system represents numbers as linear combinations of the powers of ten as shown below, e.g.

$10 = 10^1$
$20 = 2 \times 10^1$
$30 = 3 \times 10^1$
$16 = 6 + 10^1$

This is possible when the knowledge of zero is present. Several other civilizations such as the Egyptians, Chinese and Romans did not invent zero, so they used an additive decimal system. In their system, multiple repetitions of symbols were used and added to increase the magnitude. For example, in the Roman system x means 10 and xx (10 + 10) became 20, xxx (10 + 10 + 10) became 30, xvi (10 + 5 + 1) became 16. So, for want of the invention of zero they have to repeat the symbols again and again.

It is India and only India that gave the world an ingenious method of expressing all numbers by means of 10 symbols, each symbol receiving a value of the position and as well as an absolute value.

We receive first indication of decimal system from the *Yajurveda* (27.33):

एकया-दशभिः (*ēkayā-daśabhiḥ*) $1 \times 10$
द्वाभ्यां-विंशति (*dvābhyāṁ-viṁśati*) $2 \times 10$
तिसृभिः त्रिशंता (*tisṛbhiḥtriśaṁtā*) $3 \times 10$

The *Atharvaveda* (5.15.1–11) forms numbers from 1 to 1000 on the basis of decimal system:

एका च मे दश च मेऽपवक्तार ओषधे ।
ऋतजात ऋतावरि मधु मे मधुला करः ॥१॥

*ēkā cha me daśa cha me'pavaktāraōṣadhe;*
*ṛtajātaṛtāvarimadhu me madhulākaraḥ.*

[O herb used in performing *yajña* and full of juicy potentialities make us regain health if we are attacked by one disease and ($1 \times 10^1$) ten.]

द्वे च मे विंशतिश्च मेऽपवक्तार ओषधे ।
ऋतजात ऋतावरि मधु मे मधुला करः ॥२॥

*dve cha me viṁśatiśchame'pavaktāraoṣadhe;*
*ṛtajātaṛtāvarimadhu me madhulākaraḥ.*

[O herb used in performing *yajña* and full of juicy potentialities make us regain health if we are attacked by 2 diseases and ($2 \times 10^1$) twenty.]

तिस्रश्च मे त्रिंशच्च मेऽपवक्तार ओषधे ।
ऋतजात ऋतावरि मधु मे मधुला करः ॥३॥

*tisraścha me triṁśachchame'pavaktāraoṣadhe;*
*ṛtajātaṛtāvarimadhu me madhulākaraḥ.*

[O herb used in performing *yajña* and full of juicy potentialities make us regain health if we are attacked by 3 diseases and ($3 \times 10^1$) thirty.]

चतस्रश्च मे चत्वारिंशच्च मेऽपवक्तार ओषधे ।
ऋतजात ऋतावरि मधु मे मधुला करः ॥४॥

*chatasraścha me chatvāriṁśachchame'pavaktāraoṣadhe;*
*ṛtajātaṛtāvarimadhu me madhulākaraḥ.*

[O herb used in performing *yajña* and full of juicy potentialities make us regain health if we are attacked by 4 diseases and ($4 \times 10^1$) forty.]

पञ्च च मे पञ्चाशच्च मेऽपवक्तार ओषधे ।
ऋतजात ऋतावरि मधु मे मधुला करः ॥५॥

*pañcha cha me pañchāśachchame'pavaktāraoṣadhe;*
*ṛtajātaṛtāvarimadhu me madhulākaraḥ.*

[O herb used in performing *yajña* and full of juicy potentialities make us regain health if we are attacked by 5 diseases and ($5 \times 10^1$) fifty.]

षट्च मे षष्टिश्च मेऽपवक्तार ओषधे ।
ऋतजात ऋतावरि मधु मे मधुला करः ॥६॥

*ṣaṭcha me ṣaṣṭiśchame'pavaktāraōṣadhe;*
*ṛtajātaṛtāvarimadhu me madhulākaraḥ.*

[O herb used in performing *yajña* and full of juicy potentialities make us regain health if we are attacked by 6 diseases and ($6 \times 10^1$) sixty.]

सप्त च मे सप्ततिश्च मेऽपवक्तार ओषधे ।
ऋतजात ऋतावरि मधु मे मधुला करः ॥७॥

*sapta cha me saptatiśchame'pavaktāraōṣadhe;*
*ṛtajātaṛtāvarimadhu me madhulākaraḥ.*

[O herb used in performing *yajña* and full of juicy potentialities make us regain health if we are attacked by 7 diseases and ($7 \times 10^1$) seventy.]

अष्ट च मेऽशीतिश्च मेऽपवक्तार ओषधे ।
ऋतजात ऋतावरि मधु मे मधुला करः ॥८॥

*aṣṭa cha me'śītiśchame'pavaktāraōṣadhe;*
*ṛtajātaṛtāvarimadhu me madhulākaraḥ.*

[O herb used in performing *yajña* and full of juicy potentialities make us regain health if we are attacked by 8 diseases and ($8 \times 10^1$) eighty.]

नव च मे नवतिश्च मेऽपवक्तार ओषधे ।
ऋतजात ऋतावरि मधु मे मधुला करः ॥९॥

*nava cha me navatiśchame'pavaktāraōṣadhe;*
*ṛtajātaṛtāvarimadhu me madhulākaraḥ.*

[O herb used in performing *yajña* and full of juicy potentialities make us regain health if we are attacked by 9 diseases and ($9 \times 10^1$) ninety.]

दश च मे शतं च मेऽपवक्तार ओषधे ।
ऋतजात ऋतावरि मधु मे मधुला करः ॥१०॥

*daśa cha me śataṁ cha me'pavaktāraōṣadhe;*
*ṛtajātaṛtāvarimadhu me madhulākaraḥ.*

[O herb used in performing *yajña* and full of juicy potentialities make us regain health if we are attacked by 10 diseases and ($10 \times 10^1$) hundred.]

शतं च मे सहस्रं चापवक्तार ओषधे ।
ऋतजात ऋतावरि मधु मे मधुला करः ॥११॥

*śataṁ cha me sahasraṁchāpavaktāraōṣadhe;*
*ṛtajātaṛtāvarimadhu me madhulākaraḥ.*

[O herb used in performing *yajña* and full of juicy potentialities make us regain health if we are attacked by 100 diseases and ($100 \times 10^1$) thousand.]

In the above-cited verses, it is revealed that when zero is placed to the right of the numbers their power is increased by 10 times. The above verse explicitly reveals:

$1 \times 10^1 = 10$
$2 \times 10^1 = 20$
$3 \times 10^1 = 30$
$4 \times 10^1 = 40$
$5 \times 10^1 = 50$
$6 \times 10^1 = 60$
$7 \times 10^1 = 70$
$8 \times 10^1 = 80$
$9 \times 10^1 = 90$
$10 \times 10^1 = 100$
$100 \times 10^1 = 1000$

The above mantras also indicate:

That when zero is placed to the right of one (1), the power of one (1) is increased by 10 times turning 1 into 10, similarly zero makes 2 as 20; 3 as 30; 4 as 40; 5 as 50; 6 as 60; 7 as 70; 8 as 80; 9 as 90; 10 as 100 and 100 as 1,000.

Here in the above-cited verses, we find not only the numbers of two digits but also the numbers of three and four digits. By this description, it is quite clear that the rule adopted in making various numbers from 10 to 1,000 does not stop on 1,000 but also may be applied further in making the numbers of five, six digits, and further on.

Thus, from the aforementioned discussion on powers of ten, place value notions, and the base 10 decimal system, it is crystal clear that the invention of zero is a singular contribution of India to the modern world of science and technology.

## 7 Ṛgveda: Zero – a Tool to Increase the Value or Magnitude

Vedic seers found in zero a tool to increase the value or magnitude of numbers. When a zero is placed just in the right position of the units 1 to 9 increases their values ten times and the same process makes hundred and thousand from ten and hundred respectively, is described in the following verses of the *Ṛgveda* (10.51.3).

> ऐच्छाम त्वा बहुधा जातवेदः प्रविष्टमग्ने अप्स्वोधिषु।
> तं त्वा यमो अचिकेच्चित्रभानो दशान्तरुष्यादतिरोचमानम्।।
>
> *aichchhāmatvābahudhājātavedaḥpraviṣṭamagneapvotidhiṣu;*
> *taṁtvāyamoachikechchitrabhānodaśāntaruṣyādatirochamānam.*
>
> [We desire you, O Agni Jātavedas (solar energy which is known by its appearance alone), which has variously permeated into waters and plants, etc. *Yama* (your concealed nature) is recognized or perceived by your illumination, just as zero remains (*antaruśyam*) concealed into numbers and known when it increases the power of a digit by 10 times from place to place in a very interesting way.]

In the above mantra, in the following hemistich,

> दशान्तरुष्यादतिरोचमानम्।
>
> *daśāntaruṣyādatirochamānam*

the significance of zero is described. Accordingly, due to zero, the power of numbers is interestingly increased by tens from place to place.

Sāyaṇa in his commentary on the above mantra points out the fact that zero remains concealed and increases the magnitude of numbers by 10 times.

> अन्तरुष्यं गूढमावासस्थानम्। तच्च च स्थानं दशसंख्योपेतम्।
>
> *antaruṣyaṁgūḍhamāvāsasthānam;*
> *tachcha cha sthānaṁdaśasaṁkhyopetam.*

[The zero remains hidden in the numbers. It is visible when it increases the value of numbers by 10 times from place to place.]

Confirming the above fact, the *Viṣṇu-Purāṇa* (6.3.4) observes in the same manner:

स्थानात्स्थानं दशगुणमेकस्माद्गण्यते द्विज।
ततोऽष्टादशमे परार्धमभिधीयते।।

*sthānātthāsnaṁdaśaguṇame*kas*mādgaṇyatedvija;*
*tato'ṣṭādaśame parārdha mabhidhīyate.*

[O Dvija, from the place to the next in succession, the places are multiples of ten. The eighteenth one of these (places) is called *parārdha.*]

Using a simile, Vyāsa in his commentary on *Patañjali's Yoga-sūtra* (3.13) says that the same *rekhā* (digit) is termed one in the units place, ten in the tens place, and hundred in the hundreds place. The actual reference is cited below.

यथैका रेखा शतस्थाने शतं दशस्थाने दशैका चैकस्थाने ।

*yathaikārekhāśatasthāneśataṁdaśasthānedaśaikāchaikasthāne.*

[Just as a *rekhā* (a digit) connotes a hundred in the 'hundreds' place, ten in the 'tens' place and one in the 'units' place.]

We also come across a very interesting *śloka* in the *Samayochita Padma-ratna-mālikā* (*śloka* No. 48), a collection of relevant *śloka*s published in Mumbai in 1957, where it states that when zero is placed to the right of numbers their power is increased by ten times. The verse goes like this.

अंकेषु शून्यविन्यासाद्, वृद्धिः स्यात् तु दशाधिका ।
तस्माद् ज्ञेया विशेषेण, अंकानां वामतो गतिः॥

*aṁkeṣu*śūnya*vinyāsād, vṛddhiḥsyāttudaśādhikā;*
*tasmādjñeyāviśeṣeṇa, aṁkānāṁvāmatogatiḥ.*

[When zero is placed to the right of numbers their power is increased by ten times. That is why the numerals are placed from left to right.]

## 8 Nature and Philosophy of Zero

In the Vedic literature, the words used to express zero also points out its nature and philosophy as viewed by the Vedic seers.

### 8.1 *Zero Is Like Space and Brahman*

For instance, in the *Yajurveda* (40.17), the nature of zero has been defined by the term '*kha*' associating it with Brahman.

> ओं खं ब्रह्म।
>
> *Oṁkhaṁ brahma*
>
> [Brahman pervades the whole universe, just as zero permeates all the numbers.]

Brahman is all prevalent, *ananta* (infinite), *aparimita* (having no boundaries or limit), *aprimeya* (immeasurable), *asaṁkhyeya* (uncountable), and *pūrṇa* (full/absolute).

Just as Brahman pervades everything but remains invisible, similarly zero also permeates all numbers but it remains invisible. To understand the nature of zero or Brahman, we shall have to understand the *Paṇini's sūtra* (1.1.60).

> अदर्शनं लोपः।
>
> *adarśanaṁlopaḥ*
>
> [In the grammatical process, *lopa* means non-visibility, but non-visibility does not mean absence.]

Similarly, if zero is not visible in numbers written from left to right, like 1, 2, 3, 4, 5, 6, 7, 8, or 9 that does not mean that zero is missing. Zero is present in all numbers but not visible, like Brahman. The Brahman is not visible, but it never means that Brahman is missing from the creation.

The civilizations that could not identify the invisible Brahman, also could not invent zero.

Brahman is infinite, immeasurable, uncountable, and absolute, similarly zero in the Vedic parlance is not empty but infinite, immeasurable, uncountable, and absolute. The most popularly used word these days for zero is '*śūnya*' which has sprung from the Vedic word '*śūna*' meaning 'fully blown' or 'fully fledged'. Just as the space is not empty but full of all ingredients of creation, similarly zero is not empty.

Brahman is *pūrṇa* (absolute), similarly zero is absolute. The absolute value of 0 is 0.

Just as Brahman alone survives after the whole world is eliminated, similarly, when all numerals are eliminated, zero alone survives.

The famous Upaniṣadic quote reads as follows:

> ॐ पूर्णमदः पूर्णमिदं पूर्णात्पूर्णमुदच्यते ।
> पूर्णस्य पूर्णमादाय पूर्णमेवावशिष्यते ॥
>
> *pūrṇamadaḥpūrṇamidaṁpūrṇātpūrṇamudachyate.*
> *pūrṇasyapūrṇamādāyapūrṇamevāvaśiṣyate.*
>
> [Brahman is *pūrṇa* (fullness/absolute), this space is *pūrṇa* (fullness/absolute), If *pūrṇa* (fullness) is taken out of *pūrṇa* (fullness), *pūrṇa* (the fullness) does remain.]

Inversely this statement also applies to zero. Mathematically it points out $0 - 0 = 0$ or $\infty - \infty = \infty$

Just as Brahman is invisible, zero also represents invisible. In Bakhśāli Mathematics (200 AD) the invisible quantity is represented by symbol '0' which is called '*śūnya*' (zero). This is obvious from the folio 22 verso which reads as under:

> द्विगुणं द्वितीयस्य प्रथमा----तीय। प्रथमा चतुर्गुणं चैव चतुर्थे चैव दत्तवान् च शतमेकं द्वयानुगम। वदस्व प्रथमे दत्तं किं प्रमाणम्-
>
> *dviguṇaṁdvitīyasyaprathamā----tīya. prathamā chaturguṇaṁchaivacha turthechaivadattavān cha śatam ekaṁdvayānugama. vadasvaprathamedattaṁkiṁ pramāṇam-*

| | | | | | | |
|---|---|---|---|---|---|---|
| स्य... | 0<br>1 | 2<br>1 | 3<br>1 | 4<br>1 | दृश्य 200<br>(invisible 200)<br>1 | शुन्यमेकयुतं कृत्वा<br>(adding unity to invisible/ zero) |
| | 1 | 2 | 3 | 4 | ...क्षेप युत्त्या फलम्<br>(having added 1 to invisible) | |
| | 20 | 40 | 60 | 80 | | |
| | आ 20 | उ 20 | प 4 | | एवं 200<br>Thus the share is 200 | |
| | 1 | 1 | 1 | | | |

The above problem is of this type:

'A certain amount was given to the first invisible person, twice of invisible to the second, thrice to the third, and four times to the fourth. State the amount given to the invisible and the shares of others, if the total visible amount (given amount) was 200.'

In the above problem, shares are represented by 0, 1, 2, 3, and 4. (Here zero stands as a symbol for invisible). Now having added 1 to the invisible (0) the sum is 1 + 2 + 3 + 4 = 10, which shows the proper share of the first is 200/10 or 20 and the series is 20 + 40 + 60 + 80 = 200.

Here Hörnle (1983, p. 90; 1988, p. 30) and Kaye (1912, p. 357) consider the '*śūnya*' as the symbol of 'unknown'. But here the invisible person represents the invisible Brahman which is represented by *śūnya* and nothing else. The same *śūnya* has also been used for the zero for the decimal arithmetical notations. This is, indeed, its true significance. Its mention in connection with an algebraic equation, in a sense other than for arithmetical notation, is simply to indicate that the quantity which is invisible so not known. The following examples clarify this issue further.

Folio 13, verso quotes मूलं न ज्ञायते (*mūlaṁnajñāyate*)
15 verso quotes प्रथमं न जानामि (*prathamaṁnajānāmi*)
24 verso quotes पदं न ज्ञायते (*padaṁnajñāyate*)

In each case cited above, unfamiliarity with the invisible elements has been pointed out and the same invisibility has been indicated by *śūnya*.

Thus on the basis of aforesaid citations, it can safely be maintained that zero has its philosophical origin in India. Its characteristics are closely related to those of Brahman.

Like a *yogī* when keep meditating on infinite Brahman, merges with the infinite power of Brahman, in a similar manner you keep on suffixing zero to a number to increase its power to infinity (∞).

Brahman permeates all living beings and non-living objects, but His presence does not affect the individuality of a living being or non-living object, similarly zero when added or subtracted to a number it makes no difference in the individual power of a number.

When a *yogī* through *samādhi* merges with Brahman, he loses his individual identity or say as an individual he is reduced to zero and becomes one with Brahman, similarly, when zero is multiplied to a number it reduces its power to zero.

The realization of Brahman increases the power of the seeker to infinity, similarly when zero divides a number, it increases its power to infinity (∞).

### 8.2 *Zero Represents a Transitional Stage*

Zero is sandwiched between +ve and –ve powers or numbers. It is a transitional stage, when you switch over from +ve to –ve you shall have to pass through it.

This nature of zero has been cited in the following verse of the *Yogavāsiṣṭha Mahārāmāyaṇa* (Nirvāṇa Prakaraṇa, Uttarārdha, 127.20).

> पर्यन्ते तस्य नभसः स्थितं ब्रह्माण्डखर्परम् ।
> एकमूर्ध्वे परमधो गगनं मध्यमेतयोः ॥
>
> *paryantetasyanabhasaḥsthitaṃbrahmāṇḍa*kharparam,
> eka*mūrdhveparamadhogaganaṃ*madhya*metayoḥ*.
>
> [At the end of this sphere, there is the great circle of the universe; having one half of it stretching above and one below, and containing the space in the midst of them like that of zero between +ve and –ve numbers.]

This *śloka* gives the origin of zero and +ve and –ve numbers. +ve numbers are heavenly numbers and –ve numbers are infernal regions. Space representing zero is sandwiched between them.

### 8.3 *Zero Is Not a Countable Number*

The following verse of the philosophical treatise the *Yogavāsiṣṭha Mahārāmāyaṇa* (Nirvāṇa Prakaraṇa, Uttarārdha, 127.27), speaks of another characteristic of zero giving a simile of Brahman. According to it, zero inherits all countable numbers, yet it cannot be called a countable number.

नभसोऽप्यधिकं शून्यं न च शून्यं चिदात्मकम् ।
सूर्यकान्ते यथा वह्निर्यथा क्षीरे घृतं तथा ॥ 27 ॥

*nabhaso'pyadhikaṃ*śūnya*ṃnaca*śūnya*ṃcidātmakam,*
*sūryakānteyathāvahniryathākṣīreghṛtaṃtathā.*

[The Chaitanya or Brahman is more rarefied/subtle than zero or space, and yet zero or space cannot be called Brahman. This is like a sun-stone inheriting in it the fire but fire cannot be a sunstone, and the milk inheriting butter in it, but butter cannot be milk; similarly, God inherits space in it, but space cannot be God and zero inherits all countable numbers in it, but countable numbers cannot be zero.]

## 8.4 *Zero Is Contained in Zero*

At another place, the *Yogavāsiṣṭha Mahārāmāyaṇa* (Nirvāṇa Prakaraṇa, Nirvāṇa Prakaraṇa, Pūrvārdh, 2.11), speaks of yet another characteristics of zero. Accordingly, zero is contained in zero. There is no other container of zero, except zero itself.

शून्यं शून्ये समुच्छूनं ब्रह्म ब्रह्मणि बृंहितम् ।
सत्यं विजृम्भते सत्ये पूर्णे पूर्णमिव स्थितम् ॥

śūnya*ṃśūnyesamucchūnaṃ brahma brahmaṇibṛṃhitam,*
*satyaṃvijṛmbhatesatyepūrṇepūrṇamivasthitam.*

[Just as the *śūnya* (zero) is contained in zero; the vastness of Brahman is contained in the immensity of Brahman, and as truth can exist in truth; so infinity (∞) is contained in the infinity.]

## 8.5 *Addition of Infinite Numbers of Zeros Makes One Zero*

According to the *Yogavāsiṣṭha Mahārāmāyaṇa* (Nirvāṇa Prakaraṇa, Nirvāṇa Prakaraṇa, Pūrvārdh, 3.14), infinite numbers of zeros make only one zero.

यथा नानाप्यनानैव खं खे खानीति वाग्गणः ।
सार्थकोऽप्यतिशून्यात्मा तथात्मजगतोः क्रमः ॥

*yathānānāpyanānaiva*kha*ṃkhekhānītivāggaṇaḥ,*
*sārthako'pyatiśūnyātmā tathātmajagatoḥ kramaḥ.*

[Just as the one, two, or infinite numbers of zeros make one zero, although this classification of one, two, or more zeros is meaningful and justified in statements. Similarly, the infinite numbers of spaces make one space and the infinite number of *ātmans* and worlds make one Brahman.]

## 9 The Binary Number System and Zero

The binary number system is a system in which numbers are represented as linear combinations of the powers of 2.

This is a positional numeral system employing only two kinds of binary digits, viz. 0, 1. The importance of this system lies in the convenience of representing decimal numbers using a two-state system in computer technology. The system was in vogue during the Vedic period also.

TABLE 20.1 Vedic system

| | |
|---|---|
| अस्ति (*asti*) | नास्ति (*nāsti*) |
| आग्नेय (*āganeya*) | सोमीय (*somīya*) |
| सत्य (*satya*) | मिथ्या (*mithyā*) |
| ज्योतिष (jyotiṣa) | आपः (*āpaḥ*) |
| पुरुष (*puruṣa*) | प्रकृति (*prakṛti*) |
| लघु (*laghu*) | गुरू (*gurū*) |

TABLE 20.2 Modern system

| | |
|---|---|
| on | off |
| Open | closed |
| Charged | non-charged in electronic circuitry |

The discovery of binary numbers is attributed to Gottfried Leibniz (1646–1716) at the end of the seventeenth century. He is said to have come up with the idea when he interpreted Chinese hexagram depictions of Fu Hsi in I-Ching (the Book of Changes) in terms of binary code. But In India, Āchārya Piṅgal (eighth century BC as per Indian tradition/second century BC as per Western historians), the author of the *Chandaḥśāstra* and a mathematician and meterist applied binary code as many 2800 years before Christ while suggesting the *prastara* of *chandas*.

Piṅgala gives the following rule:

द्विरर्द्धे –8.28

Place two when halved

रूपे शून्यम् –8.29

When *rūpa* (unity or one) is subtracted, place zero

द्वि शून्ये –8.30

Multiply by two when zero

तावद् अर्द्धे तद् गुणितम् –8.31

Square when halved

| The Binary Code in Prastar of Gāyatrī Chanda | No | No to be placed |
|---|---|---|
| No of syllables | 6 | |
| Halve the no. | $6/2 = 3$ | 2 |
| Since 3 cannot be halved, subtract 1 | $3 - 1 = 2$ | 0 |
| Halve the number | $2/2 = 1$ | 2 |
| Since 1 cannot be halved, subtract 1 | $1 - 1 = 0$ | 0 |

Set (2,0,2,0). Let us start from last no. 0 and apply the Piṅgala rule

| 2 | 0 | 2 | 0 |
|---|---|---|---|
| $(2^3)^2 = 2^6$ | $2 \times 2^2 = 2^3$ | $2^2$ | 2 |

$$26 = 2 \times 2 \times 2 \times 2 \times 2 \times 2 = 64$$

Similarly, we can have binary code for other *chandas*

*Uṣṇika* = 7 = 0,2,0,2,0 = $2^7$ = 128
*Anuṣṭup* = 8 = $2^8$ = 256
*Bṛhati* = 9 = $2^9$ = 512
*Paṅkti* = 10 = 210 = 1028
*Triṣṭup* = 11 = $2^{11}$ = 2056
*Ati* = 12 = $2^{12}$ = 4196

## 10 Various Denominations of Zero

Zero has been represented by various denominations starting from its Vedic denomination '*kha*' to '*śūnya*'. It will be highlighted by the following citations:

### 10.1 '*Kha*', '*Śūnya*' and '*Ambara*'

We have seen while applying binary code Āchārya Piṅgala (twenty-eighth century BC traditional/200 BC modern) used for the first time the word '*śūnya*' for zero. Afterward, the word '*śūnya*' in addition to '*kha*' and its synonyms was also applied to express zero in India.

In the *Pañcha-siddhāntikā* (649 BC traditional estimate and 500 AD modern estimate), Varāhamihira addresses zero by denominations like '*kha*', '*śūnya*' and '*ambara*'. For instance in 3.2 and 3.17, '*kha*' is used to describe zero. Examples are:

एकादशाऽष्टषट् रूपोना सप्ततिः 'ख'युक्ता च। 3.2

*ēkādaśā'ṣṭaṣaṭrūponāsaptatiḥ* 'kha'*yuktā cha*। *3.2*

[The numbers of minutes to be deducted from the Sun's mean longitude are eleven (11), eight sixes (48), seventy minus one सप्ततिः रूपोना (69), and that plus zero खयुक्ता (70).]

[यम]- शिखि-गुणा-ऽग्नि-यम-शशि-वियुता सैका सरूप रूपे-का
खे-कवियुता च भानोः षष्ठिर्भुक्तिः क्रमादेवम्।। 3.17

[*yama*]- *śikhi-guṇā-'gni-yama-śaśi-viyutāsaikāsarūparūpai-kā;*
khai*-kaviyutā cha bhānoḥṣaṣṭhirbhuktiḥkramādevam.*

[The daily motion of the sun in minutes during each of 12 months amounts to sixty minus three (57), three (57), three (57), three (57), two (58), one (59), plus one (61), one (61), one (61), one (61), minus *kha* (60), one (59) in turn.]

Note: In the above *śloka* it has also been pointed out that when zero is subtracted from a number, its value does not change.

In 4.7 and 6.12, '*śūnya*' is used to denote zero. Examples are:

सेकाऽजे पंचाशत् पंचाष्टकं पंचवर्गवेदाश्च।
त्रिंशच्चतुर्भिरधिका षट्पंचशच्छराः शून्यम्।। 4.7

*saikā'jepaṁchāśatpaṁchāṣṭakaṁpañchavargavedāścha;*
*triṁśachchaturbhiradhikāṣaṭpaṁchaśachchharāḥ*śūnya*m.*

[In the first sign, the second parts are successively fifty plus one (51), five eights (40), five squared (25), thirty increased by four (34), fifty plus six (56), five (5) and śūnya (0).]

In 4.8, the denomination of '*ambara*' is used to denote zero. For example,

षट् त्रयोदशै (कोना) विंशति स्त्रयष्टकान्यतस्त्रिंशत्।
युक्ताम्बर-पंचनवा (तिज) गतिभिर्लिप्तिका वृषभे।। 4.8

*ṣaṭtrayodaśai* (*konā*) *viṁśatistrayaṣṭakānyatastriṁśat;*
*yuktāmbara-paṁchanavā* (*tija*) *gatibhirliptikāvṛṣabhe.*

[Of the signs in the second sign, the minutes' parts taking the increments in the current sign alone, are successively six (6), thirteen (13), twenty minus one (19), three into eights (24) thirty plus *ambara* (zero) (30), thirty plus five (35), and thirty plus nine (39).]

Note: The above example also indirectly informs that when zero is added to a number, its value remains unchanged.

Thus, we see in one single book three denominations were used to express zero numerals.

All the above instances occur in those sections of the *Pañcha-siddhāntikā* which deal with the teachings of Pauliśa. It seems, therefore, that such expressions are quotations from the *Pauliśa-siddhānta. Pauliśa-siddhānta* has now become extinct. Varāhamihira (649 BC as per Indian traditional chronology and 500 AD by modern scholars) in his *Pañcha-Siddhantikā* mentions five (*pañcha*) ancient *Siddhantas* arranged in order of chronology. They were the *Paitāmaha, Vasiṣṭha, Romaka, Pauliśa,* and *Sūrya*. Presently none of *Siddhāntas* has survived except *Sūrya-siddhānta*. *Sūrya-siddhānta*'s ancient version has also become extinct, only its modern version is available. From the aforementioned, we can understand the antiquity of *Pauliśa-siddhānta* (27th Mahāyuga traditional/400 AD modern) which was ancient than the ancient *Sūrya-siddhānta*. The text of *Sūrya-siddhānta* says that the *Sūrya-siddhānta* was revealed to Maya at the end of the Satyayuga of the present 28th *Mahāyuga*. As such we can safely say that *Pauliśa-siddhānta* was prevalent in this country in the hoary past. Now it has survived in the references quoted by ancient and medieval astronomers like Bhaṭotpala (305 AD traditional). It is also well known that 'word numerals' were employed by Pauliśa, so it can be safely concluded that he was conversant with the concept of the zero as a numeral.

The writings of Jinabhadra Gaṇi, a contemporary of Varāhamihira, offer conclusive evidence of the use of zero as a distinct numerical symbol. While mentioning large numbers containing several zeros, he often enumerates, obviously for the sake of abridgment, the number of zeros contained. For instance: 224,400,000,000 is mentioned as 'twenty-two forty-four, eight zeros' (*Bṛhat-kṣetra-samāsa*, 1. 69)

द्वि विंशति च चतुश्चत्वारिंशति च अष्ट शून्यानि

*dvi viṁsati ca chatuśchatvāriṁśati ca aṣṭaśūnyāni*

[Twenty-two and forty-four and eight zeros.]

Similarly, 3,200,400,000,000 is mentioned as 'thirty-two, two zeros, four, eight zeros'. (*Bṛhat-kṣetra-samāsa*, 1. 71). There are several instances of this kind in his work (1.90, 97, 102, 108, 113, 119, etc.)

At another place in the *Bṛhat-kṣetra-samāsa* (1.83) zero is mentioned in describing

$$241960\,\frac{407150}{483920} = 241960\,\frac{40715}{48392}$$

as two hundred thousand, forty-one thousand, nine hundred and sixty; removing (*apavartana*) the zeros, the numerator is 'four-zero-seven-one-five', and the denominator 'four-eight-three-nine-two'.

Here the term *apavartana* is used in the sense of the reduction of a fraction to its lowest terms by removing the common factors from the numerator and the denominator. Hence the zero of Jinabhadra Gaṇi is a specific numerical symbol used in the arithmetical calculation (Datta, 1962, p. 79).

Another contemporary mathematician, Bhaskara I in his famous text *Mahābhāskarīyam* (1.36) refers to the word '*vyoma*', a synonym of '*kha*' and '*śūnya*' and '*bindu*' (small circle/raindrop) to represent zero.

> व्योमशून्यनेत्रभाजिते फलं राशयोऽष्टभाजितेऽथ लिप्तिकाः।
> बिन्दुषड्हृते विलिप्तिका विदुः सर्वमेव योज्य गण्यते बुधः।।
>
> *vyoma*śūnya*netrabhājitephalaṁrāśayo'ṣṭabhājite'thaliptikāḥ;*
> *binduṣaḍḍhṛteviliptikāviduḥsarvamevayojyagaṇyatebudhaḥ.*
>
> [Divide *grahatanu* by 200 (*vyoma-śūnya-netra*); the result is in terms of *rāśīs*. The divide by 8, we get minutes (*liptikā*); then divide by 60 (*bindu-ṣad*), we get seconds. Adding all these (including 4 times the Sun's mean longitude as prescribed in the previous *śloka*) we get the mean longitude (*śighrochcha*) of Mercury.]

In the above *śloka*, *vyoma* and *śūnya* both have been used to represent zero (0) making the number 200. Similarly, zero of sixty (60) has been expressed by the word '*bindu*'. So we can say that '*bindu*' was an additional denomination for representing the numeral zero (0).

We come across a passage from the *Vāsavadattā* of Subandhu where the stars have been compared with '*śūnya-bindus*' (zeros). Thus, Subandhu in his book *Vāsvadattā* uses the word '*bindu*' (literally meaning small circle/raindrop, etc.) as a symbol to represent zero. The passage goes like this:

> शशिकठिनोखण्डेन तमोमशीश्यामेऽजिन इव नभसि संसारस्यातिशून्यत्वाच्छून्यबिन्दव इव.
>
> *śaśikaṭhino*khaṇ*ḍenatamomaśīśyāme'jinaivanabhasisaṁsārasyāti*śūnya *tvāchchhūnyabindavaiva.*
>
> [And at the time when the Moon rises with its blackness of night, bowing down under the guise of closing blue lotuses, the stars appear in the sky like *śūnya-bindus* (zeros).]

Subandhu was the minister of King Bindusāra (fifteenth century BC as per Indian tradition/280 BC as per European scholars) of Pāṭaliputra. The famous

Patañjali (1200 BC as per Indian tradition/150 BC as per European scholars) makes a mention of Subandhu's *Vāsvadattā*.

In his commentary on the gloss of Umāsvāti on Tattvārthādhigama-sūtra (3.11), Siddhsena Gaṇi uses, the word *śūnya* for zero while performing the square root of 3,53,44,00,00,00,00 as 1,88,00,00 by removing the four *śūnyas*' (zeros) from eight and doing the square root of remaining portion 3,53,44 into 188. According to Jain tradition, Umāsvāti lived around 200 BC. The time of Siddhsena is considered to be around the fourth century AD.

Bhāskara II and so many other scholars also extensively use the word '*kha*' as the denomination of zero. Bhāskara II uses the word *kha-guṇa* for the product of a number and zero and '*kh-hara*' for the division of a number by zero

Thus we see that '*kha*' and synonyms of '*kha*' like '*ambara*', '*vyoma*', '*ākāśa*', etc. were used as the denominations of zero right from the Vedic age till the medieval period of Indian history, but the use of '*śūnya*' was started by Āchārya Piṅgala in the post-Vedic period and became more popular now-a-days replacing all other denominations.

## 11 Origin of Zero Symbol

Throughout Vedic and post-Vedic Sanskrit literature, zero was expressed by the word '*kha*' so its symbol was also developed based upon the philosophy behind the word '*kha*'. '*Kha*' also means space and space have been characterized as '*Khagola*' meaning 'round or egg-shaped universe'. Its synonym is *Brahmāṇḍa*, which also literally means 'Egg-shaped Universe'. Since '*kha*' or the universal space as well as *chidākāśa* (space of Brahman) from which the universal space manifested, both are '*gola*' 'round or egg shaped', so zero was also represented by a 'small circle' or 'round' or 'egg shaped' figure (0). This figure is applied right from the Vedic period to date. In ancient India, zero was also denoted by the word '*Pūrṇa*' which represents *Pūrṇa Chandra* 'Full Moon'. So zero was also shaped like 'Full Moon'. In *Gopatha Brāhmaṇa* (2.2.5), we come across an example where *kha* (zero) is described as a '*chhidra*' (hole). The reference is cited below:

> मख इत्येतद्यज्ञनामधेयम्। छिद्र प्रतिषेधसामर्थ्यात् छिद्रं खं इत्युक्तम्। तस्य मा इति प्रतिषेधो मा यज्ञं छिद्रं करिष्यतीति।
>
> *makhaityetadyajñanāmadheyam. chhidraprastiṣedha-sāmarthyāt chhidraṁ khaṁ ityuktam. tasyamāitipratiṣedhomāyajñaṁchhidraṁkariṣyatīti. chhidramkhamityutkam*

> [*Yajña* is called *makha* (*ma* + *kha*). '*Ma*' represents '*Mā*' which means 'prohibition' and '*kha*' is used to represent zero (emptiness). Zero has the shape of *chhidra* (hole). So '*ma-kha*' means prohibition of *chhidra* (hole or emptiness) gap-filler. As such *yajña* fills up zero *or chhidra* (emptiness).]

The *Amarkośa* also describes zero as '*randhra*' (hole).

In addition to the above examples, we find some alterations in its shape or figure. As pointed out above there are positive paleographical shreds of evidence that zero was represented in India by a small circle in the two inscriptions of Bhojadeva (336–392 AD), and Mahipāla (Ojha, 1918, pp. 127 and Lipi-patra, 75). However, in Bakhśāli Manuscript, we find zero represented by *bindus* or solid dots (.). *Bindu* means 'rain-drop'. Raindrop is a circular shaped figure. So, zero was also represented by circular-shaped figures (o) or dots (•). Solid dots were easily scribed on birchbark instead of a circular-shaped figure (o), so the scribe would have used bold dots for the sake of convenience. However, Mukherjee (1991) identifies one particular solid dot in the manuscript as a small circle. So, using a solid dot for the sake of convenience in writing on birchbark, never means that the scribe was not aware of the circular-shaped figure of zero (o).

As already pointed out above, the *Vāsavadattā* of Subandhu in a metaphor points out *bindu* (solid dot/drop) as a representative figure for '*śūnya*' (zero) while comparing the stars with '*śūnya-bindus*' (*bindus* as figures representing zero). Thus, Subandhu in his book *Vāsvadattā* uses the word '*bindu*' (literally meaning dot, raindrop, etc.) as a symbol to represent zero. In *Naiṣada-charita* (1.21) also, Śri Harṣa (88–92 AD/Modern 606 AD) describes zero as '*śūnyabindu*' as follows:

> किमस्य लोम्नां कपटेन कोटिभिर्विधिर्न लेखाभिरजीगणद्गुणान् ।
> न रोमकूपौघमिषाज्जगत्कृता कृताश्च किं दूषणशून्यबिन्दवः ॥ २१ ॥
>
> kimasyalomnāṃkapaṭenaKoṭibhirvidhirnalekhābhirajīgaṇadguṇān;
> naromakūpaughamiṣājjagatkṛtākṛtāścakiṃdūṣaṇaśūnyabindavaḥ.
>
> [Did not the Creator reckon his merits with crores of lines, the hairs of his body? Did not the maker of the world put the pores of his skin for a small circle of zeros to indicate the absence of defects?]

Here it may be pointed out that the above examples may not be taken so seriously to decide the form of the zero symbols. *Bindu* has also been used in the sense of raindrop, spot as in *bindu-mṛga* (spotted deer) and small circle.

## 12 Paleographic Evidence of the Use of Zero Symbol

The Bakhshali Manuscript (200 AD) was found in 1881 at a place called Bakhshali, about eighty kilometers north-east of Peshawar (now in Pakistan), by a farmer who dug it up while cultivating the land. The manuscript is preserved in the Bodleian Library at Oxford University.[1] The untitled manuscript, written on birch bark, is composed in the Gāthā language (a modified form of *prākṛta*) and the scripts used are recognized as an earlier type of Śāradā, once used in the Kashmir region of India. In the seventy fragmentary leaves of the document, which is now extant, one finds numerous intricate calculations involving decimals, and place-value notation of numerals, including a solid dot and a small circle for *śūnya* (zero). The work seems to be an exercise compiled by a scribe. The manuscript records the word *śūnya* to represent invisible (zero) in calculations. For instance, on folio 56 verso, we have:

880 × गुणितं जातं (multiplied by) 964 = 848320
84 × गुणितं जातं (multiplied by) 168 = 14112
चत्वारिंश पृथक् स्थानं वर्ग (the square of 40 different places is) 1600
एष उपरापात्यशेषं (On subtracting this from the number above (numerator), the remainder is

$$848320 - 1600 = \frac{846720}{14112} = 60$$

वर्त्यजातं (on the removal of the common factor it becomes 60

For an explicit reference to zero and an operation with it, folio 22, verso has already been cited above.

On the other hand, there are positive paleographical shreds of evidence of the use of zero as a decimal digit depicted as a small circle in the two stone inscriptions of Gwalior from King Bhojadeva (336–392 AD) discovered in the nineteenth century in a temple of Lord Viṣṇu in Gwalior fort, and King Mahipāla (Ojha, 1918, p. 127 and Lipi-patra, 75).

We also find it used in Kalchuri grant dated 695 AD (Ojha, 1918, p. 115).

1 See Appendix 3 for an online petition urging the Bodleian Libraries, Oxford, UK, to take concrete steps to commission the necessary follow-up radiocarbon-dating of the Bakhshali Manuscript in the interest of scientific advancement in the field.

## 13 Journey of Zero from the East to the West

The science of Mathematics in India also goes back to Vedic times as in the case of other sciences. Algebra, Arithmetic including geometry had their origin in India. India is the inventor of many mathematical ideas. These ideas developed during the Vedic period and percolated with the passage of time to the present age. A lot of basic mathematical ideas travelled from India to other parts of the world. There are many examples that shed ample good light on the Indian influence on Arabic, Roman and Greek mathematics.

The sexagesimal place value notation of Hellenistic astronomy was a modification of Indian decimal and place value notation.

The trapezium problems in Heron can be traced to Āryabhaṭa. Brahmagupta's treatment of rational solutions of the right-angled triangle influenced Ptolemy's cyclic surds, and Euclid's quadrilaterals and influenced Diophantus' indeterminate equations of the second degree in Greece. Pythagoras' theorem can be traced back to Baudhāyana Śulvasūtras.

It is a proven fact that there was close contact between the Indian and Greek mathematicians in those old days.

The statement of a Syrian astronomer (John Leyden et al., 1921, pp. 78–79) of seventh century AD is noteworthy. He writes:

> No word can praise strongly enough the knowledge of the Hindus. If these things are known to the people who think that they alone have mastered the sciences because they speak Greek, they would perhaps be convinced though a little late in the day that other folk not only Greek but men of different tongues, know something as well as they.

Like all other Indian mathematical inventions '*śūnya*' (zero) also reached Arab and was transliterated into Arabic in Al-Khwarizmi's (780–850 CE) work *De Numero Indorum* (Hindu Art of Reckoning) as '*sifr*', meaning 'empty'. The associated Arabic root √*sfr* leads to other words of similar meaning, like '*asfara*' (meaning 'empty') or '*safir*' (meaning 'to be empty'). The slightly changed term '*sifra*' was used to denote 'zero' in '*Sefer ha misper*' (meaning, *Book of Numbers*) by Rabbi Ben Ezra (1092–1167). This one, along with many local variants like *cifra, cyfra, cyphra, zyphra*, etc., was used to denote 'zero'. However, they took a turn by the beginning of the Renaissance and one may find the Spanish word '*cifra*', derived directly from Arabic '*sifr*' being used with a changed connotation to refer to the rest of the digit numbers. Similar examples prevail in the case of other European languages as well – for instance, '*chiffre*' in French, or '*ziffer*' in German are still used to refer to digit numbers. However, the Arabic '*sifr*' was gradually latinized as '*zephyrus*' – 'the west wind' and Leonardo of Pisa, better known as Fibonacci, while writing his seminal book, *Liber Abaci* in 1202, coined the term '*zephirum*' – mere light breeze – almost nothing, to denote 'zero'. Later it got converted to '*zefiro*' in the Venetian dialect of the early Renaissance, and modern 'zero' is a mere contraction of that. The first known occurrence of the modern term 'zero' is found in an Italian book *De Arithmetica Opusculum* by Filipps Calandri, published in 1491 in Florence.

On the basis of the above discussion, we can safely say that Indian '*śūnya*' reached Arab and became '*sifir*' or 'cipher'. From Arab this '*sifir*' reached Europe following two routes – one Spanish and another Latin which is illustrated below:

TABLE 20.3 Route no. 1

| Sanskrit | Arabic | Spanish | French | English |
|---|---|---|---|---|
| शून्य<br>*śūnya* | सिफ्र<br>*sifir* | सिप्रा<br>*cifra*<br>*cyfra*<br>*cyphra*<br>*zyphra* | सिफ्रे<br>*chiffre* | सिफर<br>cypher |

TABLE 20.4 Route no. 2

| Sanskrit | Arabic | Latin | Latin | English |
|---|---|---|---|---|
| शून्य<br>*śūnya* | सिफ्र<br>*sifir* | जेफ्रम<br>*zephirum* | ज़ीफ्रो<br>*zefiro* | जीरो<br>zero |

## References

Al-Khwarizmi, Abu Jafar Muh (1990). *The Hindu Art of Reckoning*. English trans. by John N. Crossley and Alan S. Henry. *Historia Mathematica*, 17 (1990), pp. 103–131.

Agni Purāṇa (1914). Calcutta: Baṅgabāsī edition.

Arya, Ravi Prakash (1996a). *Ṛgveda Saṁhitā*. Delhi: Primal Publications.

Arya, Ravi Prakash (1996b). *Yajurveda Saṁitā*. Delhi: Primal Publications.

Arya, Ravi Prakash (1997). *Sāmaveda Saṁitā*. Delhi: Primal Publications.

Arya, Ravi Prakash (2007a). *Vedic and Classical Sanskrit*. Delhi: Indian Foundation for Vedic Science.

Aryabhaṭa (1976). *Āryabhaṭīyam*. Shukla, K. S. and Sarma, K. V. (Eds.) New Delhi: Indian National Science Academy.

Bakhshali Manuscript (1979). (Ed.) Swami Satya Prakash Saraswati, Dr. Ratna Kumari Svadhyaya Sansthana Allahabad.

*Bṛhat-kṣetra-samāsa*. Ed. with the commentary of Malayagiri. Bombay.

Clark, W. E. (1929). *Indian Studies in the honor of Charles Rockwell Lanman*. Cambridge MA: Harvard University Press.

Datta, B. (1962). *History of Hindu Mathematics*. Asia Publishing House.

Dayananda Saraswati (1961). *Yajurveda Bhāṣya*. Ajmer.

Dayananda Saraswati (1963). *Ṛgbhāṣya*. Ajmer.

Dayananda Saraswati (1984). *Ṛgvedādibhāṣvabhūmikā*, Ed. Yudhisthira Mimansaka, Bahalgarh, Sonepat.

Dvivedi, S. (1996). *Bṛhat-Saṅhitā*. Benares.

*Edinburgh Review* (1818). A Quarterly Journal (November 1817–February 1818). Vol. 29. Edinburgh and London.

*Edinburgh Review* (1820). A Quarterly Journal (January–May 1820). Vol. 33. Edinburgh and London.

Hall, F. (1859). *The Vasavadatta: A Romance*. Published under patronage of Asiatic Society of Bengal.

Halsted, George Bruce (1912). *On the Foundation and Technic of Arithmetic*. Chicago: The Open Court Publishing Company.

Hoernle, A. F. R. (1883). *Indian Antiquary*, Vol. XII.

Hoernle, A. F. R. (1888). The Bakhshali Manuscript, *Indian Antiquary*, Vol. XVII.

Hunter, W. W. (1881). *The Imperial Gazetteer of Bharat*, First Edition, 9 Vols. London: Tubner & Co.

Ifrah, Georges (2000). *The Universal History of Numbers*. John Wiley and Sons.

Kaye, G. R. (1912). The Bakhshali Manuscript: a study in medieval mathematics. *Journal of the Asiatic Society of Bengal*, Vol. VIII.

Kumar, Alok (2014). *Sciences of the Ancient Hindus: Unlocking Nature in the Pursuit of Salvation*. USA: CreateSpace.

Leyden, John and Erskine, William (1921) (Trans.) *Memoirs of Zehir-Ed-Din Muhammed Babur*, Emperor of Hindustan. Oxford: OUP.

Macdonell, A. A. (1900). *History of Sanskrit Literature*. New York: D. Appleton and Company.

Manning (Mrs.) (1869). *Ancient and Mediaeval Bharat*, 3 vols. London: Wm. H. Allen & Co.

Mukherjee, Rabindranath (1991). *Discovery of Zero and its impact on Indian Mathematics*. Calcutta: Dasgupta & Co. Pvt. Ltd.

Naishadha-charita of Shriharsha (1956). Ed. Krishna Kanta Handiqui. Poona.

Ojha, G. H. (1918). *Bharatiya PrachinLipi Mala*. Ajmer.

Pañcha-siddhāntikā by Varahamihira (1993). Ed. K. V. Sharma. Adyar, Madras: P.P.S.T. Foundation.

Piṅgala, Āchārya (1931). *Chandaḥ-Śāstram*, Ed. Sri Sitanath Sharma. Calcutta.

Raghvir (1936). *Atharvaveda*, S. V. Granthamāla. Lohore.

*Ṛgveda Saṁhitā* (1933). With the Commentary of Sāyṇāchārya, Vaidic Samshodhan Mandal. Poona.

Sachau, E. C. (1964). *Alberuni's India*. Vol. 1. Delhi: S. Chand & Company.

Samayochita Padma-ratna-mālikā (1957). Mumbai.

Sarda, *Diwan Bahadur Harbilass* (2007). Ed. Ravi Prakash Arya, *Hindu Superiority*. Delhi: Indian Foundation for Vedic Science.

Shripada Sharma (1942). *Kaṭha Saṁhitā*. Aundh.

Shripada Sharma (1943a). *Kāṭhaka Saṁhitā*. Aundh.

Shripada Sharma (1943b). *Maitrāyaṇī Saṁhitā*. Aundh.

Shripada Sharma (1943c). *Maitrāyaṇī Saṁhitā* (Brāhmaṇa portion), Aundh.

Shripada Sharma (1945). *Taittirīya Saṁhitā*. Aundh.

Tattvārthādhigama-sūtra of Āchārya Umāsvami (1926). Kapadia, H. R. (Ed.) with the commentary of Siddhasena Gaṇi. Bombay.

Weber, Albrecht (1878). *History of Indian Literature*. London: Trubner & Co. Ltd.

Williams, Monier (1875). *Indian Wisdom*. London & Co.

CHAPTER 21

# A Philosophical Origin of the Mathematical Zero

*Sudip Bhattacharyya*

## Abstract

We investigate if philosophical ideas of *nothing* could have given rise to the mathematical zero, which is distinct from *nothing* and is an independent number with a set of basic operations defined. This investigation should be complementary to other studies, e.g., those based on epigraphic and paleographic evidence in order to probe the origin of zero. First, we argue that a practical need did not give rise to the mathematical zero. Moreover, neither a place-value number system nor a placeholder in such a system was required to conceptualize zero. Therefore, we insist that zero and its basic operations were intellectually premised, which likely required a philosophical base of *nonexistence*, and not that of *emptiness* or *void* as commonly assumed. The prevalence of a philosophy of *nothing* in ancient India and the appearance of the first clear prescription of basic operations of zero in an Indian book of the first millennium CE point to a development of the concept of zero in India. We back this up with 'living evidence' of word-numerals developed in India, and propose that the comments of the Syrian Bishop Severus Sebokht, which are often used to argue against the Indian origin of zero, do not provide evidence against such an origin.

## Keywords

history – mathematics – nothing – number – philosophy – zero

## 1 Introduction

The aim of this chapter is to probe the philosophical origin of the mathematical zero and the possibility of its conception in India. At the outset, it would be useful to make a clear distinction between the idea of *nothing* and the mathematical zero. Sometimes, they are used interchangeably, which is confusing, because the origin of the idea of *nothing* and that of zero are not same. This is because zero itself is not *nothing*, it is an existing mathematical object, but it indicates or represents a category of *nothing*.

 | DOI:10.1163/9789004691568_025

A previous work classified *nothing* into four categories: *balance, absence, emptiness* and *nonexistence* (Bhattacharyya, 2021). Examples of *balance-nothing* are $m - m$ ($m$ is a nonzero number), two forces which are equal and opposite to each other, the unmanifested fundamental substance, *Prakṛti*, of the Indian Philosophical school of Sāṁkhya (Radhakrishnan, 2008b, pp. 226–307) and the fluctuating vacuum considered in modern physics (Seife, 2000, p. 172). Thus, *balance-nothing* is secondary to multiple existents. *Absence-nothing*, which implies the absence of a specific thing, is also a secondary of that *thing* which is an existent. While *emptiness-nothing* is not the absence of a particular thing, it requires a container, which is the primary and an existent. An empty space and a mathematical *empty set* are examples of *emptiness-nothing*, where space and set are the container respectively. Thus, none of these three categories of *nothing* is fundamentally nonexistent. This allows one to conceive absolute nonexistence as the fourth category of *nothing*. In this chapter, we discuss which category of *nothing* could be represented by zero.

How can one probe the origin of zero? On one hand, epigraphic and paleographic sources, as well as ancient books could be searched for and studied. On the other hand, one can look for 'living evidence' – in other words, evidence which still exists in an almost unchanged condition since ancient times. Some philosophical ideas and word-numerals are two such 'living proofs' discussed in this chapter.

First, we ask if the conception of the mathematical zero required a practical need. In section 2, we argue that neither did the progress of various civilizations require zero, nor could the place-value number system explain the origin of the mathematical zero. These point to what zero is, i.e., a representation of *nothing*, as the origin of zero. In section 3, we discuss the philosophy of *nothing* as the base of both the number zero and its basic mathematical operations. But where could this number and its operations originate? In section 4, we insist that the requisite philosophical source of zero existed in India, and discuss word-numerals as living evidence of the Indian origin of the mathematical zero. We make a few general comments on the time of origin of zero in section 5, and summarize this chapter in section 6.

## 2 Did Mathematical Zero Originate Due to a Practical Need?

### 2.1 *Was There a Need in Daily Life, Mathematics, Natural Science or Technology?*

It has been pointed out in section 1, that the mathematical zero is distinct from a broad and general idea of *nothing*. Zero is a specific mathematical existent and an independent number representing *nothing*. But it could gain the full status of a number, when at least its basic mathematical operations, such as addition, subtraction, multiplication and division, were invented. It is this mathematical zero which we consider in this chapter, and we ask if it could have originated due to a practical need.

Does one need the number zero in day-to-day activities? One does not need to consider the growth, measurement, building, buying or selling of zero amount of anything. Zero amount of knowledge and activity can more naturally be mentioned as 'no knowledge' and 'no activity' respectively. The ideas of absence and emptiness can easily be expressed by the words 'absence' and 'emptiness', and the number zero is not needed. Thus, an independent number zero representing nothing is not required in daily life, and hence it is not surprising that such a mathematical number was not conceived in every culture or civilization.

But was there a requirement of zero in mathematics? In this context, it is common practice to discuss the place-value number system and the placeholder in this system. However, we argue in section 2.2 that the place-value system and the placeholder should not have a role for the conception of zero. The idea of a reference point (e.g., for measurements) or a starting/end/boundary point could also have given the notion of the mathematical zero. But a reference/starting/end point does not necessarily imply nothing. Almost all ancient cultures and civilizations should have had the idea of a reference point. For example, the ancient Egyptians probably denoted the reference level with a symbol nfr. But almost none of these cultures could invent the mathematical zero as an independent number (e.g., Kaplan, 1999, p. 20; Seife, 2000, pp. 11–12). Note that zero was essential for the development of the current sophisticated and advanced mathematics. But one could not have foreseen this before zero's basic mathematical operations were invented, and certainly not before zero was conceived as an independent number representing nothing. Thus, a mathematical requirement might not have stimulated the invention of zero.

Finally, while many aspects of natural science and technology cannot be imagined without zero at the current time, the use of the mathematical zero in these disciplines has been a relatively recent phenomenon. For example, a number of ancient civilizations, such as those of Egyptians, Mesopotamians, Phoenicians, Greeks and Romans, made spectacular progress in technology without the mathematical zero. Thus, zero was likely conceived without a practical need.

### 2.2 *Zero, the Place-Value Number System and the Placeholder*

The most popular number system in current use is the place-value number system, in which the position of a digit in a number determines its value. It is believed that ancient Mesopotamians first invented such a system. While writing down a number using numerals, they would leave a gap or put a special placeholder symbol where the numeral zero is used today. This is why the placeholder is sometimes called the placeholder-zero and is often thought to be the first appearance of zero (Seife, 2000, p. 15). But neither was the Mesopotamian placeholder an independent number nor were its basic mathematical operations invented. Thus, it was not the mathematical zero (see section 2.1), and this zero was not invented by ancient Mesopotamians. Nevertheless, could the place-value number system and/or the placeholder have motivated the invention of the mathematical zero? This is what we examine in this section.

A connection between the place-value number system with zero is commonly made, almost implying that zero owes its existence to this system. To check if this could be true, let us consider an extreme case. Suppose each natural number is denoted by a single unique symbol (essentially implying an infinite number of such symbols). This is, while not practical, perfectly possible in principle or notionally. Note that, even in such a rudimentary number system, not only could zero be considered as an independent number with its own distinct symbol, but also its basic mathematical operations could be defined. This could be easily checked by considering just 0–9 in the current decimal place-value number system. Therefore, clearly the conception of the mathematical zero does not require the place-value system.

Here we note that, since the above-mentioned number system with a unique symbol for every natural number is not manageable in practice, various cultures invented techniques to write bigger numbers using a limited number of distinct symbols. Thus, ancient Egyptians, Greeks and Romans invented their own ways (Seife, 2000, p. 13). But the place-value system invented by ancient Mesopotamians appears to be the most compact and systematic way to write bigger numbers. Mesopotamians used a sexagesimal system with 1–59 digits (since they did not have 0). The currently used Indian decimal place-value number system has ten digits including zero, and is ideal for writing very big numbers and for performing complex calculations relatively easily. But, as argued above, such place-value systems were not required to conceive the mathematical zero, and Egyptians, Greeks and Romans could easily include zero in their number systems, had they invented this zero. All these imply that the place-value number system could not have motivated the invention of the mathematical zero.

Could at least the placeholder stimulate the idea of zero? This is unlikely, as the placeholder is required only in the place-value number system, that too only when certain numbers are written using numerals. Moreover, the placeholder did not represent the *emptiness-nothing* (see section 1) and was only a separator or a punctuation mark (Bhattacharyya, 2021) which could be understood from the following: (1) Mesopotamians never used the placeholder at the end of a number (e.g., 60) where zero should have appeared (Kaplan, 1999, p. 12); and (2) the placeholder was not used as an independent number. Note that a mere punctuation mark should not have stimulated the idea of zero. Thus, we suggest that it would be misleading to bring in the place-value number system and the placeholder in order to probe the origin of zero.

## 3 The Need of a Philosophy for the Origin of Mathematical Zero

If a practical need did not give rise to the mathematical zero (see section 2), then it must have been conceived purely intellectually. What could be the intellectual source of zero? Zero implies *nothing*, and hence it is expected that a philosophy of *nothing* would stimulate the idea of the mathematical zero.

### 3.1 *Philosophical Origin of the Number Zero*

Considering that the mathematical zero was conceptualized from a general idea of *nothing*, what category of *nothing* could it be (see section 1)? Zero is often defined by $m - m = 0$, and in such a case it would be a *balance-nothing* (section 1). But there could be two objections to this. First, $m - m$ cannot be concluded to be zero without a prior understanding of zero as *nothing*. Therefore, $m - m$ cannot define zero, as the concept of zero should be primary and $m - m = 0$ should be secondary. Second, $m$ could have any value and thus zero would be defined non-uniquely by any number, had $m - m$ defined zero. Consequently, zero would not have been an independent number. It would be like defining the number 7 non-uniquely by 1478 – 1471 and 123450 – 123443, instead of treating it as a number in its own right. It would also make natural numbers secondary to *their* subtractions, which involves a self-contradiction. Therefore, the mathematical zero should not have been conceptualized from *balance-nothing* (Bhattacharyya, 2021).

If *absence-nothing* were the source of zero, then zero would be defined by the absence of a thing (e.g., any number). Thus, again, zero would not be an independent number and would be defined non-uniquely. On the other hand, if zero would represent *emptiness-nothing*, it would be defined with respect to a container. Thus, unlike other numbers, it would not be independent. Therefore, neither *absence-nothing* nor *emptiness-nothing* should be the source of the mathematical zero (Bhattacharyya, 2021). In fact, the ideas of *balance*, *absence* and *emptiness* must have been common across cultures, and if any of these categories were a source of the mathematical zero, it would perhaps give rise to the idea of zero in multiple cultures. Hence, we suggest that a notion of the absolute nonexistence or the *nonexistence-nothing* perhaps provided the conception of zero (Bhattacharyya, 2021).

### 3.2 *Philosophical Origin of Zero's Basic Operations*

The first record of zero's basic mathematical operations is found in the book *Brāhmasphuṭasiddhānta*. Brahmagupta, an Indian mathematician and astronomer, wrote this book in 628 CE (Datta & Singh, 1962a, pp. 59, 238–244; Datta & Singh, 1962b, pp. 20–24; Kaplan, 1999, pp. 71–72). Zero attained the full status of a number due to these operations. Here, we discuss if such operations could be invented without a philosophical premise of *nothing* (Bhattacharyya, 2021).

First, we would like to point out that zero's mathematical operations are different from those of other numbers, and hence did not directly follow from such operations of nonzero numbers. This is sometimes considered an anomaly in some cultures (see Seife, 2000, pp. 19–23), which indicates that the philosophical base of zero is not perhaps fully appreciated (Bhattacharyya, 2021). Let us start with addition and subtraction. Unlike any other number, addition and subtraction of zero do not change a number. In fact, addition and subtraction of zero are also conceptually different from those of nonzero numbers, as explained in the following. Note that the notion of increase and decrease preceded and was primary to the mathematical addition and subtraction of nonzero numbers. This is because, while even non-human animals can recognize changes such as increase and decrease, the invention of mathematical addition and subtraction, which are quantitative measurements of increase and decrease, required advanced intellectual exercise during the early phase of civilization. But for zero, increase and decrease are not primary, and the rules of addition and subtraction could not be ascertained without a philosophical premise or a conception of *nothing* (Bhattacharyya, 2021). We also note that since there is no change due to addition or subtraction of zero, at the beginning, such operations might not have any practical need. This supports our conclusion in section 2.

Let us now consider multiplication. Unlike nonzero numbers, zero converts all numbers to itself by multiplication. This indicates that all numbers are treated in the same way by zero. This is true even for addition, subtraction and division, as all numbers remain unchanged due to addition or subtraction of zero and all nonzero numbers become infinity when divided by zero. This suggests that zero is profoundly different from nonzero numbers. This is to be expected if zero represents nonexistence and all nonzero numbers together imply existence (Bhattacharyya, 2021).

But why is $c \times 0 = 0$? Sometimes, this is proved in the following way: $c \times 0 = c \times (m - m) = (c \times m) - (c \times m) = 0$. But this is not proper, because $m - m$ is secondary to the concept of zero and hence does not define zero (see section 3.1). Rather, $c \times 0 = 0$ can be premised from the philosophy of *nothing* in the following way. Note that $c \times 5$ means five '*c*'s. Similarly, $c \times 0$ implies no '*c*'s, that is *nothing* or mathematically zero (Bhattacharyya, 2021). We also note that zero divided by a nonzero number is same as zero multiplied by the reciprocal of that number. Finally, a division by zero also requires a philosophical premise of *nothing* (see Bhattacharyya, 2021, for details).

## 4 Some Cases for the Indian Origin of the Mathematical Zero

If the placeholder does not have a role in the conception of the mathematical zero (see section 2.2), then it is unlikely that this zero originated in Mesopotamia or China. The ancient Egyptians and Greeks did not have the mathematical zero either (Seife, 2000, pp. 11–12). The Islamic world learned the Indian number system including zero only in the second half of eighth century CE (Sachau, 2002, pp. xxx–xliii; Bhattacharyya, 2021). On the other hand, not only were zero's basic mathematical operations first established in India (see section 3.2) but there are also several strong indications of the Indian origin of the mathematical zero available. While epigraphic and paleographic sources are important to probe the origin of zero, there are other kinds of evidence. Here we discuss two 'living proofs' – philosophical ideas and word-numerals (sections 4.2 and 4.3). But before going into these, let us examine one of the most popular objections to the Indian origin of zero.

### 4.1 *Comments of Syrian Bishop Severus Sebokht*

In 662 CE, a Syrian Bishop, Severus Sebokht, clearly annoyed by the Greeks' claim to superiority in scholarly activities, wrote that Greeks had learned science from the Babylonians; that Greeks had not invented astronomy – Syrians had; and that Hindus (implying Indians) had made more ingenious discoveries than they had. In order to establish the last point, he mentioned computations done by Hindus using *nine* signs (Kaplan, 1999, p. 46). While this is evidence of the Indian use of the decimal system, the mention of *nine*, and not *ten*, signs is sometimes thought to be evidence of the absence of the mathematical zero in India that time.

However, this alleged absence cannot be true, because the Indian mathematician and astronomer Brahmagupta had already written the basic mathematical operations of zero in his book in 628 CE (see section 3.2). Could it then be possible that Indians did not have a sign or symbol for zero in 662 CE? This is rather unlikely, because a stone inscription found near Sambor in Cambodia showed the Saka year 605 (683 CE; Datta & Singh, 1962a, p. 44) with a small dot marking the numeral zero. This was an Indian culture in Cambodia, and therefore this numeral for zero would have been in use in India long before 683 CE.

Why didn't Severus Sebokht mention *ten* signs then? Note that the Indian numbers and mathematics likely spread in his part of the world (i.e., Mesopotamia) about a century after he mentioned Indian numerals (Sachau, 2002, pp. xxx–xliii; Bhattacharyya, 2021). Hence, it is unlikely that Sebokht knew much about Indian numbers and mathematics. He had possibly vaguely heard about these numbers, but never used them himself. Therefore, having no first-hand experience, he could have had the following two strong reasons for not mentioning the Indian number zero: (1) He was not only a Christian but also an Aristotelian, and both these doctrines, particularly the latter, were not supportive of the idea of *nothing* (Seife, 2000, pp. 47–48; Bhattacharyya, 2021). Thus, he would not accept a number based on *nothing* just by vaguely hearing about it. (2) Note that his main aim was to criticize the Greeks and not to make Indian mathematics known to people. Therefore, naturally he would not mention a number based on *nothing*, because that would weaken his arguments, given that the idea of *nothing* was not accepted by the ancient Greeks (Bhattacharyya, 2021).

## 4.2 *Philosophy of* Nothing *in India*

Let us now consider the first living evidence of the Indian origin of the mathematical zero. It has been argued in section 3 that zero would not have been conceived without a prior philosophy of *nothing*. While such a philosophy was either not considered or not accepted in many ancient cultures, it was one of the primary aspects of several ancient Indian philosophical schools. Perhaps, the first instance of it is found in the 'Nāsadīya Sūkta' or 'Hymn of Creation' (Ṛg-Veda, 10:129) of the ancient Indian book Ṛg-Veda, which might have been composed before 1000 BCE. The hymn starts with, 'Then even nothingness was not, nor existence' (Basham, 2004, p. 250). While this hymn conveys philosophic doubts, it also mentions, 'That One which came to be, enclosed in nothing' (Basham, 2004, p. 250).

In a later period, multiple Indian philosophical systems, such as Upaniṣads and Sāṁkhya, had the idea of *nothing* as an important aspect. For example, *Nirguna Brahman*, or the 'Absolute without qualities' of Upaniṣads, might be somewhat close to *nonexistence-nothing*. Note that some of the Upaniṣads might have been composed in the first half of the first millennium BCE (Radhakrishnan, 2008a, pp. 106–220). On the other hand, the unmanifested fundamental substance, *Prakṛti*, of the Sāṁkhya philosophy can be an example of *balance-nothing* (see section 1).

However, the Indian philosophical idea of *nothing* was spread far and wide by Buddhism. Particularly around third century CE, the Indian Buddhist school, the Śūnyavāda, argued that everything is empty (Radhakrishnan 2008a, p. 508). Such an Indian philosophy of nothing gave birth to the *Ch'an* philosophy in China. Note that the word *Ch'an* was derived from *dhyana*, an Indian word for 'meditation'. Later, *Ch'an* spread into Japan and came to be known as *Zen* (Yu-Lan, 1976, p. 255). The Indian philosophy of *nothing* is still valued in *Zen* and some other sects of *Mahayana* Buddhism in many countries, and provides living evidence of the philosophical background of the mathematical zero.

### 4.3 *Word-Numerals: Living Evidence*

Further living evidence of the Indian origin of the mathematical zero is word-numerals. This is a way of expressing numbers by specific words, which are arranged according to the place-value system. Word-numerals were developed in India, likely in the early centuries of the Common Era (Datta & Singh, 1962a, p. 53). Many words were assigned to represent each of the numbers from 0 to 9, and sometimes even larger numbers (Datta & Singh, 1962a, pp. 54–57). Thus, *chandra* (moon) is one of the words representing 1, because moon is unique. Similarly, *pakṣa* (fortnight) is one among many words representing 2, because there are two fortnights in a month. There are a number of words to represent 0 as well, such as *śūnya* (emptiness), *kha* (sky), *bindu* (point), etc.

Why was a word needed to express a numeral? This is because, in ancient India it was customary to compose or write results and conclusions of scholarly works, including those of mathematics and astronomy, in verse (Datta & Singh, 1962a, p. 54). But why were there many words to express a single number? Perhaps, this was a result of the diversity and evolution of cultures and languages in India. Moreover, it was useful to have many different words for a single number in order to fit the meter of the poem/verse.

It should be noted that word-numerals are not words, they are actually numerals, represented by words only due to the cultural reason as mentioned above. That word-numerals are actually numerals can be found not only from their arrangement according to the place-value system, but also from their right-to-left arrangement (i.e., the least important digit is mentioned first). The latter is particularly suitable for mathematical operations because basic mathematical operations, such as addition, subtraction and multiplication, start from the least important digit. This indicates that word-numerals could be used for mathematical calculations.

An example of such an arrangement of word-numerals can be found in the ancient Indian astronomy book *Puliśa-siddhânta* (Datta & Singh, 1962a, p. 59):

*kha* (0) *kha* (0) *aṣṭa* (8) *muni* (7) *râma* (3) *aśvi* (2) *netra* (2) *aṣṭa* (8) *śara* (5) *râtripâḥ* (1) = 1582237800. Here, we note that, unlike the Mesopotamian placeholder (see section 2.2), zero was used at the end of a number, indicating that this zero was not a mere separator or placeholder. Moreover, given that there are several words to represent zero, like numbers 1–9, and that all these words for zero suggest *nothing* one way or another, one may conclude that this Indian zero is an independent mathematical number founded on the idea of *nothing*. Thus, this zero in the form of a word-numeral is the mathematical zero considered in this chapter.

The main attraction of word-numerals in order to probe the origin of the mathematical zero is that they are a living system, continuing without an interruption for almost two millennia in India (Datta & Singh, 1962a, pp. 54, 58–59). They are taught to children and used in verse even today. They have been extensively used throughout many centuries in multiple disciplines, including literature. In fact, it was rather common to use word-numerals to mention dates in poetry. For example, in the Bengali poetry *Chandimangal* written by Dwij Madhab in the sixteenth century CE, the year of composition is referred to as '*indu* (moon = 1) *bindu* (point = 0) *baan* (arrow = 5) *dhata* (god = 1)' in Śaka era (e.g., Mandal, 2021, p. 95). This implies the year 1501 Śaka or 1579 CE, as the Śaka era started in 78 CE. Thus, word-numerals for numbers including zero were popular in multiple languages in India, and are deeply entrenched in the diverse Indian culture.

We note that, while epigraphic and paleographic evidence of the origin of the mathematical zero are required for a definite proof and to pinpoint a short range of time, such evidence could be isolated and without a clear perspective. On the other hand, living evidence, like word-numerals, can provide a broader picture of cultural background. Such a background, unlike an isolated evidence, may provide a much required general confidence. The continuity of a living evidence to the present day can also make an interpretation more reliable.

## 5 Comments on the Time of the Origin of the Mathematical Zero

A reliable and precise time of the origin of the mathematical zero could be established from the dating of epigraphic and paleographic specimens (e.g., the Bakhshali manuscript; Kaplan, 1999, p. 56) in the future. However, here we make a few general comments.

First, we note that it was the invention or premise of the basic operations of zero, which made it a truly mathematical number. The first record of this was found in a book written by the Indian mathematician Brahmagupta in 628 CE (section 3.2). Thus, 628 CE should be the latest possible time for the origin of the mathematical zero, and can be considered the most important year for zero based on our current knowledge. When before 628 CE did the mathematical zero originate? Note that it is not known if Brahmagupta himself invented the basic operations of zero mentioned in his book. If he did not, then the latest possible time for zero's origin as a full mathematical number would be earlier than 628 CE. But even if he did, then zero as a mathematical number should also have been conceptualized at least several centuries before 628 CE. This could be clear from what happened in Christian Europe after the mathematical zero had formally been introduced there by Leonardo Fibonacci in early thirteenth century (Freely, 2010, pp. 132–133). Even though the immense advantage of zero in mathematical calculations, particularly for accounting in the business world, was fully known, there was a significant resistance for centuries, including by the Christian church, before it was finally overall accepted (Seife, 2000, pp. 78–84, 91–93). Therefore, one may speculate that many centuries in India might have passed before an 'unnecessary and impractical' number like zero (see section 2) – the advantage of which was not at all known (unlike in Europe about a millennium later) – was widely given the full status of a number (e.g., Kaplan, 1999, p. 71; Bhattacharyya, 2021). This strongly suggests, even without an epigraphic or paleographic evidence, that the mathematical zero must have been conceived centuries before 628 CE.

However, it is not the specific symbol of zero, but what it represents that is important. Numerals of the same decimal place-value system can be different in different languages (see Datta & Singh, 1962a, pp. 39, 121). While the hollow circle currently used to represent the mathematical zero is sometimes highlighted because it perhaps represents an *emptiness-nothing* or *śūnya*, this symbol was not unique. For example, a small dot or a point (i.e., *bindu*, perhaps representing the zero or no spatial dimension) represents the mathematical zero in the Bakhshali manuscript (Kaplan, 1999, p. 56) and the Sambor stone inscription (section 4.1). On the other hand, an inscription of 876 CE in Gwalior, India shows a hollow circle as zero. Thus, both hollow circle and point were in use to represent zero in India, which are consistent with the word-numerals *śūnya* and *bindu* respectively, representing the mathematical zero (see section 4.3). Moreover, note that a circular symbol is not very special, and many cultures had a good knowledge of a circle (Bhattacharyya, 2021). Therefore, an attempt to find the origin of the currently used symbol for zero might not be useful in probing the origin of the mathematical zero.

Nevertheless, in order to find the origin of the mathematical zero, it is customary to look for a numeral for zero and for calculations involving such a numeral in a book. However, this may not be a fair demand given the ancient Indian culture, where books used to be written typically using verses and word-numerals (see section 4.3). Even Brahmagupta did not use a symbol for zero, but used word-numerals, such as *kha* (sky), *ākāśa* (sky) and *śūnya* (emptiness) (Kaplan, 1999, p. 44). Thus word-numerals and related statements in books could be considered as a sufficient proof of the existence of the mathematical zero and its calculations. Nevertheless, as mentioned above, there were also usual numerals for zero and other numbers, as found from epigraphic and paleographic evidence. Such numerals, while not used in books in ancient times, would have been used for day-to-day calculations performed using dust spread on a board or on the floor. This is why mathematical calculations were known as *dhuli-karma* (dust-work) and arithmetic is still known as *pâṭîgaṇita* (*pâṭî* = board, *gaṇita* = science of calculation) in India (Datta & Singh, 1962a, pp. 123–124; Kaplan 1999, p. 49). Moreover, this is perhaps why ancient samples of numerals of zero are not easily found in India.

Note that, like today, different sets of numerals for the same decimal place-value number system existed in different languages in ancient India (see Datta & Singh, 1962a, pp. 39, 121). This was noted by the well-known Persian scholar Al-Bîrûnî in the first half of eleventh century CE: 'As in different parts of India the letters have different shapes, the numerical signs, too, which are called *aṅka*, differ.' (Sachau, 2002, p. 160). Al-Bîrûnî's comment clearly shows that the decimal place-value system including zero existed in India before different scripts and numerals originated or evolved independently in different parts of the country (Datta & Singh, 1962a, p. 39). This indicates the antiquity of the idea of zero and decimal place-value number system in India, and suggests how entrenched these ideas are in the Indian culture (see also section 4.3).

Finally, we note that the evolution of the number system and numerals can be a complex process in every culture, and particularly in a vast country like India. Not only may a new idea or a change take a long time to diffuse throughout the entire country, but also multiple systems, older and newer, could coexist. Thus, even when a newer and better system is fully established, the older system could still be used for certain specific purposes. A prominent example of this is the entirely outdated, inefficient and zeroless Roman number system, which is still taught in schools and is used for certain purposes throughout the world, including in India! This is particularly surprising, because even in its own time the Roman system was far less efficient than the superior Mesopotamian place-value system. Incidents of such a coexistence in ancient times could complicate an identification of the origin of the mathematical zero.

## 6 Conclusions

In this chapter, we discuss and/or conclude the following (see also Bhattacharyya, 2021):

(1) The mathematical zero, which is an independent number with its basic operations defined, is distinct from the general idea of *nothing*.
(2) *Nothing* can be classified into four categories (Bhattacharyya, 2021): *balance*, *absence*, *emptiness* and *nonexistence*.
(3) A practical need did not motivate the conceptualization of the mathematical zero, which implies that purely intellectual activities gave rise to zero.
(4) The place-value number system and the placeholder might not have a role in the invention of the mathematical zero.
(5) The invention of zero required a prior philosophical base of *nothing*, most likely the *nonexistence-nothing*. The basic mathematical operations of zero were also premised based on the idea of *nonexistence*.
(6) Comments of Syrian Bishop Severus Sebokht, which are often used to argue against the Indian origin of zero, do not provide evidence against such an origin.
(7) The philosophy of *nothing* and word-numerals are two 'living proofs' of the Indian origin of the mathematical zero.

## References

Bhattacharyya, S. (2021). Zero – a tangible representation of nonexistence: Implications for modern science and the fundamental. *Sophia, 60*, 655–676. https://doi.org/10.1007/s11841-021-00870-4.

Datta, B., & Singh, A. N. (1962a). *History of Hindu Mathematics: A source book – Part I*. Calcutta: Asia Publishing House.

Datta, B., & Singh, A. N. (1962b). *History of Hindu Mathematics: A source book – Part II*. Calcutta: Asia Publishing House.

Freely, J. (2010). Aladdin's Lamp: How Greek science came to Europe through the Islamic world. New York: Vintage Books.

Kaplan, R. (1999). *The Nothing That Is: A natural history of zero*. Oxford: Oxford University Press.

Mandal, S. K. (2021). *Bangla Sahityer Itihas: Prachin O Madhyayug*. Kolkata: Uddalak Publishing House.

Radhakrishnan, S. (2008a). *Indian Philosophy: Vol. 1*. India: Oxford University Press.

Radhakrishnan, S. (2008b). *Indian Philosophy: Vol. 2*. India: Oxford University Press.

Sachau, E. C. (2002). *Alberuni's India*. New Delhi: Rupa & Co.

Seife, C. (2000). *Zero: The biography of a dangerous idea*. London: Souvenir Press.

Yu-Lan, F. (1976). *A Short History of Chinese Philosophy*. New York: The Free Press.

CHAPTER 22

# Category Theory and the Ontology of *Śūnyata*

*Sisir Roy and Rayudu Posina*

## Abstract

Notions such as *śūnyata, catuskoti,* and Indra's net, which figure prominently in Buddhist philosophy, are difficult to readily accommodate within our ordinary thinking about everyday objects. Famous Buddhist scholar Nāgārjuna considered two levels of reality: one called conventional reality, and the other ultimate reality. Within this framework, *śūnyata* refers to the claim that at the ultimate level objects are devoid of essence or 'intrinsic properties', but are interdependent by virtue of their relations to other objects. *Catuskoti* refers to the claim that four truth values, along with contradiction, are admissible in reasoning. Indra's net refers to the claim that every part of a whole is reflective of the whole. Here we present category theoretic constructions that are reminiscent of these Buddhist concepts. The universal mapping property definition of mathematical objects, wherein objects of a universe of discourse are defined not in terms of their content, but in terms of their relations to all objects of the universe is reminiscent of *śūnyata*. The objective logic of perception, with perception modeled as [a category of] two sequential processes (sensation followed by interpretation), and with its truth value object of four truth values, is reminiscent of the Buddhist logic of *catuskoti*. The category of categories, wherein every category has a subcategory of sets with zero structure within which every category can be modeled, is reminiscent of Indra's net. Our thorough elaboration of the parallels between Buddhist philosophy and category theory can facilitate better understanding of Buddhist philosophy, and bring out the broader philosophical import of category theory beyond mathematics.

## Keywords

Cantor – contradiction – emptiness – essence – figure – functor – Nāgārjuna – natural transformation – object – property – reality – relation – set – shape – structure – structure-respecting morphism – truth value – Yoneda – zero

 | DOI:10.1163/9789004691568_026

## 1 Introduction

Buddhist philosophy, especially Nāgārjuna's Middle Way (Garfield, 1995; Siderits and Katsura, 2013), is intellectually demanding (Priest, 2013). The sources of the difficulties are many. First it argues for two realities: conventional and ultimate (Priest, 2010). Next, ultimate reality is characterized by śūnyata or emptiness, which is understood as the absence of a fundamental essence underlying reality (Priest, 2009). Equally importantly, contradictions are readily deployed, especially in catuskoti, as part of the characterization of reality (Deguchi, Garfield, and Priest, 2008; Priest, 2014). Lastly, reality is depicted as Indra's net – a whole, whose parts are reflective of the whole (Priest, 2015). The ideas of relational existence, admission of contradictions, and parts reflecting the whole are seemingly incompatible with our everyday experiences and the attendant conceptual reasoning used to make sense of reality. However, notions analogous to these ancient Buddhist ideas are also encountered in the course of the modern mathematical conceptualization of reality. These parallels may be, in large part, due to 'experience' and 'reason' that are treated as the final authority in both mathematical sciences and Buddhist philosophy. Here, we highlight the similarities between Buddhist philosophy and mathematical philosophy, especially category theory (Lawvere and Schanuel, 2009). The resultant cross-cultural philosophy can facilitate a proper understanding of reality – a noble goal to which both Buddhist philosophy and mathematical practice are unequivocally committed.

## 2 Two Realities

There are, according to Buddhist thought, two realities: the conventional reality of our everyday experiences and the ultimate reality (Priest, 2010; Priest and Garfield, 2003). In our conventional reality, things appear to have intrinsic essences. It is sensible, at the level of conventional reality, to speak of essences of objects, but at the level of ultimate reality there are no essences, and everything exists but only relationally. There is an analogous situation in mathematics. On one hand, mathematical objects can be characterized in terms of their relations to all objects, in which case the nature of an object is determined by the nature of its relationship to all objects. In a sense, there is nothing inside the object; an object is what it is by virtue of its relations to all objects. The objects of mathematics are, as Resnik (1981, p. 530) notes, 'positions in structures', which is in accord with the Buddhist understanding of things as 'loci in a field of relations' (Priest, 2009, p. 468). However, there is another level of mathematical reality, wherein we can speak of essences of objects (e.g., theories of objects; Lawvere and Rosebrugh, 2003, pp. 154–155). For example, one can characterize a set as a collection of elements or 'sum' of basic-shaped figures (1-shaped figures, where $1 = \{\bullet\}$), with basic shapes understood as essences (Lawvere, 1972, p. 135; Lawvere and Schanuel, 2009, p. 245; Reyes, Reyes, and Zolfaghari, 2004, p. 30). Similarly, every graph is made up of figures of two basic shapes (arrow- and dot-shaped figures; Lawvere and Schanuel, 2009, pp. 150, 215). This characterization of an object in terms of its contents, i.e., basic shapes or essences (Lawvere, 2003, pp. 217–219; Lawvere, 2004, pp. 11–13), can be contrasted with the relational characterization, wherein each and every object of a universe of discourse (a mathematical category; Lawvere and Schanuel, 2009, p. 17) is characterized in terms of its relationship to all objects of the universe or category (see Appendix 7.1 below). The relational nature of mathematical objects, as elaborated below, is reminiscent of the Buddhist notion of emptiness – an assertion that objects are what they are not by virtue of some intrinsic essences but by virtue of their mutual relationships.

## 3 Emptiness

According to Buddhist philosophy, everything is empty and the totality of empty things is empty. Here, emptiness is understood as the absence of essences. Things, in the ultimate analysis, are what they are and behave the way they do not because of [some] essences inherent in them, but by virtue of their mutual relationships (Priest, 2009). This idea of relational existence has parallels in mathematical practice. Mathematical objects of a given mathematical category (e.g., a category of sets) are what they are, not by virtue of their intrinsic essences but by virtue of their relations to all objects of the category. For example, a single-element set is a set to which there exists exactly one function from every set (Lawvere and Schanuel, 2009, pp. 213, 225). Note that the singleton set is characterized not in terms of what it contains (a single element), but in terms of how it relates to all sets of the category of sets. In a similar vein, the truth value set $\Omega = \{\text{false, true}\}$ is defined in terms of its relation to all sets of the category of sets. The truth value set, instead of being defined as a set of two elements 'false' and 'true', is defined as a set $\Omega$ such that functions from any set $X$ to the set $\Omega$ are in one-to-one correspondence with the parts of $X$ (ibid. pp. 339–344). To give one more example, a product of two sets is defined not by specifying the contents of the product set (pairs of elements), but by characterizing its relationship to all sets. More explicitly, the product of two sets A and B is a set A × B along with two functions (projections to the factors)

$$p_A: A \times B \to A, p_B: A \times B \to B$$

such that for every set Q and any pair of functions $q_A: Q \to A$, $q_b: Q \to B$, there is exactly one function $q: Q \to A \times B$ satisfying both the equations: $q_A = p_A \circ q$ and $q_B = p_B \circ q$, where '∘' denotes composition of functions (ibid. pp. 339–344). The universal mapping property definition of mathematical constructions brought to sharp focus the relational nature of mathematical objects. It conclusively established that 'the substance of mathematics resides not in substance (as it is made to seem when ∈[membership] is the irreducible predicate, with the accompanying necessity of defining all concepts in terms of a rigid elementhood relation) but in form (as is clear when the guiding notion is an isomorphism-invariant structure, as defined, for example, by universal mapping properties)' (Lawvere, 2005, p. 7). More broadly, Yoneda lemma (Lawvere and Rosebrugh, 2003, pp. 249–250; Appendix 7.1 below), according to which a mathematical object of a given universe of discourse (i.e., category) is completely characterized by the totality of its relations to all objects of the universe (category), is an unequivocal assertion of the relational nature of mathematical objects. Yoneda lemma, as pointed out by Barry Mazur, establishes that 'an object X of a category C is determined by the network of relationships that the object X has with all the other objects in C' (Mazur, 2008). Thus the Buddhist idea of emptiness or relational existence finds resonance in mathematical practice, especially in terms of universal mapping properties and the Yoneda lemma.

However, note that according to the Buddhist doctrine of emptiness, not only is everything empty, but the totality of empty things is also empty (Priest, 2009). In other words, even the notion of relational existence is empty, i.e., emptiness is not the essence of existence; emptiness is also empty. This idea of emptiness being empty is much more challenging to comprehend. When we say that objects are empty, we are saying that objects are mere locations in a network of relations. But when we say that the totality of empty things is empty, we are asserting that the existence of totality is also relational just like that of the objects in the totality. What is not immediately clear is how are we to think of relations especially when all we have is the totality, i.e., one object. Within mathematics, note that the totality of all objects (along with their mutual relations) forms a category. More importantly, categories are objects in the category of categories (Lawvere, 1966), and hence the totality of objects, i.e., category, is also empty or relational as much as the objects of a category. Thus the idea of śūnyata (everything is empty) resonates with the relational nature of objects and of the totality of objects (within the mathematical framework of the category of categories).

Equally importantly, Nāgārjuna's Middle Way, having gone to great lengths to distinguish two realities (conventional essences vs. ultimate emptiness) identifies the two: 'There is no distinction between conventional reality and ultimate reality' (Deguchi, Garfield, and Priest, 2008, p. 399). Contradictions (such as these) within Buddhist philosophy, on a superficial reading, are diagnostic of irrational mysticism. However, as we point out in the following, contradictions also figure prominently in the foundations of mathematical modeling of reality. In light of these parallels, 'contradiction' may be intrinsic to the nature of reality, which is the common subject of both Buddhist and mathematical investigations, and not a sign of faulty Buddhist reasoning.

## 4 Contradiction

Within the Buddhist philosophical discourse, one often encounters contradictions and these contradictions are treated as meaningful (Deguchi, Garfield, and Priest, 2008; Priest, 2014). There is an analogous situation in mathematics. Although not every contradiction is sensible, there are sensible contradictions such as the boundary of an object A formalized as 'A and not A' (Lawvere, 1991, 1994a, p. 48; Lawvere and Rosebrugh, 2003, p. 201). More importantly, within mathematical practice, it is now recognized that contradictions do not necessarily lead to inconsistency (an inconsistent system, according to Tarski, is where everything can be proved; Lawvere, 2003, p. 214). Of course, admitting a contradiction invariably leads to inconsistency in classical Boolean logic. In logics more refined than Boolean logic contradiction does not necessarily lead to inconsistency. This recognition is very important, especially since contradiction plays a foundational role in mathematical practice. Briefly, Cantor's definition of SET is, as pointed out by F. William Lawvere, 'a strong contradiction: its points are completely distinct and yet indistinguishable' (ibid. p. 215; Lawvere, 1994a, pp. 50–51). Zermelo, and most mathematicians following him, concluded that Cantor's account of sets is 'incorrigibly inconsistent' (Lawvere, 1994b, p. 6). Lawvere, using adjoint functors, showed that Cantor's definition is 'not a conceptual inconsistency but a productive dialectical contradiction' (Lawvere and Rosebrugh, 2003, pp. 245–246), which is summed up as the unity and identity of adjoint opposites (Lawvere, 1992, pp. 28–30; Lawvere, 1996).

A related notion is catuskoti, which is routinely employed in Buddhist reasoning (Priest, 2014; Westerhoff, 2006). To place it in perspective, in the familiar Boolean logic, any proposition is either true or false. Put differently, there are only two possible truth values, and they are mutually exclusive and jointly exhaustive. Unlike Boolean logic, in Buddhist reasoning more than two truth values are admissible. In the Buddhist logic of catuskoti, a proposition can possibly take, in addition to the familiar truth values of 'true' or 'false', the truth values of 'true and false', or 'not true and not false'. Given a proposition A, there are four possibilities: 1. A; 2. not A; 3. A and not A; 4. not A and not not A. Here contradiction is admissible, i.e., 'A and not A' is a possible state of affairs, which is reminiscent of the boundary operation and the unity and identity of adjoint opposites in mathematics, alluded to earlier. Moreover, double negation is not the same as identity operation as in the case of, to give one example, the non-Boolean logic of graphs (Lawvere and Schanuel, 2009, p. 355). Note that if not not A = A, then the fourth truth value of catuskoti is equal to the third.

As an illustration of how the four truth values of catuskoti could be a reflection [of an aspect] of reality, we consider the category of percepts. Perception involves two sequential processes of sensation followed by interpretation (Albright, 2015; Croner and Albright, 1999). So, we define the category of percepts as a category of two sequential functions of decoding after coding. The truth value object of the category of percepts has four truth values (Appendix 7.2 below). Thus the objective logic of perception, with its truth value object of four truth values, is reminiscent of the Buddhist logic of catuskoti (see Linton, 2005).

## 5 Indra's Net and Zero Structure

Another important concept in Buddhist philosophy is the idea of Indra's net, wherein reality is compared to a vast network of jewels such that every jewel is reflective of the entire net (Priest, 2015). In abstract terms, reality is characterized as a whole wherein every part is reflective of the whole. Admittedly, this Buddhist characterization of reality sounds mystifying, but there is an analogous situation, involving part-whole relations, in mathematics.

How can a part of a whole reflect the whole? First, note that mathematical structures of all sorts can be modeled in the category of sets (Lawvere and Schanuel, 2009, pp. 133–151). Sets have zero structure (Lawvere, 1972, p. 1; Lawvere and Rosebrugh, 2003, pp. 1, 57; Lawvere and Schanuel, 2009, p. 146). Negating the structure (cohesion, variation) inherent in mathematical objects, Cantor created sets: mathematical structures with zero structure (Lawvere, 2003, 2016; Lawvere and Rosebrugh, 2003, pp. 245–246). In comparing his abstraction of sets with zero structure to the invention of number zero, Cantor considered sets as his most profound contribution to mathematics (Lawvere, 2006). Sets, by virtue of having zero structure, serve as a blank page – an ideal background to model any category of mathematical objects (Lawvere, 1994b; Lawvere and Rosebrugh, 2003, pp. 154–155). However, structureless sets are a small part – the only part – of the mathematical universe that reflects all of mathematics. It seemed so until Lawvere axiomatized the category of categories (Lawvere, 1966; Lawvere and Schanuel, 2009, pp. 369–370). Along the lines of Cantor's invention of structureless sets, Lawvere defined a subcategory of structureless (discrete, constant) objects within a category by negating its structure (cohesion, variation; Lawvere, 2004, p. 12; Lawvere and Schanuel, 2009, pp. 358–360, 372–377). Thus, within any category of mathematical objects, there is a part, a structureless subcategory, that is like the category of sets in having zero structure, and hence serves as a background to model all categories of mathematical objects (Lawvere, 2003; Lawvere and Menni, 2015; Picado, 2008, p. 21). Modeling a category of mathematical objects requires, in addition to the subcategory with zero structure, another subcategory objectifying the structural essence(s) of the objects of the category, i.e., the theory of the given category of mathematical objects. Finding the theory subcategory also depends on the structureless subcategory, by way of contrasting or negating the structureless subcategory (Lawvere, 2007). Once we have the subcategory with zero structure and the subcategory objectifying the essence (theory) of a given category, interpreting the theory subcategory into the structureless subcategory gives us models of the given category of mathematical objects. Thus, thanks to the recognition of significance of Cantor's zero structure, every mathematical category can be modeled in any category of the category of categories.

If we compare the category of categories to Indra's net, then categories within the category of categories would correspond to jewels in Indra's net. Just as in the case of Indra's net, wherein every jewel in the network of jewels is reflective of the entire network, in the category of categories every category (part) of the category of categories (whole) reflects the whole. For example, the category of dynamical systems is a part of the category of categories. Within the category of dynamical systems, we have the constant subcategory (obtained by negating the variation) of dynamical systems (wherein every state is a fixed point), which is like the category of sets, and within which any category can be modeled. Similarly, the category of graphs is another part of the category of categories. Within the category of graphs there is the discrete subcategory (obtained by negating the cohesion) of graphs (with one loop on each dot), which is also like the category of sets, and hence can model every category. Thus, we find that within the category of categories, every part is reflective of the whole, which is reminiscent of the Buddhist depiction of reality as Indra's net: a whole with parts reflective of the whole.

## 6 Conclusion

There are similarities between Buddhist philosophy and mathematical practice, especially with regard to essence vs. emptiness, contradictions, and part-whole relations. These similarities might be a natural consequence of identical objectives – understanding reality and commitment to truth– and identical means – experience and reason – employed toward those ends. It is in this respect that the practices of the two – mathematicians and Buddhists – can be compared. Our exercise, on that score, can help better appreciate the rationality of Buddhist reasoning. Oftentimes, admission of contradiction (as in *catuskoti*) tends to be equated with irrational mysticism.

However, as we have seen, contradictions are also an integral and indispensable part of the mathematical understanding of reality. On the other hand, in drawing parallels between Buddhist thought and mathematical practice, we hope to have brought out the broad philosophical import of category theory beyond mathematics.

## 7 Appendices

### 7.1 *Yoneda Lemma*

We begin with an intuitive introduction to the mathematical content of Yoneda lemma (Lawvere and Rosebrugh, 2003, pp. 175–176, 249). With simple illustrations of figures-and-incidences (along with [its dual] properties-and-determinations) interpretations of mathematical objects, we prove the Yoneda lemma (Lawvere and Schanuel, 2009, pp. 361, 370–371). Broadly speaking, Yoneda lemma is about [properties of] objects [of categories] and their mutual determination.

First, let us consider a function

$$f: A \rightarrow B$$

We can think of the function f as (i) a figure of shape A in B, i.e., an A-shaped figure in B. For example, in the category of graphs, a map

$$d\colon \mathrm{D} \to \mathrm{G}$$

from a graph D (consisting of one dot) to any graph G is a D-shaped figure in G, i.e., a dot in the graph G. We can also think of the same function f as (ii) a property of A with values in B, i.e., a B-valued property of A (Lawvere and Schanuel, 2009, pp. 81–85).

For example, with sets, say, Fruits = {apple, grape} and Color = {red, green}, a function

$$c\colon \text{Fruits} \to \text{Color}$$

(with $c$ (apple) = red and $c$ (grape) = green) can be viewed as Color-valued property of Fruits.

Now let us consider two figures: an X-shaped figure in A

$$x_A\colon X \to A$$

and a Y-shaped figure in A

$$y_A\colon Y \to A$$

Given a transformation from the shape X to the shape Y, i.e., an X-shaped figure in Y

$$x_Y\colon X \to Y$$

we find that the X-shaped figure in Y($x_y$) induces a transformation of a Y-shaped figure in A into an X-shaped figure in A via composition of maps

$$y_A \circ x_Y = x_A$$

(where '∘' denotes composition) displayed as a commutative diagram

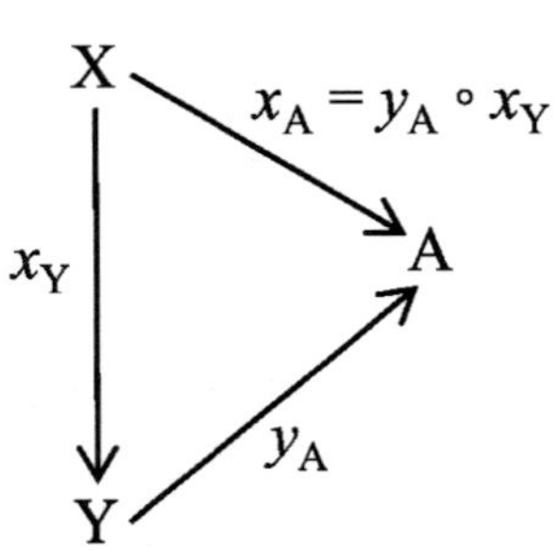

showing the transformation of a Y-shaped figure in $A(y_A)$ into an X-shaped figure in $A(x_A)$ by an X-shaped figure in $Y(x_Y)$ via composition of maps.

As an illustration, consider an object (of the category of graphs), i.e., a graph G (shown below):

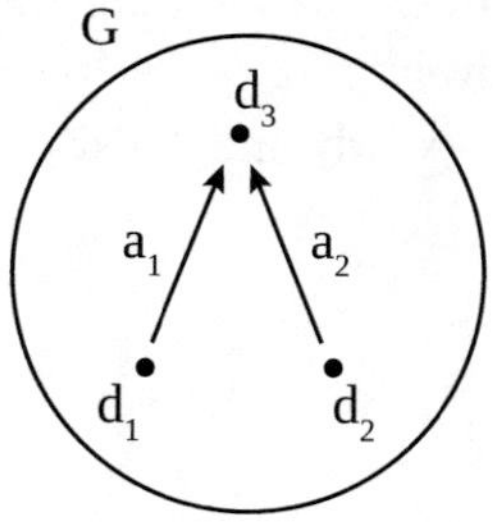


and a shape graph [arrow] A with exactly one arrow 'a', along with its source 's' and target 't', as shown

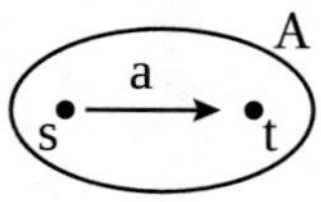


along with an A-shaped figure in G

$$a_G: A \to G$$

displayed as:

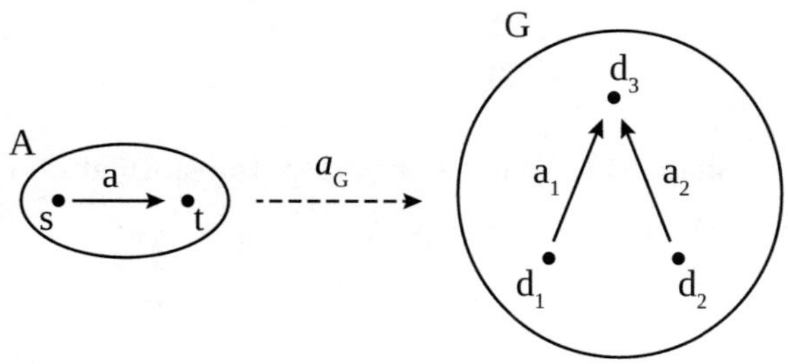


with, say,

$$a_G(a) = a_1$$

This A-shaped figure in G, i.e., the graph map $a_G$ maps the [only] arrow 'a' in the shape graph A to the arrow '$a_1$' in the graph G, while respecting the source (s) and target (t) structure of the arrow 'a', i.e., with arrow 'a' in shape A mapped to arrow '$a_1$' in the graph G, the source 's' and target 't' of the arrow 'a' are mapped to the source '$d_1$' and target '$d_3$' of arrow '$a_1$', respectively.

Next, consider another shape graph [dot] D with exactly one dot 'd'

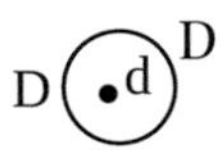


along with a D-shaped figure in A

$$d_A : D \to A$$

with

$$d_A\,(\mathrm{d}) = \mathrm{s}$$

i.e., the graph map $d_A$ maps the dot 'd' in the graph D to the dot 's' in the graph A, i.e., the source dot 's' of the arrow 'a', as shown below:

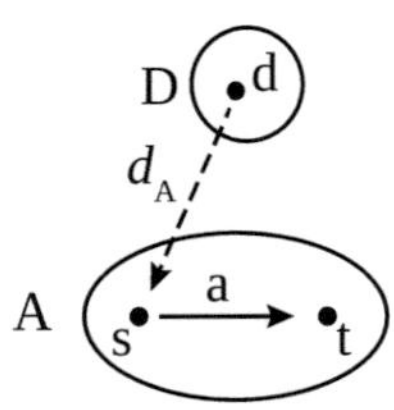


This graph map $d_A$ from shape D to shape A induces a transformation of the (above) A-shaped figure in graph G

$$a_G : A \to G$$

into a D-shaped figure in G

$$d_G : D \to G$$

via composition of graph maps

$$d_G = a_G \circ d_A$$

i.e., $d_G(d) = a_G \circ d_A(d) = a_G(s) = d_1$ as depicted below (Lawvere and Schanuel, 2009, pp. 149–150):

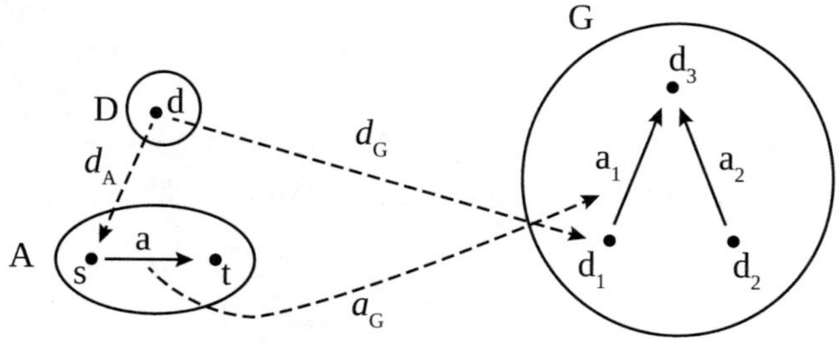


In general, every X-shaped figure in Y transforms a Y-shaped figure in A into an X-shaped figure in A, i.e., every map

$$x_Y: X \to Y$$

induces a map in the opposite direction (contravariant; Lawvere, 2017; Lawvere and Rosebrugh, 2003, p. 103; Lawvere and Schanuel, 2009, p. 338).

$$A^{x_Y} : A^Y \to A^X$$

where $A^Y$ is the map object of the totality of all Y-shaped figures in A, $A^X$ is the map object of the totality of all X-shaped figures in A, and with the map

$$A^{x_Y}$$

of map objects defined as

$$A^{x_Y}(y_A : Y \to A) = y_A \circ x_Y = x_A : X \to A$$

assigning a map $x_A$ in the map object $A^X$ to each map $y_A$ in the map object $A^Y$. Thus, the figures in an object A of all shapes (all X-shaped figures in A for every object X of a category) along with their incidences

$$A^{x_Y} : A^Y \to A^X$$

induced by all changes of figure shapes

$$x_{\mathrm{Y}} = \mathrm{X} \to \mathrm{Y}$$

(i.e., every map in the category) together constitute the geometry of figures in A, i.e., a complete picture of the object A. Summing up, we have the complete characterization of the geometry of every object A of a category in terms of the figures of all shapes (objects of the category) and their incidences (induced by the maps of the category) in the object A (Lawvere and Schanuel, 2009, pp. 370–371).

Let us now examine how figures of a shape X in an object A are transformed into figures of the [same] shape X in an object B. We find that an A-shaped figure in B

$$a_{\mathrm{B}} = \mathrm{A} \to \mathrm{B}$$

induces a transformation of an X-shaped figure in A

$$x_{\mathrm{A}} = \mathrm{X} \to \mathrm{A}$$

into an X-shaped figure in B

$$x_{\mathrm{B}} = \mathrm{X} \to \mathrm{B}$$

via composition of maps

$$a_{\mathrm{B}} \circ x_{\mathrm{A}} = x_{\mathrm{B}}$$

displayed as a commutative diagram

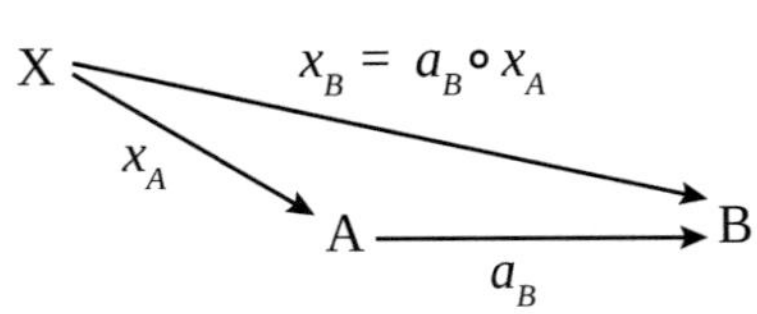


showing the transformation of an X-shaped figure in A ($x_A$) into an X-shaped figure in B ($x_B$) by an A-shaped figure in B ($a_B$) via composition of maps. Thus, every map

$$a_B: A \to B$$

induces a map in the same direction (covariant; Lawvere and Rosebrugh, 2003, pp. 102–103, 109; Lawvere and Schanuel, 2009, p. 319)

$$a_B{}^X: A^X \rightarrow B^X$$

where $A^X$ is the map object of all X-shaped figures in A, $B^X$ is the map object of all X-shaped figures in B, and with the map $a_B{}^X$ defined as

$$a_B{}^X (x_A: X \rightarrow A) = a_B \circ x_A = x_B: \text{X} \rightarrow \text{B}$$

assigning a map $x_B$ in the map object $\text{B}^\text{X}$ to each map $x_A$ in the map object $A^X$. Thus, the totality of maps $a_B{}^X$ of map objects (for all objects and maps of the category) induced by a map $a_B$ from A to B constitutes a covariant transformation of the figure geometry of object A into that of B, i.e., specifies how figures-and-incidences in A are transformed into figures-and-incidences in B.

Putting together these two transformations: (i) the covariant transformation of X-shaped figures in A into X-shaped figures in B induced by an A-shaped figure in B, and (ii) the contravariant transformation of Y-shaped figures in A into X-shaped figures in A induced by an X-shaped figure in Y, we obtain the diagram (Lawvere and Schanuel, 2009, p. 370):

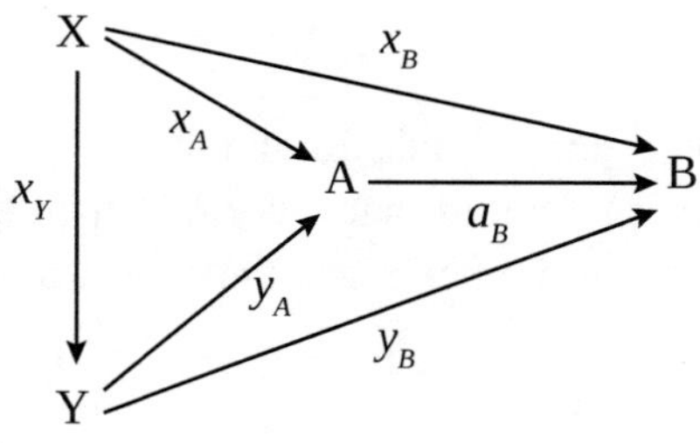


from which we notice that there are two paths to go from a Y-shaped figure in A ($y_A$) to an X-shaped figure in B ($x_B$):

Path 1. First we map the Y-shaped figure in $A(y_A)$ into an X-shaped figure in $A(x_A)$ along the X-shaped figure in $Y(x_Y)$ via composition of the maps

$$y_A \circ x_Y$$

and then map the composite X-shaped figure in $A(y_A \circ x_Y)$ into an X-shaped figure in B along the A-shaped figure in $B(a_B)$ via composition

$$a_B \circ (y_A \circ x_Y)$$

Path 2. First we map the Y-shaped figure in A ($y_\text{A}$) into a Y-shaped figure in $B(y_B)$ along the A-shaped figure in $B(a_B)$ via composition of the maps

$$a_B \circ y_A$$

and then map the composite Y-shaped figure in $B(a_B \circ y_A)$ into an X-shaped figure in B along the X-shaped figure in $Y(x_Y)$ via composition

$$(a_B \circ y_A) \circ x_Y$$

Based on the associativity of composition of maps (Lawvere and Schanuel, 2009, pp. 370–371), we find that

$$a_B \circ (y_A \circ x_Y) = (a_B \circ y_A) \circ x_Y$$

i.e., the two paths of transforming a Y-shaped figure in A

$$y_A\colon Y \to A$$

into an X-shaped figure in B give the same map

$$a_B \circ y_A \circ x_Y = x_B\colon X \to B$$

Since the associativity of composition of maps holds for all maps of any category (Lawvere and Schanuel, 2009, p. 17), we find that every A-shaped figure in B induces a covariant transformation of the figure geometry of A into the figure geometry of B.

More explicitly, each A-shaped figure in B

$$a_B\colon A \to B$$

induces a commutative diagram (of maps of map objects)

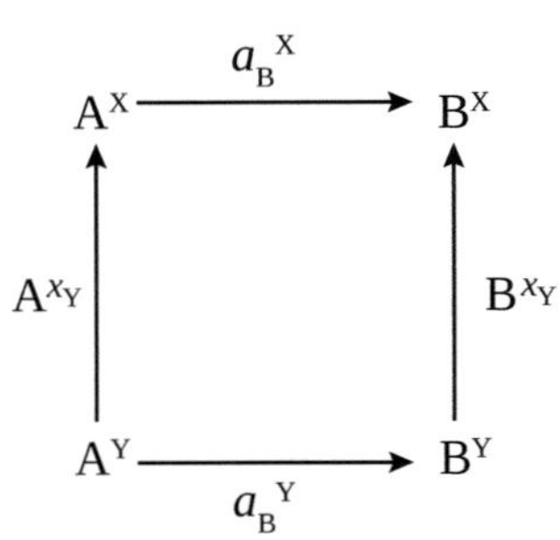


satisfying

$$a_B{}^X \circ A^{x_Y} = B^{x_Y} \circ a_B{}^Y$$

for every map in the category, and hence a natural transformation (compatible with the composition of maps) of the figure geometry of A into the figure geometry of B. To see the commutativity, consider a Y-shaped figure in A, i.e., a map $y_A$ of the map object $A^Y$ and evaluate the above two composites:

$$a_B{}^X \circ A^{x_Y}(y_A) = a_B{}^X(y_A \circ x_Y) = a_B \circ (y_A \circ x_Y)$$

$$B^{x_Y} \circ a_B{}^Y(y_A) = B^{x_Y}(a_B \circ y_A) = (a_B \circ y_A) \circ x_Y$$

Again, according to the associativity of the composition of maps

$$a_B \circ (y_A \circ x_Y) = (a_B \circ y_A) \circ x_Y = a_B \circ y_A \circ x_Y$$

and hence both composites map each Y-shaped figure in A (a map in the map object $A^Y$)

$$y_A\colon Y \to A$$

to the X-shaped figure in B (a map in the map object $B^X$)

$$a_B \circ y_A \circ x_Y = x_B\colon X \to B$$

Since we have the above commutativity for every shape (object) and figure (map), i.e., for all objects and maps of the category, we conclude that an A-shaped figure in B corresponds to a natural transformation (respectful of figures-and-incidences) of the figure geometry of A into the figure geometry of B.

Now we formally show that every A-shaped figure in B

$$a_B\colon A \to B$$

of a category C can be represented as a natural transformation

$$n^{a_B}: C(-,A) \to (-,B)$$

from the domain functor $\boldsymbol{C}(-, \mathrm{A})$ constituting the figure geometry of the object A to the codomain functor $\boldsymbol{C}(-, \mathrm{B})$ constituting the figure geometry of the object B, which is the core mathematical content of the Yoneda lemma (Lawvere and Rosebrugh, 2003, p. 249): 'Maps in any category can be represented as natural transformations' (Lawvere and Schanuel, 2009, p. 378). Since natural transformations represent structure-preserving maps between objects, the domain (codomain) functor of a natural transformation represents the domain (codomain) object of the structure-preserving map.

Let us define the (domain) functor

$$\boldsymbol{C}(-, A)\colon \boldsymbol{C} \to \boldsymbol{C}$$

as: for each object X of the category $\boldsymbol{C}$

$$\boldsymbol{C}(-, A)(X) = A^X$$

where $A^X$ is the map object of all X-shaped figures in A

$$x_A\colon X \to A$$

and, for each map

$$x_Y\colon X \to Y$$

of the category C

$$C(-, A)(x_Y : X \to Y) = A^{x_Y} : A^Y \to A^X$$

where $A^Y$ is the map object of all Y-shaped figures in A, and with the map $A^{x_Y}$ of map objects defined as

$$A^{x_Y}(y_A : Y \to A) = y_A \circ x_Y = x_A : X \to A$$

assigning a map $x_A$ in the map object $A^X$ to each map $y_A$ in the map object $A^Y$. Thus the functor

$$\boldsymbol{C}(-, A)\colon \boldsymbol{C} \to \boldsymbol{C}$$

in assigning to each map

$$x_Y = X \to Y$$

(of the domain category $\boldsymbol{C}$) its [induced] map [of map objects]

$$C(-,A)(x_Y : X \rightarrow Y) = C(-,A)(Y) \rightarrow C(-,A)(X) = A^{x_Y} : A^Y \rightarrow A^X$$

(of the codomain category $\boldsymbol{C}$) is contravariant, i.e., a transformation of a shape X into a shape Y induces a transformation (in the opposite direction) of Y-shaped figures in A into X-shaped figures in A (Lawvere and Rosebrugh, 2003, pp. 236–237).

Now, we check to see if $\boldsymbol{C}(-, \mathrm{A})$ preserves identities, i.e., whether

$$\boldsymbol{C}(-, A)(1_X: X \rightarrow X) = 1_{\boldsymbol{C}(-,A)(X)}$$

for every object X. Evaluating

$$\boldsymbol{C}(-, \mathrm{A})(1_\mathrm{X}: \mathrm{X} \rightarrow \mathrm{X}) = \mathrm{A}^{1_\mathrm{X}}: \mathrm{A}^\mathrm{X} \rightarrow \mathrm{A}^\mathrm{X}$$

at a map

$$x_A: X \rightarrow A$$

we find that

$$A^{1_X}(x_A : X \rightarrow A) = (x_A \circ 1_X) = x_A : X \rightarrow A$$

(for every map $x_A$ in the map object $A^X$). Next, evaluating

$$1_{\boldsymbol{C}(-,A)(X)} = 1_{A^X} : A^X \rightarrow A^X$$

at the map

$$x_A: X \rightarrow A$$

we find that

$$1_{A^X}(x_A : X \rightarrow A) = (x_A \circ 1_X) = x_A : X \rightarrow A$$

(for every map $x_A$ in the map object $A^X$). Since

$$A^{1_X} = 1_{A^X}$$

i.e.

$$\boldsymbol{C}(-, A)(1_X: X \rightarrow X) = 1_{\boldsymbol{C}(-,A)(X)}$$

for every object X of the category $C$, we say $\boldsymbol{C}(-, \mathrm{A})$ preserves identities.

Next, we check to see if $\boldsymbol{C}(-, \mathrm{A})$ preserves composition. Since $\boldsymbol{C}(-, \mathrm{A})$ is contravariant, we check whether

$$\boldsymbol{C}(-,A)(y_Z \circ x_Y) = \boldsymbol{C}(-,A)(x_Y) \circ \boldsymbol{C}(-,A)(y_Z)$$

where $y_Z: Y \to Z$. Evaluating

$$C(-,A)(y_Z \circ x_Y) = A^{(y_Z \circ x_Y)}$$

at any map $z_A$ in the map object $\mathrm{A}^{\mathrm{Z}}$, we find that

$$A^{(y_Z \circ x_Y)}(z_A) = z_A \circ (y_Z \circ x_Y)$$

Next, we evaluate

$$C(-,A)(x_Y) \circ C(-,A)(y_Z) = (A^{x_Y} \circ A^{y_Z})$$

also at the map $z_A$

$$(A^{x_Y} \circ A^{y_Z})(z_A) = A^{x_Y}(z_A \circ y_Z) = (z_A \circ y_Z) \circ x_Y$$

Since

$$z_A \circ (y_Z \circ x_Y) = (z_A \circ y_Z) \circ x_Y$$

by the associativity of the composition of maps, we have composition preserved

$$\boldsymbol{C}(-,A)(y_Z \circ x_Y) = \boldsymbol{C}(-,A)(x_Y) \circ \boldsymbol{C}(-,A)(y_Z)$$

Having checked that

$$\boldsymbol{C}(-,A): \boldsymbol{C} \to \boldsymbol{C}$$

with

$$\boldsymbol{C}(-,A)(X): A^X$$
$$C(-,A)(x_Y : X \to Y) = A^{x_Y} : A^Y \to A^X$$

where $A^{x_Y}(y_A) = y_A \circ x_Y$, is a contravariant functor, we consider another contravariant functor

$$C(-, B): C \to C$$

with

$$C(-, B)(X) = B^X$$
$$C(-,B)(x_Y : X \to Y) = B^{x_Y} : B^Y \to B^X$$

where $B^{x_Y}(y_B) = y_B \circ x_Y$.

With the two functors $C(-, A)$ and $C(-, B)$ representing the [figure geometry of] objects A and B, respectively, we now show that every structure-preserving map

$$a_B: A \to B$$

is represented by a natural transformation

$$n^{a_B} : C(-,A) \to C(-,B)$$

More explicitly, given a map $a_B$, we can construct a natural transformation $n^{a_B}$. A natural transformation $n^{a_B}$ from the functor $C(-, A): C \to C$ to the functor $C(-, B): C \to C$ assigns to each object X of the domain category C (of both domain and codomain functors) a map

$$a_B^X: A^X \to B^X$$

(in the common codomain category C) from the value of the domain functor at the object X, i.e., $C(-, A)(X) = A^X$ to the value of the codomain functor at X, i.e., $C(-, B)(X) = B^X$; and to each map $x_Y: X \to Y$ (in the common domain category $C$), a commutative square (in the common codomain category $C$) shown below:

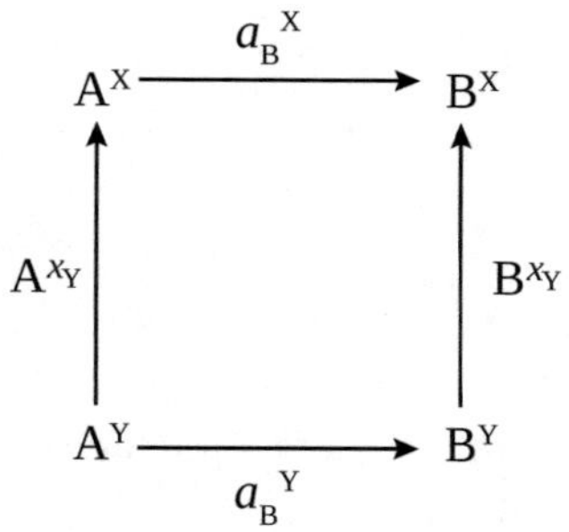


satisfying

$$a_B^X \circ A^{x_Y} = B^{x_Y} \circ a_B^Y$$

(Lawvere and Rosebrugh, 2003, p. 241; Lawvere and Schanuel, 2009, pp. 369–370). We have already seen that with the composition-induced maps (of map objects):

$$A^{x_Y}(y_A) = y_A \circ x_Y$$
$$a_B{}^X(x_A) = a_B \circ x_A$$
$$a_B{}^Y(y_A) = a_B \circ y_A$$
$$B^{x_Y}(y_B) = y_B \circ x_Y$$

the required commutativity:

$$a_B{}^X \circ A^{x_Y}(y_A) = a_B{}^X(y_A \circ x_Y) = a_B \circ (y_A \circ x_Y)$$
$$B^{x_Y} \circ a_B{}^Y(y_A) = B^{x_Y}(a_B \circ y_A) = (a_B \circ y_A) \circ x_Y$$

is given by the associativity of the composition of maps

$$a_B \circ (y_A \circ x_Y) = (a_B \circ y_A) \circ x_Y = a_B \circ y_A \circ x_Y$$

Thus, each A-shaped figure in $B(a_B)$ is a natural transformation ($n^{a_B}$; homogenous with respect to composition of maps) of the figure geometry ***C*** (–, A) of A into the figure geometry ***C***(–, B) of B.

Furthermore, we can obtain the set $|B^A|$ of all A-shaped figures in B based on the 1–1 correspondence between A-shaped figures in B and the points (i.e., maps with terminal object T of the category C as domain; Lawvere and Schanuel, 2009, pp. 232–234) of the map object $B^A$. This 1–1 correspondence, which follows from the universal mapping property defining exponentiation, along with the fact that the terminal object T is a multiplicative identity (Lawvere and Schanuel, 2009, pp. 261–263, 313–314, 322–323), involves the following two 1–1 correspondences between three maps:

$$T \to B^A$$

---

$$T \times A \to B$$

---

$$A \to B$$

Yoneda lemma says, in terms of our figures-and-incidences characterization of objects, that the set $|B^A|$ of A-shaped figures in B

$$a_B: A \to B$$

is isomorphic to the set $|C(-, B)^{C(-,A)}|$ of natural transformations

$$n^{a_B} : C(-, A) \to C(-, B)$$

of the figure geometry of A into that of B. The required isomorphism of sets

$$|B^A| = |C(-, B)^{C(-,A)}|$$

follows from the 1–1 correspondence between A-shaped figures in B and the natural transformations (compatible with all figures and their incidences) of the figure geometry of A into that of B, which we have already shown (see also Lawvere and Rosebrugh, 2003, pp. 104, 174).

Dually, a map

$$A \to B$$

viewed as a B-valued property on A induces a natural transformation

$$C(-, B) \to C(-, A)$$

of the function algebra of B into that of A (Lawvere and Rosebrugh, 2003, p. 249). Here also the proof of Yoneda lemma involves two transformations: (i) Contravariant: a map from an object A to an object B induces a transformation of properties of B into properties of A, for each type(object) of the category, and (ii) Covariant: a map from a type T to a type R (of properties) induces a transformation of T-valued properties into R-valued properties, for every object of the category. The calculations involved in proving Yoneda lemma in this case of function algebras are same as in the case of figure geometries, except for the reversal of arrows due to the duality between function algebra and figure geometry (Lawvere and Rosebrugh, 2003, p. 174; Lawvere and Schanuel, 2009, pp. 370–371). More specifically, function algebras and figure geometries are related by adjoint functors (Lawvere, 2016).

### 7.2 *Four Truth Values of the Logic of Perception*

Conscious perception involves two sequential processes of sensation followed by interpretation:

Physical stimuli → Brain → Conscious Percepts

(Albright, 2015; Croner and Albright, 1999), which can be thought of as

$$X - \text{coding} \rightarrow Y - \text{decoding} \rightarrow Z$$

and objectified as two sequential processes:

$$A - f \rightarrow B - g \rightarrow C$$

Without discounting that the processes of sensation and interpretation are much more structured than mere functions, and with the objective of simplifying the calculation of truth value object, we model percept as an object made up of three [component] sets C, B, and A, which are sets of physical stimuli, their neural codes, and interpretations, respectively, and two [structural] functions $f$ and $g$ specifying for each interpretation in A the neural code in B (of which it is an interpretation) and for each neural code in B the physical stimulus in C (of which it is a measurement), respectively (see Lawvere and Rosebrugh, 2003, pp. 114–117). The logic of [the category of] perception, whose objects are two sequential functions is determined by its truth value object (Lawvere and Rosebrugh, 2003, pp. 193–212; Lawvere and Schanuel, 2009, pp. 335–357; Reyes, Reyes, and Zolfaghari, 2004, pp. 93–107).The truth value object of a category is an object Ω of the category such that parts of any object X are in 1–1 correspondence with maps from the object X to the truth value object Ω. Since parts of an object are monomorphisms with the object X as codomain, for each monomorphism with X as codomain there is a corresponding X-shaped figure in Ω.

In order to calculate the truth value object, first we need to define maps between objects of the category of percepts. A map from an object

$$A - f \rightarrow B - g \rightarrow C$$

to an object

$$A' - f' \rightarrow B' - g' - C'$$

is a triple of functions

$$p\colon A \rightarrow A',\ q\colon B \rightarrow B',\ r\colon C \rightarrow C'$$

satisfying two equations

$$q \circ f = f' \circ p,\ r \circ g = g' \circ q$$

which make the two squares in the diagram

$$\begin{array}{ccc} A & \xrightarrow{p} & A' \\ {\scriptstyle f}\downarrow & & \downarrow{\scriptstyle f'} \\ B & \xrightarrow[q]{} & B' \\ {\scriptstyle g}\downarrow & & \downarrow{\scriptstyle g'} \\ C & \xrightarrow[r]{} & C' \end{array}$$

commute, i.e., ensure that maps between objects preserve the structural essence of the category (Lawvere and Schanuel, 2009, pp. 149–150).

Now that we have maps of the category of percepts defined, we can calculate its truth value object. The truth value object of a category is calculated based on the parts of the basic shapes (essence) constituting the objects of the category. In the category of sets, one-element set **1** (= {•}) is the basic shape in the sense that any set is made up of elements (see Posina, Ghista, and Roy, 2017 for the details of the calculation of basic shapes, i.e., theory subcategories of various categories). Since the set **1** is also the terminal object (i.e., an object to which there is exactly one map from every object; Lawvere and Schanuel, 2009, pp. 213–214) of the category of sets, and since every set is completely determined by its points (terminal object-shaped figures), we can determine the truth value object of the category of sets by determining its points, i.e., maps from **1** to the (yet to be determined) truth value object. According to the definition of truth value object, **1**-shaped figures in the truth value object are in 1–1 correspondence with parts of **1**. Since the terminal set **1** has two parts: **0** (= {}) and **1**,the truth value set has two points (elements). Thus, the truth value object of the category of sets is **2** (= {false, true}).

Along similar lines, let us calculate the terminal object of the category of percepts. Since there is only one map from any object (two sequential functions) to the object T (two sequential functions from one-element set to one-element set):

$$1 \to 1 \to 1$$

the terminal object of the category of percepts is T and since parts of the terminal object T correspond to the points of the truth value object, let's look at the parts of the terminal object. The terminal object T

$$1 \to 1 \to 1$$

has four parts:

Part 1 ($o$: $0 \to T$)

$$\begin{array}{ccc} 0 & & 1 \\ & & \downarrow \\ 0 & & 1 \\ & & \downarrow \\ 0 & & 1 \end{array}$$

Part 2 ($o_1$: $0_1 \to T$)

$$\begin{array}{ccc} 0 & & 1 \\ & & \downarrow \\ 0 & & 1 \\ & & \downarrow \\ 1 & \to & 1 \end{array}$$

Part 3 ($o_2$: $0_2 \to T$)

$$\begin{array}{ccc} 0 & & 1 \\ & & \downarrow \\ 1 & \to & 1 \\ \downarrow & & \downarrow \\ 1 & \to & 1 \end{array}$$

Part 4 ($1$: $T \to T$)

$$\begin{array}{ccc} 1 & \to & 1 \\ \downarrow & & \downarrow \\ 1 & \to & 1 \\ \downarrow & & \downarrow \\ 1 & \to & 1 \end{array}$$

These four parts correspond to the four points (global truth values) of the truth value object, which means that the component set (of the truth value object) corresponding to the stage of interpretations is a four-element set **4** = $\{o, o_1, o_2, 1\}$. Since objects in the category of perception (two sequential functions) are not completely determined by points, we look for all other basic shapes that are needed to completely characterize any object of two sequential functions. The other basic shapes, besides the terminal object T, are: domains of the parts $0_2$ and $0_1$ of the terminal object T, i.e., shape $0_2$

$$0 \quad 1 \to 1$$

and shape $0_1$

0 0 1

Since the basic shape object $o_2$ has three parts (*0, $0_1$, and 1*), there are three $o_2$-shaped figures in the truth value object, and since the object $o_1$ has two parts (*0* and *1*), there are two $o_1$-shaped figures in the truth value object, which means that the component set (of the truth value object) corresponding to the stage of neural coding is a three-element set **3** = {*0, $0_1$, 1*}, while the component set (of the truth value object) corresponding to the stage of physical stimuli is a two-element set **2** = {*0, 1*}. Putting it all together we find that the truth value object of the category of percepts is:

$$\mathbf{4} - \mathrm{j} \rightarrow \mathbf{3} - \mathrm{k} \rightarrow \mathbf{2}$$

We still have to determine the functions *j* and *k*, which can be done by examining the structural maps between the basic shapes

$$o_1 - c \rightarrow o_2 - \mathrm{d} \rightarrow T$$

which as a subcategory constitutes the theory (abstract essence) of the category of two sequential functions. More explicitly, the incidence relations between the three basic-shaped figures in the truth value object are calculated from the inverse images of the parts of the basic shapes ($0_1$, $0_2$, and *T*) along the structural maps (*d* and *c*). The inverse images of each one of the four points (*0, $0_1$, $0_2$* and *1* corresponding to the four parts of the terminal object T) along the structural maps decoding *d* and coding *c* give for each one of the four global truth values **4** = {*0, $0_1$, $0_2$, 1*} its value in the truth value sets **3** = {*0, $0_1$, 1*} and **2** = {*0, 1*} of the previous stages of neural codes and physical stimuli.

For example, the global truth value $o_2$ corresponds to the part $o_2$ of the basic shape T, and its inverse image along the structural map *d*: $o_2 \rightarrow$ T is the entire basic shape $o_2$, which corresponds to the truth value *1* (of stage **3**); and the inverse image of the entire object $o_2$ along the structural map *c*: $o_1 \rightarrow o_2$ is the entire basic shape $o_1$, which corresponds to the truth value *1* (of stage **2**). Along these lines we find that

$$j(0) = 0, j(0_1) = 0_1, j(0_2) = 1, j(1) = 1$$

$$k(0) = 0, k(0_1) = 1, k(1) = 1$$

which completely characterizes the truth value object

$$4 - j \to 3 - k \to 2$$

of the category of percepts.

## References

Albright, T. D. (2015). Perceiving. *Daedalus* 144: pp. 22–41.

Croner, L. J. and Albright, T. D. (1999). Seeing the big picture: Integration of image cues in the primate visual system. *Neuron* 24: pp. 777–789.

Deguchi, Y., Garfield, J. L., and Priest, G. (2008). The way of the Dialetheist: Contradictions in Buddhism. *Philosophy East and West* 58: pp. 395–402.

Garfield, J. (1995). *The Fundamental Wisdom of the Middle Way*, New York, NY: Oxford University Press.

Lawvere, F. W. (1964). An elementary theory of the category of sets. *Proceedings of the National Academy of Science of the U.S.A.* 52: pp. 1506–1511.

Lawvere, F. W. (1966). The category of categories as a foundation for mathematics. In S. Eilenberg et al. (eds.), *La Jolla Conference on Categorical Algebra*, New York, NY: Springer-Verlag, pp. 1–20.

Lawvere, F. W. (1972). *Perugia Notes: Theory of Categories over a Base Topos*, Perugia: University of Perugia Lecture Notes.

Lawvere, F. W. (1991). Intrinsic co-Heyting boundaries and the Leibniz rule in certain toposes. In A. Carboni, M. C. Pedicchio, and G. Rosolini (eds.), *Category Theory*, New York, NY: Springer-Verlag, pp. 279–281.

Lawvere, F. W. (1992). Categories of space and of quantity. In J. Echeverria, A. Ibarra and T. Mormann (eds.), *The Space of Mathematics: Philosophical, Epistemological and Historical Explorations*, Berlin: De Gruyter, pp. 14–30.

Lawvere, F. W. (1994a). Tools for the advancement of objective logic: Closed categories and toposes. In J. Macnamara and G. E. Reyes (eds.), *The Logical Foundations of Cognition*, New York: Oxford University Press, pp. 43–56.

Lawvere, F. W. (1994b). Cohesive toposes and Cantor's 'lauter Einsen'. *Philosophia Mathematica* 2: pp. 5–15.

Lawvere, F. W. (1996). Unity and identity of opposites in calculus and physics. *Applied Categorical Structures* 4: pp. 167–174.

Lawvere, F. W. (2003). Foundations and applications: Axiomatization and education. *The Bulletin of Symbolic Logic* 9: pp. 213–224.

Lawvere, F. W. (2004). Functorial semantics of algebraic theories and some algebraic problems in the context of functorial semantics of algebraic theories. *Reprints in Theory and Applications of Categories* 5: pp. 1–121.

Lawvere, F. W. (2005). An elementary theory of the category of sets (long version) with commentary. *Reprints in Theory and Applications of Categories* 11: pp. 1–35.

Lawvere, F. W. (2006). Why are we concerned? II, *Category Theory Post*, http://rfcwalters.blogspot.com/2010/10/old-post-why-are-we-concerned-fw.html [Accessed 10 August 2018].

Lawvere, F. W. (2007). Axiomatic cohesion. *Theory and Applications of Categories* 19: pp. 41–49.

Lawvere, F. W. (2016). Birkhoff's theorem from a geometric perspective: A simple example. *Categories and General Algebraic Structures with Applications* 4: pp. 1–7.

Lawvere, F. W. (2017) Everyday physics of extended bodies or why functionals need analyzing. *Categories and General Algebraic Structures with Applications* 6: pp. 9–19.

Lawvere, F. W. and Menni, M. (2015). Internal choice holds in the discrete part of any cohesive topos satisfying stable connected codiscreteness. *Theory and Applications of Categories* 30(26): pp. 909–932.

Lawvere, F. W. and Rosebrugh, R. (2003).*Sets for Mathematics*, Cambridge, UK: Cambridge University Press.

Lawvere, F. W. and Schanuel, S. H. (2009).*Conceptual Mathematics: A First Introduction to Categories*, (2nd ed.) Cambridge, UK: Cambridge University Press.

Linton, F. E. J. (2005). Shedding some localic and linguistic light on the tetralemma conundrums. In G. G. Emch, R. Sridharan, and M. D. Srinivas (eds.), *Contributions to the History of Indian Mathematics*, New Delhi: Hindustan Book Agency, pp. 63–73.

Mazur, B. (2008). When is one thing equal to some other thing? In B. Gold and R. A. Simons (eds.), *Proof and Other Dilemmas: Mathematics and Philosophy*, Washington, DC: MAA Spectrum, pp. 221–241.

Picado, J. (2008). An interview with F. William Lawvere. *CIM Bulletin* 24: pp. 21–28.

Posina, V. R., Ghista, D. N., and Roy, S. (2017). Functorial semantics for the advancement of the science of cognition. *Mind & Matter* 15: pp. 161–184.

Priest, G. (2009). The structure of emptiness. *Philosophy East and West* 59: pp. 467–480.

Priest, G. (2010). Two truths: Two models. In *The Cowherds, Moonshadows: Conventional Truth in Buddhist Philosophy*, New York, NY: Oxford University Press, pp. 213–220.

Priest, G. (2013). Nāgārjuna's Mulamadhyakamakarika, *Topoi* 32: pp. 129–134.

Priest, G. (2014). Beyond true and false. *Aeon*, https://aeon.co/essays/the-logic-of-buddhist-philosophy-goes-beyond-simple-truth [Accessed 13 March 2017].

Priest, G. (2015). The net of Indra. In K. Tanaka et al. (eds.), *The Moon Points Back*, New York, NY: Oxford University Press, pp. 113–127.

Priest, G. and Garfield, J. (2003). Nāgārjuna and the limits of thought. *Philosophy East and West* 53: pp. 1–21.

Resnik, M. (1981). Mathematics as a science of patterns: Ontology and reference. *Nous* 15: pp. 529–550.

Reyes, M. L. P., Reyes, G. E., and Zolfaghari, H. (2004).Generic Figures and their Glueings: A Constructive Approach to Functor Categories, Milano: Polimetrica.

Siderits, M. and Katsura, S. (2013). *Nāgārjuna's Middle Way*, Boston, MA: Wisdom Publications.

Westerhoff, J. (2006). Nāgārjuna's catuskoti. *Journal of Indian Philosophy*, 34: pp. 367–395.

CHAPTER 23

# Zero: an Integrative Spiritual Perspective with One and Infinity

*Sharda S. Nandram, Puneet K. Bindlish, Ankur Joshi and Vishwanath Dhital*

## Abstract

The world of numbers keeps us intrigued for one reason or the other. From negative numbers to imaginary numbers and from various constants to peculiar ratios, they never cease to surprise us. And our curiosity and imagination keeps on inspiring us to look for these patterns in nature. Consequently leading the way for discoveries that are important to meeting not only our material goals but also spiritual ones at the same time. However, due to an extraordinary reliance of modern applied science on numbers, we have generated more scholarly interest, especially in the last few centuries, around the application of numbers to the material realm as compared to the spiritual. Amidst our attention to the peculiar or special numbers and ratios, there are some numbers – or rather concepts – that still look in our face, inviting us to explore a whole new world of possibilities all over again. Though these concepts laid the foundation of modern day mathematics as we use it, yet at times we take them for granted and are not usually interested in revisiting our understanding about their nature, their past and future possibilities. In this chapter, we revisit three of these foundational concepts – Zero, One and Infinity (ZOI) from an integrative spiritual perspective.

## Keywords

integrative – spirituality – zero – infinity – non-duality – ancient wisdom – Hindu philosophy

 | DOI:10.1163/9789004691568_027

## 1 Introduction

> rāma nāma ko eka aṃka hai. bākī saba sādhana hai sūna, aṃka gae kachu hātha nahīṃ, aṃka rahe dasa gūna.[1]
>
> [The whole universe can become zero (i.e., all that is infinite can be reduced to zero in case of absence of the name of divinity (spiritual component). And in its presence, it becomes multi-fold.]
>
> GOSWAMI TULSIDAS

The world of numbers keeps us intrigued for one reason or another. From negative numbers to imaginary numbers and from various constants to peculiar ratios, they never cease to surprise us. And our curiosity and imagination keep on inspiring us to look for patterns based on numbers in nature. Consequently, we are leading the way for discoveries that are important to meeting not just our material goals but also spiritual ones at the same time. However, due to an extraordinary reliance of modern applied science on numbers, we have generated more scholarly interest, especially in the last few centuries, around the application of numbers to the material realm as compared to the spiritual. Amidst our attention to the peculiar or special numbers and ratios, there are some numbers – or rather concepts – like *One, Zero and Infinity*, that still look in our face, inviting us to explore a whole new world of possibilities all over again. Though these concepts laid the foundation of modern-day mathematics as we use it, at times we take them for granted and are not usually interested in revisiting our understanding about their nature, or their past and future possibilities. It would be worth exploring the questions like why One, Zero and Infinity were conceptualized in the first place? For what greater objectives? How can this understanding impact our collective future?

In our initial study, we noticed that for some civilizations, zero, as well as one and infinity, had always been a way to grapple the unknown – in other words, the phenomena beyond human experience like divinity, God, the unknown. We therefore find it more meaningful when using a spiritual lens to these concepts. Thus, in the present chapter, we will explore the spiritual dimensions of these foundational number concepts – particularly One, Zero and Infinity – from an integrative spiritual perspective of those worldviews.

1 राम नाम को एक अंक है. बाकी सब साधन है सून | अंक गए कछु हाथ नहीं, अंक रहे दस गून ||

## 2 Methodology

This is a conceptual work based on ancient and modern scholarly literature to revisit the ontology of One, Zero and Infinity (OZI) and their relevance for the field of spirituality. The key research questions are:

- How does Zero help us in understanding spirituality from an integrative perspective together with One and Infinity?
- Can a richer understanding of a spiritual perspective to OZI help infuse spirituality in other domains through holistic application of OZI in those domains?

To address these questions, we build on our previous work on the integrative perspective as a worldview. Arguments are developed to build the ontology of One, Zero and Infinity for getting an integrative grasp of its relevance for spirituality and society. The scope of concepts of spirituality and integrative as used in our research follows the following definitions:

> *Spirituality*: Any entity's quality of becoming aware of the connectedness with its existence beyond perceived existence. (Nandram, 2019)
> *Integrative*: the intent of relating all possible aspects, perspectives and purposes under holistic understanding of any context towards a coherent view among all observers of the context in focus without discarding any aspect, perspective and purpose. (Nandram et al., 2017)

This conceptual chapter acknowledges the fact that zero and its origin are complex and its understanding intersects through ancient scientific, modern scholarly literature and cultural literature of various civilizations. Therefore, we propose an integrative perspective towards our search for arguments to build the ontology of OZI for getting an integrative grasp of its relevance for spirituality and society. First, the relevant literature in modern science has been studied. In addition to the traditional review of contemporary literature, *hermeneutics of contemplation* (as compared over *hermeneutics of suspicion or faith*) (Sharma, 2001) approach has been undertaken to study selective ancient Sanskrit scriptures with the focus on zero from a spiritual perspective.[2]

---

2 *Hermeneutics of suspicion* (inspired by Marx, Nietzsche, Freud): It includes demystification of meaning presented to interpreters in the form of a disguise. It is characterized by a distrust of the symbol as a dissimulation of the real; animated by suspicion, skepticism towards the given. It is largely used by: Feminist theologians to critique traditional religious doctrines; Post-colonial scholars to highlight structural injustices; and, Religious scholars to analyze non-Western religious cultures.

During the initial phase of the study, it became clear that we would have to study not only zero but also the numbers one and infinity so that we got a better grasp of its relevance for the field of spirituality. In the next sections we provide some views on the origin of zero followed by various understandings of zero and we proceed with arguments to build the ontology of zero where we introduce different similar Sanskrit concepts. Here we draw from several Vedic scriptures and move forward to discuss not only zero but also the numbers one and infinity for getting an integrative grasp of its relevance for spirituality and society.

## 3 Background of Zero: a Complex Phenomenon

The confusion and mystery around the idea of zero has been persisting for a long time. Scholars, thinkers and philosophers have tried to provide answers based on logic, assertion and reasoning. Rotman (1987) in his book titled *Signifying nothing: The semiotics of zero* captures arithmetical, sign and monetary aspects related to zero. Also, it is important to read about the connection drawn with philosophy, culture, religion and design in connection with the idea of zero. He reviewed and studied in detail the origin of zero from Hindu systems and how it traveled to different parts of the world. He remarked that it was very clear that European consciousness took time to incorporate the sign of zero that represents 'nothing' or 'void'. The resistance to the idea was so deep that it was considered an unnecessary concept. Even the symbol used for zero faced resistance. However, later on, the idea was well received and, with the advent of binary systems and computers, the world acknowledged its importance. However, this acceptance in terms of worldly developments cannot be considered the greatest contribution. The real contribution lies in its deep spiritual significance which has taken precedence with the global interest in Yoga and Meditation.

The mathematics, science and technology that can be seen today are possible because of zero. It cannot be imagined how things would have been without zero. The concept of zero is required to define numbers. Without zero we

*Hermeneutics of faith*: It aims at the restoration of a meaning addressed to the interpreter in the form of a message. It is characterized by a willingness to listen, to absorb the message in its given form and it respects the symbol, understood as a cultural mechanism for our apprehension of reality, as a place of revelation. Romantic reclamation of tradition. (Josselson, 2004).

*Hermeneutics of contemplation*: The object is more of a philosophical exploration for truth than towards historicity. (Sherma, 2008).

cannot even imagine the calculations done today with ease and few errors (Aczel, 2015). From whichever perspective one may look, the concept of zero remained something which was not just difficult to comprehend, but also challenging to accept and incorporate in all domains like arithmetic, finance, governance, philosophy and spirituality.

The scholars and practitioners in these fields explore some ground for practical purposes and come to common ground for at least a partial understanding of concepts like zero. For instance, Barton (2020) suggests that zero is a kind of numerosity property corresponding to an absence of positive numerosity. Zero represents dual nature. On the one hand, it is a *bona fide* number to compute whereas, on the other hand, it is a representation of not being and nothingness. Also, for some, it is a number characterized by an absence property. Zero as a number has huge potential because zero offers nothing and seems to represent emptiness. It gives nothing. Strikingly, by contrast, it changes the values of other numbers when combined with them. This might look very obvious but the concept of zero is being made to convey many things (Salwi, 1988). Zero enjoys a privileged status among the numbers. It is a sign for nothing and is being represented as a vanishing point in arts and imaginary money in economic transactions (Mclennan, 1988). The acceptance of the concept of zero in these domains provides space for discussion and deliberation on the idea of zero at a more fundamental level.

### 3.1 *Some Views about Its Origins*

The idea originated in India and is known as *Śūnya* in Sanskrit as well as in Hindi. In other languages it is known by other names. For instance, it is called *Sifr* in Arabic and Cipher or Zero in English. In Medieval Latin it is called *cifra*.[3] Basically it is something which stands for naught, the absence of all quantity considered as quantity. The word 'cipher' initially had a meaning of zero but became a secret way of writing coded messages, because early codes often substituted numbers for letters. This meant that 'the key to a cipher or secret writing' was, by 1885, short for cipher key (by 1835).[4]

Zero is a unique number with special properties in calculations and is an exception to other numbers but still a number. It is the first digit in the numbers to be invented last, and the first number to be discovered last. It represents nothingness. India is the one who gave zero an arithmetical notation. To represent nothing Indians used the symbol of a dot (.) which denotes *Śūnya* – indicating emptiness. The *śūnya* represents the symbol for zero but not the

3 https://www.etymonline.com/search?q=zero.
4 https://www.etymonline.com/word/cipher.

FIGURE 23.1 Zero in different languages
Note: https://www.indifferentlanguages.com/words/zero.

number zero. This *śūnya*, when traveling to the Arab world, became *Sifr* and got translated and adapted in other languages successively (Reid, 2006).

Aczel (2015) through his inquisitiveness to find out the origin of zero has been through much research and observation for years and has credited the origin of zero to the East and falsified the claim that the concept of zero is of Western origin. The number system and zero reached the Arab world from the East and then Western countries.[5] As per *nāsadīya sūkta* in the ancient Vedic scripture *Ṛgveda*, zero in the sense of nothingness, is seen as the origin of everything. The Buddhist philosophical concept of emptiness (i.e., void) is Śūnyatā.

Aczel (2015) analyzes the meaning of infinity (*ananta*) to the conception of Indian Gods. In his travel accounts he mentions the use of numbers at various places in the East such as Khajuraho, Jaipur, Gwalior in India, and Thailand, Cambodia and Vietnam. The curiosity ended when the author found the stone K-127, finally defusing the claim of the origin of zero to the West. Apart from

5 Interestingly, the Dutch language took a slight departure on the meaning of *cijfer*. One of the meaning(s) of *cijfer* in Dutch is digits (numbers, any or all) but not zero in the sense in which it is used exactly in the parent word root. The word has an obvious influence from Latin *cifra* and/or Arabic *Sifr*, where it means zero in a sense of absence of quantity. But the meaning is not exactly the same. It would be interesting to explore whether *cijfer* philosophically was ever seen as zero in the sense of an origin of all numbers.

the archeological evidence, there are multiple references to usage of zero in scriptures in different fields.

The mathematical progress made by the countries in the world has been possible through the high mathematical acumen of India since ancient times. There is mention of mathematics in one of the most ancient Indian scriptures, Vedas. Zero also finds a place in works of Ācārya Cāṇakya also known as *Kauṭilya* and widely acclaimed for his work's, *Cāṇakya Nīti* and *Arthaśāstra*, the scriptures which deal with administration, governance, finance, management and various aspects of human life. Numbers represented as *aṇka* in India were invented before zero, which can be authenticated by the beautiful stone pillars of Aśoka, indicating that the decimal system and number symbols were present in India (273–232 BC). Chapters 121 and 122 of *Agni Purāṇa* are on *Jyotiṣa* and calculation of time. These include astronomy where study of planetary motions and time span are discussed, which is not possible without numbers. When Indians could multiply $10^{18}$, other countries could calculate only up to $10^4$. It is significant to note how the supremacy of the Indian number system was introduced to other countries. It was through the merchants, traders, travelers, scholars and soldiers that the Indian number system was introduced in far East, Arab, and European countries. Initially in the number's place zero signified nothing. Many Indian scholars worked on the operation of zero. Brahmagupta (598–660 AD) in his treatise *Brāhmasphuṭasiddhānta* and subsequently mentions the calculation operation on zero. Bhāskara (1114–1185 AD) made corrections to his work. Indian contribution in the domain of numbers, decimals and the invention of zero has simplified the calculations and has given huge assistance to the world in the field of astronomy and navigation.

The reference of zero is visible too in social science and technology. For instance, zero growth represents stability or equilibrium in social scenarios and zero defect indicates consistency of quality in any output. Though, in the sensory experienceable reality, zero does not seem to exist. For example, we cannot have zero items, we cannot practically create or be in a complete vacuum or a physical state of emptiness. But still, the concept of zero which was created in mind and exists in mind has led to profound scientific and technical advances. This has been made possible through the genius of the ancient Indian mind (Salwi, 1988). The discovery of zero freed the entire humanity from the counting board calculations (Ifrah, 1985).

Though there are various scholarly debates around the origin of zero and the journey of zero from mathematicians from one place to another, the present work does not deal with the origin of the concept of zero. It deals with the interpretations of zero from spiritual perspectives. In general, the most common conceptualization of zero is related to absence of anything, place and material. However, this is a limited understanding as even in mathematics the idea of the number line violates this basic and fundamental idea behind zero by making its place in the center of the number line and numbers of negative plane 'lesser than zero'. The concept of zero as discussed in ancient scriptures is also 'not' limited to absence or the ideas of transcendence like *nirvāṇa* or *mokṣa*. The idea of zero can be looked at from various other dimensions too.

## 4 Understanding Zero with One and Infinity

### 4.1 *Revisiting Origins as Thought Forms for Unknown*

Our understanding of the history of science from ancient times to the present informs us that all ancient civilizations have contributed in a great way to the scientific concepts and methods as we know them today. Several worldviews that influenced the evolution of science (or scientific knowledge) can be categorized into two types on the basis of the belief systems around time – linear and cyclical worldviews. The former refers to religious worldviews that believe in one life, one beginning and one end of existence. They are largely monotheistic in their faith. The latter holds the idea of multiple lives spanning a cycle of birth, death and rebirth. They are largely polytheistic or atheistic in expressions of their faith. Most of the eastern religions, especially originating from Indian civilization share this belief.

The worldview of modern science, especially that which has evolved in the past few centuries, is largely influenced by the industrial worldview, which in turn has its influence from early days of modern science. The modern scientific worldview has its genesis in the reaction towards the oppression by authority-based religious worldviews. Incidentally, or as a consequence, both worldviews share a striking similarity with regards to their predominantly linear view of the concept of time. This influenced the linear conceptualization (and subsequent material exploitation) of the concepts of One, Zero and Infinity. Quite a lot has been researched and written on those conceptualizations and their applications to material reality especially from linear worldviews. But there is still a good scope for philosophical investigation into these concepts from the perspective of cyclic worldviews. Also, it is no surprise that zero has an equal connection with every field of study including spirituality.

Before formalizing zero as a number to solve day-to-day counting or other material problems or problems in the physical realm, there were prior thought forms leading to this formalization. These formalizations should be seen as by-products that emerged from deep thinking. Our interest is to consider the end product of these thought patterns – the deep thinking. Following an integrative spiritual perspective, if we go outside the mathematical lens to see spiritual texts, we will find that the concept of zero presents itself in a variety of ways. This can be best dealt with in terms of spirituality, apart from disciplines like psychology and philosophy.

In order to appreciate this, let's revisit our usual connotation with numbers in the material realm. School children are first taught numbers through counting by relating them with material objects. However, in the traditional schools teaching ancient scriptures like *gurukula*, the first introduction to numbers begins with their connotations with relatively eternal as well as abstract objects. It is the beginning of formation of thought forms representing eternal objects, like qualities of supreme consciousness are equated with zero, one or infinity (for instance, truth is one), without naming them specifically. Other numbers are also introduced in a similar fashion. For example, a number five implying five elements (fire, water, earth, air and space), number three is introduced with some form of trinity (observer-object-observation; knower-knowable-knowledge). We find trinities in several Hindu scriptures in many expressions.

### 4.2 *Zero and Infinity*

Zero and Infinity are two concepts which signify our attempt to conceptualize the unknown. Ontologically, both concepts are dealing with something whose properties, when described under different concepts, often overlap. Therefore, the two share a deep interrelationship which may be relevant for us to examine from both material as well as spiritual ends. To delve deeper into the idea of zero and infinity from a Hindu worldview, let's look at the words associated with these concepts and ones that may have similar connotations.

- *Śūnya* denotes absence of anything. From a pure mathematics or quantification point of view, it can be interpreted as its literal meaning. Apart from the normal understanding of zero as nothing, the zero is also a number from which counting begins and is used to repeat the numbers in our commonly used decimal system.
- *Anta* or *Samāpta* means something that has been reduced to zero or end or completed. This links to the notion of infinity – also in one direction. As *Ananta*, another ontological concept for end, means something which has no end.
- *Abhāva* denotes that which does not exist or something which does not have any existence.
- *Nagaṇya* means something that cannot be counted. This also can be studied in both directions. It can be towards zero or towards infinity.

The above connotations of the concept of zero have been explored by different scientific disciplines for a variety of material applications in the past centuries. However, the aspect of zero which deserves more spiritual and philosophical

FIGURE 23.2 The idea of Subtler than the Subtlest. The biggest circle represents the subtlest (aṇu) and smallest red circle represents paramāṇu (subtler than subtlest)

attention is the fact that it represents different interpretations at the same time. Let's look at their ontological aspects more for exploring the spiritual significance of these concepts.

– *Paramāṇu* or subtler than the subtlest: Zero can represent nothing, absence or something that is smaller than the smallest (see Figure 23.2). When something is divided by this smaller than the smallest, we get infinity. Or in other words, zero can be seen as the seed of the whole tree of infinity. Upādhyāy (2018) in his study of Swami Nischalananda Saraswati's works on the philosophy of mathematics, provides a good understanding of zero, infinity and their deep interrelationship. Both of these concepts are indicative of aspects of *Brahma*, the supreme truth and innate nature of creation or growth. He quotes several verses from different Upanishadic literature: Tejobindūpaniṣad (6.87), Maitrāyaṇīya Upaniṣad (2.6, 6.38) among other quotes.

> *aṇoraṇīyānahameva tadvanmahānahaṃ viśvamahaṃ vicitram, purātano'haṃ puruṣo'hamīśo hiraṇmayo'haṃ śivarūpamasmi.*
>
> [I am more minute than the minute, I am likewise the greatest of all, I am the manifold universe. I am the Ancient One, the Consciousness and the Ruler, I am the Effulgent One, and the All-good.]
>
> JAQUES, 2022, *Kaivalya Upaniṣad*, verse 20[6]

6 अणोरणीयानहमेव तद्वन्महानहं विश्वमहं विचित्रम्। पुरातनोऽहं पुरुषोऽहमीशो हिरण्मयोऽहं शिवरूपमस्मि ॥

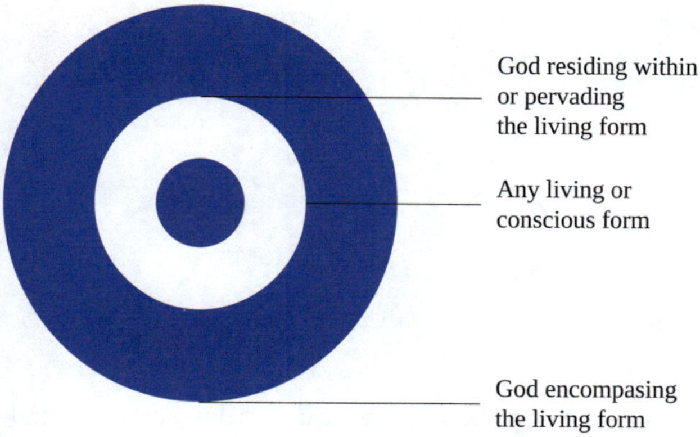


FIGURE 23.3 The idea of pervasive as well as all-encompassing nature of God

- *Ādi* and *Anādi: Ādi* signifies the fundamental thing of energy from which everything or anything originates or something which originates. Whereas *anādi means* something which has no origin. It has existed since always and hence will also not have any end. The former can be seen closer to the concept of zero and the latter to infinity. From a Hindu theological conceptualization of God, both properties are valid (i.e., from presence as well as omnipresence). In Śrīmadbhagavadgītā, the God is additionally seen as detached too from the concept of both.

> mayā tatam idaṁ sarvaṁ jagad avyakta-mūrtinā,
> mat-sthāni sarva-bhūtāni na cāhaṃ teṣvavasthitaḥ.
> na ca mat-sthāni bhūtāni paśya me yogam aiśwaram,
> bhūta-bhṛn na ca bhūta-stho mamātmā bhūta-bhāvanaḥ.
>
> [All this world is pervaded by Me in My unmanifest aspect; all beings exist in Me, but I do not dwell in them. Nor do beings exist in Me (in reality); behold My divine Yoga, supporting all beings, but not dwelling in them, is My Self, the efficient cause of beings.]
>
> SIVANANDA, 2003 p. 20 4–205, *Śrīmadbhagavadgītā*, verse 9.4, 9.5[7]

The innermost blue circle depicts God residing within or pervading the living form (represented by the white circle) and that living form is encompassed by or within God (represented by the outer blue circle).

7 मया ततमिदं सर्वं जगदव्यक्तमूर्तिना। मत्स्थानि सर्वभूतानि न चाहं तेष्ववस्थित: ॥
न च मत्स्थानि भूतानि पश्य मे योगमैश्वरम्। भूतभृन्न च भूतस्थो ममात्मा भूतभावन: ॥

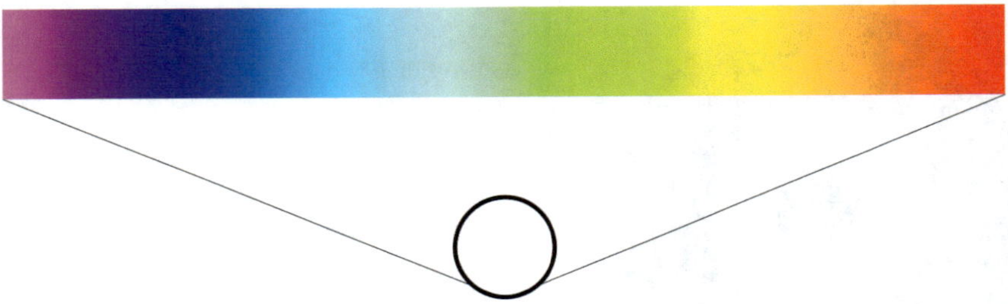

FIGURE 23.4 Depicting the idea of *Nirguṇa* and *Guṇātīta*

- *Nirguṇa and Guṇātīta: Nirguṇa* denotes something that is devoid of any qualities (or *Guṇa*), a concept that resonates with the concept of zero. Whereas *guṇātīta*, means one who is able to transcend the boundaries of qualities (or *Guṇa*) (see Figure 23.4). Another relatable concept is of *kālātita*, something which is beyond time or beyond the purview of time and something which has the capability to transcend time. Both can be equated to the idea of infinity. Interestingly, the Hindu theology of God describes it through qualities and among other qualities, two of the prominent ones are *nirguṇa* and *guṇātīta*. That means God, the supreme, is seen as someone that's subtler of the subtlest existence and also that which transcends the whole of existence, at the same time. This interrelationship of zero and infinity in the domain of unknown aspects of existence is an important element that has the potential to impact our material as well as spiritual view of existence.

The white light is the sum total of all colors and shows its constituents only when seen through a prism.

- *Nirākāra* means something without any form or finite shape (or *ākāra*). Something which is not confined to any boundaries and therefore can be zero or infinity or both. The concept is valid for either of the two or both. Here from a spiritual standpoint, zero is seen as exactly equal to infinity.

After exploring the ontology of zero from a spiritual perspective, let's now examine its implication of spiritual pursuits. An interesting context in which the discussion on spiritual implication comes in Chāndogyopaniṣad between sage Uddālaka and his son Śvetaketu, who considers himself well read and has turned arrogant after twelve years of traditional education. Uddālaka asks, 'Now that you're well read, my dear, have you ever asked for that learning by which one hears what cannot be heard, by which one perceives what cannot be perceived, by which one knows what cannot be known?'. While giving a detailed reply, which is the subject matter of that entire Upanishad, one of the verses provide the nature of zero, infinity and immanence:

*eṣa ma ātmāntarhṛdaye'ṇīyān vrīher vā yavād vā sarṣapād vā śyāmākād vā śyāmākataṇḍulād vaiṣa ma ātmāntarhṛdaye jyāyān pṛthivyā jyāyān antarikṣāj jyāyān divo jyāyān ebhyo lokebhyaḥ.*

[He who consists of the mind, whose body is subtle, whose form is light, whose thoughts are true, whose nature is like the *akasa*, whose creation in this universe, who cherishes all righteous desires, who contains all pleasant odors, who is endowed with all tastes, who embraces all this, who never speaks and who is without longing. He is my Self within the heart, smaller than a grain of rice, smaller than a grain of barley, smaller than a mustard seed, smaller than a grain of millet; He is my Self within the heart, greater than the earth, greater than the mid-region, greater than heaven, greater than all these worlds.]

AUROBINDO, 2006, *Chāndogyopaniṣad* 3.14.3[8]

The scriptures while shedding light on the ontology of these concepts, also highlight the ways to know and realize these concepts. This is important from the spirituality point of view. Kaṭhopaniṣad, for instance, hints at the futility of epistemological logic in knowing this subtler than the subtlest nature of existence.

*na nareṇāvareṇa prokta eṣa suvijñeyo bahudhā cintyamānaḥ,*
*ananyaprokte gatir atra nāsty aṇīyān hy atarkyam aṇupramāṇāt.*

[An inferior man cannot tell you of Him; for thus told thou canst not truly know Him, since He is thought of in many aspects. Yet unless told of Him by another thou canst not find thy way there to Him; for He is subtler than subtlety and that which logic cannot reach.]

AUROBINDO, 2006, *Kaṭhopaniṣad* verse 1.2.8[9]

Śrīmadbhagavadgītā provides some guidance on the ways to realize this. And Śvetāśvataropaniṣad mentions some of the outcomes of this pursuit, particularly freeing oneself from sorrows and desires.

8 एष म आत्मान्तर्हृदयेऽणीयान्व्रीहेर्वा यवाद्वा सर्षपाद्वा श्यामाकाद्वा श्यामाकतण्डुलाद्वैष म आत्मान्तर्हृदये ज्यायान्पृथिव्या ज्यायानन्तरिक्षाज्ज्यायान्दिवो ज्यायानेभ्यो लोकेभ्यः॥

9 न नरेणावरेण प्रोक्त एष सुविज्ञेयो बहुधा चिन्त्यमानः। अनन्यप्रोक्ते गतिरत्र नास्त्यणीयान् ह्यतर्क्यमणुप्रमाणात्॥

*kaviṁ purāṇamanuśāsitāramaṇoraṇīyāṁsam anusmaredyaḥ,*
*sarvasya dhātāram acintya-rūpamāditya-varṇaṁ tamasaḥ parastāt.*

[Whosoever meditates on the Omniscient, the ancient, the ruler (of the whole world), minuter than an atom (minutest), the supporter of all, of inconceivable form, effulgent like the sun and beyond the darkness of ignorance.]

SIVANANDA, 2003, p. 184, *Śrīmadbhagavadgītā* verse 8.9[10]

*aṇoraṇīyānmahato mahīyānātmā guhāyāṁ nihito'sya jantoḥ, tamakratuṁ paśyati vītaśoko dhātuḥ prasādānmahimānamīśam.*

[Subtler than even the subtlest and greater than the greatest, the Atman is concealed in the heart of the creature. By the grace of the Creator, one becomes free from sorrows and desires, and then realizes Him as the great Lord.]

TYAGISANANDA, 1949, p. 75, *Śvetāśvataropaniṣad* verse 3.20[11]

#### 4.2.1 Understanding One and Infinity

Like Zero and Infinity, One also has several connotations especially in the spiritual worldviews and philosophical schools from Indian civilizational context. One of the prominent philosophical schools, Nyāya-Vaiśeṣika philosophy, provides multiple interpretations of 'one' – in other words, one that can be divided and another undivided one. The undivided one (like sky) resembles the concept of infinity.

One holds a deep spiritual significance in the Hindu worldview. Despite different manifestations of this worldview, there are some underlying commonalities. The most prominent among them is the concept of *Ekatva* (Oneness). To see oneness (*Ekatva*) in many or diversity (*Bahutva*) and *Bahutva* in *Ekatva* is the essence of this worldview. Further, the supreme consciousness is seen as the instrumental as well as material cause behind the whole creation in its diversity (Saraswati, 2010). Therefore, the origin and ultimate destination in the journey of the soul (*Atman*) is seen as one. Saraswati (2010) quotes Śrīmadbhagavadgītā to explain this concept.

---

10 कविं पुराणमनुशासितारमणोरणीयांसमनुस्मरेद्यः। सर्वस्य धातारमचिन्त्यरूपमादित्यवर्णं तमसः परस्तात्॥

11 अणोरणीयान्महतो महीयानात्मा गुहायां निहितोऽस्य जन्तोः। तमक्रतुं पश्यति वीतशोको धातुः प्रसादान्महिमानमीशम्॥

*sarva-bhūteṣu yenaikaṁ bhāvam avyayam īkṣate,*
*avibhaktaṁ vibhakteṣu taj jñānaṁ viddhi sāttvikam.*

[That by which one sees the one indestructible Reality in all beings, not separate in all the separate beings know thou that knowledge to be Sattvic.]

SIVANANDA, 2003 p. 481, *Śrīmadbhagavadgītā*, verse 18.20[12]

*sarva-bhūta-stham ātmānaṁ sarva-bhūtāni cātmani,*
*īkṣate yoga-yuktātmā sarvatra sama-darśanaḥ.*
*yo māṁ paśyati sarvatra sarvaṁ ca mayi paśyati,*
*tasyāhaṁ na praṇaśyāmi sa ca me na praṇaśyati.*
*sarva-bhūta-sthitaṁ yo māṁ bhajaty ekatvam āsthitaḥ,*
*sarvathā vartamāno 'pi sa yogī mayi vartate.*

[With the mind harmonized by Yoga he sees the Self abiding in all beings and all beings in the self; he sees the same everywhere. He who sees Me everywhere and sees everything in Me, he never becomes separated from Me, nor do I become separated from him. He who, being established in unity, worships Me who dwells in all beings, that Yogi abides in Me, whatever may be his mode of living.]

SIVANANDA, 2003 pp. 144–145, *Śrīmadbhagavadgītā*, verse 6.2 9–31[13]

Saraswati (2010) highlights the importance of *Ekatva Darśana* (the philosophy of Oneness) inspired action by quoting verses from various ancient scriptures, especially Mahābhārata (Anuśāsanaparva, Vana Parva) and Īśāvāsyopaniṣad.

*yasminsarvāṇi bhūtānyātmaivābhūdvijānataḥ,*
*tatra ko mohaḥ kaḥ śoka ekatvamanupaśyataḥ.*

[When to a knower discovering unity, all beings become his very Self or Atman, what delusion or perplexity then (to him) and what grief or sorrow when he sees this oneness.]

HIRIYANNA, 1911, p. 14, *Iśavasya Upaniśad*, verse 7[14]

12 सर्वभूतेषु येनैकं भावमव्ययमीक्षते । अविभक्तं विभक्तेषु तज्ज्ञानं विद्धि सात्त्विकम् ॥
13 सर्वभूतस्थमात्मानं सर्वभूतानि चात्मनि । ईक्षते योगयुक्तात्मा सर्वत्र समदर्शनः ॥
यो मां पश्यति सर्वत्र सर्वं च मयि पश्यति । तस्याहं न प्रणश्यामि स च मे न प्रणश्यति ॥
सर्वभूतस्थितं यो मां भजत्येकत्वमास्थितः । सर्वथा वर्तमानोऽपि स योगी मयि वर्तते ॥
14 यस्मिन् सर्वाणि भूतान्यात्मैवाभूद् विजानतः। तत्र को मोहः कः शोक एकत्वमनुपश्यतः ॥

*Ekatva or* Integral Unity is the oneness that lies underneath and connects every form of diversity (Malhotra, 2019). This is articulated in almost all scriptures albeit at different levels of our existence and in different ways, explicit or implicit.

> *jñātā jñānaṁ tathā jñeyaṃ draṣṭā darśanadṛśyabhūḥ,*
> *kartā hetuḥ kriyā yasmāt tasmai jñaptyātmane namaḥ.*
>
> [He is the knower, the knowledge and all that is to be known. He is the seer, the (act of) seeing, and all that is to be seen. He is the actor, the cause and the effect: therefore salutation to Him (who is all) knowledge himself. Or in other words, Knower, knowledge, and to be known; Seer, observation and object; Doer, reason and action; we pray to God from whom all of these come.]
>
> BINDLISH ET AL., 2018, *Yoga* Vāsiṣṭha, 1.1.2[15]

Interestingly, a similar idea was mentioned by Richard Feynman in his Nobel Prize lecture:

> I received a telephone call one day at the graduate college at Princeton from Professor Wheeler, in which he said, 'Feynman, I know why all electrons have the same charge and the same mass.' 'Why?' 'Because, they are all the same electron!' (Feynman, 1966)

### 4.3 *Zero with One and Infinity*

When the ontology of OZI is viewed from a spiritual perspective of Hindu worldview, all the three concepts are seen as complementary, supplementary and equal at the same time. The three variations on the concept of forms from a spiritual significance are: *nirguṇa-nirākāra* (quality-less, formless), *saguṇa-sākāra* (with qualities, with form) and *nirguṇa-sākāra* (qualityless, with form). Here, zero can be seen as *nirguṇa-nirākāra* denoting the root or origin and the end of the creation; devoid of any quality or form; something denoting absence of anything. One can be seen as *saguṇa-sākāra* denoting something that's not zero; of finite existence; something between zero's two extremes; the underlying oneness. And finally, infinity can be seen as *nirguṇa-sākāra*, signifying something that transcends any qualities including material, space and time; something which is not finite or unlimited; and also the ultimate end of anything (see Table 23.1).

15 ज्ञाता ज्ञानं तथा ज्ञेयं द्रष्टा दर्शनदृश्यभूः । कर्ता हेतुः क्रिया यस्मात् तस्मै ज्ञप्त्यात्मने नमः ॥

TABLE 23.1 One, zero and infinity – an integrative spiritual perspective

| | **Quality-less** | **With quality** |
|---|---|---|
| **Form-less** | *nirguṇa-nirākāra*<br>0 | – |
| **With Form** | *nirguṇa-sākāra*<br>∞ | *saguṇa-sākāra*<br>1 |

This view also provides an inspirational understanding of the concept of deep spiritual unity among different manifestations of gods in the Hindu theology. Take for example, *Viṣṇu* and *Śiva*. Several ancient scriptures like *Purāna, Rāmāyāna, Mahābhārata, Śrīmadbhagavadgītā* that mention the two, also mention a cyclic recursive relationship between the two in several ways. Some may even depict them with no mutual difference. Taking idea from the role of *Viṣṇu* as a maintainer God for the world through infinite resources at his disposal. While Śiva, with no resources at his disposal or seen as one that transcends space and time, takes care of the dissolution of the world. These ideas have been very aptly put together in an Upaniṣad verse:

> *oṃ pūrṇam adaḥ pūrṇam idam pūrṇāt pūrṇam udacyate.*
> *pūrṇasya pūrṇam ādāya pūrṇam evāvaśiṣyate. oṃ śāntiḥ śāntiḥ śāntiḥ*
>
> [Om. That (Brahman) is infinite, and this (universe) is infinite. The infinite proceeds from the infinite. (Then) taking the infinitude of the infinite (universe), it remains as the infinite (Brahman) alone.]
>
> MĀDHAVĀNANDA, 1965, *Bṛhadāraṇyaka Upaniṣad* 5.1.1[16]

## 5 Connection with Spirituality

We explored the idea of zero in relation to one and infinity. This exploration has been drawn from scriptural references and the context in which the interconnected concepts of OZI had been used. The purpose is to advance understanding towards spirituality in different disciplines through these concepts. An integrative spiritual approach to develop understanding of interconnectedness

16 ॐ पूर्णमदः पूर्णमिदम् पूर्णात् पूर्णमुदच्यते । पूर्णस्य पूर्णमादाय पूर्णमेवावशिष्यते ॥

of OZI will be beneficial for all streams like science, mathematics, economics, psychology and spirituality as a discipline.

### 5.1 *Understanding Spirituality*

The concept of spirituality deals with our quest for the unknown realms of life and experience the associated realizations in one's known realm. From the philosophical ideas of rationality coming from the spectrum of objectivism and subjectivism to the religious and theological ideas of higher power or self, transcendence and immanence, the literature on spirituality has brought about several conceptualizations towards its understanding and application in past decades.

Amidst these debates around essentialist and existentialist, instrumental and deep, religion and secular stances on the concept of spirituality, Nandram (2019) proposed the following integrative schema approach towards enriching the field of spirituality as a discipline: *spirituality as an ability of any entity to become aware of its connectedness to its existence beyond perceived existence*. In addition to the other elements, this schema introduces the notion of *existence, beyond* and *perceived existence*. This necessitates a fresh look at the ontology of knowledge and experience of existence as such. Further, in Nandram (2022), she contributes to the field of spirituality as follows (see Figure 23.5):

In the same work, Nandram brings our attention to the analogy between the discipline of mathematics and spirituality in general and treatment of the unknown, in particular. Spirituality and mathematics as disciplines are or are becoming fundamental to other disciplines. Both have their own ways of giving place to the unknown or non-definite aspects of our reality. Mathematics has discovered several ways to incorporate several undefinable concepts such as nothing, infinity, imaginary and others. The concept of zero and the infinite symbolize the unknown in mathematics. Similarly, pure spirituality like zero and infinity together with practical spirituality like one, help us explore both the knowable and the unknowable unknown.

Historically, we have seen that these discoveries around the unknowable coming from theoretical research eventually find their way into the practical realm, dealing with definitive aspects of reality. In fact, they touch almost every aspect of mathematics. Many of these ways are still being argued towards better representations and newer applications. At the same time, mathematics is influencing other disciplines through its power of dealing with quantification of both definitive and undefinable concepts. The realm of undefinable is explored in the discipline of mathematics under a legitimate pursuit to bring them under the familiar rules governing the definitive. However, inadvertently, in that pursuit the ontology of these undefined concepts gets a reductionist treatment and thereby cripples the growth of other disciplines especially that

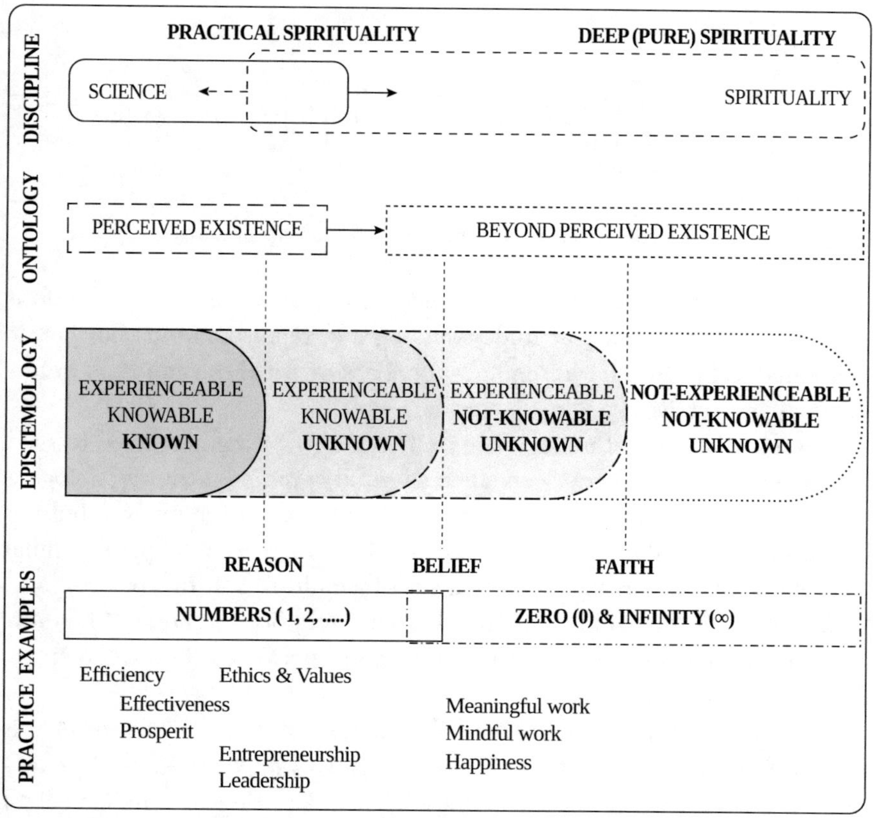


FIGURE 23.5 The ontology of discipline of spirituality (Nandram, 2022)

which concerns dealing with the unknown. For this reason, it calls for immersion of disciplines with spirituality as, historically, spirituality has been our way to deal with the unknown. This immersion in the case of mathematics implies a fresh look at the ontology of the concepts of One, Zero and Infinity besides other similar concepts. This fresh look may reveal that these concepts can have multiple connotations and these connotations may have meanings in several disciplines. This can potentially help us discover newer aspects of unknown reality in those disciplines.

## 5.2 *Revisiting Zero with One and Infinity*

In our pursuit towards a fresh ontological exploration of these concepts, revisiting the origins for their usage and connotations for material as well as spiritual aspects of reality. It is important to mention here the rationale behind the choice of discussing the three concepts *One, Zero and Infinity* together. In ancient spiritual traditions, especially Indian, these concepts are somewhere used distinctively and at times, overlapping to mean the same. Perhaps this is

FIGURE 23.6 Indian concept of self: The social and spiritual dimensions

the most fuzzy aspect of the ontology put forth by these traditions. Therefore, it requires different ways of understanding and representations not just in mathematics but in other disciplines too. Let's revisit these concepts from a spiritual perspective.

The spiritual ideas of immanence (self-contained, remained unexpressed), reality (existence) and transcendence (beyond) present a trinity analogous to *One, Zero and Infinity*. The idea of self that pervades from inner to outer is beautifully depicted in Figure 23.6 (adapted from Bhawuk, 2011, p. 72). Similar ideas have been expressed in different ways by multiple spiritual perspectives, for instance, the idea of involution and evolution by Sri Aurobindo (Malhotra, 2011). However, as earlier mentioned, these concepts are not fully understood, assuming them as strictly discrete.

The ancient scriptures talk about the oneness of reality which in turn is made up of a fabric consisting of three threads or *guṇas* (qualities), names *sattva*, *rajas* and *tamas*. *Sattva* represents true knowledge, truth and light besides other similar connotations. *Rajas* represent action orientation and *tamas* indicate darkness, inertia, ignorance (i.e., absence of *sattva*). These represent the manifestation of all life forms and other elements of our existence. Immanence and Transcendence of this type of reality is given by the concepts of *nirguṇa* and *guṇātīta* respectively. The former denotes the absence of *guṇas*, while the latter is about transcending the *guṇas*. The essence of both concepts is to highlight the higher order aspects that go beyond reality or existence comprising three *guṇas*. The idea of *nirguṇa* being connected with zero and *guṇātīta* with infinity and mutual resemblance of both phenomena have been explored further.

## 6 Conclusion

The present work has explored various dimensions of the idea of zero, a brief history and its varying interpretations. The core philosophical derivations were made to explore possible linkages based on related meanings in Hindu philosophy. The concept of zero and infinity might appear as a pair of extremes about the realm of the unknown, but Hindu philosophy provides a lens as well as space to see the oneness between these two ideas. To conclude, it can be said that these are not the extremes but a twin pair that provides intellectual stimulation to thinkers, philosophers and seekers.

Spirituality is acknowledged as a way to connect with the unknown beyond our perceived existence and hence involves dealing with the nature of the unknown. Most importantly, the conceptualizing of the unknown provides us a sense of direction in our known realm. The connotations and conceptualizations of the unknown by different worldviews and cultures shape their response to the way they explore existence and act in or with it. Therefore, with the above discussion, we have attempted to highlight the importance of qualitatively conceptualizing *One, Zero and Infinity* when building theories and models based on numbers around us. The understanding of *One, Zero and Infinity* will play an integral role in any future research on the ontology and epistemology of spirituality as a discipline. We hope that it may lead to fresh explorations and discoveries benefitting our engagement with our perceived existence too.

## References

Aczel, A. D. (2015). *Finding zero: A mathematician's odyssey to uncover the origins of numbers*. St. Martin's Press.

Aurobindo, S. (2006). *The Upanishads – II: Kena And Other Upanishads*. SriAurobindo Ashram Publication Dept.

Barton, N. (2020). Absence perception and the philosophy of zero. *Synthese, 197*(9), pp. 3823–3850.

Bhawuk, D. (2011). *Spirituality and Indian psychology: lessons from the Bhagavad-Gita*. Springer Science & Business Media.

Bindlish, P. K., Nandram, S. S., Gupta, R. and Joshi, A. (2018). How to prepare the researcher for indigenous context: an integrative approach. *International Journal of Indian Culture and Business Management, 17*(2), pp. 221–237.

Feynman, R. P. (1966). The development of the space-time view of quantum electrodynamics. *Science, 153*(3737), pp. 699–708.

Glynne-Jones, T. (2011. *The book of numbers*. Arcturus Publishing.

Hampton, A. J. (2018). Transcendence and Immanence: Deciphering Their Relation through the Transcendentals in Aquinas and Kant. *Toronto Journal of Theology, 34*(2), pp. 187–198.

Hiriyanna M. (1911). Ishavasya Upanishad with Shankara's Commentary. Sri Vani Vilas Press.

Ifrah, G. (1985). *From One to Zero*, translated by Lowell Bair. New York: Viking Penguin, pp. 428–497.

Ifrah, G. (2000). *The Universal History of Numbers*, translated from the French by David Bellos et al. Harvil, London.

Jaques, K. (n.d.). *The Philosophy of the Kaivalya Upanisad.* [online] Available at: http://www.advaita-philosophy.info/Kaivalya_Upanishad.pdf. Accessed March 10, 2022.

Josselson, R. (2004). The hermeneutics of faith and the hermeneutics of suspicion. *Narrative inquiry*, *14*(1), pp. 1–28.

Kaplan, R. (1999). *The Nothing That Is: A natural history of zero*. Oxford University Press.

Mādhavānanda, S. (1965). *The Bṛhadāraṇyaka Upaniṣad*: With the Commentary of Śaṅkarācārya. Advaita Ashrama.

Malhotra, R. (2011). *Integral Unity and Synthetic Unity: Being Different*. Harper Collins.

Mclennan, G. (1988). Brian Rotman, Signifying Nothing: The Semiotics of Zero. *Radical Philosophy*, *49*.

Mitra, V. L. (1891–93). *The Yoga-Vasishtha Maharamayana of Valmiki.* Vol. 1 [1891]; Vol. 2 [1893]. Translated from the original Sanskrit by Vihari-Lala Mitra. Reprint.

Nandram, S. S. (2019). Integrative Spirituality in the Fourth Industrial Revolution: From how we do things to why we exist. Inaugural lecture. Vrije University Press. Amsterdam. [online] Available at: https://research.vu.nl/en/publications/integrative-spirituality-in-the-fourth-industrial-revolution-from. Accessed February 27, 2022.

Nandram, S. S. (2013). Karma Theory. In Coghlan, D., & Brydon-Miller, M. (Eds.). *The Sage Encyclopedia of Action Research.* California, USA: Sage Publication, Inc., pp. 475–477.

Nandram, S. S. (2022). Spirituality: the discipline for doing business with the unknown. Inaugural lecture by Professor of Business and Spirituality at Nyenrode Business Universiteit on April 1, 2022.

Nandram, S. S., Bindlish, P. K., Keizer, W. A. J. (2017). *Understanding Integrative Intelligence*. International edition paperback, Praan Uitgeverij, Netherlands.

Reid, C. (2006). From Zero to Infinity: What makes numbers interesting. CRC Press.

Rotman, B. (1987). *Signifying nothing: The semiotics of zero*. London: Macmillan Press.

Saraswati, N. (2010). *Sanatan Siddhanta*. Puri, India: Swasti prakashan samsthan.

Salwi, D. M. (1988). *Story of Zero*. Children's Book Trust.

Seife, C. (2000). *Zero: The biography of a dangerous idea*. Penguin.

Sharma, A. (2001). The Hermeneutics of suspicion: A case study from Hinduism. *Harvard Theological Review*, *94*(3), pp. 353–368.

Sharma, G. P. (n.d.). *Tarka Sangraha of Sri Annam Bhatt.* Varanasi, India: Chaukhamba Publications.

Sherma, R. D. and Sharma, A. (Eds.). (2008). *Hermeneutics and Hindu thought: Toward a fusion of horizons* (p. 235). New York: Springer.

Sivananda, S. (2003). *The Bhagavad Gita*. India: The Divine Life Society.

Tyagisananda, S. (1949). *Shvetashvatara Upanishad*. Madras, India: Shri Ramakrishna Math.

Upadhyay, A. K. (2018). *Nature of Numbers*. Puri, India: Swasti prakashan samsthan.

CHAPTER 24

# Challenges in Interpreting the Invention of Zero

*Kaspars Klavins*

## Abstract

Attempts to explain the importance of inventing zero and the application thereof in the context of the history of culture, philosophy, and the exact sciences bring to the forefront a range of problems in relation to the overlapping of challenging issues (and terminology) in science and culture. The symbol of zero is qualified both as a scientific discovery (or invention) and an empirically found solution for the satisfaction of certain practical needs of humans. Moreover, its invention is also based on references to certain religious and philosophical teachings. All this makes the explanation of this phenomenon extremely difficult, taking into account that explanations frequently connect fundamentally different fields that actually address completely different areas of research. Furthermore, when looking more closely into the materials related to scientific-technical and religious-philosophical explanations, we see that the emergence of zero (or the idea of 'emptiness', 'nothingness') on the religious-philosophical side does not at all indicate the existence of this phenomenon in exact scientific use, and vice versa. There might be a discussion about 'emptiness' or 'nothingness' within a specific historical period in a certain society, and at the same time zero might not show up at all in the mathematical theory and practice known to that society, as was the case, for example, in medieval Europe. Likewise, the introducers of zero in mathematical practice may theoretically disagree with the idea of 'nothingness' per se at the religious-philosophical level, as was seen in the Europe of the early modern period.

## Keywords

scientific paradigm – religious mysticism – Brethren of Purity – negative theology – Meister Eckhart – *Nichts* – exact sciences – abstraction – *Śūnyatā* – Nāgārjuna

## 1 Introduction

The issue of exporting the invention of zero is extremely complex. If, based on a rather simple interpretation of information found in the history of science,

 | DOI:10.1163/9789004691568_028

we acknowledge that the idea of zero traveled from India to Europe via the Arabs and Persians, that is, Islamic civilization, that does not make the issue of 'export' simpler, because, regardless of the interrelation of Arabs/Persians and Europe to ancient Greek culture, the understanding in Islamic civilization of the interrelation between science and religion, including the understanding of mathematics, was different from that in European Christianity. Here it is useful to remember, among other things, the scientific revolution instigated by Thomas S. Kuhn regarding the interpretation that a certain discovery within a particular civilization only becomes a 'discovery' after necessary preconditions have occurred for the perception thereof. And only after becoming part of the mainstream science of that particular time and place, provided its representatives feel the need for change (that is, the existing arsenal of ideas is felt to be insufficient), does such a discovery become usable in science, namely, become a component of science. Exporting a discovery is not possible if the recipient does not have an analogous understanding of the meaning, value, and usability of such a discovery. The heliocentric model of the universe presented by the ancient Greek astronomer Aristarchus of Samos (*c.*310–*c.*230 BCE) long before Copernicus, for example, changed nothing in Europe, because the geocentric worldview for a long time remained flawless and functional in Europe (Kuhn, 1993, p. 88). Like Copernicus' 'discovery', which positioned the sun as motionless and at the center of the universe, and is incorrect from the modern perspective, it fitted well in the paradigm of early modern and modern period science, despite relative opposition to the concept from the thinkers of that time. Regarding the introduction of zero in Europe, we know that, despite information about the Hindu-Arabic numeral system (including zero) acquired by Fibonacci (*c.*1170–*c.*1240–50) from Arab sources, zero failed to gain a place in European mathematics until the seventeenth century, and even René Descartes (1596–1650), despite the introduction of zero in the center of the number line, denied in his writings the existence of emptiness (Novoszel, 2016, p. 30). Negative numbers, in turn, were not widely accepted by mainstream European mathematicians until the eighteenth century, and then only after much controversy (Shutler, 2017, p. 2). Additionally, the year zero does not exist in the Anno Domini (AD) system used to number years in the Gregorian calendar, which is still used today; likewise, Western historians have never included a year zero in the chronology of global events.

In order to provide a justified explanation for the transition of the idea of zero from Islamic civilization to Europe, a deeper, unbiased comparison of the Islamic world and the European philosophical-scientific heritage is required, at the same time repeatedly assessing the common and the different in Arabian/Persian and Indian spiritual culture. Hence, in the context of these complex issues, serious intercultural studies envisaging a range of interdisciplinary studies are of great importance, as opposed to remaining satisfied with merely some frequently repeated and simplified explanations of the exporting of the idea of zero. In this regard, difficulties may arise from the inability of representatives of the exact sciences and the humanities to cooperate in the modern, narrowly specialized world of science, yet, without this, elucidation of the genesis of zero and the integration of that phenomenon in different cultures, as well as its multifaceted manifestations, is not possible.

## 2 Challenges regarding the Genesis of the Invention of Zero

From the perspective of the genesis of the idea of zero, the Indian material provides the most convincing example of a theoretical-practical application of this symbol, and therefore the majority of scientists today acknowledge India as the birthplace of zero as it is understood by contemporary science. In addition, the spiritual heritage of ancient India – especially in the context of Buddhism – may justify the philosophical background of the genesis of zero, but only if we assume that the practical application of zero in Indian mathematics and the Buddhist concept of 'emptiness' are really related. However, a parallel discussion has already been going on for a long time among culture and science historians regarding a possible origin of the idea of zero in ancient Babylonia and ancient Greece in the context of Plato's philosophy.

We can fully agree with Peter Pesic that 'Babylonian mathematics had a special cuneiform sign used as a kind of placeholder, so that it could distinguish what we might write in modern notation as $1{,}20 = 1(60)1 + 20 = 80^o$ from $1{,}0{,}20 = 1(60)2 + 0(60)1 + 20 = 3620^o$.' (Pesic, 2004, p. 1).

At the same time not only did this cuneiform sign not look like our o, but it was notably different in significance; although it indicated positional notation, it was only a placeholder that did not stand by itself and, therefore, did not signify 'no magnitude' as such (Pesic, 2004, p. 1).

Greek mathematics, in turn, seems to not only have had no concept of zero, but such a concept would have been deeply alien to the ancient Greeks as an element of arithmetic, although Plato's philosophy does include moments worthy of attention. Of course, '... the basic Greek concept of numbers runs counter to the notion of zero', and '... the Greeks do not seem to have used what the moderns call "negative numbers", much less "imaginary numbers", nor other generalizations ...' (Pesic, 2004, p. 3). Pesic, however, provides an extremely interesting analysis of Plato's dialogue *The Sophist* (360 BCE), in which, when solving the issue of *not-being*, a version is expressed that can be clearly perceived as a contemplation of zero: '... not merely as placeholder but as the concept of non-being in number' (Pesic, 2004, p. 8). In fact, this work by Plato may be assessed as the first occasion in which zero was not merely asserted as a phenomenon of non-being but included and counted among the other counting numbers (Pesic, 2004, p. 9).

Without going deeper into Plato's dialogue and its analysis, it must first be understood that this 'discovery of zero' is subject to the already mentioned impossibility of a 'discovery' being scientifically usable in a situation where Greek science had not experienced a new paradigm in this context, and the idea of zero, even if expressed in an individual work, did not acquire a place in the mainstream scientific thought. To a certain extent this can be compared to the non-topicality of the heliocentrism idea by Aristarchus of Samos before Copernicus. Yet the interpretation of zero in the works of the Syrian Neoplatonist philosopher Iamblichus (*c.*242–327 CE) is of note, because he wrote that the series of numbers should be carried below one to zero (*οὐδέν*), which is their source (Nasr, 1964, p. 46). His commentary on the *Introduction to Arithmetic* by Nicomachus (*c.*60–*c.*120 CE) regards zero as a number. According to Pesic, Iamblichus extended multiplication to zero so that he even contemplated that the product of zero times zero equals zero. Thus he expressed the full range of operations with which modern mathematics endows the zero many centuries before these concepts were again expressed with such clarity (Pesic, 2004, p. 14). Did the neoplatonist Iamblichus indeed continue and further develop Plato's version of zero? This question cannot be answered, even more so because the concept of zero is completely absent in Nicomachus, the text to which Iamblichus refers (Pesic, 2004, p. 14).

There is another big philosophical problem. We always want to associate zero with non-existence, but there are very ancient cultures that have also influenced the genesis of zero in the European tradition – and there is no concept of non-existence in these cultures.

A distinction must be made between the philosophical meaning of zero and practical manipulations, calculations, and so on. These are completely different things.

For example, in ancient Mesopotamia (the laws of Hammurapi, eighteenth century BC), there is no concept of zero, but the interest on a loan, etc. is calculated to the third sign until the balance – equality – debt is paid off. Zero is the sign of equilibrium.

In ancient Egypt, there are both zero and negative values and signs, but ... the theoretical part is based on the assumption of the starting point – the point is absolute and everything is taken from it. The point is physical, but the dimensions can be virtual (including life, debts, treasures in life). A point or stand is fixed – a stamp, an event, a ritual, written down.

In Assyria, it is time – the starting point is time (the state of the stars) – and from it both positive and negative values.

There is something else to consider – material things never disappear. They change their shape, form, appearance, consistency – but they do not disappear. **These are not negative values**, there are no losses if the craftsman breaks the mug before the sale, they reduce future income, not losses.

However, it would be extremely interesting to assess anew the sources on the basis of which the concept of zero entered the Arab and Persian tradition, that is, Islamic civilization. In addition to the Indian influence in the genesis of the 'Arab zero' idea, we cannot rule out the influence of neoplatonism (including that of Iamblichus), considering the importance of the interpretation of neoplatonism in Islamic philosophy. This is especially true regarding the Illuminatist (*Ishrāqī*) school, for which the natural sciences and mathematics were primarily symbolic and resembled to a great extent the writings of some neoplatonists (Nasr, 2012, p. 36). This tradition was especially followed by the Brethren of Purity (*Ikhwan al-Safa*), a tenth century esoteric fraternity based in Basra and Baghdad whose members positioned themselves as 'the followers of Pythagoras and Nichomachus, especially in their treatment of numbers as the key to the understanding of nature' (Nasr, 1964, p. 37). In relation to the symbolism of zero, their encyclopedia, *The Epistles of the Brethren of Purity* (*Rasa'il Ikhwan al-Safa'*) and its sources are extremely important material because it is indicated in the esoteric version of this work (*Risālat al-jāmi'ah*) that the Infinite, or the Divine Essence, corresponds to zero: 'Zero, therefore, symbolizes the Divine Ipseity, which is above all determinations including Being (*al-wujūd*)' (Nasr, 1964, p. 46)[1] Although it is known that the Brethren of Purity were also influenced by interpretations of the achievements of Indian science, which were available to them through various Arab sources, there is no evidence that due to their activities the symbol of zero would have been promoted in the mathematics used by the Arabs. Instead, the Arabs seem to have adopted the zero symbol for practical use due to the translation into Arabic of the *Brāhmasphuṭasiddhānta* already in the eighth century and due to the activities of the Persian polymath Muḥammad ibn Mūsā al-Khwārizmī (*c.*780–*c.*850), as a result of which Hindu numerals along with the notion of zero (*śūnya*; *as-ṣifr* in Arabic) thoroughly entered the science of Islamic civilization.

We cannot explain the religious basis of zero in Islamic esotericism with the influence of Buddhism; the frequently misunderstood Buddhist concept of 'emptiness' is something completely different. Of course, we can try drawing

1 The Being in this system corresponds to one.

parallels with Hinduism in the context of *The Principal Upanishads*, keeping in mind that in the Hindu tradition 'the Supreme [we might call it God – K. K.] is both transcendental and immanent ... It is the manifest and the unmanifest, *vyaktāvyaktāḥ*, the silent and the articulate, ... the real and the unreal, *sad-asat*' (Radhakrishnan, ed., 2012, p. 83). However, regardless of certain universal analogies that can be traced in many religions around the world, we cannot consider Hinduism as a source of the understanding of God in Islam, which instead continued the traditions of Judaism and radical Christian monotheism. The idea of zero as the symbol for God postulated by the Brethren of Purity originates from the explanation of God that is acceptable to Islam, namely, as neither a material nor a spiritual phenomenon, present above all that exists, which is analogous to the 'negative' explanation of God by means of the nothingness characteristic of European Christian mysticism in the medieval and early modern period (see hereunder).

The question is, why was Islamic civilization able to accept the symbol of zero for practical use in mathematics so quickly and flexibly, unlike Europe, where it experienced extended opposition regardless of the concept of nothingness in the religious-philosophical field? In other words, why was the Indian concept of zero introduced quickly in the science used in Islamic countries while a wider integration of zero in European mathematics was accepted so much later (in the seventeenth century), despite information about the symbol of zero having arrived in Europe much earlier (in the thirteenth century)? The initial advantage of the Middle East can probably be explained by the lack of contradiction between science and religion that became the key to the Islamic civilization's past achievements, while the strict separation of these areas was for many centuries an impediment in the Western world view. The Arab philosopher, writer, philologist and zoologist al-Jahiz (*c.*776–*c.*868) once accurately explained the link between religion and science in Middle Eastern intellectual discourse in the following manner:

> Should anyone allege that belief cannot be related to the findings made in natural science, he does not sufficiently value God's omnipresence in nature (the unity of the divine and nature). Contrarily, should anyone allege that natural science can exist only through abandoning the belief in God [a higher spiritual force that humans are unable to explain – K. K.], he risks subjecting his scientific discoveries to the limits of human cognition. (Sezgin, 1975, p. 368)[2]

2 *Kitāb al-hayawān*.

## 3 A Comparison of the Concept of *Nothingness* in European Christian Mysticism and the Concept of *Emptiness* in Buddhism

Even if the introduction of the zero symbol in European science, that is, mathematics, took place late, it cannot be stated that it was related to the incompatibility of the concept of zero with the European philosophical-religious tradition or the Christian world outlook dominating in the Middle Ages and early modern period, which was extremely multifaceted and provided for many different alternatives to the official church doctrine, which was in itself not actually constant. Instead, European history showcases possibly different paths of development for the philosophical-religious world outlook on the one hand, and the methodology of the sciences on the other. The entire cultural history of Europe is a line of paradoxes. The religious thinkers who were philosophically ready to accept the concept of zero did not introduce it into science, while scholars (that is, the introducers of zero, such as René Descartes, etc.) did not accept the concept from the philosophical standpoint. Perhaps these contradictions are due to the dualism of the Western world outlook manifested as the separation of religion (ethics) from power, subject from object, ideas from things, mind from body, etc. In the field of religious thought, this dualism was manifested through opposite poles: God's children versus the devil's servants, good versus evil, beautiful versus ugly, etc. This originates from *The City of God* (*De civitate Dei*) by Saint Augustine of Hippo (354–430), who was extremely influential in Europe both in the Middle Ages and at the time of the genesis of modern science in the seventeenth and eighteenth centuries (Harrison, 2009). If, for example, the ideas of Aristotle (384–322 BCE) were relevant to the philosophy of Thomas Aquinas (1225–1274) – namely, that all scientific knowledge is good (Harrison, 2009, p. 45) – the father of the Reformation Martin Luther (1483–1546) already appealed only to 'original sin' and the 'Fall of Adam' (Harrison, 2009, pp. 53–55). A certain amount of time had to pass before European intellectual thought was able to return to a harmonious dialogue between science and religion. At the same time, religious mysticism was always present as a strong intellectual current among different strata of the population in Europe, which in some instances was able to take the 'position', while in most cases it remained in the status of 'opposition', 'heresy', or 'secret (alternative) knowledge', taking into account that representatives of both secular and spiritual power (including the Church) placed the work of thinkers under strict censorship. Once this is understood, one can also understand the separation of religious-philosophical and scientific world perceptions in Europe, which to a certain extent have left a footprint in the scientific methodology of the West to the present day.

FIGURE 24.1
Tomb of the Japanese philosopher, Nishida Kitarō. Meinolf Wewel, 1997, Tomb of the Japanese philosopher, Nishida Kitarō, 19 May 1870, Mori Kanazu (today known as Kahoku, Präfektur Ishikawa). Used with permission

Certainly, we can compare European Christian mysticism to Buddhism in the religious-philosophical context of the concept of zero. East Asian thinkers have also done so, for example, the Japanese philosopher Kitaru Nishida (1870–1945) drew parallels between the *Śūnyatā* concept in Buddhism and Meister Eckhart's (*c.*1260–*c.*1328) notion of *Nichts* (Wehr, 1989, p. 128).

However, as we will see below, such a comparison is essentially erroneous because, contrary to the ability of Buddhist thinkers to operate with the *undefined*, European thinkers from medieval times already had the tendency to precisely define the self and everything that exists, which addresses science, religion, and philosophy equally. Apparent similarities between European religious mysticism and the spiritual teachings of India and East Asia in terms of form and expression do not eliminate the cardinal differences in the very approach to reality per se between these two traditions, which has been well emphasized in the works by the contemporary Korean philosopher Byung-Chul Han (2007).

For comparison, let us look at the views of two eminent thinkers: Meister Eckhart, the most prominent representative of European medieval mysticism, and Nāgārjuna (*c.*150–*c.*250 CE), one of the most important Buddhist philosophers from India. Specifically, we can compare the explanation of nothingness (*Nichts*) in Meister Eckhart's writings with Nāgārjuna's view of the concept of emptiness (*Śūnyata*) in Buddhist philosophy. Meister Eckhart's concept of nothingness is very often used as a negative comparison of everything existing (everything created by God) with God himself, that is, in order to glorify God's being far above the world (Jung, 2014, p. 64). For Meister Eckhart, God (the intellect) stands in opposition to being as such and can thus be qualified as 'nothing' – it is the 'nothingness of being'. The intellect negates being in its entirety in order to be everything that can originate at all (Jung, 2014, p. 53). On the other hand, all that is created he calls nothing, also in comparison with God, who is the real essence of things (his creations, in turn, are non-essential):

> Alle creatûre sint ein lûter niht. Ich spriche niht, daz sie kleine sîn oder iht sîn, sie sind ein lûter niht. Swaz niht wesens hât, daz ist hint. Alle creatûre hânt kein wesen, wan ir wesen swebet an der gegenwertikeit gotes. (Pfeiffer, 1857, p. 136; Huizinga, 1924, p. 309)

At the same time, Meister Eckhart absolutely believes in the creation of the world out of nothing (*creatio ex nihilo*). For him, this nothing out of which God created the world is the divine intellect in which God designed things before he created them (Jung, 2014, p. 53). Such a perception is completely different from Buddhism, first and foremost because Buddhism is a spiritual teaching without a need for 'God' or 'belief'. The world outlook of Meister Eckhart based in stressing the opposites is actually expressly Western, if one may apply such a modern-day evaluation to a medieval European spiritual discourse. Eckhart's aim is to define God as precisely as possible, and the negative method with its use of the concept of nothingness is just a way to highlight God's mightiness and separate him from the rest of being, to stress the contrast.

Nāgārjuna, in turn, understands the concept of emptiness completely differently: 'All things are in the perpetual process of arising and passing away, ever "becoming" and thus never actually "being". Conditioned by multiple interdependent causes, all things are "empty" of any sort of independent or intrinsic nature and thus defy conceptualization' (Olendzki, 2018). Contrary to Meister Eckhart, such a position of methodologically explaining emptiness means that *emptiness itself* means observing objective reality without fixing one's attitude. If everything is empty of self-being (*svabhava*), there are no objects with ultimate reality in this sense, and, subsequently, any theory of such objects is mistaken and should therefore be abandoned (Priest, 2010, p. 36). This fully coincides with the traditional Buddhist perception of the unsubstantiality of all phenomena: 'Void is the world ... because it is void of a self and anything belonging to a self' (Thera, 2004, p. 205). Certainly, Nāgārjuna does not deny the reality of the world, he merely does not accept that it is possible and necessary to objectively explain it – and likewise with the terminology of reality or unreality. For him, according to Buddha's example, utmost denial is as futile an extreme as its opposite. From this vantage point, *Śūnyatā* follows from the interrelation of all things, which does not allow them to acquire actual independent existence; while the concept also represents reality per se in such a way that is not possible to either be perceived or explained in rational terms at the level of common consciousness (Лысенко and Терентьев, 1994, p. 279).

Is it really non-existence, emptiness of not-being as such, in the philosophy of Nagarjuna – if we are talking about India? No.

Already since the European Middle Ages, the shapelessness of reality (also Taosim in China and even Confucianism in China and Korea) is something that has been very difficult for a Western thinker to perceive, even more so where it has to be accepted without the presence of a leading and directing force (God). At the same time, however, this does not rule out the concept of nothingness, and as a result, from the philosophical-religious standpoint the acceptance of the symbol of zero in Europe and its widespread use in mathematics from the fourteenth century onward would have caused no problems. Unfortunately, the vitality of the existing scientific paradigm at that time along with the separation between empirical science/religious mysticism and the official power/church postponed this process.

The assumption existing in India and the Far East (in the teachings of Hinduism, Buddhism, and Taoism) that the nature of reality can be unaffected by our ways of trying to grasp it was extremely difficult to accept in Western mainstream intellectual discourse. Yet certain European mystics of the Middle Ages and early modern period had very closely approached such a feeling of the world and even proposed it from the standpoint of God as an imperceptible phenomenon. For example, the mystic and religious poet Angelus Silesius (*c.*1624–1677) was not afraid to write in his compilation of religious paradoxes *Cherubinischer Wandersmann* that, 'God is a complete nothing: the more you reach for it, the less it will be possible':

> Gott ist ein lauter Nichts, ihn rührt kein Nun noch Hier;
> Je mehr du nach ihm greifst, je mehr entwird er dir. (Angelus Silesius, 1935, p. 10)

Very much like the symbolic explanation of God by means of zero in Islamic esotericism, the Flemish mystic John of Ruusbroec (1293/1294–1381) also stressed God's total non-connectedness to anything actually measurable, writing that God is a 'size without measure, length without end, depth without foundation ... etc' (Ruusbroec, 1923, p. 71). This is in essence a 'negative definition' of God similar to that provided by Meister Eckhart. According to this, it is not possible to compare God with anything existent; it is only possible to stress its non-likeness to anything. Certainly, John of Ruusbroec held on to God as an absolute and permanent dominant. He stressed that God is also 'higher than anything', that is, he exists beyond nature (Ruusbroec, 1923). For European thinkers to perceive nature and reality in their constant changeability without concentrating on some mechanically separated force beyond nature (God) was as difficult as accepting the instability of reality. Philosopher-mystics, such as Jakob Böhme (1575–1624), who approached this concept, in turn, did not use the notion of nothingness as a technical term to justify their ideas. Böhme criticized the view popular in Christianity that Heaven and Earth were created out of *nothing*, because God had existed since eternity:

> Now, where nothing is, there nothing can come to be; all things must have a root, else nothing can grow. If the seven spirits of nature had not been from eternity then there would no angel, no Heaven, also no earth have come to be. (Penny, 1912, p. 191)

## 4 Conclusion

A cultural-historical survey of zero brings forth the issue of the strategy of Western science, namely, how long will it be able to live with its own fundamentalism and radicalism and fondness for accurate definitions. Phenomenology in philosophy was an attempt to reform this strategy, yet it has not essentially changed the attitude of Western science toward reality. India and East Asia, in turn, can seriously offer their own alternative only under the condition that the scientific paradigm of their region returns to its former high degree of abstraction. In the traditional communities that existed there before the region's comparatively late and dramatic modernization, people tended to think concretely – abstraction was neither acceptable nor necessary for them. For scholars to be not only introducers of technical innovations but also thinkers, they must find anew the road to the origins of their own spiritual culture in a global dialogue with colleagues who care about the synergy of theoretical and practical knowledge. If such a dialogue were to confirm the finding from ancient India about the existence of universal, omnipresent, joint knowledge that can become particular and local, that would be a huge achievement indeed.

## References

*Aus des Angelus Silesius Cherubinischem Wandersmann.* (1935). Lepzig: Insel Verlag.

Byung-Chul, Han. (2007). Abwesen: Zur Kultur und Philosophie des Fernen Osten. Berlin: Merve.

Harrison, P. (2008). *The Fall of Man and the Foundations of Science.* Cambridge: Cambridge University Press.

Harrison, P. (2009). *The Fall of Man and the Foundations of Science.* Cambridge: Cambridge University Press.

Huizinga, J. (1924). *Herbst des Mittelalters.* München: Drei Masken Verlag.

Jung, C. (2014). Die Funktion des Nichts in Meister Eckharts Metaphysik. In: *Salzburger Jahrbuch für Philosophie.*

Kuhn, T. S. (1993). *Die Struktur wissenschaftlicher Revolutionen.* Frankfurt am Main.

Nasr, S. H. (1964). *An Introduction to Islamic Cosmological Doctrines.* Cambridge, Massachusetts: The Belknap Press of Harvard University Press.

Novoszel, T. M. (2016). *Viel über das Nichts. Die historische und gegenwärtige schulische Bedeutung der Null.* Wien: Universität Wien.

Olendzki, A. What's in a Word? Emptiness. In: *Tricycle: The Buddhist Review.* https://tricycle.org/magazine/emptiness-buddhism/.

Penny, A. J. (1912). *Studies in Jacob Böhme.* London: John M. Watkins.

Pesic, P. (2004). Plato and Zero. *Graduate Faculty Philosophy Journal.*

Pfeiffer, F. (ed.). (1857). *Deutsche Mystiker des 14. Jahrhunderts. Bd. 2, Meister Eckhart.* Leipzig: Göschen.

Priest, G. (2010). The Logic of the Catuskoti. In: *Comparative Philosophy.*

Ruusbroec, Jan van. (1923). *Uut dat Boec vanden twaelf Beghinen.* Mainz: Matthias Grünewald Verlag.

Sarvepalli Radhakrishnan. (ed.) (2012). *The Principal Upaniṣads.* New York: HarperCollins.

Sezgin, H. (1975). *Geschichte des arabischen Schrifttums.* Bd. III. Leiden: E. J. Brill.

Shutler, P. M. E. (2017). A Symbolical Approach to Negative Numbers. *The Mathematics Enthusiast.*

Thera, N. (2004). *Buddhist Dictionary. A Manual of Buddhist Terms and Doctrines.* Kandy: Buddhist Publication Society.

Wehr, G. (1989). Meister Eckhart. Reinbek bei Hamburg. Rowohlt Taschenbuch.

Лысенко, В. Г. and А. А. Терентьев. (1994). Шохин В.К. Ранняя буддийская философия. Философия джайнизма. Москва: 'Восточная литература' РАН.

CHAPTER 25

# Some More Unsystematic Notes on *Śūnya*

*Alberto Pelissero*

## Abstract

This chapter is an overview of the themes regarding the concept of *śūnya* ('void') in Indian thought, mainly *mādhyamika*, with reference to *apoha* semantic theory, and in relation with another couple of problematic relationships, the metaphoric use of words (diffused in the milieu of poetics) and the paradox of ineffability (current in Vedantic Brahmanical circles). Did the mathematical, the grammatical, or the Buddhist philosophical meaning of *śūnya* come first? The chapter tries to illustrate the great deal of debate current on this question, but does not offer a unique answer.

## Keywords

*apoha* – *nirvāṇa* – Sanskrit – *śūnya* – *śūnyatā* – vacuity – void – *vyākaraṇa* – zero

## 1 Śūnya and Śūnyatā

The term *śūnya*, meaning 'void', and in mathematical literature, 'zero', derives from *śūna*, being the past passive participle of root *śvi*, 'to grow', 'to swell', according to Pāṇini (*śvi-idito niṣṭhāyām, Aṣṭādhyāyī* 7.2.14).[1] So *śūna* etymologically means swelled, swollen, increased, grown; it could seem a paradox that its semantical sphere is in fact not too far from *pūrṇa*, 'full', the key term of the *incipit* of the *Īśāvāsyopanisad, oṃ pūrṇam adaḥ pūrṇam idaṃ pūrṇāt pūrṇam udacyate / pūrṇasya pūrṇam ādāya pūrṇaām evāvaśiṣyate //*. According to *Ṛgvedaprātiśākhya* (14.2) it indicates a fault in Vedic recital, consisting in an utterance with a swollen mouth (Abhyankar, 1986, p. 393).

The term *śūnya* occurs within Upanisadic literature in the *Maitryupaniṣad* (2.4; 6.31; 7.4), together with other epithets referred to *brahman*, epithets that mean 'pure', 'clear', 'pacified' (*śuddha, pūta, śānta*). Etymologically, *śūnya*

1 The present chapter is a reworking and update of a previous work: Pelissero 2013, for a different audience and in light of more recent studies listed in the *Bibliography for further research*.

 | DOI:10.1163/9789004691568_029

should therefore mean a void space, a hole determined by a borderless opening, by an unlimited disclosing. According to lexicographers (*Amarakośa* 3.1.115), its synonyms are 'sapless', 'meaningless', 'void', 'vane', or 'hollow' (*asāra, phalgu, vaśika, tuccha, riktaka*).[2]

This kind of voidness is conceived first of all as a sort of deprivation, as we can see from a well-known literary 'good saying' (*subhāṣita*) centered around the term *śūnya*: 'Void is the house for he who is sonless, void is the time for he who is friendless, void are the four cardinal points for he who is silly, void is the whole world for he who is poor' (Śūdraka, *Mṛcchakaṭikā* 1.8).[3] The reference to the cardinal points (*diś*) is not at all a trivial one, because it explains why the term *śūnya* could be made synonymous with 'ethereal space', 'atmospheric space', 'heaven' (*ākāśa, kha, vyoman*). The accepted synonymity of *śūnya* with *kha* could be considered a crucial turning point in the elaboration of the concept of the mathematical zero.

The abstract derivate *śūnyatā* is recorded in Buddhist literature, mainly of the *mahāyāna* type, first of all in the *mādhyamika* school headed by Nāgārjuna, as meaning 'voidness', 'the fact of being void', and even (though this kind of translation is *sub judice*) as 'vacuity', 'emptiness' (Streng, 1967; De Jong, 1972–1974; Huntington Jr., 1989). The mathematical zero cannot be compared to any other number, being their very precondition, and it is tendentially compared with the concept of infinity (*ananta*).

The concept of zero grade is fundamental within Indian grammatical tradition (*vyākaraṇa*), but in fact the term *śūnya* is actually never employed in this context. Phonic zero, intended as absence of any sound whatsoever, to be found in alternation with sound, especially within vocalic gradation (apophony), is widely known and used as the apophonic grade. But we must note that the grades known in Western use as normal or full grade and extended or lengthened grade, both correspond to a Sanskrit technical term, respectively *guṇa* (a, e, o) and *vṛddhi* (ā, ai, au), which among other things can be taught as a replacement for a, i, and u, respectively (A 1.1.3 *iko guṇa-vṛddhī*).

By contrast the grade that we call weak or reduced or zero grade does not correspond to a univocal Sanskrit technical word, because it is treated exactly as the other zero-replacements of phonemes and, what is most important, it never takes the name of *śūnya*. It is not mere chance that what we call zero grade is not described by Indian grammarians in positive words, but only as an exception, subject to specific rules of application, to *guṇa* and *vṛddhi*

2 (Amarakośa 3.1.115) asāraṃ phalgu śūnyaṃ tu vaśikaṃ tuccha riktake.

3 This passage is echoed by *Cāṇakyaśataka* 47, quoted in Vatsyayan et al., 1992, p. 400. Translations from Sanskrit are by the author, unless otherwise indicated.

grades (e.g., A 1.1.5 *kṅiti ca* suffixes with K and Ṅ markers): it is impossible to describe an absence, a deprivation, a limitation in positive terms. The technical term used in such cases is *lopa* (e.g., 1.1.4 *na dhātu-lopa ārdhadhātuke*; 1.1.62 *pratyaya-lope pratyaya-lakṣaṇam*), a name given to the meaning of *adarśanam* (non-perception) by means of a metarule (*paribhāṣā*) (*Mahābhāṣya* ad 1.1.60 *tatra adarśanam lopaḥ iti lopasañjñā prāpnoti*).

Which of the three main shades of meaning of *śūnya* first suggested the other two? Did the mathematical, the grammatical, or the Buddhist philosophical meaning come first? There is a great deal of debate on this question.

Therefore, even if we cannot rule out the possibility that the apophonic zero could be the base of the mathematical zero, it is only the latter that takes the name of *śūnya*. The doubt whether or not the philosophical use could precede the mathematical one still remains. Even in the field of architecture, the value of the void asserts itself: it is sufficient to think of what we define as the *sanctum* (Indians call it *garbhagṛha*, 'house of the embryo') in the sanctuary of Śiva Naṭarāja in Chidambaram, enclosing the signum 'made of space' (*ākāśaliṅga*), technically a void space, that represents the icon being worshipped by the devotees.

Within the mathematical field, zero is the base of the system of numerical positional notation on the decimal scale: it is the void space that permits the passage from units to tens and so on. The *Yajurveda* (*Vājasaneyisaṃhitā* 17.2) enumerates the names of the powers of 10 starting from 100 *eka* up to 1012 *parārdha*. The synonyms of zero to be found in mathematical, astronomical, and astrological texts (*jyotiḥśāstra*), are all specifications of a semantic field that generally covers the concept of 'space'. But it is a large sphere that combines different notions, and is variously declined as ethereal space, surrounding space, void space, or atmospheric space (*ākāśa*, *ambara*, *kha*, *gagana*). Other kinds of synonyms are more interesting, because they range from an apparent antonym meaning 'full' (*pūrṇa*), to the term for 'point' (*bindu*), up to the little circle used in writing as a sign for zero (*chidra*, *randhra*, both words meaning 'hole'). It is possibly not a mere chance that the first quotation of zero as a mathematical symbol is to be found within a metrical text (Piṅgala, *Chandaḥsūtra* 8.2831).[4] Obviously, quotations from mathematical literature are numerous.[5]

---

4 Thus Sarma, 1992, pp. 400–11, but see Bronkhorst, 1994, for a radical criticism of arguments not grounded on solid chronology with reference to mathematical and prosodical sources.

5 See e.g., Āryabhaṭīya 1.2; Pañcasiddhāntikā 1.17; Bṛhatkṣetrasamāsa 1.69–71; Tattvārthādhigamasūtra 3.11.

At least in the Vedic period, within priestly circles the value of fullness (*pūrṇatā*) and full (*pūrṇa*) prevails, for example in passages such as 'full that, full this, from the full this full is born, having taken the full from the full, full only remains' (*Īśāvāsyopanisad, incipit* quoted above; *Bṛhadāraṇyaka-upaniṣad* 5.1.1). This primacy of fullness does not entail any sort of undervaluation of voidness, because without the void the full itself could not hold ('in the beginning indeed this was not being, from this the being is based [...] who could live, who could breathe, if within space [*ākāśa*] there was not bliss?' (*Taittirīyopaniṣad* 2.7 *asadvā idamagra āsīt* [...] *ko hyevānyātkaḥ prāṇyāt / yadeṣa ākāśa ānando na syāt*). This fact entails a twofold consequence. First of all, being and not being (*sat, asat*), full and void, are complementary entities: each is indispensable to the other. In some way, each is the matrix of the other (*Ṛgvedasaṃhitā* 10.129.1–4).[6]

Two considerations should be taken into account in this regard. First, thanks to the doctrine of the different levels of truth (*saṃvṛtisatya paramārthasatya*), each one can be derived from the other. Secondly, bliss is associated with the void. Buddhism will value both concepts highly.

In the Buddhist field, *śūnya* and its abstract *śūnyatā* cannot be considered as signs of a nihilistic doctrine. Vedantic doxographical tradition will put a conscious distortion into practice that rejects the 'emic' denomination of Nāgārjuna's school, 'followers of the middle path' (*mādhyamika*), preferring an ambiguous term (doomed to a certain degree of success), that is nihilists (*śūnyavādin*); in fact *śūnya* cannot be considered as a *vāda*, a valid doctrine, from Nāgārjuna's point of view, but only a convenient dialectical device. Void and voidness only signify the mere negation of every possible sort of positive assessment within the field of experiential reality. It is not proper to ascribe the status of doctrine (*vāda*) to *śūnya* and *śūnyatā*.[7]

The concept of vacuity or emptiness, *śūnyatā*, so relevant in Nāgārjuna, has been fully fledgedly theorized in the literary genre of the 'transcendent gnosis', *prajñāpāramitā*, where we can find different lists, ranging from four to

6 *nāsád āsī̱n no sad ā̍sīt ta̱dānī̱ṁ nāsī̱d rajo̱ no vyómā pa̱ro yat |*
*kim āvárīva̱ḥ kuha̱ kasya̱ śarma̱nn ambha̱ḥ kim ā̍sī̱d gahánaṁ gabhī̱ram* || RV 10.129.1
*na mṛ̱tyur ā̍sīd a̱mṛta̱ṁ na tarhi̱ na rātryā̱ ahná āsīt prake̱taḥ |*
*ānī̍d avā̱taṁ sva̱dhayā̱ tad eka̱ṁ tasmā̍d dhā̱nyan na pa̱raḥ kiṁ ca̱nāsá* || RV 101.29.2
*tamá āsī̱t tamásā gū̱ḻham agré 'prake̱taṁ sáli̱laṁ sarvám ā i̱dam |*
*tu̱cchyenā̱bhv apíhita̱ṁ yad āsī̱t tapása̱s tan máhi̱nājā̍ya̱taikám* || RV 10.129.3
*kāma̱s tad agre̱ sam ávarta̱tādhi̱ manáso̱ retáḥ pratha̱mamaṁ yad āsī̍t |*
*sa̱to bandhu̱m asáti̱ nir ávindan hṛ̱di pra̱tīṣyā̍ ka̱vayó manī̱ṣā* || RV 10.129.4.

7 For this debate see Ruegg, 1986.

20 elements.[8] The list including 18 terms, the most widely accepted one, considers the following varieties:

1. vacuity relative to the interior realm (*adhyātmaśunyatā*), where the six awarenesses (*vijñāna*, five related to the senses and the sixth a mental one) are revealed as empty;
2. vacuity relative to the exterior realm (*bahirdhāśūnyatā*), where both sensory and mental objects are revealed as empty ones;
3. vacuity relative to interior and exterior (*adhyātmabahirdhāśunyatā*), where the very same distinction between interior and exterior is revealed as empty;
4. vacuity of vacuity (*śūnyatāśūnyatā*), where the very same notion of vacuity is revealed as empty;
5. great vacuity (*mahāśūnyatā*), where space is revealed as empty;
6. vacuity of absolute reality (*paramārthaśūnyatā*), where transcendent reality is revealed as empty;
7. vacuity of all composite entities (*saṃskṛtaśūnyatā*), where every compounded entity is revealed as empty, because it depends on causes and conditions;
8. vacuity of non-composite entities (*asaṃskṛtaśūnyatā*), where every non-compounded entity is revealed as empty, beginning with *nirvāṇa*;
9. final vacuity (etymologically 'vacuity beyond the limit', *atyantaśūnyatā*), where the very same border dividing permanence and destruction is revealed as empty;
10. beginningless and endless vacuity (*anavarāgraśūnyatā*), where the whole cycle of transmigration (*saṃsāra*) is revealed as empty;
11. vacuity of what is not subject to scattering (*anavakāraśūnyatā*), where *nirvāṇa* is revealed as empty;
12. vacuity of the object principle (*prakṛtiśūnyatā*);
13. vacuity of all phenomena whatsoever (*sarvadharmaśūnyatā*);
14. vacuity of what is self-defining (*svalakṣaṇaśūnyatā*), where what is self-defining (*svalakṣaṇa*), i.e., the point-instant (*kṣaṇa*), is revealed as empty;
15. vacuity of what is not known (*anupalambhaśūnyatā*), where all events, considered as cut off from any reference with the time in which they take place (past in the future, future in the past, present in the future and in the past), are revealed as empty;
16. vacuity of the absence of one's own mode of being, or of not existence (*asvabhāvaśūnyatā*, or sometimes *abhāvaśūnyatā*), where the very same

8 For a general outline of this field see Venkata Ramanan, 1966.

absence of one's own mode of being (*svabhāva*) of the *dharmas* is revealed as empty;

17. vacuity of one's own mode of being (*svabhāvaśūnyatā*);
18. vacuity of non-existence and of one's own mode of being (*abhāvas-vabhāvaśūnyatā*), where the very same distinction between real and unreal is revealed as empty.

As we can see, the path toward theologizing and hypostatizing the concept of vacuity has already been outlined. It will be one of the main concerns of Nāgārjuna, as a champion of the anti-intellectual trend of the followers of the middle path, to deconstruct the huge doctrinal building of the followers of the transcendent gnosis, in view of a strong antimetaphysical vision.

Nāgārjuna's thought is particularly reluctant to be pigeonholed within discursive categories. It is characterized by a background of anti-intellectualism, and it makes use of eristical techniques that are most useful in debate. It will be sufficient to mention here the doctrine of the double truth, absolute and wordly (*paramārthasatya, saṃvṛtisatya*); the use of the logical tetralemma (*catuṣkoṭi*) in order to defeat any metaphysical assessment whatsoever; the concept of vacuity or emptiness (*śūnyatā*) as a category in effect identical with the process of conditioned coproduction (*pratītyasamutpāda*); the dialectical use of the *reductio ad absurdum* (*prasaṅga*); and the concept of insubstantiality (*niḥsvabhāvatā*) (see Ruegg, 1977–1978; Westerhoff, 2006; Sprung, 1973; Huntington, 1983; Lindtner, 1981; Eckel, 1987).

The current Vedantic interpretation (first of all in such works as the *Sarvadarśanasaṃgraha*) of *mādhyamika* (or *madhyamaka*) doctrine as *śūnyavāda*, 'doctrine of the void', a term generally rendered with 'nihilism', is based on a great equivocation and causes a gnoseological misunderstanding.[9] Following Mādhava's statement of *śūnyavāda*, we will be confronted with a position according to which the gnoseological triad, formed by the knowing subject, the known object and knowledge, effectively amounts to fully interdependent elements. In such a case, the reality of each and every element of the triad depends on the reality of the other two elements. So it will be sufficient to prove the falseness of one single element in order to deduce the falseness of the other two. When we erroneously perceive the snake instead of the rope, the snake is no doubt false; so, even the subject perceiving a false object, and the very same knowledge deriving from such a perception, are equally false. Thus reality withdraws more and more, until it wholly disappears, and universal falseness can be translated as void, *śūnya*, or as an abstract principle such as vacuity, emptiness, insubstantiality, *śūnyatā*. First of all, this is a conscious

9 For an epistemological evaluation of this work see Pelissero, 2020.

doxographical distortion of *mādhyamika* thought. In fact, the *mādhyamika* school refers to the 'middle path' evoked by the Buddha, the path standing as intermediate between two opposite conceptions, eternalism and annihilationism (or, less properly, nihilism) (*śāśvatavāda, ucchedavāda*). If nihilism is one of the two risks that must be avoided, it is not possible to attack the school with the charge of nihilism. Secondly, *mādhyamika* thought does not deny reality at all (for this would amount to accepting a metaphysical position); rather, it criticizes from its very root the concept of substantiality of the phenomenal world, of what is apparent within the domain of the senses and of the mind. Beyond the phenomenal world there is no substance whatsoever, mental or extramental, there is only the void, *śūnya*. And the void, in turn, cannot be made into a substance. The term 'void', or 'vacuum', *śūnya*, covers two entirely different shades of meaning: it may indicate the phenomenal world as it is 'void of one's own nature' (*svabhāvaśūnya*), or it may indicate the absolute reality as it is 'void of the manifoldness of manifestation' (*prapañcaśūnya*). So the term may be rendered both as 'void, vacuum' and as 'devoid, deprived'. This last meaning restates the character of (inter)relation that is typical of *śūnya*: it always means a deprivation of something else, it is not a self-contained term, it cannot be reified within positive self-sufficient terms.[10]

Candrakīrti will declare that vacuity, *śūnyatā*, acts for intellectual activity as a purge acts for the body, purifying it and being expelled together with the pathogenic factors carried away by its action. That being the case, vacuity cannot cling to the intellect as a conceptual construction, but it must be carried away in its turn when it has carried out its goal. Otherwise, it will be a cause of further problems, in the same way as if a purge could not be carried away, flowing out of the body. In these conditions, the description of reality can be made only in negative terms, it can only negate substantiality, and this negation does not equate at all with the same conceptual predicament as if it could positively assess unsubstantiality. The real nature of objects and of the world cannot be ascertained; it can only be described in negative terms such as through the category of void, an intrinsically empty predicament. In fact, what is real must be wholly independent from any other element, in order that it may be described in conceptual terms. But universal interdependence, deriving from *pratītyasamutpāda*, negates every sort of independence of anything whatsoever. So the very same reality of the world is negated. But this does not mean that the world can be described as unreal, because what is unreal (the aerial city of the heavenly musicians on the clouds, *gandharvanagara*) never comes into being. On the contrary, the world does come into being, it is manifest, and

10 Following the proposal of Sharma, 1960, p. 86 f.

everyone can testify to this fact. Neither can we say that the world is real and unreal at the same time, nor that it is neither real nor unreal. This eristic or dialectical practice applies a fourfold negation to a thesis A in four steps (not A, not non-A, not either-A and-non-A, not neither-A-nor-non-A). It is mainly applied to the category of being, *sat*, as not being, not non-being, not either being-and-non-being, not neither-being-nor-non-being, so as to thwart any ontological claim. It will take the name of tetralemma (*catuṣkoṭi*), and it will become the main device of negative dialectic (eristic) of the *mādhyamika* school, being able to deconstruct any possible conceptual construction whatsoever, with rigid and pitiless grace. An adequate description of reality being impossible, its best approximation is the term vacuity (*śūnyatā*), that has in itself its own antibodies, being able to prevent any possible reification of itself in terms of a substance. The 'things' of the world appear, are manifest to us, but when we try to analyze them, they escape any possible definition in terms of reality, non-reality, either-reality-and-non-reality, or neither-reality-nor-non-reality. In the *Madhyamakakārikā* the method of the tetralemma is successfully applied to vanify concepts such as motion, causation, and so on. This deconstruction definitely proves the inadequacy of common sense and of trivial thought, which cannot efficiently and authentically describe the complexity of phenomena. Interdependence affirms itself as the preferential entry to reality, even if it is an apophatic and limited approach. Interdependence and vacuity are the same thing, they are synonyms: *pratītyasamutpāda*, interdependent coproduction, and *śūnyatā*, vacuity or emptiness, are two different ways to describe the same situation. There is no feature (*dharma*) that might be considered independent, so every feature is intrinsically devoid of one's own nature. One's own way of being, the essential nature (*svabhāva*) of a thing (*dharma*), its being non-factitious (*akṛtrima*), not dependent on anything whatsoever for its own being, is not ascertainable in any way at all. So universal interdependence, say emptiness, is just as well rendered as absence of one's own nature (*niḥsvabhāvatā*). Indeed, the concept of emptiness, vacuity (*śūnyatā*), in its extreme shade of meaning, refers properly to the absence of one's own way of being (*svabhāva*). It can therefore be indicated as *niḥsvabhāvatā*, absence of one's own nature or of one's way of being. This does not only concern the empirical reality, but the whole of the *dharmas*. Each and every *dharma* is conceived as devoid of one's own nature, we could even say that each and every *dharma* is unsubstantial (but this could be interpreted as a tendentially metaphysical assessment). Following Candrakīrti *ad* Nāgārjuna, the term 'one's own way of being' can be understood in three ways: 1) essential property, e.g., heat in the case of fire; 2) essential feature of a single *dharma*, absolute specificity, *svalakṣaṇa*; and, finally 3) existence that is not dependent upon other factors. In this last meaning, *svabhāva* indicates an absolute non-subjugation to change in the past, present or future, independence from causes and conditions, unborn and unproduced nature. Nāgārjuna's criticism hits precisely this last shade of meaning, disclosing every object whatsoever as being void, unsubstantial. Nāgārjuna is fully aware of the risk, that is of transforming the doctrine of vacuity, originally a mere dialectical device, into a new sort of substantialism: he strongly reaffirms the emptiness, the ineliminable self-contradictory nature of any thesis (*pratijñā*) whatsoever, positive or negative (See Fenner, 1990).

## 2 Nirvāna

Nevertheless, an underlying reality is beyond the range of phenomena, a reality that is no less authentic, even if our power of understanding and our expressive capacity fail to grasp it firmly. It is the reality of extinction (*nirvāṇa*), proclaimed by the Buddha: the extinction of pain, of the grief bound to human worldly experience. If this extinction were unreal, there would be no use in undertaking the way doomed to lead to it; the Four Noble Truths would be overcome, together with the Noble Eightfold Path. The unknowability of the real through ordinary gnoseological tools does not clash with this underlying reality. Contradiction is avoided through the medium of a doctrine of double truth. There is indeed a trivial, empirical, temporary, conditioned truth, the truth of worldly affairs, *saṃvṛtisatya* (etymologically 'covering truth'), and in addition to it, an underlying, eternal, unconditioned truth, the absolute truth, *paramārthasatya*. The distinction between *saṃvṛtisatya* and *paramārthasatya* probably arises from a dichotomy present in the *sarvāstivāda*, where there is the opposition between what is existent *qua* substance, *dravyasat*, and what is existent *qua* convention, *prajñaptisat*, even if the two sets of criteria do not completely overlap. According to *mādhyamika* thought, the very same reality that from the ordinary point of view appears as interdependence, transmigration, *saṃsāra*, if it is observed from the point of view of the higher truth, is manifested as extinction, *nirvāṇa*. So *saṃsāra* and *nirvāṇa* are the two faces of the same coin. The description of *nirvāṇa* can be attempted only in apophatic terms; the same is true for the description of the 'so gone' or 'he who comes and goes in the same way', *tathāgata*, the being who has fully realized in himself this kind of truth. The silence of the Buddha with reference to metaphysical questions can thus be explained as a programmatic silence, full of meaning, the silence of he who declares that some questions are undecidable; they cannot be worked out in conceptual and verbal terms, but only with apophatic tools. *Mādhyamika* apophatism has often been compared with

Upanisadic and Vedantic apothatism (mainly with reference to Gauḍapāda and Śaṅkara), giving life to a sort of game in the looking glass, involving mutual charges of cryptovedantism and cryptobuddhism (with reference to Nāgārjuna and Śaṅkara). In particular the incipit of *Madhyamakakārikā* can be read as an attempt to reinterpret Nāgārjuna's terminology in Vedantic terms. But the direction of the loan is not clear at all: who lends, who borrows, within this hermeneutical circle? The *incipit* of the *Madhyamakakārikā* contains, with reference to *pratītyasamutpāda*, a series of epithets clearly of Vedantic appearance. First of all, we have the eight negations relative to cessation, birth, annihilation, eternity, unity, multiplicity, going and coming (*anirodha, anutpāda, anuccheda, aśāśvata, anekārtha, anānārtha, anāgama, anigama*), and subsequently more specific epithets, such as 'appeasement of the display of discursive thought' and 'kindly disposed' (*prapañcopaśama, śiva*), which are present in the *Māṇḍūkyopaniṣad*. So the Upanisadic references present in the text, which show a sort of similarity in the concept of void with terms such as 'pure', 'appeased', 'full', and 'bliss' (*śuddha, śānta, pūrṇa, ānanda*), acquire a new meaning: they are appropriated from a Buddhist point of view.[11]

The process of conditioned coproduction (or interdependent co-origination), *pratītyasamutpāda*, is a good hermeneutical tool, but it cannot be reified in its turn. This reification would involve the violation of unsubstantiality: it is not possible to consider any concept whatsoever a substance. Even the concept of the self is critically deconstructed: there is no substance that is independent from its qualities; there is no self that is independent from the states of consciousness; an agent exists only with reference to action, and action with reference to an agent, but in absolute terms neither of the two exists as a substance. The very same knowledge is a logical impossibility: it cannot exist, because it depends entirely on difference. For instance, a cow is not defined in positive terms, but only negatively, as not non-cow, that is as different, for example, from a horse or a sheep (foreshadowing the *apoha* theory): the cow exists in so far as it is not existent as a horse or a sheep, as it is not existent as something different from it. Any knowledge whatsoever is relative, *rectius*, it is relational. Thought cannot grasp the world, even less could it grasp itself. Truth can be compared with silence, the semantically pregnant silence of the Buddha with

11 Compare *Māṇḍūkyopaniṣad* 7: *nāntaḥprajñaṃ na bahiḥprajñaṃ nobhayataḥprajñaṃ na prajñānaghanaṃ na prajñaṃ nāprajñam | adṛṣṭam avyavahāryam agrāhyam alakṣaṇam acintyam avyapadeśyam ekātmapratyayasāraṃ prapañcopaśamaṃ śāntaṃ śivam advaitaṃ caturthaṃ manyante | sa ātmā sa vijñeyaḥ ||* with *Madhyamakakārikā* 1.1–2: *ani rodhamanutpādamanucchedamaśāśvatam / anekārthamanānārthamanāgamamanirga mam //yaḥ pratītyasamutpādaṃ prapañcopaśamaṃ śivam / deśayāmāsa saṃbuddhastaṃ vande vadatāṃ varam //.*

reference to vain questions. True knowledge consists in the awareness of the impossibility of knowing anything with discursive and argumentative tools, because the world does not fit logical criteria. Indeed, not even the Buddha, the Tathāgata, exists: from the absolute point of view there is no distinction whatsoever between truth and error, *saṃsāra* is the same as *nirvāṇa,* there is neither death nor birth, neither unity nor multiplicity.[12]

## 3 Apoha

So we can now examine the last term that can help us in our attempt to understand the subject-matter of the chapter. The term has already been incidentally introduced. Its use and diffusion are due to the great master of the so-called logico-epistemological school of Buddhism, Dignāga (or Diṅnāga). The term is *apoha*, 'exclusion'.

Dignāga reduces the field of application of perception to the particular (*svalakṣaṇa*, 'one's own feature') and the field of application of inference to the universal (*sāmānyalakṣaṇa*, 'common feature'), and in so doing he negates a pillar of the *nyāya* (logical) school, maintaining the suitability of different means of knowledge to the very same object (*pramāṇasamplava*). The *svalakṣaṇa* is endowed with causal efficiency (*arthakriyā*, the power to produce its object), different from any other object, inexpressible in words, unknowable with the help of signs different from itself, being able to determine a difference in the form manifested into cognition: a real fire is a fire being able to illuminate and to heat; a fire that cannot illuminate nor heat is an unreal one. *Svalakṣaṇa* is a key term, not entirely defined by Dignāga, being glossed and interpreted by Dharmakīrti and Mokṣākaragupta. Real, unique, determined by the time, space, and form that are peculiar to it, *svalakṣaṇa* is also an undivided instant (*kṣaṇa*). Perception is free from conceptual constructions (*kalpanāpoḍha,* where *kalpanā* means a cognition being associated with a linguistic expression), and is error-proof (*abhrānta*). It comprehends sensorial perception, but in addition to it, even mental awareness of perceptions and emotions, and the supernatural perception typical of the *yogin* (*yogipratyakṣa*). Existence is equated with efficiency, and efficiency is change; what is changeless is not efficient, therefore it is unreal, because reality is change, instantaneous, perpetual kinesis, according to the theory of universal flux. All the same, causality is not a sort of functional production, but merely and exclusively interdependence.

---

12 For *apoha* theory see Arnold, 2006; Gupta, 1985; Katsura, 1991; Patil, 2003; Payne, 1987; Siderits, 1985.

The cause cannot produce the effect because it does not have not the necessary time to do so; it is instantaneous like any other thing. All we can say is that the cause precedes the effect, and the effect follows the cause. Existence is efficiency, and efficiency itself is the cause. As *mādhyamika* thought had already demonstrated, things cannot arise from themselves, from 'non-themselves', from both or from neither. Effects are not at all produced, they only functionally depend upon their causes. Therefore, all *dharma*s are inactive, forceless (*nirvyāpāra, akiñcitkāra*), there is no efficiency beyond existence, existence itself equates with causal efficiency (*sattvaiva vyāpṛti*), and causality is only invariable precedence (*ānantaryaniyama*), the fact that the cause invariably precedes the effect. In other words, the effect is distinguishable from the cause only because it invariably follows the cause, it cannot come before it. Causality is the determination of the successive states through the preceding states. The universal, being the object of inference, is conceived in strictly apophatic terms, in order to avoid its reification current within Brahmanical schools, and is thus defined as 'exclusion of other' (*anyāpoha*).

The linguistic theory of exclusion (*apoha*) can now be entirely spelled out. A word expresses its objects only through a negative way, denying other meanings, different from it. The semantical sphere of each and every word is obtained by cutting out every other possible extraneous meaning, i.e., it is not possible to delineate any meaning in positive terms. So, for example, the horse will be described as a living being (excluding non-living beings), four-footed (excluding two-footed beings, snakes and so on), not endowed with horns (excluding cows and so on), a non-cloven-footed quadruped (excluding cloven-footed animals such as pigs and so on), cutting out the semantic sphere of the term by progressively excluding further specifications, through a series of successive negations. Language is the sole tool of communication for discursive thought; it does not express universals that are present in particular objects in an undivided way. This last position is maintained by the followers of epistemological realism (first of all, the *nyāya* school). Language for Dignāga expresses only a 'difference', *bheda*, through the exclusion of what is different from itself, of everything that, in its extreme variety, can be combined into a whole only by the fact that it is different from something else, that it has different effects from something else. The own way of being of a thing, its intrinsic nature, is something inexpressible, it is a mere difference. The only common feature is the exclusion of what is different from itself: this feature in fact replaces the notion of universal, because it takes on such attributes as unity, permanence, invariable presence with reference to its particulars. The denotation of the meaning of a word is inseparable from the exclusion of other meanings. Difference and exclusion are not separable, either from a gnoseological or from a conceptual point of view. Difference indeed cannot be considered as a real existing thing, a *vastu*, and the same is true for exclusion. It is only a relative, interdependent term, it has no reality in and of itself. Only the form, *rūpa*, should be considered as existent in itself (but the interest of the school lies in epistemology, not in ontology): but the word is not concerned with form, *rūpa*, rather only with difference, *bheda*.

The three main positions of the school with regard to *apoha* can be summarized as follows. The word expresses: 1) only negation (Dignāga, Dharmakīrti); 2) firstly a positive entity, secondly, only by implication, the exclusion of all other entities (Śāntarakṣita, Kalamaśīla); 3) a positive entity, being qualified by the exclusion of all other entities (Jñānaśrīmitra, Ratnakīrti). A word denotes only the 'portion' of the thing corresponding to the exclusion of all the other. So the 'thing' perpetually escapes the word, it is continually elusive. A word may only aspire to remove the causes of misunderstanding of the thing, but it can never grasp its essence, its intimate core.

Śāntarakṣita and Kalamaśīla subsequently elaborate more and more upon the notion of exclusion, *apoha*, noting the existence of two types of negation: relative negation or exclusion properly, *paryudāsa*, and absolute negation or denial, *niṣedha*. Relative negation is itself divided into two types: due to a conceptual difference (*buddhyātma*), and due to an objective difference (*arthātma*). Things are in and of themselves reciprocally different one with respect to the other, but due to a limited power (*niyataśakti*) some of them develop some sort of conception of similarity. On the basis of such a conception, a kind of reflection (*pratibimbaka*) arises. This reflection is erroneously conceived as an 'object'. *Apoha* is only the conception of such a reflection. The denotative function of the word is only the production of such a reflection, the exclusion of other objects from its own semantic sphere. Relative negation is known directly, whereas absolute negation is known through implication. In order to better understand the distinction between the two types of negation, we may remember that it dates back to grammatical thought, where we find the distinction of two different shades of application of the negative particle *na*. This happens through the dichotomy between *paryudāsa* and (*prasajya*)*pratiṣedha* (this last term being replaced with *niṣedha* within Buddhist epistemology). The process can be summarized as follows: the composition of *na* with a noun (*abrāhmaṇa*, one who is not a priest) in order to indicate a negation; and composition of *na* with a verbal form, in order to indicate a prohibition. Negation being limited to a nominal form, it can generate a certain amount of semantical choice. See for example the phrase 'this is polite': when verbally negated it becomes 'this is not polite'; when nominally negated it becomes 'this is impolite'. But when we apply the double negation we can generate such a phrase as 'this is not impolite'. This phrase is not entirely semantically equivalent with 'this is polite', because nominal negation (*paryudāsa*) does entail the possibility of a choice. The way is paved for the possible existence of a kind of behavior that is intermediate between polite and impolite. The object cow and the object no-cow are both well delimited, their respective boundaries are clear. But the word cow is unreal, because it depends on the will of those who use it in speech. Words do not know external objects, only their reflections. Due to ignorance, we take these internal reflections as if effectively representing external objects. But no word can ever directly 'touch' (or enter in contact with) its own object: a word is always a mere verbal abstraction. The word 'cow', definable as 'not no-cow', is entirely different from the object 'no-cow', it never enters in contact with it. A word is imprisoned within the net of its own conceptual definitions; it never succeeds in tearing this net, in order to directly enter in contact with reality. Reality is not negated by this school, it is merely revealed as not accessible to our understanding; it is understandable only through the severe rigid mediation of language. From the epistemological point of view, the *apoha* theory corresponds to the same extension of the mere absence of observation (*adarśanamātra*) of that which is not denoted by the semantical sphere, and in so doing it is eventually elevated to the rank of inference. Inference is that special type of exclusion that excludes from itself the object not denoted by it (*vyavacchedānumāna*). From the point of view of the different levels of reality, we may conclude that Dignāga equates the level of perception with the dependent level, *paratantra*, the level of inference with the imaginary level, *parikalpita*, and the level or pure consciousness with the absolute level, *parinispanna*.

## 4 Conclusion

As a conclusion, we may note that Indian mystics tried to follow three different paths in order to express what is ineffable.[13] The first path is the path of poetic language through the use of metaphor, the main way, the other two being subordinate to it. The limits of expressibility are transcended by poetic language, and mystics use this tool of communication in order to reach what is ineffable. The second path is paradox, the use of contradictory predicates in order to characterize the specific experience we want to communicate. Within Jainism, this path is present in the type of the systematic simultaneous application of predicates that are authentical but at the same time mutually

13 Following Matilal, 1990, p. 151 ff.

contradictory, with reference to a metaphysical assessment. Thus, from a certain point of view it is possible to assert a proposition, and from another point of view it is possible to negate the same proposition. For example, the phrase 'John does not drink' may be intended as valid with reference to the fact that John does not drink spirits, but it is invalid with reference to the fact that John does effectively drink water (it is not possible to conceive the hypothesis that John does not drink any liquid substance whatsoever). The third path is apophatism, negative dialectics, according to the Upanisadic dictum 'not so, not so' (*Bṛhadāraṇyakopaniṣad* 2,3,6 *neti neti*), further elaborated by *mādhyamika* eristic into an antisubstantialist device. To assign any possible predicate to a mystical object is but a vain venture, every sort of predication being successively negated. But if we are able to reiterate this process more and more, progressively negating every variety of possible descriptions, we will be able to convey the desired meaning in a negative way.

To sum up, to the question 'are there precursor/allied concepts that can be associated with the mathematical idea of zero?', our tentative answer could be as follows: the connection between *śūnyatā* and mathematical zero cannot be denied, but it is very difficult, indeed impossible, at present to detect a genetic or at least chronological relation between them, in order to say which one of the two comes before the other one. As far as the second relation is concerned, the one linking *śūnyātā*, through *apoha*, to the concept of ineffability, we should conclude that probably the deep need to find a way to tell with words the impossibility to describe with accuracy *brahman*, prompted the *kevalādvaitavāda* school of *vedānta* to reelaborate the buddhistic idea of *apoha* in order to adapt it to the (logically impossible) description of *brahman* in apophatic terms. This is perhaps the most amazing link between the different facets of *śūnya*: the 'metaphysical' one (*śūnyatā*), the linguistic one (*apoha*) and the mathematical one.

## References

Abhyankar, Kashinath Vasudev. (1896). *A Dictionary of Sanskrit Grammar*. Baroda: Oriental Institute, p. 393, s.v. *śūna*.

Arnold, Daniel A. (2006). On Semantics and Saṃketa: Thoughts on a Neglected Problem with Buddhist Apoha Doctrine, *Journal of Indian Philosophy* 34(5), pp. 415–478.

Bronkhorst, Johannes. (1994). A Note on Zero and the Numerical Place Value System in Ancient India, *Asiatische Studien* 48(4), pp. 1039–42.

De Jong, Jan Willem. (1972–1974). The Problem of the Absolute in the Madhyamaka School, *Journal of Indian Philosophy* 2, pp. 1–6. Id., Emptiness, *Journal of Indian Philosophy* 2, pp. 7–15.

Eckel, Malcolm David. (1987). *Jñānagarbha on the Two Truths*. Albany/New York: State University of New York Press.

Fenner, Peter. (1990). *The Ontology of the Middle Way*. Dordrecht: Kluwer.

Gupta, Rita. (1985). Apoha and the Nominalist/Conceptualist Controversy, *Journal of Indian Philosophy* 13, pp. 383–398.

Huntington Jr., C. W. (1983). The System of the Two Truths in the Prasannapadā and the Madhyamakāvatāra: A Study in the Mādhyamika Soteriology, *Journal of Indian Philosophy* 11, pp. 77–106.

Huntington Jr., C. W. (1989). *The Emptiness of Emptiness*. Honolulu: University of Hawaii Press.

Katsura, Shoryu. (1991). 'Dignāga and Dharmakīrti on Apoha', *Studies in the Buddhist Epistemological Tradition*, Proceedings of the Second International Dharmakīrti Conference, Wien, 11–16 June, 1989. Wien: Österreichischen Akademie der Wissenschaften. Ed. by Ernst Steinkellner, pp. 129–146.

Lindtner, Christian. (1981). Atiśa's Introduction to the Two Truths, and its Sources, *Journal of Indian Philosophy* 9, pp. 161–214.

Matilal, Bimal Krishna. (1990). *The Word and the World, India's Contribution to the Study of Language*. Oxford: Oxford University Press.

Patil, Parimal G. (2003). On What it is that Buddhists Think about: Apoha in the Ratnakīrti-nibandhāvali, *Journal of Indian Philosophy* 31(1–3) pp. 229–256.

Payne, Richard K. (1987). The Theory of Meaning in Buddhist Logicians: The Historical and Intellectual Context of Apoha, *Journal of Indian Philosophy* 15, pp. 261–284.

Pelissero, Alberto. (2013). *Much Ado about Nothing: Unsystematic Notes on śūnya*. In: Signless Signification in Ancient India and Beyond. London: Anthem Press. Ed. by Tiziana Pontillo and Maria Piera Candotti, pp. 17–31.

Pelissero, Alberto. (2020). The Epistemological Model of Vedantic Doxography According to the *Sarvadarśanasaṃgraha* for the Study of Indian Philosophy, *Annali di Ca' Foscari* Serie orientale Vol. 56, pp. 245–272.

Ruegg, David Seyfort. (1977–1978). The Uses of the Four Positions of the Catuṣkoṭi and the Problem of the Description of Reality in Mahāyāna Buddhism, *Journal of Indian Philosophy*, 5, pp. 1–71.

Ruegg, David Seyfort. (1986). *Does the Mādhyamika have a Thesis and Philosophical Position?* In: Buddhist Logic and Epistemology, Studies in the Buddhist Analysis of Inference and Language. Dordrecht: D. Reidel Publishing Company. Ed. by David Seyfort Ruegg and Bimal Krishna Matilal, pp. 229–237.

Sarma, S. R. (1992). '*Śūnya*: Mathemathical Aspect.' In: *Kalātattvakośa. A Lexicon of Fundamental Concepts of the Indian Arts*. Ed. by Kapila Vatsyayan. New Delhi: Indira Gandhi National Centre for the Arts and Motilal Banarsidass, vol. 2 *Concepts of Space and Time*. Ed. by Bettina Bäumer, pp. 400–11.

Sharma, Chandradhar. (1964). *A Critical Survey of Indian Philosophy*. Delhi: Motilal Banarsidass [Rider, London 1960].

Siderits, Mark. (1985). Word Meaning, Sentence Meaning, and Apoha, *Journal of Indian Philosophy* 13, pp. 133–151.

Sprung, Merwyn. (ed.) (1973). *The Problem of Two Truths in Buddhism and Vedānta*, Dordrecht: Reidel.

Streng, Frederick J. (1967). *Emptiness. A Study in Religious Meaning*. New York: Abingdon Press.

Vatsyayan, Kapila, S. R. Sarma and G. C. Pande (1992). '*Śūnya/śūnyatā*.' In: *Kalātattvakośa. A Lexicon of Fundamental Concepts of the Indian Arts*. Ed. by Kapila Vatsyayan. New Delhi: Indira Gandhi National Centre for the Arts and Motilal Banarsidass, vol. 2 *Concepts of Space and Time*. Ed. by Bettina Bäumer pp. 399–428.

Venkata Ramanan, Krishniah. (1966). *Nāgārjuna's Philosophy as Presented in the Mahā-Prajñāpāramitā-Śāstra*. Tokyo: Charles E. Tuttle Company.

Westerhoff, Jan. (2006). Nāgārjuna's catuṣkoṭi, *Journal of Indian Philosophy* 34(4), pp. 367–395.

## Further Research

Arnold, Dan. (2006). On Semantics and *saṃketa*: Thoughts on a Neglected Problem with Buddhist *apoha* Doctrine, *Journal of Indian Philosophy* 34(5), pp. 415–78.

Bäumer, Bettina and Dupuche, John R. (eds.) (2005). *Void and Fullness in the Buddhist, Hindu and Christian Traditions. Śūnya—Pūrṇa—Plerôma*. New Delhi: D. K. Printworld.

Bhatt, Siddheshwar Rameshwar (ed.) (2019). *Quantum Reality and Theory of Śūnya*. Singapore: Springer Nature.

Bhattacharyya (Chakrabarti), Bhaswati. (1979). The Concept of Existence and Nāgārjuna's Doctrine of *śūnyatā*, *Journal of Indian Philosophy* 7(4), pp. 335–44.

Bugault, Guy. (2000). 'The Immunity of *śūnyatā*: is it possible to understand *Madhyamakakārikās* 4, 8–9?, *Journal of Indian Philosophy* 28(4), pp. 385–97.

Dargyay, Lobsang. (1990). What Is Non-existent and What Is Remanent in *śūnyatā*, *Journal of Indian Philosophy* 18(1), pp. 81–91.

De Jong, Jan Willem. (1972–1974). Emptiness, *Journal of Indian Philosophy* 2(1), pp. 7–15.

Duckworth, Douglas D. (2010). De/limiting Emptiness and the Boundaries of the Ineffable, *Journal of Indian Philosophy* 38(1), pp. 97–105.

King, Richard. (1989). *Śūnyatā* and *ajāti*: Absolutism and the Philosophies of Nāgārjuna and Gauḍapāda, *Journal of Indian Philosophy* 17(4), pp. 385–405.

Lopez, Donald S. Jr. (1988). Do *śrāvakas* Understand Emptiness?, *Journal of Indian Philosophy* 16(1), pp. 65–105.

Sen, Syamal K. and Agarwal, Ravi P. (eds.) (2016). *Zero. A Landmark Discovery, the Dreadful Void, and the Ultimate Mind*. London: Elsevier.

Siderits, Mark. (2004). Causation and Emptiness in Early Madhya-maka, *Journal of Indian Philosophy* 32(4), pp. 393–419.

Frederick J. Streng. (1970–1972). The Buddhist Doctrine of the Two Truths as Religious Philosophy, *Journal of Indian Philosophy* 1(3), pp. 262–71.

Tola, Fernando and Carmen Dragonetti. (1981). Nāgārjuna's Conception of 'Voidness' (*śūnyatā*), *Journal of Indian Philosophy* 9(3), pp. 273–82.

CHAPTER 26

# Much Ado about Nothing or, How Much Philosophy Is Required to Invent the Number Zero?

*Johannes Bronkhorst*


## Abstract

This article argues that the search for a philosophy that gave rise to the number zero is misguided. No philosophy is required to invent this number. The article further shows that there are good reasons to accept that Buddhism did not play a role in this invention. It further points out that the notion of number as developed in Indian philosophy had no place for zero.



## Keywords

zero – empty – philosophy – numerical place-value – Buddhism – numbers in Indian philosophy


## 1 Introduction

The question what philosophy made the invention of zero possible has often been raised. A question less often raised is whether any philosophy at all is required for this invention. This article will raise this question and find that it is far from obvious that philosophy has to play a role here. It will subsequently look for whatever evidence there might be for a connection between Buddhist philosophy and the invention of zero and repeat the, by now, well-known conclusion that there is none. It will also briefly survey what ideas about numbers were produced in Indian philosophy and point out that these ideas could not possibly give rise to the number zero.

 | DOI:10.1163/9789004691568_030

## 2 Zero and Philosophy

One of the uses of a written numerical place-value system – perhaps the most important one – is that it facilitates arithmetical operations that would be complicated without it. The contrast between Roman and Hindu-Arabic numerals illustrates this sufficiently:

> The addition 123 + 234 (= 357) is easily carried out with Arabic numerals by adding the digits in each of the three columns, but becomes complex with Roman numerals: CXXIII + CCXXXIV = CCCLVII. Subtraction and multiplication become even more complex.

Let us assume, for argument's sake, that certain cultures have adopted a written numerical place value system at least in part for this reason: to facilitate arithmetical operations. If so, they cannot have done without a way of indicating that certain places are 'empty'. Consider the following:

> $123 = 1 \times 100 + 2 \times 10 + 3 \times 1$. The value of each digit ('1', '2', or '3' in this case) depends on its position.

In the analysis of 103, there must be a way to indicate that the second 'column', which represents the value 10, is empty. This can be done by leaving that place open: 1/ /3. It will be less confusing to put some kind of visible marker at that place, for example 1/*/3, or simply 1*3. Having such a marker does not imply that one has accepted the number 0, far from it. However, in specifying how addition, subtraction, and multiplication can be carried out, rules about how to deal with that marker will be necessary. Consider 103 + 234 = 337. Our hypothetical mathematician does not have the number 0. To him this equation looks like this: 1*3 + 234 = 337. There is yet no doubt about the outcome of this addition. But how can the digits of the second column be added if there is there no digit, as in 1*3? How can the known result (viz. 337) be obtained?

Clearly the digits in the '1 category' must be added, and the sum of this addition will occupy the '1 category' of the result: -3 + -4 = -7. The same rule applies for the '100 category': 1- + 2- = 3-. But how to deal with the '10 category', which is empty in one case? The answer is straightforward. It cannot but be: -*- + -3- = -3-.

An analysis of other cases, including subtraction and multiplication, will reveal that the empty placeholder * follows rules that may be expressed as follows:

X + * = X
X − * = X
X × * = *

Note that these rules can be obtained without the help of a zero *concept*. Perhaps one should say: *Zero needs no concept in order to be useful in arithmetic*. Anyone who knows the rules can calculate with zero and profit from its usefulness.

## 3 Origin

When did the written numerical place value system arise in India?[1] In a short article that came out many years ago (Bronkhorst, 1994), attention was drawn to a passage in Vasubandhu's *Abhidharmakośa Bhāṣya* that is ascribed to a certain Bhadanta Vasumitra. This same passage occurs in the Chinese translations of the *Mahāvibhāṣā* and the *Vibhāṣā*. These texts may have been composed during the reign of Kaniṣka, the *Vibhāṣā* presumably somewhat earlier. All this supports the claim that Vasumitra is to be dated at that same period. Well, the passage attributed to him illustrates his position with the help of a *vartikā* that in the unit position has the value of a unit, in the hundreds position that of a hundred, and in the thousands position that of a thousand. This was presented as evidence for the existence of a written numerical place value system during the early centuries of the Common Era. This conclusion seems no longer valid. Dominik Wujastyk (2018, p. 41) rightly makes the following observation: 'Vasubandhu's description may refer not to writing but to placing a strip or tube on a marked board, perhaps analogous to an abacus. The word *vartikā* that he used means a wick, stalk, paintbrush, or twist of cloth. It is not clear what Vasumitra was describing.' In short, this passage provides no proof that a written numerical place value system existed at that time.

However, a variant of Vasubandhu's passage occurs in the *Yogaśāstra* on sūtra 3.13. It uses *rekhā* ('line, scratch') instead of *vartikā*, which could suggest some kind of writing. 'This passage has been cited as the earliest unambiguous description of written place-value notation using digits, and Patañjali's version

1 Lam (1986; 1987; 1988) argues that a written numerical place value system with zero existed in China, too. Martzloff (1995) shows that these conclusions have to be looked at with suspicion.

is datable to the period 375–425' (Wujastyk, 2018: p. 41). This would then be evidence for a written place value system that existed around 400 CE.[2]

A century after Patañjali, the author on Yoga, the situation becomes clearer. A number of texts from the middle of the first millennium onward present clear evidence for a written place value system (See, for example, Gāṅguli, 1932; 1933). One of these is Varāhamihira's *Pañcasiddhāntikā*, which 'incidentally states two fundamental arithmetical operations by the zero … viz. addition and subtraction, in more than one place' (Datta, 1926, p. 451). Other texts, too, show acquaintance with both the written numerical place value system and with the rules how to manipulate zero.

## 4 Zero and Philosophy Again

Is zero anything more than this? Consider the following observation:

> If zero merely signified a magnitude or a direction separator, the Egyptian zero, **nfr**, dating back at least four thousand years, amply served these purposes. If zero was merely a placeholder symbol, then such a zero was present in the Babylonian positional number system before the first recorded occurrence of the Indian zero. If zero was represented by just an empty space within a well-defined positional number system, such a zero was present in Chinese mathematics a few centuries before the beginning of the Common Era. The dissemination westwards of the Indian zero as an integral part of the Indian numerals is one of the most remarkable episodes in the history of mathematics … the Indian zero was a multi-faceted mathematical object: a symbol, a number, a magnitude, a direction separator, and a placeholder, all in one operating within a fully established positional numeration system. (Joseph, 2008, pp. 37–38)

This is not the place to evaluate these claims. However, it is far from obvious where – in the development of these aspects of zero – philosophy comes in. To my eye, these features considerably facilitated arithmetic operations and were

2 Certain Jaina texts were apparently acquainted with a written numerical place value system. It is unfortunately extremely difficult (if not impossible) to date these texts, so that it is impossible to draw chronological conclusions. The *Lokavibhāga* (Joseph, 2016, p. 105) might be an exception. The fact that it only survives in a Sanskrit translation that is younger than the original text makes it, once again, difficult to draw chronological conclusions.

a plausible consequence of the introduction of a written numerical place value system, and an intelligent extension of it. *But where is the philosophy*?

Philosophy did, of course, play a role in the history of zero, if not in its invention. As is well known, zero did not find a warm welcome in Europe when it was introduced there in the thirteenth century. Aristotelian philosophy had no room for empty space, and the arrival of the number zero was felt as a threat by many a churchman, delaying its full establishment by centuries. Lots of philosophy here, but philosophy that stood in the way of zero, not philosophy that promoted its invention or use.

## 5 Buddhism and Zero

Returning to India, no one has yet shown that there is any connection between the Buddhist philosophy of emptiness and the invention of zero. Scholars go on and on about it, no doubt because that philosophy frequently uses the term *śūnya*. That same term is used in mathematical literature to designate zero. By itself, this proves nothing. The word *śūnya* means 'empty'. 'Empty' is a common word and can be applied to numerous altogether different situations, both in India and in the West. No arguments to support the connection between Buddhist emptiness and zero are known to me, apart from pure (one would almost say 'empty') speculation.

The lack of evidence for a link between Buddhism and zero is not surprising. Elsewhere attention has been drawn to the virtual non-existence of Buddhist treatises on astrology, astronomy, and mathematics. This non-existence, it was there proposed, is due to the fact that Buddhism in classical India had taken the position that there were occupations that were best left to Brahmins, and these included astrology, astronomy, and mathematics.[3]

Consider at this point some of the other words that were used in mathematical texts to refer to zero; these include *kha*, *ambara*, *antarikṣa*, *gagana*, *abhra*, *viyat*, *nabhas*, *ākāśa*. These words have one philosophically loaded meaning in common: 'sky, ether'. One might think that empty space is meant, but this would not be correct. Ether, in Indian philosophical thought, is not empty. It is an omnipresent element and therefore an existing element. This is even so in Buddhist scholasticism, where ether is an unconditioned element (*asaṃskṛta dharma*) and therefore an existing 'thing'.

3 Bronkhorst, 2011, section 3.1, esp. p. 109 (but cf. Bronkhorst, 2018, p. 319).

## 6 Numbers in Indian Philosophy

No one seems to have paid attention to what Indian philosophers actually thought about numbers and related matters. The texts that inform us about zero and the numerical place value system are, most of them, Brahmanical texts. No Buddhist texts betray acquaintance with them. It will therefore be appropriate to look at what Brahmanical thinkers thought about the nature of numbers.

Vaiśeṣika is the school to look at. Its vision of the world influenced other schools of thought. Vaiśeṣika had sophisticated ideas on the nature of numbers from *one* onward. These numbers were thought of as qualities that inhere in substances. 'One tree', for example, refers to a tree that has the quality 'one'; 'two trees' refers to two trees that share the quality 'two'; etc. The story as to how the numbers from two onward come about is complicated and involves the observer. It is not necessary to deal with it in detail.[4] More important for us at present is that these numbers (1, etc.) cannot exist independently of the substances in which they inhere. What is more, this understanding of numbers has no place for zero.

Incidentally, Vaiśeṣika had no room for infinity either. According to its classical text (called *Padārthadharmasaṃgraha* or *Praśastapādabhāṣya*), there is a highest number, called *parārdha*.[5]

---

4 By way of example, here is a description of how the number two ('duality') and its cognition come about according to the fourteenth century *Compendium of All Philosophies* (*Sarvadarśanasaṃgraha*), ed. Abhyankar p. 221: (1) First there is connection between sense and object. (2) It gives rise to knowledge of the universal one-ness. (3) After this a combining cognition comes about. (4) From this duality arises. (5) This gives rise to knowledge of the universal duality-ness. (6) It gives rise to knowledge of the quality duality. (7) After this the idea 'these are two substances' comes about. (8) This gives rise to a mental trace.

5 Bronkhorst & Ramseier, 1994: xxx: *ekādivyavahārahetuḥ saṃkhyā/ sā punar ekadravyā cānekadravyā ca/…/ anekadrvavyā tu dvitvādikā parārdhāntā/.* Ganeri (2001, p. 435 n. 6) makes the following perceptive observation about *parārdha*: 'Praśastapāda states that, after one, the numerical qualities run from two to a large but finite number *parārdha*. The number *parārdha* is mentioned in many texts as the highest decimal place name. The precise value of *parārdha* varies: in the *Taittirīya Saṃhitā* and the *Maitrāyaṇī Saṃhitā* it is given as $10^{12}$, while the *Kāṭhaka Saṃhitā* records both $10^{12}$ and $10^{13}$. Among the mathematicians, it is always $10^{17}$. There are names for higher decimal powers in Buddhist and Jaina texts. Only the Nyāya-Vaiśeṣika takes *parārdha* to be the highest number, and not merely the highest named place value. There is no room for the idea of a maximal finite number if one thinks of the number series as generated by recursive application of the successor function, but among the Nyāya-Vaiśeṣika authors, only Bhāsarvajña attempts so to construct the number series. Within a conception of number as qualities of substances, indeed, it seems that there has to be a largest number, if the number of things in the cosmos is finite.' '*Parārdha* is not a number *per se*, but the number of substances' (Lysenko, 1994, p. 786).

Some years ago, there was occasion to observe that early mathematical texts in India betray no acquaintance with philosophical literature, even though their acquaintance with grammatical literature is strong and evident. The earliest mathematical work of some length, Bhāskara's commentary on Āryabhaṭa's *Āryabhaṭīya*, 'often cites grammatical and generally linguistic texts … astronomical texts, some religious and literary treatises, but not a single philosophical work' (Bronkhorst, 2001, p. 64). The Vaiśeṣika ideas about numbers existed well before the texts that provide evidence for numerical place value and zero. It is yet possible to exclude that our mathematicians drew inspiration from these ideas; if they had, they would not have been able to develop their new methods. Influence in the opposite direction did not take place either. Vaiśeṣika remained unperturbed by the new arithmetic and held on to its understanding of the nature of numbers.[6]

## 7 Conclusion

It seems highly unlikely that there was any link between the invention of zero and Indian philosophy in any of its forms. The importance of zero in the history of mathematics (and much more) cannot be underestimated, but it is important to remember that this discovery did not need philosophy. Worse, when philosophy got involved – i.e., when zero and the written place value system reached Western Europe – it stood in the way of this marvelous invention and unnecessarily delayed its adoption by several centuries.

## References

Bronkhorst, Johannes. (1994). A note on zero and the numerical place-value system in ancient India. *Asiatische Studien / Études Asiatiques* 48(4), pp. 1039–42.

Bronkhorst, Johannes. (2001). Pāṇini and Euclid: Reflections on Indian geometry. *Journal of Indian Philosophy* 29(1–2), pp. 43–80.

Bronkhorst, Johannes. (2011). *Buddhism in the Shadow of Brahmanism*. Leiden–Boston: Brill.

6 Cp. Ganeri (2001, p. 430): 'The theory of number developed by the philosophers is not based on any great awareness of the work of mathematicians, nor is its purpose to study the philosophical foundations of mathematics or scientific practice.' Lysenko (1994, p. 784): '[W]e cannot find in our [Vaiśeṣika] texts any account of mathematical operations like addition, subtraction, multiplication and division.'

Bronkhorst, Johannes. (2018). Were Buddhist Brahmins Buddhists or Brahmins? *Mitrasampradānam: A collection of papers in honour of Yaroslav Vassilkov*. Ed. M. F. Albedil and N. Yanchevskaya. Saint Petersburg: Mae Ras. pp. 311–323.

Bronkhorst, Johannes and Ramseier, Yves. (1994). *Word Index to the Praśastapādabhāṣya*. Delhi: Motilal Banarsidass. Reprinted 1999.

Datta, Bibhutibhusan. (1926; 1931). Early literary evidence of the use of the zero in India. *American Mathematical Monthly* 33(9), pp. 449–54 and 38(10), pp. 566–72.

Ganeri, Jonardon. (2001). Objectivity and proof in a classical Indian theory of number. *Synthese* 129, pp. 413–437.

Gāṅguli, Sāradākānta. (1932; 1933). The Indian origin of the modern place-value arithmetical notation. *American Mathematical Monthly* 39(5), pp. 251–56; 39(7), pp. 389–93; 40(1), pp. 25–31; 40(3), pp. 154–57.

Joseph, George Gheverghese. (2008). A brief history of zero. *Tārīkh-e 'Elm: Iranian Journal for the History of Science* 6, pp. 37–48.

Joseph, George Gheverghese. (2016). Indian Mathematics: Engaging with the world, from ancient to modern times. London etc.: World Scientific.

Lam, Lay-Yong. (1986). The conceptual origins of our numeral system and the symbolic form of algebra. *Archive for History of Exact Sciences* 36(3), pp. 183–95.

Lam, Lay-Yong. (1987). Linkages: Exploring the similarities between the Chinese rod numeral system and our numeral system. *Archive for History of Exact Sciences* 37(4), pp. 365–92.

Lam, Lay-Yong. (1988). A Chinese genesis: Rewriting the history of our numeral system. *Archive for History of Exact Sciences* 38(2), pp. 101–08.

Lam, Lay Yong and Ang, Tian Se. (1992). Fleeting Footsteps: Tracing the conception of arithmetic and algebra in Ancient China. Singapore: World Scientific.

Lysenko, Victoria. (1994). 'Atomistic mode of thinking' as exemplified by the Vaiśeṣika philosophy of number. *Asiatische Studien/Études Asiatiques* 48, pp. 781–806.

Martzloff, Jean-Claude. (1995). Review of Lam & Ang 1992. *Historia Mathematica* 22, pp. 67–73.

Wujastyk, Dominik. (2018). Some problematic Yoga Sūtra-s and their Buddhist background. *Yoga in Transformation*. Ed. Karl Baier, Philipp A. Maas & Karin Preisendanz. Göttingen: Vienna University Press; V&R Unipress. pp. 21–47.

CHAPTER 27

# From Emptiness to Nonsense: the Constitution of the Number Zero (for Non-mathematicians)

*Erik Hoogcarspel*


## Abstract

The subject is introduced by a reflection on the different aspects of numbers and a review of a paper written by Andreas Nieder: *Representing Something Out of Nothing: The Dawning of Zero*. This is followed by a reflection on the precise nature of the transformation of zero from a signifier of quantity to a number. Edmund Husserl called this the constitution (*Stiftung*) of the zero as a new ideal object and he described the process in *The Origin of Geometry*. Zero as a number is an element of a self-contained symbolic deductive system that does not refer to the life world and has no sense at all; it is, in the words of the analytic philosophical tradition, nonsense. This implies that no meaning can possibly be transferred from zero as a quantity to zero as a number. The history of zero as a number coincides with the history of abstract mathematics and the development of calculation systems, the tremendous usefulness of which is precisely based on its very meaninglessness. All speculation about any eventual progress of the concept of zero as a quantity to zero as a number is therefore unsubstantiated. Marc Richir showed that the constitution of an ideal object is a leap; it leaves a gap between its new sense and its historical basis. This chapter will conclude with the question whether there is a link between this historical basis of zero as a quantity and the concept of emptiness in Buddhism; this supposed relation has been and still is the pet of quite a few Buddhologists.



## Keywords

natural number – placeholder – whig history – emptiness – meditation – logical theory – Buddhism


## 1 Introduction

In this article I will argue that emptiness as understood in the Buddhist tradition has no bearing on the mathematical concept of the number zero. I will

 | DOI:10.1163/9789004691568_031

start with an overview of the conception of zero as a number on the basis of natural numbers, placeholders and the set theory. Through a short review of an article of Ralf Nieder I will argue that the history of numbers is *a posteriori*, but their use and significance are *a priori*. The gap between both is one between numbers having a conventional meaning and numbers being meaninglessness or nonsense and therefore exceptionally useful. Finally, I will argue that the concept of emptiness in the Buddhist tradition has no bearing whatsoever on the concept of zero, nor on its history.

## 2 Number or Quantity?

Is black a color? Not according to the Rolling Stones,[1] but it is according to the shop assistant who is trying to sell you a pair of black pants. Is zero a number? This question might be the subject of heated discussions, but the answer remains a matter of opinion. In the nineteenth century natural numbers were defined as numbers that were the result of counting. It is obvious that zero is not the result of counting. When children learn to count, they start from one and when we learn a foreign language, we are learning the numbers in the same way: *un*, *deux*, *trois*, *quatre*, and so on. Children only at a later stage become familiar with the concept of zero; in fact it is only when they go to school and learn arithmetic that they discover that zero is a number. So zero is not a natural number, it is the result of subtraction of equal numbers, the result of an arithmetical operation. From this it becomes clear that there is a difference between zero as a number and zero as a quantity. The threshold from quantity to number is crossed when numbers are no longer conceived as the result of counting, but as elements in an independent system of possible operations.

Children who have never heard of the zero are not puzzled when they see their cup is empty or when they discover that the candy is finished. The concept of emptiness is quite natural; it naturally comes to mind when something is missing. You are looking for something, but it is not there, so you have to look somewhere else or find a replacement. Emptiness is not, however, a number, because when you are missing something, you are not counting. Counting starts when we are confronted with things to count. Mathematicians conceived the numbers that were the result of counting as the ordered system of natural numbers. They included the zero in their set of natural numbers (usually symbolized by the sign N), not because we start counting with zero,

---

1 One of their famous songs is *Paint it Black* It contains the line: 'No colors anymore, I want them to turn black'.

but for their own convenience. In order to lift the natural limits of subtraction, the negative numbers were introduced. Without negative numbers there is no answer to the question, what is '2 – 5?' When numbers are allowed to become negative, there is an answer: '–3'. This implies that the set of natural numbers has been expanded to the larger set of integers, symbolized by Z. The answer to the question, 'what is 2 – 2?' needs yet another kind of number that is not negative and not positive: the zero. Zero is an even number, because it is in between 1 and –1, which are both uneven, and even and uneven always alternate. It can be divided by any number and the result is always equal to itself. You cannot divide by zero, because there is no result (it is called infinity, symbolized by ∞; this is not a number). The most useful quality of zero is that it is neutral in respect to addition and subtraction. When you add or subtract zero to any number, the number does not change. The number 1 is the neutral element in respect to multiplication and division, because a number does not change when multiplied or divided by 1. Neutral elements can be very useful in applied mathematics, for instance when describing a card game where one is allowed to pass, or throw a card without value, or when describing traffic, where cars have to wait for traffic lights. In mathematical systems they are often essential.

## 3 Sets and Places

In the beginning of the twentieth century, the set theory in mathematics was proposed by John von Neumann. Counting could be represented by means of sets. The most basic set was defined as the empty set, the set without any elements, in symbols written as {} or ø. The second set was represented as the set you get when you make a set of the empty set, so it has only one element: the empty set. This is written as {{}} or {ø}. Therefore, the first set is the zero, the second one is the set that contains the zero, etc. It is possible to repeat this process *ad libitum* and the result is a series of sequential sets of which each one is the exclusive element of the next one. The set of all these sets makes a set that has a structure identical to the set of natural numbers including zero, which is the only real element, and the ordinals make the set of natural numbers without the zero.[2]

The empty set is an abstraction, of course; we may experience absence sometimes when we look into our wallets or check our bank account at the end of the month, but we do not think of it as a set. The bank account is already an abstraction in itself. When it shows a zero, it merely symbolizes the practical

2 I thank Professor Jeffrey A. Oaks for his corrections.

situation that for us cash machines are blocked for the time being. Our wallet is more concrete: we peek into it and we do not see anything there, not even zero or an empty set. Absence is not a number and the trick with zero is that it is a number made from absence. Absence is not a natural phenomenon; it is only possible if someone expects something to be there, and we expect by convention to find certain things in certain places. A wine shop never runs out of bread and a bakery never runs out of wine. No one walks into a bakery asking 'Where is the Chianti?' and no wine merchant will point to an empty shelf and say to a customer 'I am sorry sir, but we ran out of rye bread today'. However, when you start to count things and also take things away sometimes, you will sooner or later end up with the quantity zero.

So how do we come by the idea of zero? Professor Andreas Nieder has written a detailed description of its history (Nieder, 2016). He explained that there are four steps in the evolution of the concept. It starts when people simply take notice of something not being there; this is the awareness of nothing. Subsequently comes the thought of missing something. The third step is taking notice of the fact that one expects something to be there and, finally, one associates the number zero with the situation.

A very interesting discovery in the history of mathematics is the value position system. This is still clearly visible nowadays in the way people were used to multiplying large numbers before the introduction of pocket calculators. Say you wanted to multiply 367 by 891 the old-fashioned way. The result is shown in Figure 27.1. The zeros are not real numbers, they are placeholders. They do not have any meaning other than to show the place they are holding must be ignored; it is very much possible to use another sign instead of zero. I used to have a friend at school who had learned to write dots instead of zeros, but this did not have any influence on his calculations; they were never different from mine. Placeholders have the same function in large numbers. The number 30,507, for instance, means three ten-thousands and five hundreds and seven. Without the zeros to fill in the fixed places, it would be written like '357' and that could mean a lot of numbers, such as 30,057, 35,700, 305,070, and so on. There are other possible conventions, of course; one could write it like 3*5*7 or 3tt5h7. The use of the zero, however, has turned out to be convenient. So the function of a placeholder is, as it were, a side job of the zero and has no consequences for its sense or reference.

According to Nieder the first culture to use a placeholder was the Babylonian, attested by texts from around 500 BCE. They used a wedge, not a zero. A century later the Greeks started to use a circle instead. It is interesting and perhaps of some influence that the widespread use of money had been introduced a century earlier and that most calculations in those days were done in

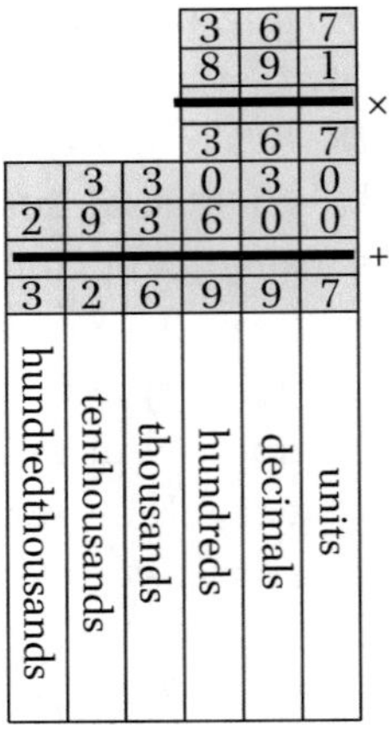


FIGURE 27.1
The zero as a placeholder

bookkeeping. The fact that a placeholder other than the zero was used in those days did not, however, affect the accuracy of the arithmetic.

## 4 Toward Becoming a Number

A significant step toward the zero becoming a number was the conception of a certain instrumentality of numbers in general, when they were taken out of the direct context of counting and quantities. This happened first when conventional values were conferred to numbers, such as being lucky or ominous. In China, the number 4 *sì* (四) is considered ominous, because it sounds like *sǐ* (死), 'death' in Mandarin. In the West we have a fear of the number 13 and a complete system of values is associated with numbers in the Kabbalistic literature. It is as if the number appears as meaningless for just a moment and people immediately hasten to fill the gap with a new, this time, conventional meaning. This was probably predated by the instrumentalization of language in rites. In India this happened quite early (at least before 1200 BCE, when the Vedas were consolidated) when mantras were introduced, meaningless words that were supposed to be heard by the gods or have physical effects when pronounced, just because of their very sound. This became acceptable through the assumption that a certain language, Sanskrit in this case, is the only proper language and even the physical structure of the universe. Language exists in itself, is eternal, according to the Vedas, it speaks itself, and it is not a human institution (Ruegg, 1959, p. 13).

The first time the zero was considered to be a real number was according to Nieder in the seventh century in India. The Indian astronomer Brahmagupta presented in his book *Brāhmasphuṭasiddhānta* (*The Explanation of the Brahma System of Astronomy*, 628 CE) a set of rules prescribing how to deal with the number zero. On a temple wall in Gwalior (central India) there is even an inscription to be found where the zero was written for the first time in a modern fashion; this inscription is dated 876 CE. The zero was called *śūnyam*, 'empty'. The Arabs, who took over a lot of Indian mathematics, translated this word to *as-sifr* (the empty), which became *cifra* or *cephirum* in Latin. The gloom of secrecy that hung over the zero in those days inspired the English expression 'to decipher a code'.

We are not born with the idea of zero present in our mind, we have to learn it and Nieder reports that research has shown that children between the ages of six and nine are increasingly capable of understanding the rules Brahmagupta discovered a long time ago: zero is smaller than any positive number, zero plus any number is that number, and a number minus zero is again that very number. Apparently, children at that age are sufficiently capable of thinking in concepts and become aware of our number system and the function of the zero. There is no way to tell whether animals have any idea of zero, concludes Nieder, and that should not surprise us, I would like to add, because zero is a convention. Zero is not a thing existing in itself out there waiting to be discovered; it is something that humans invented, just like Santa Claus or democracy.

No paper these days is complete without a jingle about the brain. So Nieder (who happens to be a professor in animal physiology at the University of Tübingen) tries to tell us how difficult it is for the neurons to represent 'nothing, empty sets, or the number zero'. Nieder does not think so because his neurons have been complaining to him, but because he thinks that evolution has predisposed humans only to look for something present. I am not sure how you can be aware of something present without being aware of other things absent, but I am not an expert. I presume, however, that were Nieder to skip his dinner, his neurons would convince him otherwise, like they do in all kinds of animal brains. Anyway, to locate the exact place of the zero in the brain is a job for biologists or medical experts. A philosopher cannot rule out the possibility that one day an alien from outer space with a totally different set of neurons and a brain that is in no way comparable to ours, will partake in the discussion about the zero and teach us humans some new insights. In philosophical language, the concept of zero is *a priori* and not *a posteriori*, so numbers are not supposed to have a place anywhere, neither in the human brain nor in a calculator or a computer.

So, what triggered the introduction of the zero as a number? The jump from zero as a quantity to zero as a number is a huge one that took, according to Nieder, more than a millennium. These days we use both concepts of zero alternatively without even being aware of the difference. When we are solving a Sudoku puzzle, we treat the digits (OK, there is no zero in Sudoku) as numbers, mere elements of a fixed set of formal elements (the set of the natural numbers between zero and ten). This set we have to spread out in every row in such a way that every element of the set appears precisely once. When we inspect a cash-register slip on leaving a supermarket, we regard the digits as quantities. In other words, the huge difference is that a zero as a quantity refers directly and naturally to a counted set of objects in the world, while a zero as a number does not, it is more abstract. Other examples of formal use of numbers are number puzzles or the use of numbers as a code. In many practical situations numbers are used as an ordering principle, for instance in competitions where the best performer is called the first one, the next best the second, and so on.

## 5 Crisis

Money links numbers with value, because everybody loves money. There is a Buddhist story about a monk who found a precious gem. He finally decided to give this gem to a rich person, because he needed it the most. This is a way to express the fact that things, the gem in this case, have no value in themselves, they are associated with values because people value them. For the monk the gem had no value because he did not need any money, in contrast to the rich person. Most people do; this is the reason why they are curious about the digits on cash-register slips. The introduction of money spread a veil of values over everything in the world. Some people are so obsessed with value that they no longer perceive things in a natural way. They buy lots of stuff not because they need it, but because of the price.

The philosopher Edmund Husserl stated in his lecture and book *Die Krisis der europäische Wissenschaften und die transzendentale Phänomenologie* (The crisis of the European sciences and transcendental phenomenology, Husserliana VI) that science has thrown a veil of ideas over the world. The veil of ideas and numbers hides the way we experience things; the veil of science hides our day-to-day experiences under a cover of scientific expressions and formulas. What did he mean? Let us take an example. If we want to paint a wall, we want to know how much paint we need, and in order to do that we have to measure it. Let us presume the wall measures 3.2 meters by 5.6 meters. Subsequently, we reduce the wall to a geometrical shape, a rectangle, and calculate the surface of this rectangle by multiplication of the numbers. This is 17.92 $m^2$, so now we know how big a can of paint we need to buy. We have gained this knowledge by throwing a veil of ideas and numbers over the world.

Note that we associated numbers to the wall by measuring it. The numbers we measure are not the sizes of the wall we perceive. We cannot tell the measurements of the wall just by looking at it; we need a tool that tells us how many conventional standard sizes, meters or feet, width and height the wall is. Having done this, we reduce the shape of the wall to a rectangle. The shape of the wall is not exactly identical to a rectangle, its edges are not smooth and bend slightly, perhaps even slant. This does not matter, because we do not need to know exactly how much paint we need; we merely need to know whether we need a 10 $m^2$ or a 20 $m^2$ can of paint. So after our calculation is finished, we jump back to the practical level of our daily life. While painting the wall and afterwards we forget the veil of ideas and numbers; we experience the surface of the wall, first its roughness and later the beauty of the result of our work.

The jump from zero as a quantity to zero as a number is similar to our example. First of all, the jump did not take place in Brahmagupta's book, because he presented a set of rules that describe how to deal with the zero as a number. You do not make rules for something no one knows how to do. So, apparently people were beginning to regard the zero as a number and ran into difficulties. Brahmagupta told how to avoid these. Besides, Brahmagupta did not deal only with the zero, he also presented rules for manipulating both negative and positive numbers, a method for computing square roots, methods of solving linear and some quadratic equations, and rules for summing series. As in our example, there already had to be a more or less complete mathematical system of numbers, in which the zero is now included and can be used. This mathematical system is not part of our life world, just as the rectangle is not the wall and the wall has not the exact shape of a rectangle. A mathematical system is more like the rules of a Sudoku game; it is completely abstract. A formula like $\sin^2 x + \cos^2 x = 1$ does not mean anything; it is the expression of a set of rules made up by human beings. Formulas can receive meaning, however. The formula we use to calculate the surface $S = l \times b$, has a minimal sense, restricted to the calculation of surfaces, but it can be used to calculate the surfaces of all kinds of rectangles, ranging from very tiny to very large. If a meteorologist states that the temperature is 12°C, I know I need a coat if I want to go for a walk. If I read on a lightbulb that it is 12 V, the number 12 has a totally different meaning. Both volts and degrees Celsius are not the result of counting. They are measurements taken out of the laboratory into the life world, and we have grown so accustomed to them that we use them without realizing what they really mean.

In other words, if the zero is inserted into a mathematical system, it means that it has reached a higher level of abstraction and is no longer a quantity. This event is not just a change in the understanding of zero, it is a transformation of the mathematical system of numbers as a whole. It implies the constitution of a calculating system, a structure of possible calculations, as a tool to explain and predict events in the life world. The addition of the zero to this calculating system reinforces it, but it does not change its character.

To say that the mathematical system is meaningless does not mean that it is useless; on the contrary, it can be applied in a great variety of practical situations. In the case of Brahmagupta, it was probably used for astronomical calculations, and perhaps by others at the time in bookkeeping.

We have seen that the zero did not reach a higher level of abstraction just by itself; its level of abstraction depends on that of the system it is part of. So when Brahmagupta described the rules for handling the zero, he implied the existence of an abstract mathematical system of numbers, one in which the zero became just an element. So the zero has not gained significance; on the contrary it has lost all of it. It no longer means anything; it is a sign that is completely defined by its position in the system. In this way it has become very useful and has made it possible to develop the different calculating systems that are the functional heart of all sciences.

## 6 Constitution

How does such a transition proceed? In *The Origin of Geometry*, Edmund Husserl tried to analyze this process. In Brahmagupta's time, mathematics was not yet the heart of science – the fusion of both took place in Europe – but was introduced in the work of Galileo Galilei (1564–1642). The use of zero as a number is made possible by a constitution (*Stiftung*) of it, by making it a cultural or scientific fact. This is a purely mathematical development by which numbers have become elements in their own independent field. Husserl thought this new development could not logically be deduced from external factors.

> Aus all dem ist nun jedenfalls zu erkennen, daß der Historismus, der das historische bzw. Erkenntnistheoretische Wesen der Mathematik von Seiten der magische Bewandtnisse oder sonstigen Apper/zeptionsweisen eines zeitgebundenen Menschentums aufklären will, ganz prinzipiell verkehrt ist. (Husserliana VI: pp. 385–86)

> [From this it is clear anyway that historicism is totally wrong if it tries to explain the historical or rather epistemological essence of mathematics by magical affairs or other such perceptions of an historical mankind.]

Like Nieder, Husserl tried to describe the development of a constitution in different steps: five in his case. These steps are different from the four steps of

Nieder, because the latter is convinced that he has to explain how we become aware of the zero as a quantity, which is a natural consequence of the fact that have expectations. Husserl wanted, however, to clarify how an ideal or cultural object like the number zero is introduced. It started, in his view, when at a certain moment someone suddenly experienced a change in perspective and saw clearly that zero was, in fact, just a number like all the others. She did not forget this moment and started to realize that this was a new point of view. This new discovery became a fixed idea that she remembered while practicing mathematics, and in this way it became a more or less solid theory. Now the discovery was ready to be discussed with colleagues. In the case of the zero one of those would have been Brahmagupta, who wrote it down. By this the constitution was completed (Husserliana VI, p. 370); alternatively it could become formalized as part of an oral tradition, as was the case with earlier Indian mathematics.[3] One of the consequences of the constitution was that the environment that made the opportunity of the constitution was covered up (Richir, 2011, p. 72). In other words, every time we treat the zero as a number it stops being a quantity.

In other words, the use of the zero as a number today is due to two elements: it is fixed in writing, and furthermore the knowledge and capability exists to understand the writing and to put it into practice (philosophers of science would speak of the transmission of a paradigm). Both elements are essential if we want to make sure that students do not have to repeatedly discover that zero can be treated as a number. They need textbooks and they have to go to college.

If the zero has lost sense, what sense did it have before, and apart from the use as a number in mathematics? As already mentioned, zero means absence, but absence of what? If we run out of bread, there is zero bread. In this case zero makes no sense by itself. It is possible, however, to summarize all these situations where we have run out of something, by saying that zero is the situation of absence of that which we are counting. In other words, zero as a quantity is the absence of something we expect to find, it is emptiness. It is not nothingness, because it is impossible to become aware of nothingness. Consciousness is always conscious of something; if it becomes conscious of nothing, it turns into a lack of consciousness. When you are thinking about something and you suddenly realize there was a moment of nothingness, you have a gap in your memory. If not and you can remember that moment, there

3 The first mathematical text that is preserved dates from 499 CE, but there were probably other texts in use before that time. In the early days of the Christian era many texts that until then had been orally transmitted were gradually put into writing.

has to be something to remember; if there is not, the obvious conclusion is that you dozed off.

The words 'zero' and 'emptiness' have no meaning by themselves; when we say that a room is empty, we mean to say that there is nobody there, or that there is no furniture. We do not necessarily mean to say that there is no carpet on the floor or no painting on the wall, or that there is a vacuum in the room. Besides, if emptiness were a quality of the room, it would reveal itself to someone who is just admiring the room. Emptiness only means absence to someone who is expecting to find something there. Jean-Paul Sartre mentioned a famous example of the absence of Pierre in the café where they had an appointment. Pierre was late, the café was empty of Pierre. The funny thing is that the day before Sartre was also in the café, but at that moment the café was not empty of Pierre (Sartre, 1943, p. 44). So if someone says a room is empty, different persons may have different ideas about what is meant. Perhaps this has caused the mistaken idea that there is something like emptiness in itself. Another source of confusion is the identification of emptiness with nothingness, the idea of absence of everything whatsoever.

The word 'emptiness' has made a stunning career in Buddhism, but from early on the texts warned not to interpret it in the sense of zero as a number or a quantity. Edward Conze's translation of probably the oldest text about emptiness, the *Prajnaparamita Ratnagunasamcayagatha*, dated at least a century BCE, says:

> What is this wisdom, whose and whence, he queries,
> And then he finds that all these dharmas (phenomena) are entirely empty.[4]

and:

> to imagine these skandhas (kinds of phenomena) as being empty;
> Means to course in the sign, the track of non-production ignored.
>
> (in other words, someone who merely imagines emptiness is just aware of signs and does not see that phenomena never arise)
> and:
> All dharmas are not really there, their essential original nature is empty.
> To comprehend that is the practice of wisdom, perfection supreme
> …

4 vyupariksate punar ayam katar'esu prajna kuto va imi sunyaka sarva-dharmah (verse 9).

> And just that merit is declared to be just worthless,
> And likewise empty, insignificant, void, and unsubstantial.[5]

The word 'empty' occurs 15 times in the text, so it is a very important concept. The last quotation by the way, is also the answer the Buddhist saint Bodhidharma is supposed to have given around the end of the fifth century to the Chinese emperor Wu of Liang, who asked how much merit he had gathered by his patronage of Buddhism.[6] Anyway, what is the meaning of this term 'emptiness' that is so important for the Buddhist redemption? For starters, it is called wisdom, so it is not just information, it is something that one must ponder on and understand. It is called the highest wisdom: *prajñapāramitā*. Someone who understands it will find that 'all these *dharmas* are entirely empty', so once emptiness is understood, it becomes obvious that all *dharmas* are empty. 'All *dharmas*' means all phenomena, everything that appears to us. Normally phenomena do not look very empty: they are material, or effects of solid matter. The chair you are sitting on is not empty, nor is your body that rests on it. So how can a kind of wisdom suddenly evaporate all this solid matter? It is also said that the phenomena are originally empty, so they may not appear to be empty now, but they do if they appear in their original nature. So this wisdom comprehends phenomena in their original nature. It is not just by imagination that it does this. Someone who imagines phenomena to be empty, for instance by considering that they are composed of electromagnetic strings, which is the current scientific model, is missing the point. The *skandhas* are the five classes of phenomena: physical appearances or impressions, feelings, perceptions, inclinations, and consciousness. They are the components of a person. It is important not to forget that a person and his world are two sides of the same coin in Buddhism. As in Western phenomenology, a person is not existing in his world like a car is parked in a garage. The world is not a location, but the very stuff of existence. Someone who is just imagining emptiness is restricted to signs, to conventions, and language. The emptiness of phenomena is beyond representation, it has to take you by surprise. And if it does, you realize that the phenomena 'are not really there'. Well then, if you have parked your car in a garage and find that somebody stole it, you would not say that it is not really there. You would say that the garage is empty and the car has gone. If you were to say that your car is empty, it would still have to be there. How else could it be empty? If saying something is empty amounts to

---

5 https://archive.org/details/Astasahasrika and http://www.dharmafellowship.org/library/texts/practice-of-holy-transcendental-wisdom.htm.

6 https://en.wikipedia.org/wiki/Bodhidharma.

saying it is not there, you could also say that the moon is empty because that is not in your garage either. So emptiness means that phenomena are there, but not really; they just appear to be there. They are 'nothing but phenomena', as the French philosopher Marc Richir called it. Perhaps we should conceive of this kind of emptiness as an ontological transparency.

Much of the discussion about emptiness is, however, not about the tradition of the *prajñapāramitā* texts, (several texts by this name exist, ranging from 20 to 100,000 lines), but about the work of the Indian philosopher Nāgārjuna, who lived about 250 years later. His most famous work is the *Mūlamadhyamakakārikāh*, (*Verses about the Root of the Doctrine of the Middle*). It is curious, however, that Nāgārjuna did not refer to the *prajñapāramitā* tradition, but used the word emptiness (*śūnyatā*) 20 times and the word 'empty' 30 times. Moreover, he put the word between brackets: 'All that is dependently originated we call emptiness. This is a figure of speech. This indeed is the middle way' (Chapter 24, verse 18, Hoogcarspel, 2005: p. 68). That Nāgārjuna agreed with the teaching of the *Ratnagunasamcayagatha* is obvious, and he too denied that it could be called a doctrine, a conceptual structure of propositions that are to be explained and justified in words by means of logic: 'The Victorious Ones have declared emptiness to be the transcendence of all doctrines. They have said, however, that those who adhere to a doctrine of emptiness are incurable (Chapter 13, Verse 8; ibidem: p. 39)'. He also explained the danger: 'Emptiness that has been misunderstood can cause the downfall of an unintelligent person, like a wrongly pronounced magic spell or a snake grasped at the wrong end' (Chapter 24, verse 11; ibidem: p. 68). Nāgārjuna gave another indication: he said that phenomena are empty of substance (*svabhāvaśūnyatā*, ibidem: pp. 7, 8, 25). What appears in dependency on something else and does not exist by itself is like a shadow that disappears when you change the light. The causes that make phenomena appear to most people as if they are substantial things are said to be twofold: it is the lack of wisdom and the involvement in life or in the world, which has in Western philosophy been called the will to live (Arthur Schopenhauer) or life as self-affection (Michel Henry). This has been called 'thirst' by the Buddha. The *prajñapāramitā* tradition goes back to the Buddha himself, as is testified by his adherence to the principle of dependent existence,[7] by the unanswerable questions[8] and by his rejection of the concept of being.[9]

7 Assutavā Sutta, Samyutta Nikāya 12.61.

8 Aggi-Vacchagotta Sutta, Majjhima Nikāya 72.

9 Kaccayanagotta Sutta, Samyutta Nikāya 12.15.

So the million dollar question is: where does the zero come in? I do not want to spoil the party, but I cannot find any single connection. Both in the *Ratnagunasamcayagatha* and in the *Mūlamadhyamakakārikāh* the words 'emptiness' and 'empty' are not used in the common way. They do not refer to an absence of things, nor to a quantity, and certainly not to a number. Both sources stressed that it is not a concept, nor even something that is imaginable or representable. Emptiness is not a doctrine; it transcends both logic and language, it is a necessary metaphor for a meditation experience. The emptiness of the *prajñapāramitā* tradition and the *Mūlamadhyamakakārikāh* is not a thing among other things, nor even a phenomenon among other phenomena, it is phenomenality itself, the way phenomena phenomenalize.

All those experts who wrote about emptiness and the tetralemma as an advanced kind of logic, and sometimes even linked it to quantum mechanics,[10] are secret Platonists who believe that what they conceive as the metaphysical foundation of the life world is a timeless truth, that every culture in every historical period in vain has been trying to discover and that science now has delivered. This is a mistake that has been part and parcel of the dominant tradition in Western philosophy since Parmenides in the fifth century CE. Parmenides claimed that thinking is connected with being and being is equal to truth, the latter being totally different from experience. In Western philosophy it is not uncommon to presume that logic is the art of right thinking. Logic is, in this view, the royal way that leads us to the truth. Aristotle firmly believed that logic reflects the structure of the universe, like the Brahmin priests believed that Sanskrit does. At the same time, it is clear from the history of logic that it is invented, just like mathematics for instance. Both are conventional, formal, self-contained, deductive symbolic systems and even mutually related. Following the rules, the logician makes a logical reasoning that is valid, they are not delivering any truth. Truth is the result of a right interpretation. The reasoning 'if A then B, A is the case, so B must be the case too' is valid. A valid reasoning may result in truth if the premises are true; for instance, the conclusion that Jane is a witch might be true if it is established that only witches can fly on a broomstick and Jane is flying on one. Truth is a relation to the world; logic does not need any world.

Graham Priest made a great analysis of the tetralemma (Priest, 2010, pp. 24–54). He showed that it can be formalized fully only in paraconsistent logic. I am not an expert in the field, but it seems to me that he has done a great job. The only problem I have with his article is that the early pioneers of paraconsistent logic lived in the late nineteenth century, so it is impossible

---

10 http://www.chinabuddhismencyclopedia.com/en/index.php/Nagarjuna and Kohl.

that this analysis makes anything clear about the intentions of the author of the *Mūlamadhyamakakārikāh*.

Bruno Latour gave a wonderful example of this problem in his article *On the Partial Existence of Non-Existing Objects* (Latour, 2000, pp. 247–269). In 1976 the mummy of Ramses II was investigated, and the conclusion was that the cause of death had been tuberculosis. Latour asked in his article how was it possible that someone could die from a disease that came into existence only 3,000 years later, because tuberculosis was not discovered until the second half of the nineteenth century. The answer is obviously that this is impossible unless the cause of the disease has been around all the time, waiting to be discovered. This would be the usual materialistic explanation; it just ignores history. Nieder seems to follow the same drift, suggesting that the zero has been there all the time, just waiting to be discovered and given its rightful function. The complete history of the world is, in his eyes, not much more than the advent of Western technology, the discovery of the number zero being a decisive axial point. The problem is that there is no guarantee that new discoveries in the future will turn things upside down again and show that the present state of science has it all wrong.

The answer to the enigma whether present scientific models and culture were true facts in the past or not, is to be found in the fact that developments like tuberculosis as well as the zero have been constituted, although perhaps not both quite in the same way, because the disease was preceded by a lot of research and the construction and invention of a lot of tools and machines.

Marc Richir noticed a few interesting consequences of any constitution (Richir, 2011, p. 73). To begin with, a constitution is a discontinuity in history, it is an invention, something new that could not be foreseen. Furthermore, it is a contingent conceptualization of the phenomenal world; therefore it is not right or wrong, it just might be more or less successful. Finally, it has no beginning. After the constitution has been achieved, it appears as though things have never been different. Both tuberculosis and the number zero have achieved the status of a timeless Platonic *eidos*; they are conceived as having been true facts of our world all along, although it was only at a certain point in history that they entered the scientific discourse. Latour expressed this in a sweet and short way: after the constitution of tuberculosis we became inhabitants of a new world. The problem of both Nieder and Priest is that they try to live in two worlds at the same time, mixing up both and acting as a ventriloquist using historical figures such as Brahmagupta and Nāgārjuna. This has happened frequently, because specialists in exotic languages sometimes neglect the historical and philosophical background of the authors they are translating.

## 7 From Onto-Logic to Phenomeno-Logic

Nāgārjuna was not aware of paraconsistent logic, but what was he up to and why did he use the tetralemma? Let us take an example. The first verse of the *Mūlamadhyamakakārikāh* goes:

> Never has there been anything whatsoever, that is arisen without cause, by itself, by something else or by both.[11]

This is a clear example of the *catuṣkoti*, or tetralemma. Something has a cause or not, there is no alternative, yet Nāgārjuna mentions four possibilities: arisen by a cause, without one, both, and neither. In fact, this happens to be a complex *catuṣkoti*, because first there is the dilemma of things having a cause or not, and the things that do have a cause are said to have three options: to have itself as a cause, something else, or both. The fourth option would be to have neither itself nor something else as a cause, which amounts to having no cause at all, and that has been covered by one of the options of the first dilemma. Anyway, the first dilemma already says it all: things both have and do not have causes, a flagrant rejection of the famous law of non-contradiction. Not only did Nāgārjuna ascribe the same quality, having a cause, to one thing while simultaneously denying it, on top of that he was talking about all things. Therefore, having and not having a cause is something all things have in common. Aristotle pondered a long time on the question whether general propositions about all things are possible and finally his answer was: yes, these propositions are ontological, because all things are. Nāgārjuna did not just tell us that some things both have and do not have causes, he said this is the way all things exist. Now we also know that he quoted the Buddha's rejection of being and non-being. These are extremes that are wrong according to the Buddhist view; the right way is the middle way, this is what the word *Madhymaka* means. The middle way between being and non-being is precisely just appearing. Things appear as having a cause and not having a cause simultaneously. In an ontological philosophy this is impossible because, as we saw before, this would violate the law of non-contradiction. According to phenomenology, the philosophy of appearance, however, this makes perfect sense. The chair I am sitting on needs nothing else to be what it is; even if it were brought into orbit around Jupiter, it still would be this chair. On the other hand, if it did not have causes,

---

11 10na svato nāpi parato na dvābhyāṃ nāpy ahetutaḥ | utpannā jātu vidyante bhāvāḥ kvacana kecana ||

it could never have been produced or destroyed, nor used for new purposes. In ontology, both dependency and independency are separated; in phenomenology they are combined and the result is a philosophy of pure or transparent appearance where phenomena appear as nothing but phenomena.

Therefore, Nāgārjuna's dilemmas and tetralemmas are not just logical tricks as Richard Hayes would have it (Hayes, 1994, p. 299), they are meaningful phenomenological descriptions. The unthinkability of them in ontological logic is not a flaw, it gives them extra rhetorical force that helps to liberate the reader from the shackles of traditional logic. Priest's analysis may be a fine job for a logician, but it misses the point entirely. In other words, it is a whig historical cover-up of what Nāgārjuna was trying to say. There are several verses that give away that Nāgārjuna was not working out a logical theory, but instead wanted to pave the way to *nirvāṇa*, for instance 5.8:

> However, the simpletons seeing existence and non-existence,
> do not see what there is to see: the blissful appeasement of things.[12]

Or 25.24

> The appeasement of all objects is the blissful appeasement of uncontrolled thought.
> The Buddha never taught any teaching to anyone.[13]

For those who maintain the idea that Nāgārjuna was trying to put forward a new doctrine about the nature of emptiness or zero, there is the following verse 13.8:

> The Victorious Ones have declared emptiness to be freedom of all doctrines. They have also said that adherers to a doctrine of emptiness are incurable.[14]

It all boils down to the idea that emptiness is not nothingness, it is phenomenality, the open happening of phenomena, which are nothing but phenomena. Nāgārjuna called it dependent arising, and an illusion, because every

12 astitvaṃ ye tu paśyanti nāstitvaṃ cālpabuddhayaḥ | bhāvānāṃ te na paśyanti draṣṭavyopaśamaṃ śivam ||

13 sarvopalambhopaśamaḥ prapañcopaśamaḥ śivaḥ | na kva cit kasyacit kaścid dharmo buddhena deśitaḥ ||

14 śūnyatā sarvadṛṣṭīnāṃ proktā niḥsaraṇaṃ jinaiḥ | yeṣāṃ tu śūnyatādṛṣṭis tān asādhyān babhāṣire ||

phenomenon pretends to be there all by itself, while being fully dependent on everything else. In fact, Nāgārjuna's emptiness is short for the expression 'being empty of existence by itself' (*svabhāvaśūnyatā*); it is not the emptiness of your wallet at the end of the month. Nāgārjuna's emptiness is an unavoidable metaphor, because the experience of the common emptiness, which is missing something you expect to find, is remotely similar to the experience of the phenomenality of phenomena.

## 8 Conclusion

My findings can be briefly summarized. The zero started out as a sign referring to a quantity, and it still is in many of our dealings with it. At a certain moment it was introduced as a placeholder, but that did not affect its sense, nor the arithmetic, so this is of minor importance. The big leap is the constitution of calculating systems, where numbers are no longer quantities, but elements whose meaning is defined by their place in the system. Calculating systems were made complete when the zero was inserted and mathematicians became aware of its special function in the system.

Zero as a quantity is associated with emptiness, the experience of missing something you expect to find. Zero as a number can only be associated with emptiness by crossing the gap and going back to zero as a quantity. In a calculating system every number is empty, because it pretends to have a value of its own, while being totally dependent on the other numbers. There is no doubt – it can be inferred from several verses he wrote – that Nāgārjuna was not involved in mathematics, but practiced Buddhist meditations in the *prajñapāramitā* tradition, where the experience of phenomenality as emptiness is an important step to *nirvāṇa*.[15] So any supposed link between the development of calculating systems and Nāgārjuna's metaphor of emptiness is non-existent, and the idea is a combination of superficial whig history and metaphysical speculation.

15 This misunderstanding, by the way, could be caused by some Chinese Buddhist texts, where Nāgārjuna's metaphor of emptiness is often mixed up with the older concept of emptiness in Daoism, though both have different names in Chinese. Daoist emptiness is wù (無). In the Dàodéjīng this kind of emptiness or nothingness is mentioned as the empty space in a window or between the spokes of a wheel. This is clearly no metaphor. The word for emptiness in Buddhist texts is often translated in Chinese as kōng (空), which also means free or open. When taken out of the context of meditation, especially in discussions about the influence of Zen Buddhism or Chán Buddhism on art, both conceptions of emptiness have however not been kept entirely apart.

## References

Access to Insight. https://accesstoinsight.org/tipitaka/index.html (for all Buddhist texts).

Conze, Edward. http://lit.lib.ru/img/i/irhin_w_j/prajnaparamita108/ratnagunasamcayagatha.pdf.

Brahmagupta. (1966). *Brāhmasphuṭasiddhānta*. Indian Institute of Astronomical and Sanskrit Research.

Hayes, Richard P. (1994). Nāgārjuna's Appeal, in *Journal of Indian Philosophy* 22: 299–378.

Hoogcarspel, Erik. (2005). *The Central Philosophy, Basic Verses*. Olive Press, Amsterdam.

Husserl, Edmund. (1976). *Die Krisis der europäischen Wissenschaften und die transzendentale Phänomenologie, Husserliana VI*. Martinus Nijhoff, Den Haag.

Kohl, Christian Thomas. (2007). Buddhism and Quantum Physics: A strange parallelism of two concepts of reality, in *Contemporary Buddhism* 8 (1): 69–82, https://philarchiv.org/rec/THOBAQ.

Latour, Bruno. (2000). On the Partial Existence of Existing and Non-existing Objects 1996, in *Biographies of Scientific Objects*, Lorraine Daston (ed.), Chicago University Press, 247–269.

Nieder, Andreas. (2016). Representing Something Out of Nothing, The Dawning of Zero, in *Trends in Cognitive Science* 20/11 November 2016, 831–42.

Priest, Graham. (2010). The Logic of the Catuskoti. *Comparative Philosophy* 2, 24–54.

Richir, Marc. (2011). Über die Phänomenologische Revolution: einige Skizzen, in *Phänomenologie der Sinnereignisse*. Wilhem Fink Verlag, München.

Ruegg, David Seyfort. (1959). *Contributions de la Philosophie Linguistique, Indienne*, E. de Boccard, Paris.

CHAPTER 28

# The Fear of Nothingness

*John Marmysz*

## Abstract

The Western fear of nothingness can be traced back to Thales, Anaximander, and Anaximenes, three ancient Presocratic thinkers who comprised the first school of Western philosophy: the Milesian School. Despite the varied and ephemeral nature of the world's appearances, the Milesian School suggested that there remains something stable, permanent, and dependable beneath it all. Whether it be Thales' claim that 'all is water,' Anaximander's claim that the universe arises from *apeiron*, or Anaximenes' claim that 'all is air,' the strategy pursued by these ancient Greek thinkers served to offer comfortable assurance that our cosmos has a steady and knowable foundation. The universe ultimately rests on one 'thing' rather than on nothing at all.

In setting this precedent, the Milesian School influenced later Western philosophers whose concerns concentrated on establishing fixed and substantial foundations for the world while also repudiating systems of thought emphasizing the primacy of nothingness. Such systems came to be criticized as 'nihilistic'; a moniker intended to highlight the negativity and meaninglessness of nothingness.

This chapter examines the logic of the Milesian thinkers in order to highlight the basic assumption shared by these first philosophers: nothing comes from nothing. This negative view of nothingness may help to account for why it is that the number zero was initially discovered in the East but rejected in the West.

## Keywords

Anaximander – Anaximenes – ancient philosophy – Milesians – nihilism – nothingness – Presocratics – Thales

## 1 Introduction

The fear of nothingness has deep roots in the West. Whereas Eastern 'emptiness' is commonly associated with spiritual peace and creative potential, in the West nothingness is more commonly associated with complete non-existence,

 | DOI:10.1163/9789004691568_032

oblivion, and the extinction of all value and meaning. In this regard, Westerners have traditionally conceived of nothingness as a dreadful and terrifying lack: something to be overcome and defeated rather than something to be embraced.

Michael Novak has observed that while the 'experience of nothingness' is itself a universal human concern, in the West it has tended to be solidified into a dark philosophical perspective called nihilism (Novak, 1970: 10). The first uses of the word nihilism date back only to the eighteenth century – when European authors such as Jacob Hermann Obereit, Daniel Jenish, and Friedrich Heinrich Jacobi used it as a criticism of Kantian philosophy's reduction of the 'real' world to 'nothing' – but the more general condition that the word describes can be traced back much further, to the very beginnings of ancient Greek Presocratic thought. With the Milesians, the first Presocratic school of philosophy, we find a group of thinkers who recoiled from the experience of nothingness rather than embracing it, thus initiating not only the study of physics, but also the advent of Western nihilism.

In what follows, we shall examine the nature of Milesian thought with an eye toward highlighting a basic assumption established by the ancient Greek thinker Thales and later reinforced by his students Anaximander and Anaximenes, philosophers who, although quite different in many ways, nonetheless rejected nothingness while holding that the universe is rooted in one physical 'thing' rather than in nothing at all. Insisting that the multiplicity of the world's appearances must be grounded by an underlying, unitary substance, these first philosophers established a perspective that Western philosophy has tended to adopt ever since; a perspective that, in its rejection of nothingness, may also, as Charles Seife suggests, be responsible for the ancient Greek rejection of the number zero (Seife, 2000, p. 40).

## 2 Something from Nothing

It is commonplace for those of us who have been raised and educated in the West to presume as a matter of common sense that, as Lucretius insisted, 'nothing can come from nothing' (Lucretius, 2003, p. 19). This is a first principle so seemingly unquestionable that its denial strikes many people as plainly absurd. Declaring that the various tangible and perceptible properties of our empirical world originated from nothing would be tantamount to resigning ourselves to the belief that there is no final and ultimate explanation for all that we see, taste, touch, and feel. After all, if our world arose out of nothing, then any attempt to trace the universe's chain of cause and effect back to its origin would leave us empty-handed, and some of the most profound questions that philosophy asks would fall flat. There would be no reason, ultimately, for anything being the way that it is, and as a result we would be reduced to silence when trying to articulate the origins of our universe. Traditionally for Westerners, this has been considered an unacceptable prospect, thus constituting a sufficient reason for rejecting the absurd power of nothingness.

Nevertheless, prior to the advent of the Presocratics, many ancient thinkers did attempt to convey how it is that something could come from nothing. For instance, we find in the mythic accounts of the Babylonians, the Hebrews, and Hesiod bold efforts to articulate the generation of the universe out of a void. Normally, these accounts rely on anthropomorphism, characterizing the origins of things in terms of human-like actions carried out by supernatural entities that either arise out of a chaos of nothingness or that manipulate that chaos and form it into something. The problem with these sorts of explanations is clear, for none of them tell us: 1) where the supernatural entities themselves come from; or 2) by what mechanism nothingness became something. If a myth does not tell us where the first supernatural entities came from, then it does not really provide an account for the origin of *all* things. And if it can't articulate the logic or the process by which nothing became something, then a myth does not really answer the question of *how* the world came to be. Consequently, a full account of the primal origin of all things remains concealed and unanswered in traditional mythology. The best that can be offered by these accounts is that, in the beginning, something happened that exceeds our capacity for logical understanding. Nothingness somehow gave birth to our world, but why or how this occurred must remain mysterious.

Take, as illustration, the opening lines from the English Standard Version of the *Book of Genesis*:

> In the beginning, God created the heavens and the earth. The earth was without form and void, and darkness was over the face of the deep.
> And the Spirit of God was hovering over the face of the waters.
> And God said, 'Let there be light,' and there was light. And God saw that the light was good. And God separated the light from the darkness. God called the light Day, and the darkness he called Night.
> And there was evening and there was morning, the first day. (Genesis 1:1–1:5)

In this, the Hebrew creation myth, we find an already existing God presiding over a dark, formless void. The inexplicable nature of the void is reified as deep water – presumably because readers would find it impossible to picture sheer nothingness – while God's power of creation is represented in terms of speech. His words bring light and day into existence and, as the story of creation progresses, God continues to speak, while the void yields to His commands. Yet, nowhere do we find an explanation of where God Himself came from, nor do we find an explanation of what it is about His words that are potent enough to heave a universe out of nothing at all. It may be that creation *ex nihilo* is here intended to convey the infinity of God's power, yet it still does nothing to articulate the mechanism that makes this possible. Creation is explained as an act of willfulness, lacking a logic or a strategy that can be understood by the human mind.

This creation myth is Eastern in origin, dating back at least to the sixth century BCE, though probably relying on a much older tradition (Porter, 1993, p. 140). In it, the fertile potential of nothingness is emphasized, and yet this fertility requires the intervention of an extant being in order for it to be teased out and developed. The genesis of such a being is not itself explained. God is portrayed as existing contemporaneous with the void. It is not that God is dependent on the void, nor that the void is dependent on God. Rather, the two exist together, in contrast with one another, as Being stands to nothingness. This same relationship can be found in many ancient creation myths, illustrating the conundrum faced by anyone attempting to describe the origin of all things out of nothing. To even begin the account, there must be something that already exists in order to coax something from nothing. But this leaves open the question as to where that already existing 'something' came from.

Dissatisfaction with this kind of open question is precisely what motivated the Presocratics to rebel against mythic thinking in favor of an approach that has ever since influenced the development of Western philosophy and science. With the first of the Presocratics, the beginning of the universe was explained not by reference to something coming from nothing, but by all things coming from some pre-existent substance: one 'thing' that underlies and supports all other things that we can see, taste, touch, and feel. The world was thus not underpinned by a void of nothingness, but sat stable and secure upon an eternal, substantial foundation.

## 3 The Presocratics

Traditionally regarded as the first Western philosophers, the Presocratics established a precedent that has guided the development of philosophy in the West ever since. Their unique blend of materialism and logic encouraged a rebellion against myth-making, initiating an early step in the development of the natural sciences. In the process, the example they set may also have introduced the fear of nothingness into Western philosophy.

The evidence used to reconstruct Presocratic philosophy is fragmentary, and there is quite a bit of creative interpretation involved in attempts to understand their doctrines and beliefs. Much of what we know about the earliest Presocratic thinkers comes to us second-hand, through the writings of Plato and Aristotle, who of course had their own philosophical agendas. But despite the challenges involved in trying to understand them, there does seem to be a general consensus on some of the basic ideas that made the Presocratics revolutionary thinkers.

The first of these ideas is Presocratic materialism. Instead of supernatural explanations, these first philosophers sought to account for the origins of things in terms of material causes and effects. Whereas mythic thinkers told stories about gods and goddesses who willfully brought the universe into existence through magical or coercive means, the Presocratics made appeals to material elements and their natural qualities. Water, air, fire, or earth replaced the gods as the source of what we see, taste, touch, and hear. And since material elements follow impersonal patterns of cause and effect, the world itself came to be viewed as predictable and mechanical, rather than as capricious and arbitrary.

Materialism may have served as a way for the Presocratics to avoid the unpredictability and instability of explanations relying on the void of nothingness. Nothingness has no predictable patterns or governing laws, and so if the universe arose out of nothing, it too must retain an element of volatile mystery. Material substances, on the other hand, possess knowable and tangible qualities that are relatively stable and regular. If the world is made of such substances, then it too must possess those qualities, and thus humans should be able to comprehend and potentially control that world. By replacing the void with matter, the Presocratics shifted emphasis away from nothingness as the generative source and directed it toward a 'thing' that was substantial, tangible, and perceptible. Nothing was replaced with something and, for many of the most influential of these thinkers, that something was one thing. Instead of zero being the starting point, all things were thought to begin in one material substance.

### 3.1 *Thales of Miletus*

Thales of Miletus is a perfect illustration of this materialist emphasis in ancient Greek Presocratic thought. Although Thales never wrote anything, Aristotle tells us that his importance lies in the assertion that 'all is water' (Aristotle, *Metaphysics* 983b, pp. 6–11). Strange as this may sound, its significance is that water is a natural, material element whose known qualities can be appealed to in order to explain the transformations occurring in our empirical world.

Water can take on differing forms, such as liquid, solid, and gas. If it is frozen it becomes hard. If it is boiled, it vaporizes. These differing empirical states mask an underlying unity of substance, and that single, underlying substance, ruled by the laws of nature, systematically and predictably responds to the forces that are applied to it. To say 'all is water' is to say that the world is, in its deepest essence, one thing, and that this one thing has knowable properties and qualities that are subject to natural laws. Whereas the nothingness of the myth-makers cannot accurately be visualized or imagined, a person can picture a single watery universe in the imagination. This image can be held in the mind, offering a kind of stable foundation that is at once comprehensible and comforting.

Following from this materialist principle, is the implication that the universe is not a vast irrational, incomprehensible place, but one that is governed by causal patterns that adhere to a consistent logic. Unlike with supernatural entities, the objective laws governing material substances are not haphazard or subject to moods, emotions, or passing whims. Water doesn't get angry, sad, or feel jealous. It always behaves predictably, freezing or boiling at precise temperatures. Thus, in positing water as the underlying substance out of which everything else arises, Thales also posits logical predictability as a natural part of our world. Reality unfolds according to rules and regularities. It is, in a word, rational. Both Bertrand Russell and Wallace Matson highlight this rational and logical aspect of Thales' philosophy as perhaps his most important contribution to the West (Russell, 1945, pp. 28–29; Matson, 1987, p. 10). With it, the world starts to take on the characteristics of an object rather than the characteristics of a human. It becomes a thing rather than a person, and more importantly, it becomes a thing that explains all other things.

Water, according to Thales, did not spring out of nothingness, but is rather the eternal, uncreated, and indestructible first source out of which all things arise and return. It is something like what followers of the Abrahamic religious tradition call God, but with a difference. Whereas the God of the Jews, Christians, and Muslims is a willful, caring, and good creator who generates the universe out of a void of nothingness, for Thales, water is neither willful, nor caring, nor good. It simply is. Crowding out the nothing, water replaces the void as the singular raw material from which all things emerge by necessity rather than by deliberate intention. In the Abrahamic tradition there is an opposition between the singular God and the void: a dualism between Being and nothingness. For Thales, on the other hand, there is a thoroughgoing monism that dispenses with everything except the first principle of water. There is no 'nothing' for Thales, only Being.

This, of course, runs us into the same problem that we encountered in myth-making. Just as myth-makers could not give an adequate account of the origin of the creator god(s), so it seems Thales cannot give an explanation of water's origin. In this way, we might conclude that his account is no more or less satisfying than those formulated by the Babylonians, the Hebrews, or Hesiod. In the end (or the beginning) we are left mute when it comes to the fundamental question concerning the origin of *all* things. Perhaps Thales is no different from any other myth-maker, except that he replaces God with water.

Friedrich Nietzsche nevertheless insists that Thales was in fact different, and rightfully called the first philosopher, although not due to his materialism or his use of logic. According to Nietzsche, while Thales' materialist explanation of reality makes him a natural scientist, what really makes him the first philosopher, and thus different from previous myth-makers, is his claim that 'all things are one' (Nietzsche, 1994, p. 39). This claim is rooted in his assertion that all is water, but it also goes beyond it in drawing a non-empirical generalization concerning cosmic unity. His speculation about the world's watery nature may still be hampered by a kind of ancient picture-thinking – probably influenced by earlier myths that also picture the cosmic origin as a watery chaos – but it also initiates a movement away from picture-thinking by leaping into complete monism. It is his monistic view, conceiving of the cosmos as one 'thing,' that initiates the Western philosophical tradition. In his attempt to conceptualize Being as a singular whole, rather than as an opposition between two things (Being and nothingness), Thales becomes the author of a perspective endorsed by an entire tradition of thinkers up through present times. As Giovanni Reale states, this perspective is concerned with knowing 'the universal in which all particular things are subsumed, that is, that universal which gives meaning to the particulars, unifying them' (Reale, 1987, p. 307). This monistic perspective insists that the universe is rooted in some underlying, stable substance tying everything together into a single whole that can be grasped and understood by the human mind. And while it still leaves open the question of how and why this singularity came into existence, it nonetheless reduces reality to one thing rather than two, and in the process this one thing crowds out nothingness altogether.

Thales is regarded as the founder of the Milesian school; a way of thinking about the universe that is commonly regarded as initiating the Western tradition in both philosophy and science. While it is certainly not the case that mythic, religious, or other forms of thought died with the advent of the Milesians, it is clear that their influence has been tremendous, priming many of those who came after Thales to adopt his basic assumptions and then work through the implications of those assumptions in their own unique ways.

### 3.2 *Anaximander of Miletus*

One of the most significant of Thales' followers was Anaximander, a man whose actual words comprise the oldest surviving fragment of Western philosophy (Barnes, 1987, p. 74; Heidegger, 1975, p. 13):

> Whence things have their origin, there they must also pass away according to necessity; for they must pay penalty and be judged for their injustice, according to the ordinance of time. (Heidegger, 1975, p. 13)

This fragment, taken out of context, appears mysterious, strange and perhaps incomprehensible. What could Anaximander possibly mean when he wrote about 'penalty' and 'injustice' in connection with the origin and the passing away of things? What could these moral concepts have to do with a world that is generated and that decays according to laws of 'necessity'? If we understand him as a student who was influenced by Thales, however, what we find here can be seen as illustrating the next step in the Greek resistance to nothingness. In fact, according to Nietzsche this passage 'takes two steps beyond' Thales by addressing, first, the question of how the many can be generated from the one and, second, the question as to why it is that the one has not already passed away into nothingness (Nietzsche, 1994, p. 49). In tackling these questions, Anaximander further entrenches himself in the conviction that the universe is ultimately one thing that is self-generating and self-sustaining, with no room for non-existence.

If all things come from one thing, then this implies that the multitude of things that now exist must somehow be capable of resolution back into the one thing from which they arose. So, if we accept Thales' speculation that 'all is water', it should follow that the multitude of things in our world – air, fire, and earth among them – should be harmonious and continuous with water. They must, in some sense, be capable of coexisting with their watery source. While it is unclear as to whether or not Thales himself considered this inference (Lloyd, 1970, p. 19), it was suggested by Aristotle that Anaximander was inspired by just this thought, and came to the conclusion that Thales' teachings were inadequate to account for it: 'air is cold, water moist, fire hot; if one were infinite, the others by now would have ceased to be' (Aristotle, *Physics* 204b, pp. 26–28). Water, in other words, cannot possibly be the one single substance out of which all things arise since water possesses properties that are incompatible with other elements that also exist. The 'self-contradictory, self-consuming and negating character of the many' (Nietzsche, 1994, p. 49) led Anaximander to the conclusion that Thales was wrong. If water was the origin of all things, then nothing other than water would now exist. Infinite water would have long since extinguished all fire, dampened all air, and dissolved all earth. Everything would have collapsed back into a watery abyss. Yet things other than water do in fact exist. Thus, water cannot be the singular origin of everything.

From this line of reasoning, one might plausibly conclude that all things do not, in fact, arise from one source at all, but from many. Yet Anaximander recoils from this conclusion, instead retaining his teacher's assumption that there must be a single, unitary, underlying substance that grounds all of reality. Why retain this assumption? Aristotle gives us a clue. If there is, in fact, a finite diversity of contradictory (or contrary) elements in existence, then they would have to 'always balance' (Aristotle, *Physics*, 204b, p. 15) in order to avoid the sort of eradication discussed previously. A tip in any direction – more water, more fire, more air, more earth – would result in the most common element 'prevailing' over the others. But, if the elements are all finite and balanced, then it is impossible for them to fill infinity, since there would always be a remainder of nothingness left over. A finite number of elements cannot fill an infinite universe. Therefore, all things cannot come from numerous finite elements, nor can they come from one of the elements in infinite quantity.

The three assumptions that guide this line of reasoning are: 1) The universe is infinite; 2) Earth, air, fire, and water exist as discrete, yet opposed elements; and 3) Nothingness does not exist. Anaximander might have abandoned any one of these assumptions, but chose instead to retain them all, going on to reason that in a world thus constituted, there must be one infinite, invisible substance, not opposed to any of the finite visible elements, but which supports and underlies them all. Since such an infinite substance would have to be capable of sustaining opposing, seemingly contradictory elements, it must be ambiguous in its make-up; neither fully earth, air, water, nor fire. Anaximander called this substance *apeiron*: 'the boundless', 'the unlimited', 'the infinite', or 'the indeterminate'. *Apeiron* is eternal, uncreated, and immortal. It never came into existence and it cannot go out of existence. It is the singular source of all things. Anaximander's speculation addresses how it is that the many might be generated from the one. He suggests that *apeiron* is in motion, and that due to its natural movements, the various elements it supports become separated from one another, as in a centrifuge, with the lightest elements (like air and fire) being drawn outwards, while the heavier elements (like earth and water) are drawn inwards. This is why water lies on the surface of the earth, while air fills the skies and fire shines from the sun and the stars. Each element inhabits a differing region of Being, and so they all are able to coexist in the same world at the same time. This avoids the sort of inconsistency found in the account of

Thales, who insisted not only that all things are composed of water, but that seemingly contradictory elements coexist with one another in the same place and time.

Anaximander's speculation also addresses why it is that all of the seemingly opposed things in our world have not passed away into nothingness. Since the infinity of the universe is grounded not by any single element possessing a determinate or particular set of qualities, at the most fundamental level, there is no strife or discord in the heart of Being itself. *Apeiron* is not in conflict with anything and so, considered from the grand perspective, it does not wither or decay. In its movements there is a never-ending generation and decline of apparent phenomena, but in its hidden essence it always remains what it is: unbounded and indeterminate. Like whirlpools in a lake, the phenomena of existence make their appearance, but the underlying substance of the universe, like lake water, remains firm and established.

Another question remains, however. What is it that Anaximander intends when he claims that the things in the world 'must pay penalty and be judged for their injustice, according to the ordinance of time'? According to Heidegger, this is in fact the most important part of the fragment, constituting the only 'immediate, genuine words of Anaximander' (Heidegger, 1975, p. 30). But it is also the most puzzling, as, in contrast to the materialism and naturalism found in the rest of the passage, these words introduce an ethical/moral dimension into Anaximander's thought. Many interpreters are quick to dismiss this part of the passage as metaphorical or merely poetic. However, it may be that something more than mere style is underscored here, revealing aspects of Anaximander's own assumptions concerning non-being in an especially dramatic manner.

If we look at the structure of Anaximander's fragment, it appears to have the form of an argument. The word 'for' (after 'necessity') seems to act as a premise indicator. If that is the case, then the second part of the fragment – the part making apparent moral claims – constitutes the premise, and the first part of the fragment – the part making claims about physics – constitutes the conclusion. If this is correct, then Anaximander is arguing that the reason why things must pass back into that from which they take their origin is *because* those same things are unjust and must be judged to pay a penalty. The existence of things is a corruption and the necessary penalty is reabsorption into the infinite. So what is the injustice here? What is wrong with things that emerge out of *apeiron*? Why must they be judged and returned to their origins?

Nietzsche suggests that the 'injustice' Anaximander is calling our attention to is the 'fall from being to coming-to-be' (Nietzsche, 1994, p. 50). As things separate out from the infinite unity of *apeiron*, they fall from Being, moving from the realm of the eternal and stable into a state of impermanence and flux. This 'fall' constitutes an 'injustice' that must be penalized precisely because it is a degradation of the sublime nobility of *apeiron*'s magnificent unity. In other words, the preferred, most noble mode of existence according to Anaximander is the stable, unified presence of Being itself, not the ephemeral, unstable state of flux and change. To 'be' in the fullest, most legitimate sense is to perdure, to remain stable and fixed, like *apeiron* itself. The various temporary things in our world, then, are 'unjust' and must be 'judged' due to their ephemerality and lack of permanence. The 'penalty' for this lack of permanence is to pass back into the eternal, leaving the realm of the temporary and unstable behind. Reabsorption into the infinity of *apeiron*, then, is the ultimate form of justice, which makes recompense for the instability of impermanent existence.

But this raises the question as to why stability is more 'just' than instability and change. What is so great about unchanging Being, and what is so wrong with impermanence? Here we encounter the crux of the issue. Western philosophy, beginning at least with the Milesians, stands firm in the belief that to be is better than not to be, and so that which exists in the most stable and permanent fashion is better (and more just) than that which exists in a less stable or less permanent fashion. This appears to be the case because stability, on the one hand, is equated with Being while change, on the other hand, is equated with nothingness. Nothingness is the real enemy here. Standing in opposition to the nobility of Being, nothingness steals something away from it; it degrades and erodes its integrity, acting as a negative force of destruction. Thus, whatever allows the scourge of nothingness to creep into our world acts as an accomplice in a crime against Being. And this is what temporary, changing and ephemeral 'things' do. While they partially exist, things that come into and go out of existence also partially do not exist. They hover in an indeterminate state between reality and illusion, opening a door to the void and inviting non-being to make an appearance. For Anaximander, this is an unjust insult that must be punished.

Anaximander's thought continues the tradition of his teacher, Thales, working out the implications of a monistic metaphysics: a metaphysics rooted in the assumption that the universe must find its origin in one thing rather than in nothing at all. In doing so, he 'set the major questions for later pre-Socratics' (Graham, 1999, p. 29) while also establishing the goals for Western philosophy to come. After Anaximander, Presocratic thought (and Western thought in general) presumes that the presence of things in our world implies that there must be a pre-existing foundation from which those things emerged. You can't get something out of nothing, as Lucretius asserts, and so the existence of anything at all already suggests the reality of an eternal, stable, and fundamental cosmic foundation that lies beyond appearances. Additionally, Anaximander imports an ethical assumption into this monistic worldview that also continues to affect the development of Western thought: to be is better than not to be. Nothingness is an 'injustice' that must be conquered and eradicated by pure Being itself, which, in its battle against nothingness, constantly reasserts its dominance through its own eternal presence. While particular things come and go, Being remains, absorbing and reconciling the contradictions and inconsistencies found in the 'unjust' world of mere becoming. Being endures these changes and forever resolves them into a superior, overarching unity. The task of philosophy since Anaximander has been the contemplation of this noble unity.

With Thales, the West was first introduced to a thoroughgoing monism that established Being as one thing out of which all other things emerge. Water, eternal and indestructible, replaced the gods and nothingness, forming a cosmic unity with no place for the void. Anaximander, in the course of criticizing his teacher, advanced Thales' monism, giving a logical account for how it is that one eternal element, *apeiron*, accounts for the existence of the diverse and contradictory sorts of objects and qualities that make an appearance in our empirical world. But he also introduced a moral judgment into his cosmology that ranks the underlying unity of Being as superior to its various empirical manifestations. These two Milesians together established a new worldview, one in which the cosmos is understood as comprising an unseen moral unity. It was up to the third and final member of the Milesian school of philosophy, Anaximenes, to formulate a quantitative account of the mechanisms that operate to fracture this moral unity into a diversity.

### 3.3 *Anaximenes of Miletus*

Anaximenes' speculations are often deemed less innovative, and thus not as important, as the contributions of Thales or Anaximander (Graham, 1999, p. 29; Reale, 1987, p. 46). Whereas Thales proposed water as the singular first principle and Anaximander proposed *apeiron*, Anaximenes tells us that the first principle is actually air. Superficially, it may appear that what Anaximenes has done here is simply to switch out one element for another, not really contributing much to the development of the Milesian worldview beyond that. However, if we ask ourselves *why* he felt it necessary to introduce air as the first principle, his contribution to this tradition – and his influence on Western thinking as a whole – becomes much more notable.

While Anaximander did offer an account of how the one becomes the many, he did so in a way that makes reference only to the apparent *qualities* of *apeiron*'s inner properties. According to Anaximander, the four elements – earth, air, fire, and water – are separated out from an indefinite mix by the motions of the cosmos, coming to rest in their own separate regions of the universe and thus forming the world of impermanence and change that we experience through our senses. It seems that in Anaximander's conception, all of the differing qualities of these elements (wetness, dryness, hotness, coldness) were in some sense already there, embedded in the one undivided, boundless substance that, in its unified, ambiguous aggregate, is the only truly noble reality. Those qualities make their appearance because they become separated through motion, but what is it about motion that makes these differing qualities manifest? What is the precise mechanism through which the hot and cold, the wet and dry become differentiated from one another out of the elemental stuff that is *apeiron*?

The problem with positing *apeiron* as the first, singular cosmic principle is that it is so occult that its inner workings must forever remain mysterious. Recall that the reason Anaximander initially proposed it as the primal source was to account for the existence of the various conflicting qualities that appear in our world. If all things come from one thing, and if the qualities of the various elements conflict with one another, then all things cannot come from any one of the elements. They must, instead, come from an ambiguous substance that encompasses and contains all of the contradictory qualities in existence. But if this is so, then there arises a problem with articulating the nature of this first principle fully and in a rational way. To say that *apeiron* is both hot and cold, or wet and dry, is like saying I am both here and there, or that an apple is also an orange, or that squares are also circles. Such assertions defy logic. Declaring that *apeiron* is the origin of all conflicting qualities, thus, relegates it to the mystical: a realm in which language and human comprehension cease to function. To say that everything originates from the undivided boundlessness of *apeiron* is not much different from saying that all things originate from God, since both God and *apeiron* work in mysterious ways. In this sense, Anaximander's *apeiron* also seems to share a similar nature with the Tao of the *Tao Te Ching*, which cannot truly be spoken of or rationally understood. (*Tao Te Ching*, 1). Both substances are, in their essence, utterly uncanny.

Anaximenes may have regarded this as a problem and so he began a search for some other, less mysterious substance out of which the empirical world may have emerged. Note that he never departs from the Milesian assumption that all things must come from one thing. Like his predecessors he is a monist, assuming that all things ultimately are unified. His own innovation was to suggest that the one substance underlying all others is air in varying *quantities* of rarefaction and condensation.

All living things need air to survive, and so Anaximenes' inference that everything is made of air was perhaps related to Thales' own thought process in speculating that everything is made of water: an element that is also required for life. However, unlike Thales, who used water's *qualitative* changes from liquid to solid to gas in order to account for the various things that exist in our world, Anaximenes instead suggested that the *quantity* of air packed into a given area gives an explanation for why things exhibit the qualities that they do. So, for instance, when air is densely packed together, it exhibits the qualities of solidity and coldness. When it is loosely packed together, it exhibits the qualities of slackness and heat. This explains how the apparently conflicting elements of earth, water, and fire might be formed from the same underlying substance by making an appeal not to the empirical *qualities* of the substances themselves, but by appealing to the *amount* or *quantity* of a fundamental substance that is present in a particular space. Plutarch suggests that Anaximenes may even have offered experimental evidence in support of his theory by observing 'the breath is cooled when it is compressed and condensed by the lips, but when the mouth is relaxed and it is exhaled it becomes hot by reason of its rareness' (Barnes, 1987, p. 79). Compress air tightly in your mouth then blow it out vigorously onto the back of your hand and it feels cold and hard, like earth. Allow rarefied air loosely to flow out of your mouth onto the back of your hand and it feels hot and slack, like fire.

Leaving aside the obvious problems with such an experiment, the general speculation that Anaximenes offers is one that has continued to be very influential in the history of Western philosophy and science: namely that the qualities we experience in our world can be explained and understood in terms of quantifiable differences arising in some sort of underlying substance. In proposing this, Anaximenes' system finally gives a consistent articulation of the dynamics of the universe based on number and amount, producing a naturalistic philosophy 'fully coherent with its premises' (Reale, 1987, p. 47). All things are really one thing, and appearances to the contrary are best explained by more or less of this one thing being packed into the various regions of Being.

Nothingness does not exist for Anaximenes. Our universe consists of a more or less densely packed 'something' that crowds out the void and occupies the infinity of Being. The universe is based in the one, and the many are just multiplications of this one, leaving no room for zero.

## 4 The Fear of Nothingness

The Milesian thinkers Thales, Anaximander, and Anaximenes are customarily regarded as comprising the very first school of Western philosophy: a school of thought that has had a major influence on the development of Western philosophy and science up to the present day. Among their major contributions to the tradition are the introduction of materialism, naturalism, and the use of reason and logic in explaining the world. But perhaps even more influential is their monistic approach to reality. As Nietzsche states, the Milesian school begins with faith in the contention that 'all things are one' (Nietzsche, 1994, p. 39), and it is this, he claims, that makes these early Western thinkers truly philosophical. But there is a cost that goes along with the adoption of such a thoroughgoing philosophical monism. If all is one, then there is no such thing as nothingness, and if there is no such thing as nothingness, then the empirical world of impermanence, which is infected by nothingness, is an illusion that deludes us about the real nature of the universe. By tracing all things in the cosmos back to one thing, the Milesians set the stage for a perspective that regards the only true reality as that which is permanent, present, stable, and never changing. If this is so, then there is only one real thing, and that one thing is something that can be known only in pure thought, if it can be known at all.

The equation of Being with unitary, eternal, noble stability, on the one hand, and of nothingness with unjust impermanence, on the other, initiated Western philosophy's fear of nothingness. The Western tradition generally holds that Being is better and more just than non-being precisely because it is only Being that is stable and present. But why are stability and presence better than non-stability and non-presence? Why not value non-existence over existence? For Westerners, the answer to this question has traditionally been found in the presumed negativity and destructiveness of non-existence. Whereas Eastern thought has tended to find something creative and productive in nothingness, Westerners have instead tended to see it as a dead end, a dreadful lack out of which nothing at all can be generated. The assumption that nothingness is inevitably partnered with oblivion seems to be at the root of the common Western contention that out of nothing comes nothing. The fear of nothingness grows out of this assumption. Nothingness is fearful because is it at odds with, and thus threatens, the integrity of Being.

Nishida Kitaro writes: 'I think that we can distinguish the West to have considered being as the ground of reality, the East to have taken nothingness as its ground' (Carter, 2013, p. 36). In the West, philosophies that embrace nothingness have tended to be disparaged as 'nihilistic', suggesting not only that they perversely value non-existence over existence, but that because of this they are also permeated with a dark negativity that is hostile to life, creativity, and constructive insight. If nothing comes from nothing, then to embrace nothingness, according to this way of thinking, is to embrace absolute oblivion and extinction. It is to abandon all hope for progress or positive creation.

When we encounter impermanence in the empirical world, our first inclination may be to despair over the ephemerality of existence. Because they come and go, ephemeral things might seem not fully 'real', and thus strike us as valueless. For this reason, we may be tempted to denigrate them, coming to the conclusion that no particular thing is important or of value. This is a negative conception of nothingness that Nishitani Keji calls 'relative nothingness', a nothingness that is discovered through the negation of things (Carter, 2013, p. 115). This form of nothingness is 'relative' because we come to understand it only in relation to the fleeting existence of beings. In the West, it has tended to be assumed that this form of nothingness exhausts the concept altogether. Nothingness in this sense consists only in the absence of all things.

But consider the *Dhammapada*, in which Siddhartha Gautama states: 'When you have understood the destruction of all that was made, you will understand that which was not made' (*Dhammapada* XXVI). Here it is proposed that in thinking the obliteration of all particular things, a path is opened toward a different conception of nothingness; one that constitutes the underlying ground out of which all ephemeral things come and go. In this we find an affirmative understanding of nothingness. In order for things to come into being, according to this tradition, there must already exist an emptiness within which those things can make their appearance. This emptiness is the backdrop against which the entirety of the empirical world unfolds. It is not something to fear or something from which we should recoil precisely because it is not something that is 'unjust' or corrupt. It is, rather, a powerfully fecund 'hollow' (Nishitani, 2006, p. 143) or a 'field' out of which 'every thing arises, in its suchness' (Carter, 2013, p. 118).

Nishitani Keji calls this productive form of the nothing 'absolute nothingness'. Unlike relative nothingness, it exists apart from particular things, and so is not the same as the nothingness understood merely by clearing all beings away. It is, instead, a positive potentiality that is the precondition for the existence of beings. In absolute nothingness, the ground of all beings is revealed not to be a thing in itself, but a 'place where beings "be"' (Carter, 2013, p. 115). Just as emptying a room of all of its furnishings does not eliminate the room itself, so it is that in thinking beyond the existence of all particular beings, we do not necessarily decimate Being altogether, but rather find that Being itself is just the emptiness of nothingness. In this sense, Being is more like a cavity than a foundation; an open space waiting to be populated by beings. Without this hollow, there would be no place for beings to exist, and so the emptiness of nothingness becomes the positive precondition for all things. In this positive understanding, nothingness is not so much a lack as it is a constructive potential. It is not the oblivion of beings, but the necessary requirement for their existence in the first place. This Eastern concept of a positive nothingness is called *Śūnyata*, or 'emptiness', and it offers a productive counterpoint to the Western conception of nihilism. As authors such as D. T. Suzuki and Huston

Smith point out, nothingness thus conceived is neither negative nor unjust, but indicative of 'boundless life itself' (Smith and Novak, 2003, p. 52). It is a far cry from the nothingness that was resisted, and feared, by the Milesians.

## 5 Conclusion

Western philosophy is traditionally thought to begin with the Milesian thinkers of the sixth century BCE. Their influence is evident in the common Western assumption that nothing can come from nothing, and thus that the existence of our universe must be grounded in an eternal, singular substance. The assumption that all things must emerge from one thing may in fact help to explain the ancient Greek rejection of the number zero. In their fear of the destructive power of the void, and in their consequent embrace of materialist monism, the Greeks may have been predisposed to ground their thinking in terms of singularity rather than nothingness. As Charles Seife writes, among the Greeks '... zero was inexorably linked with the void – with nothing. There was a primal fear of void and chaos. There was also a fear of zero' (Seife, 2000, p. 9).

However, in the East, where nothingness tended to be associated with creative potential, this same fear did not exist. With *Śūnyata*, or creative emptiness, Eastern philosophers conceived of the nothing not in negative, destructive, and thus threatening terms, but in positive terms, as a ground that allows for the unfolding of empirical reality. This, in turn, may have primed them to be more open to the usefulness of the number zero, and so perhaps it is no coincidence that its discovery occurred in the East rather than in the West.

## References

Aristotle. (1941). *Physics. The Basic Works of Aristotle*. New York: Random House.

Aristotle. (1941). *Metaphysics. The Basic Works of Aristotle*. New York: Random House.

Barnes, Jonathan (ed.). (1987). *Early Greek Philosophy*. New York: Penguin Books.

Carter, Robert E. (2013). *The Kyoto School: An Introduction*. Albany: SUNY Press.

Gautama, Siddhartha. (2000). *Wisdom of the Buddha: The Unabridged Dhammapada*. F. Max Müller (trans.) Mineloa: Dover.

Graham, Daniel W. (2005). Anaximander. *The Cambridge Dictionary of Philosophy*. Cambridge: Cambridge University Press.

Heidegger, Martin. (1975). *Early Greek Thinking*. New York: Harper and Row.

Lao Tzu. (1988). *Tao Te Ching*. Stephen Mitchell (trans.) New York: HarperCollins.

Lloyd, G. E. R. (1970). *Early Greek Science: Thales to Aristotle*. New York: W. W. Norton.

Lucretius. (2003). *De Rerum Natura*. London: Psychology Press.

Matson, Wallace. (1987). *A New History of Philosophy: Ancient and Medieval*. New York: Harcourt Brace Jovanovich.

Nietzsche, Friedrich. (1994). *Philosophy in the Tragic Age of the Greeks*. Washington, DC: Gateway Editions.

Nishitani, Keji. (1990). *The Self-Overcoming of Nihilism*. Albany: SUNY Press.

Nishitani, Keji. (2006). *On Buddhism*. Albany: SUNY Press.

Novak, Michael. (1970). *The Experience of Nothingness*. New York: Harper Torchbooks.

Porter, J. R. 1993. Creation. *The Oxford Companion to the Bible*. Oxford: Oxford University Press.

Reale, Giovanni. (1987). *From the Origins to Socrates*. Albany: SUNY Press.

Russell, Bertrand. (1972). *A History of Western Philosophy*. New York: Simon and Schuster.

Seife, Charles. (2000). *Zero: The Biography of a Dangerous Idea*. New York: Penguin Books.

Smith, Huston and Philip Novak. (2003). *Buddhism: A Concise Introduction*. New York: HarperOne.

Suzuki, D. T. (1964). *An Introduction to Zen Buddhism*. New York: Grove Press.

CHAPTER 29

# The Concept of Naught in Jewish Tradition

*Esti Eisenmann*

## Abstract

The article looks at the concept of zero, from the philosophical and mathematical sides, in the Jewish tradition from the Bible to the modern age. With regard to philosophy, three aspects will be considered: (1) ex nihilo creation; (2) the definition of the deity as the Naught and the creation out of the Naught; (3) the existence of the vacuum in the material world after the Creation. Two mathematical aspects will be examined: (1) the Hebrew system for designating numbers (numerals) from the Bible until the adoption of the Arabic numerals; (2) the notion of zero as a numerical value. The sources for both discussions are mainly the Bible, rabbinic texts of the Talmudic age, *Sefer ha-Yezirah*, the medieval philosophic literature (including the work of Abraham Ibn Ezra, the very first Hebrew author to deal with zero), and medieval Kabbalah.

## Keywords

Judaism – Jewish philosophy – Bible – Sefer Yezira – infinity – creation ex nihilo – God as null – Gematria – Ibn Ezra – zifr

## 1 Naught in the Jewish Philosophical Tradition

### 1.1 Ex nihilo *Creation*

Do the Bible's famous opening words report that the world was created out of nothing? Was there naught before then? Most biblical scholars do not think that Genesis postulates *ex nihilo* creation. First, they say, *pace* the standard translations, the opening verse does not mean 'in the beginning God created the heavens and the earth', but rather (in Robert Alter's rendering) '*when God began to create* the heavens and the earth, the earth then was welter and waste and darkness over the deep'. That is, the creation of the universe began with the creation of light, the separation of the waters, the heavens, and dry land. This is how God overcame the chaos of darkness and water that preceded

 | DOI:10.1163/9789004691568_033

these creations.[1] Second, even if we do accept the traditional translation, it is not clear whether it means that in the beginning God created the heavens and the earth *ex nihilo* – or that He did so from pre-existing matter that is simply not mentioned. In other words, the first chapter of Genesis does not necessarily support the idea of *ex nihilo* creation or the concept of nothingness (Kister, 2007). Nor do any other verses in the Bible state unequivocally that God created the heavens and the earth out of nothingness. Scholars tend to relate the biblical narrative to the Babylonian creation story in which creation is the imposition of order on confusion rather than the production of something from nothing.

By contrast, the Septuagint does translate Genesis 1:1 as Ἐν ἀρχῇ ἐποίησεν ὁ θεὸς τὸν οὐρανὸν καὶ τὴν γῆν, which matches the traditional rendering. Here too, however, there is no way to know whether this means that there was absolutely nothing before creation or, instead, that the universe was formed from primordial matter. In contrast to the vagueness of the Septuagint, Philo of Alexandria (first half of the first century CE) was a vigorous opponent of the assertion that the universe is eternal. He maintained that God created the logos from His mind, after which He created the hylic matter from which, by means of the logos, he created everything all at once. Philo adds, though, that the six days of Creation are not a chronological description but a logical account of the order in which things came into being.

The Church fathers, perhaps influenced by Philo, tended to the idea of *ex nihilo* creation (Wolfson, 1948). But the Mishnaic and Talmudic sages in Palestine and Babylonia, who shaped the concepts of rabbinic Judaism starting early in the Common Era, were not familiar with Philo's work (though they may have been exposed to it indirectly, through their disputes with Christians). They also confronted various positions they knew in the ancient world, ranging from the Greek philosophers to the Gnostics. In general, many of the sages had reservations about discussing the origins of the universe. Their reticence ranged from the simple statement, 'you may not inquire what is above, what is below, what before, what after' (*b Hagigah* 11b), to a ban on public discussion of 'the work of Creation' (*m Hagigah* 2:1). It is evident, however, that some of them did not accept the idea that God had created the universe from pre-existing matter, as in the well-known midrash (*Gen. Rabbah* 1:9):

---

1 I do not wish to delve into the sense of the words *tohu*, *bohu*, and *tehom*. According to biblical scholars, the creation story in Genesis incorporates elements of the Babylonian myth of the sea goddess Tiamat (to whom *tohu* or *tehom* may allude), and perhaps also the Phoenician goddess of darkness, Baau. Whatever the case, in later strata of the Bible *tohu* and *bohu* are understood to be void and chaos, reflecting the primordial confusion that the Babylonian myth associated with the battle against Tiamat, whose body consists of water.

> A certain philosopher asked R. Gamaliel, saying to him: 'Your God was indeed a great artist, but surely He found good materials which assisted Him?'
>
> 'What are they?' said [R. Gamaliel] to him.
>
> 'Tohu, bohu, darkness, water, wind, and the deep,' he replied.
>
> 'Woe to that man,' [R. Gamaliel] exclaimed. 'The term "creation" is used by Scripture in connection with all of them.'

The sage's response clearly rejects the notion that the universe was created from pre-existing matter. We do not know who the philosopher might have been, but Urbach believed that he must be a Jewish sectarian, because he tried to support the concept of primordial matter from Scripture (Urbach, 1979, p. 188). On the other side, there were sages who accepted that the universe was created from primeval substance.

In contrast to the vagueness about creation in the Jewish tradition of antiquity (the Bible, Mishnah, and Talmud), the rabbis of the Middle Ages took an unequivocal stand in favor of *ex nihilo* creation. Saadia Gaon (882–942), who later in life lived in Babylonia where he was head of one of the major Talmudic academies, came out against the idea, which he knew from his reading of Greek philosophers in Arabic translation, that the universe as a whole, or at least the hylic matter from which everything derived, was not created from absolute nothingness and is eternal. He wrote explicitly:

> After these considerations, which proved to me the tenability of the hypothesis that a thing could create itself and which led to the necessary conclusion that someone outside of itself must have created it, I have inquired by means of the art of speculation whether its Maker had created it out of something else or, [whether He had done it] out of nothing, as it has been transmitted in the Holy Scriptures. (Saadia Gaon, 1948, 1:47)

It was clear to Saadia that the Bible posits *ex nihilo* creation, which he supplemented with intellectual proofs. But not all Jewish philosophers of the Middle Ages agreed with him. Some accepted the Greek rejection of the vacuum as contrary to logic, and all the more so creation *ex nihilo*, and implicitly accepted Parmenides' idea that 'whatever is is, and what is not cannot be'. These philosophers rejected the idea that 'something' was created from 'nothing' and insisted that creation must have been 'of something from something'; that is, that God merely imposed the fixed order of matter that we see in nature.

As part of their rejection of the idea that something could proceed from nothing, many Jewish philosophers adopted, in part or in whole, the Neoplatonist

Plotinus' doctrine of emanations. According to Plotinus (205–279), the material world emanates from the perfect and nonmaterial Godhead, in a slow and gradual process that includes intermediate spiritual entities whose perfection and spirituality steadily decrease, and of which most important are the mind (*nous*) and the soul. This concept reached medieval Jewish thinkers in paraphrase. For some of them, the idea of the original creation is compatible with Plotinus' theory of emanations; they held that the biblical account of creation describes the emergence of material substance from spiritual substance.

Despite this multiplicity of opinions about *ex nihilo* creation, in the later Middle Ages and modern times, and especially after the emergence of Jewish Orthodoxy, Jewish tradition has increasingly held to the idea that the biblical position binding on every Jew is that of total and absolute *ex nihilo* creation. There is also a tradition that appears from time to time, going as far back as the Talmudic age (Genesis Rabbah 3:7), that at the end of days the universe will be destroyed and the absolute void will be restored as it was before. Then there will be a new divine creation and a new universe, it too created *ex nihilo*, like the process that led to the creation of our current universe (see, for example, Harvey, 1998, pp. 13–22).

### 1.2 *The Definition of the Godhead as a Void*

The medieval discussions of the origins of the universe – whether from absolute nothingness or primordial matter – were conducted mainly by Jewish thinkers who were trying to reconcile the Jewish religion with the dominant Aristotelian science, which they knew in Arabic translation. They had to deal with the issue of *ex nihilo* creation because Greek philosophy rejected the concept of the void in general and of *ex nihilo* creation in particular.

Alongside these philosophical discussions, medieval Judaism also spawned mystical streams of thought that were peculiar to Judaism, and especially what is known as Kabbalah, which delved into the origin of the universe with no conscious or overt link to Greek philosophy. It is true that Kabbalah drew some inspiration from Plotinus' emanations; but, unlike that doctrine, it asserted that the intermediate entities that emanated from God and stand between Him and the other worlds that He emanated from within Himself are the ten spiritual entities known as the *sefirot*. The *sefirot* are vessels, measures, and God's mode of action, including *din*, *ḥesed*, and *raḥamim* – judgment, love, and compassion. Seeking to combine their doctrine of emanation with the idea of *ex nihilo* creation, many Kabbalists described the process of emanation from God as the true meaning of *ex nihilo* creation. They seem to have been relying on a particular interpretation of the philosopher and poet Solomon Ibn Gabirol, who, like many Jewish thinkers, maintained a variant of the Plotinian theory of emanation. In one of his most famous poems, *Keter malkhut* ('Royal Crown') he wrote:

אַתָּה חָכָם וּמֵחָכְמָתְךָ אָצַלְתָּ חֵפֶץ מְזֻמָּן / כְּפוֹעֵל וְאֻמָּן,
לִמְשֹׁךְ מֶשֶׁךְ הַיֵּשׁ מִן הָאַיִן / כְּהִמָּשֵׁךְ הָאוֹר הַיּוֹצֵא מִן הָעַיִן.

> [You are wise, and from Your wisdom, You emanated the pre-ordained will, like a craftsman and an artist,
> In order to make flow the flow of existence – from nothingness, like the flow of light that issues from the eye.[2]]

These lines can be read to mean that for Ibn Gabirol, what exists flows and emanates from a higher entity – 'nothingness'. Similarly, the Kabbalists referred to the first *sefirah* that emanates from God (frequently referred to by them as *keter* (crown) or *razon* (will) as 'nothingness'). In their interpretation, the creation of something from 'nothing' is merely a slow and gradual emanation of what exists from a spiritual entity (*sefirah*) whose name is 'nothingness.'

To sum up, even according to this interpretation of creation, there was no concept of absolute nothingness, because what exists was created from a perfect entity, the Godhead. Paradoxically, then, in their theory of emanations the Kabbalists, who frequently asserted that they and not the philosophers were the standard-bearers of authentic Judaism, rejected the concept of *ex nihilo* creation in its plain sense (an idea that, as noted, later generations consecrated as the obligatory orthodox view) and denied the possibility of the void of nothingness.

After the expulsion from Spain in 1492, few Jews remained in Europe who could be exposed to the revolutionary ideas of the emerging new science. Only after the Emancipation, in the eighteenth century, were some Jews exposed to its ideas. Then, however, when Jews discussed the concept of the naught (as a mathematical term with implications for philosophy), they did so from a pure scientific perspective and made no attempt to reconcile these theories with their Jewish beliefs and traditions. They wrote in the vernacular and addressed their books to a general audience. Some were assimilated and denied their Jewish origins. This is why the present article must end its survey of the Jewish

2 The sense of these lines is unclear and ambiguous; I have provided my own translation in keeping with how I understand it, rather than relying on any of the many that have been produced in the past. For our purposes, the main message of *Keter malkhut* is the emanation of what exists from nothingness. Thus we can say that for Ibn Gabirol, Nothingness designates the perfect existent from which everything emerged, or the will, as I explain below.

concept of naught at the end of the Middle Ages, around the turn of the sixteenth century; from that time on the discussion did not have a particular Jewish nature, neither in its content (the link between the idea of nothingness and Jewish tradition) nor its form (the target audience and language).

### 1.3 *The Definition of God as Infinite*

If we linger briefly in the Middle Ages we can say that despite their implicit rejection of the concept of the absolute nothingness (*ayin*), the Kabbalists did not reject its 'twin' – infinity (*ein-sof*). Whereas the Greeks had rejected both the void and actual infinity as illogical, for the Kabbalists the Godhead from which the first *sefirah* emanated was the *Ein-Sof* or Infinite. Following Plotinus, they held that God transcends any verbal description; any description reduces Him and mars His perfection. For them, the fact that He exceeds our comprehension stems from His infinite and unlimited essence. In this way the concept of actual infinity, rejected on logical grounds by Greek philosophy, took root in Kabbalistic theology and was identified with God, who stands opposite to finite and limited human beings.

After discussing the concept of infinity in the medieval mystic tradition, we turn to the infinity of Jewish philosophers, who, as noted, were seeking to reconcile Aristotelian science with Jewish tradition. Echoing Aristotle, most Jewish thinkers distinguished a theoretical and future infinity, which is possible only with regard to sequential elements that follow one another – such as time, motion, and the repeated division of matter into smaller parts – and an actual infinity of things that coexist simultaneously, such as the size of the universe. Most of them rejected the latter outright.[3] Hasdai Crescas (1340–1410/11) was the first medieval Jewish philosopher who did not do so (Harvey, 1998, pp. 8–9). Denying many of the basic postulates of Aristotelian physics, he held that the universe is of infinite magnitude and that there could be an infinite number of worlds. But his revolutionary ideas did not leave a substantial mark on the subsequent course of medieval Jewish thought.

### 1.4 *The Existence of a Vacuum in the Post-creation Material World*

Thus far we have discussed the void that, according to some Jewish systems, existed before creation, and the infinite and transcendent Godhead that is

3 Unlike the Kabbalists, they did not assign the term infinite to God (except when they wished to give a positive form to the negative statement that God is not finite and limited). They also grappled with the problem of infinite time, which necessarily entails the idea that the world is eternal and uncreated (and not even derived from primeval matter). But we will not expand on this here.

separate from the material world. Now we move on to consider how traditional Jewish thinkers related to the possibility of the vacuum. This topic is absent from the ancient Jewish texts of the Bible, Mishnah, and Talmud and comes up only in medieval Jewish philosophy, which, as noted, had to grapple with the relationship between Greek science and the Jewish Torah. Overall we can say that Jewish philosophers in the Middle Ages rejected outright the concept of the vacuum and employed the proofs adduced by Aristotle in *Physics* IV to demonstrate its impossibility. Here again Crescas was the most prominent Jewish thinker of the later Middle Ages who demurred from this consensus.[4] On the contrary, Crescas explained that this concept was essential for defining place and asserted that place is the empty space in which bodies are located. For him, space is quite independent of physical objects (Harvey, 1998, p. 6). He defined it as infinite (from which we see that his acceptance of the vacuum, devoid of matter, was linked to his acceptance of its twin, actual infinity). As Zev Harvey has shown, his acceptance of the vacuum and of a universe of infinite magnitude, mentioned above, was directly or indirectly influenced by the ideas of Nicole Oresme (1323–1382) (Harvey, 1998, pp. 22–29). Via Crescas, Oresme influenced Benedict Spinoza (1632–1677), who had read Crescas' *Or Hashem* and defined space as extended and infinite substance (*substantia extensa*) (Harvey, 1998, pp. 29–30). However, Crescas had virtually no influence in this issue on the Jewish philosophical tradition of the late Middle Ages.

## 2 The Mathematical Naught – Zero – in Jewish Tradition

### 2.1 *Methods of Counting and Denoting Numbers in Antiquity, from the Bible through the Mishnah and Talmud*

The biblical system of numeration is decimal, but based on the numbers from one to ten and has no zero. The fact that the biblical term for eleven is 'one-ten' (*aḥad-asar*) supports the idea that the first in the series of numbers is one and the last is ten; the next set of ten begins with 'one plus ten'. Similarly, the word 'hundred' does not reflect a decimal system of counting, because the Hebrew *me'ah* has no etymological connection with either *eḥad* or *asarah*. The same applies to *elef* (thousand) and *revavah* (ten thousand). The Bible is also replete

4 So did the Iraqi Jew who converted to Islam late in his life, Abu al-Baraka al-Baghdadi (1080–*c.*1165). But his ideas about physics, including the vacuum, exerted no real influence on the annals of medieval Jewish philosophy, whose center was in Spain and Provence. Hence it is not clear whether Crescas had any knowledge of his arguments.

with typological numbers; ten is one of them, but no different from any of the others.

The Bible has no concept of numerals. Numbers are indicated only in words: 'twelve' rather than 12; 'one thousand one hundred and twenty-five' rather than 1,125. The sages of the Mishnah and Talmud, who had been exposed to Greek culture, adopted the Greek system of denoting numbers by means of letters. In this system, the first ten letters represent the numbers one through ten, the next eight letters represent the eight decades from 20 through 90, and the last four represent 100, 200, 300, and 400. (Because the Hebrew alphabet has only 22 letters, 400 was the maximum value that could be indicated this way.) To denote the number 118 one sets down, from right to left (the Hebrew reading order), the letter *qof*, the nineteenth letter in the alphabet, followed by the letter *yod*, the tenth letter, and then the letter *ḥet*, the eighth letter. That there was no way to indicate zero in the system did not matter, given that the very concept of zero was unknown to the Talmudic rabbis.[5]

A curious phenomenon is that this method of numeration inspired the sages to move in the opposite direction: if numbers could be denoted by letters, words could be broken down into their letters to calculate a numerical value from which various implications could be extracted. This led to the well-known system of *gematria*, in which every word or phrase has a numerical value that may indicate its links with others that have the same numerical value. For example, the sages interpreted Deuteronomy 33:4, 'Moses commanded us the Torah', to mean that at Mount Sinai Moses taught the Israelites exactly 611 precepts – because the numerological value of *torah* is 611.[6] The Talmudic rabbis did not make extensive use of *gematria*. But the mystics of the Middle Ages extended its scope so far that for R. Judah the Pious (1150–1217), who lived in Germany, the addition or omission of a single word in the prayers upset the numerical equilibrium of the entire world. This idea drew not only from the Talmudic *gematria* but also from an idea found in *Sefer Yeṣirah*, which we will discuss below.

As long as Jews used the Hebrew alphabetic system of numeration they found it difficult to denote large numbers. If they wanted to divide a book into

---

5 Although the Bible could have been cognizant, from the Babylonian numeration system, of the concept of the zero as a placeholder, we find no evidence of such awareness. As noted, the Bible expresses numbers only in words, never in symbols.

6 At Sinai, according to tradition, the Israelites heard the first two commandments ('I am the Lord' and 'You shall not have') directly from God, and not from Moses. The sages added those two to the 611 extracted from the numerological value of *torah* to produce the classic 613 precepts of Judaism.

numbered sections, they were liable to need a very long string of letters. In general they resolved this problem by restarting the count at one in each new chapter or topic. Another method adopted was to combine the alphabetic and biblical systems. For example, they could denote 5,000 by writing the letter *heh*, which represents five, followed by the word *alafim* (thousands). *Sefer Ḥasidim*, by Judah the Pious, is the only extended work with continuous numbering of its sections from start to end.[7] In some editions it has as many as 2,000 sections, the last of which are denoted by a string of seven letters or signs – almost twice as many as required by the use of Arabic numerals.

Mathematical computations are extremely complex when numbers are represented by letters, so we may assume that in antiquity Jews generally employed an abacus to calculate, like the societies around them. Because in ancient times Jews were not interested in pure mathematics, they rarely had to perform difficult calculations. They engaged in more intricate computations only to deal with practical matters, such as the astronomical calculations required to determine the Hebrew calendar, or various geometrical problems.[8]

### 2.2 *Numbers in* Sefer Yesirah

*Sefer Yeṣirah*, an extremely enigmatic work written in Hebrew, which scholars date to between the second and seventh centuries CE, begins by setting the number ten as superior to all other numbers. It asserts that the universe was created by the 22 letters of the Hebrew alphabet as well as by ten *sefirot* (that is, ten different units), for a total of 32 'vessels' or ways employed by God to create the universe. For *Sefer Yeṣirah*, the number ten has cosmogenic and cosmological significance, meanings that are manifested in the dimensions of time, space, and the morality (good and evil) of the universe, as well as in the materials of which it is constructed. *Sefer Yeṣirah* emphasizes that the number ten is not random but precise: 'ten spheres and not nine; ten and not eleven'. It seems to be impressed by the fact that in a decimal system one can continue to infinity: 'ten *Sefirot*, ... their measure is ten that has no end.'

*Sefer Yeṣirah* also divides the 22 letters into groups: 12 regular letters, seven letters that have a double pronunciation in Hebrew, and three fundamental

7 It is unclear when this numbering was assigned to the text. It seems unlikely that Judah the Pious himself was responsible for it.

8 The Talmudic sages needed to calculate the minimum size of a round sukkah (b *Sukkah* 7b), after the Mishnah defined it only as a rectangle with walls of minimum length (with no explicit statement of the area). Hence they had to derive the minimum area and calculate the minimum diameter of a round sukkah that would produce the same: R. Johanan uses the value of 3 for *pi*, even though contemporary Greek science knew better.

letters.[9] These three groups, too, are manifested in the world, both on the cosmological plane (for example, there are seven planets, 12 zodiacal constellations, and three material elements – air, water, and fire) and in the dimension of time (for example, there are seven days in the week, and 12 months and three seasons – summer, winter, and temperate – in the year), and in the human body (for example, there are seven organs in the face – including two ears; 12 paired organs in the rest of the body, such as the hands; and three segments of the body – the head, abdomen, and heart). As can be seen, none of these groups consists of ten items. The only numerical link between the letters and the number ten to which the book may allude is that the 22 letters of the alphabet equal the sum of the ten digits on the hand, the ten digits on the feet, the tongue (the organ of speech, which lies between the hands), and the penis[10] (the organ of generation, which lies between the feet) – a total of 22 items. In any case, the number ten has little importance in the numerical division of the letters, so it is hard to understand from its text how the author of *Sefer Yeṣirah* conceived of the link between letters and numbers as the implements employed by God to create the world.

We can see that the numerological element is dominant in *Sefer Yeṣirah*, and that ten holds a place of honor. Here too, however, as in the Bible, the decimal system runs from one to ten: there is no zero. Nor is there any uniformity in how numbers are represented: the author of *Sefer Yeṣirah* intermingles the use of words alone, as in the Bible, with the alphabetic numeral system. For people today the idea that the decimal system is based on only ten units but is infinite may seem a trivial matter. For the author of *Sefer Yeṣirah*, for whom letters did not have a symbolic function from which the decimal system could be derived, the revelation that counting is infinite led to joy and astonishment.

### 2.3 *The Mathematical Zero in Medieval Jewish Culture*

People in the Middle Ages, especially philosophers who had been introduced to Greek science, were extremely interested in mathematics. This is demonstrated by the dozens of medieval manuscripts of Euclid's *Elements* in Hebrew translation, as well as six manuscripts of a Hebrew abridgment of the *Arithmetic* by Nicomachus of Gerasa. The fact that these works were not aware of the Indo-Arabic system of numerals and stuck to the system of the Hebrew letters

9 Scholars are not sure why the letters *alef*, *mem*, and *shin* are defined as fundamental, aside from the fact that they may represent the Hebrew words for the three elements: *alef* as the first letter of *avir* (air); *mem* as the first letter of *mayim* (water); and *shin* as the last letter of *esh* (fire) (its initial letter, *alef*, having already been assigned to air).

10 In Hebrew, *millah* (word) and *millah* (circumcision) are homonyms.

required verbal presentations of extremely complex mathematical problems and their solutions, which naturally made it harder to understand them.

Even though the Indo-Arabic numerals were known in the Muslim world in the early Middle Ages and were used by Muḥammad ibn Musa al-Khwarizmi (780–850), another three centuries passed before the system was adopted, and then only partially, by the Jews. *Sefer ha-Mispar* (The Book of the Number) by Abraham Ibn Ezra (1080–1164) (Sela, 2003), was the first work written in Latin Europe that offered a detailed explanation of the use of Arabic numerals, anticipated Leonardo Fibonacci (1170–1240) by a century.[11] Ibn Ezra's treatise, which seems to have been written in Lucca, Italy, between 1142 and 1145 (Sela and Freudenthal, 2006), is a full and systematic treatment of arithmetic.[12]

In his introduction, Ibn Ezra explained that the system he presented was of Indian origin and that he would adopt it in a Jewish version. 'Therefore, the wise men of India denoted all their numbers by nine and devised shapes for the nine digits. I wrote instead of them [the first nine Hebrew letters].' Because there were only nine numbers in this system, he added: 'For nine is the end of every counting and these are called the units.'[13] He also associated the number nine with the number of orbs posited by the astronomy of his day.

Ibn Ezra continued by introducing the sign 'zero', explaining that it served as a placeholder. Because of its shape he called it 'wheel', (*galgal*) and said that this was in fact a metaphor: 'It rolls the way straw does in the wind' (Sela, 2003, p. 22). He also noted that its Arabic name was *sifra*.[14] In order to teach his readers how to use this system of numeration, he called the units 'items' or 'ones' (פרטים, אחדים), used 'multiple' (or 'groups' or 'collection': כלל) for the tens, hundreds, and so on, and explained how each was written in a fixed place or 'rank':

> If you have a number of units before the multiples, which are tens, always write first the number of the units, then the number of the multiples. If there is no number in the units, but there is a number in the second rank, which is the tens, put a shape of a wheel first, to indicate that there is no number in the first rank, and write the number that he has in the tens afterwards. If its multiple is in the hundreds and the tens, write a wheel

11 According to Sela, 2003, pp. 21–22, it can be seen as Ibn Ezras first scientific treatise.

12 There is a debate about whether Ibn Ezra was familiar with al-Khwarizmis work. See Aradi, 2013, pp. 238–39.

13 The translation is modified from that on the internet at https://mispar.ethz.ch/wiki.

14 Because the primary sense of the Arabic *zifr* is emptiness or nothing.

> in the first [rank], then the number of the tens in the second [rank] and the number of hundreds in third [rank].

We see that Ibn Ezra did not distinguish 'numeral' from 'number', which is why there are only nine 'numbers' in the *Book of the Number*. Because one and ten are similar, the latter is not considered to be a number; or as he wrote, 'because ten resembles one and twenty resembles two'. The wheel/zero was merely a placeholder and certainly not considered to be a number. What is more, we can see that Ibn Ezra related to all the numerals, including the zero, as a concise and practical system of notation that is convenient for algebra. But he did not employ it to denote quantities; for that he always used alphabetic numerals.

Despite the idea that there are only nine fundamental numbers, Ibn Ezra remained faithful to the concept of the Bible, and even more so of *Sefer Yeṣirah*, that there are ten fundamental numbers. In his long commentary on Exodus, written about a decade after the *Book of the Number*, he asserted: 'All the numbers are nine in one way and ten in another way' (on 3:15). Even though there are nine orbs (the Hebrew *galgal* means both 'wheel' and 'orb'), he said, one must take account of an additional orb, the tenth spiritual.[15] In the commentary on Exodus he did not provide a full description of the Indian numerals or mention the wheel/zero. Had he done so, perhaps the concept would have spread among the Jewish community at large, and not only among intellectuals, because his Bible commentaries were extremely popular.

Nevertheless, Ibn Ezra as a mathematician influenced educated Jews, who knew what he had written about the 'wheel'. In subsequent generations they referred to zero only as a convenient tool for calculation and did not see it as a number. Gersonides (1288–1344), Immanuel Bonfils (*c.*1340–1377), Isaac Ibn al-Ahdab (fourteenth century), Joseph Moses Zarfati (fl. 1422), Jacob Caphanton (died *c.*1439), Mordecai Comtino (1402–1482), Isaac ben Moses Eli (fifteenth century), Elijah Mizrahi (*c.*1450–1526), and Gad Astruc (fifteenth to sixteenth century), all of whom wrote mathematical treatises and mathematical textbooks in Hebrew for Jews, discussed the topic. Some of them called it '*zefer/zifr/sifra*,' while others used 'wheel'. Jacob Caphanton presented a table of the numbers one through nine, parallel to the first nine letters of the Hebrew alphabet, but in practice employed the Arabic numerals instead of the Hebrew letters. Comtino, who lived in Byzantium, mentioned that in Greek the 'wheel' is *ouden*, and in Latin, *nulla*. After him, Elijah Mizrahi explained that the word *sifra* means 'naught'.

15 On the spiritual orb, see Eisenmann, 2011.

### 2.4 *Zero as a Numerical Value*

As we have seen, Jews in the Middle Ages and early modern era considered zero to be purely a symbol and not a number. As mentioned above, Jews were not full partners to the discoveries of the New Science in Europe. When they again found a place in European society they learned the concept of zero as a number. By that time, the links between religion and science and between theology and philosophy had weakened and they felt no need to write about mathematical ideas in Hebrew for specifically Jewish audiences. Hence we cannot speak about zero as a numerical value in the Jewish tradition, because there is nothing in texts by Jews with any particular Jewish character, neither in content nor in form.

## 3 Conclusions

Even though the decimal system was in common use from biblical times on, in Antiquity and the Middle Ages the Jews, like many other cultures, did not employ 'zero' as a placeholder digit or number. This was because the Bible always sets out quantities in words and never employs numerals. From the early centuries of the Common Era, Jews adopted the Greek system of alpha-numerics, using the Hebrew alphabet, but the attendant inconveniences kept it from being a practical solution and we are unaware of the use of algorithms to perform mathematic computations with it. It was only in the twelfth century that Abraham Ibn Ezra adopted the zero from Arabic civilization, along with the use of only nine additional symbols, in order to facilitate mathematical computations. But he continued to employ the first nine letters of the Hebrew alphabet for this, rather than the Arabic glyphs. He conveyed the use of the zero to the Jewish world, whose learned circles came to use it extensively. Note that zero was used exclusively as a placeholder and not seen as a numerical value. It was only when Jews entered the world of the secular universities in modern Europe that they began to consider zero to be as a true number, but this had no impact on Jewish tradition per se.

Jewish philosophy and science were also marked by the absence of the concept of null. Jews who were exposed to Greek thought adopted its conclusions and rejected the possibility of the void as absurd. This was expressed, inter alia, in the conclusion that the vacuum did not exist. One Jewish tradition that became influential mainly in the Middle Ages and modern era proposed that the material world was preceded by an absolute void. But except for this concept of ex nihilo creation – which was rejected by many Jewish thinkers, explicitly or implicitly – Jewish tradition tended to reject the possibility of a vacuum in our post-Creation world.

## References

Aradi, Naomi. (2013). An Unknown Medieval Hebrew Anonymous Treatise on Arithmetic. *Aleph*, 13(2), pp. 235–309.

Eisenmann, Esti. (2011). On the Heaven in Jewish Middle Ages Philosophy. *AJS Review*, 35(2), Hebrew section, pp. 1–20.

Harvey, Warren Zev. (1998). *Physics and Metaphysics in Hasdai Crescas*. Amsterdam: J. C. Gieben.

Kister, Menahem. (1996). Tohu wa-Bohu, Primordial Elements and Creatio ex Nihilo, *JSQ*, 14(3), pp. 229–56.

Saadia Gaon. (1948). *The Book of Beliefs and Opinions*, translated by Samuel Rosenblatt, New Haven: Yale University Press.

Sela Shlomo and Gad Freudenthal. (2006). Abraham Ibn Ezra's Scholarly Writings: A Chronological Listing, *Aleph*, vol. 6, pp. 13–55.

Sela, Shlomo. (2003). Abraham Ibn Ezra and the Rise of Medieval Hebrew Science. Leiden: Brill.

Urbach, Ephraim E. (1979). *The Sages: Their Concepts and Beliefs*, translated by Israel Abrahams, Jerusalem: Magnes Press.

Wolfson, Harry A. (1948). The Meaning of Ex Nihilo in the Church Fathers, Arabic and Hebrew Philosophy, and St. Thomas. In *Mediaeval Studies in honor of J. D. M. Ford*, eds. U. T. Holmes and A. J. Denomy, Cambridge, MA: Harvard University Press, pp. 355–70.

CHAPTER 30

# How Does Tom Tillemans Think?

*Erik Hoogcarspel*

## Abstract

How the zero eventually developed from the concept of *śūnyatā* and if it developed from there at all, may still be open to discussion. Certain, however, is that any such development has to cross the line between intuition and formalization and this crossing can even go both ways. It is, for instance, possible that the Hindu concept of *karma* has developed from the practical relations and registration of debt. The formalization of this with the introduction of bookkeeping and money might in its turn have led to a formalization and dogmatization of the concept of *karma*. Did such a border crossing also happen in the case of zero? Tom Tillemans has argued that it did, which means that the emptiness intuition of Nāgārjuna has been formalized (or has lost a dimension, as I will argue) into a logical structure that might have instigated the institution of zero. I will argue that we need to practice phenomenology if we want to recover the other side of the border of formalization and understand the nature of border crossing.

Tom Tillemans (TT) is a historian and philologist of Buddhist texts, who has an impressive record of publications. He has done intensive research into the Buddhist philosophy of emptiness called the Mādhyamika. He also conducted several discussions with colleagues about the subject. This resulted in a series of publications, of which the most important have recently been published in two books. In this article I want to review the most important problems with which TT et al. were confronted and suggest a way out which in my opinion has been overlooked due to self-imposed philosophical restrictions.

## Keywords

logic – catuskoṭi – intuition – emptiness – phenomenology – depth – history

 | DOI:10.1163/9789004691568_034

## 1 Introduction

To begin with, it is not my intention to read Professor Tillemans' mind. The title of this article refers to a book by the title 'How do Mādhyamikas think?' in which TT and some of his colleagues try to elucidate a few important historical and comparative aspects of the Buddhist Mādhyamika philosophy. The book contains several contributions from TT, one of them having the same title as the book and there is even a follow up. TT also does not pretend to possess a crystal ball by which he can read others' minds in the past, so the title must be short for 'How does TT think Mādhyamikas think?'. In other words, the question in the title of the book brings up the question in the title of this article. We can't possibly know what the Mādhyamikas think if we don't know how TT thinks, at least during the time when he wrote his book. I respect Professor Tillemans very much for his expertise of Buddhism and his knowledge of Buddhist texts. Besides, he is a very reasonable and rational person. In other words, he may be right about many things, but I nevertheless think that some critical notes are in place.

One important problem is that TT thinks from within a bubble: the circle of analytic philosophers who focus on logic and language and are all admirers of the Buddhist logicians Dignāga (480–540 CE) and Dharmakirti (sixth to seventh century). Both logicians were following the Yogācāra philosophy, which entails a rejection of the Mādhyamika way of thought and proposes a kind of idealism or cognitivism. Everything we perceive is said to be nothing but cognition. This is somehow combined with an atomic realism, that represents the ultimate reality. Although the name Yogācāra means 'practitioner of yoga', the followers did not show much proficiency in spiritual exercises, but merely wrote bulky books. The philosophy was particularly dominant in the Northwest of India around the city of Gandhara (now Kandahar), where also the founders Vasubandhu and Asanga lived during the fourth century. Mādhyamikas, like Nāgārjuna (second century), Candrakirti (600–650) and Atiśa (982–1054), mainly lived in the East of the Indian continent.

The logicians had a profound influence on Buddhist philosophy, for instance on the work of later Mādhyamikas like Śāntarakṣita (755–788) and Kamalaśīla (740–795), who called themselves Yogācāra-mādhyamikas. This is a typical example of inclusive thinking, a kind of Hegelianism Indian style. Yogācāra was supposed to be a good start for dummies and Mādhyamika the real stuff for advanced or clever people. In the same drift, the teachings of the historical Buddha were described in the Lotus Sutra (second century, Gandahar) as no more than a pedagogical introduction, for those simpletons who could not understand the true teaching that was supposed to have been presented by the metaphysical Bollywood Buddhas of the Mahāyāna movement.

## 2 Quietism

A real puzzle for modern analytical logicians like TT et al. appears to be the philosophical quietism of the Mādhyamikas, i.e., their abstaining from any philosophical claim or thesis. Logic is the science of deduction, which is mainly the procedure by which new true propositions are produced from accepted true propositions, and this is exactly what the Mādhyamikas refuse to do. Their one and only purpose is to show the inconsistencies of accepted true propositions and representations.

TT compares this attitude with the philosophy of Wittgenstein, but this philosopher was still in search of the truth. The obvious Western counterparts are of course the Skeptics. It is telling that TT does not pay much attention to them, in spite of the fact that one of the most influential Skeptics, Sextus Empiricus, lived in the same period as Nāgārjuna and both share a considerable number of arguments. Sextus even mentions the famous logical anomaly: the *tetralemma* or *catuṣkoti*. The most important fact that TT overlooks, however, is that both schools aim for some sort of inner peace, called *ataraxia* by the Skeptics, and *nirvāṇa* by the Madhyamikas. TT's 'quietism' is however meant to be logical or epistemological, not existential. In ancient Greece the tradition of inner transformation started with Socrates, who in his Apology said that he did not care about worldly concerns like most people, but took care of himself (*epimeleia heautou*). Socrates, as is well known, also claimed that he did not know anything and that he did not have anything to say about the truth. He simply followed the instructions of the god Apollo. The Skeptics too were not so much trying to establish yet another kind of metaphysics, they sought inner transformation, through *époche* (distancing) and *isostates* (indifference). Nāgārjuna, likewise, did not try to build a new metaphysical or ontological system. He just wanted to reinstate what, in his view, was the real teaching of the Buddha, a teaching that he feared was about to be forgotten. In several passages of the *Pali Canon* the Buddha affirmed that the one and only subject of his teaching was *nirvāṇa*, nothing else mattered to him. Again, it is all about personal transformation, and this explains the Socratic irony as well as the Nāgārjunian rejection of taking a stand or making any claims. In Wittgenstein's view, these are the sort of things about which philosophy had to remain silent, but philosophy should not remain silent about the rest, but be outspoken and clear.

## 3 Flatland

Recently, scientists have been developing the thesis that our three-dimensional world is a projection of a two-dimensional essentially true reality, a Flatland

(Scargill, 2020). TT is a reasonable and rational man, he is very able to reason, so he avoids leaps, discontinuities, or transformations. That is why he thinks according to laws of Flatland, a place without depth, without involvement, where philosophers consider themselves to be strictly separated from their thoughts. Flatland is true reality, its inhabitants survey the world from above, avoiding every hint of a point of view. To them everything is transparent and within the limits of the two dimensions true and false. No one can live in Flatland because soccer is boring, the beer is flat and women have no sex appeal. Therefore the Flatlanders continuously cross borders in order to enjoy their family life and their guilty and innocent pleasures.

Inside Flatland, one speaks Flattish, or Ontologese, a language without inclinations for time and space, without emotions and without perspective. In Flattish the word *physical* has nothing to do with our bodies, 'to naturalize' means to rob something of its nature and bring it under the Flat laws. Common sense consists of Flat patriotic thought that has been approved by the FPC, the Flattish Party Committee. Finally, humor is unknown, nor poetry, anyone who commits such a felony definitely loses his or her citizenship. It is the paradise of the Theory of Everything (ToE) that explains everything one can think of, except itself. In other words, it does not explain what it has to explain, how it explains and why. In fact, the ToE is the naturalized Platonic Form of the Supreme Good, saved from oblivion by monotheism. In Flatland there are no hiding surfaces, everything is clear and unchanging, even dying is impossible. Discontinuities and transformations are unthinkable, surprises and contingencies have been outlawed. It is a haven of security.

It is important to understand that with the reduction of three dimensions to two, depth has become impossible. Depth is the experience of our physical involvement in the world. It makes movements possible. It is not just the experience of an arrangement in perspective. The laws of perspective, which have been developed during the Renaissance, are a deception because they suggest a personal point of view through the implementation of an impersonal mechanical construction. That is why a painting is often more realistic than a photograph, in spite of being blurry and unclear. When we look at things, we move our eyes and our heads, and sometimes even our entire body. Because of this, space and the things in it are a kind of living presence to us, and we feel like moving around the things or picking them up. Registering objects by a video camera is a continuous solid process, seeing objects requires discontinuities and emptiness. Our experience of the world is always distorted, just because we are part of it, says the phenomenologist Marc Richir. The security and clarity of Flatland is therefore nothing but a dream.

## 4 How Many Truths?

It looks like Nāgārjuna was playing in the same league as Socrates, but how can TT naturalize him? Well, a lot of work had already been done by the Indian logicians and TT happily quotes them when they claim that Nāgārjuna was not denying that things do not exist. He conveniently forgets the passages where Nāgārjuna calls the world an illusion, like a mirage or a magic show. In other words, Nāgārjuna was ontologized and taken to be not very different from a Hindu or from a Yogācārin as if he was exercising the very same hobby as traditional Western philosophers from Plato to Heidegger, separating truth from falsehood. The logicians suggest that Nāgārjuna only criticized the false imputation of a substance (*svabhāva*, self being) upon things. As a consequence there are innocent things and apart from those there are other things that are wrongly considered to be substances, because of our symbolic institution or system of conventions (TT calls them 'customs'). Anyone who is the least familiar with the history of Western philosophy should remember that we have been at this point before, when David Hume expressed his amazement when he discovered that causality is not to be found anywhere in the world, but still seems to be self evident by all people in all walks of life – yes even by little children and some animals. Immanuel Kant saw the problem and decided that causality is part and parcel of the way we all experience things. That is why little children know how to blow out candles on a birthday cake. TT et al. seem to overlook this, because they do not mention the aftermath of Kant's philosophy and the problems and solutions that came up in due course.

Every perception is guided by concepts. Concepts without perception are empty, while perceptions without concepts are blind. If this is true, 'innocent things' which are not reified or part of the causal network are impossible. According to TT Candrakirti and the Buddhist logicians as well as Śāntarakṣita and Kamalaśīla, but also Tibetan philosophers like Tsongkapa, these innocent things are basically familiar to all of us. This can only be the case if we understand them in a way and know them through concepts. If this is true, how can these very innocent things not be reified? TT himself does not see any problem and thinks this is a great idea. He assumes that Tibetan meditations are designed to sift the malicious reified things from the innocent ones. It is telling that he mentions a meditation on the ego as an example, because in doing this he confuses psychology with philosophy. Psychology is a science *a posteriori*, it is about living persons reifying things, by a wrong mental habit for instance. It does not ask the question whether they have sound arguments to do so.

The advantage of TT's assumption of innocence is that the Mādhyamika philosophy has instantly become harmless and receives a passport for Flatland. Suddenly things are real again, and a transcendent truth of emptiness is introduced apart from the vulgar one. The transcendent truth is among Tibetans said to be a secret shared by the elite class of lamas. Some dummies still reify things, but they can be cured if they develop enough devotion for the lamas and learn some tricks from them. Those tricks or meditations are described by TT as a kind of phenomenology without *époche*, i.e., without distancing oneself from the life world. He does not realize that this *époche* is not very different from the practice of *samatha* or *zhi nas*, which is part and parcel of every Buddhist meditation.

However, we are confronted with a problem at this point, because TT seems to be a bit confused. He quotes, agreeing wholeheartedly, the conclusion of the Japanese Buddhologist Teruyoshi Tanji that 'emptiness is the nature of all things'. How should we understand this? Apparently all things have a nature, which is emptiness. That which has emptiness as its nature must be empty. If not, what would be the point? That which is empty can in no way cause anything, it lacks any causative force, can not be distinguished from any other thing and therefore it can not possibly have a 'nature'. 'Innocent things' that are not reified, could therefore neither be causes nor results, and would therefore have to be considered as pure fiction. Teasing out reification would amount to the introduction of a universal fictionalism, a philosophy that considers everything to be merely made up (we may think for instance of Nietzsche), which TT finds ridiculous.

On a second thought, TT thinks that deflationism might save the day. This theory of truth wants to do away with the concept of truth itself, because this is assumed to be redundant. To say 'it is true that snow is white' just means that snow is white. Apart from the problem that many flavors of this theory are competing for the truth, it is easy to see what a Mādhyamika's answer would be: the proposition 'snow is white' is not analytical, because in that case it would be identical with 'white is white' and 'snow is snow'. It has to be synthetical, which means that you need one self being that makes snow to be snow and another one that makes snow to be white. Aristotle did not see this problem because he took over the inexplainable Platonic concept of participation, but self being, or substances, can not have contrasting attributes. Both substances will therefore always exist independently, so they can not be related to each other. Anyone who wants things to be otherwise has to accept their emptiness. In other words, deflationism only works in Flatland and on top of that it opens a Pandora's box of problems, because the many types of deflationism all have their specific drawbacks and the discussions are still going on. TT is however looking for allies because he is afraid that the sophisticated modern scientific theories will be sacrificed on the altar of the higher truth, therefore he places his bets on Tsongkapa, in his view one of the most sophisticated ontologists. In other words, he wants two truths, both neatly separated from each other.

## 5 Set Theory

A new element in the discussion about the two truths is a proposal from Graham Priest and Jay Garfield. They suggest that it is all about a kind of Russell's paradox. Bertrand Russell discovered a flaw in the set theory: the set of all sets would be a member of itself, which is a contradiction. In the case of the Mādhyamika philosophy, the set of all empty things would itself according to Priest and Garfield 'have the nature of being empty' and therefore be a member of itself. This does not mean that the set of all empty things itself would be empty, because this amounts to saying that no empty things exist. It is its nature that is empty. In other words, the concept 'set of all empty things' suddenly is thought to have a nature, as if it were a thing by itself. This is a clear case of Flattish reification. This reification has nothing to do with the Mādhyamika philosophy, but everything with the reification of sets in the set theory. After all, several sets have this problem, for instance the set of all concepts and the set of all things made up are also their own member. It is just the consequence of Frege's rule that sets are not allowed to contain themselves. In most such cases, the theory is modified, for instance when the square root of -1 appeared to be impossible, imaginary numbers were introduced. So what is the problem here? Besides, having an empty nature, or any nature at all, is not a necessary property of sets, like for instance being made up is. Sets are analytical, they are not real. It looks like we can also pull off this argument if we would try to prove that the set of red things is itself red. This brings us to the second problem, which is that a set does not have a nature at all, it is a mere concept. If I wear a pair of black shoes, the pair is not black. If I would paint one of my shoes white, the pair does not change color, it does not become gray or black and not black. If I lose one of my shoes, the pair suddenly does not exist anymore. This is because the pair is empty as a way of being but not by nature, it exists solely by implication (*pratītya samutpāda*). But it certainly does not have an empty nature. An empty nature is something we experience when we see that synthetic things dissolve as a unity. Anything that has an empty nature, like a shoe for instance, is perceived and considered to be a thing. It is a part of our life world. I can put it on and it makes walking comfortable for me. It has an empty 'nature' because in order to know what a shoe is and how to put it on, I have to take a multitude of non-existing things into consideration. I have to understand how the shoes were made, how to use them and where I have stored them, these are events in the world that do not exist anymore. I have to know when they have to be polished or repaired, or whether they are impermeable in case it is going to rain. These are events in the world that do not yet exist.

What appears, but does not exist, has an empty nature, and we can understand now how existing things are kept together by emptiness. A concept does not appear, it can be thought and mentioned, but it never enters our life world. Perhaps one may try to move the shoes to Flatland, where time does not exist and sets and members are made identical. In that case we would have to decide what color the shoes are. The conclusion is inevitable: sets exist on a meta level and members of sets on an object level. Both levels are characterized by two different kinds of emptiness, analytical and synthetic and these should not be mixed or confused.

## 6 The Historical View

TT does not agree with the analysis of Priest and Garfield. He advances a strong argument saying that this would take the very force out of Nāgārjuna's rhetoric, which is the very force and meaning of the Mādhyamika movement itself. He wants to be fair and respect the ambition of the movement and this seems only possible by accepting their quietism. In due time, however, this quietism has vanished and the Mādhyamika philosophy has changed. We can understand this from new sociological developments in Buddhism and its place in Indian society. The bulky commentaries, the discussions, the ever larger monasteries with ever more political ties and interests, the competition between different abbots, all this made the Mādhyamika school into an institution and this changed the very philosophy itself. It became sophisticated, perhaps even middle class *salonfähig* and the meditation, the *epimeleia heautou* became an item for one's CV. Abbots were supposed to be great yogis, but no one ever had to prove anything, it was supposed to be an inevitable result of scholarship. Mādhyamika became reified, turned into books and words.

## 7 Phenomenology

Is there a better way to understand emptiness? A way that does take the *epimeleia heautou* into consideration? I think there is. TT refers to it occasionally, but he does not understand enough to see the impact it could make on the understanding of emptiness. When Husserl developed his phenomenology as a new kind of philosophy, he was reflecting about the nature of logic and mathematics. What interested him more than anything else was the nature of logical and mathematical thought. What happens in the mind of the mathematician? How is evidence achieved? He had learned from his teacher Franz Brentano that a thought cannot appear as anything else but itself. So evidence happens if something merely appears. Husserl called these appearances 'phenomena' and the study of them 'phenomenology'. He adopted the Kantian structure of the transcendental ego as a necessary principle of unity of the structure of the world. Perhaps this was the result of the old Platonic bias of the appearances to be a chaotic bunch, a kind of school class without teacher. A phenomenon is anything that appears, a perceived image, a sound, an emotion or a thought. All phenomena refer directly or indirectly to the world. The world as such is left out of consideration, but it remains the point of reference for all phenomena. That is why Husserl affirms that in the phenomenological reduction nothing is lost. The world does not disappear, nor is it excluded, it is only not considered to exist on its own. The reductions are a major ontological clean up.

For those who study the works of Husserl it is not very easy to become clear about the reductions, because the major developments in the work of Husserl appear only in the unpublished manuscripts, which have now almost all been published in the series of Husserliana. One of such developments is the discovery of the transcendental reduction, which is described by Eugen Fink, Husserl's assistant, in his 'Sixth Cartesian Meditation'. Husserl has been looking over Fink's shoulder when he wrote it and he has agreed with the contents. When one focuses on that which appears, one sees that every appearance refers to other appearances. It is possible to investigate this structure of reference by using the eidetic reduction, leaving out everything that is coincidental. It is however also possible to leave out all reference and just focus on the process of appearing, the phenomenality. Phenomena appear in this way as nothing but phenomena. They do not exist, they merely appear, in such away that appearing and disappearing are both identical. This new way of understanding phenomena changes everything.

Suddenly Buddhism emerges at the horizon, because phenomenality has the very same characteristics as is described by the Buddha when he was talking about the world (*tilakkhana*): phenomena are impermanent, selfless and unsatisfactory. They are impermanent because they merely appear while vanishing, they are selfless because they are not caused by anyone or anything and they are unsatisfactory because every phenomenon refers to other phenomena, it is in itself unfulfilled. The transcendental reduction leaves behind any trace of an ontology, so there is nothing to claim, no argumentation is possible. Phenomena in themselves do not mean anything, meaning is just the illusion of phenomenal interference. If we try to understand Nāgārjuna's philosophy from the point of view of the transcendental field, everything falls into place. The *samvṛtisatya*, the conventional truth is the ontological one, the *paramārthasatya* is the transcendental one. Both truths are not transcendent to one another, they cannot be separated. The transcendental truth is the very necessary condition for the conventional one, albeit one that is usually forgotten or overlooked. That is why the conventional truth only can give access, by reduction, to the transcendental one. We have to accept the conventional truth from the life world, not because it is the infallible and absolute truth, but because it is the very necessary condition for all language and understanding. It is also the truth of our body and we need our body to meditate and think. TT is puzzled by the acceptance of the conventions by early Mādhyamika, but Husserl thought along the same lines in his speech and study called '*Die Krise der Europäische Wissenschaften und die transcendentale Phänomenologie*'. The 'sophisticated' science and logic hide the life world under a veil of ideas and because of this they lack any foundation. Some scientists even have suggested that the whole universe could be an illusion, not realizing that this includes their own thinking.

Another important reason for the phenomenological approach of Mādhyamika is the similarities between Buddhist meditation and phenomenological and transcendental reductions. What could be more realistic, more *yathābhūta*, than seeing phenomena as phenomena? It is finally beyond the scope of this article to dive too deep into recent research into transcendental phenomenology as has been conducted by philosophers like Marc Richir and Alexander Schnell, but in a sense it is all about *dhamma*, which is one of the possible Buddhist Pali translations of the word 'phenomenon', but also means truth and teaching.

## 8 How Does TT Really Think?

We have seen several attempts to reconstruct or explain the philosophy of the Mādhyamikas from within the limits of the tradition of analytical philosophy. Their overall failure should not surprise the attentive reader, because both philosophical traditions have their own play field and therefore overlap only partially. They have a different history and do not share the same values and purposes. TT, being an historian, is definitely aware of this fact, perhaps even more than he realizes. This might also be the reason why he keeps on reminding the reader that he or she should be fair and invoke the principle of charity in the evaluation of the Mādhyamika philosophy. TT does however not put his money where his mouth is, because he tries to explain the Mādhyamika philosophy through comparing it with the ideas of Wittgenstein, without any consideration of the historical gap that separates both. This is a returning flaw of analytical philosophers – they elevate themselves to a metaphysical level far beyond history with the excuse that they have the copyright to absolute truth, because they are the only ones who do justice to logic, science and mathematics. However, this also entails a self restriction, because they forbid themselves to reflect on the ontology that is uncritically assumed by logic, science and mathematics. TT should have been aware of this, because of his familiarity with historical impermanence, but he is not only a reasonable but also a very modern thinker. He is a staunch believer in the progress and sophistication of modern science. This is a metaphysical dogma, which does not go well with the intuition of emptiness and relativity which is at the heart of the Mādhyamika philosophy. It is this caveat which returns again and again in his books and articles, in several disguises.

Both the quietism and the acceptance of conventional opinions by the early Mādhyamikas have been explained by TT et al. in terms of logic, as also was done by the Buddhist logicians and the *svatantrikas*. TT believes in progress, so he insists it is a development of sophistication. It may, however, also be an indication that something important was lost, perhaps because of the increasing importance of debate and competition. The famous debate between sudden and gradual enlightenment in the Tibetan monastery of Samye, for instance, which TT also mentions, looks more like an Italian opera than like a philosophical discussion. The event took several months, people got killed, reputations were shattered and nothing really changed. This should not have surprised anyone, because what was at stake was not research into the nature of Buddhahood, but simply to determine once and for all who was right and who was wrong.

If Nāgārjuna really was picking up from the rejection of the Buddha of both being and non-being, as he states himself in his writings, than quietism and accepting conventions are the obvious and necessary elements of Nāgārjuna's philosophy. The tetralemma can in such a situation only be a rhetorical move and not a higher form of logic. In case of an absence of both being and non-being there neither can be truth nor untruth and the philosopher has freed himself from the shackles of logic.

It is not fitting to blame TT for his adventures with fictionalism and deflationism – after all, he admits himself that his expertise is mainly philology and history. On the contrary, the reader can only be impressed by his vast knowledge and clear expositions of the Mādhyamika texts. All this opens new discussions and horizons of further research. Without this, my observations would also have been impossible.

## References

Garfield, J. L., & Priest, G. (2003). Nāgārjuna and the Limits of Thought. *Philosophy East and West, 53*(1), 1–21. http://www.jstor.org/stable/1400052.

Garfield, J. L., Tillemans, T. J. F. and D'Amato, M. (ed.) (2009). *Pointing at the Moon: Buddhism, Logic, Analytic Philosophy*. Oxford: Oxford University Press.

Scargill, J. H. C. (2020). *Existence of life in 2 + 1 dimensions, arXiv.org*. Available at: https://arxiv.org/abs/1906.05336.

Tillemans, T. J. F. (2016). *How do Mādhyamikas think? And Other Essays on the Buddhist Philosophy of the Middle*. Wisdom Publications.

CHAPTER 31

# Overhauling the Prevailing Worldview: an Essay

*Peter Gobets*

## 1 Introduction

> Champollion deciphered the wrinkled granite hieroglyphics. But there is no Champollion to decipher the Egypt of every man's and every being's face. Physiognomy, like every other human science, is but a passing fable. If then, Sir William Jones, who read in thirty languages, could not read the simplest peasant's face in its profounder and more subtle meanings, how may unlettered Ishmael hope to read the awful Chaldee of the Sperm Whale's brow? I but put that brow before you. Read it if you can. – Moby Dick by Herman Melville

My name is not Ishmael, but we are going on a whale-hunt all the same. In fact, we have even bigger fish to fry. The 'Universe' no less and the prevailing worldview based upon it, both in bad need of overhaul. Take heed, we will not play by the rules ...

The account you are about to read aims to consider and reject both the prevailing worldviews of an objective Reality 'out there' and of a subjective Reality 'in here'. Each position is deemed to be untenable on closer analysis, which has momentous consequences for the sciences.

Both ideologies have served their purpose in practice but have outlived their theoretical usefulness, even become counterproductive by eclipsing alternative worldviews. Consequently, the sciences have landed in crisis, unable any longer to account for certain cutting-edge laboratory results.

A viable option is offered to surmount the philosophical-logico-linguistic hurdles encountered to usher in a new worldview and the foundationless 'Nonoverse'. And with it, at last, a way out of 'the belly of the whale', to use Melville's Moby Dick-simile, swallowing up all mortal creatures. As it turns out, zero is the key to unlocking the new worldview.

 | DOI:10.1163/9789004691568_035

## 2 Secular Priesthood

Look at your teachers, don't listen to them, to paraphrase Woody Allen.

Barring rare exceptions, the 'group think' in mainstream academia has taken on preposterous proportions, with, as consequence, tunnel vision on a massive scale. This myopia is fostered by the culture of specialization rather than interdisciplinary generalization coupled to the peer review / publication / funding regimes in place to reinforce the *status quo* over the centuries, given Western/ Eurocentric cultural domination.

The main objection to this trend is that the prevailing worldview eclipses all rival options to the detriment of science in particular and society as a whole. In the words of Paul Davies, who wrote the Introduction to Werner Heisenberg's *Physics and Philosophy* (Heisenberg, 1958):

> Einstein's opinions are labelled 'dogmatic realism', a very natural attitude, according to Heisenberg. Indeed, the vast majority of scientists subscribe to it. They believe that their investigations actually refer to something real 'out there' in the physical world and that the lawful physical universe is not just the invention of scientists. ... So natural science is actually possible without the basis of dogmatic realism.

Rationalism has demystified the world, reduced now to what is by and large the mainstream scientific narrative of our contemporary 'Secular Priesthood', preaching with Missionary Zeal their Gospel of a Material, Mechanical Universe – humans as machines/computers – replete with Unquestioned Dogma, tacit assumptions that in the final analysis do not stand up to scrutiny.

The converts to this global, institutionalized Secular Church are legion among professionals and public alike.

None of this is to deny the breathtaking scientific advances *in practice*, leading to breakthroughs in technology and the whole range of consumer products, creature comforts, etc.

Rather, the thrust of this exploration is to challenge the wholesale adoption and the internalization of received 'wisdom' over the past centuries in the form of often times outdated and thread-worn Judeo-Christian-Neoplatonist-Greek concepts that have led to this 'progress'.

## 3 Harpoon

We will arm ourselves on this whale-hunt with the barbed harpoons of modern linguistics crafted by de Saussure, Wittgenstein et al, of language, words, concepts not *corresponding* to some presumed Objective Reality 'out there', but deriving their meaning from the *internal coherence* within the language system concerned: be this natural or artificial human languages, including mathematics/computer coding and thus by extension the entire realm of modern mathematics and physics.

And with this lethal harpoon, we will puncture the 'Universe' and deflate it into the 'Nonoverse' by dismissing Realism and Idealism in favor of 'nonism' as ontological point of departure: no objects, no subjects, no self, and lastly no no-self. For by this tour de force we arrive at the *tabula rasa*, the clean slate, to enable us to assess afresh our abode in the Trackless Void.

It will turn out that the supposed Laws of Nature are *not* fundamental as presumed but subordinate. These Laws of Nature are 'emergent', supervening upon the subject – object split, read: the observer and observed Universe of science and citizens alike.

## 4 Target Practice

Let's take as example the late-Stephen Hawking, brilliant scientist, no doubt, but deplorable philosopher, who in *A Brief History of Time*, concluded:

> the people whose business it is to ask why, the philosophers, have not been able to keep up with the advance of scientific theories. In the eighteenth century, philosophers considered the whole of human knowledge, including science, to be their field and discussed questions such as: Did the universe have a beginning? However, in the nineteenth and twentieth centuries, **science became too technical and mathematical for the philosophers, or anyone else except a few specialists.** Philosophers reduced the scope of their inquiries so much that Wittgenstein, the most famous philosopher of this century, said 'The sole remaining task for philosophy is the analysis of language.' What a comedown from the great tradition of philosophy from Aristotle to Kant! (Hawking, 1988, emphasis added)

What unadulterated rubbish! In challenging prevailing scientific theory or theories, and to show up Hawking, we state the obvious here, namely that the concepts that constitute the backbone of the sciences such as mass, energy, atoms, molecules, subatomic particles, quarks, time, space, entropy, causation, information are all *useful anthropomorphic fictions* on the one hand, while that of the presumed *autonomous* self or observer – inside as well as outside of the lab – is no less of a fiction, a conventional construct, convincing as this may seem to the gullible (i.e., High Priest Hawking the First and the entire capped and gowned Guild of Clergymen and women and their hapless students).

According to Paul Davies' reading of Heisenberg (Heisenberg, 1990):

> What then is an electron, according to this point of view? It is not so much a physical thing as an abstract encodement of a set of potentialities or possible outcomes of measurements. It is a shorthand way of referring to a means of connecting different observations via quantum mechanical formalism. But the reality is in the observation, not the electron.

## 5 Conundrums

What we observe in the lab in practice are blips and beeps, dials and gauges, computer printouts – on that we can agree. But what these evident signals in practice signify in theory – that, epistemologically, is anybody's guess.

What to make of the particle/wave dualism? In the old days it would have been dismissed out of hand owing to the Laws of Thought and the Excluded Middle (Aristotle). Particle or Wave, but not both. And for some time it was looked at askance, far-fetched as it seemed. How could a wave suddenly turn into a particle and *vice versa*? Meanwhile it is common currency, bandied about by every Tom, Dick and Harry as Unquestioned Dogma.

Erstwhile Certainty led to Uncertainty; Determinacy to Indeterminacy and all it implies in terms of the human psyche. Who can imagine Einstein's 4-dimensional spacetime, let alone the 11-dimensions of String Theory? Who can imagine a Black Hole? Is this science or science fiction?

We have meanwhile bred generations of 'Sleepwalking Selves' lost in Scientific and Academic Slumberland.

## 6 Beware of Greeks Bearing Concepts

Nobel Laureate Stephen Weinberg once commented in an interview that we had to 'unlearn' much of what the Greeks had taught us. Yet in another interview he looked back with nostalgia to the Hellenist Greek Golden Age as source of all knowledge. A blatant instance of Eurocentric astigmatism since the Hellenists borrowed heavily from the Mesopotamians and Egyptians. Let's not forget the Chinese or Indians, who were no slouches either in donating generously to Europe's scientific kitty.

But a little knowledge is a dangerous thing: Nothing cannot be Something, decreed Aristotle (again) and with him the whole rest of Europe almost two millennia hence! Yet the Big Bang Theory tells us that Nothing exploded into Everything. Everybody and his/her uncle and/or aunt quotes Heisenberg: 'Nothing' in physics is not our usual 'nothing'. How can that possibly be?! Was Aristotle wrong after all?

You see, the 'Nothing' in physics fluctuates! The quantum vacuum spontaneously generates virtual particles. And what of the Big Bang? Fluctuations all the way down! Aha, now we understand! Or do we?

## 7 Cul-de-Sac

Not only is our language misleading, but so is our traditional twofold, either/or logic that still dominates our thinking today as it did in Aristotle's day. It seems that you cannot have Something *without* Nothing – Nothing is thrown into the bargain free of charge; each other's obverse. The concept of Complementarity looms once such *traditional* Greek logic is overcome. And with it the particle/wave dualism and other enigmas become palatable.

Wise words come from theoretical physicist, emeritus professor Raphael D. Sorkin of the Perimeter Institute (Canada), in an article entitled *To What Type of Logic Does the 'Tetralemma' Belong?* (Sorkin, 2010). In it he analyzes a *non-traditional* fourfold logic extant among skeptical circles in Southeast Asia 2500 years ago. As an aside, Pyrrho of Ellis was said to have accompanied Alexander the Great on his incursion into India, inspiring him to introduce his own brand of skepticism in the form of a threefold logic – unfortunately never to become mainstream in the West:

Sorkin takes a stab at taking up Ishmael's challenge to read Moby Dick's wrinkled brow:

> Considered from the standpoint of classical logic, the fourfold structure of the so-called tetralemma (catuṣkoṭi) appears to be irrational, and modern commentators have often struggled to explain its peculiar combination of alternatives ... A possible answer comes from quantum mechanics, where certain alternative logics have been proposed as **a solution to the paradoxes that arise in the attempt to describe subatomic reality**. In the early proposals of this sort, known collectively as "quantum logic", the laws for combining propositions were modified in such a way that the distributive law no longer holds. More recently though a different type of

> logical structure has been put forward in which the rules for combining propositions are classical ones but what changes are the rules of inference. It is these "anhomomorphic" logics, I would suggest, that hold the key to understanding the catuṣkoṭi form. (Emphasis added)

One of the key lessons to be drawn from Sorkin's article, which is not made explicit by him, is that the so-called 'classical homomorphic logic' referred to is Aristotelian 2-value logic, a legacy of the Greeks, that while useful under certain restricted circumstances, hamstrung Western thinkers as an insidiously tacit assumption depriving us of a range of options by warping our take on things.

Earlier the Japanese symbolic logician, H. Nakamura, had already concluded that the fourfold logic was tantamount to the definition of the null class (i.e., 0) in modern set theory. Nakamura's conclusion – when corroborated independently – would constitute further circumstantial evidence contextualizing the invention/discovery of zero to philosophy, linguistics, logic – and conversely to zero's absence when the cultural context was insufficiently evolved (Nakamura, 1954).

Recall here that it was on Aristotle's authority that the Greeks rejected the concept of 'Nothing' that, by the way, precluded the invention of zero as number in the West.

This same predisposition against the concept of 'emptiness', militated against the notion of a physical vacuum, which was also dismissed out of hand – again on the authority of Aristotle: Nature abhors a vacuum (*horror vacui*).

It took Torricelli and Pascal in the seventeenth century, who had meanwhile been exposed to the decimal system plus zero and as such had a powerful symbol at their disposal to contemplate emptiness and as such the vacuum, to eventually overturn the earlier Greek decree against the vacuum. As Brian Rotman put it:

> Pascal, Torricelli, Newton and others did not, after all, exist in a semiotic void with regard to 'nothing'. They had at their disposal, and were in fact immersed in, the whole practice and mode of discourse about 'nothing' built into the Hindu numeral system … But, as will be obvious by now, the mathematical infinite was the fruit of the mathematical nothing: it is only by virtue of zero that infinity comes to be signifiable in mathematics. (Rotman, 2004, p. 117)

## 8 Recap

Thus to recap, modern physics has meanwhile ground to a halt, *theoretically* speaking, by furthermore running head-on into the intractable and insurmountable 'measurement problem', where somehow the observer influences the observation; and the instrumentation manifests irksome *artefacts* that cannot be factored out as these are integral to the measurement process – *both in lab instruments as well as in the observer – that human instrument*!

In practice this fact does not pose a problem, under the dictum: 'Shut up and calculate'. Don't ask thorny questions, dear students. Only the results count. But theory balks at the 'bit' – pun intended.

All this has led to the most far-fetched notions in a desperate attempt to make sense of lab results, most notably entanglement/non-locality/the Many (read: Infinite) Worlds Theory and the rather tautological Anthropic Principle to account for our own world and discount the Infinity Minus-1 Many Other Worlds. Where is Occam's Razor and Strop when you need it?

## 9 Second Copernican Revolution

Meanwhile the much-vaunted Theory of Everything (ToE) remains as elusive as ever, defying 'Unification' despite the massive application of brainpower over the past one hundred years from Einstein to Hawking and ever since. Both men invoked 'God' and as such reveal their monotheistic leanings. Einstein's Old One, but then as Pantheist God of Spinoza; and Hawking's Mind of God that he sought to know by dint of formulating *God's ToE*. Nobel Prize assured! But hold on a minute, what if a ToE is ruled out in advance? The Nobel Prize will stay on the shelf for a long time to come.

Science is in a quandary, theory all akimbo, but everyone pretends it's business as usual. According to his biographer, Abraham Pais, Einstein held the view that quantum mechanics was consistent but incomplete; while a hundred years on, Nobel Laureate Sir Roger Penrose in a recent interview with Jordan Peterson stated that quantum mechanics is inconsistent. And in another interview with *New Scientist* exclaimed with reference to singularities: "It's worse! Quantum mechanics breaks down." (Brooks, 2022).

And so what we see happening is that just as in the days of Ptolemy, we keep piling on epicycles upon epicycles to save the *outdated prevailing scientific worldview* until it collapses under its own weight. Must we wait for another 100 years to conclude that unification is a pipedream? Time therefore to unleash a second Copernican Revolution.

After Copernicus flipped the geocentric planetary model of Aristotle and Ptolemy to a heliocentric model, the new worldview flips the Universe in favor of the Nonoverse. The scientific, empirical enterprise begins only after self and Universe arise co-dependently from the Nonoverse. And we applaud its achievements – let there be no doubt. We pronounce here only on the unwitting transgression of the boundaries of empiricism.

Thus the old incongruous Greek metaphysical concepts, the measurement problem referred to above and the ToE-quandary are also potentially helpful, hopeful hints that diehard scientists may heed rather than gloss over or dismiss out of hand at their own peril. As so often in the past, it takes a crisis to drive progress. After all, scientists find themselves in a box, theoretically, stymied at this juncture in history. Something has to give.

Consider for a moment: Could it be the case that the wild-goose ToE-chase is the predictable manifestation of the drive towards Unification of that deceptive Fictious Self (i.e., the observer) projected upon a presumed Objective Reality? Could it be that there is no Unity 'out there', but manufactured 'in here'?

To quote Sir Arthur Stanley Eddington, astronomer, physicist, mathematician and writer on science and the philosophy of science, who says in his book *Space, Time and Gravitation* (Eddington, 1920):

> All through the physical world runs that unknown content which must surely be the stuff of consciousness. Here is a hint of aspects deep within the world of physics, and yet unattainable by the methods of physics. And, moreover, we have found that where science has progressed the farthest, the mind has but regained from nature that which the mind has put into nature. We have found a strange footprint on the shores of the unknown. We have devised profound theories, one after the other, to account for its origin. At last, we have succeeded in reconstructing the creature that made the footprint. And lo! it is our own.

The relentless human drive towards unity (1) is not surprising, when seen as the obverse of the sense of Self, emerging from the mists of history and fastidiously cultivated in the socialization process from birth onwards to produce responsible, law-abiding members of society: our conventional, work-a-day world.

In his *Mysticism and Logic*, Bertrand Russell (1988) writes:

> That the things which we experience have the common property of being experienced by us is a truism from which obviously nothing of

> importance can be deducible ... **The generalisation of the second kind of unity, namely, that derived from scientific laws, would be equally fallacious**, though the fallacy is a trifle less elementary [emphasis added].

Or take philosopher-logician W. V. Quine when in the preface to his *Theories and Things*, he justifies taking Objective Reality seriously:

> Our talk of external things, our very notion of things, is just a conceptual apparatus that helps us to foresee and control the triggering of our sensory receptors in the light of previous triggering of our sensory receptors. The triggering, first and last, is all we have to go on.
>
> In saying this I too am talking of external things, namely people and their nerve endings. Thus what I am saying applies in particular to what I am saying, and is not meant as skeptical. There is nothing we can be more confident of than external things – some of them, anyway – other people, sticks, stones. **But there remains the fact – a fact of science itself – that science is a conceptual bridge of our own making**, linking sensory stimulation to sensory stimulation; there is no extrasensory perception [emphasis added].

Quine queries perception but not *introspection* and *the illusory sense of Self co-created by perception*, just as Russell, too, takes the Self and Objective Reality as given.

Herein lies the crux of the issue, *the reciprocal relationship between the sense of self, perception of an object, and ultimately the Universe itself.*

## 10 Bold Step

The final bold step would be to posit that if a Universe does not 'exist' in the strict sense, that is as Objective Reality, then there is no question of an origin, and as such we must conclude that supposed signs of an origin are misleading, such as the Cosmic Microwave Background Radiation (CMBR). CMBR must be 'instrument artefact' – that is, the combined impact/outcome of both observer and observed, as indeed applies more generally to what is commonly referred to as 'human experience'.

Nor can CMBR be factored out of the equation as it is inherent in the observation process, and so may be expected to be 'smooth' (uniform) across the sky. At both ends of the spectrum, macro- and micro-level, ultra large and ultra small, access is prohibited, as it were, wrested from our grasp, and so beyond the empirical realm of science-proper. To overreach would smack of 'scientism'.

Furthermore, the so-called Constants of Nature would not be universal, but vary from galaxy to galaxy. The only requirement imposed by the Nonoverse is that in all cases their Conservation Laws sum to zero.

We are caught in a language (thought) bubble and so is our reading of laboratory instruments that serve as extensions of our senses. *Ergo: one can never think* (*or, indeed, measure*) *one's way out of 'the belly of the whale' – only further in.*

Puncture the bubble and both observer and observed, Self and Universe, vanish into thin air. But to give it a name and designation anyway, we refer to the Nonoverse, nonism and nontology in order to communicate this insight in the context of social convention within polite society – our human condition.

This realization must have dawned across human civilizations – privy to shamans, mystics, prophets and seers – since time immemorial. Priest castes and their scribes must have recognized the predictive power in manipulating numbers, applied in service of the rulers of the day.

What appears to us today as the straightforward science of metrology in Mesopotamia (standardized lengths, surfaces, volumes and weights, for example), must have seemed to the elite as uncanny control over the lives of the populace during the urban revolution; pyramid construction/architecture generally in ancient Egypt; altar construction in India; natural phenomena and affairs of state in China; astronomy in all.

As such, the business of manipulating numbers that led to the innovation of the placeholder and the number zero is of particular interest in surfacing in mathematics since zero is more than a metaphor that may be dismissed out of hand, as some do religion or philosophy, for example (Hawking) – mere words. Hence the significance of zero's emergence in the historical record since zero has proven so utterly indispensable.

Future research may bear out a likely link between the priest castes, religion, numerology/numbers and, as such, suggest a likely relationship with the respective civilizations' Creation Myths and the propensity for the eventual signification of Nothing as Something.

## 11 From 1 to 0

Note, then, the universal starting point of all counting systems across civilizations from 1 onwards – and not from zero! Counting was apparently a very concrete, practical affair in the lives of common people, who derived their sense of Self from the 'objects' in their lives – in other words, interaction with the environment.

It was only after tens of thousands of years of making tally marks on sticks and stones and bones, followed by 'cipherization', that a sophisticated concept of emptiness arose to finally facilitate the Great Leap from 1 to 0, a feat universally hailed as one of the greatest innovations in human history.

Our whale-hunt is not an anti-science tirade but nautical chart out of 'the belly of the whale', read: the 'Universe', of current scientific theorizing in which generations have got lost at sea, and indeed perished.

As far as that is concerned, compare Melville's Moby Dick to the biblical Jonah being swallowed up by a 'big fish' and the myth of Gilgamesh of a comparable thrust – apparently all in quest of immortality. Or take the perspicuous Polynesian proverb: *"Standing on a whale, fishing for minnows."*

The 'passing fables' of human science bravely undertook to study in minutes detail the interior of the 'belly of the whale', i.e., the Universe that swallows us all up.

## 12 Out of the Belly of the Whale

À la Eugene Wigner, leading lights in the sciences ask why mathematics is so 'unreasonably effective', but find it not worth noting that human language itself is at least as unreasonably effective.

Ask anyone directions how to get from A to B, follow the directions, and presto!, you found your way from A to B. What a miracle! Follow a mathematical formalism like Schrödinger's superposition wave equation to predict by statistical probability (not Certainty!) experimental outcomes, and by golly, it works. The latter is unreasonably effective; the former not? After all, both literacy and numeracy are aspects of the same neural network in the same brain.

Alternatively stated, physics is *prescriptive*, not as is commonly assumed *descriptive* [Michel Bitbol, philosopher of science, CNRS, Paris]. We are told by math/physics how to arrive at likely outcomes, statistical probabilities, not to conjure up 'a picture of the world as it really is' – in other words, the Einstein–Bohr dispute.

## 13 Modern Science as Sophisticated Form of Divination?

Manipulating squiggles (i.e., numbers) on paper to predict the future – something that Homo sapiens have always done by other means and other media like stargazing, tea leaves, cracked tortoise shells etc.

Time for Rationalism to be re-Mystified, at least to some extent. Our pretensions not withstanding, we cannot know what is in the offing, but are able to make best-guess estimates. It is all that humans can do. And admittedly we've got pretty good at it.

Modern linguistic analysis serves well to resolve the issue of how we were led astray for millennia. A few hundred thousand years of cultural development equipped us with a impressive linguistic toolkit to tackle sundry jobs (Wittgenstein, Mr. Hawking!), including how to get from A to B, indeed from the Earth to the Moon and beyond. But language *mesmerizes* speaker and listener alike, and therein lay the danger of *reification*. Fictions are turned into factoids in support of our prevailing worldview with subscribers worldwide.

But modern science has now run up against the limits of language to describe the subatomic world of quantum weirdness, entanglement, non-locality etc. etc. The fly in the ointment, proves to be our naïve and literal interpretation of *internal* language (including mathematics and physics) as standing in one-to-one correspondence to *external* objects. Rather, meaning is derived from coherence of signs and syntax within the semiotic system concerned (i.e., human language, mathematics).

Nor is this a plea for Idealism (subjectivity), as argued. Idealism, too, must be dispensed with, just as objectivity in the sciences, if we are to leave behind 3000 years of futile philosophic fussing. Both objectivity and subjectivity assume there to be a self, which on closer examination vanishes into thin air – as does the air itself. While the neurosciences produce 'neural correlates' galore these days, 'consciousness' remains a mystery; no self to be found anywhere, and hence no observer either, leaving the sciences bereft and in bewilderment. Are we a 'brain-in-a-vat' or a computer simulation?

Nor, we hasten to add, is this proleptical rant either a Meaningless Nihilist or Self-Absorbed Solipsist Apologia, as will be shown below by dispensing with the tacit assumption of a Self at all; and therewith obviate Idealism as final refuge. There is no place to run, no place to hide.

## 14 Constants of Nature as Fudge Factors

The common assumption today still is that classical physics does give the observer an accurate *description* of the world, but may perhaps not apply to the quantum realm. Yet the quest for a ToE persists, just in case QM and Gravitation may be reconciled. The assumption at work is that classical physics (macro-world) gives us *exact* answers, while quantum physics (micro-world) can only provide *statistical probabilities*.

Grey-mopped Einstein was simply wrong when he said that the most incomprehensible thing about the universe is that it is comprehensible. Wishful thinking by old Albert. It is sooner the other way around once we comprehend *why* the universe is incomprehensible.

Rather, I submit, even Newton's and Einstein's classical physics are no less approximate than quantum physics is probabilistic. The 'sleight of hand' that hoodwinks even today's slick 'mathemagicians' themselves, is the introduction of the so-called 'physical constants of nature'. But these 30+ constants of nature are 'cooked', they are 'fudge factors' to compensate for all the myriad interdependent phenomena left out of consideration, which impinge on both the experimenter and experiments performed to verify their veracity.

No phenomenon arises in isolation but is integral to the entire web of the universe; and with it engenders the juxtaposed sense of self that seemingly 'observes' the phenomenon under scrutiny 'subjectively'. These fudge factors club together the sum of myriad disparate effects of impinging forces to enable us to arrive at seemingly simple but deceptive formulations that mislead us into thinking that the universe is indeed comprehensible and our formulas exact (deterministic). Divination masquerading as sophisticated science.

Not only that but in concert with each other the Bureaus of Standards (both national and international) 'conspire' to annually filter out reported divergent lab data in measuring these physical constants to have them appear constant in theory – rather than admit fluctuations in practice.

Inconstant constants may otherwise raise suspicions and undermine the Universal Church of Science and with it the Power Hierarchy of the Secular Priesthood. Billions in funding are at stake, not to mention status, money, power. It is this that accounts for the inertia of the prevailing worldview.

Our universe is a Seamless Whole, or rather a Seamless Hole, Nothing manifesting as Something, rendered by conventional human agency in dissecting it into bite-size chunks so as to digest and assimilate experience for human consumption (read: Descartes' Method).

But in so doing, we shatter the (W)Hole, forever after stuck with the shards that cannot be glued together again: Humpty Dumpty-like after his Great Fall (a suggestive reference to the Fall from Grace in Christianity that expelled us form the Garden of Eden forever – read: today's Scientific Labyrinth). The rationale is fairly obvious. According to Barrow (1992):

> The current breed of candidates for the title of a 'Theory of Everything' hope to provide an encapsulation of all the laws of nature into a simple and single representation. The fact that such a unification is even sought tells us something important about our expectations regarding the Universe. These we must have derived from an amalgam of our previous experience of the world and our inherited religious beliefs about its ultimate Nature and significance. **Our monotheistic traditions reinforce the assumption that the Universe is at root a unity** … [emphasis added]

*Ergo*: God's ToE – in other words, Unification – is ruled out in advance. At root there is Nothingness, Emptiness, as far as the empty eye can see. It is a vision as daunting as Moby Dick himself must have been to Melville's Captain Ahab: the *Mysterium Tremendum*.

Nothingness, Emptiness at bottom, too, when serious-minded scientists contemplate the Holographic Principle or the Simulation Hypothesis.

What the Holographic Principle contends, among others, is that 'Reality' may be a projection from a 2-dimensional plane onto our classical work-a-day world of 3-dimensions. It is furthermore contended that as the quantity of data/computing power required to model volume would be too great, and in fact unnecessary, projection of bounding surfaces on a plane suffices.

But equally this fanciful notion implies a form of 'emptiness' at the core, itself the likely inspiration for the invention/discovery of zero as number.

Something similar holds in the case of the Simulation Hypothesis presently in vogue, namely that 'Reality' as we experience it in daily life, could conceivably be a computer simulation designed by an advanced alien civilization, turning us humans into the characters in a computer game, unable to tell whether we are 'Real' or not.

But wouldn't such characters be 'empty'? Grating indeed in the context of the prevailing worldview, yet the natural outcome once overhauled in the second Copernican Revolution. The writing has been on the wall for over a century now. In the words of Herman Melville quoted in the opening: *I but put that brow before you. Read it if you can.* To paraphrase Melville: *all human sciences are passing fables.*

Each alteration of earlier prevailing worldviews was fought tooth and nail by the established order of the day and their vested interests. One is reminded of William James' witticism, quoted by Hazel E. Barnes, in the translator's introduction to Sartre's *Being and Nothingness*, when she noted:

> Any new theory, said James, first 'is attacked as absurd; then it is admitted to be true, but obvious and insignificant; finally it is seen to be so important that its adversaries claim that they themselves discovered it.'

It was the plight of not only Copernicus, but no less, too Galileo, Pascal and many others, as their correspondence on the subject with, among others, Church authorities bears out. Today it is the Secular Priesthood and their coterie with whom we have to contend.

## 15 The Culprit

The 'Devil' at work is our dominant Monistic scientific worldview, functional in the conventional work-day-day world, but misleading as compass out of the Labyrinth. Cartesian Dualism and Idealist Monism have served their purpose well in getting us this far. But Dualism (2) and Monism (1) need to be augmented in order to *overhaul the prevailing worldview*.

To that end we propose to introduce into the vocabulary of the traditional range of ontologies the 0-based ontology of Nothingness or Emptiness, by coining the neologist 'Nonism' in tandem with Monism and Dualism. *Nonism, the notion that the world, the universe, is not constituted of any essence, substance or principle.*

Just as in the thirteenth century, our numeral system had to be augmented by the number zero (0) that entered Europe via the Islamic world, which adopted it from India in the late-eighth century, just so our ontologies need to be augmented as well to comparable salubrious effect. Never forget that the adoption of zero and its undergirding concept was what eventually would lead to cutting-edge science and concomitant scientific theory – but severed by then from its philosophical moorings.

As such we are able to finally shift supplely from the 'Universe' (1) to the 'Nonoverse' (0), our Home in the Trackless Void, that is, our Cosmic Abode. The erstwhile Universe turns out to be a fiction, a tacit assumption, like so much else that we were spoon-fed. Talk about thinking out of the box.

Subjectivity does no better than objectivity when the tacit assumption of the sense of self is teased apart into its various components, which in turn vanish. This might well have been anticipated by the sciences when presumed objects are found to be composed of atoms, which are composed of subatomic particles, which finally vanish into what are supposed to be rarefied fields that spontaneously generate virtual particles. Read: a modern-day Monist Creation Myth that holds Believers in thrall as did so many other Pluralist Creation Myths in their day.

Since modern science evolved out of the Western tradition based on ancient Greek metaphysical misconceptions that dismiss the concept of Nothingness out of hand, such a outmoded mindset maybe expected.

## 16 Conclusion

High time, therefore, to overhaul the prevailing worldview with an extra degree of freedom to toy with both objectivity as well as subjectivity, as desired, but without buying into either.

'What is to be gained by overhauling the prevailing worldview?', the diehard skeptic may and should ask. As things stand, the prevailing worldview in the sciences and society generally tends to eclipse what may be even more profound insights.

Intoxicating as the Universe-hypothesis is – more of an a-scientific tacit assumption than anything else since we cannot step outside the Universe to observe it – the Nonoverse is even more breathtaking once comprehended in its profundity. Emptiness as far as the empty eye can see. Everything arising out of, and vanishing into, Nothingness: Baron von Munchhausen's bootstrapping, but then without the bootstraps.

The invention or discovery of zero marked the moment in human history when Nothing was signified as Something in mathematics. The number zero would in due course unleash revolutions in physics that finally led to our assumption of the existence of a Universe, with Nothing at both ends of the spectrum: the macro- and micro-levels, the ultra large and ultra small.

The proper question to ask then is not, 'Why is there Something rather than Nothing' – the mantra of modernity – but to realize, as Bohr et al. did, that these are complementary aspects characteristic of a non-traditional logic. Aristotle finally side-lined. You cannot have Something without Nothing; and *vice versa*.

See there the novelty of the Nonoverse, of Nothing manifesting as Something, but that Something being devoid of fundamental essence, substance, selves, just as nonplussed scientists are finding when probing matter at both extremes.

Nor does it prevent R&D, in pursuit of more and more practically useful data and relata, but then disabused of the obsession for Unification. Science prescribes, telling us how to best calculate outcomes, predict the future. It may be a bitter pill to swallow but science does not describe, nor has recourse to, any such thing as Reality.

To paraphrase novelist and New Yorker editor, the late-Peter de Vries (1984), has one guests at a posh cocktail party say to another in a snatch of conversation for the reader's delectation: We used to think that life was an illusion, but we now think that it only seemed that way.

## References

Barrow, John D. (1992). *Theories of Everything*. Vintage Edition.

Brooks, M. (2022). Roger Penrose: 'Consciousness must be beyond computable physics'. *New Scientist*, 14 November 2022.

Eddington, A. S. (1920). *Space, Time and Gravitation*. Cambridge University Press.

Hawking, Stephen. (1988). *A Brief History of Time*. New York: Bantam Dell Publishing Group.

Heisenberg, Werner (1958). *Physics and Philosophy: The Revolution in Modern Science*. New York: Harper.

Heisenberg, Werner (1990). *Physics and Philosophy: The Revolution in Modern Science*. London: Penguin Books.

Nakamura, H. (1954) *Journal of Indian and Buddhist Studies* (*Indogaku Bukkyogaku kenkyu*), vol. III, no. 1, September 1954, pp. 223–231.

Quine, W. V. (1981). *Theories and Thing*. Belknap Harvard.

Rotman, B. (2004). Signifying Nothing: The Semiotics of Zero (1987). In Heath, S., MacCabe, C., and Riley, D. (Eds). *The Language, Discourse, Society Reader*. Palgrave Macmillan.

Russell, B. (1988). *Mysticism and Logic. Mysticism and Logic*, Rowman & Littlefield Publishers.

Sartre, J.-P., trans. Hazel E. Barnes. (1993). *Being and Nothingness*. Washington Square Press.

Sorkin, Rafael D. (2010). To What Type of Logic Does the 'Tetralemma' Belong? *Physics* Paper 7, http://surface.syr.edu/phy/7.

de Vries, P. (1984). *Slouching Towards Kalamazoo* (London: Penguin Books).

# PART 2

## *Zero in the Arts*

# Introduction to Part 2

The reader may have gathered that the earliest signification of the empty space or placeholder on Mesopotamian clay tablets, illustrated in the cover photograph, is of utmost historic importance as marking the earliest conscious recognition of the underlying medium itself as integral to written numeric notation.[1]

Prior to that, however, the substrate of material concerned – be it stone, bone, wood, papyri, paper, for example – had always already played an indispensable role as 'contrast-gainer' in epigraphy, albeit it intuitively, to distinguish sequential mathematical signs; or even instinctively as in the space between tally marks in the Neolithic. Comparably 'absence', as meaningful category of thought, fulfilled the same role across a range of cultural disciplines, including in the Arts (blank canvas, formless clay or other materials, papyri, paper, and so on). Even as the silence between spoken words is indispensable to meaningful communication – or the silence between notes in music.

Our brief foray into Zero in the Arts focuses exclusively on the work of one visual artist. The Selected Works by ANISH KAPOOR that follow – inspired by the Indian concept of 'Shunyata' or 'emptiness', where emptiness is form and form is emptiness – exemplify the universal phenomenon more generally.

As George Gheverghese Joseph puts it in *The Crest of the Peacock: Non-European Roots of Mathematics*:

> Śūnyata is recommended in writing poetry, composing a piece of music, producing a painting, or any activity that comes out of the mind of the artist. An architect was advised in the traditional manuals of architecture (the Silpas) that designing a building involved the organization of empty space, for 'it is not the walls that make a building but the empty spaces created by the walls' … The mathematical correspondence was soon established. 'Just as emptiness of space is a necessary condition for the appearance of an object, the number zero being no number at all is the condition for the existence of all numbers.'[2]

---

1 See chapter by Brian Rotman, 'On The Semiotics of Zero' and Andreas Nieder, 'The Unique Significance of Zero: A Sense of "Nothing"'.

2 George Gheverghese Joseph, *The Crest of the Peacock: Non-European Roots of Mathematics* (Princeton and Oxford: Princeton University Press, 2011), pp. 344–5. See also Roger Lipsey (Ed.), *Coomaraswamy: Vol 2 – Selected Papers Metaphysics* (Princeton: Princeton University Press, 1977); Chapter 20, Devangana Desai, 'Sunya in the Context of Temple Art' and Chapter 21, Prem Lata Sharma, 'Sunya in the Indian Tāla System' in A. K. Bag and S. R. Sarma (Eds.), *The Concept of Śunya* (Indira Gandhi National Centre for the Arts, Indian National Science Academy and Aryan Books International, 2003).

 | DOI:10.1163/9789004691568_036

CHAPTER 32

# Selected Works by Anish Kapoor

*Peter Gobets*

The Volume Editors are grateful to include the following selected works by Anish Kapoor: Descent into Limbo, 1992; Untitled, 1990; Turning the World Inside Out II, 1995; At the Edge of the World, 1998; Untitled, 1996; Untitled, 1996; Descension, 2014; My Body Your Body, 1993.[1]

The following excerpts provide contextual information for the eight artworks:

> If [Yves] Klein's art was informed by the occult Rosicrucian belief in space as 'Spirit in its attenuated form', rather than an empty void, applicable to Kapoor's art is the Buddhist philosophy that the void is a plenum rather than a vacancy … The energy of this void, which Buddhists describe as *shunyata*, cannot be paraphrased; it can only be approached through the indexical directions and approximations of a visionary poetics, examples of which include the paradoxical songs of the Siddha adepts and the riddling *kōans* of the Zen masters. In the same spirit, Kapoor's sculptures are not objects so much as propositions, staging complex reconfigurations of space and perception.[2]
>
> The void works are for me a poetic and spiritual concept. A void object is not an empty object; its potential for generative possibility is ever present. It is pregnant. The void returns the gaze. Its blank face forces us to fill in content and meaning. Emptiness becomes fullness. Things are turned upside down. This must reveal our phantasies about voiding ourselves. Somehow, we must avoid our own death. We are not done, finished, ended. Contradiction is its essential truth. A thing and its opposite.[3]

1 All images © Anish Kapoor. All rights reserved DACS/ PICTORIGHT.

2 Nancy Adajania, Essay: 'The Mind Viewing Itself', Anish Kapoor. Available at: https://anishkapoor.com/459/the-mind-viewing-itself-by-nancy-adajania.

3 Anish Kapoor, Collège de France lecture, 23 June 2016.

© PETER GOBETS, 2024 | DOI:10.1163/9789004691568_037

FIGURE 32.1 Descent into limbo, 1992. Concrete, stucco and pigment

FIGURE 32.2 Untitled, 1990. Fiberglass and pigment, dimensions variable
PHOTO: J FERNANDES & S DRAKE

FIGURE 32.3 Turning the world inside out II, 1995. Chromed bronze, 180 × 180 × 130 cm
PHOTO: ATTILO MARANZANO, SIENNA. COLLECTION: FONDAZIONE, PRADA, MILAN

FIGURE 32.4 At the edge of the world, 1998. Fiberglass and pigment, 500 × 800 × 800 cm
PHOTO: JOHN RIDDY

FIGURE 32.5 Untitled, 1996. Concrete
PHOTO: GERRY JOHANSSON

FIGURE 32.6 Untitled, 1996. Wood and pigment, dimensions variable
PHOTO: GERRY JOHANSSON

FIGURE 32.7 Descension, 2014
PHOTO: ELA BIALKOWSKA

FIGURE 32.8 My body your body, 1993. Fiberglass and pigment, 248 × 103 × 205 cm
PHOTO: DAVE MORGAN

# PART 3

## *Zero in Mathematics and Science*

# Introduction to Part 3

ANDREAS NIEDER commences the exploration of zero in mathematics and the sciences, in his chapter, 'The Unique Significance Of Zero in Thinking: A Sense Of "Nothing"' and, from a perspective of cognitive neuroscience, both behavioral and neural processing, Nieder considers the four stages in the emergence of zero – from the mental/neural resting state lacking a specific signature to the extension of empty-set representation to finally becoming the number zero and a combinatorial number of the system of signs used for calculation and mathematics, telling us a great deal about the mind in the process. MARINA VILLE addresses the question, 'Can We Divide by Zero?' arguing that division by zero implies enlarging the family of numbers and making space for an infinite number, ∞, as in the Riemann sphere, or even many infinite numbers, as the founders of calculus envisioned.

In 'Division By Zero (*khahara*) In Indian Mathematics', AVINASH SATHAYE proposes that from Brahmagupta in the seventh century, the reciprocal 1/0 of the number 0 was used in mathematical operations. Sathaye analyzes how the ideas were not limited to the notions of limits in calculus but appear to be unusual algebraic constructs without parallel in the history of mathematics, using ideas from modern Algebra and discussing similar structures in modern mathematics. MAYANK N. VAHIA and UPASANA NEOGI consider the versatility of zero – as a representation of the absence of items in arithmetic, the absence of sound in the merging of words, to the zero as a number in arithmetical operations – considering how culture drives the human response to derivative ideas, with particular reference to European and Indian cultures and how Indian mathematicians explored zero much earlier than their European counterparts in 'Zero: In Various Forms.

MARCIS AUZINSH analyzes the deep and complicated concept of Nothing and its relation to the real physical world and empty space or quantum vacuum in 'Nothing, Zeno Paradoxes and Quantum Physics'. Auzinsh thus explores Nothing from the perspective of Quantum Physics, Zeno's paradoxes of motion as an analysis of the possibility of infinite division of time and space intervals, the ultraviolet catastrophe, and Einstein's special theory of relativity.

From the perspective of conservation laws, symmetry and non-linear dynamics, JOSEPH A. BIELLO and R. SAMSON consider 'The Significance to Physics of the Number Zero' – in particular, the role of zero as an anchor point or eye at the center of the hurricane. In accessible language, Biello and Samson demonstrate that although many mathematical models are so complex that they defy even the most powerful computers, exact or approximate

 | DOI:10.1163/9789004691568_038

results deriving from conservation or equilibrium reveal many characteristics of these complex systems. Finally, in 'A World Without Zero', R. SAMSON draws the interdisciplinary treatment of zero to a close, addressing the question of what the world would be like if the number zero had never been invented. This anti-historical question is dodged by asking a slightly less difficult question: what would the world have been like if the so-called complex numbers (ubiquitous in many fields of science and engineering) had never been invented?

CHAPTER 33

# The Unique Significance of Zero in Thinking: a Sense of 'Nothing'

*Andreas Nieder*

### Abstract

Zero is a magic number. It represents emptiness, nothing – and yet it is considered one of the greatest cultural achievements of mankind. For a brain that has evolved to process sensory stimuli ('something'), conceiving of empty sets ('nothing') as a meaningful category demands high-level abstraction. Recent studies of cognitive neuroscience now provide an insight into how an abstract concept like zero can emerge. In both behavioral and neural processing, the emergence of zero passes through four stages. In the first stage, the absence of a stimulus, 'nothing', corresponds to a (mental/neural) resting state lacking a specific signature. In the second stage, the absence of stimulus is grasped as a meaningful behavioral category, but is still devoid of quantitative relevance. In the third stage, 'nothing' acquires a quantitative meaning and is represented as an empty set at the low end of a numerical continuum or number line. Finally, the empty-set representation is extended to become the number zero, thus becoming part of a combinatorial number of the symbols system used for calculation and mathematics. The concept of zero shows how our minds and brains originally evolved to represent stimuli and detach from empirical properties to achieve the ultimate in abstract thinking. Because of this, the story of zero tells us a great deal about the mind, leaving empirical grounds behind and raising it to new intellectual heights.

### Keywords

development – phylogeny – neural activity – brain – cognition – animals – precursors of zero – numerical competence

## 1 Introduction

Zero is considered one of the greatest cultural achievements of humankind, integral to numerous breakthroughs in science and mathematics. However,

 | DOI:10.1163/9789004691568_039

zero is a most abstract and difficult concept. It took a long stretch of human history for zero to be recognized and appreciated. Children show a delayed understanding of numerosity zero, long after they comprehend positive integers. Nonhuman animals with whom we share a nonverbal quantification system exhibit a rudimentary grasp of zero numerosity. For a brain that has evolved to process sensory stimuli ('something'), conceiving of empty sets ('nothing') as a meaningful category demands high-level abstraction. It requires the ability to represent a concept independently of experience and beyond what is perceived. The brain needs to interpret 'nothing' as 'something'.

Until recently, the mental and neural foundations of the understanding of zero were unknown. This chapter traces the different prerequisites and precursors of an understanding of zero on four interrelated levels. It will first briefly recount the discovery of the number zero in human history, then follow its development in human ontogeny, next trace its evolution throughout animal phylogeny, and finally elucidate how the brain transforms 'nothing' into an abstract zero category. The emergence of zero passes through four corresponding stages in all of these realms (Nieder, 2016). At the most primitive stage, sense organs simply register the presence of stimuli. At the next higher level, 'nothing' is conceived as a behaviorally relevant category. After that, a set containing no elements is realized as an empty set or a null set. At the final and highest stage, zero becomes the number 0 used in number theory and mathematics.

## 2 Zero as a Latecomer in Human History

As shown throughout this book, the number zero is a surprisingly recent development in human history and mathematics. There was a time, not too long ago, when mankind only had positive natural numbers. The word, the symbol, and the very concept of zero had not yet been discovered or invented, respectively. Zero was first used as a sign to indicate an empty place in the positional notation system, also called the place value system. The discovery of zero as a positional placeholder is therefore entwined with the invention of positional notation systems. Zero as a sign for an empty column in positional notations appears to have first been used around 400 BCE in ancient Mesopotamia by the Babylonians, who used two slanted wedges as a placeholder (Ifrah, 2000). 'Nothing', the absence of a numeral in a positional notation system, was now realized as a meaningful category for the first time – a void placeholder, denoted by a sign.

Importantly, however, these signs for zero invented by the Babylonians did not have their own numerical value. However, once mankind had a sign to denote emptiness in a notation system, it was only a small step to realize that this sign represents an empty set. This concept of an empty set is important, because only with it does zero acquire a quantitative meaning. Zero becomes a set that is empty and is adjacent to a set that contains only one element. Now this sign denotes zero as a numerical member on a number line. When exactly this discovery was made in history is difficult to determine, but it marks a turning point in thinking, because it requires the insight that even if a set is empty, it still is a quantitative set.

Once zero was associated with a quantitative meaning, with time it turned into a real mathematical concept: the number zero. Zero first became associated with an elementary concept of number in India around the seventh century. The first written record of the use of zero as a number in its own right comes from the Indian mathematician Brahmagupta (Ifrah, 2000). The numeral for zero as we know it today (0) first appeared in India on the wall of a temple in Gwalior (central India), dated 876 CE (Menninger, 1968). A small circle in engraved numbers denotes the measures of a flower garden made as a gift to the temple. Now, zero was no longer a humble empty set, but a number that became part of a complex number theory and a combinatorial symbol system, enabling arithmetical operations. The number zero arrived circa 1200 CE in the West along with the rest of the Indo-Arabic numerals 1 to 9 and a base–10 (decimal) positional numeral system.

## 3 Development of Zero-Like Concepts in Children

Relative to positive integers, the grasp of zero as a number is protracted in developing children. Newborns and infants have the capacity to recognize the number of items in a set (Izard et al., 2009) and to solve basic addition and subtraction operations when making sense of the appearance and disappearance of puppets in puppet shows (Wynn, 1992). However, eight-month-old infants fail in operations with empty sets. According to Wynn and Chiang (1998), when infants saw a 'magical appearance' versus an 'expected appearance' (1 − 1 = 1 versus 0 + 1 = 1), they did not discriminate between these two conditions. This has been interpreted as the inability of infants to conceive of a null quantity, even though they are able to understand the absence of entities (Liszkowski et al., 2009). Mirroring the cultural invention of zero, children during their first years of development pass the same stages: first zero as 'nothing' versus 'something', next as an empty quantity (empty set), and finally as a number (Wellman and Miller, 1986).

Around three years old, children start to realize that 'nothing' can be a meaningful category that differs from all other categories comprising things. In a counting task in which children are asked to count backward as long as possible from a given number of cubes, they comprehend that the condition arrived at by taking away the last cube is 'none', 'nothing', or 'zero'. Here, zero becomes a sign for absence. Importantly, however, at this stage zero is not yet integrated in their quantitative knowledge of other small integers. For example, when asked, 'Which is smaller, zero or one?' children often insist that 'one' is smaller (Wellman and Miller, 1986). In a study in which children were asked to distribute cookies, then note the quantity of cookies on a sticky note, and finally read their notation later, a similar finding was reported (Bialystok and Codd, 2000). In this situation, the children's most common way of denoting zero was to leave the sticky note blank. When asked why they left the paper blank, children often said 'it means no cookies', or 'because there are no cookies', It is obvious that the blank paper serves as a sign of absence, much like the empty column denoting an empty position in a positional notation system.

An important leap forward for children is to understand that zero is a quantity, and as such can be placed on a number line along with other positive integers. When faced with tasks employing number words and numerals, children realize that zero is the smallest number in the series of positive integers by about age six (Wellman and Miller, 1986). This is rather late in development and may have to do with their problem with dealing with symbolic number tasks in general, rather than an actual lack of understanding of zero.

However, children younger than age six can comprehend empty sets as a void quantity if they don't have to use symbols in such tasks. This was indeed confirmed in a more direct, nonverbal numerosity ordering task in which four-year-old children were tested on numerosities represented by the number of dots in displays (Merritt and Brannon, 2013). Here, children were required to select the quantitatively smaller of two numerosities by touching it on a touchscreen with a finger (Figure 33.1A and B).

As was expected, based on earlier studies, it was found that children were more accurate at ordering countable numerosities compared with ordering pairs that contained an empty set. More importantly, however, they also found evidence that the children had a rudimentary grasp of empty sets. To figure out whether children treat empty sets as quantitative stimuli, rather than a category beyond numbers, the numerical distance effect was exploited. The numerical distance effect describes the well-known psychophysical finding that two numerosities can be more easily discriminated the greater the

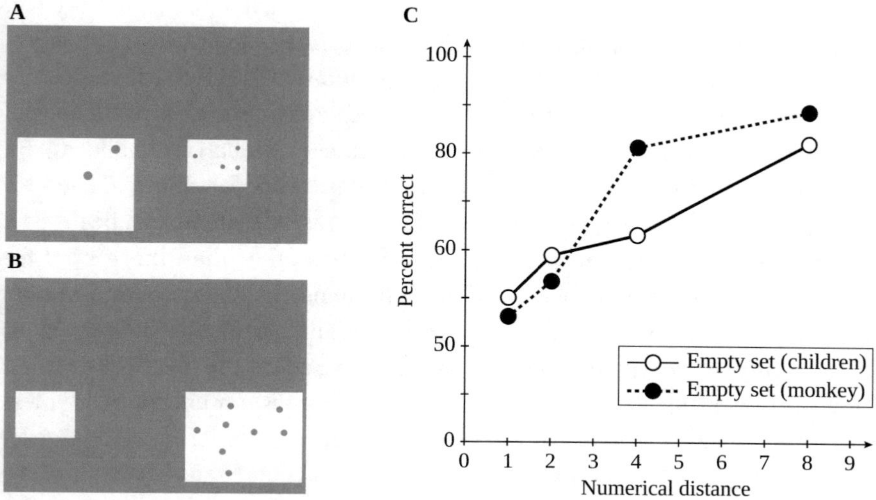


FIGURE 33.1 Four-year-old children and monkeys represent empty sets. (**A**) Children and monkeys learned in standard trials to order numerosities (dot collections on a white background) in ascending order by selecting them on a touch-sensitive screen. In this example image, they first had to select numerosity 2 (*left*) and then numerosity 4 (*right*). (**B**) In intermingled empty-set probe trials, one of the sets contained no item. Again, both sets had to be touched in ascending order. (**C**) The performance accuracy of children and two rhesus monkeys for empty sets as a function of numerical distances of 1, 2, 4, and 8. For instance, accuracy at distance 1 represents performance for empty sets and numerosity 1, accuracy at distance 2 represents performance for empty sets and numerosity 2, and so on (after Merritt et al., 2013)

magnitude difference between them. This means that if children define empty sets as having a quantitative meaning in relation to countable numerosities, they should have the most difficulty in discriminating between empty sets and the smallest numerosity, 1. Put differently, they should mix up the empty set with sets of one more often than with other sets. Indeed, this is what happened: the children who could reliably order countable numerosities exhibited a distance effect with empty sets (Figure 33.1C). The result of this study indicates that around four years of age, children begin to include a non-quantitative representation of 'nothing' as an empty set into their mental number line.

The final and most advanced stage in development is reached when children around the ages of six to nine acquire the concept of zero as a number. At that time, they are increasingly able to correctly affirm and deny simple algebraic rules, particularly if they involve zero. For instance, they know that 'If you add 0 to a number, it will be that number'. From age seven onward, children typically understand three general rules, namely: $0 < n$, $n + 0 = n$, and $n - 0 = n$. They are able to justify and reason about such rules, again with zero. It has therefore been concluded that young elementary school children possess some understanding of simple algebraic rules, and that zero holds a special status in fostering their reasoning about such rules (Wellman and Miller, 1986).

No doubt, learning the meaning of zero is hard work for children. Even for adults, zero remains a special number. Behavioral experiments indicate that the representation of zero is based on principles other than those used for positive integers. For instance, while adult humans' reading time for numbers from one to 99 grows logarithmically with the number magnitude, zero takes consistently more time to read than expected based on the logarithmic function (Brysbaert, 1995). In parity judgment tasks, zero is not judged as a typical even number and probably not part of the mental number line (Fias, 2001; Nuerk et al., 2004). When elementary school teachers were tested on their knowledge of zero, they exhibited confusion as to whether or not zero is a number, and showed stable patterns of calculation error using zero (Wheeler and Feghali, 1983). This clearly shows that zero is a very special number and stands out among other integers, even in adulthood.

## 4 Zero-Like Concepts in Animals

If zero-like concepts emerge early in human development, maybe primordial concepts of zero are also phylogenetically present in the animal kingdom. After all, research in recent decades has shown that humans and animals share an approximate number system for processing a numerical quantity of countable items (Nieder, 2005). Today, we know that animals indeed also show zero-like concepts, but it is important to take a closer look. Similar to the developmental progression in children, it is necessary to classify the types of zero precursors in animals. Before animals understand that zero refers to the null quantity of an empty set, they need to grasp that 'nothing' is a behavioral concept that opposes 'something'.

Animals can be trained to report not only the presence, but also the absence of stimuli ('nothing'). Rhesus monkeys, for instance, can be taught to press one of two buttons to indicate the presence or absence of a light touch (de Lafuente and Romo, 2005), or to report the presence or absence of a faint visual stimulus (Merten and Nieder, 2012). Clearly then, animals are able to recognize 'nothing' not just as the absence of a stimulus, but also as a behaviorally relevant category.

Can animals also ascribe a quantitative meaning to 'nothing' as an empty set? In many earlier studies, the situation is far from clear. Animals have been trained to associate set sizes with visual or vocal labels, including a sign for 'no item', in an attempt to mimic 'symbolic' number processing. An African grey parrot named Alex used human speech sounds to report the absence or presence of same/different relationships between objects (Pepperberg, 1988). For instance, when asked 'What's different?' when faced with two identical objects, he correctly responded with 'none' (Pepperberg and Gordon, 2005). However, Alex failed to respond similarly when asked how many items were hidden under two empty cups in a follow-up study (Pepperberg, 2006). Maybe the parrot used 'none' to signify the absence of object attributes rather than the absence of the objects themselves. Alternatively, Alex's response may have simply indicated a failed search. In the absence of reports of a numerical distance effect, it would be premature to attribute a quantitative meaning to the parrot's response.

Similar interpretational limitations arise in studies in which nonhuman primate studies were trained to associate sets with specific visual signs. A female chimpanzee learned to match a 0 sign with an empty tray (Boysen and Berntson, 1989). When the chimp was shown a pair of numeral signs, she was able to select the sign that indicated the arithmetic sum of the two signs (e.g., shown 0 + 2, she selected 2). Importantly, however, the chimp would also have succeeded if she had simply ignored the 0 sign that denoted nothing. Whether she learned to interpret the zero sign in a quantitative way therefore remains an open question. In another study, squirrel monkeys were trained to choose between all possible pairs of numerals 0, 1, 3, 5, 7, and 9 to receive that many peanuts in return (Olthof et al., 1997). The monkeys always chose the larger sum between two sets of numeral signs (e.g., 1 + 3 versus 0 + 5). However, the easiest interpretation of why the monkeys succeeded in this task is that they chose the sign that was associated with the greater amount of reward and therefore went for a hedonic, not numerical value. The simplest interpretation of this behavior is that the 0 sign most likely meant 'nothing' to the squirrel monkeys rather than the null quantity. As a final example, a female chimpanzee learned to successfully discriminate between numerosities and signs associated with numerical values (Matsuzawa, 1985). She had also been trained to match certain numbers of dots to signs using a computer-controlled setup (Biro and Matsuzawa, 2001). Eventually, she mastered matching blank squares containing no dots with the sign for zero. However, she failed to transfer the zero sign from the matching task (cardinal domain) to the ordinal task without further training (Biro and Matsuzawa, 1999). If she had learned a concept of zero quantity, she should have been able to immediately transfer it to a related numerical task. In conclusion, the aforementioned studies allow only limited conclusions about whether animals have zero-like concepts. First, it cannot be ruled out that animals associate the 0 sign with 'nothing' instead of with 'null quantity'. Second, even though animals learn to associate the absence of items with arbitrary shapes, their ability to transfer that to novel contexts (e.g., from cardinal to ordinal tasks) is severely limited.

However, more recent studies with rhesus monkeys showed that these primates recognize empty sets quantitatively with other numerosities along a numerical continuum by exhibiting a numerical distance effect with empty sets (Merritt et al., 2009). The monkeys were trained to touch pairs of variable dot numerosities shown on a touchscreen in ascending order; for instance, first two dots, then five dots (Figure 33.1A). After they managed this task, transfer trials with empty sets as an occasional member of the numerosity pairs were presented to the monkeys (Figure 33.1B). Because they were arbitrarily rewarded for responses to numerosity pairs containing empty sets, they did not have a chance to learn the correct response. Still, the monkeys were immediately able to order empty-set stimuli of variable appearance as the smallest numerosity, thus indicating a conceptual understanding of null quantity (Figure 33.1C). Later, the numerical distance effect with empty sets was confirmed in monkeys trained to match numerosities including empty sets (Ramirez-Cardenas, 2016). Both monkeys tested in this study mistakenly matched empty sets to a numerosity of one more frequently than to larger numerosities.

One may expect that nonhuman primates, our closest animal cousins on the tree of life, grasp the concept of an empty set. However, even honeybees, a small insect on a branch very remote from humans on the animal tree of life, belong to the elite club of animals that comprehend the empty set as the conceptual precursor of zero (Nieder, 2018). Honeybees can rank numerical quantities according to the rules of 'greater than' and 'less than', and they extrapolate the 'less than' rule to place empty sets next to a set of one at the lower end of their mental number line (Howard et al., 2018). In these experiments, the bees were rewarded for discriminating between displays on a vertical rotating screen that showed different numbers of items, ranging from zero to five. First, the bees were trained to rank two numerosity displays at a time. One group of bees was rewarded with a sweet sugar solution whenever they flew to the display showing more items, thereby following a 'greater than' rule. The other group of bees was trained on the 'less than' rule and was rewarded for landing at the display presenting fewer items. The bees learned to master this task with displays consisting of between one and four items, not only with familiar numerosity displays, but also for novel displays. Surprisingly, the bees obeying

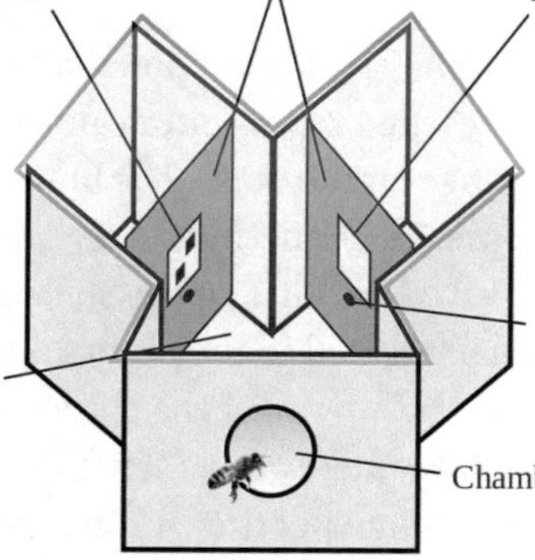


FIGURE 33.2 Honeybees shown an understanding of empty sets. Top view of the testing apparatus known as Y-maze (covered by a Plexiglas ceiling) used in one of the experiments. A bee enters through the small hole into the decision chamber where it is presented with two numerosity options and must make a decision on which pole to land to potentially collect a reward for a correct choice. Bees trained on the 'less than' rule spontaneously chose the empty-set display over all displays showing items (after Scarlett et al., 2018)

the 'less than' rule spontaneously landed upon the occasionally inserted and unrewarded displays showing no item (i.e., an empty set). When entering the test apparatus (Figure 33.2), they found it hard to judge whether the empty set was smaller than 1, but were progressively better when they had to compare 2, 3, or larger numbers with empty sets. In doing so, bees understood that the empty set was numerically smaller than sets of one, two, or more items. With this, the bees demonstrated the numerical distance effect with empty sets, a hallmark of number discrimination. These series of experiments therefore demonstrate that bees grasp the empty set as a quantitative concept.

Collectively, these studies provide clear evidence that animals, when faced with a range of numerosities, can judge the empty set as a quantity that is related to other numerical values. However, there is no evidence that any animal will ever transcend empty-set representation in order to arrive at the last and most advanced stage of zero-like concepts to understand zero as a true number symbol. The number zero therefore will remain a uniquely human concept. Still, it is fascinating to see that the earlier and more basic concepts of empty sets we first need to understand as children are already within the reach of animals. The foundations or precursors of a zero concept can therefore be investigated from a evolutionary point of view in the brains of monkeys.

## 5 Neuronal Representations of 'Nothing' and Empty Sets

But how does the brain and its anatomical and functional units, the neurons, deal with zero-like concepts? Clearly, when representing 'nothing' empty sets, or the number zero, must be a challenge for the brain. The brain and its sensory neurons have evolved to always represent something – an obstacle, a predator, a mate; all of them can be perceived because of the sensory energy these objects radiate can be detected and processed in the brain. In the same way, a collection of items is represented by the energy of the stimuli that constitute this set. Neurons signal such stimuli by increasing their firing rates, i.e., the number of electrical impulses they generate per time. In the absence of stimulus energy, neurons are inactive and in a resting state.

Because cognitively advanced animals can learn that the absence of stimulation is also a meaningful condition and relevant for behavior, the lack of sensory stimulation can also become a meaningful category and thus can be encoded by neurons. Indeed, we found evidence that the absence of a stimulus is actively represented by neurons in the prefrontal cortex of rhesus monkeys (Merten and Nieder, 2012). At the behavioral level, we made sure that 'nothing' was an important category by training rhesus monkeys to report both the presence and the absence of a stimulus. This was done in a visual detection task in which half of the trials were visible stimuli that were flashed on a monitor, whereas in the other half of the trials no stimulus at all was presented. The intensity of the stimuli was varied so that the faint ones were barely visible. This was a difficult task for the monkeys, and they were often uncertain whether a stimulus was shown or not. If a sensory stimulus is faint enough that neurons have difficulty sensing it, the monkeys are forced to make up their minds and subjectively judge whether they had or had not seen a barely visible stimulus. As a result, the monkeys respond from trial to trial in opposite ways to a faint stimulus that had the same physical energy. The internal status of the monkeys determined whether they judged a stimulus as being present or absent, allowing us to investigate how neurons encode both the 'present' and 'absence' categories regardless of the stimulus' energy, which was identical in these situations.

When we recorded the electrical impulses of single neurons in the prefrontal cortex while the monkeys performed this task, we found two groups of neurons that signaled the monkeys' decisions about the stimulus. One group of neurons increased their firing rates and thus encoded the monkeys' 'stimulus-present' decision whenever the monkeys later reported to have seen the stimulus. Such a response might be expected for neurons that respond to the energy from a present stimulus, and similar responses in the frontal lobe have also been reported for monkeys responding to touch (de Lafuente and Romo, 2006).

Surprisingly, however, a second group of neurons increased their impulse rates whenever the monkeys decided that they had *not* seen a stimulus, but such neurons remained silent whenever the monkeys voted for a 'presence' (Merten and Nieder, 2012). The neuronal responses of such neurons were highly correlated with the monkeys' subjective reports. Analyses of trials in which the monkeys failed to detect a faint stimulus showed that the activity of these neurons predicted the monkeys' judgments even before a response could be planned. This finding suggests that behaviorally relevant 'stimulus absent' decisions are not encoded by resting-state neuronal responses. Rather, the brain seems to internally translate the lack of a stimulus into a categorical and active 'stimulus absent' representation signified by increased firing. In other words, 'nothing' as a meaningful condition does excite the brain.

This kind of neuronal activity does not have to bear any quantitative meaning. As a signature for a quantitative representation, neurons are expected to represent not only the absence of stimulus but also signal empty sets as a quantity at the lower end of a numerical continuum. We therefore recorded from monkeys that treated empty sets as a quantity in a numerosity matching task (Ramirez-Cardenas et al., 2016). Previous studies has shown that 'number neurons' in the prefrontal cortex (PFC) and the ventral area of the intraparietal sulcus (VIP) of primates are tuned to preferred numerosities by maximum activity and show such a distance effect at the neuronal level by a progressive decline of activity relative to the preferred numerosity (Nieder, 2016) (Figure 33.3A). Indeed, we found populations of single neurons tuned to empty sets, just as other neurons were tuned to countable numerosities, and therefore treating them as conveying a quantitative null value.

However, the way such neurons encoded empty sets differed in interesting ways between the VIP in the parietal lobe and the PFC in the frontal lobe. Neurons in the VIP, at the input to the cortical number network, responded vigorously to empty sets but not at all to other countable numerosities; in other words, they encoded empty sets as categorically different from all other numerosities (Figure 33.3B). For VIP neurons, the empty set obviously was not part of a numerical continuum, but simply 'nothing' as opposed to the something of all countable numerosities.

Higher up the cortical hierarchy, however, PFC neurons were also tuned to empty sets, but they treated zero numerosity as more similar to numerosity one than to larger numerosities; in other words, PFC neurons exhibited a numerical distance effect that is indicative of empty sets being incorporated

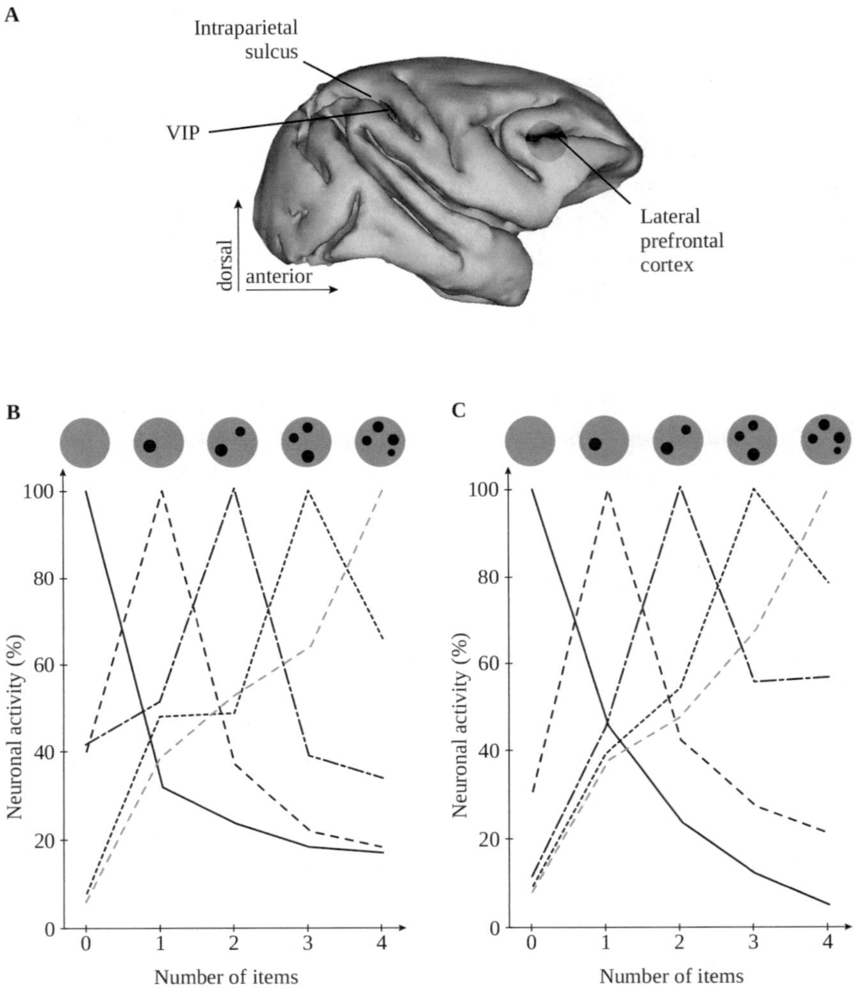


FIGURE 33.3 Number neurons in the frontal and parietal association cortices of monkeys represent zero quantity. (**A**) A lateral view of a monkey brain shows the areas in the VIP and PFC from which empty-set representations were recorded. (**B, C**) Besides neurons that preferred countable numerosities from 1 to 4, some neurons in the ventral intraparietal area (**B**) and prefrontal cortex (**C**) were also tuned to zero, i.e., empty sets, and had quantity zero as preferred numerosity. While empty-set neurons in the ventral intraparietal area responded categorically (**B**), empty-set neurons in the prefrontal cortex showed a graded response characteristic of a neuronal distance effect (from Ramirez-Cardenas et al., 2016)

together with countable numerosities along the number line (Figure 33.3C). Moreover, prefrontal neurons represented empty sets abstractly and irrespective of stimulus variations. Compared to the VIP, the activity of numerosity neurons in the PFC also better predicted the successful or erroneous behavioral outcome of empty-set trials.

These findings indicate a cortical hierarchy of processing from the VIP to PFC. While the VIP is still discriminating between 'nothing' and 'something', empty sets are steadily detached from visual properties and gradually positioned in a numerical continuum on their way to the highest cortical integration center, the PFC. Interestingly, this internal sequential process in the brain neatly mirrors the timeline of the cultural and ontogenetic advancement described above. The brain transforms the absence of countable items ('nothing') represented in brain areas lower in the hierarchy, like the VIP, into an abstract quantitative category ('zero') in areas higher in the hierarchy, such as the PFC.

Understanding the physiological mechanisms behind empty-set representations is challenging. In contrast to countable numerosities, which are represented spontaneously (Viswanathan and Nieder, 2013), coding the absence of stimuli and null quantity requires explicit training. As a behaviorally relevant category, zero-like representations need to develop over time as the result of trial-and-error reinforcement learning.

This neuronal scenario outlined above indicates an effortful cognitive and neuronal process to arrive at a concept of the number zero. The studies mentioned in this chapter from human history, developmental psychology, animal cognition, and neurophysiology provide evidence that the emergence of zero passes through four stages (Figure 33.4). In the first and most primitive stage, the absence of a stimulus ('nothing'), corresponds to a (mental/neural) resting state lacking a specific signature. In the second stage, the absence of a stimulus is grasped as a meaningful behavioral category, but its representation is still devoid of quantitative relevance. In the third stage, 'nothing' acquires a quantitative meaning and is represented as an empty set at the lower end of a mental number line. Finally, the representation is extended to become the number zero, thus becoming part of a combinatorial number symbol system used for calculation and mathematics.

Since cognitive capabilities originate from the workings of neurons, the historical and ontogenetic struggle of mankind to arrive at a concept of zero may at least partly, and in addition to socio-cultural factors, be a reflection of the neurobiological challenge for the brain to leave the sensory world behind

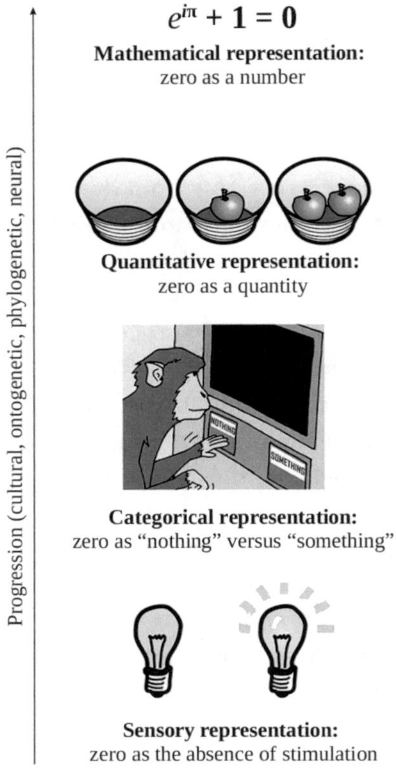


FIGURE 33.4
The four stages of zero-like concepts appearing in human culture, ontogeny, phylogeny, and neurophysiology. At the most primitive stage (bottom), sense organs register the presence of stimuli, such as light. In the absence of stimulation, sense organs are in an inactive resting state. At the next higher level, 'nothing' is conceived as a behaviorally relevant category, as exemplified by a monkey judging the absence of a stimulus. With the advent of quantitative representations, a set containing no elements is realized as an empty set or a null set. Finally, zero becomes the number 0 used in number theory and mathematics. For instance, 0 is the additive identity in Euler's identity. Each higher representation encompasses the previous, lower one. The conception of zero as a number requires a quantitative understanding of empty sets, which in turn necessitates a grasp of 'nothing' as an abstract category (from Nieder, 2016)

and start pondering about concepts that can no longer be experienced. The concept of zero shows how our brains, originally evolved to represent sensory objects and events, detach from empirical properties to achieve ultimate abstract thinking. The story of zero, therefore, shows how the mind and brain leave empirical grounds and rise to new intellectual heights.

## References

Bialystok, E., Codd, J. (2000) Representing quantity beyond whole numbers: Some, none, and part. *Canadian Journal of Experimental Psychology* 54: 117–128.

Biro, D., Matsuzawa, T. (1999) Numerical ordering in a chimpanzee (Pan troglodytes): Planning, executing, and monitoring. *J. Comp. Psychol.* 113: 178–185.

Biro, D., Matsuzawa, T. (2001) Use of numerical symbols by the chimpanzee (Pan troglodytes): Cardinals, ordinals, and the introduction of zero. *Anim. Cogn.* 4: 193–199.

Boysen, S. T., Berntson, G. G. (1989) Numerical competence in a chimpanzee (Pan troglodytes). *J. Comp Psychol.* 103: 23–31.

Brysbaert, M. (1995) Arabic number reading – on the nature of the numerical scale and the origin of phonological recoding. *J. Exp. Psychol. Gen.* 124: 434–452.

De Lafuente, V., Romo, R. (2005) Neuronal correlates of subjective sensory experience. *Nat. Neurosci.* 8: 1698–1703.

De Lafuente, V., Romo, R. (2006) Neural correlate of subjective sensory experience gradually builds up across cortical areas. *Proc. Natl Acad. Sci. USA* 103: 14266–71.

Fias, W. (2001) Two routes for the processing of verbal numbers: Evidence from the SNARC effect. *Psychological Research* 12: 415–423.

Howard, S. R., Avarguès-Weber, A., Garcia, J. E., Greentree, A. D., Dyer, A. G. (2018) Numerical ordering of zero in honey bees. *Science* 360: 1124–1126.

Ifrah, G. 2000. Universal history of numbers: from prehistory to the invention of the computer. John Wiley & Sons Inc., New York.

Izard V., Sann C., Spelke E. S., Streri A. (2009) Newborn infants perceive abstract numbers. *Proc. Natl Acad. Sci. USA* 106: 10382–5.

Liszkowski, U., Schäfer, M., Carpenter, M., Tomasello, M. (2009) Prelinguistic infants, but not chimpanzees, communicate about absent entities. *Psychol. Sci.* 20: 654–60.

Matsuzawa, T. (1985) Use of numbers by a chimpanzee. *Nature* 315: 57–9.

Menninger, K. 1969. Number words and number symbols. MIT Press, Cambridge, MA.

Merritt, D. J., Brannon, E. M. (2013) Nothing to it: precursors to a zero concept in preschoolers. *Behav. Processes* 93: 91–7.

Merritt, D. J., Rugani, R., Brannon, E. M. (2009) Empty sets as part of the numerical continuum: Conceptual precursors to the zero concept in rhesus monkeys. *J. Exp. Psychol. Gen.* 138: 258–269.

Merten, K., Nieder, A. (2012) Active encoding of decisions about stimulus absence in primate prefrontal cortex neurons. *Proc. Natl Acad. Sci. USA* 109: 6289–94.

Nieder, A. (2005) Counting on neurons: the neurobiology of numerical competence. *Nat. Rev. Neurosci.* 6: 177–90.

Nieder, A. (2016) The neuronal code for number. *Nat. Rev. Neurosci.* 17: 366–82.

Nieder, A. 2016. Representing Something Out of Nothing: The Dawning of Zero. *Trends Cogn. Sci.* 20: 830–842.

Nieder A. (2018) Honey bees zero in on the empty set. *Science* 360: 1069–1070.

Nuerk, H.-C., Iversen, W. Willmes, K. (2004) Notational modulation of the SNARC and the MARC (linguistic markedness of response codes) effect. *Quarterly Journal of Experimental Psychology: Human Experimental Psychology* 57: 835–863.

Olthof, A., Iden, C. M., Roberts, W. A. (1997) Judgments of ordinality and summation of number symbols by squirrel monkeys (Saimiri sciureus). *Journal of Experimental Psychology: Animal Behavior Processes* 23: 325–339.

Pepperberg, I. M. (1988) Comprehension of 'absence' by an African grey parrot: Learning with respect to questions of same/different. *Journal of the Experimental Analysis of Behavior* 50, 553–564.

Pepperberg, I. M. (2006) Grey parrot (Psittacus erithacus) numerical abilities: Addition and further experiments on a zero-like concept. *Journal of Comparative Psychology* 120: 1–11.

Pepperberg, I. M., Gordon, J. D. (2005) Number comprehension by a grey parrot (Psittacus erithacus), including a zero-like concept. *Journal of Comparative Psychology* 119: 197–209.

Ramirez-Cardenas, A., Moskaleva, M., Nieder, A. (2016) Neuronal Representation of Numerosity Zero in the Primate Parieto-Frontal Number Network. *Curr. Biol.* 26: 1285–94.

Wellman, H. M., Miller, K. F. (1986) Thinking about nothing: Development of concepts of zero. *Br. J. Dev. Psychol.* 4: 31–42.

Wheeler, M., Feghali, I. (1983) Much ado about nothing: preservice elementary school teachers' concept of zero. J. *Res. Math. Educ.* 14: 147–155.

Wynn K. 1992. Addition and subtraction by human infants. *Nature* 358: 749–50.

Wynn, K., Chiang, W. C. (1998) Limits to infants' knowledge of objects: the case of magical appearance. *Psychol. Sci.* 9: 448–455.

CHAPTER 34

# Can We Divide by Zero?

*Marina Ville*

## Abstract

Not dividing by zero is one of the strongest prohibitions that we learn as children. Yet, as we grow in age and in competence, we realize that many mathematical constructions would be much simpler if we could divide by zero. But this implies enlarging the family of numbers and making space for an infinite number, $\infty$, as in the Riemann sphere, or even many infinite numbers, as the founders of calculus envisioned.

**Keywords**

zero – infinity – complex numbers – projective geometry – Riemann sphere – differential and integral calculus – infinitesimals

## 1 Do Not Divide by Zero

Careful! Beware! Do not divide by zero!! This prohibition is repeated so much in school that for many students, division by zero is not a mathematical impossibility but an illegal activity like driving through a red light. But the reasons for red lights are much better understood than the impossibility of dividing by zero.

Take a number, say 100 and divide it by 4, we get 25; then multiply 25 again by 4 and we get back to 100. In math symbols we write this:

$$\frac{100}{4} = 25, \text{ then } 25 \times 4 = 100$$

So in a sense, multiplication and division are inverse to one another.

But this does not work with 0. Indeed, suppose we can divide 100 by 0 and get a regular number which we write ⟐ since we do not know what it is. We would have

 | DOI:10.1163/9789004691568_040

$$\frac{100}{0} = \boxed{?}$$

Then we should have

$$0 \times \boxed{?} = 100$$

which is impossible since any number multiplied by zero becomes zero, thus

$$0 \times \boxed{?} = 0$$

## 2 An Infinite Number?

If we still want to divide by zero, we now know that the result will not be a usual number (the technical term is real number); so what will it be?

To find a good candidate for $\frac{1}{0}$ we experiment. We divide 1 by by numbers which get closer and closer to zero:

$$\frac{1}{0.1} = 10 \quad \frac{1}{0.01} = 100 \quad \frac{1}{0.001} = 1000 \quad \frac{1}{0.00000001} = 100000000$$

etc. So $\frac{1}{0}$ would be a number larger than any real number; we would say "$\frac{1}{0}$ is infinity" and write:

$$\frac{1}{0} = \infty$$

Granted, $\infty$ would not behave like any other number, for example $\infty + 4 = \infty$, yet for any real number, say $x$, $x + 4$ is different from $x$; but the number 0 also has its idiosyncracies.

Let us indicate how $\infty$ could be a concrete number. Write words in the 26 letters of the alphabet without requiring them to have a meaning in any language or even to be easy to pronounce – thus allowing words like kkghketranvgfooapf. How many words can we write using 2 letters? Answer: 26 times 26. How many with one million letters? The math savy reader will answer $26^{1000000}$ but the non math savy reader will also see that it is a very large number. It is, however, a finite number: it would take time but we could write them all. For any given length, we have a finite number of words of that length. But if we do not specify the length and just ask how many words we can write, the answer is infinitely many.

To describe one context where ∞ was welcomed into the family of numbers, we first need to enlarge this family very much.[1]

## 3 Complex Numbers

Today we do not think twice about a $-5°$ temperature, a $-400€$ bank balance or a $-1$ button in an elevator, but negative numbers had a long history before becoming a matter of course. In the sixth and seventh centuries, accountants in India began using them to denote debts, similarly to the overdraft in our bank statements today (Guedj, 1996). In the end of the fifteenth century negative numbers reached Western Europe and became essential tools for computing and solving equations; for example, $-3$ is the solution of

$$x + 3 = 0$$

Despite their widespread use, even the seventeenth century scientist and philosopher René Descartes (1596–1650) was uncomfortable with negative numbers and called them false solutions of equations. Finally, in his book Algebra (1685), the English mathematician John Wallis (1616–1703) described the real line that we now know and use:

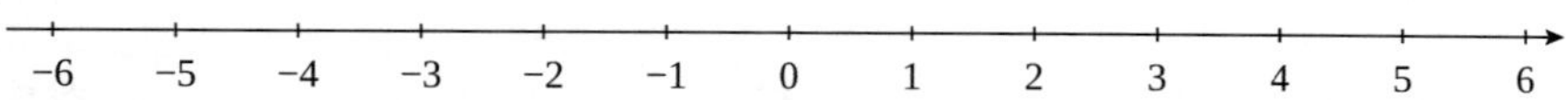


FIGURE 34.1 The real line

Wallis explained: if I stand at the number 3 on this line, I get $3 + 4$ by walking 4 steps to the right and $3 - 4$ by walking 4 steps to the left.

But negative numbers were not enough to solve all algebraic equations because of another prohibition that we learn in school do not take the square root of a negative number! Indeed, the squares of positive numbers are positive and so are the squares of negative numbers

$$3 \times 3 = (-3) \times (-3) = 9$$

---

1 In this chapter, we touch upon several topics in the history of mathematics; the reader interested in deeper expositions can look at Kline, 1972; Stillwell, 2002; Struik, 1987 (not an exhaustive list, by any means).

So there is no number which has –9 as its square; in technical terms, –9 has no square root.

Yet it would be nice to have square roots of negative numbers; consider the equation with unknown x

$$x^2 + 1= 0$$

This is a very simple equation; if we try and solve it, we can add –1 to both sides and get

$$x^2 = -1$$

and this is crying for a square root of –1. This square root appeared in sixteenth-century Italy, in the search for solutions of cubic algebraic equations (cubic=with a term in $x^3$. It is a colorful story replete with friendships, secrets and treasons. Gerolamp Cardano (1501–1576), in his 1545 *Ars Magna*, was the first one to have the audacity of using roots of negative numbers, although he was uncomfortable with them and called them impossible.

What we now call complex numbers were born: they are the sum of a real number and the root of a negative number, for example:

$$10+7\sqrt{-1},\ 3.5+4\sqrt{-1},\ -5+2\sqrt{-1},\ 4\ -9\sqrt{-1}\ \text{etc}$$

To add complex numbers, just add the real part and the multiple of $\sqrt{-1}$ part (called imaginary), e.g., $5-2\sqrt{-1}+4+5\sqrt{-1}=9+3\sqrt{-1}$.

Mathematicians began energetically computing with these strange new numbers which proved an incredible bonanza! Not only could they give solutions to all quadratic and cubic equations, but it turned out that all polynomial equations have solutions which are complex numbers. For example, letting my imagination fly, I name the first polynomial equation which comes to my mind:

$$0.3x^{68} - 8.2x^{41} + 5.1x - 236 = 0$$

and I know it has 68 solutions which are complex numbers: this is called the fundamental theorem of algebra.

For more than a century, complex numbers were widely used and very fruitfully so; after calculus took off, some of its objects and methods were extended to complex numbers. Yet, these continued to be viewed with unease and disbelief and Descartes called them imaginary. They became legitimate only when the mathematical giant, Carl-Friedrich Gauss (1777–1855) represented them by points in the plane; Gauss is also the one to have coined the term complex numbers. Note that a Norwegian cartographer, Caspar Wessel (1745–1818), and a French bookseller, Jean-Robert Argand (1768–1822) had discovered this representation shortly before Gauss, but their ideas did not have the echo they deserved.

The real numbers are a special case of complex numbers and the real line described by Wallis sits as a horizontal line inside the plane of complex numbers.

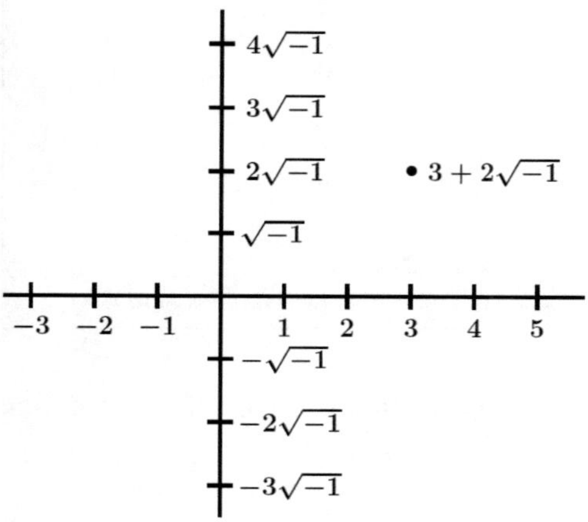


FIGURE 34.2
The plane of complex numbers

The curious reader may ask how addition and multiplication work for two points A and B of the plane.

To find the sum A + B, draw a line starting at A which is
parallel to the line between 0 and B
of length the distance between 0 and B
Then the other end of that line gives us A + B.
Note that we can get the same point by switching the points A and B

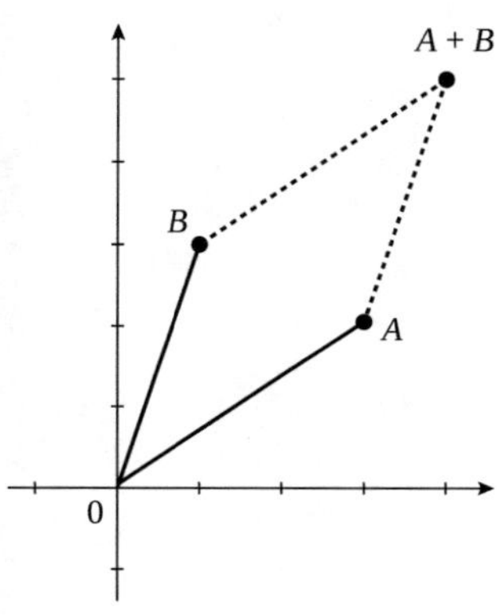


FIGURE 34.3
The sum of two complex numbers

In geometric language, A + B is the 4-th vertex of the parallelogram of vertices A, 0 and B.

To get the product AB of A and B,

multiply the lengths of the segments OA, OB (these are called the magnitudes of the complex numbers)

add the angles between OA and OB and the real axis (called the arguments of the complex numbers)

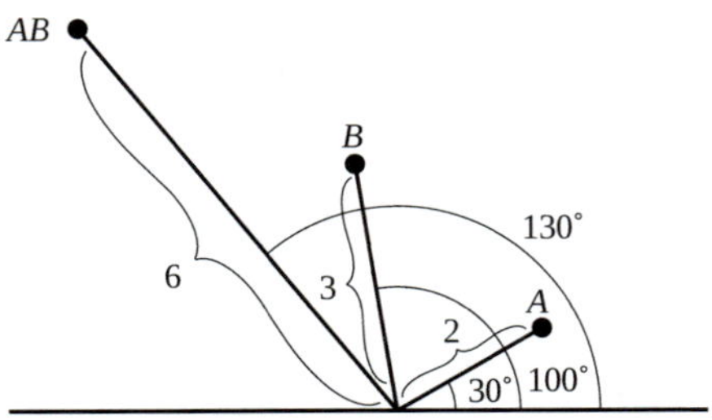


FIGURE 34.4
The sum of two complex numbers

Thus numbers have become geometric objects; to get to the infinite number, we recall some more geometry.

## 4 Going to Infinity: the Projective Space

For classical Euclidean geometry, two parallel lines never meet – and two non-parallel ones meet exactly once (Figure 34.5).

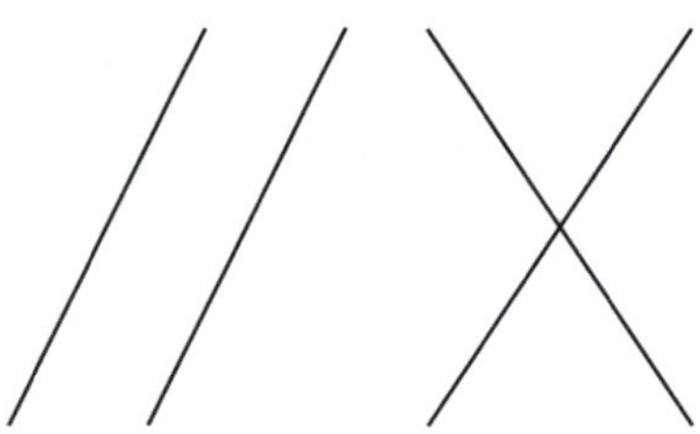

FIGURE 34.5
Lines in the plane

FIGURE 34.6
Train tracks

Yet, look at these parallel train tracks (Figure 34.6), do they not seem to meet at a point at the horizon? They do meet, but in a geometry different from the Euclidean one and called projective geometry. In projective geometry, points at infinity are added to the usual space, like the meeting point of the train tracks.

Let us go in search of points at infinity. On the real line, if you want to walk very far away from zero, you have two possibilities, either go far in the positive direction or go far in the negative direction. In the plane, where the complex numbers live, you can walk away from zero, i.e., towards infinity, by going in infinitely many directions (Figure 34.7a below). But we will only add a single point at infinity, denoted ∞, so all the points walking away from 0 will actually go towards a single point (Figure 34.7b below). When we attach ∞ to the plane, the result will look like a sphere (Figure 34.8 below).

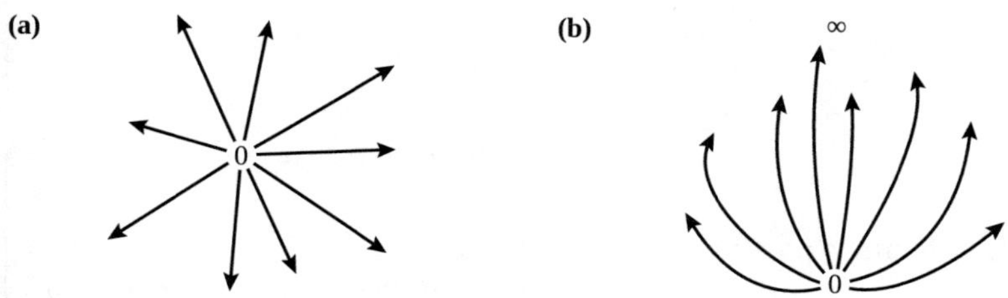


FIGURE 34.7 Points going to infinity

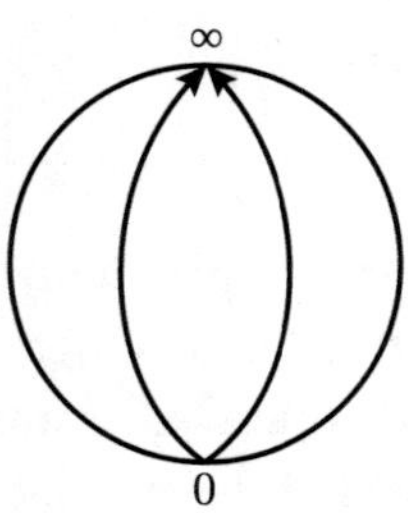


FIGURE 34.8
The Riemann sphere

To recap: the complex numbers live on a plane and when we add ∞ to the family of complex numbers, the new family lives on a sphere, called the Riemann sphere after Bernhard Riemann (1826–1866), another nineteenth century mathematical giant.

Putting points of the plane and points of the sphere in correspondence is something cartographers have been doing for centuries; one of their classical constructions, called the stereographic projection tells us how to relate a complex number viewed as a point in the plane to the same complex number viewed as a point in the sphere.

Look at the sphere in Figure 34.9 below: it has a North Pole, identified with ∞, a South Pole, identified with 0, as in Figure 34.8 above. It also has an equator represented on the figure by a horizontal circle. We take the plane going through this equator – it is a horizontal plane – and map the points of the sphere, except for the North Pole, to this plane.

Take a complex number which corresponds to a point Â in the plane, draw the line between A and ∞, a.k.a. the North Pole. This line meets the sphere at a point A which is the point on the sphere corresponding to the point A on the plane.

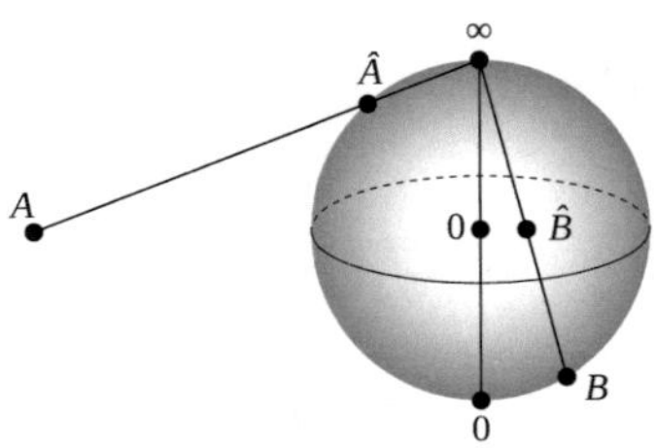


FIGURE 34.9
The stereographic projection

Looking at Figure 34.9, the reader will also notice:

- The number 0 on the sphere, a.k.a. the South Pole, corresponds to 0 in the complex plane
- Small complex numbers correspond to points on the sphere close to 0 on the sphere (look at B and $\hat{B}$); large complex numbers correspond to points close to ∞ (look at A and $\hat{A}$).

This construction of adding an infinite number to the plane of complex numbers has revolutionized the studies of curves and other objects defined by algebraic equations. This is the topic of algebraic geometry which now plays an important role in cryptography, despite being one of the most abstract fields of mathematics.

So, this number ∞ is incredibly useful, but is it all that we need? We can do some computations, such as $\infty + 4 = \infty$, $\infty \times 4 = \infty$, but what about what about $0 \times \infty$, $\frac{\infty}{\infty}$ or for that matter $\frac{0}{0}$? For these to make sense, having a single zero and a single infinity is not enough; we need several of them and that is what we now explore.

## 5 Calculus

We experiment again and replace 0 by numbers which get closer and closer to 0. Take 0.0000001 and 0.00000000000001 two numbers very close to 0. Then

$$\frac{0.0000001}{0.0000001} = 1 \quad \text{and} \quad \frac{0.00000000000001}{0.00000000000001} = 1 \text{ but}$$

$$\frac{0.0000001}{0.00000000000001} = 10000000 \text{ which is very large}$$

$$\frac{0.00000000000001}{0.0000001} = 0.0000001 \text{ which is very small}$$

So the quotient of two very small numbers can be small, large, or any real number: the size of this quotient depends on how small the two numbers are relatively to one another. Similarly the readers can check themselves that the product of a very large and a very small number can be anything, large, small, regular.

Thus we cannot specify a value for $\frac{0}{0}$ and for $0 \times \infty$.

Instead of talking of a single 0 and a single ∞, we look at infinitely small and infinitely large quantities and investigate quantities of the form

$$\text{infinitely small} \times \text{infinitely large} \quad \text{and} \quad \frac{\text{infinitely small}}{\text{infinitely small}} \tag{1}$$

Such questions are at the core of calculus which started in the seventeenth century but they go back to the Antiquity. Without going deeply into calculus, let us look at a couple of examples to see how expressions of the form (1) naturally arise.

## 6 Examples

### 6.1 *Length of a Curve*

If we have a ruler, we can measure the length of a straight line and also the length of a curve made of straight lines as in Figure 34.10.

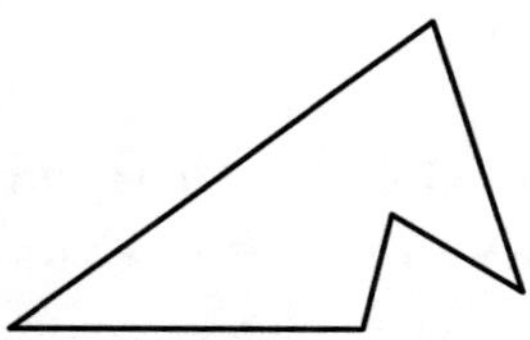

FIGURE 34.10
Length of a curve made of straight lines

What about a more general curve as in Figure 34.11a? We approximate it with straight lines as in 11b and measure the length of these lines. To get a better approximation, we take smaller lines, see 11c. For better and better approximations, take more and more segments of smaller and smaller lengths. Ideally, we should take infinitely many segments, each of them infinitely small; thus the length of the curve will be something of the form $0 \times \infty$!

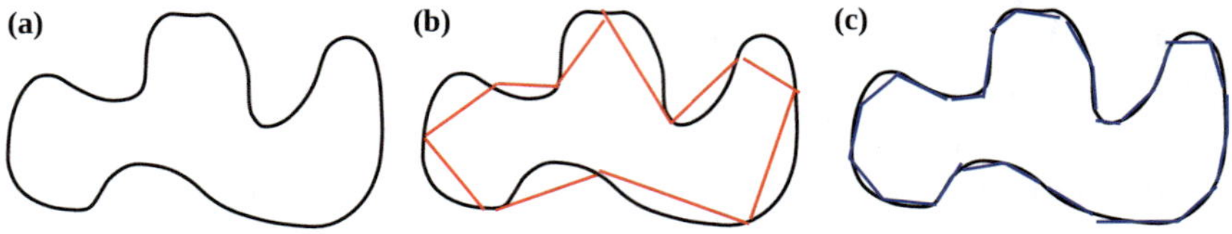

FIGURE 34.11 Length of a curve

The Greeks already had an inkling of this kind of method which blossomed with the advent of calculus in the seventeenth century: we call it integration.

## 6.2 *Speed of a Car*

If we get into a car at 9 AM, drive 140 kilometers and stop at 11 AM, our average speed will be $\frac{140}{2} = 70$ kilometers per hour. We can also measure the speed every quarter of an hour, for example if we drive 15 km the first quarter of an hour, that will give us a speed of $\frac{15}{0.25} = 60$ km per hour. We can improve our estimates by measuring speed at closer and closer time intervals. Clearly we will have the best estimate if we take measurements every instant. Ultimately we would measure the distance driven – an infinitely small distance – during an infinitely small amount of time. So finding the instantaneous speed of the car means dividing one infinitesimally small quantity by another one.

The reader probably feels, and rightly so, that it is not necessary to have such a precise information about the speed of the car. But this is an example of a rate of change of a phenomenon (here the distance on the road) which varies continuously. Mathematicians call these rates of change derivatives. Nature's laws, heat, electromagnetic forces, the movement of planets etc, are all written as equations involving derivatives (called differential equations).

Integration and derivation are the two main constructions of calculus; they are inverse of one another, but this is a different story.

## 6.3 *Formal Definitions?*

Infinitely small quantities or infinitesimals as they were called, were the main building blocks of calculus when it started and first grew, in the seventeenth and eighteenth centuries. The results were numerous and spectacular but by the nineteenth century many were feeling a lack of fully rigorous foundations for calculus. These foundations were finally given by Augustin-Louis Cauchy (1789–1857) and later Karl Weierstrass (1815–1897), but they took a different route, now called the epsilon and delta approach, which did not rely on infinitesimals. These became relegated to mathematicians' scrap paper but do not appear in the final "official" proofs of their theorems.

Roughly one century after Weierstrass, infinitesimals were finally given a rigorous definition by Abraham Robinson (1918–1974) who used the heavy machinery of twentieth century logic. He enlarged the set of numbers by introducing new non standard numbers, for example he defined

a number e is infinitesimal if, for every positive number a, $-a < e < a$

Robinson's non standard numbers, although very useful, did not have the success they may have deserved, probably because the definitions in the spirit of Weierstrass were by then completely classical and established.

## 7 Conclusion

Today we take the number 0 for granted but it took a long time to understand it as a bona fide number. Indeed, we can represent 2 by 2 apples, 3 by 3 apples, but representing 0 by no apple is more abstract. Moreover, 0 is not a number like the other ones, since $\frac{1}{0}$ is not a real number. If we want to consider $\frac{1}{0}$ as a number – and that is important for calculus and real life applications – we need to treat infinity, or $\infty$, as such. This has been achieved by the Riemann sphere which contains all the complex numbers and $\infty$. But even there, $\frac{0}{0}$ has no meaning, and neither has $\frac{\infty}{\infty}$ or $0 \times \infty$. Calculus addressed this quite successfully by introducing infinitely many zeroes and infinitely many infinities, or rather gradations of infinitely large quantities or infinitely small ones, these latter ones being called infinitesimals. But these were sacrificed on the altar of mathematical rigour, despite Abraham Robinson's valiant attempt to salvage them by giving them mathematically correct foundations. Infinitesimals remain essential behind the scenes, though.

There are certainly other approaches and there will be more in the future. We borrow the final words from a paper by Abraham Robinson (1973):

> Number systems, like hair styles, go in and out of fashion, it's what's underneath that counts. (…) The collection of all number systems is not a finished totality whose discovery was complete around 1600, or 1700 or 1800, but that it has been and still is a growing and changing area, sometimes absorbing new systems and sometimes discarding old ones and relegating them to the attic.

## References

Guedj, D. (1996). *L'empire des nombres*. Gallimard.

Kline, M. (1972). *Mathematical thought from ancient to modern times*. Oxford: Oxford University Press.

Robinson, A. (1973). Numbers – What are they and what are they good for?, *Yale Scientific Magazine* 47.

Stillwell, J. (2002). Mathematics and its history, Undergraduate texts in mathematics. New York: Springer-Verlag.

Struik, J. (1987). *A concise history of Mathematics*, Fourth revised edition. Dover.

CHAPTER 35

# Division by Zero (*khahara*) in Indian Mathematics

*Avinash Sathaye*

## Abstract

Since Brahmagupta (seventh century), the reciprocal 1/0 of the number 0 was used in mathematical operations. The ideas were not limited to the notions of limits in calculus but appear to be unusual algebraic constructs which have no parallel in history of mathematics. We present an analysis using ideas from modern Algebra. We also discuss numerous similar structures in modern mathematics.

## Keywords

Indian algebra – division by zero – infinity – Bhaskaracharya – Brahmagupta

## 1 Introduction

The ideas of zero and negative numbers are, of course, old and, by now, quite familiar to everybody. One has thus become used to these numbers called the 'integers' which include zero and are unbounded on both ends. Most of the world routinely uses the decimal number system with its ten digits 0, 1, 2, 3, 4, 5, 6, 7, 8, 9 which change value according to their place in the displayed expression of the number. This makes it easy to build arbitrarily large numbers and use them with ease.

Thus, even a single digit, say 2, can be repeatedly used as 2222 to produce the number $2000 + 200 + 20 + 2$ and clearly there is no limit on the size of numbers needed. Any number can be written as $a_0\, 10^n + a_1\, 10^{n-1} + \cdots a_{n-1}\, 10 + a_n$ where all the $a_i$ are among the 10 digits and $n$ is as large as desired.

This is the so-called Hindu Arabic system, developed in India and propagated through the Arabic mathematical sources across Europe.

It is interesting to observe that Archimedes had thought about how to represent large numbers and had suggested a myriad ($10^8$) based system, but

 | DOI:10.1163/9789004691568_041

because of the huge base and lack of the digit zero, it did not develop into a system with useful systematic calculation techniques.[1]

The Babylonians, using a sexagesimal (base 60) system did develop the idea of a positional system where numbers multiplied by larger powers of 60 would be written further to the left.

Empty spaces would be left for missing powers of 60 and later they even invented a place marker for the empty space, to clarify the arrangement. But none of these systems led to the creation of a number 0 eligible for arithmetic operations. The first documented place in literature is Brahmagupta in his *Brāhmasphuṭasiddhānta*. He not only declared zero as an algebraic quantity but went on to define operations involving zero and any number.

Infinity is a more complicated story. It is easy to imagine bigger and bigger numbers, but one does not work on the resulting 'infinity' as a number by itself. The symbols $\infty$ or $-\infty$ routinely used in modern calculus are simply descriptions of behavior of functions as their arguments get bigger and bigger, or smaller and smaller, or get closer and closer to some number.

Mathematicians have overcome the difficulty of deciding the existence (or its lack) of a limit to a sequence of numbers. Yet, one does not do algebra with the limits if they are $\pm\infty$.

In fact, the usual commandment in mathematics is that 'Thou shalt not divide by zero!' We will first discuss the explicit intentional break of this commandment in Indian mathematics. We propose a modern explanation of the process.

Afterwards, we will discuss several concepts in higher mathematics which also circumvent this rule and build a logically sound system where division by zero makes sense.

In the algebra books of ancient India, we find that 0 is treated like any other number, accepting six operations of addition, subtraction, multiplication, division, squaring and square root.

Bhāskarācārya's books *Līlāvatī* and *Bījagaṇita* include some rather intriguing exercises based on these ideas. Of course, one needs to be aware of the pitfalls of dividing by indiscriminately.

---

1 Archimedes' work *Sand-Reckoner* is available online. This number $10^8$ is called *daśakoṭī* (10 crores) in Sanskrit and is often used as the term for a very large number. Indian mathematics has names for larger powers of 10, commonly up to $10^{18}$ but in Buddhist literature the list goes on, well beyond. See History of Large Numbers, Wikipedia.

We shall discuss their definitions and precautions to avoid these pitfalls.

If we simply follow the modern mathematical conventions, then these appear to be nonsensical operations and both Brahmagupta and Bhāskarāchārya have been criticized in modern times for leaving such material in their otherwise excellent treatises.

I have discussed the details of these exercises together with a modern justification in 'Bhāskarācārya's Treatment of the Concept of Infinity' (Sathaye, 2015).

I will present a modified explanation as well as discussion of the curious assertion by Brahmagupta in *Brāhmasphuṭasiddhānta* (ch. 18, v. 34) that $0/0 = 0$.

A modern work by James Anderson et al. (2007) has sought to give a full axiomatic system which extends the real numbers to 'transreals' by adding $\pm\infty$ as well as an additional incarnation of zero which he denotes by a symbol looking like $\Phi$. In his system, $0/0$ is proved to be this new incarnation $\Phi$ of zero.

I am not sure if his system can help develop properties of our Bhāskara ring constructed below. One important difference is that in our model only one infinity is needed, the other being its formal negative.

An explanation is also discussed in section 3.

In spite of my modern justification, it is not my claim that these justifications were widespread in Indian mathematics. The ideas certainly never got elaborated in either the original mathematical texts or their subsequent commentaries.

The first of the three problems being presented here is from *Līlāvatī* and it can be comfortably explained using the natural modern ideas of limits. It was also declared to be of great use in astronomical and other calculations by Bhāskarāchārya himself and discussed extensively by commentators.

The next two problems are from *Bījagaṇita*. They fail to yield an analysis of limits and indeed may be declared as erroneous.

These problems, however, can be explained by a system of idempotents (numbers equal to their own squares) and I hereby produce such an analysis.

## 2 Basic Definitions

Here is the summary of declarations about zero and its operations. We shall give references from

- *Bīja*: Bhāskarācārya's book *Bījagaṇita* on Algebra
- *Līlā*: Bhāskarācārya's book *Līlāvatī* on Arithmetic

The verse numbers are approximate due to a large number of varied editions.

### 2.1 *Zero Operations*

1. By multiplication with zero, the answer is zero, whereas *khahara* results when a number (*rāśi*) is divided by zero. (*Bīja* 18).
*Comment*: Usually, *rāśi* refers to a positive integer (or at least a positive real number), although it may also be negative when declared as such. However, 0 is not among the plausible interpretations here and so $0/0$ is not formally defined here. Here *khahara* (zero-divided) is a technical term, meaning something with denominator zero. It is simply a new number denoted by $x/0$ and does not produce an ordinary numerical value right away.
2. In *Līlā* 46, 47, there is additional information. Zero plus (minus) zero is zero and powers of zero are zero. A number divided by zero is *khahara* (zero-divided) and one multiplied by zero is *khaguṇa*. It evaluates to zero, however, one should consider it as a zero-multiplied number during the rest of the operations till the final answer. When, after zero being a multiplier, if subsequently it is divided by zero, then the original number is unchanged. Similarly, a number with zero added to or subtracted from stays unchanged.
*Comment*: For $a \neq 0$, we may deduce that $(a \cdot 0)/0 = a = a \cdot (0/0)$. Thus, this quantity $0/0$ must behave like a multiplicative identity, so naturally we expect it to be 1.
3. Moreover, in *Bīja* 20, we have the following: In this *khahara* number (*rāśi*) there is no change by adding or subtracting any numbers, just as living beings entering during creation or leaving during final destruction do not affect the infinite stable state.
**Comment:** This is often explained as 'infinity plus any (finite) number is infinity'. I would suggest that it means a *khahara* stays a *khahara* and indeed the same *khahara* before the addition. This can be explained as

$$\frac{a}{0} + b = \frac{a}{0} + \frac{0.b}{0} = \frac{a+0}{0} = \frac{a}{0}$$

This raises the question why we did not wait to evaluate $0 \cdot b$ as 0, as suggested above. Our explanation developed below would give an answer. I would also suggest that, perhaps, Brahmagupta deduced $\frac{a}{0} + b = \frac{a}{0}$ by the same calculation using $0/0 = 0$.

The numbers which are *khahara*, *khaguṇa* and the usual numbers are in three separate compartments, and, at one time, we combine only numbers of the same type. In this view, the sum $\frac{a}{0} + b$ is simply disallowed. Only multiplication or division by zero allows one to go from one compartment to another.

We will give a formal proof of this principle after we construct the ring $B(x)$ below.

## 3 Formalism

For convenience, let us make our own formal definitions using modern terminology. It is clear that if we want to multiply or divide by zero then we need a formal place holder symbol for it, so that it does not get evaluated to zero until the formal rules have been properly applied.

So, we choose a distinctive symbol.

*Definition*: Let $\Re$ stand for the usual set of real numbers and let $\epsilon$ stand for the multiplier zero. If a is any number in $\Re$, then by a$\epsilon$ we shall denote the corresponding *khaguṇa*. The set of all *khaguṇa* can be denoted as $\Re\epsilon$. Similarly, we write $\infty$ to denote the specific *khahara* 1/0. Naturally, in our formalism this could also be defined as $1/\epsilon$.

*Comment*: We note that an exception needs to be made for $0 \cdot \epsilon$. In our final analysis, this will reduce to an ordinary real zero. Similarly, $0/\epsilon$ will also reduce to a real zero after our analysis. This is another instance of 0/0 coming out 0.

Thus, every *khahara* can be represented as a$\infty$ as a varies over $\Re$. Thus, the set of *khaharas* can be written as $\infty$.

We now have three kinds of numbers: ordinary ($\Re$), *khaguṇa* ($\Re\epsilon$) and *khahara* ($\Re/\epsilon$ or $\infty$). We will discuss a formal explanation of the operations by constructing the Bhaskara ring $B(x)$.

### 3.1 *The Three Problems*

We list the three problems involving *khahara* and *khaguṇa*, using modern notation. Full details can be found in Sathaye, 2015.

1. The first problem is from *Līlā* 48: What is a number $t$ such that

   $$\epsilon \to \cdot\ (3(t + t/2))/\epsilon = 63?$$

   If we interpret $\epsilon$ as a quantity tending to 0, then the limit makes sense even in modern conventions and we get as ordinary equation $3(t + t/2) = 63$. The answer is clearly

   $$t = 14$$

2. The second is from *Bīja* 120. What is a number $t$ such that

$$\epsilon((t\infty \pm 1000000)^2 + (t\infty \pm 1000000)) = 90?$$

This is supposed to reduce to $t^2 + t = 90$ and hence $t = 9$ ($-10$ is not used since $t$ is a *rāśi*).

3. The third is from *Bīja* 121. What is a number $t$ such that

$$[((t + t/2)\epsilon)^2 + 2\,((t + t/2)\epsilon)]\ \infty = 15?$$

This is supposed to reduce to $9t^2/4 + 3t = 15$ and leads to $t = 2$ if we discard $t = -10/3$ by the convention about the term *rāśi* again.

### 3.2 *Formal Model*

We take a little diversion into algebra to make a useful model for our purpose. Let $\Re$ be the set of all real numbers.

Consider the set of all Laurent polynomials with real coefficients which means all expressions of the form

$$f(x) = a_0 x^n + a_1 x^{n-1} + \cdots + a_n + b_1 x^{-1} + b_2 x^{-1} + \cdots + b_m x^{-m}$$

where $0 \neq a_0, a_1, \ldots, a_n, b_1, \ldots, b_m \in \Re$

**Note:** Despite the display of the formula, $n$ can be zero or negative. In case $n < 0$, we assume that all coefficients are various $b_i$.

Let us denote the set of such Laurent polynomials by $L_{\Re}(x)$.

*Important convention*: In case $f(x)$ is just a zero polynomial we just write $f(x) = 0$. As illustrative examples, consider:

$$g_1(x) = x^3 + 3x + 4$$
$$g_2(x) = 2x^2 - 5\text{x} + 3 + 4x^{-2}$$
$$g_3(\text{x}) = 3x^{-1} - 3x^{-2} - 3x^{-5}$$
$$g_4(\text{x}) = 4 + \text{x}^{-2} + \text{x}^{-3} \text{ etc.}$$

We shall define $\deg f(x)$ to be $d$ if $d$ is the highest power of $x$ appearing in $f(x)$. Thus $\deg g_1(x) = 3$, $\deg g_2(x) = 2$, $\deg g_3(x) = -1$ and $\deg g_4(x) = 4$. $\deg f(x)$ is undefined if $f(x)$ is the zero polynomial. Thus, if $\deg g(x) = d$, then we have $g(x) = cx^d + h(x)$ where $0 \neq c \in \Re$ and either $h(x) = 0$ or $\deg h(x) < d$.

Moreover, it is useful to define a simplification operator $S$ such that

If $d > 0$, then $S(g(x)) = cx$
else if $d = 0$, then $S(g(x)) = c$
else if $d < 0$ then $S(g(x)) = cx^{-1}$.
If $f(x)$ is the zero polynomial, then we define $S(f(x)) = 0$.

*Note*: Thus, $S(f(x))$ is always a single term with degree 1, 0 or −1.
Thus, we see that $S(g_1(x)) = x$, $S(g_2(x)) = 2x$, $S(g_3(x)) = 3x^{-1}$ and finally, $S(g_4(x)) = 4$

1. Let us consider the set of all simplifications of Laurent polynomials, i.e., $S(L_{\Re}(x)) = B(x)$. We shall call it the Bhāskara ring. Note that every element of $B(x)$ is of one of the three types:
   $cx$ where $0 \neq c \in \Re$
   $cx^{-1}$ $0 \neq c \in \Re$
   $a \in \Re$
   If we denote $x^{-1}$ by the symbol $\epsilon$ and $x$ by $\infty$ then we get a complete model for our calculations with the three problems.
2. The operations in the Bhāskara ring $B(x)$ may be summarized thus: $f(x) + g(x) = S(f(x) + g(x))$, and $f(x)g(x) = S(f(x)g(x)$.
   In other words, do the operations followed by simplification.
   We may explain Bhāskarācārya's calculations in *Bīja* 120 above by proposing the following rule:
   $a\infty + b = a\infty$ for any $0 \neq a, b \in \Re$
   We called this the 'winning rule' for infinity in Sathaye, 2015.
   In the ring $B(x)$, this is simply a natural operation: $S(ax + b) = ax$ which becomes the winning rule.
3. We also described a corresponding winning rule applicable for a *khaguṇa* thus for
   $0 \neq \text{a}, \text{b} \in \Re$
   $a + b\epsilon = S(a + bx^{-1}) = a.$
   Likewise $S(ax + b + cx^{-1}) = ax$ if $a \neq 0$, so infinity wins over everything else.
   Thus, when a *khahara* is added to any non-*khahara* number, then only the *khahara* survives!
4. Coming back to the 0/0 problem, We note that in our simplification process $S(0x) = 0 = S(0x^{-1})$ since 0 is the only number which has no degree and these are all the numbers which lack a degree. Thus, 0 may be thought of as $0x$ or $0x^{-1}$ but in either case it evaluates to 0.
   We can now justify Brahmagupta's assertion $0/0 = 0$ by thinking of the expression $0/0$ as $0 \cdot \infty = 0x$ and, as already observed, this number has no degree, so it is the real 0. Of course, traditionally, Brahmagupta's assertion is declared incorrect and $0/0$ is declared to be 1 since this is necessary for Bhāskarāchārya's problems.
   The reasoning for this alternate value of Bhāskarāchārya may come from the observation that the expression $S(x \cdot x^{-1})$ must be 1 in $\Re$.
   Of course, neither should be regarded as more correct. Since the only true 0 lives in $\Re$, this is bound to happen.

Perhaps, this is the reason why Andersen et al. (Anderson, Völker and Adams, 2007) created a special Φ symbol for zero, keeping it separate from zero. His system only proves that $0/0 = \Phi$.

*Comment*: Our representation of $\Re x$ as all the *khahara* and $\Re x^{-1}$ as all the *khaguṇa* may be compared with the thoughts of Ramanujan expressed during private discussions with P. C. Mahalanobis (the 'Father of Indian Statistics'). Here is a summary of their reported conversation (Ranganathan, 1967):

'Ramanujan spoke of zero as representative of the Absolute (*nirguṇa brahma*), something which has no attributes and no description. Infinity, on the other hand, was the totality of all possibilities capable of being manifest in reality. Further, the product of zero and infinity would supply the whole set of finite numbers.'

5. Note that during simplification $S(x^2) = x$ and also $S(x^{-2}) = x^{-1}$. Thus, in $B(x)$ both $x$ and $x^{-1}$ are idempotents (meaning quantities equal to their own squares).

Usually, in a ring only 0 and 1 are idempotents. Our ring has two more: $x$ and $x^{-1}$.

Now we shall take up the discussion of three 'exercises' from Bhāskarācārya's works to illustrate how our structure of $B(x)$ helps us get the alluded answers. We will also clarify why our added structure is necessary.

## 4 The Justifications

4.1 ***Problem 1.*** The problem is developed thus:

(a) Start with an unknown number $t$.

(b) Multiply by 0 i.e., make $tx^{-1}$

(c) Combine with its half, namely $(1/2)tx^{-1}$ to yield $tx^{-1} + (1/2)tx^{-1} = (3t/2)x^{-1}$.[2]

(d) Multiply by 3 to yield $3(3t/2)x^{-1}$.

(e) Divide by zero, i.e., multiply by $x$. Now we are multiplying a degree $-1$ and a degree 1 terms, so the answer will have degree zero: $3(3t/2)x^{-1}x = 3(3t/2)$. Equate this to 63, thus: $3(3t/2) = 63$ which leads to the solution $t = 14$.

(f) As explained, this does not need the 'winning rules' at all.

2 Note that these operations are all within *khaguṇa* set and hence simplification is automatic.

4.2 *Problem* 2. The problem is developed thus:
(a) Start with an unknown $t$.
(b) Divide by zero to yield $tx$.
(c) Add or subtract 10,000,000. We get $tx \pm 10{,}000{,}000$ which simplifies to $tx$ since the degree of the expression is 1 and lower degree terms are dropped under the simplification operation $S$. (We are assuming that our final answer is non-zero).
(d) Square it. Thus, we now have $S(tx \cdot tx) = S(t^2x^2) = t^2x$.
(e) Augment by its own square root. We get $t^2x + tx$.
This deserves an explanation. What is the square root of $t^2x$? Note that $S(tx)^2 = S(t^2x^2) = t^2x$.
Note that the same argument works for $-tx$. However, our $t$ is intrinsically positive. At least the convention that $\surd$ yields a positive number will give $(t^2 + t)x$
(f) Multiply by 0, i.e., by $x^{-1}$ and equate to 90.
This is $(t^2 + t)xx^{-1} = t^2 + t = 90$.
We solve to get $t = 9$.
As already explained, we have to discard the negative root $t = -10$.
If we were to allow $t$ to have either sign, then the equation becomes $t^2 + |t| = 90$. This will allow $t = -10$ as a solution as well!

4.3 *Problem* 3. The problem is developed thus:
(a) Start with an unknown number $t$.
(b) Add in its half. This is $t + t/2 = 3t/2$.
(c) Multiply by 0, thus, we now have $3t/2x^{-1}$.
(d) Square and add its root twice. Thus, we have $[3t/2]^2x^{-2} + 2(3t/2)x^{-1}$ where, we have assumed $3t/2$ to be positive. As before, simplification leads to $(9t^2/4) + 3t)x^{-1}$.
(e) Finally, divide by 0 (and simplify) to get $9t^2/4 + 3t = 15$.
As before, we accept the positive solution $t = 2$, but discard the negative root due to the square root operation.
If we let $t$ be positive or negative, then the modified equation will be $9/4t^2 + 12|t| = 60$ and $t = -30$ gives a valid solution also.

## 5 Modern Analogues

In Sathaye, 2015, I gave numerous illustrations of structures in modern mathematics which show that the new ideas have many precedents. I present some further parallels here.

Here are the main departures from standard practices which are noticeable in Bhāskarācārya's methods.

### 5.1 *Breaking the Rule of Not Dividing by Zero*

The reason for the rule is to avoid mistakes in manipulation of rational expressions. If you live in the world of real numbers only, then $1/0$ cannot be equated to any real number $a$.

This is because $1/0 = a$ leads to $1/0 = a/1$ or $1 \cdot 1 = 0 \cdot a = 0$. Thus, it leads to a false claim $1 = 0$.

The reason is that if we permit this, then for any $c$, we get $c = c \cdot 1 = c \cdot 0 = 0$. Clearly this is untenable!

A little further analysis shows a similar objection will result if $c$ is a number such that $c \cdot d = 0$ with $d \neq 0$, then we derive a contradiction by observing that $\frac{1}{c} = \frac{d}{dc} = \frac{d}{0}$. Thus, unless $d = 0$ we have a problem with writing $\frac{1}{c}$.

In modern algebra, one gets around the difficulty as follows: Choose a set $A$ of $c$-values to invert (i.e., make $1/c$ valid).

Collect the set $B$ all $d$ such that $cd = 0$ for some $c \in A$.

If every member of $B$ is made zero in a new place, then in that place all $c$'s in $A$ are invertible.

One standard example is the ring $Z_6$ which consists of integers whose addition and multiplication is defined by doing the operation and then replacing the answer by its remainder modulo 6.[3]

For example, in this ring, we have **after reducing modulo** 6

$$6 = -12 = 0, \quad 2 \cdot 3 = 4 \cdot 3 = 0, \quad 5 \cdot 5 = 1$$

So, in this ring, neither $1/3$ nor $1/2$ are valid. Note that, $1/5 = 5$ is valid here as observed.

However, we can just 'go modulo 3', which means that we take our numbers into $Z_3$ where we reduce modulo 3 and thus the image of 3 is zero. So, inverting 2 is not a problem. Indeed $2 \cdot 2 = 1$ modulo 3.

If, we choose to go modulo 5 in $Z_6$, then both 5, 6 become zero and hence so does 1.

So, we are left with only one element, namely zero.

3 This gives an interesting parallel to our 'simplification' operator $S$. Here $S$ corresponds to reducing the answer modulo 6.

### 5.2 *Simplification of Expressions after Operations*

We can consider the ring $K[X_1, \cdots, X_m]$, the polynomial ring in $m$ variables over a field $K$ (for example $K = \Re$). Then we have a degree $\deg(f)$ for each polynomial $f$. We further set $S(f)$ to be the polynomial obtained from $f$ by only keeping the highest degree terms in $f$.

This is analogous to our simplification $S$.

A polynomial $f$ is said to be homogeneous of degree $d$ if all its terms have the same degree $d$ and we write $\deg(f) = d$.

The zero polynomial 0 is not assigned a degree, but defined to be homogeneous and

$S(0) = 0$.

Observe that a polynomial $f$ is homogeneous if and only if $S(f) = f$. It is easy to see that $S(fg) = S(f)S(g)$ if $S(f) \neq 0 \neq S(g)$.

For addition, we get

If $\deg(f) > \deg(g)$, then $S(f+g) = S(f)$.
If $\deg(f) < \deg(g)$, then $S(f+g) = S(g)$.
If $\deg(f) = \deg(g)$, then $S(f+g) = S(f) + S(g)$ unless $S(f) + S(g) = 0$.

If $f, g$ are homogeneous of equal degree, then $S(f+g) = S(f) + S(g)$ where both sides may be zero polynomials.

This discussion shows that if we consider the set of *homogeneous* polynomials, then

$$S(fg) = S(f)S(g) \text{ and } S(f+g) = S(f) + S(g).$$

We note that this gives us a model of different layers (by degree) and shows how the 'winning rule' naturally works.

The ring $B(x)$ is a slightly more general version of such homogeneous rings since it also includes inverses of monomials.

### 5.3 *The Bhāskara Ring*

The ring $B(x)$ is derived from the set of homogeneous polynomials by three modifications.

- We start by taking only one variable $x$. Then all homogeneous polynomials are either single monomials or the zero polynomial.
- We invert the element $x$, to supply all negative powers of $x$.
- Finally, we use a stronger form of simplification so that all positive powers of $x$ are reduced to one and all negative powers are reduced to minus one.

### 5.4 *Infinity as a Number*

The idea of treating infinity as an additional number and using it in formal algebraic calculations is widespread in modern mathematics. The idea is similar to our discussion. Supply the missing number as a new object and extend the Algebraic operations. The hard part is to make the resulting structure so that it behaves similar to ordinary algebraic operations.

One of the well-known examples is that of complex numbers C. We start with two copies of C, say $A$ and $B$. Then all elements of $A - \{0\}$ are matched with all elements of $B - \{0\}$. If we were to simply match equal complex numbers, then we run into the usual problems with handling infinity. But if we agree to map $z \in A - \{0\}$ with $1/z \in B - \{0\}$, then those difficulties go away.

Under this model, we take the usual addition in the $A - \{0\}$ part, but in $B - \{0\}$, we define the sum of $p$ and $q$ by a matching addition denoted as $\bigoplus$ so that $p \oplus q = \frac{(pq)}{(p+q)}$. The idea of this complicated formula is simple.

The addition in $B$ is defined by outsourcing it to the $A$ part.

Thus we take these steps:

- Go to $A$ i.e., take $1/p$, $1/q$.
- Add the elements in $A$ as usual: $1/p + 1/q = \frac{pq}{p+q}$.
- Take the number in $B$ matching this answer, i.e., take $\frac{pq}{p+q}$.

The 0 in $B$ is now formally paired with 0 in $A$ and it is the necessary $1/0$ element.

Higher mathematics has examples of adding several infinities as well as adding towers of infinities, but that would be too much diversion.

## 6 Conclusion

Bhāskarāchārya evidently had a novel algebraic idea in his three problems.

Why did he not pursue them further? One possibility is that after the first problem, his second and third problems do not seem to have a 'practical' application in his astronomical endeavors. So, he may not have had time to pursue his abstract algebra.

Perhaps, there are more unusual ideas and arguments in the immense body of Indian mathematics. I would like to find a few more such gems.

## References

Anderson, J. A. D. W., Völker, N., and Adams, A. A. (2007). Perspex Machine VIII: axioms of transreal arithmetic. *Vision Geometry XV*, Proceedings of SPIE, vol. 6499.

Bhāskarāchārya, *Bījagaṇita* Numerous editions including e-versions with commentaries'.

Bhāskarāchārya, *Līlāvatī*. Numerous editions including e-versions with commentaries.

Sathaye, A. (2015) Bhāskarācārya's treatment of the concept of Infinity. *Gaṇita Bhāraī*, *37*, 1–2, pp. 5567.

Brahmagupta, *Brāhmasphuṭasiddhānta*. Numerous editions including e-versions with commentaries.

Archimedes, *Sand-Reckoner*, trans.

Mendell, H. (1967), Chapter 4. Available at http://www.bit.ly/Sand-Reckoner. (Website includes numerous translated Greek works.)

Ranganathan S. R. (1967) Ramanujan: The Man and the Mathematician.

History of Large Numbers, Wikipedia, https://en.wikipedia.org/wiki/History_of_large_numbers.

CHAPTER 36

# Zero: in Various Forms

*Mayank N. Vahia and Upasana Neogi*

## Abstract

Zero is a very versatile entity, making its appearance in a variety of environments. It can appear as a representation of the absence of items in arithmetic to the absence of sound in the merging of words, to the zero as a number in arithmetical operations. All these aspects of zero are fascinating in their own right. However, while the idea of arithmetical zero came to the human mind very early in its intellectual journey, the human response to derivative ideas has been driven by cultures. While, on the one hand, European cultures came to the conclusion that the universe exists and therefore, by definition, a non-existent entity cannot represent reality; on the other hand, the Indian philosophers, used to the ideas of non-existent entities in other contexts, especially in the grammar of conjunction of words and sounds, were fascinated by the null and assumed that it was the null that held the key. They therefore traced the idea of the origin of the universe to this null entity, which by definition is not knowable since it is formless. This divergence meant that intellectual perspectives and studies were driven by this differing approach to zero. It allowed Indian mathematicians to explore zero much earlier than their European counterparts. Modern science, having embraced the zero in all its glory, then went on to twist its existence in different ways. In the present article we discuss some of the aspects of this journey of zero.

## Keywords

zero – Indian ideas of null – zero in cultural context – zero and large number of obsessions in ancient India – zero of mathematics – zero of computers

## 1 Introduction

### 1.1 *The Sources of the Zero Symbol*

Zero made its first appearance as a placeholder in Babylonian account-keeping, which used a place value system, around the second millennium BCE.

 DOI:10.1163/9789004691568_042

The Babylonians used a sexagesimal (base 60) system. In that, the value of a digit depends on the location where the number appears. For example, a number, say, 3 in the left-most column meant 3 while in the second column it meant 30, and so on. This was a significant invention.

The Babylonians started using two slanted wedge shapes to represent an empty space or empty column on the abacus and as a separator in bilingual texts. In 331 BCE, the Babylonians were invaded by the Greeks under Alexander. During that time, the Greeks might have discovered the crucial role of zero in counting and adopted zero from the Babylonians. They employed zero to the Greek astronomical papyri. The symbol for zero used by the Greeks was the lowercase omicron 'o', which looks like the modern-day zero. Omicron was probably used to refer the first letter of the Greek word for nothing, i.e., *ouden*. But the Greeks rejected zero because it conflicted with the two fundamental philosophical beliefs of the West – the void and the infinite.

In the Indian context, the idea of zero is found in the Rigveda (*c.* 1200–1500 BCE). The Vedic literature mentions numbers in powers of 10 (ten, hundred, thousand). This appears in reference to prayers asking for riches in multiples of 10. However, it is clear that the Vedic people did not just mention the powers of ten but had an idea about the place value system with references such as RV 1.54.9, from *The Hymns of the Rigveda* (translated by Ralph T. H. Griffith, 2nd edition, Kotagiri (Nilgiri) 1896):

> With all-outstripping chariot-wheel, O Indra,
> thou far-famed, hast overthrown the twice ten Kings of men,
> With sixty thousand nine-and-ninety followers,
> who came in arms to fight with friendless Susravas.

Note that in mentioning 60,990 as the number of followers there is an implicit understanding of the place value zero. There is also a hint of the idea of multiplication with references such as RV, 1.10.10 (*ibid.*):

> We know thee mightiest of all,
> in battles hearer of our cry.
> Of thee most mighty we invoke the aid that giveth thousandfold.

The Indians went on to using the idea of null in a variety of contexts such as language, grammar, philosophy, mathematics, and science. We will see examples of this in the next section.

Another independent origin of the concept and sign of zero is found in the Mayan civilization, which flourished from 300 BCE to 900 CE in the Yucatan peninsula. The representation of a tattooed man in a necklace with his head thrown back was used by the Mayans as the symbol for zero. But the Mayans sometimes used faces, full figures, half-flowers, snail-shells, or something unnamed (?) as the glyphs for zero.

As zero evolved, some sort of ornamented bars always appeared on it. Around 150 CE, Ptolemy used 'O' with a fancy bar over it to indicate the absence of some kind of measure, such as degrees or minutes or seconds in his *The Almagest* (The Greatest Synthesis). The symbol 'O' is also seen on the Darius vase from the fourth century BCE, found in Apulia. It depicts the signs for monetary values of the tribute paid by the conquered nation which the royal treasurer is seen calculating with his representative crouched before him. One of the signs among the monetary values is 'O', the Boeotian for '*obol*', i.e., a coin of no worth.

The idea of zero as a null entity in a wider context as well as the arithmetic and algebraic calculations regarding zero was developed in India. Around 200 BCE, the Indian scholar Pingala used the word '*śūnya*' to refer to zero. Aryabhata, the Indian mathematician-astronomer, used the word '*kha*' for *śūnya* (empty), *ambara* (sky) and *akasa* (atmosphere), which later became zero's common names. Another Indian astronomer, Varahamihira used the name '*bindu*' (dot) for zero, but no symbol. The Jain Tirthankara Mahavira, in his book *Ganita-Sara-Sangraha*, used zero extensively by the name *kha*, but without any symbol. *Lokavibhaga*, a Jain text dated to 458 AD, is probably the earliest text to use a decimal place value system with a zero, referring to it as *śūnya*. Different Indian schools of philosophy called *śūnya* a *Nirguna Brahman* (attribute-less truth) or *Poojyam* (the object of worship). Around 628 CE, the numeral zero with the arithmetic rules and negative numbers was first used by Brahmagupta in India. *Brāhmasphuṭasiddhānta* is the first work that contains the rules governing the use of zero.

The Arabs adopted Hindu-Arabic numerals along with zero. The Arabic word '*sifr*' or '*as-sifr*' is the translation of the Indian *śūnya*, which means 'void'. The word *sifr* describes an empty state.

Probably, the first recorded representation of zero has been found on a stone tablet, dated to 683 AD, in the Cambodian jungle by the mathematician Amir Aczel. The stone inscription means '*The Chakra era has reached 605 on the fifth day of the waning moon*', in Old Khmer language. The zero was represented as a single dot chiseled into stone. Another symbol of zero could be seen as '∘' on a stone inscription of 876 AD, found at the Chaturbhuja temple in Gwalior, India, which shows the measurement of a garden.

### 1.2 *The Importance of Zero*

The idea of zero as a symbol of non-existent entity caught the imagination of Indian thinkers in a variety of ways and they extended the concept of zero to a variety of fields. However, zero now has various manifestations worldwide in different subjects and formats. We list some of the different forms of zero in Indian grammar, philosophy and mathematics, and modern-day science and electronics below.

## 2 Sanskrit Grammar

In *Sanskrit Grammar* by Panini, zero or the loss of sound or sounds (adarshanam) can mean any of the following – *aśravanam* (non-hearing), *anuccāraṇan* (non-pronunciation), *anupalabdhih* (non-availability), *abhāva* (non-existence), and *varnaviṇāśa* (loss of sound). Panini used zero to signify '*lop*', i.e., the disappearance of some sounds when words are conjugated, verbs are converted, or tense change occurs.

For a word, rules are defined about how conjugations happen and these are applied by replacing removed sounds and indicating *lop* or 0. The new words are created by not reading the zeros. Apart from *lop* there is also *apratyaksha*, a sound whose existence is implied. For example, *panditasyputraha* becomes *panditaputraha* and *asy* is *apratyaksha.*

The closest example of this in English is as follows. 'Two times' becomes 'twice'. Here, in Panini's terms 'o' is zero'ed and through a series of conversion rules 'times' becomes 'ice'. The rule is valid for 'three times' becoming 'thrice', but breaks down at the next level. 'Four times' does not become 'frice' and 'five times' does not become 'fice' or something. Poorly defined rules can also become a source of puns, for example in the pun 'if 20 dogs run after one dog, what time is it?' where the answer is 'twenty past one'. Here the word 'dog' is zero'ed and in the absence of a clear rule, in the answer of the pun that zero, by instinct is replaced by 'minutes' and 'o'clock', based on our daily experience. It is for these reasons that a language needs to properly define its grammar.

## 3 Philosophy

The invention of zero shook the foundations of Western philosophy. The Greeks were afraid to accept the concept of zero because the pillar of their philosophy was 'there is no void'. Whereas Indian philosophy had the concept of void or nothingness for a long time, which led them to the invention of the numeral zero. The *Nasadiya Sukta* (not the non-existent) in the Rigveda describes the beginning of all creation from nonexistence in the following way:

> At first was neither non-being nor being. There was not air nor yet sky beyond.
> What was its wrapping? Where?
> In whose protection? Was water there, unfathomable and deep?
> There was no death then, nor yet deathlessness; of night or day there was not any sign.
> The One breathed without breath, by its own impulse.
> Other than that was nothing else at all.
> Darkness was there all wrapped around by darkness, and all was water, indiscriminate.
> Then, that which was hidden by the void,
> That One, emerging stirring, through power of ardor, came to be.
> Rigveda X, 129, 1–3

## 4 Mathematics

In mathematics, zero is the integer immediately preceding 1, but it is neither positive nor negative. Zero is both a natural number and an even number. Zero in mathematics is not a null element; rather, it signifies a definite element. A set with zero as an element is not a null set. It is also a specific solution to an equation where entities cancel each other, and in some sense represents the center of the number system.

The true credit for understanding the complexity of mathematical zero should go to Brahmagupta, who attempted to give the rules for arithmetic involving zero and negative numbers in the seventh century. He explained that if we subtract a given number from itself, we obtain zero. He gave the following rules for addition that involve zero:

1. the sum of zero and a negative number is negative;
2. the sum of a positive number and zero is positive;
3. the sum of zero and zero is zero;
4. a negative number subtracted from zero is positive;
5. a positive number subtracted from zero is negative;
6. zero subtracted from a negative number is negative;
7. zero subtracted from a positive number is positive;

8. zero subtracted from zero is zero (Brahmagupta then said that any number when multiplied by zero is zero, but struggled when it comes to division);
9. a positive or negative number when divided by zero is a fraction, with the zero as denominator;
10. zero divided by a negative or positive number is either zero or is expressed as a fraction, with zero as numerator and the finite quantity as denominator;
11. zero divided by zero is zero (Brahmagupta struggled in suggesting the result if a number 'x' is divided by 0).

As a consequence of Brahmagupta's division rule, it was formulated that x/0 is undefined because 0 has no multiplicative inverse (no real number multiplied by 0 produces 1).

## 5 Science

Different branches of science use zero differently. In physics, the value of zero plays a special role. Absolute zero, considered as 0, is the lower limit of the thermodynamic scale, a state at which the enthalpy and the entropy of a cooled ideal gas reaches its minimum value. Absolute zero is taken as −273.15° on the Celsius scale, which equates to −459.67° on the Fahrenheit scale. Vacuum is another entity that is close to zero and it is one of the most complex concepts of physics where material and energy may be spontaneously created.

The most common call to zero in physics (and astronomy) is the call to singularity. It is where some quantities become zero and partner quantities become infinite. For example, the density at the center of the black hole is infinite because a finite mass is compressed to a zero volume. The initial singularity is hypothesized as a singularity of infinite density containing the whole mass and space-time before the expansion through quantum fluctuation to create the present-day universe.

In Einstein's Special Theory of Relativity, the rest mass of light is zero but it contributes to the inertia of any system containing it. Zero gravity or zero-g, is defined as weightlessness, i.e., the absence of weight. A body feels weightless and experiences no g-force acceleration while freefalling in a uniform gravitational field that does not cause stress or strain itself.

The hypothetical element tetraneutron is supposed to have a zero atomic number because it has no proton or charge in the nucleus. Physicists have shown that a cluster of four neutrons may be stable enough for considering it as an atom in its own right.

## 6 Digital Electronics

The nature of zero shows a huge transformation when it comes to the field of computers and digital electronics. Here, zero is not always used to represent the absence of anything. The use of zero in the binary number system is very significant. The development of the binary system dates back to the second century BCE, when the Indian scholar Pingala applied it to describe prosody, that is the study and use of rhythmic structure of a verse or lines in verse. Pingala's binary system was in the form of short and long syllables where the length of the long syllable is equal to twice that of the short.

The Chinese also started using binary notation from the ninth century BCE. The modern binary number system was invented by Gottfried Leibniz in 1679, where he used only 1 and 0. The two-symbol system, 1 and 0, is used to form binary code to represent text, computer processor instructions or other data. In 1847, George Boole published a paper describing an algebraic system of logic where 1-0 means a yes-**no**, on-**off** function for the operation of basic logic gates. These functions of 0 are used to signify the 'low' voltage state of an electronic circuit or no flow of electricity.

In mathematics, there is no distinction between a positive zero and a negative zero, i.e., $+0 = -0 = 0$. However, in computing $+0$ and $-0$ are regarded as equal numerically, but with different behaviors in particular operations. With the advancement of computer science, 0 is used as a parity bit's value, if the count of 1s in a given set of bits is even. Parity bits are used as the simplest form of error-detecting code to check the accuracy of transmission of information. In computing, a null pointer is a pointer that does not refer to any objects or functions. 0 is used to refer a null pointer in C programming because 0 is converted into the null pointer at the time of compilation when it appears in a pointer context.

Digitalization has a major role in evolving the concept of zero. In daily life, a leading zero before a number is no longer useless in many cases. A leading zero has a symbolic value when used as in a password or credit/debit card number. '02160' is not the same password as '2160' for any kind of account (mail, bank, etc).

In telephony also, 0 is used before dialing a phone or mobile number from one state to the other. This preceding 0 is used to tell the operator that the number we are dialing belongs to the same country.

One arbitrary use of zero in science is in defining 'zero' level, which can lead to some interesting errors. For example, we define temperature in terms of degree Celsius. So one place may have a temperature of 15°C, while another may be at 30°C. However, this does not mean that the second place is *twice* as hot. The absolute scale of temperature should be Kelvin, which comes to −273°C. So in reality, the first location is at a temperature of 288 K and the second one at 303 K, which is a difference of 5%. It is just that the centigrade scale feels more natural to us. But arbitrary zero can create serious problems in interpretation.

## 7 Conclusion

Nothing is the origin of everything. According to some physicists, everything might end into nothing again. The numeral representation of this nothing or void is '0'. Zero is the root of all puzzles in science and philosophy. Zero, which lies in between the positive and negative numbers, is the origin of the coordinate system. But zero is causing a problem for the starting of the calendrical system. There is no specified zero year between 1 BCE and 1 CE. The existence of year zero is still a matter of controversy.

The search for nothing is never-ending. Modern science is using zero in various forms. The universe is nothing without the nothing.

## Acknowledgments

We want to acknowledge the role of the Zero Project for giving us an opportunity to think about the fascinating zero. We also wish to thank V. Nandagopal for his precious comments on the manuscript.

## References

Crew, Bec. (2018). *Search for the World's First Zero leads to the Home of Ankor Wat*, https://www.smithsonianmag.com/history/origin-number-zero-180953392/.

Hockney, Mike. (2014). *The Mathmos*, Hyperreality Books.

Kaplan, Robert. (2000). *The Nothing that is: A Natural History of Zero*, Oxford University Press.

Seife, Charles. (2000). Zero: The Biography of a Dangerous Idea, Penguin Books.

Wikipedia. 0, https://en.wikipedia.org/wiki/0.

CHAPTER 37

# Nothing, Zeno Paradoxes and Quantum Physics

*Marcis Auzinsh*

## Abstract

When the concept of Zero is analyzed, it is frequently related to another equally deep and complicated concept – Nothing. If in science, Zero is closely associated with mathematics, then Nothing is primarily related to the material world. The question of whether Nothing can be found in the real physical world is often discussed in relation to the vacuum which, at some point in the history of the development of physics, is thought about as an empty space. Nevertheless, theories of modern physics have demonstrated that even in a vacuum, when all the matter is evacuated from a certain region of space, and this region is isolated from ordinary fields – gravitation field, electromagnetic fields etc – there still remain physical fields called vacuum zero fluctuations that can not be removed in principle. These vacuum fields are properties of space. In this chapter I will analyze another physics theory – namely, Quantum Physics – and will consider its less discussed relations to the concept of Nothing. One of the very first Greek philosophers, Thales (624–548 BCE) from Miletus, declared that Something can not appear from Nothing and it can not disappear, turning into Nothing. This idea, in some aspects, was further developed by Zeno from Elea (495–430 BCE) when he created his paradoxes of motion. These paradoxes can be understood as an analysis of the possibility of infinite division of time and space intervals.

When we come to Quantum Physics, the very first appearance of the idea of the smallest possible portion of energy, quantum, was stimulated by the analysis of the spectrum of radiation of heated body – black-body radiation. German physicist Max Planck (1858–1947) demonstrated in 1899 that if we assume that there exists a smallest portion of radiation energy that can not be further divided into smaller parts, we arrive at the description of spectrum of the black-body radiation that perfectly matches the observable reality. In contrast, if we assume that radiation energy can be divided an infinite number of times and there does not exist the smallest possible portion, we arrive at the radiation spectrum which is in a sharp contradiction with observable reality. This contradiction is so strong that it has acquired a special term – the ultraviolet catastrophe. Certain parallels in Zeno's reasoning and approach used in Quantum Physics can clearly be seen and will be discussed. In the conclusion, I will briefly touch on a similar situation in another field of physics when at the other extreme, departure from the idea of the possibility of the infinitely large quantities, this

 | DOI:10.1163/9789004691568_043

time velocities of motion, gave birth to another theory of modern physics – Einstein's theory of special relativity.

**Keywords**

nothingness – vacuum – vacuum zero field – Zeno paradoxes – Quantum Physics – divisibility of the energy – Planck's hypothesis – light quantum – black body radiation

## 1 Introduction

Nonexistence, nothing, has always been an object of fascination and excitement for philosophers, mathematicians, religious thinkers, mystics and, when physics was established as an independent research field, for physicists likewise. New concepts often undergo a similar evolvement in the public's perception – initially, they face resistance, struggle to be accepted, and then gradually, as soon as they are adopted in the everyday discourse, people simply use the concepts as something which is a given, obvious, without devoting serious consideration to their origin and deeper meaning. I think that we can rightfully consider that this is also the case regarding the concept of nothing.

The development of the concept of nothing is not easy to trace – both due to scarcity of documented historical evidence and as a consequence of the complexity and the many different aspects of the concept itself.

Viewed from a certain angle, concept of nothing can be related to the concept of zero. The first traces of the usage of zero, which probably at that point could not even be called a number, but rather marked as a placeholder, a vacant slot in accounting documents to indicate that some items are not present, can be found as early as several hundred years before the Common Era in Babylon (Barrow, 2001).

From a different perspective, the concept of nothingness can be related to Buddhist notion of *Śūnyatā*, which has been translated into English by various authors in different ways. The most frequently used forms are emptiness or voidness (*Middle Length Discourses*, 1995). The Buddhist concept of emptiness should be considered as old as Buddhism itself. It was used massively in the earliest Buddhist texts – discourses or *Nikayas* in the Sutta part of Pali Canon. Notions of emptiness probably can be traced further back to a Hindi worldview. One of the difficulties in translating the Buddhist term *Śūnyatā* into modern European languages is related to the complexity of the term itself. *Śūnyatā* does not mean that nothing at all exists. This concept is often explained as impossibility of *independent* existence. It is a bit risky to try to interpret even this last statement, but it can be extended as an idea that in the complex non-dual universal existence it is impossible to separate individual entities. All is interdependent, all is one.

This is one of the aspects where the ancient Buddhist concept of nothingness comes close to the ideas of modern physics, namely, with the notion of vacuum. The parallel between the Indian Buddhist notion of *Śūnyatā* and the vacuum of modern physics in the contemporary research literature is discussed to some extent, see, for example, (Roy, 2019). In physics, the idea of empty space or vacuum has traveled a long way before it reached the present-day understanding. To make this introductory discussion reasonably brief, I will not touch on ancient Greek and medieval thinkers, but will say a few words about the experimental development of the idea of empty space in physics. Initially, around the mid-seventeenth century, Evangelista Torricelli in Italy and Blaise Pascal in France independently conducted similar experiments with a mercury barometer, where in a closed glass tube above the mercury column a presumably empty space without air was created. Lately, when the fields – electric, magnetic and gravitational – were discovered, it became clear that, if all the matter was evacuated from the region of space, there still remained fields. To have a true vacuum, this region should be isolated from fields, as well. We will not be discussing the technicalities of accomplishing this, but let us assume that it can be done. Will we have a totally empty space?

The answer of modern physics to this question is at least twofold. It states that – yes, in this case, when matter and external fields are not present in a certain region of space, we will say that there is complete vacuum in this region. However, at the same time, according to the theories of present-day physics, we know that in this region of space there still exists vacuum zero field fluctuations. According to the Heisenberg uncertainty relations, one of which states

$$\Delta E \cdot \Delta t \geq h/4\pi, \tag{1}$$

that energy uncertainty $\Delta E$ times time uncertainty $\Delta t$ can not be smaller than Planck's constant $h$ divided by $4\pi$ (Heisenberg, 1927). In slightly simplified terms, it means that for a sufficiently short time intervals we can expect the energy fluctuations, which satisfy the inequality shown above. These energy fluctuations are assumed to be in the form of electromagnetic radiation. The existence of these fluctuations has very profound consequences. Without these fluctuations, free atoms that have been excited would remain in the excited state forever. The vacuum field zero fluctuations are coupling excited-state atoms to the ground state and initiating decay of excited atoms. Secondly, it is possible to make calculations and subsequently to carry out experimental tests to see how energy levels of free atoms in vacuum are shifted by vacuum fluctuations – Lamb shift. Needless to say, these calculations and experimental results agree perfectly. And finally, special experiments can be designed and conducted to test the action of zero field fluctuations on, for example, conducting metal plates placed close to each other in the vacuum – Casimir effect. These plates attract each other exactly as predicted (Klimchitskaya, Mohideen and Mostepanenko, 2009).

It is not difficult to see the parallels between emptiness – *Śūnyatā* in Buddhist texts and vacuum in the modern physics. Both are complicated realities, and in both cases, there is a certain manifestation of these realities in our everyday life.

## 2 Zeno's Paradoxes and Birth of Quantum Physics

There exists another parallel worth exploring between the ancient concept of zero as the representation of nothing, and the material world described by physics. This simile seems to be less investigated. A good starting point to describe this parallel is the paradox devised by the ancient Greek philosopher Zeno from Elea (495–430 BCE). We know about Zeno's paradoxes from Aristotle's 'Physics' (Barnes, 1984), where he describes and analyzes these paradoxes. Although it is known that the Greeks did not have zero as a part of their counting system, the paradox is strongly related to the idea of nothing or zero. One of Zeno's paradoxes, which is known as 'Achilles and the tortoise' in modern rephrasing (Gill and Pellegrin, 2009) states: that the fastest runner in Greece, Achilles, is competing with a tortoise. Achilles graciously lets tortoise to start running some distance ahead of him. Zeno's paradox asserts that Achilles will never catch the tortoise. The reasoning is the following: when the race starts, Achilles must run to the position taken by the tortoise initially. However, during this run, the tortoise will have advanced a little bit. Now Achilles would need to reach the new position of the tortoise. Yet, in the meantime, the tortoise will have advanced a bit further. And so, the reasoning can continue forever. No matter how fast Achilles reaches the new position of the tortoise, it will have advanced a distance again – no matter how little.

There are several ways to look at this paradox. The simplest mathematical one is to consider all the time intervals that Achilles is spending to reach the new position of the tortoise and to notice that, mathematically speaking, these time intervals form an infinite decreasing geometrical series. We know that if we are taking the sum of all the terms in such a series, despite the fact that the number of terms is infinite, the sum is finite. This sum of successive time intervals that is required for Achilles to reach the previous position of the tortoise in case of Zeno's paradox about Achilles and the tortoise equals exactly to the time that is required in our ordinary understanding for Achilles to catch tortoise, if we know $l$ the distance how much ahead of Achilles the tortoise starts the race and the speed at which Achilles $v_A$ and the tortoise $v_t$ move. Namely,

$$t = \frac{l}{v_A - v_t}. \tag{2}$$

The mathematical tools to solve Zeno's paradox about Achilles and the tortoise as presented above, were developed only in the late nineteenth century. Nevertheless, even after that, it is still clear that Zeno's paradoxes have a much deeper meaning (Gill and Pellegrin, 2009). It is not only a mathematical problem about the sum of infinite series. Before and after the mathematical solution of the paradox it was discussed by such great minds as Aristotle, Archimedes, Thomas Aquinas, Hermann Weyl, Henri Bergson, Bertrand Russel and many more. In modern days, an extension of Zeno's paradoxes known as the quantum Zeno's effect is actively discussed and even more, tested in experiments (Schäfer et al., 2014).

In this essay, I would like to touch upon one more aspect of Zeno's paradox. It is a question: if we divide some quantity an infinite number of times, what will remain? If we are approaching this problem from the perspective of an experiment, it is legitimate to ask: is there the smallest portion of the quantity that we are dividing which, when reached, means that it will not be possible to divide the quantity any further? Consequently, the process of the division would ultimately end.

This type of questions have played an important role in our understanding of physical world again and again. One example that I would like to mention here is the discussion about the structure of matter, which commenced at the end of the nineteenth century and ended in 1905, when Albert Einstein (1905) explained the Brownian motion of particles. In his explanation, he used the concept of atoms as the smallest part to which a matter can be divided, at least until it loses its chemical identity. Before that, even such prominent scientists as the Riga-born chemist, Nobel Prize winner Wilhelm Ostwald (1853–1932) were convinced that there did not exist any smallest portion of matter, which could not be divided further. In other words, Ostwald was advocating the viewpoint that the matter was continuous in a sense that it could be divided an infinite number of times. After Einstein provided his explanation of Brownian motion, Ostwald publicly announced the change of his opinion and accepted the theory of the atomic structure of matter (Kim, 2006).

Meanwhile, the quantities that were considered as continuous still continued to exist in physics. One of those quantities was energy, in particular, the energy of electromagnetic radiation – light. What does it mean in practice? Let us assume that we have a light beam and that we are placing an ideal half-silvered mirror in front of it. Such a mirror reflects one half of the light and lets through its other half. One can now continue the division by placing another half-silvered mirror in front of the transmitted portion of the light, and so on. And again, the question is, will there be the last portion of light that cannot be divided further, or will we arrive at the final portion (quanta) of light that cannot be divided further?

In the history of physics, two theories of light have been proposed – the corpuscular, which assumed that a light beam consisted of many small particles, and the wave, which held that light was a wave that consequently could be divided in parts an infinite number of times. Incidentally, Isaac Newton passionately advocated the corpuscular theory of light. Nevertheless, starting from the early nineteenth century, when interference and diffraction properties of light were firmly established, the wave theory overwhelmingly dominated in physics. The reason is obvious, – both interference and diffraction are the properties of waves and cannot be explained, if light is to be considered as a flow of particles (Simonyi, 2012).

Zeno's paradox, in this context, can be considered as a logical examination of a problem of divisibility and a quest for an answer to the question: what will be left after an infinite number of divisions. If we were to pose the question radically: can we, after the infinite number of divisions of something, arrive at nothing at all?

In the case of light (electromagnetic radiation), can we relate this problem to something that can be measured and thus tested in an experiment? At the turn of centuries – the end of the nineteenth century and the beginning of the twentieth century, a problem emerged: what radiates from a heated body? We all know from our everyday life, that if we heat any body to a very high temperature, it will start to glow. It appears that if the experiment is carefully conducted and radiation from this heated body reaches a thermodynamic equilibrium with the environment, the spectrum of this radiation depends only on the temperature of this body, but does not depend upon the type of material this body is made of. In physics, this radiation is called 'black body radiation'. The challenge that theoreticians faced was to explain the spectrum of this radiation from the first principles of physics. This problem arose very prominently when in 1899 two German physicists Lummer and Pringsheim (1899) measured the spectrum of this radiation with great accuracy. This was a problem that German physicist Max Planck (1899) tried to solve on Christmas eve of 1899.

It is not trivial to decide how to approach this problem. One of the ways is the following: we can perform a two-step calculation – in the first step, we can assume that to achieve the thermodynamical equilibrium, the heated body is kept inside the box, which is at the same temperature as the radiating body. Then we further assume that we do not need any specific object inside the box. The box itself can serve as a black body which radiates and an enclosure that ensures that radiation is in the equilibrium of environment. If this model is assumed, then at the first step we can calculate how many different types of radiation (modes) per unit volume $V$ and related to small frequency interval $d\nu$ can exist inside this box. In the second step of the calculation, the average energy of each radiation type that exists inside the box should be calculated, and this leads us to the spectrum of the black body radiation.

In this essay, I will not be going into the details of these calculations, but will present only the main results. After the first step, we can find that the small number of radiation modes $dN$ per unit volume and per small frequency interval can be found as

$$\frac{dN}{V \cdot d\nu} = \frac{4\pi\nu^2}{c^3}, \tag{3}$$

where $c$ is the speed of light in vacuum. In the second step, one can calculate the average energy of one mode. In classical physics, it is Boltzmann distribution that relates the probability of certain mode of radiation to the energy of this radiation mode. Boltzmann states that at a certain temperature $T$ probability $W$ of a system (here mode of radiation) to have energy $E$ is proportional to

$$W \propto e^{-\frac{E}{k_B T}}, \tag{4}$$

where $k_B$ is one of the fundamental constants of nature – Boltzmann constant. If we find the average energy $\bar{\epsilon}_{Kl}$ of the radiation mode with Boltzmann distribution, according to classical physics, we arrive at

$$\bar{\epsilon}_{Kl} = \frac{\int E e^{-\frac{E}{k_B T}} dE}{\int e^{-\frac{E}{k_B T}} dE} = k_B T. \tag{5}$$

If we put together equations (3) and (5), we arrive at the well-known contradiction or paradox in classical physics, known as the ultraviolet catastrophe. We see that the classical physics predicts that every object that has a certain temperature radiates electromagnetic waves with spectrum, which is

$$\rho_{Kl} = \frac{dN}{V \cdot d\nu} \cdot 2\,\overline{\in}_{Kl} = \frac{8\pi\nu^2}{c^3}\,k_B T. \tag{6}$$

Even those who are not enthusiasts of mathematics can easily see that, according to this equation, at any temperature $T$ the higher the radiation frequency $\nu$, the higher the energy of black body radiation $\rho_{Kl}$. If this were true, it would be impossible for any biological life forms as we know them today to exist. A higher radiation frequency means more energetic radiation. According to this black body radiation formula provided in the framework of classical physics, the environment would be filled with intense x-ray radiation and even more intense gamma radiation. Clearly, in such an environment, life would not survive.

This appears to be a direct consequence of the infinite divisibility of the energy. If, similarly to Zeno's paradox of Achilles and the tortoise, where time intervals can be divided an infinite number of times and in some other Zeno's paradoxes where space can be divided an infinite number of times, energy can be divided an infinite number of times, we arrive at a contradiction – ultraviolet catastrophe.

Nature itself forces us to conclude that in practical terms, when dividing the energy of radiation, we cannot start with a certain amount of energy, divide it into parts an infinite number of times, and arrive at nothing or zero amount of energy.

Max Planck tried to explore this conclusion suggested by Nature. He assumed that the smallest portions of energy of radiation that could not be divided further were

$$\Delta E = h\nu, \tag{7}$$

where, again, $h$ is one of the most prominent fundamental constants of Nature – Planck's constant. If this assumption is adopted, then the average energy with the same Boltzmann distribution must be calculated in a slightly different way. One must keep in mind that not all energies are allowed, but only the multiples of smallest energy portion – multiples of energy quanta $E = nh\nu$. The average energy of the radiation mode under this assumption is similar to (5), only continuous integration is substituted by discrete summation

$$\bar{\epsilon}_{KM} = \frac{\sum nh\nu e^{-\frac{nh\nu}{k_B T}}}{\sum e^{-\frac{nh\nu}{k_B T}}} = \frac{h\nu}{e^{\frac{h\nu}{k_B T}} - 1}. \tag{8}$$

Nevertheless, with this seemingly small and technical difference, we arrive at completely different results for average energy. With this average energy calculated according to Planck's assumption about minimum energy of quanta, we can calculate the spectrum of black body radiation as

$$\rho_{KM} = \frac{8\pi h\nu^3}{c^3} \cdot \frac{1}{e^{\frac{h\nu}{k_B t}} - 1}, \tag{9}$$

which, when tested experimentally, perfectly agrees with what can be observed when spectral distribution of the radiation from heated object is measured.

## 3 Analysis and Conclusions

What does this result allow us to conclude conceptually? Probably one of the obvious conclusions is that Zeno's paradox and the possibility to arrive from something to nothing, or an amount of energy that could be described as zero, at least in this particular case is impossible. Nature itself in this situation imposes restrictions as to how close we can approach zero in the physical world. I think that it would not be an overstatement to say that there exist many more similar situations in physics, especially modern physics, when Nature restrains us from closely approaching zero.

Specifically, Zeno's paradox speaks about the division of time and space, not energy. So, the question of how close we can arrive to zero, when we are dividing space and time, remains an open question. We know that one of the ways to approach this question is by combining fundamental constants of physics to obtain a quantity, which represents length and time (Wilczek, 2005). These quantities are the so-called Planck's time

$$T_P = \sqrt{\frac{\hbar G}{c^5}} \approx 5.391 \times 10^{-44} s \tag{10}$$

and Planck's length

$$L_P = \sqrt{\frac{\hbar G}{c^3}} \approx 1.616 \times 10^{-35} m. \tag{11}$$

In both formulas, $G$ is another fundamental constant of Nature that quantifies gravitational interaction – the universal gravitation constant. It may seem a bit artificial to obtain some quantities by simply combining fundamental constants of Nature in the way to obtain quantities with certain dimensions like time or length. These units were introduced by Max Planck himself (1899). To justify this approach, two things must be said. Firstly, there exists only one way how, from the fundamental physics constants, the quantity of required dimension can be obtained. Another fact is even more important. Historically, we have been defining our fundamental constants in the most convenient way for practical use. As a result, we do not have expressions and formulas in physics which have very large or very small numerical coefficients in them. As a consequence, the approach of estimating some quantity simply by combining constants that are important for a particular problem proves to be surprisingly efficient. This essay is probably not the right place to expand on this, but I would like to demonstrate one example here.

Even if we do not know much about quantum physics, this approach allows us to estimate, for example, the energy of hydrogen atoms. Precisely, it can be calculated only by solving the main equation of quantum physics – the Schrodinger equation, which is a rather lengthy procedure and, in the first place, requires us to know this equation itself. However, we can estimate this energy even without knowing the Schrodinger equation simply by applying dimension analysis. We can easily tell that, as along as a hydrogen atom is composed from proton, which attracts an electron, the charge of an electron and proton $e$ should be important. We can assume that the massive proton is at rest, and the electron moves. Thus, the mass of the electron $m$ should play a role. It is a quantum problem, so Planck's constant should be present. Combining these constants, we can construct a quantity with the dimension of energy in one unique way. Namely,

$$E = \frac{e^4 m}{\hbar^2} \,. \qquad (12)$$

If compared with the energy of the hydrogen atom obtained from the precise solution of the Schrodinger's equation, we will see that we are off only by a factor of 2 (see, for example, Auzinsh, Budker and Rochester, 2014). The correct ground state energy of hydrogen atom is the quantity given by Eq. (12) divided by 2.

This not only *allows* us but, I would say, even *forces* us to think very seriously about Planck's length and time in the context of Zeno's paradoxes and possibility or impossibility to have in principle zero time interval and zero space interval in the real physical world. At this stage, we must accept that the smallness of Planck's length and time does not allow physicists to test their existence experimentally, at least directly.

## 4 Final Remarks

There seems to be one interesting difference in the ways in which the investigation of zero and also its counterpart – infinity – influenced the development of mathematics and physics. When zero was introduced into mathematics, it led to the development of calculus, because it allowed mathematicians to think meaningfully about rationing two infinitesimally small quantities (to divide zero by zero) and to ascribe to this ratio a well-defined value. To put this in simple terms, it facilitated the realization that zero divided by zero can have a well-defined value and, as a result, enabled discussion about concepts such as instantaneous velocities, instantaneous acceleration, etc. At the opposite end, the concept of infinity opens the doors to new fields of mathematics and allows the introduction of ideas stating that not all infinities are equal. In some infinite sets, the number of elements can be infinite times larger than in other sets with the infinite number of elements. To examine this in greater detail, the reader must be referred to the works of Georg Cantor (1845–1918), who was the first mathematician to introduce the notion of the power of a set of infinite elements (Dauben, 1990).

In physics, as it was shown in this essay through a case of one particular example – the birth of quantum physics, the process went the opposite way. Until we realized the impossibility of zero energy for the electromagnetic wave, we had classical physics. With the realization of existence of the smallest portions of this energy, quanta, appeared a new physics – quantum physics.

Similar situations can also be identified concerning very large quantities. When physics did not impose any restrictions on how fast an object could move, we again had classical physics' Newtonian mechanics. This changed when, in 1905, Albert Einstein understood that there was a fundamental limit to how an object could move in the real world. In principle, it is impossible for any object to move faster than the speed of light in a vacuum. Infinite velocity in a physical world is impossible in principle. Of course, one should be cautious and add – 'as we think at this stage of development of physics'. This limitation gave birth to another fundamental branch of modern physics – Einstein's theory of relativity.

## References

Auzinsh, M., Budker, D., Rochester, S. (2014). Optically Polarized Atoms: Understanding Light-Atom Interactions. Oxford University Press.

Barnes, J. (ed). (1984). *Aristotle: Physics. The revised Oxford Translation.* vol. 1, Bollingen Series LXXI. Princeton University Press.

Barrow, J. D. (2001). The Book of Nothing. Vintage Books.

Dauben, J. W. (1990). *Georg Cantor His Mathematics and Philosophy of the Infinite.* Princeton University Press.

Einstein, A. (1905). Über die von der molekularkinetischen Theorie der Wärme geforderte Bewegung von in ruhenden Flüssigkeiten suspendierten Teilchen [On the Movement of Small Particles Suspended in Stationary Liquids Required by the Molecular-Kinetic Theory of Heat]. *Annalen der Physik* (in German), *322* (8): 549–560.

Gill, M. L., Pellegrin, P. (eds.). (2009). A Companion to Ancient Philosophy. Wiley-Blackwell; 1st edition.

Heisenberg, W. (1927). Über den anschaulichen Inhalt der quantentheoretischen Kinematik und Mechanik, Zeitschrift für Physik (in German), *43* (3–4): 172–198.

Kim, M. (2006). Wilhelm Ostwald (1853–1932). *International Journal for Philosophy of Chemistry, 12* (1): 141.

Klimchitskaya, G. L., Mohideen, U., and Mostepanenko, V. M. (2009). The Casimir force between real materials: Experiment and theory. *Rev. Mod. Phys.*, 81.

Lummer, O., and Prinsheim, E. (1899). Die Verteilung der im Spektrum des schwarzen Körpers und des blanken Platins. *Verh. Dt. Phys. Ges.*, 215–235.

*The Middle Length Discourses of the Buddha: A Translation of the Majjhima Nikaya* (*The Teachings of the Buddha*). Wisdom Publications; new edition (1995). MN 121 Cūḷasuññata Sutta. The Shorter Discourse on Voidness; MN 122 Mahāsuññata Sutta. The Greater Discourse on Voidness.

Wilczek, F. (2005). On Absolute Units, I: Choices. *Physics Today*, American Institute of Physics, *58* (10): 12–13.

Planck, M. (1899). Über irreversible Strahlungsvorgänge. Sitzungsberichte der Königlich Preußischen Akademie der Wissenschaften zu Berlin (in German), *5*: 440–480.

Roy, S. (2019). Intrinsic Property, Quantum Vacuum, and Śūnyatā. In: Bhatt, S. (ed.). *Quantum Reality and Theory of Śūnya.* Springer Singapore.

Schäfer, F., Herrera, I., Cherukattil, S. et al. (2014). Experimental realization of quantum Zeno dynamics. *Nat. Commun. 5*, 3194.

Simonyi, K. (2012). *A Cultural History of Physics*. Kramer, D. (transl.). CRC Press.

CHAPTER 38

# The Significance to Physics of the Number Zero

*Joseph A. Biello and R. Samson*

## Abstract

Although the numeral zero is ubiquitous in physics, it takes pride of place there in certain environments. This is illustrated by focusing on conservation laws, symmetry and non-linear dynamics. In all of these subjects, the zero fulfills the role of an anchor point, the eye at the center of the hurricane, so to speak. We try to make these connections clear in a language that is accessible to non-experts. We hope to make clear that although many mathematical models are so complex that they defy even the most powerful computers, exact or approximate results deriving from conservation or equilibrium reveal many characteristics of these complex systems.

## Keywords

conservation laws – non-linear dynamics – symmetry – Noether's Theorem – equilibrium versus chaos – weather forecasting – non-Cartesian logic

## 1 Introduction

What is the significance of the number zero for physics? At first sight this may seem like a rather silly question. Viewing physics as a branch of applied mathematics,[1] the latter as the business of dealing with numbers, and zero as the origin of the real numbers, it's obvious that zero is the alpha and the omega of physics. Take zero away and what's left of physics is like taking the oxygen out of the air and asking what's left of life on Earth. Not much, and certainly not very much exciting stuff.

---

1 Here, we're taking at least three unforgiveable shortcuts in one sentence. Most physicists will certainly not agree with the statement that physics is 'merely' a branch of applied math; applied math has many more nuts to crack than just numbers, and there's much more interesting stuff in numberland than only the integer or real numbers. That said, it's still abundantly clear that without the integer or real numbers and particularly their seminal origin, the number zero, most of the natural sciences would be in a sorry state.

 | DOI:10.1163/9789004691568_044

So, why write a paper on a subject that is so thin and worn-out? Well, consider the following. In physics, as in George Orwell's famous novel *Animal Farm*, not all zeros are necessarily of the same standing. Some zeros may be seated higher on the Olympian mountain than others.

This statement demands elucidation. In the next section we shall discuss the so-called conservation laws of mathematical physics. In the context of these laws, we shall demonstrate that the zeros that feature in these laws rightfully belong to the physics hall of fame; they definitely have a higher standing than any other odd zero (if this crippled figure of speech be pardoned).

One qualifier must be made at this point. This chapter is aimed at readers with little or no background in either mathematics or the natural sciences. It is our goal to explain to such an audience what role the number zero plays in the working practice of a professional physicist.

### 1.1 *Conservation Laws*

Natural phenomena – as they present themselves to our observing eye or to our measuring instruments – either seem to be seething with movement (the flowing of a river, the incessant movement of the clouds against the sky, the burning of trees, and other organic material in a wood fire, etc.), or else to be unshakably and majestically still (a mountain, the stars in the night sky, etc.). A layperson might think that a physicist who wants to study a subject in the category 'seething with movement' would dive headlong into the exciting business of chaos and violent upheavals. Nothing could be further from the truth. To the layperson's astonishment, the physicist's first and most important questions have to do with stillness. Typically, the physicist will ask the following questions:

1. In the midst of all the furious change, are there things that *don't* change, ever?
2. If so, what are those things? How are they defined?
3. Can we somehow understand *why* they don't change?

At this point, it's best to become concrete so that the discussion does not bleed to death in abstractions. It would be beautiful if we could give an example from a real life-and-blood situation, like the flowing river we mentioned before. For pedagogical reasons, however, it's better to take a far simpler example, something not quite as complicated as a river.

Consider a pistol being fired and follow, in your mind's eye, the trajectory of the bullet exiting the pistol. Under real-life conditions, the bullet's trajectory is not that simple because of the influence of gravity, of air resistance and of sudden and unpredictable gusts of wind. To make life really, really simple, let's assume that we could somehow magically eliminate gravity and the influence of the air molecules. In that case, the bullet's trajectory would be a perfectly straight line. The bullet would neither accelerate nor decelerate from the moment it leaves the gun; it would just keep on flying at the same speed in a straight line until doomsday. Things don't get sweeter (and more boring) than this.

Now we can answer questions 1 and 2. The thing that doesn't change is the velocity of the bullet: both its direction and its magnitude. Question 3 is a bit trickier. Do we understand why it doesn't change? Different generations of physicists have given (slightly) different answers to this question. Isaac Newton (1643–1727), the father of classical physics, would probably answer roughly as follows. 'Effects' (read: 'changes') in Nature, are the outcome of 'causes' (read: 'disturbing influences'). Without causes, no effects. By eliminating gravity and air drag, we have in fact eliminated all the (extraneous) causes. Hence: no effects; that is, there is no reason for the velocity of the bullet ever to change. In later centuries, some scholars would adapt the wording of this argument somewhat, but the basic idea didn't really change.

This apparently simple-minded physical experiment is a special case of something much more general that we mentioned before: it describes a so-called conservation law. In this particular case, the conservation law tells us that there's something (to wit: the velocity of the bullet) that doesn't change as long as there are no disturbing factors (no 'forces'). In physics, there are many such conservation laws. All such laws, expressed in a mathematical formula, have the form

$$\text{“A”} = 0 \tag{1}$$

We still haven't specified what 'A' is. Whatever 'A' may be, the thing that all equations expressing conservation laws have in common is the fact that on the right-hand side of the equation, we see a zero. This zero tells us that 'something' does not change. In the case of the flying bullet, the physicist would say that the rate of change of the momentum vector is equal to zero. (For reasons that we don't need to go into here, physicists prefer talking about conservation of momentum, rather than about conservation of velocity. In a popular treatise like the present one, it boils down to the same thing).

Physicists are absolutely mad about these conservation laws. The latter are truly the cornerstones of physics. Above, we've given one particular example: conservation of momentum. Likewise, there is a plethora of other physical quantities that obey similar conservation laws. The most important among these – apart from linear momentum – are: energy, angular momentum (a quantity that is of importance in rotational motions), and electrical charge. These laws are always expressed in the same form as Equation (1) above: 'something' (the rate of change of a quantity) is equal to zero.

Besides conserving 'something' (momentum, energy, etc.), these laws are intimately related to another central concept of physics: symmetry. In the next subsection, we shall further discuss this subject.

## 2 Symmetry

A layperson leafing through a text book on modern physics might be amazed at how often the word 'symmetry' is used and she might well wonder what it is about symmetry that so fascinates physicists. One of the reasons for her disconnect is that laypersons and scientists do not assign precisely the same meaning to the word symmetry. For laypersons, symmetry is a concept that has meaning exclusively in relation to forms and shapes. 'His face is almost perfectly symmetric on the left and right', etc. Although the physicist also knows and uses this conceptual meaning, she tends to use the word in a rather more abstract context as well. Scientists talk about symmetry (or rather about *symmetry operations*) as actions that, when performed on objects (geometrical or mathematical objects), leave them unchanged.

By way of example, take an equilateral triangle (one of the delights of stuffy high school teachers: the type of triangle with three sides of equal length). Now, rotate this triangle 120 degrees around its central point (the point in its middle that is equally far removed from the three vertices). This is illustrated in Figure 38.1. What has happened in this rotation? In fact: nothing at all (zero). In order to show the reader that we're not cheating and that we didn't sit on our hands instead of giving the triangle a 120-degree spin, we marked the three vertices by a black disk, a yellow square, and a red mini-triangle. Comparing the first and the second large triangles in Figure 38.1, we see that the markers (disk, square, and mini-triangle) executed a musical-chairs dance. But apart from the markers, nothing else was affected by the rotation. So, if we imagine that there are no markers, we wouldn't be able to tell the difference between the triangle that had been left alone (triangle 1), the one that was rotated 120 degrees (triangle 2) and the one that was rotated 240 degrees (triangle 3). We call this the rotational symmetry (or rotational invariance) of the equilateral triangle.

An interesting way to generalize this concept is by considering a geometrical figure of which not just three points, but *every* point is equidistant from a center point: the circle. This figure has the marvelous property of being

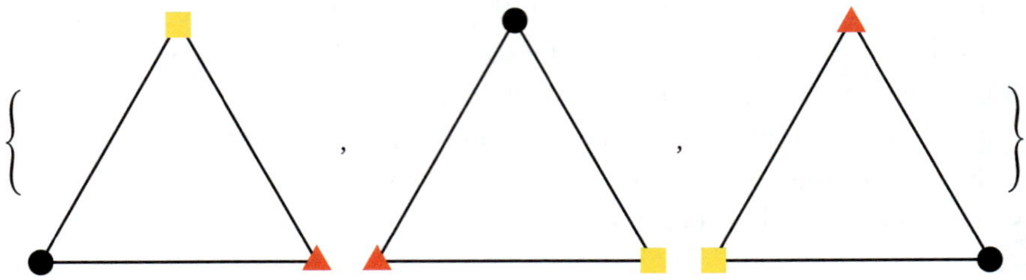

FIGURE 38.1 A snapshot of an equilateral triangle, after rotating it by 0, 120 and 240 degrees, demonstrating its rotational symmetry

symmetrical under rotation over *any* arbitrary angle around the center. In this way, we have found a *continuous* analog of the previous case (the equilateral triangle) and its corresponding symmetry group.

Roughly around the time that quantum theory knocked at the door of physics (shortly after 1900), physicists started to realize that there was a deep connection between symmetry and physics that badly needed further study. This realization did not come out of thin air. It was preceded by more than a century of feverish developments in mathematics starting in the early 1800s. Great nineteenth century mathematical geniuses such as Joseph-Louis Lagrange (1736–1813), Niels Henrik Abel (1802–1829) and Evariste Galois (1811–1832), discovered that there was a world of incredibly exciting math buried in the systematic study of symmetry operations, such as the rotations of the triangle and the circle that we looked at. This activity gave birth to a new branch of mathematics called group theory.

For a long time, group theory had a reputation among physicists of being of no practical use, until in the late 1920s Eugene Wigner (1902–1995) and others saw how group theory fit the quantum theory of atoms and molecules hand-in-glove. Physicists who learned their trade after 1930 simply cannot imagine physics without group theory.

Here's a layperson's one-paragraph summary of what physicists actually do with group theory. Suppose that a molecule existed of three identical atoms that are arranged in an equilateral triangle as shown in Figure 38.1 (to the best of the authors' knowledge such a molecule does not exist, at least not as a stable entity). As soon as a physicist knows the structure of such a molecule, she has an enormous amount of extra information at her fingertips. She knows, for example, that the interactions between this molecule and light (or any other form of electromagnetic radiation) must obey the specific dictates of the triangular symmetry group. The theory tells us which spectral lines will or will not be emitted or absorbed by this molecule. In other words, the confluence of quantum mechanics and group theory leads to a marvelously rich and powerful toolkit for the practical physicist.

At this point in the story, the reader may wonder if the authors have not strayed miles away from their topic of the zero in physics and conservation laws. We shall revert to our major topic momentarily, but not until after we have indulged in the following.

## 3 Intermezzo – the Life and Times of Emmy Noether

In a popular treatise such as the current one, the temptation to interrupt the story with a biography of one of the main persons in our story is irresistible. If we were looking for a Hollywood-style plot full of sound and fury, our choice – without a doubt – would be to relate the life of Evariste Galois, one of the main founders of group theory and modern algebra. His biography is almost beyond belief. Somebody ought to write an opera about this man.

However, we will resist this temptation and choose someone else whose life was considerably less action-packed, but whose biography gives us pause to consider some serious 'issues' in the academic community and, indeed, in society in general. The person we refer to is Emmy Noether (1882–1935). Many renowned mathematicians and scientists (among them Einstein) have called her the most important woman in the history of mathematics. In 1915, she was invited by David Hilbert to join his group at the University of Göttingen. For those not in the know, in 1915 practically every living mathematician considered David Hilbert to be the most important mathematician alive. In Hilbert's group, Emmy Noether was definitely in the 'place to be'.

Although her academic output was rich and far-reaching, here we will consider only one of her many achievements: the theorem that is now generally known as Noether's theorem. She conceived of it in 1915 and published it in 1918. Einstein, after reading the paper, wrote to Hilbert: *'Yesterday I received from Miss Noether a very interesting paper on invariants. I'm impressed that such things can be understood in such a general way. The old guard at Göttingen should take some lessons from Miss Noether! She seems to know her stuff.'*

In popular terms, the content of Noether's theorem boils down to the following: 'Between conservation laws and symmetries there is a close correspondence: to each conservation law corresponds one specific continuous symmetry group and vice versa.'

Let's try to understand this theorem in a specific case: the case of the flying bullet that moves at a constant velocity in the absence of any perturbing forces – the case we discussed in Section 2. As stated earlier, the conservation law at work here is conservation of momentum. What is the corresponding symmetry law?

To answer this question, we'll ask the reader to suspend disbelief while we present the following – at first sight, completely mad – argument. Suppose we could travel to a vantage point outside our universe. Assume further that we had a gigantic machine at our disposal, whereby we could move the entire universe one meter to the left (or up or down: it doesn't matter in which direction, as long as it's not to the right). The question we now ask is: 'Did anything change, anything at all, inside the universe, as a result of it having been shifted by a meter?' Although we cannot actually do the experiment, physicists are convinced that the answer is negative. Nothing changed, nothing at all. We cannot prove this, but at least we can provide plausible arguments for our claim. Say we perform an experiment in a laboratory, then we shift our lab table to another location and redo the test. Unless there are very trivial circumstances that changed (e.g., that our lab table was perfectly horizontal in the first location but somewhat tilted in the second location) we will always find the same outcome in the two experiments. Of course, this is not scientific proof, but at least it lends some credence to our hypothesis.

By use of this strange argument, we have arrived at a new symmetry law: the law of translational invariance. Noether's theorem states that conservation of momentum and translational invariance are two members of a conjoined twin: one implies the other and vice versa.

According to Noether this correspondence holds for every conservation law: for energy (which is a conserved quantity, just like momentum), the corresponding symmetry is time invariance (invariance under time shifts); for angular momentum, the twin sister is rotational invariance; etc.

Noether's theorem has become one of the most important concepts in mathematical physics. Its importance can hardly be overstated.

Before returning to the main thrust of our discussion, let's stop for a second and ask ourselves why it is that Noether's name is so little known to the general public? It's hard to avoid the conclusion that this has everything to do with the fact that she was a woman. In her day, for a woman to seriously make a career (in science or in any other field) took almost superhuman efforts. The prejudices of society and the obstacles to be overcome were beyond belief.

Here's one particular example, not referring to Emmy Noether but to another eminently gifted woman: Fanny Mendelssohn. Fanny's younger brother Felix Mendelssohn (1809–1847) became world famous as the composer of numerous great musical works. The Mendelssohn family was probably one of the most open and liberal group of people in the Germany of their time. Like Felix, Fanny (1805–1847) was a huge musical talent, both as composer and as pianist. Nevertheless, this is what their father wrote to her in a letter in 1820:[2] *'Music will perhaps become his [Felix's] profession, while for you it can and must be only an ornament.'* Although Felix was privately broadly supportive of her as a composer and a performer, he was cautious (professedly for family reasons) of her publishing her works under her own name. He wrote:

> From my knowledge of Fanny I should say that she has neither inclination nor vocation for authorship. She is too much all that a woman ought to be for this. She regulates her house, and neither thinks of the public nor of the musical world, nor even of music at all, until her first duties are fulfilled. Publishing would only disturb her in these, and I cannot say that I approve of it.

With supporters like this, who needs adversaries?

Back to Emmy Noether. Although eventually she found her place in the academic world, she had to endure tribulations that few male scientists – let alone, male scientists of her stature – would have accepted. Before her appointment at Göttingen, she worked for seven years at the University of Erlangen without pay. Once accepted at Göttingen, she did not lecture under her own name, but under Hilbert's name. Finally, in 1919 her habilitation was accepted so that she could carry the title of *Privatdozent* (private lecturer), which certainly did not do justice to her eminent academic achievements.

## 4 Non-linear Dynamics:[3] Equilibrium, Chaos and Everything in Between

In this section we shall discuss another subject in which the numeral zero plays a central role, namely, 'non-linear dynamics'. First, we need to explain the relevance of non-linearity in physics. Most objects studied by physicists are aggregate systems that can be regarded as a collection of smaller subsystems. A gas, for example, is usually regarded as an assembly of molecules. The relevant

2 https://en.wikipedia.org/wiki/Fanny_Mendelssohn.

3 This subject – or parts of it – is sometimes referred to under a number of different names: Chaos Theory, Theory of Fractals, Catastrophe Theory, etc. In this section, we lean heavily on Strogatz (1994).

questions in this context are: Can a composite system be regarded as the sum of its parts? If so, is this an approximate or an exact statement? For linear systems, the answer is: Yes. The whole is exactly equal to the sum of its parts. This is a great bonus in practice. If one wants to understand the properties of a gas, one studies one single molecule and 'somehow' adds them all up. Of course, this only works if the molecules have no or only very weak mutual interactions. In practice, this is usually not the case. This is where all the complexities of non-linearity enter the story. Non-linear systems do not allow us to simply disassemble the whole into parts, and are thus considerably more complex than their linear counterparts.

Although the subject of non-linear dynamics has some very significant antecedents in the nineteenth and early twentieth century, it really only took off after the advent of the digital computer; that is, after World War II. The solutions to most of the problems in non-linear dynamics require massive computing power. The calculations are often so complex, that even modern supercomputers fail to do the job. The two best known problems in this category are meteorology and fluid mechanics.

In the paragraphs below we shall attempt to describe some of the problems solved in non-linear dynamics and mention the relevance of zero to this subject.

Before attempting to tackle headlong the mathematical equations of non-linear dynamics, the mathematician usually first takes a step back and tries to understand the *general, qualitative* (rather than the *detailed, quantitative*) features of the equations first. A very powerful tool in this endeavor is the use of graphical methods. In trying to understand the overall graphical shape of certain characteristic mathematical expressions relevant to this problem, the mathematician gets a bird's-eye view of the gross features of the problem, even if they do not have precise knowledge of all the details. At this stage, they look at the graphs in much the same way that a geographer might look at an altitude map of an irregularly shaped mountain landscape: 'Aha ... here's a deep gorge. A bit further to the northeast is a huge mountain top. At the far edges, the whole landscape seems to be surrounded by a high wall.' This metaphor has its limitations because the kinds of structures that the mathematician finds in such maps may be richer, more varied, and more fantastic than the geographic offerings of Mother Nature.

The type of mathematical expression that yields the information about the valleys and rifts, etc., invariably looks like this:

$$\text{“B”} = 0 \tag{2}$$

Here, 'B' is an expression that is derived from the full mathematical statement of the problem, with the important qualifier that solving Equation 2 is many, many times simpler than solving the original equation. Quite often, solving Equation 2 is the type of calculation that mathematicians like to call a 'back-of-the-envelope' calculation (sometimes tongue in cheek).

The attentive reader who remarks that Equation 2 looks just like our earlier Equation 1, which we saw in the section on conservation laws, is quite right. The point, however, is that the significance of 'B' is quite different from our earlier 'A'. In the present case, Equation 2 does not express the existence of a conserved quantity but 'something else'.

The solution of Equation 2 in fact provides us with a map of the most important features of the graphical landscape. Among these, we mention real and metastable equilibrium points (roughly comparable to the lowest point in a valley and the highest point on a mountain top); saddle points (points that are stable in one direction and unstable along another direction); centers (points around which the system gets trapped in a periodic orbit); isoclines (closed or open orbits along which some quantity remains constant); etc. Once the system gets trapped in an equilibrium point, it will remain there for the rest of its life. The metastable equilibrium points may look deceptively like equilibrium points, but even the slightest upset of the system (the proverbial fluttering of a butterfly on a distant continent) is enough to send the system crashing down the mountain slope.

Often this is sufficient information for the mathematician to be able to predict the behavior of the system. Sometimes, however, one cannot avoid going into more detail (e.g., in weather forecasting) and then one just has to bite the bullet and solve the system as best as one can by brute-force computation.

Let us look at a number of practical examples to get a better feel for the gist of the story.

## 5 The Pendulum

Consider the case of a pendulum. A pendulum is a mass fixed to one end of a rigid rod. At its other end, the rod is attached to a fixed point in space (the vertex) but is otherwise free to rotate around that point. The location of the mass is measured by the angle between its location and an imaginary line pointing straight downward from the vertex; positive angles are to the right of the imaginary line, negative angles to the left. This is a simple physical system where both notions of zero (conservation as well as equilibrium) are manifest. Since the behavior of the pendulum does not depend on the point in time when we set it in motion (yesterday, today, or tomorrow), it follows that the system has time invariance and, by Noether's theorem, its total energy is conserved. We write:

The change in total energy of the pendulum = 0.

Figure 38.2 below depicts the iso-energy curves of the pendulum (lines of constant energy).

We also have the other notion of zero, that of equilibrium. It is obvious that if the mass is left undisturbed at angle zero (pointing straight downward) it will remain there for all time. At the antipode of this location (the mass being straight above the vertex of the pendulum), there's also an equilibrium point, albeit a precarious one. In theory, the mass could also remain at the antipodal point forever; however, even the smallest perturbation will cause it to topple over. This location represents a metastable equilibrium point.

The horizontal axis measures the angle (in degrees) of the pendulum measured from the downward direction. The vertical axis measures the angular velocity, or the speed with which the pendulum goes around. The curves are constant energy levels and, therefore, the trajectory of the pendulum in this 'phase space' must lie along one of these curves. The dots (at angle = 0 and +/− 180 degrees) denote the downward and upward pointing equilibria, where no motion is expected to occur.

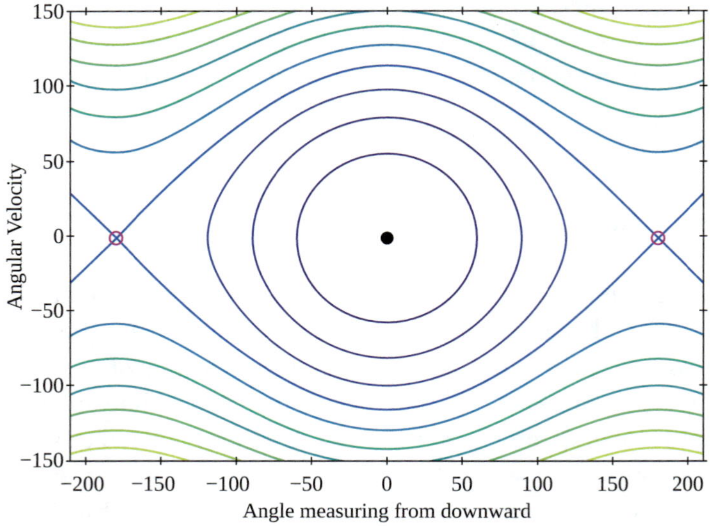


FIGURE 38.2 Phase diagram of the pendulum

## 6 Population Dynamics

Strogatz (1994) calls this the 'rabbits versus sheep' problem. This is a problem in animal population dynamics. The (highly simplified) mathematical model assumes that two biological species, rabbits and sheep, compete for a limited food supply. The model ignores all other extraneous influences – predators, diseases, climatic conditions, human intervention, etc. The graphical 'altitude map' in Figure 38.3 shows the conclusions of the model at a glance:

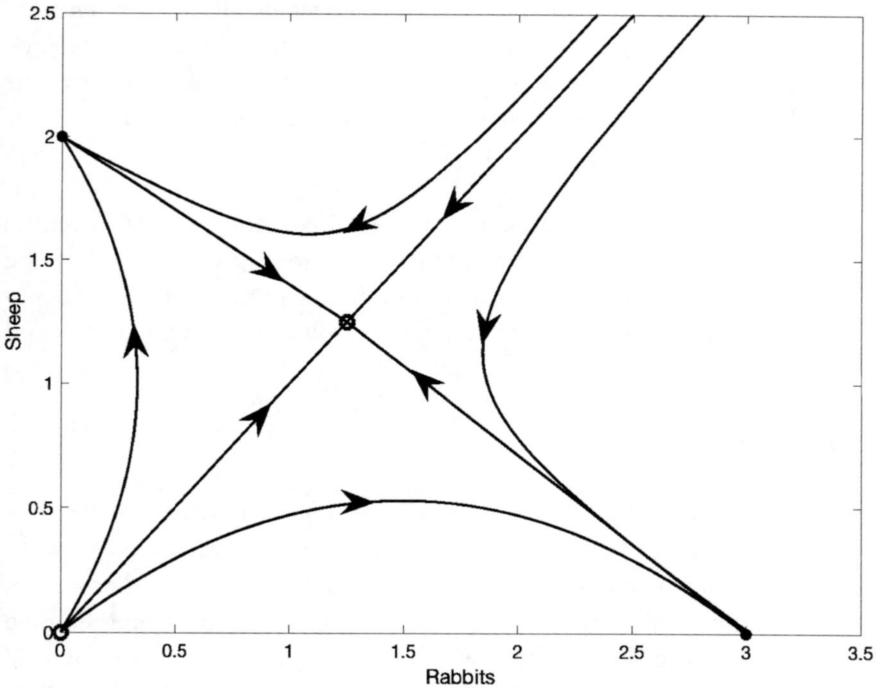


FIGURE 38.3 Graphical representation of the 'rabbits versus sheep' problem

The black dots on the horizontal (rabbits) and the vertical (sheep) axes represent two stable equilibria. These are stable situations in which one species survives only at the expense of the other. The two white circles are two unstable nodes. The arrowed lines depict possible time trajectories. The flow is always from a chosen initial point in the direction of the arrow. All scenarios eventually end in the extinction of one of the species. All lines below the line labeled as 'separatrix' end in the survival of rabbits only; conversely, all lines above the separatrix lead to survival of sheep only.

The model doesn't give rise to any form of permanent peaceful co-existence. This, of course, is not 'the truth', but the result of the raw simplifications of the model. The most characteristic feature of this problem is that all flow lines ultimately converge in a smooth fashion on one of the two stable points.

## 7 Meteorology

The following problem in non-linear dynamics instantly became a *cause célèbre* in the scientific community upon the publication of a paper on meteorology in 1963 by E. N. Lorenz (1917–2008) (Lorenz, 1963). The effect discovered by Lorenz came to be known by the nickname 'butterfly effect': an extreme dependence of the outcome of meteorological calculations on the input data.

We quote the following from Wikipedia:[4]

> In 1961, Lorenz was running a numerical computer model to redo a weather prediction from the middle of the previous run as a shortcut. He entered the initial condition 0.506 from the printout instead of entering the full precision 0.506127 value. The result was a completely different weather scenario.

Lorenz wrote:

> At one point I decided to repeat some of the computations in order to examine what was happening in greater detail. I stopped the computer, typed in a line of numbers that it had printed out a while earlier, and set it running again. I went down the hall for a cup of coffee and returned after about an hour, during which time the computer had simulated about two months of weather. The numbers being printed were nothing like the old ones. I immediately suspected a weak vacuum tube or some other computer trouble, which was not uncommon, but before calling for service I decided to see just where the mistake had occurred, knowing that this could speed up the servicing process. Instead of a sudden break, I found that the new values at first repeated the old ones, but soon afterward differed by one and then several units in the last decimal place, and then began to differ in the next to the last place and then in the place before that. In fact, the differences more or less steadily doubled in size every four days or so, until all resemblance with the original output disappeared somewhere in the second month. This was enough to tell me

4 https://en.wikipedia.org/wiki/Butterfly_effect.

> what had happened: the numbers that I had typed in were not the exact original numbers, but were the rounded-off values that had appeared in the original printout. The initial round-off errors were the culprits; they were steadily amplifying until they dominated the solution. (Lorenz, 1993, p. 134)

One meteorologist remarked that if the theory were correct, one flap of a seagull's wings would be enough to alter the course of the weather forever. The controversy has not yet been settled, but the most recent evidence seems to favor the seagulls. Following suggestions from colleagues, in later speeches and papers Lorenz used the more poetic butterfly.

Far from being a pathological case, the situation that Lorenz describes here has turned out to be almost the rule rather than the exception for many problems of practical interest. The reader should understand that this is very often not a problem that can be fixed by having more powerful computers. Of course, more computing power is always welcome. However, the demand (by society, by industrial users, etc.) for more mathematical accuracy can easily overwhelm the tempo in which computers are being improved.

Strogatz gives the following illustration of this problem. Suppose there is a meteorological service company that delivers 24-hour predictions of the weather with a degree of accuracy that satisfies the customers. The meteorologists now investigate to what degree they would need to improve their input data in order to deliver 3-day forecasts with the identical degree of accuracy as their current 24-hr forecasts. The answer is a staggering factor of one-hundred-million (100,000,000)!! It goes without saying that this is an unrealistic demand, not just with respect to the computers, but more especially also with respect to the equipment that measures the *current* weather conditions, needed as input for the weather forecast. In other words: high-accuracy, long-time weather forecasting is likely to remain pie-in-the-sky, not just in the near future, but probably for as long as we can foresee. It is a melancholy thought that the most important conclusion of the expansion of our scientific knowledge is that we get a sharper focus on the limits of our knowledge and indeed of what is and what is not knowable.

## 8 A Philosophical Aside

### 8.1 *A Closer Look at Noether's Theorem*

Conservation laws are basically statements about material objects: bullets, sound waves, electrical discharges, etc. These laws say things like 'The bullet flies at constant speed', or 'The ESA satellite orbits around the earth at constant angular momentum', or 'Electrical charges are neither destroyed nor created'. These are statements about things we can touch, things that can hurt us when they collide with us, etc.

Symmetry laws are things of an entirely different nature; they are statements about space and time. They tell us that the universe can be shifted in space with impunity, or that the universe has no North Pole or any other identifiable 'special' direction, or that there is no universal clock, etc. These are abstract concepts: things we cannot touch or collide with.

If attributes of matter in some way follow from properties of space and time, this raises the question: *what is the nature of matter*? Is it correct to endow matter with an independent, autonomous existence? Or is matter 'merely' an ephemeral by-product of space and time?

These questions are not new, of course. Both within the confines of physics, but more especially in philosophy, many brilliant thinkers have reflected on these questions for the past 4,000 years.

Einstein, in his theory of general relativity, came to the conclusion that gravity is not a property of material objects (the Earth, the Sun, stars, etc.), but of time and space itself (or rather, of a new entity that he called space-time: space and time rolled into one). He thereby 'solved' one of the most vexing philosophical questions of Newtonian mechanics: how do celestial bodies attract each other when there is no tangible medium (an 'ether') that guides the force from one to the other?

In quantum field theory, one encounters another exotic property of space. When quantum physicists study the properties of empty space, they come to the conclusion that there is no such thing as 'empty space'. They observe that empty space spontaneously emits energetic particles ('something coming out of nothing'), thereby destroying its own emptiness. Instead of emptiness, they find a furiously boiling soup of particles and anti-particles that continuously destroy each other and/or spew forth new particles. This observation effectively disrupts the neat distinction between space (or space-time) and matter. Where does one end, and where does the other begin?

## 9 Ancient Indian Views

Some branches of philosophy take an even more radical point of view by denying or questioning the reality of the material world altogether. Probably no branch of philosophy went further in this than the Nagarjuna school of

Buddhism in ancient India (*c.*150–*c.*250 CE). Nagarjuna's tetralemma (the so-called fourfold negation) has few competitors in its ruthless denial of reality.

In this lemma, Nagarjuna considers the validity ('true' or 'false') of the four following scenarios for a proposition 'P':

1. 'P' itself
2. the denial of 'P'
3. any logical combination of #1 and #2 (#1-and-#2, #1-or-#2)
4. the denial of #3

In this school of philosophy, whenever the proposition 'P' is applied to *any* aspect of our tangible ('real') world, it is stated that each and every one of the four scenarios is false.

This, of course, runs squarely against the spirit of traditional Western – Cartesian – logic in which a proposition is either true or false (the logic of the 'excluded third'). Physics, along with the rest of Western science and humanities, has always operated strictly within the confines of Cartesian logic. Nagarjuna's tetralemma feels like a merciless public flogging of René Descartes (1596–1650), 1,400 years before the poor chap was even born.

We have discussed two concepts that are products of ancient Indian culture: the number zero and the tetralemma. One wonders whether there might be a causal connection between the two concepts: could one have led to the other and, if so, which gave rise to which? In order to say something sensible about this subject, the relative chronological priority is important: which preceded which? About the dating of the tetralemma there is not much uncertainty: it must have coincided with Nagarjuna's life (roughly between 150 and 250 CE). Unfortunately, the same is not true for the number zero: its 'birth date' (in India) is the subject of wildly varying speculation (between roughly 200 BCE and 600 CE). As a result, nothing can be said with any degree of accuracy about this subject until further research is conducted.

The eminent scholar of Buddhism, D. S. Ruegg (2010), stated that he saw no evidence that the mathematical zero had any influence on the conception of the tetralemma. Although there is also no solid basis for claiming the reverse connection (tetralemma leading to zero), there is at least one scholar[5] who did argue that the tetralemma had an influence on the birth of the so-called 'null class' in modern mathematical set theory.

5 Hajime Nakamura, *Buddhist Logic Expounded by Means of Symbolic Logic*, written in the 1950s and referred to by Ruegg in the above reference.

## 10 Bridging the Gap between East and West

In this chapter we have mentioned many instances in which modern physics is indebted, so to speak, to the numeral zero. One wonders if there's possibly more that the scientific community could learn from zero. Physicists are proud to think of themselves as hard-headed and down to earth, and not inclined to mystical hocus-pocus. Although the common scientific method (empiricism, coupled to Popper's falsification principle) has been unbelievably successful, there have been areas of science where progress has been less spectacular. Is it foolish to ask whether or not this can be related to the fact that the scientific community is hostage to its own Cartesian dualistic logic? In India and elsewhere, there exist many competing (non-binary) logical systems that could provide a different viewpoint than the conventional Cartesian one.

Returning to Western science and the question of what it could learn from ancient Eastern philosophical systems, a relevant case is the discussion in the early days of quantum mechanics whether electrons (and other sub-microscopic particles) should be regarded as particles or as a wave-like entities. The realization that it could be both, depending on the type of experiment to which the electron is subjected, caused (and still causes to this day) endless discussions in the physics community. One wonders whether a culture that is imbued with non-binary logical thinking would not have accepted this concept with greater ease. (A sophisticated recent discussion of the relation between the paradoxes of quantum mechanics and Buddhist philosophy is given by M. Bitbol (2019).)

Efforts to couple non-binary logical systems to modern science, if they have been undertaken at all,[6] have probably been few and far between and very tentative. One can only wonder whether there is not room for improvement here. Who knows? And who knows whether the zero could not serve as a welcome pointer in this endeavor: a bridge between two contrary, but possibly supplementary, thought-systems, as well as a bridge between East and West.

## 11 Conclusion

Although the numeral zero is as ubiquitous in physics as the oxygen molecule in the Earth's atmosphere, there are certain environments in physics where zero takes pride of place. As an illustration of this, we concentrated on conservation laws. The latter occupy a central position in physics. We discussed the

6 One interesting attempt can be found in Sorkin (2010).

links between conservation laws, symmetry and Noether's theorem: all central concepts in modern physics.

We also reviewed the subject of non-linear dynamics. Many non-linear mathematical models are so complex that they defy even the most powerful computers. Physicists studying these complex systems always look for the calm in the eye of the tornado, which is the zero on the right-hand side of Equation 2. Understanding the geometrical structure (stable and unstable nodal points, etc.) of the system makes the taming of these otherwise unassailable systems slightly more feasible.

The current discussion has shown that the ubiquitous numeral zero occupies a central role in physics. We have argued that zero, being the signpost of conservation and equilibrium, fulfills the role of an anchor point in the description of complex physical systems.

## Acknowledgments

René Samson gratefully acknowledges many stimulating discussions with Peter Gobets on Nagarjuna's philosophy and its potential significance for modern science.

## References

Bitbol, M. (2019). Two aspects of Śūnyatā in quantum physics: relativity of properties and quantum non-separability, in S. R. Bhatt (ed.): *Quantum Reality and Theory of Śūnya*. Springer Verlag.

Lorenz, E. N. (1963). Deterministic nonperiodic flow. *Journal of Atmospheric Sciences*, 20, pp. 130–141.

Ruegg, David Seyfort. (2010). The Buddhist Philosophy of the Middle, Essays on Indian and Tibetan Madhyamaka. Boston: Wisdom Publications, p. 90.

Sorkin, Rafael D. (2010). *To What Type of Logic Does the 'Tetralemma' Belong?* Physics, Paper 7. http://surface.syr.edu/phy/7.

Strogatz, Steven H. (1994). Non-linear Dynamics and Chaos (with Applications to Physics, Biology, Chemistry and Engineering). Perseus Books Publishing.

CHAPTER 39

# A World without Zero

*R. Samson*

## Abstract

In this chapter, the question is addressed what the world would be like if the number zero had never been invented. This anti-historical question is dodged by asking a slightly less difficult question, namely: what would the world have been like if the so-called complex numbers (ubiquitous in many fields of science and engineering) had never been invented?

## Keywords

zero – metaphor – complex numbers – electromagnetism – double-entry accounting – religious rituals

## 1 Introduction

What would the world be like if the number zero had never been invented?

In more than one sense this is an extremely silly question. Similar questions of a much more down-to-earth nature are likewise next to impossible to answer. What would it be like to live in Upper Mongolia? What would it be like to be colorblind? What would it be like to be an octopus? Unless one is blessed with an extraordinary imagination or one belongs to a very specific subspecies (Mongolian, colorblind, octopussian) almost nobody can answer such questions off the cuff with any degree of accuracy.

Even so, let's give it a shot and see how far we get. Before going off the deep end, let's try to answer a much more modest question (more modest in the sense that it refers to a more recent but comparable development in human history than the invention of zero): what would it be like if there were no 'complex numbers'? For the mathematically uninitiated reader, a complex number is nothing more or less than an ordered pair of two 'normal' numbers. The concept of complex numbers arose in sixteenth-century Italy. For several centuries, complex numbers were considered a rather esoteric toy for mathematicians

 | DOI:10.1163/9789004691568_045

with no use in the 'real world'. However, all of this changed in the nineteenth century when suddenly all kinds of applications in physics and engineering presented themselves. One example is in the field of electromagnetism. Through the work of a couple of great British physicists (Faraday, Maxwell), this field developed enormously during the nineteenth century. Much to their own surprise, the physicists discovered that the description of electromagnetic phenomena appeared to cry out for the use of complex numbers. Even for a layperson it's not difficult to understand why that was the case. Imagine an electrical engineer – let's call him Max – who is studying alternating currents (this is the kind of electricity that power lines deliver to your house and that makes your lamps glow; it's the AC in AC/DC). Max might visualize the electrical current as a sine curve on the screen of an oscilloscope (see Figure 39.1). Such a curve (which shows, for example, how the electrical current is changing as a function of time) naturally needs two numbers (as shown in Figure 39.1) for every point on the graph. It shouldn't come as a surprise that complex numbers seem to come in quite naturally here. Electromagnetism is the hand that had a glove (complex numbers) waiting for it, made 300 years earlier.

Now let's ask Max what his calculations would look like if complex numbers were not available. His facial expression would probably betray his intense distaste. He might mumble something like, 'Well, in that case we would simply have to invent them.' Or – if we would block out this possibility – he might admit with great displeasure that work-arounds would have to be invented but none would be quite as powerful and compact as complex numbers.

So much for complex numbers. Let's now consider the 'real thing': the 'zero'. What would the world have been like without it?

Once upon a time, in deep prehistory, early humans' only use for numbers probably consisted of the numerals 1 through 10, not accidentally coinciding with the number of our fingers (or perhaps 20, taking into account that shoes were still not around so that toes were helpful for counting too). It's obvious why a hunter-gatherer would feel the need to count. Which hunter should be given the biggest meal tonight? Well, he who shot the largest number of game today. There you go.

Very soon in history the numbers would increase beyond 10 or 20. A well-to-do shepherd might have many more than 20 goats. Once barter and trade developed, people would have to invent ways of adding and subtracting numbers of traded objects. Trade inevitably implies not only credits but debts as well. How does one keep track of all the mutations? Probably, people first experimented with systems in which credits and debts were recorded separately, but this soon became annoying and eventually somebody invented the concept of the double-entry accountancy system or some ultra-primitive precursor of it.

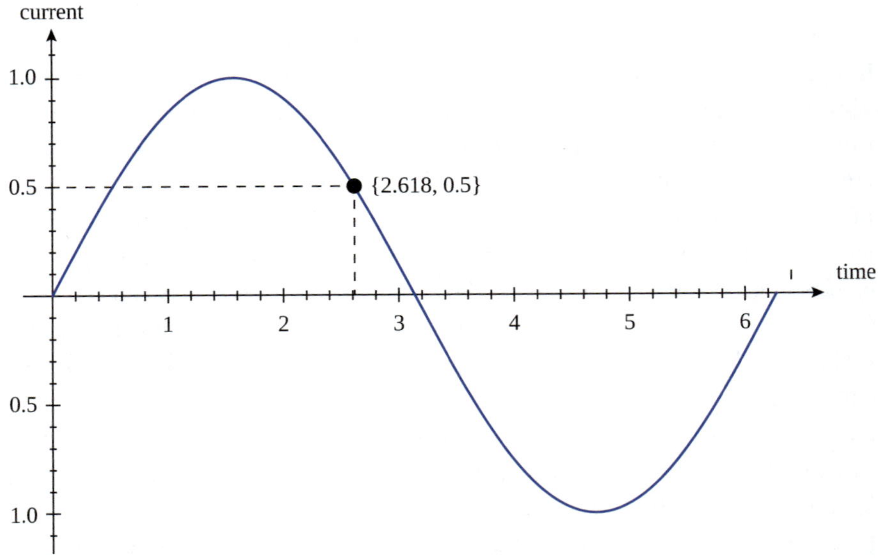


FIGURE 39.1 Alternating electricity depicted as a simple sine curve. The current in the power cable goes alternatingly forwards and backwards. Two numbers (one referring to the horizontal axis – time, one to the vertical axis – current) are needed to characterize every point on this curve

As soon as one starts thinking about debts as – somehow – the mirror image of credits, the question arises: but what's exactly in the middle between credits and debts and how shall we call that center point? This, in the context of the practical life of early humans, is the birth of the concept of zero.

Now, the story above is completely fictional and it's almost certainly *not* how it went. It is much more likely that there were many, many different strands that all contributed after many millennia of human development to the birth of zero. Besides mercantile influences (as described above), it's quite likely that religious rituals, astrology, astronomy, architecture, agriculture, warfare, etc., all played a role in this multifaceted development. In the above, I have deliberately demystified the story of zero's birth and accentuated purely utilitarian aspects. This is most probably a gross oversimplification. It is quite likely that religious and philosophical systems played an enormous role in the adaption of the concept of zero. It is quite conceivable that ancient India with its strong consciousness of 'emptiness' greeted the zero as a natural consequence of their philosophy, while medieval Christian Europe with its strong rejection of the vacuum (and related concepts) had a natural distaste for the zero.

To come back to our original question: What would the world be like if the number zero had never been invented? If I had to answer that question, I'd pull the same face – except worse, much worse – than Max pulled when asked what it would be like to manage without complex numbers. Our world, our hi-tech twenty-first-century world, is simply unimaginable without the number system that we have nowadays, in which the zero takes pride of place. Can we imagine modern science, high-power computing machines, mobile phone technology, internet, etc., without a well-oiled decimal numerical system including the zero? Hardly. Like Max, one can think of workarounds but I'm sure that each one has some pretty unsavory disadvantages. We can think of better ways to put our brainpower to good use.

## 2 Deeper Lessons

What are the deeper lessons this story teaches us? Let's try to itemize some salient points:

2.1 The reasons for the birth of the zero were probably many-faceted, both utilitarian, as well as philosophical/mystical.

2.2 It probably had a lot to do with the 'birth' of the concept of negative numbers. Once you have 1 and minus 1, it's natural to ask: is there something right in between those two?

2.3 The 'decision' to accept 'zero', to give it a name and to consider it a candidate for all the well-known mathematical operations (addition, multiplication, etc.) that were accepted for the natural numbers (1, 2, 3, etc.) was by no means trivial and would indeed lead to some astounding surprises (for example: division by zero is 'unacceptable').

2.4 In talking about zero it is necessary to make a distinction between the use of zero as a 'placeholder'[1] and the acceptance of zero as a fully fledged number with the same status as other natural numbers. It may well be that in some ancient civilizations the zero appeared only as a placeholder, while in some others it appeared both as a placeholder and as a fully fledged numeral. *This is a major point that merits further research.*

## 3 Conclusion

At the risk of stating the obvious, the number zero has become so central to our modern digital world, that even conceiving of a world without it is beyond our powers of imagination.

1 Zero as a placeholder refers to the use of zero in, e.g., the decimal numbers to keep a position open for 'empty' quantities. For example, the 0 in 101 denotes the absence of multiples of 10.

# Epilogue

*Peter Gobets*

## 1 Questions

The Zero Project does not *a priori* assume that the numeral zero (0) was invented or discovered in India. It is precisely the question to be addressed.

That the *modern* zero, worldwide in use, comes from India is well-known, as is the route it travelled. But just when and where the numeral zero was invented or discovered remains shrouded in mystery to this day. Was it indigenous to India or transmitted to India from points east or west? Was zero independently co-invented elsewhere? And as important to a better understanding would be to learn what cultural influences facilitated zero's invention/discovery.

Questions abound that warrant future research to shed light on what has been universally hailed as among the greatest innovations in human history. We have included a list of such challenge questions compiled by the contributors (appended below) to be taken up in due course by the *Expertise Center on Zero* (XCZ) – meanwhile launched by the founding signatories across Asia, Europe and North Africa.

Such questions requiring attention may contribute to a better understanding of the significance of the invention/discovery of zero (IDZ). The Zero Project and this monograph on zero are intended as stepping stones *en route* to systematically addressing these and related questions.

## 2 Answers?

Nor is the issue of zero's invention a matter of 'mere' historical interest. Could it be the case that we still have not exhausted zero's utility and versatility? As far as that is concerned, multifaceted zero and its unique significance mathematically as well as meta-mathematically, raise, among others, issues related to the underlying philosophy and non-traditional logic that may have facilitated its emergence as a number; and whether this may have had a knock-on effect in the sciences at the foundational level?

A number of articles in the monograph conduce well with the Zero Project's working hypothesis, suggesting more generally – as their authors do – that the evolution of a sophisticated concept of nothingness or its derivative emptiness

 | DOI:10.1163/9789004691568_046

may have been a necessary (if not necessarily sufficient) prerequisite for the invention of zero. But were there other factors involved?

The mere functional adoption elsewhere in the world of the transmitted mathematical zero as part of the decimal system that reached Europe via the Islamic world may, by 'reverse semiotic mechanism of transmission', may have hastened the dawning of its underlying concept, perchance even facilitating the modern scientific worldview of uncertainty, indeterminacy, complementarity, indeed of the physical vacuum, the quantum vacuum and the Big Bang itself: Something emerging out Nothing, previously deemed unthinkable in the West.

Given that after reaching points west the modern zero, severed from its philosophical moorings, went unrecognized in its most important aspect as a number – rather than just as a placeholder – for centuries on end, it may not be too far-fetched to further investigate whether a related *ontological* myopia may also plague us today still? After all, we speak only of monism (1), dualism (2) and pluralism (3 or more substances or principles).

## 3 Proposal

But what of a 0-based ontology? Just as the number zero before it, a 0-based ontology is still missing in the West, other than in a 'trivial' form as Nihilism, deracinated of its creative, profounder aspect of potentiality.

Perhaps, then, the question worth posing following the range of interdisciplinary chapters collected in the monograph on zero, may be: 'Has the time come to adopt a 0-based ontology in tandem with the number zero?' To that end we have proposed just such an ontology coined here as 'nonism' (in tandem with monism, dualism and pluralism).

Hence the thrust of the monograph on zero is to call for further research into these and other challenge questions, not only of finding fresh evidence of zero's invention/discovery in mathematics, but also of the axial significance of the underlying concepts of nothingness and the derivative 'emptiness'.

## 4 Epicenter?

We now indulge in an admittedly speculative account to cap the book project and to 'kickstart' requisite research under the *Expertise Center on Zero*.

What follows conduces well with our recommended methodology in tackling the thorny issue of finding fresh evidence of zero's invention by taking a cross-cultural view of commonly understood, plausible historical developments that does justice to all actors in the zero-story.

We take as point of departure the thesis advanced by Thomas McEvilley (*The Shape of Ancient Thought*) in which he refers to the cradle of Indo-European civilization in Central Asia steeped in Bronze Age mythology. According to McEvilley's bird's-eye view – spanning over six millennia – from at least the sixth century BCE onwards, the overriding influence of knowledge flowing from the East (Mesopotamia, China, India) and the South (Egypt) into Europe served as catalyst for the advent of the Golden Age of Ancient Greece, including the introduction of the philosophic concept of ontological monism, which was to become the mainstay of the Presocratics.

This heady mix injected into the intellectual climate of the day resulted in the flow of ideas and knowledge from diverse tributaries at the confluence of the cultural torrent during that most memorable historic moment.

In due course, McEvilley argues, this exposure to Eastern knowledge led Greek thinkers from Parmenides onward to eventually 'crack the code of language' that produced the Age of Rationalism – based on epistemic logical analysis rather than mythology. It forms still the backbone of science/scientific theory today and our concomitant worldview, which is argued elsewhere, requires urgent overhaul.

Several centuries later, following the incursions of Alexander the Great into Asia, the rich and fruitful interaction between all these cultures in ancient Bactria/Gandhara once more became the hotbed of cross-fertilization. It was through this interaction that Buddhist thinkers were exposed to Greek thought under the reign of 'Greco-Indian' kings, and adopted and adapted the Greek twofold logic of their interlocutors to engage in public debates.

As an example, we may refer to the famous questions of King Milinda (Milindapañha), in which the Buddhist monk Nagasena (*c.* second century BCE) engaged in such debates on, among other things, the concept of Emptiness (*Śūnyatā*), which apparently convinced King Milinda, who was said to have subsequently subscribed to Buddhist philosophy. Nagasena's, it may be noted, was a fourfold logic (tetralemma).

Of particular significance was the simile of the chariot and personal identity, both said to be composites of elements, each of which in turn were composites of other elements – which, when teased apart, vanished altogether. Other than the ancient Greeks, there was no 'essence' or 'substance' to be found at the core. In a word, Emptiness.

It is quite conceivable therefore that Gandhara may have been the spawning ground of the roughly contemporaneous Mesopotamian – and for that matter the Chinese – placeholder zero (as sign or empty place/column). And that, as such, Gandhara may have been prelude to the transition in Indian mathematics from their non-positional Brahmi numeral system without zero to the decimal system with zero – that is, not only as a placeholder but as a number in its own right.

The reader may recall that the earliest appearance of the 'dot-zeroes' in the Bakhshali Manuscript were encountered in the general vicinity of ancient Gandhara. As such, ancient Gandhara should be one of the focal points for further research to find fresh evidence of zero's emergence in the historical record.

The monograph chapters by Alexis Lavis, Fabio Gironi, Kaspars Klavins and Alberto Pelissero, in particular, explore this lead both pro and con.

## 5 Disruptive Moment

If our tentative conclusion based on the working hypothesis of the book project and online event warrants the assumption that the number zero slowly evolved over the centuries into what *today* we consider to be the number zero, then the truly disruptive moment in the history of mathematics was the introduction of the *placeholder*.

Subsequent interaction between that placeholder and the other numbers (via signs and syntax) would eventually lead to zero as a number alongside the other numbers – within the respective notational semiotic system concerned (decimal, vigesimal or sexagesimal).

And indeed the innovation of the placeholder in ancient Mesopotamian mathematics/astronomy in the early second millennium BCE (and a millennium or so later in China) marked just that disruptive moment in the evolution of consciousness itself when Nothing (absence) became 'a meaningful quantitative category of thought', that is, when Nothing was first signified. And once signified, it could evolve further.

'Gandhara' as crossroads between these ancient cultures may then have accounted for the spurt that zero took in ancient Indian mathematics (in the words of monograph contributor Andreas Nieder, 'a leap of abstraction'), which would in due course lead to the adoption of the decimal system worldwide – with, at its center, zero's indispensable role eventually in modern mathematics.

By the same token, a similar argument may be advanced, *mutatis mutandis*, for closer scrutiny of Hellenist Alexandrian mathematicians/astronomers, known to have adopted [note: not invented/discovered] the Mesopotamian placeholder during roughly the same historical timeframe.

Jeff Oaks' chapter however, calls into question whether the Hellenists understood zero as a number, as several other scholars with whom we had corresponded on the subject had contended (see Introduction, footnote 2). Oaks' exhaustive documentation of medieval Arabic mathematics showed that in fact zero was explicitly denied to be a number in the Islamic world.

The issue of 'semantics versus syntaxis' (what people say about numbers and what people do with numbers) to define the distinction that we 'moderns' make anachronistically between placeholder and number will top the list of priorities to be taken up in due course by the *Expertise Center on Zero* (XCZ).

The case of the Mayan zero should occasion further study in the absence of sufficient information regarding its independent invention/discovery – the more remarkable for its geographical isolation by contrast to the 'Asian zeroes'. The Zero Project explored various avenues to identify and invite a scholar on ancient Mayan mathematics, which in the end proved unsuccessful.

In both cases then, of the placeholder as well as of the number zero, their common or separate origin remains an irksome enigma to this day. We hope that this monograph on zero may serve as spur to the proposed interdisciplinary research program in the spirit of scientific advancement in order to rise to the challenge posed.

## 6 Disclaimer

The Zero Project's late chairperson, René Samson (PhD Weizmann Institute/ Postdoc MIT) had cautioned at the outset in 2015 that 'we will be spared no humiliation' in pursuit of the Zero Project's objective, given the vested nationalist interests as well as academic reputations at stake.

His prediction was not far wrong as over the years since then we have been charged with 'political motivation' and/or 'academic ineptitude' by several publishers as well as academics, despite the fact that all ancient civilizations are fairly represented in the monograph.

Samson himself set the right example by exploring within his own field of expertise the question as to what our world would be like today without zero; as well as a second chapter, coauthored by his good friend and collaborator, Joseph Biello, on the pivotal role of zero in the Conservation Laws.

Criticasters may bear in mind that as far as we are aware no scholar in the history of mathematics these past centuries has ever proposed to conduct such an in-depth interdisciplinary research project into the origin of zero. Targeted literature search by the Zero Project alone has already overturned the tacit assumption of 'scholarly consensus' that zero was understood as a number in addition to its placeholder function. Who knows what else zero holds in store?

## 7 Optimistic Note

On an optimistic note, we were heartened by the fairly recent discovery of previously unknown manuscripts bearing on early appearances of zero. The first was disclosed in a paper by Bill Mak to which we were alerted by Kim Plofker (The Date and Nature of Sphujidhvaja's Yavanajātaka Reconsidered in the Light of Some Newly Discovered Materials. *History of Science in South Asia* 1 (2013); 1–20); and the second in a private collection yet to be published by Ingo Strauch, to which we were referred in email correspondence by Bodleian Libraries, Oxford. Who knows what other evidence may be waiting to be found?

But these and all like them were serendipity finds and not the outcome of systematic, targeted research as the Zero Project proposes. Ideally therefore such concerted efforts may be made by dedicated teams in all regions of the world where local actors played a role in the invention/discovery of zero (IDZ). High time to get started.

## 8 Interdisciplinary Research Questions

We conclude the book project and online event with a preliminary set of questions proposed by contributors – in the order received – to be pursued by the *Expertise Center on Zero* (XCZ), so as to realize the main objective of shedding light on the origin of the number zero and/or its significance in the field of expertise of the contributor concerned:

1. First and foremost, to find fresh evidence – if any – of the otherwise inexplicable emergence of zero as a mathematical numeral in the historical record (as distinct from the placeholder zero that had been known for centuries).
2. Conversely, to find fresh evidence of why the emergence of zero as numeral may have been inhibited in some cultures.
3. Scrutinize possible transmission/diffusion channels among the ancient civilizations concerned.
4. Methodological question on evidence: which kind of evidence can be accepted as incontrovertible proof for the existence of which kind of zero.
5. Completion of the test cases for the widest possible range of literate societies.

6. More complete textual instances, e.g., your example, of Egyptian tableaux with empty cases.
7. Work on initial zeros in cuneiform astronomical sources, etc.
8. Cross-cultural comparative work on zeros in astronomical tables with positional notation (how did these tables deal with zeros across Mesopotamian, Egyptian, Greek, Arabic and medieval European examples).
9. Whether the numeral zero was invented or discovered (IDZ). This is an unresolved issue.
10. Year 0 in early calendrics.
11. Plot geographical coordinates of early appearances of the numeral zero worldwide and their dates on a map; connect the dots to find the 'epicenter' of zero's emergence.
12. Data-mine digitized archives of ancient manuscripts for 'tell-tale' indications of early zero or its possible cultural antecedents.
13. Draft a roadmap to conduct fresh research with a view to also answering the following question: Was the invention or discovery of the numeral zero a gradual evolutionary process from the placeholder or was it the stroke of genius that it is widely held to be?
14. Relation between zero and infinity – its cultural and mathematical context. This is an interesting question as $1/0 = \infty$. In mathematics there are many different infinities, depending on one's axioms for set theory. Are there many different zeros? For example, ℵ0 is often taken to denote the countable infinity, i.e., the cardinality of any set in one-to-one correspondence with the integers. ℵ1 is sometimes defined as the cardinality of all subsets of ℵ0, etc. Is $1/\aleph 0 = 1/\aleph 1$? Has anyone made sense of this? It calls to mind the zero's of non-standard analysis. It would be interesting to have a knowledgeable mathematician contribute something on this.
15. Were cultures able to make correct calculation by operating on their respective placeholder zero (in the absence of the numeral zero) or were correct results obtain by 'ignoring' the placeholder as 'nothing there'?
16. What were the respective concepts of number in diverse cultures (quantities or abstract mathematical objects)?
17. How the concepts of nothingness and emptiness are related to the invention of zero.
18. The profound nature of human response to zero – in the sense that there was resistance (active or passive) to accepting the concept of zero; active in Europe, passive in China and none in the Arabic world, while embraced in India.
19. Inductive and deductive thinking/logic and its response to zero.

20. Relation between numbers, counting and algebra – and development of these subjects with the use (or without the use) of formal zero.
21. Representation of zero in the arts.
22. Movable zero – in physical frames of reference and in analytical studies.
23. Zeros that are real – absolute zero of temperature etc. and zeros that are relative or arbitrary.
24. Relation between zero and infinity – its cultural and mathematical context.
25. Identity of zero – the identity as in number systems but its loss of uniqueness, say in passwords, credit card numbers etc. where leading zeros have a 'meaning'.
26. Textual studies concerning the concept of empty space in ancient cultures and the concept of infinity of space.
27. Precursors in ancient science or proto-science like mathematics but also astronomy.
28. The concept of emptiness and nothingness and introspection, a form of meditation after all.
29. Whether a phenomenology of emptiness is possible.
30. The function of zero in metaphysics.
31. The ways in which zero might function as a metaphor.
32. The mystic significance of zero.

We avail of this opportunity to invite the reader to join the *Expertise Center on Zero* (XCZ) either in his or her personal capacity or in the context of the associated university, by signing the Memorandum of Understanding (see Appendix 1).

To provide the XCZ the wherewithal to commission future research in collaboration with signatories, considerable funding is required. We therefore appeal to donors with vision to enable the XCZ to delve deeper into the urgent question of the cultural context that gave rise to the invention/discovery of zero and its significance across a range of interdisciplinary fields.

Finally, in order to attain our objective of finding fresh evidence of zero's origin, nothing less would be required than organized campaigns in each region of the world that contributed to the invention/discovery of zero (IDZ) – be it anthropological, religious, philosophical, linguistic, archaeological, epigraphic or any other relevant field. Naturally the Zero Project will do what it can and invite fellow travelers along on the journey.

APPENDIX 1

# Expertise Center on Zero Memorandum of Understanding

between Zero Project Foundation of the Hague, the Netherlands
and ____________________ (university/academic institute),
hereinafter referred to as the "Parties".
Recognizing that the Zero Project Foundation wants to further explore new evidence on the umbrella theme of discovery/invention of the number Zero;
Recognizing that ____________________ is academically interested in conducting research on topics related to the said umbrella theme;
Further recognizing that besides Parties, there are other universities and institutions who are currently or in the future interested in participating in the research endeavour;
Believing that such research will further deepen the understanding of the contribution of the number zero in amongst others mathematical, philosophical, linguistical and cultural context;
Desiring therefore to work together and with other universities, academic and research institutions in a multi-disciplinary approach
Agree as follows:

## Article 1: Objective of the Memorandum

The aim of this MOU is to further discuss, explore and establish the principles and guidelines of research projects as outlined in this MOU, both bilaterally between Parties and multilaterally with other institutions. The outcome will be an agreement on the participation of the Parties in the establishment of an *Expertise Center on Zero* which will foster and develop scientific, academic and research cooperation as well as encouraging student and academic exchange between them on the basis of equality and mutual benefit.

## Article 2: Areas of Cooperation

Parties, within their available resources, will cooperate in the following areas:

a. joint research projects and publications;
b. exchange of academic materials and other information;

 | DOI:10.1163/9789004691568_047

c. participation in seminars, conferences, scientific activities and academic meetings;
d. any other areas mutually agreed upon by the Parties.

### Article 3: Exchange of Publications and Academic Documents

The Parties shall exchange their publications and periodicals under the intellectual property laws and regulations enforced in their respective countries.

### Article 4: Exchange of Support

The Parties shall undertake to support participants in the work programs by providing the information and facilities required for the cooperation and by settling problems related to organizing issues, in accordance with the enforced regulations in both countries.

### Article 5: Establishing the *Expertise Center on Zero*

The cooperation between Parties will ultimately lead to the establishment of the *Expertise Center on Zero*, as part of the Zero Project Foundation, but academically governed by the participating universities and institutions.

### Article 6: Dispute Settlement

Any dispute that may arise between the Parties regarding the interpretation or the implementation of this MOU shall be settled amicably by direct consultation and negotiations without involving to any third Party.

### Article 7: Final Provisions

a. This MOU shall enter into force on the date of signing and shall remain in force for a period of one (1) year.
b. Thereafter, it shall be automatically extended for a further period of one (1) year.
c. Notwithstanding anything in this Article, either Party may terminate this MOU by notifying the other Party in writing at least three (3) months prior to termination term.

d. The MOU will also terminate once Parties agree to continue their cooperation as mentioned in Article 5 of this MOU.

e. The termination of this MOU will not affect any ongoing activities until their completion.

f. Either Party may amend, change or add any item or article of this MOU by mutual written consent. These amendments, or changes or additions will enter into force according to the same procedure mentioned in paragraph (a) of this Article, and it will be considered as an integrated part of it.

g. Each Party may nominate a coordinator (or a management committee) for the execution of the MOU within a maximum period of three months from its entry into force.

h. This MOU does not give any Party the right to be a procurator or a representative of the other Party, and does not either constitute a joint venture partnership or an official business of either of the Parties.

i. The execution of this MOU is subjected to the enforced laws and regulations in the countries where the Parties are based.

Drawn up and signed in ____________________(location) on ____________________ (date) in two original copies in English.

____________________ For Zero Project Foundation, the Hague, the Netherlands, Chairperson: Mr. Eric F. Ch. Niehe

____________________ For ____________________ (university/academic institution) (Representative)

APPENDIX 2

# Online Presentations

Many of the contributors took part in online presentations which are available to readers of this volume, providing a valuable resource for ongoing reference and research.

**3 October 2021**

Eric Niehe, Dr. Robert Lawrence Kuhn, Prof. Sharda Nandram, Wende Wallert, Debra G. Aczel, Dr. Miriam R. Aczel
https://bit.ly/Opening_Niehe_Kuhn_Nandram_Wallert_Aczels

**10 October 2021**

Dr. Bhaswati Bhattacharya, Prof. Paul Ernest
https://bit.ly/B_Bhattacharya_Ernest

**24 October 2021**

Dr. John Marmysz, Dr. Erik Hoogcarspel
https://bit.ly/Marmysz_Hoogcarspel

**31 October 2021**

Dr. Marina Ville, Prof. Avinash Sathaye
https://bit.ly/Ville_Sathaye

**7 November 2021**

Prof. Parthasarathi Mukhopadhyay
https://bit.ly/Mukhopadhyay

**14 November 2021**

Prof. Alexis Lavis
https://bit.ly/Alexis_Lavis

 | DOI:10.1163/9789004691568_048

**21 November 2021**

Prof. Beatrice Lumpkin, Prof. Anupam Jain
https://bit.ly/Lumpkin_Jain

**28 November 2021**

Prof. Joseph A. Biello
https://bit.ly/Joseph_Biello

**12 December 2021**

Dr. Célestin Xiaohan Zhou
https://bit.ly/Celestin_Zhou

**19 December 2021**

Dr. Esti Eisenmann
https://bit.ly/Esti_Eisenmann

**2 January 2022**

Dr. T. S. Ravishankar, Dr. Solang Uk
https://bit.ly/Ravishankar_UK

**16 January 2022**

Prof. Jeffrey A. Oaks, Jonathan J. Crabtree
https://bit.ly/Oaks_Crabtree

**30 January 2022**

Prof. Alberto Pelissero, Prof. Mayank N. Vahia, Upasana Neogi
https://bit.ly/Pelissero_Vahia_Neogi

**6 February 2022**

Debra G. Aczel and Dr. Miriam R. Aczel
https://bit.ly/aczels

**13 February 2022**

Prof. Jim Ritter, Prof. Friedhelm Hoffmann
https://bit.ly/Ritter_Hoffmann

**20 February 2022**

Prof. Mayank N. Vahia
https://bit.ly/vahia

**27 February 2022**

Prof. Sisir Roy, Venkata Rayudu Posina
https://bit.ly/Roy_Posina

**6 March 2022**

Prof. Jim Ritter, Prof. Friedhelm Hoffmann, Prof. Jeffrey A. Oaks, Prof. Mayank N. Vahia, Prof. Avinash Sathaye
https://bit.ly/Ritter_Hoffmann_Oaks_Vahia_Sathaye

**13 March 2022**

Prof. Sharda S. Nandram
https://bit.ly/Nandram

**20 March 2022**

Prof. Kaspar Klavins, Prof. Marcis Auzinsh
https://bit.ly/Klavins_Auzinsh

**27 March 2022**

Prof. Sudip Bhattacharyya
https://bit.ly/Sudip_Bhattacharyya

**15 May 2022**

Dr. Esther Freinkel Tishman
https://bit.ly/Esther_Freinkel_Tishman

APPENDIX 3

# Status Update/Petition on the Bakhshali Manuscript

The Zero Project (www.TheZeroProject.nl) has been in contact with the Bodleian Libraries, Oxford, UK, regarding the importance of radiocarbon-dating of the Bakhshali Manuscript since 2016. It would be remiss of the Zero Project not to pursue this issue vigorously as it bears directly on the central objective of facilitating in-depth academic research to find fresh evidence of the invention of the mathematical number zero.

In 2017, the Bodleian Libraries conducted radiocarbon-dating of the manuscript. However, since the results were inconclusive, scholars from around the world (including most of the forty contributors to this volume and associated online event) have urged the Bodleian Libraries to have follow-up verification tests conducted in the interest of scientific advancement in the field.

In addition, offers of technical assistance and full funding for follow-up radiocarbon-dating, available through our extensive international network, have been formally communicated to the Bodleian Libraries.

We regret that despite repeated urgent pleas to the Bodleian Libraries over the years to have follow-up radiocarbon dating conducted, this has not yet been undertaken. Nor did the Bodleian Libraries respond to our invitation to contribute a chapter on the Bakhshali Manuscript to this volume or even provide a 'status update' of their plans for the Bakhshali Manuscript for inclusion in this book.

On the eve of submission of the manuscript to the publisher, we finally received a 'status update':

> While an exceptional case was made in 2016 for the Bakhshali Manuscript, the Bodleian exercises a policy of non-destructive analysis.

In a word, the Bodleian Libraries will not conduct follow-up radiocarbon dating of the Bakhshali Manuscript.

Readers of this collected volume are cordially invited to sign the online Petition to encourage the Bodleian Libraries to have follow-up radiocarbon dating conducted: https://www.petitions.net/radiocarbon-dating_of_the_bakhshali_manuscript. Once you have signed it, please share widely.

 | DOI:10.1163/9789004691568_049

# Index

Printed in the United States
by Baker & Taylor Publisher Services